Early Evolutionary History of the Synapsida

Vertebrate Paleobiology and Paleoanthropology Series

Edited by

Eric Delson
Vertebrate Paleontology, American Museum of Natural History,
New York, NY 10024, USA
delson@amnh.org

Eric J. Sargis
Anthropology, Yale University
New Haven, CT 06520, USA
eric.sargis@yale.edu

Focal topics for volumes in the series will include systematic paleontology of all vertebrates (from agnathans to humans), phylogeny reconstruction, functional morphology, Paleolithic archaeology, taphonomy, geochronology, historical biogeography, and biostratigraphy. Other fields (e.g., paleoclimatology, paleoecology, ancient DNA, total organismal community structure) may be considered if the volume theme emphasizes paleobiology (or archaeology). Fields such as modeling of physical processes, genetic methodology, nonvertebrates or neontology are out of our scope.

Volumes in the series may either be monographic treatments (including unpublished but fully revised dissertations) or edited collections, especially those focusing on problem-oriented issues, with multidisciplinary coverage where possible.

For other titles published in this series, go to
www.springer.com/series/6978

Early Evolutionary History of the Synapsida

Edited by

Christian F. Kammerer

Division of Paleontology and Richard Gilder Graduate School, American Museum of Natural History, New York, NY, USA

and

Museum für Naturkunde, Leibniz-Institut für Evolutions- und Biodiversitätsforschung an der Humboldt-Universität zu Berlin, Berlin, Germany

Kenneth D. Angielczyk

Department of Geology, Field Museum of Natural History, Chicago, IL, USA

Jörg Fröbisch

Museum für Naturkunde, Leibniz-Institut für Evolutions- und Biodiversitätsforschung an der Humboldt-Universität zu Berlin, Berlin, Germany

Springer

Editors
Christian F. Kammerer
Division of Paleontology and Richard Gilder Graduate School
American Museum of Natural History
New York, NY
USA

and

Museum für Naturkunde
Leibniz-Institut für Evolutions- und Biodiversitätsforschung an der Humboldt-Universität zu Berlin
Berlin
Germany

Kenneth D. Angielczyk
Department of Geology
Field Museum of Natural History
Chicago, IL
USA

Jörg Fröbisch
Museum für Naturkunde
Leibniz-Institut für Evolutions- und Biodiversitätsforschung an der Humboldt-Universität zu Berlin
Berlin
Germany

ISSN 1877-9077 ISSN 1877-9085 (electronic)
ISBN 978-94-007-6840-6 ISBN 978-94-007-6841-3 (eBook)
DOI 10.1007/978-94-007-6841-3
Springer Dordrecht Heidelberg New York London

Library of Congress Control Number: 2013943718

Cover illustration: An articulated skeleton of the Middle Permian dicynodont *Eosimops newtoni* (BP/1/6674) set against South African outcrop. Drawing by Alix Lukas, photo courtesy Christian F. Kammerer.

Printed on acid-free paper

Springer is part of Springer Science+Business Media (www.springer.com)

Preface

The Cenozoic radiation of mammals is one of the major events in vertebrate evolution, and has been a subject of inquiry for as long as the science of paleontology has existed. However, the Cenozoic is not the first time that members of Synapsida, the clade that includes all extant mammals and their extinct relatives, were the most conspicuous component of terrestrial ecosystems. Between the Late Carboniferous and the beginning of the Late Triassic, a great diversity of nonmammalian synapsids evolved, and although some include the ancestors of mammals, the great majority are parts of now completely extinct lineages. Among their numbers were the largest animals of their time, some of the first tetrapod herbivores, species that flourished in the aftermath of the largest mass extinction in Earth history, and a mixture of morphotypes that are unlike any seen among crown-group mammals (e.g., sail-backed carnivores; tusked, beaked herbivores) as well as some that re-evolved numerous times in synapsid history (e.g., saber-toothed carnivores). Their fossil record consists of tens of thousands of specimens, they have been key to understanding the evolutionary history of many of the distinctive features of mammals, and they have been studied by iconic paleontologists such as Sir Richard Owen, Edward Drinker Cope, Robert Broom, Alfred Romer, and Everett Olson. Yet even among vertebrate paleontologists, nonmammalian synapsids are relatively obscure, and the nonmammalian synapsid most widely known to the general public, *Dimetrodon*, frequently is mistaken for a dinosaur.

The first edited volume on nonmammalian synapsids, *The Ecology and Biology of Mammal-like Reptiles* by Hotton et al. (1986), appeared a quarter of a century ago. In the time since, and particularly in the past decade, there has been a noteworthy increase in research on nonmammalian synapsids. Much of this work has had a taxonomic and/or phylogenetic focus, but a variety of other topics are represented as well, including biostratigraphy, patterns of diversity and faunal turnover, bone histology and growth, functional morphology, the causes and effects of the end-Permian mass extinction, paleoecology, and the description of exciting new specimens from areas and times that have historically received little attention. This book grew out of a symposium ("New Perspectives on the Early Evolutionary History of the Synapsida"; organized by C.F.K. and K.D.A) at the 69th Annual Meeting of the Society of Vertebrate Paleontology held in Bristol, England, and our intention for the symposium was to highlight some of the current range of research on nonmammalian synapsids. Our goal for this book is similar, and the topics covered include morphological description, systematics, functional morphology, terrestrial paleoecology, the end-Permian extinction, paleopathology, biostratigraphy, biogeography, and diversity through time. Research on synapsid bone histology and its implications are not included here because that topic is well covered by *Forerunners of Mammals* (Chinsamy-Turan 2012). When taken together with *Forerunners of Mammals*, we hope that readers will gain a sense of why it is an exciting time to be studying nonmammalian synapsids, both in terms of advances that are being made as well as the many outstanding questions that remain to be investigated. We are particularly pleased that the majority of chapters in our book and in Chinsamy-Turan (2012) focus on nonmammalian synapsids for their own importance and interest, as opposed to only what these fascinating animals can tell us about the evolution of mammals.

Early Evolutionary History of the Synapsida includes 14 chapters that are organized into four thematic parts, each with a short introduction that helps to place the specific contributions into the broader context of synapsid research. The first part features four chapters on basal synapsids, or "pelycosaurs", with an introduction by Robert Reisz. The chapters consist of descriptions of new material of the synapsids *Oedaleops* and *Aerosaurus*, a new member of the famed Early Permian upland fauna preserved at the Bromacker Locality, and a reconsideration of the long-standing hypothesis that *Ophiacodon* was semi-aquatic. The second part includes three chapters on anomodont therapsids, including a review of dicynodonts from Zambia, an investigation of cranial variability in *Lystrosaurus*, and an examination of pathologies preserved in Permian and Triassic dicynodont fossils. Jörg Fröbisch provides the introduction to this section. The third part consists of five chapters on gorgonopsians, therocephalians and cynodonts, with an introduction by Christian Kammerer. Two of these chapters focus on gorgonopsians, providing much-needed insight into their early radiation and phylogeny, whereas the remainder consider the phylogeny and biostratigraphy of Triassic gomphodont cynodonts and bauriamorph therocephalians. The two chapters of the final part focus on patterns of diversity through time and the end-Permian mass extinction in the Karoo Basin of South Africa, and are introduced by Kenneth Angielczyk. We consider each of these 14 contributions to be an interesting and important contribution to research on nonmammalian synapsids, and we hope that this collection of works will be a useful tool for synapsid researchers in the years to come.

As in any undertaking of this kind, there are many people to thank. First and foremost, we thank our many colleagues and friends for their contributions to *The Early Evolutionary History of the Synapsida*. Their hard work and inquisitive research are on display on its pages, and we appreciate their patience as we assembled the book. We also thank the many individuals who served as peer-reviewers for the chapters, who are also acknowledged in each chapter as appropriate. The book in its present form would not be possible without their efforts.

Alix Lukas drafted the cover illustration of a nearly complete skeleton of the dicynodont *Eosimops newtoni* (BP/1/6674).

Pete Makovicky made the initial suggestion of publishing the volume in the Vertebrate Paleobiology and Paleoanthropology Series. We also thank Eric Delson and Eric Sargis, Series Editors for Vertebrate Paleobiology and Paleoanthropology, and Judith Terpos, Springer's senior assistant publishing editor for earth and environmental sciences, for their aid in bringing this volume to publication.

Christian F. Kammerer
Kenneth D. Angielczyk
Jörg Fröbisch

References

Chinsamy-Turan, A. (Ed.). (2012). *Forerunners of mammals: Radiation histology biology*. Bloomington: Indiana University Press.

Hotton, N., III, MacLean, P. D., Roth, J. J., & Roth, E. C. (Eds.) (1986). *The ecology and biology of mammal-like reptiles*. Washington, DC: Smithsonian Institution Press.

Contents

Contributors

Note * indicates preferred address for correspondence

Fernando Abdala
Evolutionary Studies Institute and School for Geosciences, University of the Witwatersrand, Private Bag 3, Johannesburg, Wits 2050, South Africa
nestor.abdala@wits.ac.za

Kenneth D. Angielczyk
Department of Geology, Field Museum of Natural History, South Lake Shore Drive, Chicago, IL 60605, USA
kangielczyk@fieldmuseum.org

David S Berman
Section of Vertebrate Paleontology, Carnegie Museum of Natural History, 4400 Forbes Avenue, Pittsburgh, PA 15213, USA
bermand@carnegiemnh.org

Jennifer Botha-Brink
*Department of Karoo Palaeontology, National Museum, Bloemfontein, South Africa
Department of Zoology and Entomology, University of the Free State, Bloemfontein, South Africa
jbotha@nasmus.co.za

Anusuya Chinsamy
Department of Zoology, University of Cape Town, Private Bag X3, Rondebosch, Cape Town 7701, South Africa

Michael A. Cluver
Iziko South African Museum, PO Box 61, Cape Town 8000, South Africa

Ryan N. Felice
Department of Biological Sciences, Ohio University, Athens, OH 45701, USA
rf273509@ohio.edu

Jörg Fröbisch
Museum für Naturkunde, Leibniz-Institut für Evolutions- und Biodiversitätsforschung an der Humboldt-Universität zu Berlin, Invalidenstr. 43, 10115, Berlin, Germany
joerg.froebisch@mfn-berlin.de

Eva V. I. Gebauer
Senckenberg Research Institute, Natural History Museum, Senckenberganlage 25, 60325 Frankfurt, Germany
Eva.Gebauer@senckenberg.de

Amy C. Henrici
Section of Vertebrate Paleontology, Carnegie Museum of Natural History, 4400 Forbes Avenue, Pittsburgh, PA 15213, USA
henricia@carnegiemnh.org

James A. Hopson
Department of Organismal Biology, University of Chicago, 1027 East 57th Street, Chicago, IL 60637, USA
*3051 Piney Ridge Road, Ludington, MI 49431, USA
jhopson@uchicago.edu

Adam K. Huttenlocker
Department of Biology, University of Washington, Seattle, WA 98195, USA
huttenla@u.washington.edu

Tea Jashashvili
*Institute for Human Evolution, University of the Witwatersrand, Private Bag 3, Johannesburg, Wits 2050, South Africa
Anthropological Institute and Museum, University of Zurich, 8057, Zurich, Switzerland
Department of Geology and Paleontology, Georgian National Museum, 0105 Tbilisi, Georgia
tjashashvili@yahoo.fr

Sandra C. Jasinoski
Department of Zoology, University of Cape Town, Private Bag X3, Rondebosch, Cape Town 7701, South Africa
sandra.jasinoski@uct.ac.za

Christian F. Kammerer
Division of Paleontology and Richard Gilder Graduate School, American Museum of Natural History, Central Park West at 79th Street, New York, NY 10024, USA
*Museum für Naturkunde, Leibniz-Institut für Evolutions- und Biodiversitätsforschung an der Humboldt-Universität zu Berlin, Invalidenstr. 43, 10115 Berlin, Germany
christian.kammerer@mfn-berlin.de

Jun Liu
Key Laboratory of Evolutionary Systematics of Vertebrates, Institute of Vertebrate Paleontology and Paleoanthropology, Chinese Academy of Sciences, 100044 Beijing, People's Republic of China
liujun@ivpp.ac.cn

Michael W. Maisch
Staatliches Museum für Naturkunde, Rosenstein 1, 70191 Stuttgart, Germany
michael.maisch@smns-bw.de

Thomas Martens
Abteilung Paleontology, Museum der Natur, PSF 217, 99853 Gotha, Germany
martens@stiftung-friedenstein.de

Sean P. Modesto
Department of Biology, Cape Breton University, Sydney, Nova Scotia, Canada
seanmodesto@yahoo.ca

Valerie Pelletier
Department of Biology, California State University, 5500 University Parkway, San Bernardino, CA 92407, USA
uraprimate@msn.com

B. Daya Reddy
Centre for Research in Computational and Applied Mechanics, University of Cape Town, Private Bag X3, Rondebosch, Cape Town 7701, South Africa

Robert R. Reisz
Department of Biology, University of Toronto Mississauga, 3359 Mississauga Road N., Mississauga, ON L5L 1C6, Canada
robert.reisz@utoronto.ca

Bruce S. Rubidge
Evolutionary Studies Institute and School for Geosciences, University of the Witwatersrand, Private Bag 3, Johannesburg, Wits 2050, South Africa
bruce.rubidge@wits.ac.za

Christian A. Sidor
Burke Museum and Department of Biology, University of Washington, Seattle, WA 98195, USA
casidor@u.washington.edu

Roger M. H. Smith
Department of Karoo Palaeontology, Iziko South African Museum, Cape Town 8000, South Africa
rsmith@iziko.org.za

Jean-Sébastien Steyer
Department of Earth Science, CNRS, Muséum National d'Histoire Naturelle, 8 rue Buffon, CP 38, Paris 75005, France
steyer@mnhn.fr

Stuart S. Sumida
Department of Biology, California State University, 5500 University Parkway, San Bernardino, CA 92407, USA
ssumida@csusb.edu

Stephen Tolan
Chipembele Wildlife Education Centre, Malama Road, Chowo Site, PO Box 67, Mfuwe, Zambia
info@chipembele.org

Juri van den Heever
Department of Botany and Zoology, University of Stellenbosch, Private Bag X1, Matieland, Stellenbosch 7602, South Africa
javdh@maties.sun.ac.za

Cristina Silveira Vega
Departamento de Geologia, Centro Politécnico, Universidade Federal do Paraná, Caixa Postal 19.001, Curitiba, PR 81531-990, Brazil
cvega@ufpr.br

Robin L. Whatley
Department of Science and Mathematics, Columbia College Chicago, 600 South Michigan Avenue, Chicago, IL 60605, USA
rwhatley@colum.edu

Part I
"Pelycosaur"-Grade Synapsids

Chapter 1
"Pelycosaur"-Grade Synapsids: Introduction

Robert R. Reisz

The Permo-Carboniferous fossil record is characterized by a rich, diverse fauna of amniotes, represented by two major clades, the synapsids and the reptiles. Synapsids appear to dominate the earliest stages of terrestrial vertebrate evolution, with abundant fossil remains and numerous taxa (Reisz 1986). They first appear in the fossil record at Joggins, Nova Scotia (311–314 Ma, Moscovian Stage, Pennsylvanian Period) in the same *Sigillaria* stumps as the oldest known reptile, *Hylonomus* (Carroll 1964). Although the precise phylogenetic relationships of the earliest synapsid from this locality, *Protoclepsydrops*, are uncertain, its remains clearly belong to a "pelycosaur"-grade (i.e., non-therapsid) synapsid. In the slightly younger *Sigillaria* stumps in Florence, Nova Scotia (307–306 Ma), there is already evidence of at least five synapsid taxa and one reptile (Carroll 1969; Reisz 1972). In another Carboniferous locality near Garnett, Kansas (305 Ma, Kasimovian Stage, Pennsylvanian Period), the synapsid fauna is represented by at least six taxa (Reisz and Berman 1986), while only one reptilian body fossil is present (Kissel and Reisz 2004). By this stage in synapsid evolution, we already have evidence of three of the major "pelycosaur" clades (see below). In the Early Permian, synapsid taxa greatly outnumber reptiles and are widely distributed throughout the equatorial regions of Pangaea. Permo-Carboniferous synapsids also encompass a much broader size range and morphological diversity than coeval reptiles.

"Pelycosaur"-grade synapsids were first discovered in the mid-nineteenth century, and their history is intertwined with the discovery of dinosaurs and other Mesozoic reptiles. These discoveries were summarized by Romer and Price (1940) and Reisz (1986). Leidy (1854) described the first "pelycosaur", *Bathygnathus*, from Prince Edward Island, but identified it as an archosaur; its true identity as a sphenacodontid synapsid was only recognized in 1905 (Case 1905). Most of the early work on basal synapsids was done by the great paleontological antagonists E. D. Cope and O. C. Marsh. In typical fashion, they competed at naming new taxa, with Cope naming *Clepsydrops, Dimetrodon*, and *Edaphosaurus* in 1875, 1878 and 1882, while Marsh named *Ophiacodon* and *Sphenacodon* in 1878. Important subsequent studies of these early synapsids were published by E. C. Case, including the pivotal "Revision of the Pelycosauria of North America" in 1907. He recognized the biological significance of the dental patterns in these early synapsids and erected three families; this structure was generally followed by subsequent workers, including Williston (1912) and Romer and Price (1940), the latter using the taxonomic categories of Ophiacodontia, Sphenacodontia, and Edaphosauria in their "Review of the Pelycosauria". This large monograph set a new standard for a thorough, detailed review not only of these early synapsids, but for all Paleozoic tetrapods. The "Review of the Pelycosauria" continues to be useful today, even though we have changed dramatically the way we analyze evolutionary relationships and history.

Phylogenetic analyses from the last three decades have shown that Permo-Carboniferous synapsids form six distinct clades. These clades also represent differing feeding strategies. Two of the most interesting clades, Caseidae and Edaphosauridae, include taxa ranging in size from small to very large animals (Olson 1968; Mazierski and Reisz 2010). Both caseids and edaphosaurs are generally considered to be high fiber herbivores, as indicated by a variety of osteological features, including rib cage morphology, cranial anatomy, and dental anatomy (Modesto and Reisz 1992; Sues and Reisz 1998). Interestingly, caseids have "leaf-shaped" teeth reminiscent of extant herbivorous iguanids (Olson 1962; Maddin et al. 2008), while edaphosaurids have massive palatal and mandibular dental batteries for extensive oral processing (Modesto 1995).

The other four clades of early synapsids are all faunivorous, presumably feeding on other vertebrates and arthropods. One clade, the Eothyrididae, are currently only known

R. R. Reisz (✉)
Department of Biology, University of Toronto Mississauga, 3359 Mississauga Road N, Mississauga, ON L5L 1C6, Canada
e-mail: robert.reisz@utoronto.ca

Christian F. Kammerer, Kenneth D. Angielczyk and Jörg Fröbisch (eds.), *Early Evolutionary History of the Synapsida*, Vertebrate Paleobiology and Paleoanthropology, DOI: 10.1007/978-94-007-6841-3_1,

from the fragmentary remains of two taxa, *Oedaleops* and *Eothyris* (Reisz et al. 2009). These are generally considered small predators, possibly insectivorous, and are characterized by having greatly enlarged caniniform teeth in an otherwise small, low skull. Eothyridids are by far the most poorly known early synapsids, and in this volume, Sumida and colleagues present valuable new information on the postcranial anatomy of *Oedaleops*. This is a particularly significant contribution because the skeletal anatomy of this group is expected to help define the primitive condition for the clade (Sumida et al. 2013).

The remaining three faunivorous clades among basal synapsids are the Varanopidae, Ophiacodontidae, and Sphenacodontia. In many features, these three clades represent very different body designs and different dentitions. Varanopids are relatively lightly built predators with strongly recurved dentition. Ophiacodontids are awkward looking, large headed predators with numerous teeth. Sphenacodontians include the top predators of the Early Permian, with massive heads and very strongly developed incisiform and caniniform teeth.

Varanopid synapsids have attracted a lot of attention in the last few decades (Berman and Reisz 1982; Dilkes and Reisz 1996; Reisz and Laurin 2004; Maddin et al. 2006; Botha-Brink and Modesto 2009; Campione and Reisz 2010). These are relatively rarely-found small to medium-sized, gracile predators, with labio-lingually compressed, strongly recurved teeth. Frequently, the smaller members of this clade have been misidentified as diapsid reptiles (Reisz and Berman 2001; Reisz et al. 2010), partly because these two clades are characterized by similar teeth (Reisz and Modesto 2007). Despite the low number of specimens that have been collected, their fossil record extends from the Late Pennsylvanian (Dilkes and Reisz 1996) to the Middle and Late Permian (Anderson and Reisz 2004; Modesto et al. 2011). Their skeletal remains have been found not only in North America but also in Russia and South Africa, and include evidence for the early evolution of parental care in amniotes (Botha-Brink and Modesto 2007, 2009).

In this issue, there are two papers dealing with this fascinating group of synapsids, both involving larger members of the clade. These large varanopids have dramatically modified their cranial architecture when compared to their smaller relatives, and have developed a strongly sloping occiput and very long temporal fenestrae. One of the papers in this volume is particularly exciting, because it describes a new large varanopid from the Early Permian of Germany, thus providing the first evidence that large varanodontine varanopids were present outside of North America (Berman et al. 2013). The second paper on varanopids herein (Pelletier 2013) employs excellently preserved fossil remains from the Early Permian of New Mexico to reconstruct the postcranial skeleton of *Aerosaurus wellesi*.

Members of the Ophiacodontidae characteristically have very tall, narrow skulls, large skull to trunk ratios, and relatively short, broad limbs. Their marginal dentition has a very high count and is tightly packed in the jaws, and the individual teeth are generally slightly recurved and with anterior and posterior cutting edges. Even in the oldest known member of the clade, *Archaeothyris* (Reisz 1972), the snout is not only elongated but also quite tall and slender. This feature of the ophiacodontid skull is exaggerated further in the Early Permian members of the clade (Reisz 1986; Berman et al. 1995). The biology of these strange early synapsids is poorly understood, but it has been suggested that *Ophiacodon* may have had an amphibious lifestyle. This hypothesis, however, had never been tested quantitatively prior to work in Felice and Angielczyk (2013) utilizing a morphometric perspective.

The most spectacular and largest predators of the Early Permian were members of the sphenacodontian clade, which includes the Sphenacodontidae and their close relatives ("haptodonts") as well as the Therapsida. The sphenacodontids include four relatively well-known taxa (*Ctenospondylus, Dimetrodon*, *Secodontosaurus*, and *Sphenacodon*) ranging in size from moderate to very large (~4 m length) terrestrial predators with massive skulls, relatively elongate limbs, and neural spine elongation of varying degrees. For nearly a century, a close relationship between sphenacodontids and therapsids has been recognized, but the details of this relationship have been difficult to resolve because of the perceived temporal and geographic separation between the two groups. That gap has been bridged by recent discoveries and studies, but much remains to be done. The evolutionary history of the sphenacodonts remains controversial, and a review of the phylogenetic relationships of the taxa that are loosely arranged within the larger Sphenacodontia is required. Several recent papers have reignited the controversy surrounding the origins of therapsids, with the description and reinterpretation of various forms from China and North America that have been considered early, basal therapsids (Liu et al. 2009; Amson and Laurin 2011). While the relationships between the various sphenacodontids and their close "haptodont" relatives like *Haptodus, Cutleria,* and *Pantelosaurus* are not specifically discussed in this volume, extensive emphasis has been placed on therapsids, which are examined in detail in subsequent sections.

References

Amson, E., & Laurin, M. (2011). On the affinities of *Tetraceratops insignis*, an Early Permian synapsid. *Acta Palaeontologica Polonica, 56*, 301–312.

Anderson, J. S., & Reisz, R. R. (2004). *Pyozia mesenensis*, a new, small varanopid (Synapsida: Eupelycosauria) from Russia. *Journal of Vertebrate Paleontology, 24*, 173–179.

Berman, D. S, & Reisz, R. R. (1982). Restudy of *Mycterosaurus longiceps* (Reptilia, Pelycosauria) from the Lower Permian of Texas. *Annals of Carnegie Museum, 51*, 423–453.

Berman, D. S, Reisz, R. R., Bolt, J. R., & Scott, D. (1995). The cranial anatomy and relationships of the synapsid *Varanosaurus* (Eupelycosauria: Ophiacodontidae) from the Early Permian of Texas and Oklahoma. *Annals of Carnegie Museum, 58*, 99–138.

Berman, D. S, Henrici, A. C., Sumida, S. S., Martens, T., & Pelletier, V. (2013). First European record of a varanodontine (Synapsida: Varanopidae): Member of a unique Early Permian upland paleoecosystem, Tambach Basin, central Germany. In C. F. Kammerer, K. D. Angielczyk, & J. Fröbisch (Eds.), *Early evolutionary history of the Synapsida* (pp. 69–86). Dordrecht: Springer.

Botha-Brink, J., & Modesto, S. P. (2007). A mixed-age classed 'pelycosaur' aggregation from South Africa: Earliest evidence of parental care in amniotes? *Proceedings of the Royal Society, Series B, 274*, 2829–2834.

Botha-Brink, J., & Modesto, S. P. (2009). Anatomy and relationships of the Middle Permian varanopid *Heleosaurus scholtzi* based on a social aggregation from the Karoo basin of South Africa. *Journal of Vertebrate Paleontology, 29*, 389–400.

Campione, N. E., & Reisz, R. R. (2010). *Varanops brevirostris* (Eupelycosauria: Varanopidae) from the Lower Permian of Texas, with a discussion of varanopid morphology and interrelationships. *Journal of Vertebrate Paleontology, 30*, 724–746.

Carroll, R. L. (1964). The earliest reptiles. *Zoological Journal of the Linnean Society, 45*, 61–83.

Carroll, R. L. (1969). A Middle Pennsylvanian captorhinomorph and the interrelationships of primitive reptiles. *Journal of Paleontology, 43*, 151–170.

Case, E. C. (1905). *Bathygnathus borealis* Leidy and the Permian of Prince Edward Island. *Science, 22*, 52–53.

Case, E. C. (1907). *Revision of the Pelycosauria of North America*. Washington, DC: Carnegie Institution of Washington.

Dilkes, D. W., & Reisz, R. R. (1996). First record of a basal synapsid ('mammal-like reptile') in Gondwana. *Proceedings of the Royal Society of London B, 263*, 1165–1170.

Felice, R. N., & Angielczyk, K. D. (2013). Was *Ophiacodon* (Synapsida, Eupelycosauria) a swimmer? A test using vertebral dimensions. In C. F. Kammerer, K. D. Angielczyk, & J. Fröbisch (Eds.), *Early evolutionary history of the Synapsida* (pp. 25–51). Dordrecht: Springer.

Kissel, R., & Reisz, R. R. (2004). Synapsid fauna of the Upper Pennsylvanian Rock Lake Shale near Garnett, Kansas and the diversity patterns of early amniotes. In G. Arratia, R. Cloutier, & M. V. H. Wilson (Eds.), *Recent advances in the origin and early radiation of vertebrates* (pp. 409–428). München: Verlag Dr. Friedrich Pfeil.

Leidy, J. (1854). On *Bathygnathus borealis*: An extinct Saurian of the New Red Sandstone of Prince Edward's Island. *Proceedings of the Academy of Natural Sciences of Philadelphia, 8*, 449–451.

Liu, J., Rubidge, B. S., & Li, J. (2009). New basal synapsid supports Laurasian origin for therapsids. *Acta Palaeontologica Polonica, 54*, 393–400.

Maddin, H. C., Evans, D. C., & Reisz, R. R. (2006). An Early Permian varanodontine varanopid (Syanpsida: Eupelycosauria) from the Richards Spur locality, Oklahoma. *Journal of Vertebrate Paleontology, 26*, 957–966.

Maddin, H. C., Sidor, C., & Reisz, R. R. (2008). Cranial anatomy of *Ennatosaurus tecton* (Synapsida: Caseidae) from the Middle Permian of Russia and the evolutionary relationships of Caseidae. *Journal of Vertebrate Paleontology, 28*, 160–180.

Mazierski, D. M., & Reisz, R. R. (2010). Description of a new specimen of *Ianthasaurus hardestiorum* (Eupelycosauria: Edaphosauridae) and a re-evaluation of edaphosaurid phylogeny. *Canadian Journal of Earth Sciences, 47*, 901–912.

Modesto, S. P. (1995). The skull of the herbivorous synapsid *Edaphosaurus boanerges* from the Lower Permian of Texas. *Palaeontology, 38*, 213–239.

Modesto, S. P., & Reisz, R. R. (1992). Restudy of Permo—Carboniferous synapsid *Edaphosaurus novomexicanus* Williston and Case, the oldest known herbivorous amniote. *Canadian Journal of Earth Sciences, 29*, 2653–2662.

Modesto, S., Smith, R., Campione, N., & Reisz, R. R. (2011). The last "pelycosaur": A varanopid synapsid from the *Pristerognathus* Assemblage Zone, Middle Permian of South Africa. *Naturwissenschaften, 98*, 1027–1034.

Olson, E. C. (1962). Late Permian terrestrial vertebrates, U.S.A. and U.S.S.R.. *Transactions of the American Philosophical Society, New Series, 52*, 1–224.

Olson, E. C. (1968). The family Caseidae. *Fieldiana: Geology, 17*, 225–349.

Pelletier, V. (2013). Postcranial description and reconstruction of the varanodontine varanopid *Aerosaurus wellesi* (Synapsida: Eupelycosauria). In C. F. Kammerer, K. D. Angielczyk, & J. Fröbisch (Eds.), *Early evolutionary history of the Synapsida* (pp. 53–68). Dordrecht: Springer.

Reisz, R. R. (1972). Pelycosaurian reptiles from the Middle Pennsylvanian of North America. *Bulletin of the Museum of Comparative Zoology, Harvard, 144*, 27–62.

Reisz, R. R. (1986). Pelycosauria. In P. Wellnhofer (Ed.), *Handbuch der Paläoherpetologie* (Vol. 17A). Stuttgart: Gustav Fischer Verlag.

Reisz, R. R., & Berman, D. S (1986). *Ianthasaurus hardestii*, a new edaphosaurid mammal-like reptile (Synapsida, Pelycosauria) from the Late Pennsylvanian Rock Lake Shale near Garnett, Kansas. *Canadian Journal of Earth Sciences, 23*, 77–91.

Reisz, R. R., & Berman, D. S (2001). The skull of *Mesenosaurus romeri*, a small varanopseid (Synapsida: Eupelycosauria) from the Upper Permian of the Mezen river basin, northern Russia. *Annals of Carnegie Museum, 70*, 113–132.

Reisz, R. R., & Laurin, M. (2004). A re-evaluation of the enigmatic Permian synapsid *Watongia* and of its stratigraphic significance. *Canadian Journal of Earth Sciences, 41*, 377–386.

Reisz, R. R., & Modesto, S. P. (2007). *Heleosaurus scholtzi* from the Permian of South Africa: A varanopid synapsid, not a diapsid reptile. *Journal of Vertebrate Paleontology, 27*, 734–739.

Reisz, R. R., Godfrey, S., & Scott, D. (2009). *Eothyris* and *Oedaleops:* Do these early Permian synapsids form a clade? *Journal of Vertebrate Paleontology, 29*, 39–47.

Reisz, R. R., Laurin, M., & Marjanovic, D. (2010). *Apsisaurus witteri* from the lower Permian of Texas: Yet another small varanopid synapsid, not a diapsid. *Journal of Vertebrate Paleontology, 30*, 1628–1631.

Romer, A. S., & Price, L. I. (1940). Review of the Pelycosauria. *Geological Society of America Special Papers, 28*, 1–538.

Sues, H.-D., & Reisz, R. R. (1998). Origins and early evolution of herbivory in tetrapods. *Trends in Ecology and Ecology, 13*, 141–145.

Sumida, S. S., Pelletier, V., & Berman, D.S (2013). New information on the basal pelycosaurian-grade synapsid *Oedaleops*. In C. F. Kammerer, K. D. Angielczyk, & J. Fröbisch (Eds.), *Early evolutionary history of the Synapsida* (pp. 7–23). Dordrecht: Springer.

Williston, S. W. (1912). Primitive reptiles. *Journal of Morphology, 23*, 637–666.

Chapter 2
New Information on the Basal Pelycosaurian-grade Synapsid *Oedaleops*

Stuart S. Sumida, Valerie Pelletier, and David S Berman

Abstract The Early Permian amniote *Oedaleops* is generally considered to be one of the basalmost pelycosaurian-grade synapsids. Thus it occupies a key position for understanding the phylogenetic relationships of basal synapsids specifically and basal amniote interrelationships more generally. This assessment has until now been based almost exclusively on the remains of a single skull from the Lower Permian Cutler Formation of north-central New Mexico. The identification of additional cranial as well as numerous postcranial elements of at least three additional individuals now permits a more complete understanding of its anatomy and allows the first attempt at a partial body reconstruction of this basal pelycosaurian-grade synapsid. *Oedaleops* is confirmed as an extremely basal synapsid taxon, but the addition of postcranial data from *Oedaleops* to data matrices of earlier phylogenetic analyses unexpectedly weakens, as opposed to strengthens, support for the hypotheses of a monophyletic Eothyrididae.

Keywords Eothyrididae • Caseasauria • Cutler Formation • Lower Permian • Arroyo del Agua • New Mexico

S. S. Sumida (✉) · V. Pelletier
Department of Biology, California State University San Bernardino, 5500 University Parkway, San Bernardino, CA 92407, USA
e-mail: ssumida@csusb.edu

V. Pelletier
e-mail: uraprimate@msn.com

D. S Berman
Section of Vertebrate Paleontology, Carnegie Museum of Natural History, 4400 Forbes Avenue, Pittsburgh, PA 15213, USA
e-mail: bermand@carnegiemnh.org

Introduction

The advent of cladistic analysis has brought greater clarity to our understanding of phylogenetic relationships while simultaneously demonstrating the difficulty of describing and defining primitive, basal members of any natural group. The basalmost members of Synapsida are no exception. Commonly and historically referred to as Pelycosauria, it is clear that they represent a primitive grade of evolution within Synapsida. Whereas the term basal, pelycosaurian-grade synpasid is an accurate characterization of those taxa, they are alternatively referred to hereafter simply as "pelycosaurs" for the sake of simplicity.

In a series of phylogenetic analyses that have stretched over the past 30 years (Reisz 1980, 1986; Berman et al. 1995; Reisz et al. 2009), *Oedaleops* and *Eothyris*, comprising the Eothyrididae, have consistently been shown to be basal and generalized members of the Caseasauria, sister taxon to the Eupelycosauria. Significantly, eothyridids offer a much more generalized model of basal "pelycosaurian" characteristics than do caseids. The latter, while members of the basal Caseasauria, possess a variety of autapomorphic and highly derived features including extremes in body size and head:body proportions. Hopson (1991) stressed the importance and utility of using basal, as opposed to highly derived, members of clades as representatives in phylogenetic analyses. Unfortunately, until now, *Oedaleops* and *Eothyris* have been of little use for offering basal or outgroup data for postcranial characters, leaving much larger and more extremely shaped taxa such as caseids or non-synapsid diadectomorphs as the only alternatives. Thus any new insight on the morphology of *Oedaleops*, particularly for postcranial characters, sheds welcome light on the earliest stages of synapsid evolution and biology.

Oedaleops campi was described initially on the basis of a single skull and a few tentatively associated postcranial elements by Langston (1965). Subsequent restudy of the Eothyrididae (Reisz et al. 2009) focused primarily on the

Christian F. Kammerer, Kenneth D. Angielczyk and Jörg Fröbisch (eds.), *Early Evolutionary History of the Synapsida*, Vertebrate Paleobiology and Paleoanthropology, DOI: 10.1007/978-94-007-6841-3_2,

same skull and continued the practice of excluding the postcranial material originally reported by Langston from any analysis. The holotypic skull, University of California Museum of Paleontology (UCMP) specimen 35758 was recovered in 1928 from the University of California's "Camp Quarry" of Arroyo del Agua in Rio Arriba County, north-central New Mexico. Langston (1953, 1965) placed the Camp Quarry in the Early Permian Cutler Formation, which is Wolfcampian in age. An earliest Wolfcampian designation was subsequently suggested by Berman et al. (1987) and Eberth and Miall (1991). More recently, Lucas et al. (2005) re-elevated the Cutler to Group status in central New Mexico, including within it the older El Cobre Canyon Formation spanning the Pennsylvanian-Permian boundary and the overlying Arroyo del Agua Formation. Placing the Camp Quarry in the upper part of the El Cobre Canyon Formation Lucas et al. (2005) reiterated an early Wolfcampian interpretation for the Camp Quarry.

Independent study of postcranial material of the varanopid "pelycosaur" *Aerosaurus* by Pelletier (2013), which is also known from the Camp Quarry, indicated that in fact three different pelycosaurian-grade taxa were present in the blocks recovered from there (Sumida et al. 2009). A series of five large blocks (UCMP 40093, 40094, 40095, 40096 and 40097) with scattered remains of at least three different synapsids were examined, resulting in the identification of *Aerosaurus*, a large sphenacodontid eupelycosaur, and numerous scattered elements of a much smaller primitive amniote represented by a small amount of isolated cranial material as well as numerous additional postcranial elements. Modesto and Reisz (1992) reported on an edaphosaurid eupelycosaur, and Berman (1993) also reported *Edaphosaurus* and the ophiacodont eupelycosaur *Ophiacodon* from the Camp Quarry, but no materials attributable to those taxa are evident on the blocks examined in this study. The isolated cranial elements recovered from UCMP blocks 40093 to 40097 include dentigerous elements and dermal roofing elements that may be assigned to *Oedaleops* with confidence. (*Oedaleops* is a monospecific genus, and the genus and species names are used interchangeably here.) Individual elements prepared away from the main blocks and given separate specimen numbers include a number of appendicular elements and a partial skull UCMP 69679 (Fig. 2.1). The skull may be only tentatively assigned to *Oedaleops*, but following Reisz et al. (2009) it is considered here to be of the appropriate size for the collection of synapsid taxa found in the Camp Quarry blocks.

Although other vertebrate taxa are known from the Camp Quarry (Langston 1953), *Oedaleops* is the only other tetrapod from the Camp Quarry that matches these materials in size and it was deemed unlikely that a new fourth taxon so similar to *Oedaleops* in size might have been present. The elements found appear uniformly well ossified and complete in their development, so it is highly unlikely that these smaller materials might represent immature individuals of *Aerosaurus* or the sphenacodontid found at the Camp Quarry. Further, the remains of at least one partially associated skeleton on UCMP block 40094 includes both pectoral and pelvic girdles, and scattered limb elements that are clearly associated with UCMP 67225, a right dentary of *Oedaleops*. These materials are also associated with a left maxilla tentatively assigned to *Oedaleops* by Langston (1965).

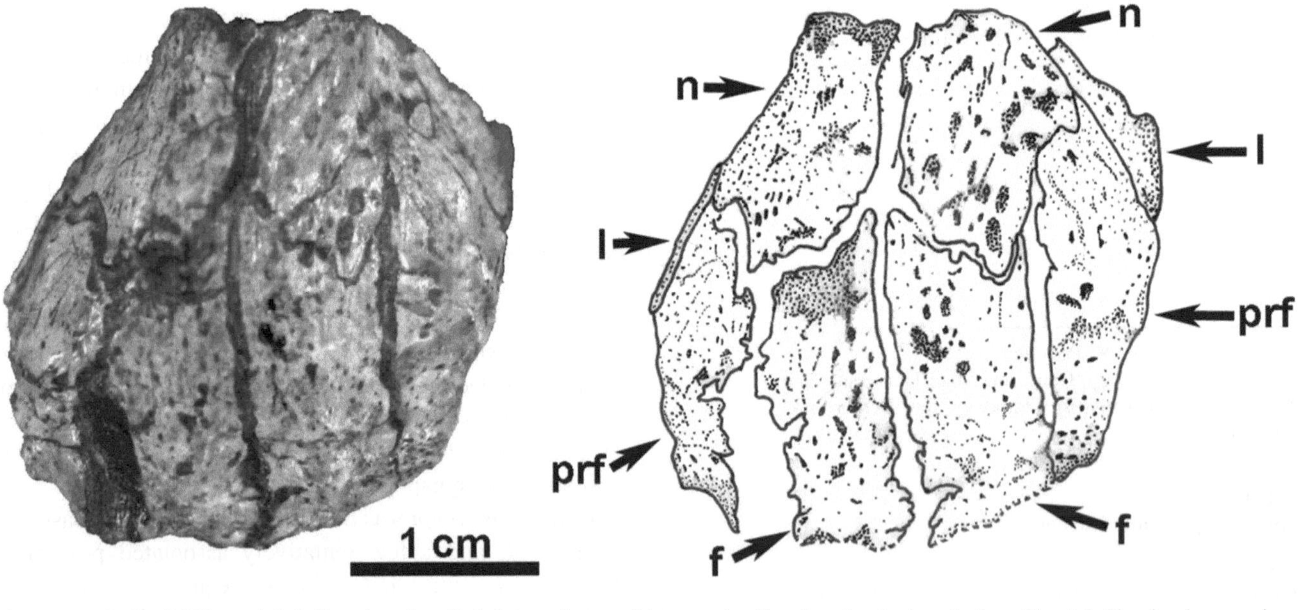

Fig. 2.1 UCMP 69679, partial skull assigned to *Oedaleops*, photo and interpretive line drawing in dorsal view. *f* frontal, *l* lacrimal, *n* nasal, *prf* prefrontal

Although the skull of *Oedaleops* remains best represented by the holotype, UCMP 35758, the isolated materials identified here do add information to our understanding of the genus. Wherever possible, both photographic and reconstructed illustrations are provided for newly described materials. Additionally, reconstructions are included in comparative illustrations to facilitate anatomical and phylogenetic comparison with other synapsids, basal amniotes, or their near relatives.

New Material Assignable to *Oedaleops*

Neither Langston (1965) nor Reisz et al. (2009) described the premaxilla in *Oedaleops* specifically, other than indirectly in the former's reference to the shape of the narial opening and the procumbent nature of the rostrum. Two isolated premaxillae are preserved on UCMP block 40096. The better-preserved right side confirms a count of three teeth, all of which are approximately equivalent in size. There is no evidence of an external narial shelf. Langston (1965) suggested that the ventral naris may have been floored by a broad shelf comprised of inward reflections of the premaxilla and maxilla. The isolated premaxilla does show some development of a dorsally directed surface internal to the lateral exposure of the element. Langston's description and reconstruction may have been what prompted Reisz et al. (2009) to code primitive conditions for the septomaxilla in *Oedaleops* despite that none are preserved. The structure of the element here neither confirms nor contradicts their speculation that the septomaxilla was a curled structure in the external naris with a lateral sheet-like exposure.

A possible left nasal is exposed in internal view on UCMP block 40096 (Fig. 2.2a) and the caudal portions of both nasals are present in UCMP 69679 (Fig. 2.1). The isolated nasal is about 20 % larger than those of the holotypic skull. Its interpretation as a nasal is prompted by facets for articulation with the premaxilla rostromedially, a smooth margin interpreted as the margin of the left naris, and a posterolateral notch interpreted as the articulation with the prefrontal. UCMP 69679 confirms the presence of distinct caudal and posterolateral processes of the nasal as separated by the rostral extension of the prefrontal. This distinction is even greater in UCMP 69679 than it is in the isolated UCMP 40096 element.

As with the nasals, UCMP 69679 preserves partial paired frontals. An isolated left frontal on UCMP block 40094 (Fig. 2.2b) is almost exactly rectangular in outline, with little evidence of the lateral lappet reported to enter the orbital margin by Langston (1965) and Reisz et al. (2009). The isolated nature of the better preserved left frontal might suggest that this small lappet could have been broken off when the skull became disarticulated, or it might alternatively suggest that there was some degree of variation in this character in *Oedaleops*. Another rectangular element from UCMP block 40097 may be an additional frontal. That element appears to show what might be an extremely short smooth edge along its lateral surface. Notably, Reisz et al. (2009) point out that the lateral frontal lappet, while present in the closely related *Eothyris*, is miniscule in its degree of entry into the orbit. On the other hand, the posterolateral process of the frontal, while short, is quite distinct and of proportions very similar to that of the holotype. Rostrally, the isolated nature of the frontal reveals that it underlaid the posterior exposure of the nasal in a well-developed lap joint. The paired frontals of UCMP 69679 do not preserve the region of a possible lateral lappet. However, their articulation with the nasal is virtually identical to the outline of the exposed dorsal margin of the nasal articulation in UCMP 40097. Both Langston (1965) and Reisz et al. (2009) noted that the dermal skull roof in the holotype was sculptured and the isolated and paired frontals confirm this clearly. However, it again suggests some degree of variation; whereas previous reports of the sculpture describe it as an essentially ridge-and-furrow pattern, the isolated element possesses a more radiate pattern of sculpture.

The prefrontals in UCMP 69679 are generally triangular as in the descriptions of Langston (1965) and Reisz et al. (2009). As mentioned above, this specimen demonstrates clearly the interpolation of this element between the caudal and posterolateral processes of the nasal (Fig. 2.1).

Both right and left lacrimals have been slightly crushed and pushed mediolaterally in UCMP 69679, however, their caudal margins are preserved well enough to demonstrate two distinct foramina for the lacrimal canal on the smooth orbital margin of the element.

UCMP block 40096 includes an isolated postorbital (Fig. 2.2c) reasonably assignable to *Oedaleops*. As in the holotypic skull, it is a subcrescentic element. The ventral process is even more elongate than represented in Langston's (1965) reconstruction of the holotype or the specimen drawing of Reisz et al. (2009). However, its ventral tip is unsculptured so this exaggerated length may be due to its having underlain the dorsal process of the jugal. More dorsally, the portion that bounds the posterodorsal corner of the orbit is not as thickened as in the holotype, but this could be due to aggressive preparation. As in the holotype the orbital rim is paralleled by a series of pits. Sculpture is evident though not as marked on the posterior process.

An isolated element measuring approximately 2 cm in transverse width that may be identified as a single left pterygoid is visible in dorsal view on UCMP 40095 (Fig. 2.2d) and is tentatively assigned to *Oedaleops* here.

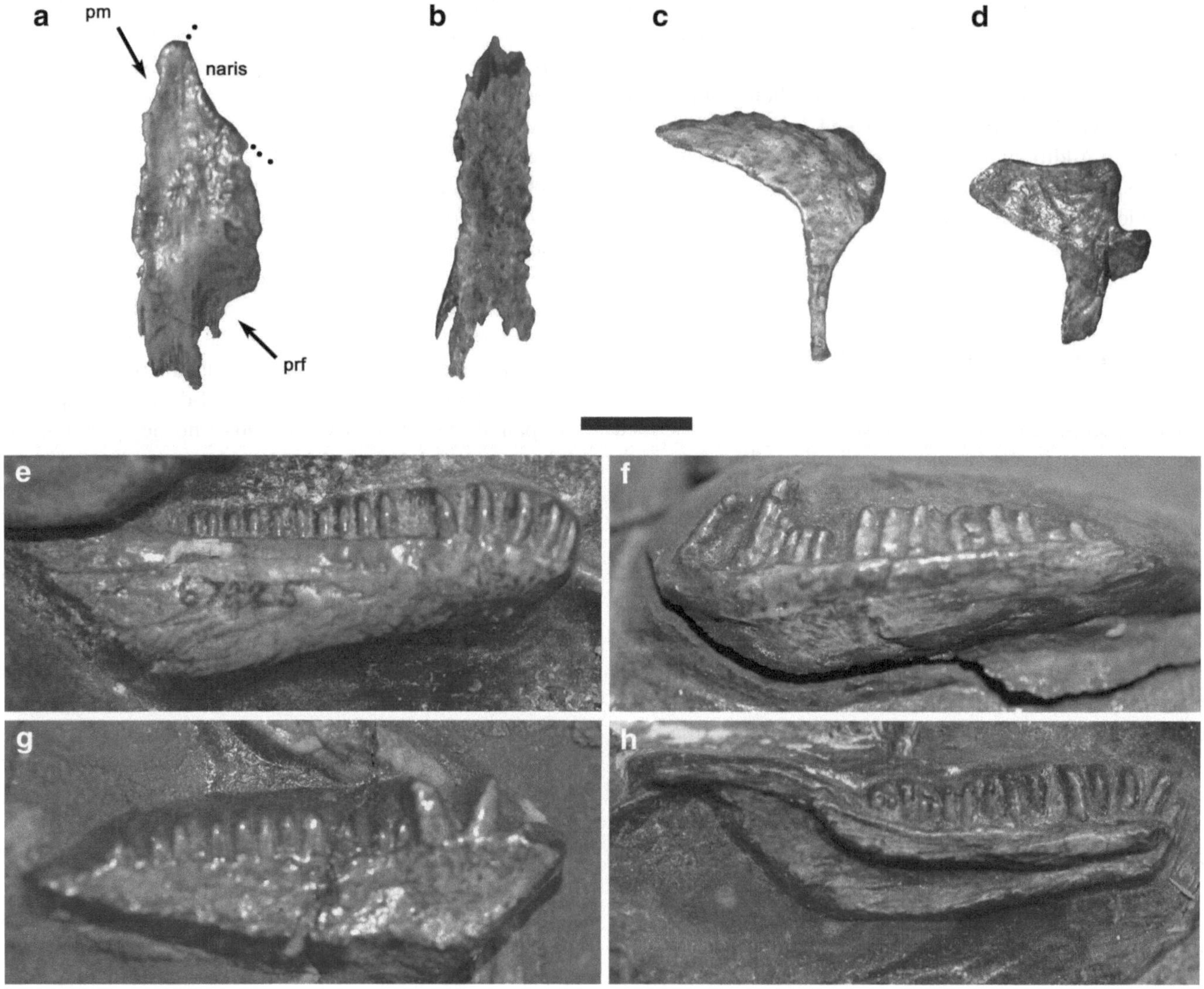

Fig. 2.2 Isolated cranial elements assigned to *Oedaleops*. **a** Isolated left nasal part of UCMP 40096, internal view; rostral is to *top* of page; **b** isolated right frontal part of UCMP 40094, dorsal view; rostral is to *top* of page; **c** right postorbital part of UCMP 40096; **d** left pterygoid part of UCMP 40095, dorsal view; **e** UCMP 67325, right dentary in lateral view; **f–h** isolated elements from UCMP 40095: **f** left dentary in lateral view; **g** right dentary in lateral view; **h** left dentary in medial view. Scale bar equals 1 cm

The palatal process is broad, and the notched nature of the rostral end of the element indicates it may not be preserved completely. The basisphenoid articulation is shallow but distinct, marked by a surface that is an elongate, partial oval. Due to the disarticulated nature of the element, the exact direction of the quadrate process cannot be determined with confidence, though it appears to have been oriented nearly directly caudally.

Numerous isolated dentaries are present (Figs. 2.2e–h, 2.3) showing both lateral and medial views. In overall outline, the dentary is nearly triangular in shape, though the dental margin does dip mesially to accommodate the overhanging premaxilla of the rostrum. Its sculpture is a combination of both ridges and grooves, and mostly irregular pits. In at least one specimen (UCMP 67325; Fig. 2.2e) the pits are arranged as a single line of foramina that parallel the dental margin. Lingually, a distinct and well-developed ridge of bone parallels the dental margin and underlies the teeth (Fig. 2.2h). On one specimen (UCMP 40095), the surangular may be seen to continue this shelf in a gently sigmoidal fashion. In UCMP 67325 seventeen tooth positions are clearly visible with a single tooth missing at position six. Other specimens exhibit tooth numbers ranging from 15 to 17. Notably, two specimens (both part of UCMP block 40095; Fig. 2.2f, g) do demonstrate some degree of caniniform tooth development at what appear to be the third and fourth tooth positions. Maxillary and dentary dentitions of these elements are not sufficiently different from the descriptions of Langston (1965) and Reisz et al. (2009) to warrant further description.

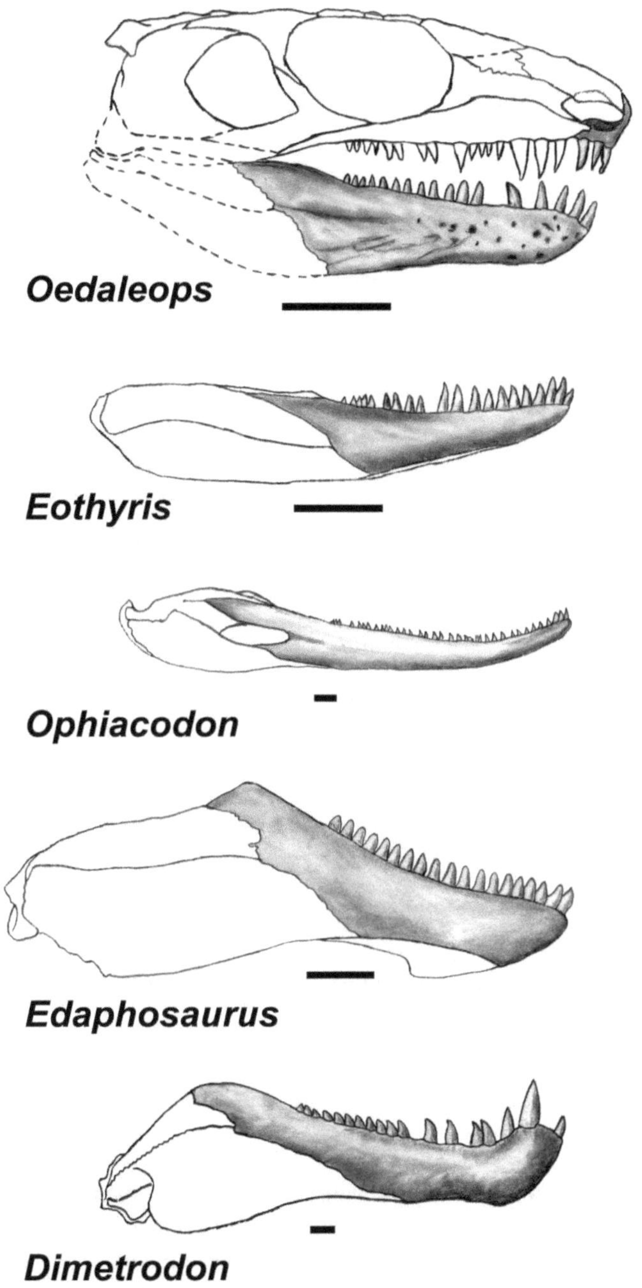

Fig. 2.3 Comparative reconstructions of lower jaws highlighting dentary in *Oedaleops* and the dentary of other pelycosaurian-grade synapsids in right lateral view. Note dermal sculpture in dentary in *Oedaleops*; skull reconstruction after Langston (1965) in part. *Eothyris* in part after Reisz et al. (2009). *Ophiacodon*, *Edaphosaurus*, and *Dimetrodon* in part after Romer and Price (1940), and *Edaphosaurus* in part after Modesto and Reisz (1992)

Numerous vertebrae appropriate to the size of cranial materials of *Oedaleops* are scattered about the five UCMP blocks examined, but no articulated strings are present, thus there is no way to know the exact number of presacral vertebrae. They match neither the vertebrae clearly associated with *Aerosaurus* (Pelletier 2013), nor are of the size appropriate to the much larger sphenacodontid from the Camp Quarry. Of other taxa known from north-central New Mexico, they lack the blade-like neural spines of ophiacodonts and are much too small to be those of an edaphosaurid. They could be argued to be similar to those of a protorothyridid (e.g., *Protorothyris* in Fig. 2.4), but it is deemed unlikely that so much postcranial material attributable to *Oedaleops* should be associated with strictly protorothyridid vertebrae and a coincident lack of any other protorothyridid material.

All of the vertebrae visible in the UCMP blocks examined are preserved lying on their lateral surfaces, so exact dimensions of the centra are difficult to discern. However, cranio-caudal length estimation is straightforward, with most centra measuring approximately 0.75–0.80 cm. The neural arches are smooth and slightly concave, with no evidence of the lateral expansion or "swelling" characteristic of many other basal amniote groups (Fig. 2.4) or the ophiacodont eupelycosaur *Varanosaurus* (Sumida 1989a). There are no excavations on the lateral surface of the neural arches as in varanopids and sphenacodonts (Romer and Price 1940; Maddin et al. 2006; Campione and Reisz 2010; Pelletier 2013). The neural spines are well developed but relatively short. Their smoothly finished distal tips are nearly quadrangular in shape, and are just slightly greater in measure than the maximum height of the centra. Numerous scattered ribs may be found throughout the five UCMP blocks examined. Their scattered and disarticulated nature does not allow detailed description. However, their proportions do not appear to suggest a broadly barrel-shaped trunk, and given the small size of *Oedaleops* more likely are indicative of a typically generalized insectivore. If a generalized presacral count of 24–27 vertebrae (Romer and Price 1940) is assumed then the presacral vertebral column would have measured approximately 19–20 cm.

No cleithrum has been identified with confidence. However, a clavicle, an interclavicle and at least three well-preserved scapulocoracoids (Fig. 2.5a–c) may be seen. A left clavicle (Fig. 2.5c) is well preserved in UCMP 40095 and another is present but in poorer condition on UCMP block 40094. In comparison to known "pelycosaurian" clavicles, it is most similar to the condition seen in ophiacodontids in its retention of a relatively narrow dorsal process and ventral plate. The dorsal process terminates in well-developed striations that parallel its longitudinal axis. Thus, although no cleithrum has been found for *Oedaleops*, it seems quite certain one was present. Immediately ventral to the grooved depression, the smooth lateral surface of the clavicle begins as a posteroventrally directed ridge. The ventral plate is somewhat distorted by postmortem flattening. Despite that, it remains clear that the ventral plate is not dramatically expanded in the manner of caseids,

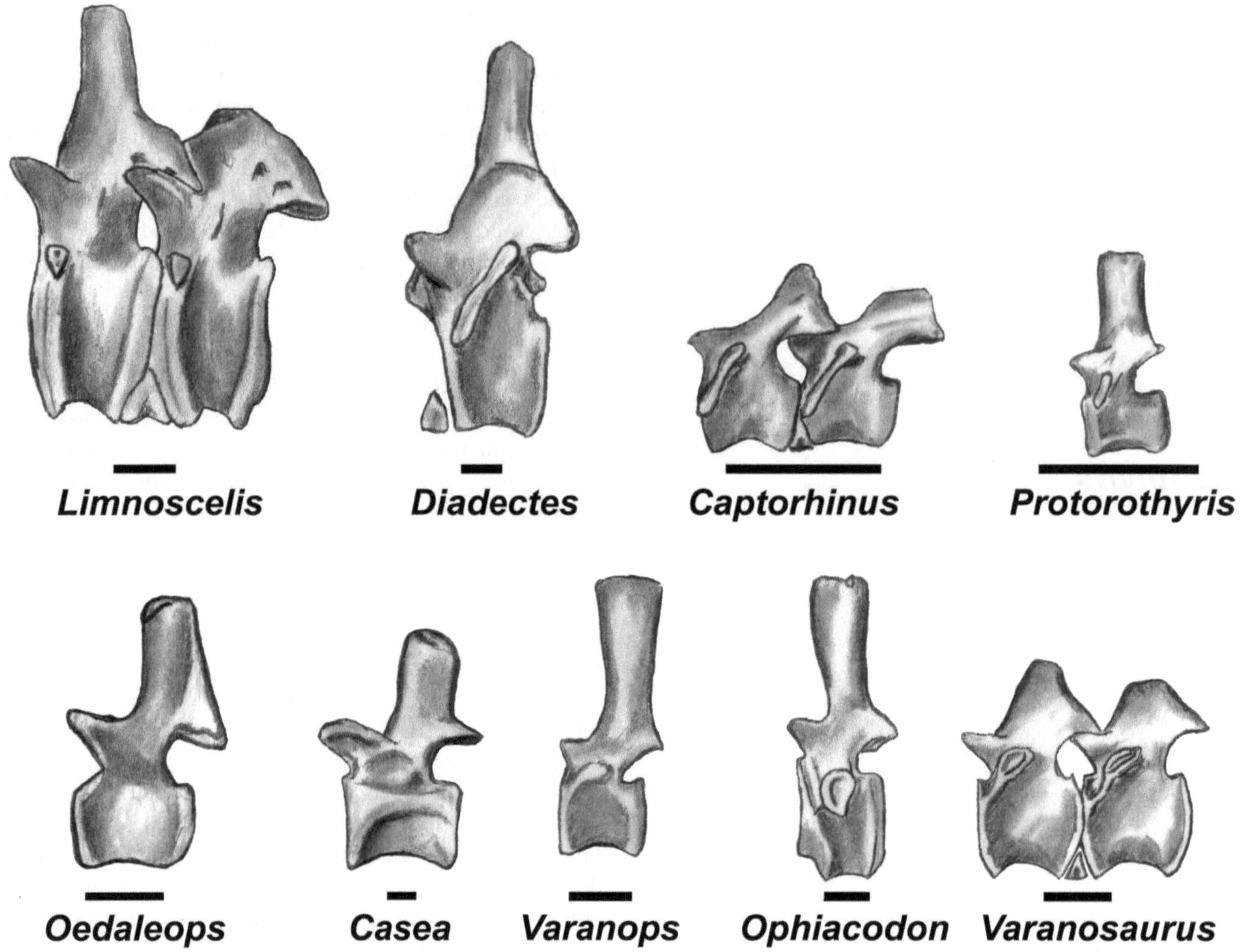

Fig. 2.4 Comparative reconstructions of dorsal vertebrae in left lateral view of the diadectomorphs *Limnoscelis* and *Diadectes*, the eureptiles *Protorothyris* and *Captorhinus*, and pelycosaurian-grade synapsids including *Oedaleops*. Reconstructions of *Limnoscelis*, *Captorhinus*, and *Varanosaurus* shown as pairs to highlight variability in neural spine construction. *Varanosaurus* after Sumida (1989a); all other except *Oedaleops* after Sumida (1997). Scale bars equal 1 cm

varanopids, sphenacodontids, or edaphosaurs (Romer and Price 1940; Pelletier 2013) (Fig. 2.6).

The interclavicle (Fig. 2.5a) is a "T-shaped" element with no significant cranially directed process evident, though this may be due to postmortem damage. Lateral processes of the interclavicle come to a blunted pointed terminus while the midline shaft of the element is approximately 50 % longer than the cranio-caudal length of the coracoid region of the scapulocoracoid.

Though Langston focused his description of the scapulocoracoid on UCMP 40282 (Fig. 2.5b) three additional right scapulocoracoids (Fig. 2.5b; Table 2.1) help to fill out a description of the element. It is much more rounded and curved in outline than the more angular condition common to other "pelycosaurs" and basal amniotes (Fig. 2.6). This condition is confirmed by another specimen in UCMP block 40097. No supraglenoid foramen is visible, though the supraglenoid buttress is well developed. The glenoid fossa itself is "screw-shaped" in the manner typical of primitive tetrapods. Immediately caudal to the glenoid a triceps tubercle is very well developed. Langston (1965) reported and illustrated a well-developed coracoid foramen in UCMP 40282, but re-examination of the specimen shows no evidence of it.

At least two humeri can be confidently attributed to *Oedaleops* from the materials examined. Langston (1965) focused on UCMP 40283, but another from UCMP block 40097 (Fig. 2.5d) adds significantly to our understanding. Notably, the deltopectoral crest in UCMP 40097 is extremely robust. Whereas it is not as strongly developed as in limnoscelid and diadectid diadectomorphs, the deltopectoral crest is as well developed, if not more-so, than in other "pelycosaurs" (Fig. 2.7). The humeral shaft is distinct, and although *Oedaleops* shows the robust development of the

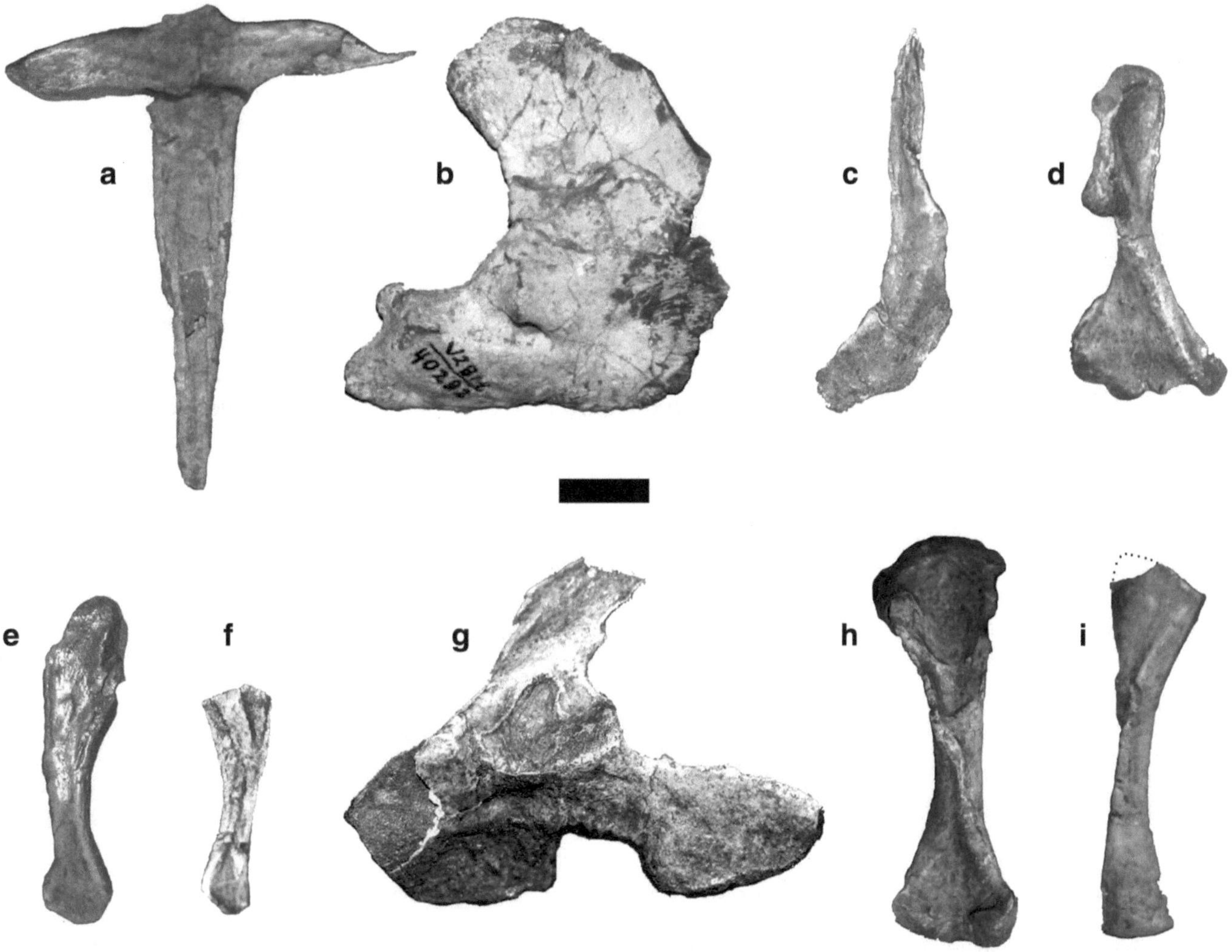

Fig. 2.5 Appendicular structures of *Oedaleops campi*. **a** UCMP 40094, interclavicle in ventral view; **b** UCMP 40282, right scapulocoracoid in lateral view; **c** UCMP 40095, right clavicle in lateral view; **d** UCMP 40094, right humerus; **e** UCMP 40096, left ulna in medial view; **f** UCMP 40095, right radius in medial view; **g** UCMP 40094, nearly complete left pelvic girdle in lateral view; **h** UCMP 40094, right femur in ventral view; **i** UCMP 40095, left tibia in ventral/medial view. All elements are isolated elements on larger UCMP blocks also containing *Aerosaurus* and/or sphenacodontid materials. Scale bar equals 1 cm

ectepicondyle and entepicondyle typical for "pelycosaurs", they do not take up as much of the overall length of the humerus as in caseids or other more derived "pelycosaur" families. The entepicondylar foramen is an elongate slit very close to the margin of the process. With the exception of the construction of the deltopectoral crest, in overall shape, the humerus in *Oedaleops* is more similar to that in basal eureptiles than to caseids or basal eupelycosaurian families.

Langston (1965) mentioned the ulna only briefly and did not illustrate it. Despite its small size, the element preserved in UCMP block 40096 (Figs. 2.5e, 2.8) has a well-developed olecranon process that projects well above the trochlear notch. Immediately distal to the olecranon head is a small but distinct process. Holmes (1977) suggested that this was, at least in part, a point of attachment for the anconeus muscle. A distinct ridge runs along the dorsolateral surface of the ulna in a manner more similar to that of sphenacodontids than more basal eupelycosaurs or caseids (Fig. 2.8). A radius has also been recovered from UCMP 40096 (Fig. 2.5f). It is of approximately the length expected were it paired with the ulna described above. It is exposed in medial view and possesses a strongly sigmoid ridge in the manner characteristic of most "pelycosaurs" (Romer and Price 1940).

Both pelves are preserved in block 40094. The right is visible only partially in medial view, presenting a smooth, unremarkable surface. However the left pelvis (Fig. 2.5g, UCMP 40094; Fig. 2.9) is well preserved. It is virtually complete with the exception of the dorsalmost tip of the ilium and a portion of the ischium at its ventral border with the pubis. The leading edge of the pubis is also missing, but

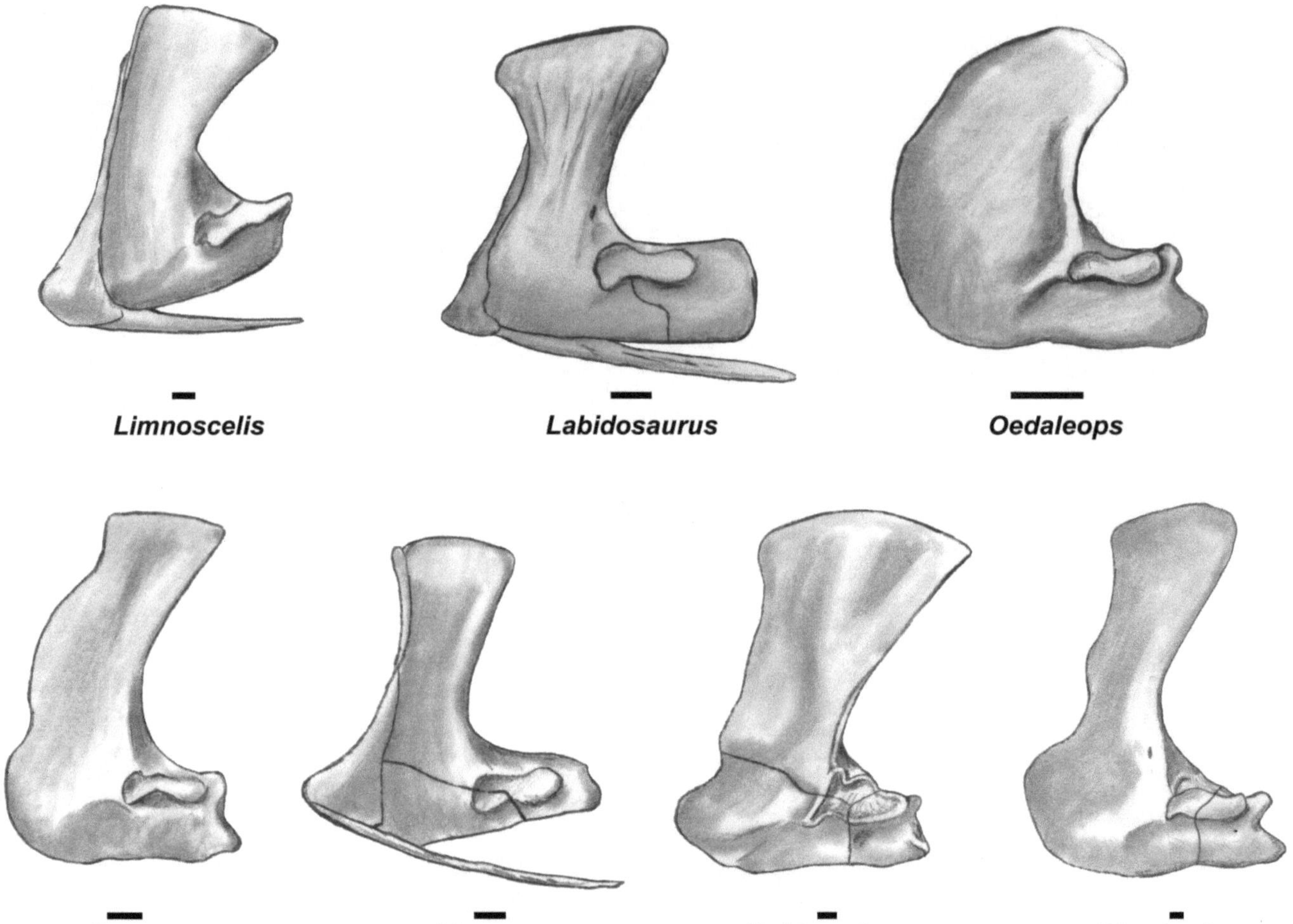

Fig. 2.6 Comparative reconstructions in lateral view of left pectoral girdles of the diadectomorph *Limnoscelis*, the eureptile *Labidosaurus*, and pelycosaurian-grade synapsids including *Oedaleops*. *Labidosaurus* after Sumida (1989b); all others except *Oedaleops* after Sumida (1997). Although four different right scapulocoracoids are assigned to *Oedaleops*, it is reconstructed in left lateral view here to facilitate comparison. Note relatively more rounded and curved outline of scapulocoracoid in *Oedaleops* when compared to those of other "pelycosaurs" and basal amniotes. Scale bars equal 1 cm

its outline is well preserved as an impression. Sutures between the girdle's component parts are not visible with the possible exception of the ventralmost suture between the ischium and pubis. Despite its missing tip, it is clear that the ilium tapered distally as in ophiacodonts and varanopids (Romer and Price 1940; Maddin et al. 2006; Pelletier 2013; Berman et al. 2013) and did not expand broadly distally as in caseids, sphenacodontids, and edaphosaurids (Romer and Price 1940; Sumida 1997). The pubis is not as acuminate as in other pelycosaurian-grade genera; rather it is more rounded in a manner similar to that of closely related outgroups such as diadectomorphs and basal eureptiles (Fig. 2.9). However, as in many "pelycosaurs", the ischium is essentially a broad plate. A gently arching ridge spans nearly the entire length of the puboischiadic plate, presumably marking the margin of attachment of the puboischiofemoralis externus muscle. As in most other "pelycosaurs" and some diadectomorphs (Fig. 2.9) the acetabulum is not a simple oval in outline. A strongly developed posterodorsal embayment together with a very shallow embayment cranially, results in a triradiate acetabular outline. The medial surface of the girdle is not visible in the specimen as preserved.

Langston (1965) illustrated a right femur, UCMP 40284. That element is lightly built with an unusually elongate intertrochanteric fossa and therefore relatively short adductor crest. UCMP block 40094 preserves another right femur (Figs. 2.5h, 2.10) still embedded in matrix, exposing the ventral surface only, but providing considerably more detail than that illustrated by Langston (1965). The femur in *Oedaleops* has a distinctly cylindrical shaft offsetting expanded proximal and distal heads. The proximal articular surface is gently curved, terminating in a robust internal trochanter. The intertrochanteric fossa is clearly developed,

Table 2.1 Cranial and appendicular elements assigned to *Oedaleops* for determination of minimum number of individuals present at the Camp Quarry locality

Specimen number	Element	Right/left	Minimum number individuals by element
UCMP 40096	Nasal	Right	3
UCMP 69679	Nasal	Right	
UCMP 69679	Nasal	Left	
UCMP 35758	Nasal	Right	
UCMP 35758	Nasal	Left	
UCMP 40094	Frontal	Right	3
UCMP 69679	Frontal	Right	
UCMP 69679	Frontal	Left	
UCMP 35758	Frontal	Right	
UCMP 35758	Frontal	Left	
UCMP 69679	Prefrontal	Right	2
UCMP 69679	Prefrontal	Left	
UCMP 35758	Prefrontal	Right	
UCMP 35758	Prefrontal	Left	
UCMP 40096	Postorbital	Right	2
UCMP 35758	Postorbital	Right	
UCMP 35758	Postorbital	Left	
UCMP 40095	Pterygoid	Left	1
UCMP 40094	Jugal	Left	2
UCMP 35758	Jugal	Right	
UCMP 35758	Jugal	Left	
UCMP 67222	Maxilla	Right	2
UCMP 40094	Maxilla	Left	
UCMP 35758	Maxilla	Right	
UCMP 67222	Dentary	Right	2
UCMP 40095	Dentary	Left	
UCMP 40095	Dentary	Right	
UCMP 40095	Dentary	Left	
UCMP 40094	Interclavicle	–	1
UCMP 40095	Clavicle	Right	1
UCMP 40282	Scapulocoracoid	Right	4
UCMP 67248	Scapulocoracoid	Right	
UCMP 40094	Scapulocoracoid	Right	
UCMP 40097	Scapulocoracoid	Right	
UCMP 40094	Humerus	Right	2
UCMP 40283	Humerus	Right	
UCMP 40096	Ulna	Right	1
UCMP 40094	Pubis	Right	1
UCMP 40094	Pubis	Left	
UCMP 40094	Femur	Right	2
UCMP 40284	Femur	Right	
UCMP 40095	Tibia	Left	1

Data from type specimen UCMP 35758 (Langston 1965) that overlap with elements described here are included for completeness

slightly squared off, and not as elongate as described by Langston. It delimits clearly the insertion of the puboischiofemoralis externus muscle. On its ventral surface, the shaft is characterized by a distinctly sigmoidal adductor ridge, which continues all the way to the distal fibular condyle. Approximately one-third to one-half the distance

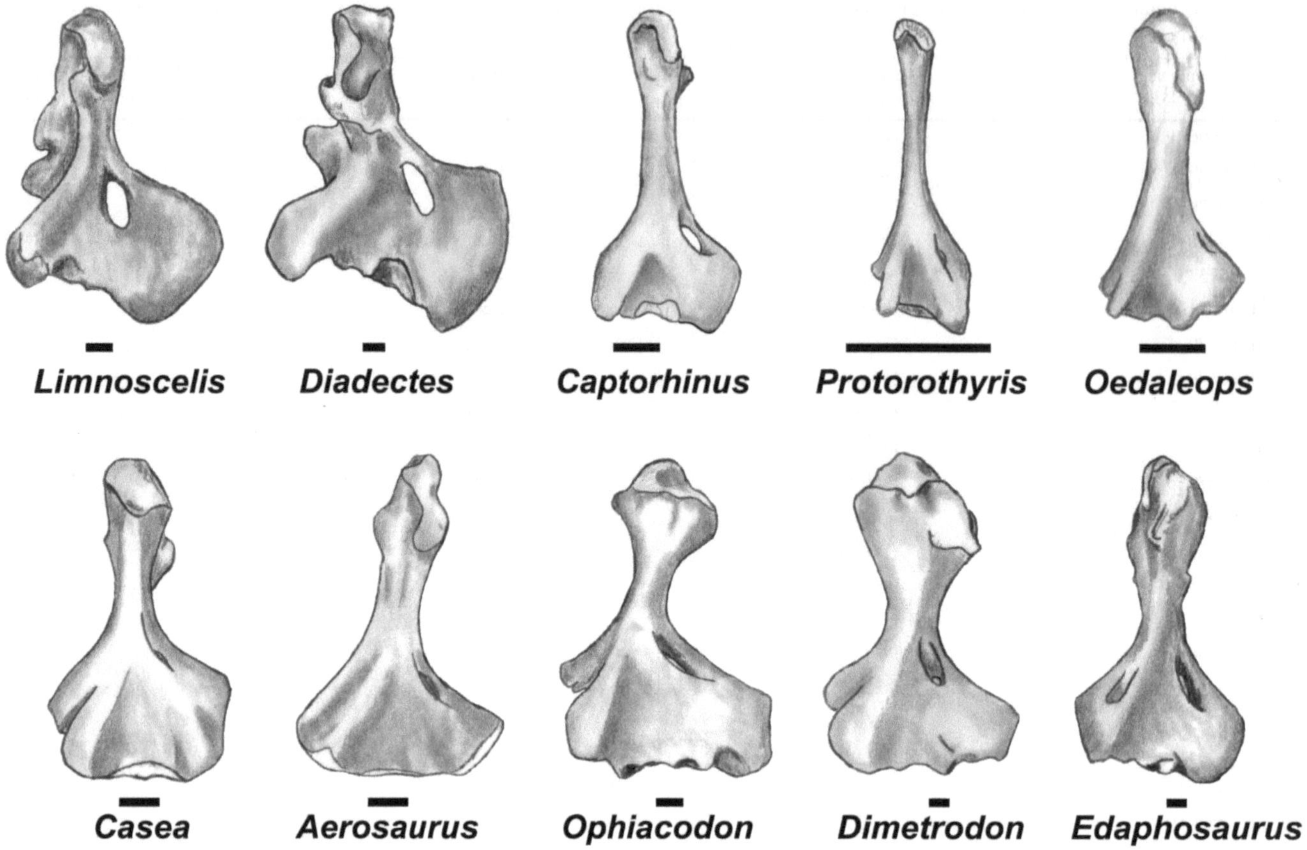

Fig. 2.7 Comparative reconstructions of the left humerus in distal ventral aspect of the diadectomorphs *Limnoscelis* and *Diadectes*, the eureptiles *Captorhinus* and *Protorothyris*, and pelycosaurian-grade synapsids including *Oedaleops*. *Ophiacodon*, *Dimetrodon*, and *Edaphosaurus* after Romer and Price (1940); all others except *Oedaleops* after Sumida (1997). Note the well-developed deltopectoral crest in *Oedaleops*. Scale bars equal 1 cm

along the femoral shaft and at the most pronounced curvature within the ridge, a fourth trochanter is moderately well developed for attachment of the caudifemoralis muscle. The adductor ridge terminates at the distal fibular condyle which is just slightly more rounded and heavily developed than the tibial condyle.

A number of elements are present that are likely tibiae assignable to *Oedaleops*, the best of which may be seen on UCMP block 40095 (Figs. 2.5i, 2.11). Like those of other "pelycosaurs", the tibia is slightly curved and has a moderately well developed ridge running the length of its ventral surface. However it is relatively more slender than those of most other "pelycosaurs". As preserved, its articular surfaces do not show significant complexity, but there is no way to know if this is due to the degree of maturity or a genuine character difference.

A small number of phalangeal elements may be found scattered about the five UCMP blocks examined in this study. Most are disarticulated, and it is not possible to determine if they are manual or pedal elements. Thus, no clear indication of manual or pedal shape or counts may be offered. However, there are a number that are claw-shaped and can be identified clearly as unguals. This includes one short articulated string of three phalangeal elements on UCMP block 40096 (Fig. 2.12), as well as a number of scattered, isolated distal claws on UCMP blocks 40096 and 40097. While otherwise unremarkable, the clearly claw-like condition of the distal elements is consistent with the suggestion of Maddin and Reisz (2007) that true claws are characteristic of Amniota, here confirmed by *Oedaleops* as one of its basalmost members.

Reconstruction

A partial skeletal reconstruction of *Oedaleops* is now possible (Fig. 2.14). A summary of all elements to the description of *Oedaleops* here (Table 2.1) indicates a minimum number of four individuals including the type skull. No proportional changes to the cranial reconstruction of Langston (1965) are proposed, but a revised interpretation of the surface texture of the premaxilla and dentary is now offered (Fig. 2.3). Vertebral elements presumed to belong to *Oedaleops* are disarticulated and scattered across the five

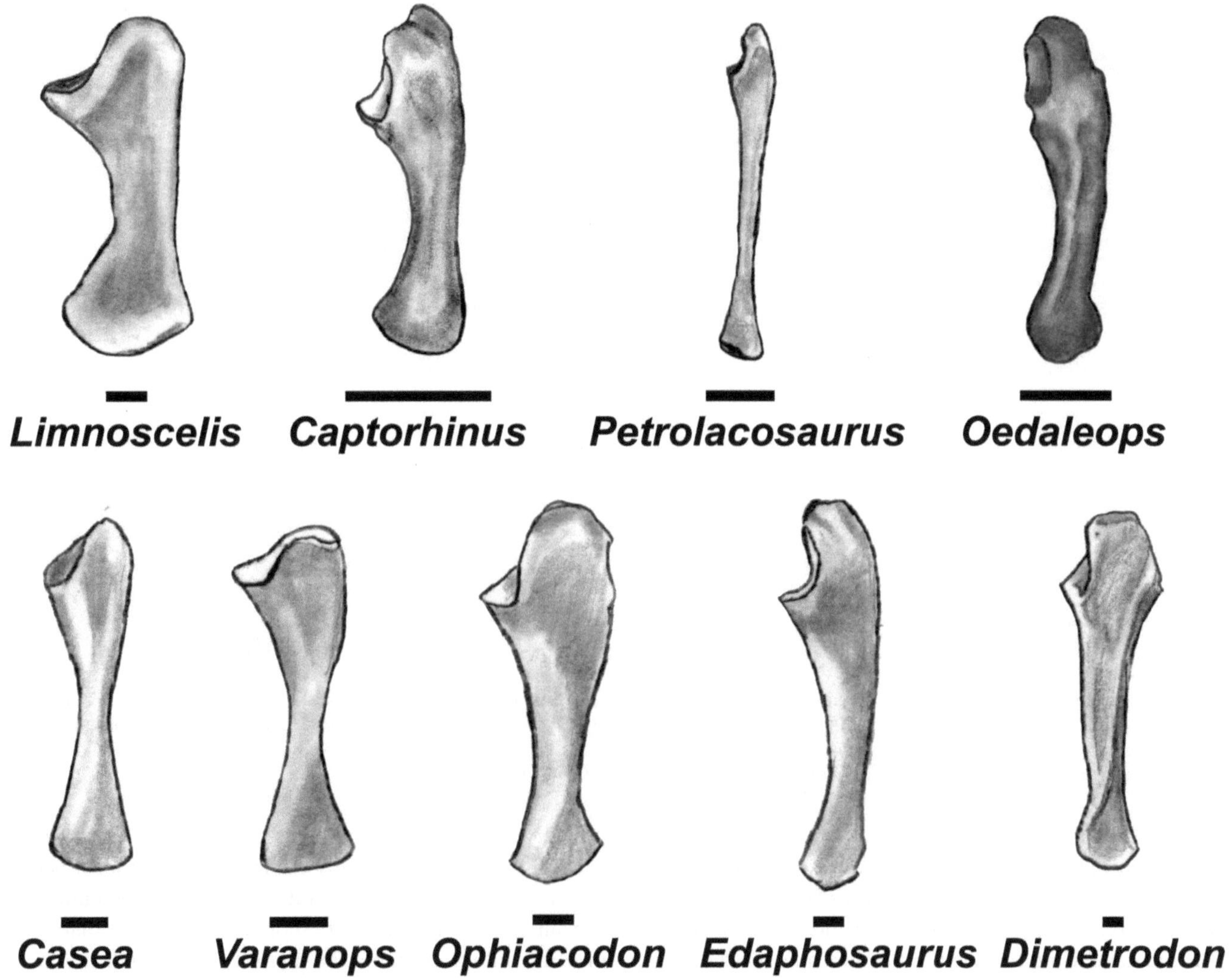

Fig. 2.8 Comparative reconstructions of the left ulna of the diadectomorph *Limnoscelis*, the eureptile *Captorhinus*, the basal araeoscelidian reptile *Petrolacosaurus*, and pelycosaurian-grade synapsids including *Oedaleops*. All nonsynapsid taxa after Sumida (1997). All synapsid taxa with the exception of *Oedaleops* after Romer and Price (1940). Scale bars equal 1 cm

different blocks that comprised this study. As mentioned above, the average cranial-caudal length of the vertebrae examined is approximately 0.75–0.80 cm. Romer and Price (1940) suggested a typical primitive vertebral count as 27 in what are now considered basal eupelycosaurian ophiacodonts. Romer and Price (1940) and Reisz (1986) suggested that caseids probably possessed only about 24 presacral vertebrae. Greater clarity will be brought to this question with the more complete description of a primitive caseid from the Lower Permian Bromacker locality of central Germany (Berman et al. 2009; Reisz et al. 2010). Until then, a conservative range for primitive synapsid presacral count of 25–27 vertebrae is assumed, then the snout-vent length of *Oedaleops* as reconstructed would be approximately 25 cm and an axial column length of approximately 19–20 cm. Although ribs are scattered amongst the five UCMP blocks examined, none are definitively associated with individual vertebrae; thus is it not possible to determine trunk shape or volume precisely. The vertebral column is reconstructed as sloping slightly ventrally from cranial to caudal. This interpretation, which follows the strategy for a number of pelycosaurian-grade taxa by Romer and Price (1940), is of course subject to alternative interpretation. Ribs are not included in the reconstruction because the scattered nature of the ribs does not allow association of any single rib with any single vertebral element with any confidence. However, until completely articulated specimens demonstrating definitive proportions of both girdles and fore- and hindlimbs of a single individual are discovered, we here take the conservative position of following the orientation offered in Romer and Price (1940). Limb orientation was presumed to be in a sprawling position as in other "pelycosaurs" and basal amniotes (Romer and Price 1940; Holmes 1977; Heaton and

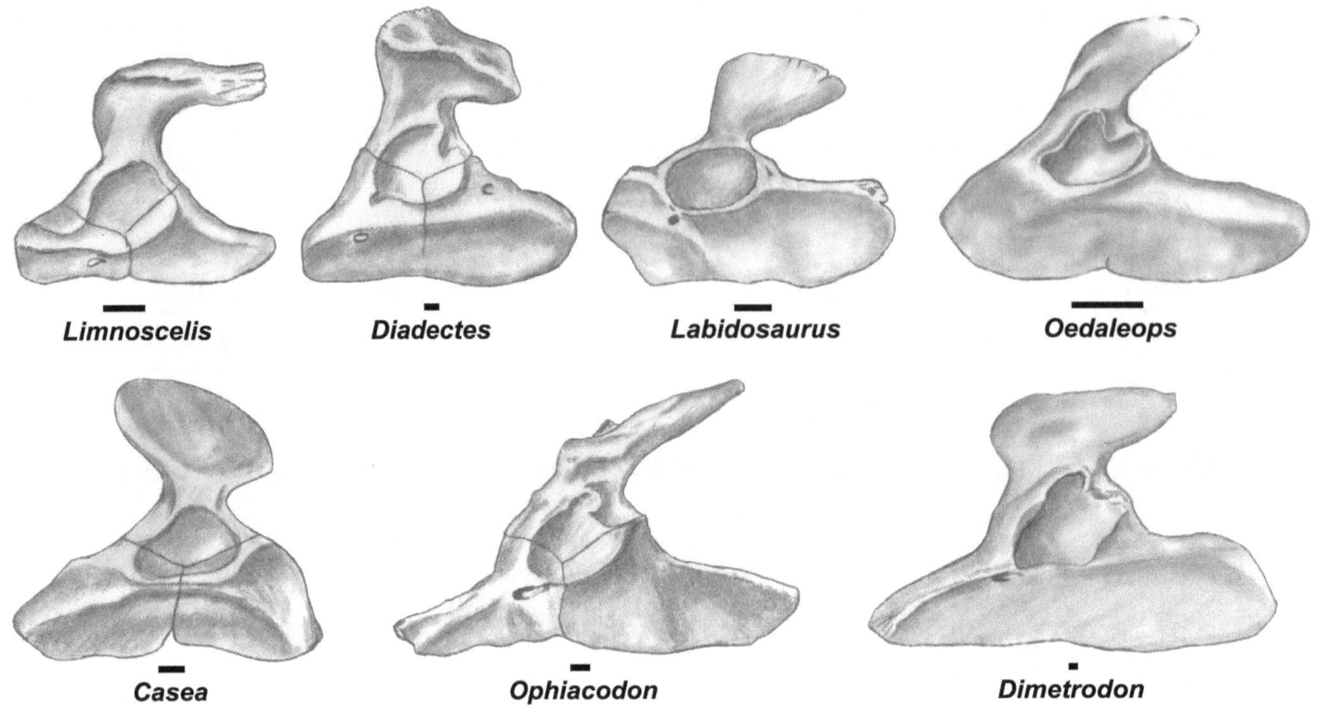

Fig. 2.9 Comparative reconstructions of pelvic girdles in left lateral aspect of the diadectomorphs *Limnoscelis* and *Diadectes*, the eureptile *Labidosaurus*, and pelycosaurian-grade synapsids including *Oedaleops*. *Labidosaurus* after Sumida (1989b), *Limnoscelis* and *Diadectes* after Sumida (1997) all others except *Oedaelops* after Romer and Price (1940). Note distinct posterodorsal embayment of the acetabulum in *Oedaleops* as well as the eupelycosaurians *Ophiacodon* and *Dimetrodon*. Scale bars equal 1 cm

Reisz 1980; Sumida 1997). Admittedly, there is a necessary level of speculation when proposing a reconstruction based on dissociated elements from numerous individuals. However, only data that could be derived directly from observable elements were included in the data matrix used in the phylogenetic analysis below.

The lengths of ribs observed do not suggest a large barrel-shaped herbivore and the generalized tooth structure most likely suggests an insectivorous habitus. Conjecture as to the diet of other Paleozoic amniotes or their near relatives has focused predominantly on tooth structure and body shape as suggested by rib structure. The ribs assigned here to *Oedaleops* are not elongate enough to suggest a large, barrel-shaped body as in caseid synapsids or diadectid diadectomorphs (e.g., Berman et al. 2004). The upper dentition of *Oedaleops* occupies approximately 51 % of the length of the upper jaw as measured from the posterior part of the jaw articulation. This is not as great a percentage as that of (an as yet unnamed but) definitively insectivorous parareptile (65 %) reported by Modesto et al. (2009) from the Lower Permian (Artinskian) of Oklahoma. In the latter case the remains of chitinous arthropod exoskeletal material confirms insects as at least part of the parareptile's diet. The relatively extensive post-dentition portion of the jaw might suggests that in addition to insects *Oedaleops* could have fed on relatively recalcitrant food, including other arthropods and possibly taking in plant matter as well. However, beyond the suggestion that insects likely comprised a significant part of the diet of *Oedaleops*, other interpretations of potential diet remain speculative at best, though it is likely that plant material did not make up the majority of the diet.

Phylogenetic Analysis

Until now, the most recent phylogenetic analysis incorporating *Oedaleops* is that of Reisz et al. (2009), but as noted earlier, the data for *Oedaleops* were essentially the same as those in use since Langston's (1965) original description. It is now possible to incorporate both a small number of new cranial features, as well as postcranial data as well.

The study of Reisz et al. (2009) included both *Oedaleops* and *Eothyris* in its analysis, but a wider and more recent study of "pelycosaurian" interrelationships is that of Campione and Reisz (2010). That study focused primarily on the interrelationships of the family Varanopidae, but because it was both the most recent available, and incorporated characters from a variety of earlier phylogenetic analyses of basal synapsids and other basal amniotes, it was initially considered as a context in which to reconsider the new data from *Oedaelops*. However, during the compilation

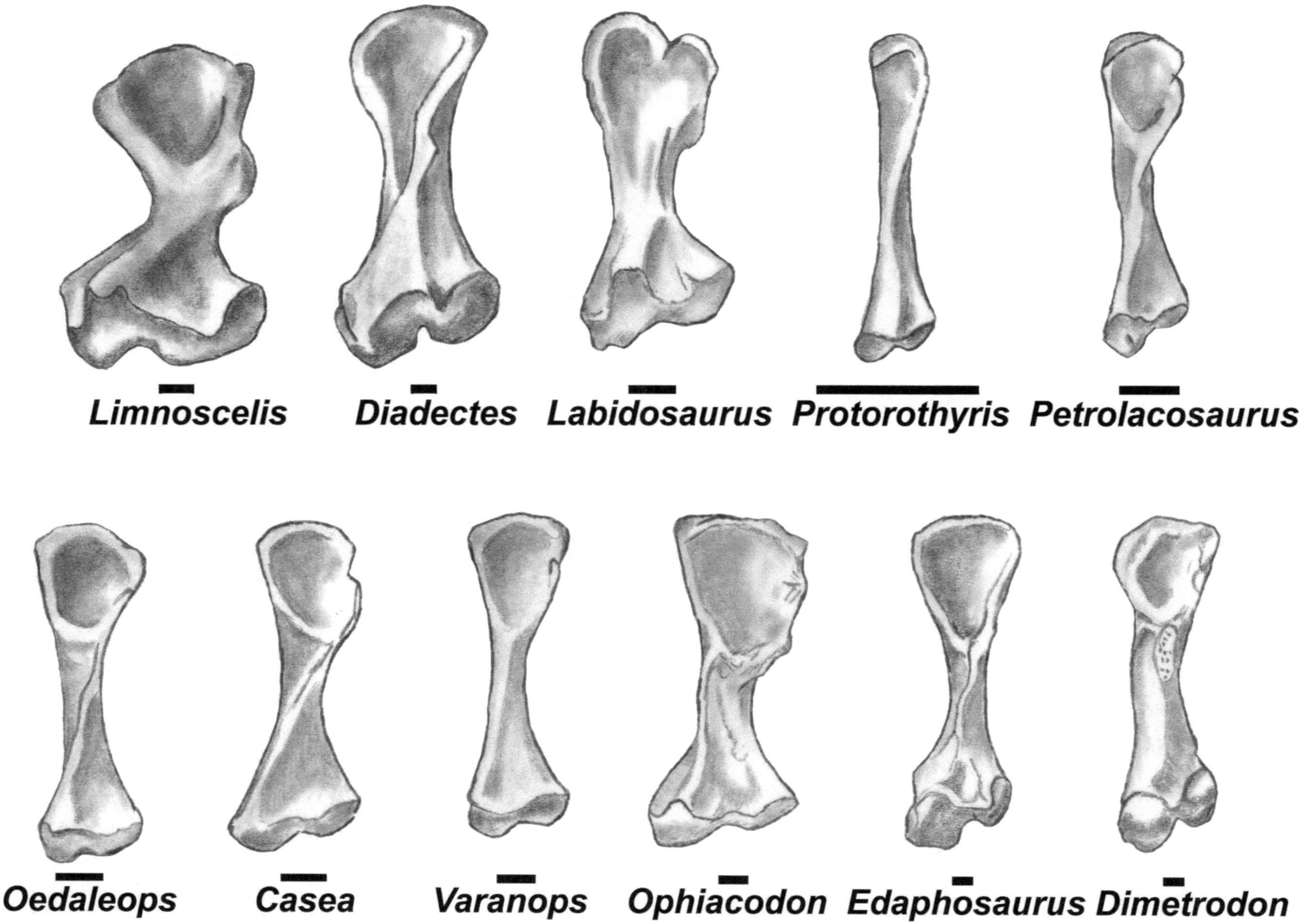

Fig. 2.10 Comparative reconstructions of the left femur in ventral view of the diadectomorphs *Limnoscelis* and *Diadectes*, the eureptiles *Labidosaurus* and *Protorothyris*, the basal araeoscelidian reptile *Petrolacosaurus*, and pelycosaurian-grade synapsids including *Oedaleops*. Note development of fourth trochanter along adductor ridge in *Oedaleops*. *Varanops* and all nonsynapsid taxa after Sumida (1997). All other synapsid taxa with the exception of *Oedaleops* after Romer and Price (1940)

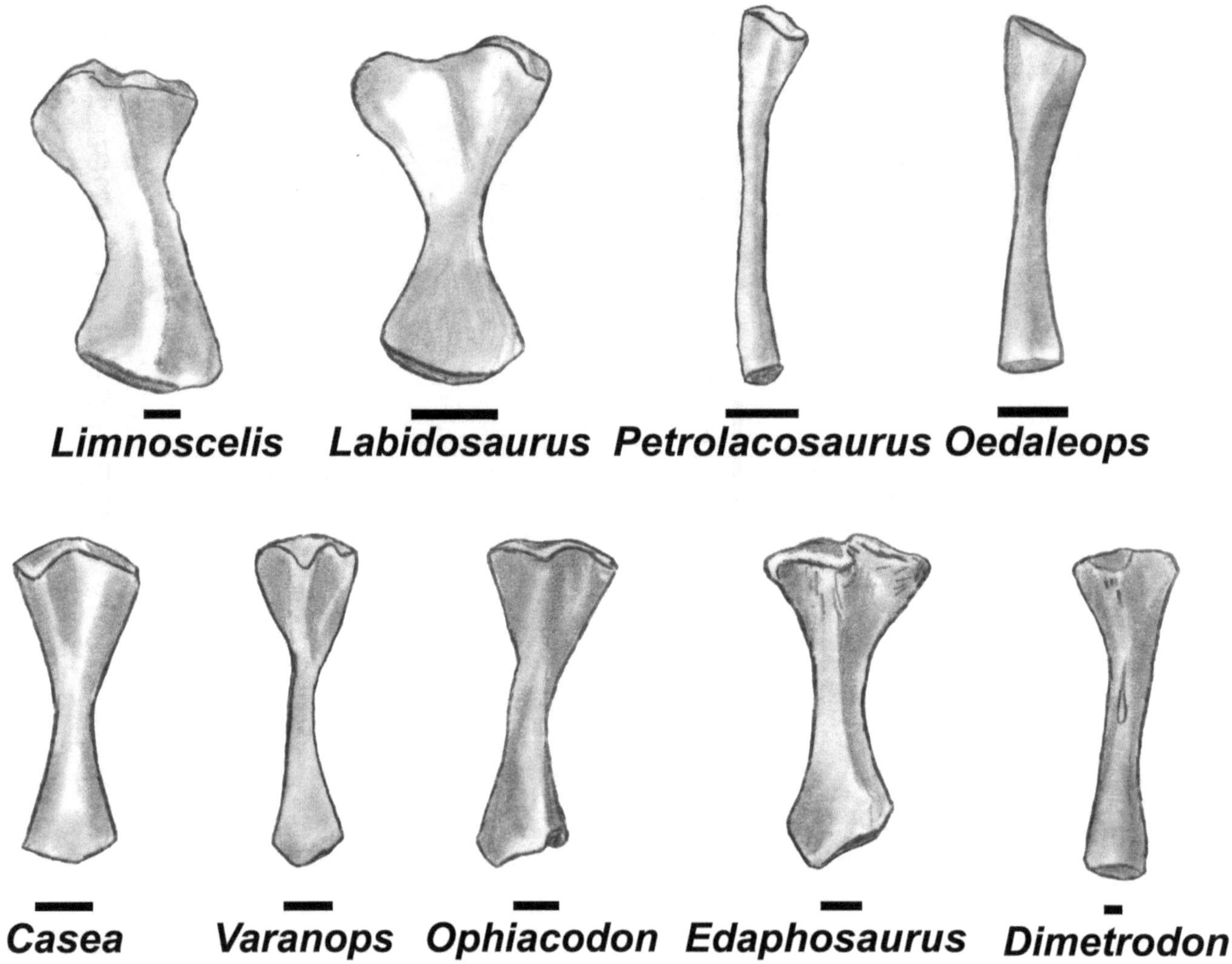

Fig. 2.11 Comparative reconstructions of the left tibia in ventral view of the diadectomorph *Limnoscelis*, the eureptile *Labidosaurus*, the basal araeoscelidian reptile *Petrolacosaurus*, and pelycosaurian-grade synapsids including *Oedaleops*. *Ophiacodon* and all nonsynapsid taxa after Sumida (1997). All other synapsid taxa with the exception of *Oedaleops* after Romer and Price (1940)

of data from this study, as well as those of Berman et al. (2013) and Pelletier (2013) some disagreements with character-state coding and interpretation, particularly for *Cotylorhynchus*, were discovered. Thus, here we have reverted to the character-state coding of Maddin et al. (2006) with some minor exceptions. The Richards Spur varanopid and *Pyozia* from the Russian Mezen fauna were omitted from this analysis, as was the Bromacker varanopid described elsewhere in this volume (Berman et al. 2013) whereas *Oedaleops* was added to the data matrix. Characters 45 and 60 were dropped from the analysis, thus characters 45–58 here (Appendix 2.1) are equivalent to characters 46–59 of Reisz et al. (2010). Minor changes in coding include changing the state of character 52 (number of sacrals) to (1) for *Aerosaurus* and (2) for *Cotylorhynchus*. Maddin et al. (2006) code the equivalent of characters 57 (4th metacarpal/ulna length ratio) as (0; less than 0.5) and 58 (distal phalanx length/width ratio) as (0; less than 1.5) for *Aerosaurus*, but the work of Pelletier (2013) indicates both characters should be coded as (1).

A maximum parsimony analysis was performed in PAUP*4.0b10 (Swofford 2002) using a branch-and-bound search with the following character treatments: multistate taxa were considered as polymorphic, and undetermined character states were treated as missing data. All characters were unordered with equal weight. A bootstrap analysis to estimate clade support was also performed, with the same character treatment, using a branch-and-bound search with 1000 replicates. Reptilia was used as the outgroup. Figure 2.13 is the maximum parsimony tree with the consensus of 1000 bootstrapped replications above the branches.

The results provided both expected and unexpected components. Interestingly, *Oedaleops* is positioned as the sister group to *Eothyris*+*Cotylorhynchus*. The basal position

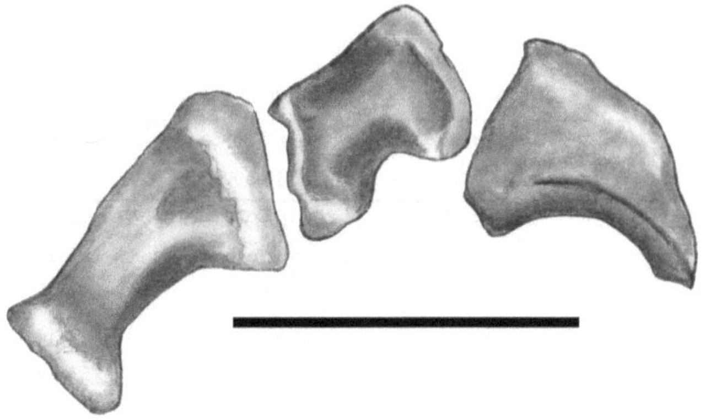

Fig. 2.12 UCMP 40096, phalangeal elements provisionally assigned to *Oedaleops*. Scale bar equals 1 cm

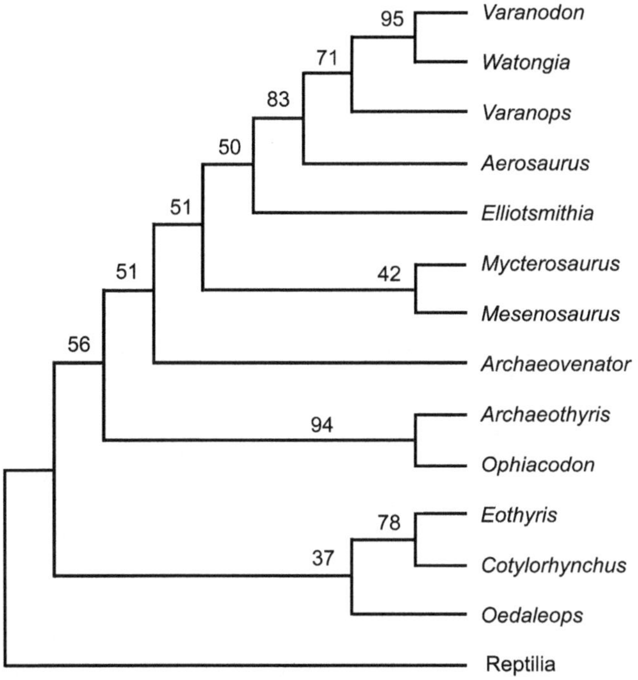

Fig. 2.13 Phylogenetic tree illustrating hypothesis of relationships of basal pelycosaurian-grade synapsids. Hypothesis is a maximum parsimony tree with the consensus of 1000 bootstrapped replications above the branches based on the data matrix in Appendix 2.1 (See text for details of this analysis)

of *Oedaleops* and the monophyly of the caseasaurian taxa included here are not surprising. However, separating *Oedaleops* and *Eothyris*, generally considered to comprise the family Eothyrididae (Reisz et al. 2009), was not expected. This result is only a suggestion of an alternative hypothesis, and is not meant to suggest the definitive dismantling of the family Eothyrididae. However, it is clear that the inclusion of postcranial characters weakens as opposed to strengthens the hypothesis of a robust Eothyrididae as suggested by Reisz et al. (2009). Notably, support for the monophyly of the Caseasauria is low when

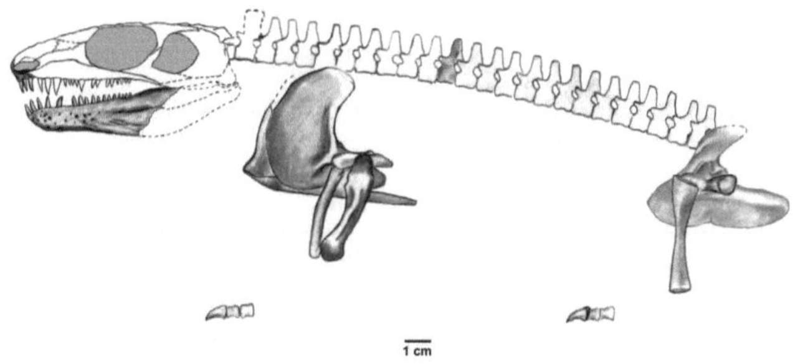

Fig. 2.14 Partial skeletal reconstruction of the basal caseasaurian pelycosaurian-grade synapsid *Oedaleops campi* in left lateral view. *Shaded areas* indicate new interpretations, *unshaded areas* are after Langston (1965), and *dashed areas* are those that remain uncertain

compared to that for the ophiacodontids included here (*Ophiacodon* + *Archaeothyris*) or the varanopid subfamilies. This could be due in part to the use of the derived *Cotylorhynchus* as representative of caseids. Utilization of a more basal representative caseid (Berman et al. 2009; Reisz et al. 2010) might clarify relationships amongst more basal taxa of the analysis somewhat. On the other hand, inclusion of *Oedaleops* does not change substantially the hypothesis of relationships of the more derived taxa from the analyses of Maddin et al. (2006), Campione and Reisz (2010), or Berman et al. (2013). What is clear is that although the phylogenetic data bearing on basal amniote interrelationships are dominated by cranial characters, the inclusion of postcranial data can have a significant impact on the results generated. Of course the recovery and analysis of additional, more complete materials, including postcrania of both *Oedaleops* and *Eothyris*, would be most welcome. Until such time, any additional phylogenetic analyses of pelycosaurian-grade synapsids should: (1) combine characters derived from as many independent phylogenetic analyses as possible (e.g., Reisz 1986; Berman et al. 1995; Reisz et al. 1992, 2009; Maddin et al. 2006; Campione and Reisz 2010); (2) utilize as many basal synapsid taxa as possible (Hopson 1991); (3) utilize both cranio-dental and postcranial characters; and (4) use multiple outgroups including basal eureptiles and diadectomorphs.

Acknowledgments The authors are grateful to Drs. Pat Holroyd and Kevin Padian who facilitated the examination and loan of specimens and encouraged both the *Aerosaurus* and *Oedaleops* studies included in this volume. Drs. Hillary Maddin and Robert Reisz provided thoughtful and measured reviews that helped improve the final version of this study markedly. Particular thanks are due Dr. Reisz regarding the potential diet of basal amniotes. The study of these specimens was made possible by a UCMP Welles Fund Award (to VP). Dr. Anthony Metcalf of the Department of Biology at California State University San Bernardino (CSUSB) provided invaluable help with the phylogenetic analysis. The Professors Across Borders Program at CSUSB provided support to SSS in his initial presentation of these data. SSS thanks Dr. Elizabeth Rega for leading enough interference with Darwin and Owen Sumida to allow the completion of his share of this project. SSS also thanks Kathleen Devlin for critical computer imaging and other logistic support. VP thanks David Pelletier for unflagging support.

Appendix

Appendix 2.1 Character-taxon matrix used for phylogenetic analysis of relationships of basal pelycosaurian-grade synapsids

	1	1111111111	2222222222	3333333333	4444444444	55555555
	1234567890	1234567890	1234567890	1234567890	1234567890	12345678
Reptilia	0000100000	0000000100	00000?000?	0010?00000	0000?00000	00000000
Eothyris	0001110000	0010000000	0000001000	0100100?00	0000000???	????????
Oedaleops	0000000?00	0000?00001	00000?000?	??0???????	?0010???00	0?0????0
Cotylorhynchus	0000110010	0010000000	0000101001	0100100000	0000011000	02000110
Archaeothyris	0101?0??00	0001001101	010000?100	0?10000?00	0010001?10	10??????
Ophiacodon	01?1000000	0001001101	0100000100	0110000000	0010301010	10100000
Mycterosaurus	1011001?11	0000111111	011010?110	0010?0011?	?111201?11	11?????1
Mesenosaurus	1011001?11	0000111111	0110101110	0010110110	1111210111	010?0001
Aerosaurus	1001001110	0?00201111	11?121?111	1101011211	1210110111	11001011
Varanops	11?1001110	1100201111	110121?110	11011?1211	1210110112	12011010
Varanodon	1101001110	2200201111	111121?111	11011???11	1210101121	1211111?
Archaeovenator	0100101100	0000011111	010010?111	00?0000000	0210210?10	010???0?
Watongia	1?0???????	??????11??	??1??????1	1?????????	????11??21	1?11111?
Elliotsmithia	1?1???????	??0??11111	0111101110	01110?????	1??11101??	????????

Data matrix is primarily that of Maddin et al. (2006) with the exception that characters 45 and 60 have been dropped; thus characters 45–58 here are equivalent to characters 46–59 of Reisz et al. (2010). *Oedaleops* has been added and certain characters have been recoded as discussed in the text

References

Berman, D. S (1993). Lower Permian vertebrate localities of New Mexico and their assemblages. *New Mexico Museum of Natural History and Science Bulletin, 2*, 11–21.

Berman, D. S, Reisz, R. R., & Eberth, D. A. (1987). *Seymouria sanjuanensis* (Amphibia, Batrachosauria) from the Lower Permian Cutler Formation of north-central New Mexico and the occurrence of sexual dimorphism in that genus questioned. *Canadian Journal of Earth Sciences, 24*, 1769–1784.

Berman, D. S, Reisz, R. R., Bolt, J. R., & Scott, D. (1995). The cranial anatomy and relationships of the synapsid *Varanosaurus* (Eupelycosauria: Ophiacodontidae) from the Early Permian of Texas and Oklahoma. *Annals of Carnegie Museum, 64*, 100–133.

Berman, D. S, Henrici, A. C., Kissel, R. A., Sumida, S. S., & Martens, T. (2004). A new diadectid (Diadectomorpha), *Orobates pabsti*, from the Early Permian of central Germany. *Bulletin of Carnegie Museum of Natural History, 35*, 1–36.

Berman, D. S, Henrici, A. C., & Sumida, S. S. (2009). Pelycosaurian-grade synapsids from the Lower Permian Bromacker locality, central Germany. *Journal of Vertebrate Paleontology, 29*, 62A.

Berman, D. S, Henrici, A. C., Sumida, S. S., Martens, T., & Pelletier, V. (2013). First European record of a varanodontine (Synapsida: Varanopidae): Member of a unique Early Permian upland paleoecosystem, Tambach Basin, central Germany. In C. F. Kammerer, K. D. Angielczyk, & J. Fröbisch (Eds.), *Early evolutionary history of the Synapsida* (pp. 69–86). Dordrecht: Springer.

Campione, N. E, & Reisz, R. R. (2010). *Varanops brevirostris* (Eupelycosauria: Varanopidae) from the Lower Permian of Texas, with discussion of varanopid morphology and interrelationships. *Journal of Vertebrate Paleontology, 30*, 724–746.

Eberth, D. A., & Miall, A. D. (1991). Stratigraphy, sedimentology and evolution of a vertebrate-bearing, braided to anastomosed fluvial system, Cutler Formation (Permian-Pennsylvanian), north-central New Mexico. *Sedimentary Geology, 72*, 225–252.

Heaton, M. J., & Reisz, R. R. (1980). A skeletal reconstruction of the Early Permian captorhinid reptile *Eocaptorhinus laticeps* (Williston). *Journal of Paleontology, 54*, 136–143.

Holmes, R. (1977). The osteology and musculature of the pectoral limb of small captorhinids. *Journal of Morphology, 152*, 101–140.

Hopson, J. A. (1991). Systematics of the non-mammalian Synapsida and implications for patterns of evolution in synapsids. In H.-P. Schultze & L. Trueb (Eds.), *The origin of higher groups of tetrapods: Controversy and consensus* (pp. 635–693). Ithaca: Cornell University Press.

Langston, W., Jr. (1953). Permian amphibians from New Mexico. *University of California Publications in Geological Sciences, 29*, 349–416.

Langston, W., Jr. (1965). *Oedaleops campi* (Reptilia: Pelycosauria) new genus and species from the Lower Permian of New Mexico, and the family Eothyrididae. *Bulletin of the Texas Memorial Museum, 9*, 1–47.

Lucas, S. G., Harris, S. K., Spielman, J. A., Berman, D. S, Henrici, A. C., Heckert, A. B., et al. (2005). Early Permian biostratigraphy at Arroyo del Agua, Rio Arriba County, New Mexico. In S. G. Lucas, K. E. Zeigler, & J. A. Spielman (Eds.), *The Permian of Central New Mexico. New Mexico Museum of Natural History and Science Bulletin, 31*, 163–169.

Maddin, H. C., & Reisz, R. R. (2007). The morphology of the terminal phalanges in Permo-Carboniferous synapsids: an evolutionary perspective. *Canadian Journal of Earth Sciences, 44*, 267–274.

Maddin, H. C., Evans, D. C., & Reisz, R. R. (2006). An Early Permian varanodontine varanopid (Synapsida: Eupelycosauria) from the

Richards Spur locality, Oklahoma. *Journal of Vertebrate Paleontology, 26*, 957–966.

Modesto, S., & Reisz, R. R. (1992). Restudy of Permo-Carboniferous synapsid *Edaphosaurus novomexicanus* Williston and Case, the oldest known herbivorous amniote. *Canadian Journal of Earth Sciences, 29*, 2653–2662.

Modesto, S. P., Scott, D. M., & Reisz, R. R. (2009). Arthropod remains in the oral cavities of fossil reptiles support inference of early insectivory. *Biology Letters, 5*, 838–840.

Pelletier, V. (2013). Postcranial description and reconstruction of the varanodontine varanopid *Aerosaurus wellesi* (Synapsida: Eupelycosauria). In C. F. Kammerer, K. D. Angielczyk, & J. Fröbisch (Eds.), *Early evolutionary history of the Synapsida* (pp. 53–68). Dordrecht: Springer.

Reisz, R. R. (1980). The Pelycosauria: A review of phylogenetic relationships. In A. L. Panchen (Ed.), *The terrestrial environment and the origin of land vertebrates* (pp. 553–592). London: Academic Press.

Reisz, R. R. (1986). Pelycosauria. In P. Wellnhofer (Ed.), *Handbuch der Paläoherpetologie* (Vol. 17A). Stuttgart: Gustav Fischer Verlag.

Reisz, R. R., Berman, D. S, & Scott, D. (1992). The cranial anatomy of *Secodontosaurus obtusidens*, an unusual mammal-like reptile (Synapsida: Sphenacodontidae) from the Lower Permian of Texas. *Zoological Journal of the Linnean Society, 104*, 127–184.

Reisz, R. R., Godfrey, S. J., & Scott, D. (2009). *Eothyris* and *Oedaleops*: Do these Early Permian synapsids from Texas and New Mexico form a clade? *Journal of Vertebrate Paleontology, 29*, 39–47.

Reisz, R. R., Fröbisch, J., Berman, D. S, & Henrici, A. C. (2010). New Permo-Carboniferous caseid synapsids from North America and Europe and their evolutionary significance. *Program and Abstracts, Society of Vertebrate Paleontology Annual Meeting* (150A) Pittsburgh, Pennsylvania.

Romer, A. S., & Price, L. W. (1940). Review of the Pelycosauria. *Geological Society of American Bulletin, 28*, 1–538.

Sumida, S. S. (1989a). Reinterpretation of vertebral structure in the Early Permian pelycosaur *Varanosaurus acutirostris* (Amniota, Synapsida). *Journal of Vertebrate Paleontology, 9*, 451–458.

Sumida, S. S. (1989b). The appendicular skeleton of the Early Permian genus *Labidosaurus* (Captorhinomorpha, Captorhinidae) and the hind limb musculature of captorhinid reptiles. *Journal of Vertebrate Paleontology, 9*, 295–313.

Sumida, S. S. (1997). Locomotor features of taxa spanning the origin of amniotes. In S. S. Sumida & K. L. M. Martin (Eds.), *Amniote origins: Completing the transition to land* (pp. 353–398). San Diego: Academic Press.

Sumida, S. S., Pelletier, V., Berman, D. S, & English, L. (2009). New information on the basal pelycosaurian-grade synapsid *Oedaleops*. *Journal of Vertebrate Paleontology, 29*, 188A.

Swofford, D. L. (2002). *PAUP*: Phylogenetic Analysis Using Parsimony (*and other methods). Version 4*. Sunderland, MA: Sinaeur Associates.

Chapter 3
Was *Ophiacodon* (Synapsida, Eupelycosauria) a Swimmer? A Test Using Vertebral Dimensions

Ryan N. Felice and Kenneth D. Angielczyk

Abstract *Ophiacodon,* a Permian synapsid, has been hypothesized to be semi-aquatic. This interpretation is based on a range of evidence, including observations of histology, phalangeal morphology, dentition, and taphonomy. However, many of these data are inconclusive or have been reinterpreted. Here we investigate whether the morphology of the axial skeleton in *Ophiacodon* displays specializations for aquatic locomotion. Qualitative and quantitative comparisons of *Ophiacodon* to extant terrestrial and semi-aquatic tetrapods demonstrate that the distribution of centrum lengths in its vertebral column is similar in some ways to those of extant semi-aquatic reptiles. However, other basal synapsids that are widely regarded as terrestrial show comparable patterns, and the correlation between swimming style and vertebral morphology in extant semi-aquatic tetrapods may be weaker than previously thought. Therefore, vertebral proportions provide little support for a semi-aquatic lifestyle in *Ophiacodon*. Given that most lines of evidence are equivocal at best, we suggest that future studies that consider the ecology of *Ophiacodon* use a terrestrial lifestyle as a null hypothesis.

Keywords Permian • Carboniferous • Centrum length • Limb length • Aquatic tetrapods • Ophiacodontidae

R. N. Felice (✉)
Department of Biological Sciences, Ohio University, Athens, OH 45701, USA
e-mail: rf273509@ohio.edu

K. D. Angielczyk
Department of Geology, Field Museum of Natural History, 1400 South Lake Shore Drive, Chicago, IL 60605, USA
e-mail: kangielczyk@fieldmuseum.org

Introduction

The eupelycosaurian synapsid *Ophiacodon* (Fig. 3.1) is a common component of Late Carboniferous and Early Permian terrestrial ecosystems. Six species of *Ophiacodon* are currently recognized: the type species *Ophiacodon mirus* Marsh, 1878; *O. uniformis* (Cope, 1878) Romer and Price, 1940; *O. navajovicus* (Case, 1907) Romer and Price, 1940; *O. hilli* (Romer, 1925) Romer and Price, 1940; *O. retroversus* (Cope, 1878) Romer and Price, 1940; and *O. major* Romer and Price, 1940. All are found in North America (New Mexico, Texas, Utah, or Kansas) in terrestrial deposits (Marsh 1878; Reisz 1986; Eberth and Berman 1993; Eberth and Miall 1991), with the exception of *O. hilli*, which is known from marine limestones in Kansas (Romer 1925) and a fragmentary dentary from the Kenilworth Breccia in the United Kingdom that Paton (1974) referred to the genus. *Ophiacodon* was one of the larger animals in these communities, ranging from around 160 to 300 cm in total length (Reisz 1986) and between 26 and 230 kg in mass (Romer and Price 1940).

Ophiacodon has long been interpreted as a semi-aquatic piscivore (e.g., Case 1907; Romer and Price 1940; Kemp 1982; Reisz 1986; Germain and Laurin 2005) and several lines of evidence have been cited in support of this hypothesis. For example, Romer and Price (1940) suggested that the flattened unguals of *Ophiacodon* were indicative of aquatic habits, and its small, sharp teeth resemble those of extant crocodiles, perhaps indicating a piscivorous diet (Kemp 1982). Unique among Permian synapsids, the skeleton of *Ophiacodon* tends to be poorly ossified, particularly the limb bones (Reisz 1986). This morphology is thought to suggest a slowing of ossification in *Ophiacodon* relative to other taxa such as *Dimetrodon* (Brinkman 1988), and is reminiscent of the condition commonly seen in secondarily aquatic tetrapods (Romer 1948). The bone microanatomy of *Ophiacodon* also has been interpreted as consistent with a semi-aquatic lifestyle (Germain and Laurin 2005). Finally, a

Christian F. Kammerer, Kenneth D. Angielczyk and Jörg Fröbisch (eds.), *Early Evolutionary History of the Synapsida*, Vertebrate Paleobiology and Paleoanthropology, DOI: 10.1007/978-94-007-6841-3_3,

Fig. 3.1 Photograph of *Ophiacodon mirus* (FMNH UC 671) on display at the Field Museum of Natural History

single specimen, the holotype of *Ophiacodon hilli*, was found in a marine limestone (Romer 1925), although the majority of specimens occur in more terrestrial sediments that also produce other basal synapsids (Romer and Price 1940). Although these observations are consistent with an amphibious lifestyle, none are completely unequivocal and little detailed analysis of the semi-aquatic *Ophiacodon* hypothesis has been performed.

Here we test whether *Ophiacodon* was likely semi-aquatic by examining the morphology of the axial skeleton, specifically whether proportions of its vertebrae are comparable to other swimming tetrapods. The relationship between axial morphology and locomotor style in swimming tetrapods is well documented (Fish 1984, 1994; Ritter 1996; Buchholtz 1998, 2001a, b; Pierce et al. 2011), with the degree of adaptation for aquatic locomotion reflected in the flexibility of the vertebral column. Vertebral column flexibility depends in part on the proportions of the individual centra: long, thin, spool-shaped centra indicate greater flexibility, whereas short, wide, disc-shaped centra imply greater stiffness (Buchholtz 1998, 2001a, b). In addition, semi-aquatic and aquatic tetrapods have peaks or plateaus in centrum length corresponding to the point(s) of maximum undulation in the vertebral column, which are not found in terrestrial tetrapods (Buchholtz 1998, 2001a). We compared vertebral proportions in *Ophiacodon* to several extant and extinct tetrapods with known locomotor patterns to test whether its vertebral morphology was consistent with its hypothesized semi-aquatic lifestyle. In addition, we investigated the disparity in hind limb and fore limb lengths noted by Romer and Price (1940) and Kemp (1982) to test the hypothesis that the limb proportions of *Ophiacodon* were most similar to those of semi-aquatic tetrapods. A more decisive determination of the lifestyle of *Ophiacodon* will not only provide a better foundation for interpreting the distinctive features of its anatomy, but will also facilitate an improved understanding of the paleoecology of late Paleozoic Euramerican terrestrial communities.

Institutional abbreviations: FMNH, Field Museum of Natural History, Chicago, IL, USA; UCMP, University of California Museum of Paleontology, Berkeley, CA, USA.

Review of Evidence for a Semi-aquatic Lifestyle in *Ophiacodon*

The hypothesis that *Ophiacodon* was semi-aquatic has a long history, starting with Case's (1907) suggestion that ophiacodontids (Poliosauridae in Case 1907) were probably aquatic. However, Case provided no evidence in that monograph to support this assertion, and it seems to have received a mixed reaction. For example, Williston (1911, p. 80) stated that "the assumption that the animals were probably aquatic is evidently wrong," and did not include *Ophiacodon* among the taxa he discussed in his book on aquatic reptiles (Williston 1914). Likewise, Williston and Case (1913, p. 59) concluded that *Ophicaodon* was not a swimming animal, particularly noting that its slender tail "would have been of no use in the water in propulsion." Case (1915) himself is somewhat equivocal, commenting that ophiacodontids do not show aquatic adaptations of the feet or tail, but nevertheless proposing that they frequently took to the water to hunt and/or escape danger. Romer (1925) downplayed the significance of the discovery of the holotype of *Ophiacodon hilli* in a marine limestone, speculating that it was a terrestrial animal that had been washed out to sea after its death. In contrast, Romer and Price (1940) reasserted that *Ophiacodon* was likely semi-aquatic, and provided various lines of evidence from skeletal anatomy and taphonomy in support of the hypothesis. The assumption of a semi-aquatic ecology for *Ophiacodon* also played a key role in Romer's (1957, 1958) hypothesis that the amniotic egg evolved in the context of aquatic or semi-aquatic adults venturing onto land to reproduce. Most subsequent authors who have considered the subject cite Romer and Price's (1940) arguments directly, or use very similar lines of evidence.

Several morphological features of *Ophiacodon* have been used to infer that it was semi-aquatic. For example, *Ophiacodon* displays unusually poor ossification of the endochondral elements of the skull. The sutures between the elements of the braincase of other basal synapsids are normally indistinguishable, but they are all unfused and discernible in *Ophiacodon* (Reisz 1986). The joint surfaces of the long bones also appear to have remained cartilaginous throughout a longer period of ontogeny in *Ophiacodon* than in *Dimetrodon*, with the ulna and femur never reaching the degree of ossification seen in adult *Dimetrodon* specimens (Brinkman 1988). The sutures in the pelvis also remain poorly fused (Romer and Price 1940; Olson 1941). Romer and Price (1940) cited this poor ossification as evidence suggesting that *Ophiacodon* was semi-aquatic, and noted that this feature is well defined even in *O. major*, the largest species in the genus. Olson (1941) carried this argument further, using the lack of fusion of the pelvis in *O. major*, along with the species' expanded anterior ribs, tall caudal neural spines, and relatively late stratigraphic occurrence to propose that *Ophiacodon* became progressively more adapted to a semi-aquatic lifestyle over the course of its history. Whereas reduced ossification could be diagnostic of juvenile individuals with extremely rapid growth rates, specimens of *Ophiacodon* show little intraspecific size variation and are interpreted as adults (e.g., Romer and Price 1940; Ricqlès 1989). Reduced ossification of long bone articular surfaces and open sutures are widespread among secondarily aquatic tetrapods, and may be related to heterochronic changes in other aspects of gross morphology and microanatomy that frequently accompany secondary adaptation to semi-aquatic or aquatic habits in tetrapods (e.g., Ricqlès 1989; Ricqlès and Buffrénil 2001), so this evidence is consistent with a semi-aquatic lifestyle in *Ophiacodon*. At the same time some parts of the skeleton, such as the manus and pes, tend to be well ossified in *Ophiacodon* so the pattern is not perfect.

The skull and tooth morphology of *Ophiacodon* have been cited as evidence of a semi-aquatic lifestyle (Romer and Price 1940; Kemp 1982), primarily because they were interpreted as consistent with a piscivorous diet. Romer and Price (1940) noted that the skull of *Ophiacodon* is relatively tall and narrow compared to most other pelycosaur-grade synapsids, and that the jaws are relatively slender. They considered these features to translate into a mechanically weak skull that would be poorly suited to dealing with large terrestrial prey. This issue is difficult to address rigorously because the mechanical responses of "pelycosaur" skulls to loading have not received the same degree of attention as some therapsids (Jenkins et al. 2002; Jasinoski et al. 2009, 2010a, b, 2013; Jasinoski and Chinsamy-Turan 2012), and various lines of evidence suggest different conclusions about the mechanical properties of the skull of *Ophiacodon* compared to similarly-sized sphenacodontids such as *Dimetrodon*. For example, factor analysis indicates that its dimensions are generally similar to those of sphenacodontids (Gould 1965), primarily because of their proportionally longer snouts than herbivorous taxa such as *Edaphosaurus*. The large palatines of *Ophiacodon* also brace the maxillae in a manner similar to that described for *Dimetrodon* (Thomason and Russell 1986), which likely helped strengthen the snout against the bending and torsional loads expected when dealing with large prey. Additional evidence that the skull of *Ophiacodon* was likely resistant to such loads can be found in the fact that relatively tall, narrow skulls (i.e., oreinirostral skull morphologies) have been found to resist bending and torsion better than flatter (platyrostral) skulls in dinosaurs and crocodilians (McHenry et al. 2006; Rayfield et al. 2007; Rayfield and Milner 2008). However, the skull bones of *Ophiacodon* are relatively thin (Williston and Case 1913), which would weaken the skull even if it was similarly constructed.

The teeth of *Ophiacodon*, especially in the mandible, are smaller and more numerous than those of other large carnivorous pelycosaur-grade synapsids, and vary in shape from conical and relatively straight to slightly recurved. Serrations or distinct carinae are absent, although very weak ridges are sometimes present near the base of the crown. Strong wear features are not apparent, perhaps because of relatively frequent replacement (Reisz 1986). These features are not inconsistent with a piscivorous diet [e.g., the relative size and morphology of the teeth are similar to those in Massare's (1987) Pierce II and General guilds of marine reptiles, both of which were inferred to have a fish component in their diets], although the teeth of *Ophiacodon* are less numerous and more generalized than the slender, uniform, sharply pointed teeth of the closely related *Varanosaurus* (Berman et al. 1995). Indeed, Romer and Price's (1940) description of ophiacodontid teeth, which they use to argue for a piscivorous diet for members of the group, more closely resembles the dentition of *Varanosaurus* than *Ophiacodon*.

Romer and Price (1940) suggested that the relatively broad, flat unguals of *Ophiacodon* were indicative of a semi-aquatic lifestyle because they were unlike the laterally compressed, sharply pointed unguals of other basal synapsids. However, Maddin and Reisz (2007; also see Berman et al. 2004) recently examined changes in ungual morphology in basal synapsids and diadectids, and found evidence of an evolutionary trend from broad, flat unguals in the most basal synapsids and the outgroup to taller, more strongly curved, laterally compressed unguals in more crownward taxa, irrespective of inferred diet. In this context, the ungual morphology of *Ophiacodon* seems most likely a reflection of its relatively basal phylogenetic position than a proxy for its ecology.

Romer and Price (1940) and Kemp (1982) considered the disparity in limb lengths observed in *Ophiacodon* (hind limbs longer than fore limbs) as evidence of semi-aquatic habits. Little justification was provided for why this disparity would be consistent with such an ecology, aside from Romer and Price's statement that similar proportions are found in the aquatic mesosaurs. Differences in fore and hind limb proportions as adaptations to semi-aquatic or aquatic lifestyles have received attention in mammals, particularly in the context of whale evolution (e.g., Thewissen and Fish 1997; Gingerich 2003). Among mammals, limb proportions are useful in distinguishing hind limb- and fore limb-powered paddlers (Gingerich 2003). However, most of the changes in limb length occur in the manus or pes, particularly in the length of the metacarpals or metatarsals and the phalanges (Thewissen and Fish 1997; Gingerich 2003; Samuels and Van Valkenburgh 2008), reflecting the fact that the optimal shape for a drag-based propulsor is a tall triangle (Webb 1988; Thewissen and Fish 1997). The manus and pes of *Ophiacodon* do not show these sorts of proportional changes, however. Although the pes is larger than the manus (e.g., Romer and Price 1940), neither is especially triangular or elongate. Likewise, the length of the third and fourth metacarpals and metatarsals do not differ greatly from the lengths of the second and third metacarpals and metatarsals (e.g., Fig. 52 of Romer and Price 1940), similar to terrestrial mammals sampled by Thewissen and Fish (1997), but not semi-aquatic mammals with paddle-like feet.

In addition to gross morphology, bone histology has been used to infer a semi-aquatic lifestyle for *Ophiacodon*. Enlow and Brown (1957) and Ricqlès (1974) both noted that the cortices of the *Ophiacodon* specimens they sectioned were unlike those of other pelycosaur-grade synapsids they sampled in being highly vascularized. Enlow and Brown (1957) interpreted this as evidence that *Ophiacodon* was fast growing, but Ricqlès (1974) interpreted this observation, along with the primarily longitudinal orientation of the vascular canals and the lack of a clear distinction between the cortex and spongiosa as indicating a semi-aquatic lifestyle. In their quantitative analysis of bone cross-sectional area, Germain and Laurin (2005) explored the link between the lifestyle of tetrapods (including *Ophiacodon*) and several parameters related to the compactness of the radius. The compactness (density) of limb bones of tetrapods generally relates to whether the organism is terrestrial, amphibious, or aquatic. Semi-aquatic tetrapods, as well as some aquatic forms, typically have more compact bone as an adaptation to reduce buoyancy, whereas highly pelagic taxa have reduced skeletal density to achieve near neutral buoyancy (e.g., Taylor 1994; Madar 1998; Ricqlès and Buffrénil 2001; Laurin et al. 2004; Germain and Laurin 2005; Houssaye 2009; Northover et al. 2010; although see Nakajima 2010). Germain and Laurin (2005) took cross-sections of the radii of 46 species of tetrapods whose lifestyles were known and three taxa (including *Ophiacodon*) whose lifestyles were unknown, and calculated several parameters for a model fit to the compactness profile of each taxon. The low overall compactness (*C* in their model; also see Girondot and Laurin 2003) they report for *Ophiacodon*, along with its high *S* and *P* values (implying a slow transition between the medullary cavity and cortex, with the transition point between the two zones being located relatively far from the center of the bone), are consistent with Enlow and Brown's (1957) and Ricqlès' (1974) observations, and are most similar to the parameters of extant aquatic tetrapods. However, *Ophiacodon* was an outlier in a linear discriminant analysis of the data, falling outside of the ranges occupied by extant terrestrial, semi-aquatic, and aquatic taxa, leading Germain and Laurin (2005) to tentatively conclude that *Ophiacodon* was likely semi-aquatic. Although they noted that additional sampling of basal synapsids and sauropsids would be necessary to provide context for understanding the bone histology of *Ophiacodon*, the taxon was not included in similar analyses of the humerus and tibia (Kriloff et al. 2008; Canoville and Laurin 2010), making it uncertain whether the taxon's bones consistently show compactness profiles expected for a semi-aquatic tetrapod, as well as how they compare to coeval taxa such as *Dimetrodon* or *Captorhinus*.

Finally, taphonomy has been used to argue that *Ophiacodon* was semi-aquatic. As noted above, Romer (1925) documented the occurrence of *Ophiacodon hilli* in a marine limestone, but speculated that the specimen had been washed into that depositional environment after its death. Romer and Price (1940) ascribed greater significance to this occurrence, and included it among the evidence they cited in support of a semi-aquatic lifestyle for the taxon. While it is true that *Ophiacodon* is not represented in the most terrestrial Lower Permian fossil localities (Sullivan and Reisz 1999; Martens et al. 2005; Voight et al. 2007; Evans et al. 2009; Berman et al. 2013), and that marine occurrences are very unusual for basal synapsids, it is not unheard of for terrestrial tetrapod fossils to be preserved in marine rocks. More importantly, the vast majority of *Ophiacodon* fossils have been collected in rocks that represent terrestrial lowland floodplain, lacustrine, and deltaic environments (Hentz 1988, 1989; Eberth and Miall 1991), so it seems unlikely that the marine record of *Ophiacodon* provides particular insight into its ecology.

Taken together, these observations paint an equivocal picture of potential aquatic adaptation in *Ophiacodon*. Although some data (e.g., bone histology, reduced ossification of limb elements) are certainly very suggestive of a semi- aquatic way of life, much of the other evidence (e.g., limb proportions, ungual morphology, taphonomy) are either contradicted by more recent studies or are anecdotal at best. Therefore, it is necessary to bring new data to bear on the semi-aquatic *Ophiacodon* hypothesis.

Vertebral Anatomy and Aquatic Locomotion in Tetrapods

Many semi-aquatic and aquatic tetrapods, such as whales and their ancestors, ichthyosaurs and their ancestors, crocodilians, and even semi-aquatic mammals like otters or minks, move with axial undulatory locomotion (Buchholtz 1998, 2001a, b). Undulatory swimming involves waves of motion passing repeatedly along a propulsor, in this case the vertebral column (Buchholtz 1998, 2001a, b). Various parts of the vertebral column are reinforced and stiffened in order to increase the wavelength and decrease the amplitude of undulations (Buchholtz 1998, 2001a), which may be laterally- or dorsoventrally-directed. The degree of undulatory flexibility is controlled by regional variation in the axial musculature, soft tissues, and skeleton (Buchholtz 2001a). If *Ophiacodon* was indeed semi-aquatic, it likely swam in an undulatory manner analogous to crocodilians and other semi-aquatic tetrapods, and thus may show similar modifications of the vertebral column.

Vertebral proportions serve as an osteological correlate for regional flexibility in the vertebral column, and the relationship between vertebral morphology and locomotor style in swimming tetrapods is well known (Fish 1984, 1994; Ritter 1996; Buchholtz 1998, 2001a, b; Pierce et al. 2011). Long, narrow, spool-shaped vertebrae are at the center of the undulatory wave (Buchholtz 2001a), with the elongated shape increasing the displacement of the intervertebral joints, and therefore axial flexibility. Conversely, short, disk-shaped vertebrae impart less flexibility (Buchholtz 2001a) because the disk shape increases the articular surface of the centrum, limiting angular displacement of the intervertebral joint (Buchholtz 1998, 2001a). These differences in individual vertebral proportions have been proposed to translate into functionally significant patterns that can be recognized across the vertebral column. For example, depending on speed, the trunks of lizards with robust limbs bend into either a standing or traveling lateral wave during terrestrial locomotion (Ritter 1992). The location of this undulation is reflected by a distinct peak in centrum length between the girdles, with the relatively longest vertebrae corresponding to the peak of the (standing) wave (Buchholtz 1998). The tails of extant semi-aquatic reptiles such as the marine iguana or gharial typically move in traveling waves of lateral undulation, and in these taxa a plateau of relatively long centra are present in the tail reflecting where series of vertebrae undergo similar patterns of displacement (Buchholtz 1998). Peaks in centrum length also are found in the locations of undulatory waves in extant and fossil semi-aquatic and aquatic mammals, with the position and size of the peaks depending on the details of swimming style (Buchholtz 1998).

These observations allow us to make predictions about the distribution of vertebral proportions across the column of *Ophiacodon* if it was indeed a swimmer. Given that *Ophiacodon*, like most basal synapsids, possessed limb and vertebral morphologies that would correspond to a sprawling posture and symmetrical gait (e.g., Romer and Price 1940; Kemp 1982, 2005; Carroll 1986; Panko 2001), we predict that *Ophiacodon* engaged in lateral undulation during terrestrial locomotion to increase its stride length and perhaps aid in force transmission, although the degree of undulation may be less than observed in modern reptiles (e.g., Kemp 1982, 2005; Hunt and Lucas 1998, although see Carpenter 2009) and may have included a larger rotary component (Sumida and Modesto 2001). A corollary of this prediction is that *Ophiacodon* should show a peak in centrum length in its trunk region corresponding to the peak of its undulatory wave, much like extant sprawling tetrapods. Similarly, because the locomotor style of *Ophiacodon* was most likely more similar to extant reptiles than extant mammals, we also predict that if *Ophiacodon* was a swimmer then it should show a plateau in centrum length in its caudal series that is comparable to that observed in extant swimming reptiles.

Materials and Methods

We examined a total of 50 specimens from the Field Museum of Natural History and two from the University of California Museum of Paleontology (Table 3.1), representing at least 28 fossil and extant species. Our fossil sample consisted of 19 specimens representing at least 12 species of pelycosaur-grade synapsids belonging to several synapsid taxa: Caseidae (*Angelosaurus, Casea, Cotylorhynchus*), Varanopidae (*Aerosaurus*, *Varanops*), Ophiacodontidae (*Ophiacodon*), and Sphenacodontidae (*Dimetrodon, Sphenacodon*). All of the specimens represent articulated or semi-articulated individuals for which vertebral positions were known or could be determined. In some cases the specimens were mounted for display and included reconstructed vertebrae, but we excluded the reconstructions from

Table 3.1 Values for range (R), polarization (C), irregularity (Cm), concentration ($E1$), and smoothness ($E2$) metrics, and fore and hind limb lengths, for measured specimens

Taxon	Specimen	R (length)	C (length)	Cm (length)	$E1$ (length)	$E2$ (length)	R (height)	C (height)	Cm (height)
Angelosaurus romeri	FMNH UR 827	0.491	0.196	0.004	0.294	0.193	–	–	–
Casea broilii	FMNH UC 656	0.488	0.191	0.001	0.297	0.191	–	–	–
Casea sp.	FMNH UC 960	0.432	0.204	0.002	0.228	0.202	0.372	0.154	0.001
Cotylorhynchus hancocki	FMNH UC 581	1.307	0.779	0.002	0.528	0.777	–	–	–
Cotylorhynchus hancocki	FMNH UR 272	1.441	0.634	0.002	0.808	0.631	–	–	–
Ophiacodon major	FMNH UC 1638	0.502	0.137	0.001	0.365	0.136	0.627	0.238	0.002
Ophiacodon mirus	FMNH UC 671	0.499	0.204	0.002	0.294	0.202	0.941	0.276	0.003
Ophiacodon retroversus	FMNH UC 458	0.649	0.178	0.002	0.471	0.176	0.576	0.278	0.001
Ophiacodon retroversus	FMNH UC 709	0.063	0.030	0.001	0.032	0.029	–	–	–
Ophiacodon uniformis	FMNH UC 690	0.358	0.136	0.004	0.222	0.132	–	–	–
Varanops brevirostris	FMNH UC 644	0.782	0.275	0.001	0.508	0.274	–	–	–
Varanops sp.	FMNH UR 607/UR 616	0.250	0.136	0.004	0.114	0.132	0.295	0.169	0.006
Varanops sp.	FMNH P 12841	0.264	0.124	0.003	0.140	0.121	0.311	0.131	0.004
Aerosaurus wellesi	UCMP V2814/40096	0.196	0.096	0.006	0.100	0.090	0.271	0.171	0.008
Dimetrodon giganhomogenes	FMNH UC 112	0.275	0.145	0.003	0.130	0.142	0.340	0.157	0.002
Dimetrodon loomisi	FMNH UC 1322	0.959	0.447	0.002	0.512	0.446	1.116	0.651	0.003
Dimetrodon sp.	FMNH UC 1758	0.861	0.336	0.003	0.525	0.333	–	–	–
Sphenacodon ferox	FMNH UC 35	0.634	0.248	0.002	0.386	0.246	1.048	0.431	0.002
Sphenacodon ferox	UCMP V3529/34226	0.442	0.218	0.002	0.224	0.216	0.884	0.422	0.003
Amblyrhychus cristatus	FMNH 22213	0.423	0.172	0.001	0.251	0.172	0.658	0.239	0.001
Iguana iguana	FMNH 22085	0.916	0.293	0.001	0.623	0.292	1.191	0.513	0.001
Varanus bengalensis nebulosus	FMNH 22495	0.626	0.352	0.001	0.273	0.352	0.796	0.383	0.002
Varanus bengalensis nebulosus	FMNH 211856	0.560	0.271	0.001	0.289	0.270	0.753	0.380	0.001
Varanus dumerili	FMNH 228151	0.473	0.283	0.001	0.190	0.282	0.588	0.314	0.001
Varanus exanthematicus	FMNH 228398	0.735	0.369	0.001	0.366	0.368	0.619	0.308	0.001
Varanus komodensis	FMNH 22197	0.936	0.460	0.000	0.476	0.459	1.437	0.753	0.001
Varanus rudicollis	FMNH 98947	0.606	0.299	0.000	0.307	0.299	0.851	0.441	0.001
Varanus salvator	FMNH 22204	1.251	0.481	0.000	0.770	0.481	1.552	0.771	0.001
Varanus salvator	FMNH 98866	0.536	0.194	0.001	0.341	0.194	0.557	0.243	0.001
Varanus salvator	FMNH 195576	1.019	0.338	0.001	0.681	0.337	1.210	0.505	0.001
Varanus salvator	FMNH 211938	0.707	0.404	0.001	0.303	0.403	0.661	0.381	0.001
Caiman crocodilus	FMNH 9150	0.265	0.082	0.001	0.183	0.081	1.240	0.449	0.001
Caiman crocodilus	FMNH 13062	0.505	0.136	0.001	0.369	0.135	0.862	0.331	0.001

(continued)

Table 3.1 (continued)

Taxon	Specimen	*R* (length)	*C* (length)	*Cm* (length)	*E*1 (length)	*E*2 (length)	*R* (height)	*C* (height)	*Cm* (height)
Caiman crocodilus	FMNH 217159	0.240	0.087	0.001	0.153	0.087	1.012	0.410	0.001
Melanosuchus niger	FMNH 218507	0.331	0.128	0.000	0.204	0.127	1.002	0.373	0.001
Crocodylus rhombifer	FMNH 34677	0.653	0.202	0.001	0.451	0.202	1.575	0.689	0.001
Gavialis gangeticus	FMNH 82681	0.444	0.204	0.001	0.241	0.203	1.478	0.573	0.001
Chironectes minimus	FMNH 60576	1.705	0.971	0.001	0.734	0.970	0.993	0.443	0.002
Chironectes minimus	FMNH 60578	1.730	0.910	0.001	0.820	0.908	1.198	0.542	0.001
Castor canadensis	FMNH 18525	1.406	0.628	0.001	0.778	0.627	1.304	0.611	0.001
Castor canadensis	FMNH 134455	1.414	0.640	0.001	0.774	0.639	1.701	0.669	0.002
Castor canadensis	FMNH 141992	1.286	0.614	0.001	0.672	0.613	1.498	0.613	0.001
Lontra canadensis	FMNH 60655	0.585	0.310	0.001	0.274	0.309	0.906	0.274	0.001
Lontra canadensis	FMNH 160115	0.505	0.257	0.001	0.249	0.256	0.483	0.192	0.001
Lontra canadensis	FMNH 175293	0.514	0.229	0.001	0.285	0.228	0.912	0.285	0.001
Lontra canadensis	FMNH 180757	0.864	0.301	0.001	0.563	0.300	1.386	0.326	0.001
Neovison vison	FMNH 59739	1.047	0.392	0.002	0.655	0.390	0.762	0.228	0.001
Neovison vison	FMNH 59776	0.799	0.330	0.001	0.468	0.329	0.672	0.196	0.002
Neovison vison	FMNH 59793	0.817	0.402	0.002	0.415	0.401	1.061	0.269	0.001
Neovison vison	FMNH 112982	0.828	0.348	0.001	0.480	0.347	0.904	0.232	0.002

Taxon	Specimen	*E*1 (height)	*E*2 (height)	*R* (width)	*C* (width)	*Cm* (width)	*E*1 (width)	*E*2 (width)	Fore limb length	Hind limb length
Angelosaurus romeri	FMNH UR 827	–	–	–	–	–	–	–	–	–
Casea broilii	FMNH UC 656	–	–	0.862	0.297	0.001	0.565	0.296	15.10	12.61
Casea sp.	FMNH UC 960	0.219	0.152	0.648	0.279	0.002	0.369	0.277	–	–
Cotylorhynchus hancocki	FMNH UC 581	–	–	–	–	–	–	–	61.50	62.50
Cotylorhynchus hancocki	FMNH UR 272	–	–	–	–	–	–	–	46.50	43.50
Ophiacodon major	FMNH UC 1638	0.389	0.236	0.563	0.223	0.002	0.340	0.221	–	–
Ophiacodon mirus	FMNH UC 671	0.664	0.273	0.533	0.223	0.002	0.310	0.222	19.87	21.61
Ophiacodon retroversus	FMNH UC 458	0.298	0.277	0.284	0.103	0.003	0.181	0.100	–	–
Ophiacodon retroversus	FMNH UC 709	–	–	0.226	0.153	0.003	0.073	0.150	–	–
Ophiacodon uniformis	FMNH UC 690	–	–	0.262	0.144	0.009	0.119	0.135	–	–
Varanops brevirostris	FMNH UC 644	–	–	1.400	0.646	0.002	0.754	0.644	13.28	17.08
Varanops sp.	FMNH UR 607/UR 616	0.125	0.163	0.388	0.238	0.010	0.150	0.228	–	–
Varanops sp.	FMNH P 12841	0.181	0.127	0.155	0.061	0.003	0.094	0.058	–	–
Aerosaurus wellesi	UCMP V2814/40096	0.100	0.163	–	–	–	–	–	9.30	11.14
Dimetrodon giganhomogenes	FMNH UC 112	0.183	0.155	0.397	0.163	0.003	0.234	0.160	–	–
Dimetrodon loomisi	FMNH UC 1322	0.465	0.648	–	–	–	–	–	–	–

(continued)

Table 3.1 (continued)

Taxon	Specimen	*E*1 (height)	*E*2 (height)	*R* (width)	*C* (width)	*Cm* (width)	*E*1 (width)	*E*2 (width)	Fore limb length	Hind limb length
Dimetrodon sp.	FMNH UC 1758	–	–	–	–	–	–	–	–	–
Sphenacodon ferox	FMNH UC 35	0.617	0.429	0.790	0.389	0.002	0.400	0.387	25.50	27.87
Sphenacodon ferox	UCMP V3529/34226	0.462	0.419	–	–	–	–	–	19.10	22.40
Amblyrhychus cristatus	FMNH 22213	0.419	0.238	0.673	0.294	0.001	0.379	0.293	12.46	13.90
Iguana iguana	FMNH 22085	0.678	0.512	1.665	0.453	0.001	1.212	0.452	15.10	18.22
Varanus bengalensis nebulosus	FMNH 22495	0.412	0.382	0.719	0.383	0.001	0.336	0.382	9.11	10.91
Varanus bengalensis nebulosus	FMNH 211856	0.372	0.38	0.791	0.400	0.001	0.392	0.399	8.28	9.90
Varanus dumerili	FMNH 228151	0.274	0.313	0.505	0.292	0.001	0.212	0.292	7.92	8.97
Varanus exanthematicus	FMNH 228398	0.311	0.307	0.654	0.342	0.001	0.312	0.341	9.85	10.74
Varanus komodensis	FMNH 22197	0.685	0.752	1.189	0.677	0.001	0.513	0.676	28.97	31.99
Varanus rudicollis	FMNH 98947	0.411	0.440	1.031	0.393	0.001	0.638	0.391	14.97	16.53
Varanus salvator	FMNH 22204	0.781	0.771	1.355	0.764	0.000	0.590	0.764	19.18	19.77
Varanus salvator	FMNH 98866	0.314	0.242	0.562	0.329	0.001	0.233	0.328	15.96	17.82
Varanus salvator	FMNH 195576	0.704	0.504	1.080	0.432	0.001	0.649	0.431	21.50	21.17
Varanus salvator	FMNH 211938	0.280	0.380	0.672	0.403	0.001	0.269	0.402	11.16	12.74
Caiman crocodilus	FMNH 9150	0.791	0.448	1.672	0.670	0.001	1.002	0.669	–	–
Caiman crocodilus	FMNH 13062	0.531	0.330	1.128	0.475	0.001	0.653	0.474	14.51	17.54
Caiman crocodilus	FMNH 217159	0.602	0.409	1.304	0.471	0.001	0.833	0.470	–	–
Melanosuchus niger	FMNH 218507	0.629	0.372	1.175	0.485	0.001	0.690	0.484	–	–
Crocodylus rhombifer	FMNH 34677	0.885	0.688	2.298	0.883	0.001	1.416	0.881	18.10	26.95
Gavialis gangeticus	FMNH 82681	0.905	0.572	1.655	0.658	0.001	0.997	0.656	12.23	15.15
Chironectes minimus	FMNH 60576	0.550	0.442	1.025	0.378	0.002	0.647	0.376	10.33	11.57
Chironectes minimus	FMNH 60578	0.656	0.541	1.440	0.451	0.001	0.990	0.449	10.91	12.29
Castor canadensis	FMNH 18525	0.694	0.609	1.108	0.349	0.001	0.758	0.348	17.58	21.90
Castor canadensis	FMNH 134455	1.032	0.667	1.569	0.467	0.002	1.102	0.465	–	–
Castor canadensis	FMNH 141992	0.885	0.612	1.280	0.352	0.001	0.928	0.351	17.74	22.72
Lontra canadensis	FMNH 60655	0.632	0.273	1.233	0.400	0.002	0.833	0.398	15.00	17.14
Lontra canadensis	FMNH 160115	0.292	0.190	0.680	0.267	0.001	0.413	0.267	15.10	17.20
Lontra canadensis	FMNH 175293	0.627	0.283	1.563	0.500	0.001	1.062	0.499	16.35	18.27
Lontra canadensis	FMNH 180757	1.060	0.325	2.110	0.686	0.002	1.424	0.685	16.21	17.85
Neovison vison	FMNH 59739	0.534	0.227	1.371	0.679	0.001	0.693	0.677	8.27	9.58
Neovison vison	FMNH 59776	0.476	0.194	0.847	0.399	0.002	0.448	0.397	8.52	9.95
Neovison vison	FMNH 59793	0.792	0.268	1.622	0.710	0.001	0.912	0.709	10.42	11.81
Neovison vison	FMNH 112982	0.672	0.230	1.946	0.710	0.002	1.235	0.709	8.44	9.72

See text for details on calculation of the metrics. Fore limb length is calculated as the sum of humerus and radius length; hind limb length is calculated as the sum of femur and tibia length. Metrics for FMNH UC 581 are based on measurements from Olson (1962) our measurements. We only measured limb lengths (see below) for *Aerosaurus* because most of the vertebrae in the specimen are not exposed in a manner that allows centrum dimensions to be measured.

We also measured 31 specimens representing 16 species of extant semi-aquatic and terrestrial reptiles and semi-aquatic mammals. The extent reptile taxa measured were *Amblyrhynchus cristatus* (marine iguana), *Caiman crocodilius* (spectacled caiman), *Gavialis gangeticus* (gharial), *Iguana iguana* (green iguana), *Melanosuchus niger* (black caiman), *Varanus bengalensis* (Bengal monitor), *Varanus dumerilii* (Dumeril's monitor), *Varanus exanthematicus* (savannah monitor), *Varanus komodoensis* (Komodo dragon), *Varanus salvator* (water monitor), and *Varanus rudicollis* (roughneck monitor). The mammalian taxa were *Castor canadensis* (American beaver), *Chironectes minimus* (water opossum), *Lontra canadensis* (North American river otter), and *Neovison vison* (American mink). *Iguana iguana* and all of the *Varanus* species except for *V. salvator* were chosen as representatives of moderately-sized terrestrial reptiles that likely have a locomotor pattern comparable to that of basal synapsids; the other species were chosen because they are small- to medium-sized semi-aquatic reptiles and mammals that can provide insight into the range of variation in vertebral proportions found in semi-aquatic tetrapods with lifestyles that may be analogous to that of *Ophiacodon*. *Crocodylus rhombifer* is of interest because it spends more of its time on land than many other extant crocodiles and possesses specializations for terrestrial locomotion such as reduced webbing of the feet and powerful hind limbs (Braziatis 1973). Only specimens that had multiple centra in sequence were selected, and then only if the positions of the centra along the column could be determined.

We measured centrum length, width, and height for each available centrum except the atlas and axis, with all measurements made to the nearest 0.1 mm using either digital or dial calipers (Fig. 3.2). Measurements for FMNH UC 581 are taken from Olson (1962). To obtain a graphical overview of each specimen's vertebral profile that could be easily compared, we plotted centrum length against vertebral position (Figs. 3.3, 3.4, 3.5, 3.6, 3.7, 3.8, 3.9, 3.10). We also measured lengths of a femur, tibia, humerus, and radius for each specimen when these bones were available. Limb bone lengths represent the longest dimension of the bone.

In addition, we quantified the properties of the vertebral columns of the extant and fossil taxa using the series of metrics developed by McShea (1992; also see McShea 1993). Each metric (*R, C, Cm, E*1 and *E*2, see below) was calculated for each vertebral parameter (centrum length, width, and height) of each specimen. For each parameter for a given specimen, the series X_1 through X_N represents all

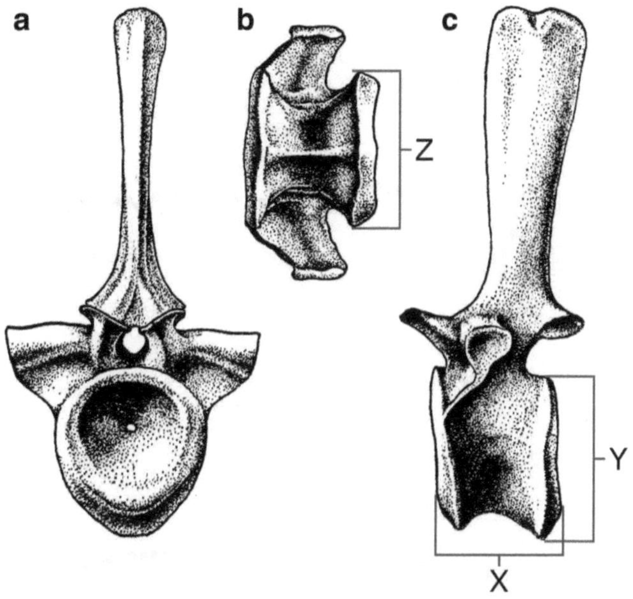

Fig. 3.2 Dorsal vertebrae of *Ophiacodon mirus* in anterior (**a**), inferior (**b**) and left lateral (**c**) view. Centrum width (X), centrum height (Y), and centrum width (Z) were evaluated as illustrated. Drawings from Romer and Price (1940)

the measurements along the vertebral column in order. The range, *R*, of that series is simply:

$$R = \log(X_{max} - X_{min}).$$

Polarization, *C*, which considers the degree to which the vertebral column is divided into sections with dimensions clustered at high and low values, is calculated as:

$$C = \log\left(2\sum_{i=1}^{N}(X_i - \overline{X})/N\right).$$

Irregularity, *Cm*, which quantifies the average difference in dimensions between successive vertebrae across the column, is calculated as:

$$Cm = \log\left(\left(\sum_{i=1}^{N-1}(X_{i+1} - X_i)\right)/(N-1)\right).$$

Related to polarization is *E*1, or concentration, which measures the degree to which centrum dimensions are concentrated at the mean or at the extremes, is calculated as:

$$E1 = R - C.$$

Related to irregularity is *E*2, or smoothness, which measures the degree to which successive vertebrae are constrained to have similar dimensions, is calculated as:

$$E2 = C - Cm.$$

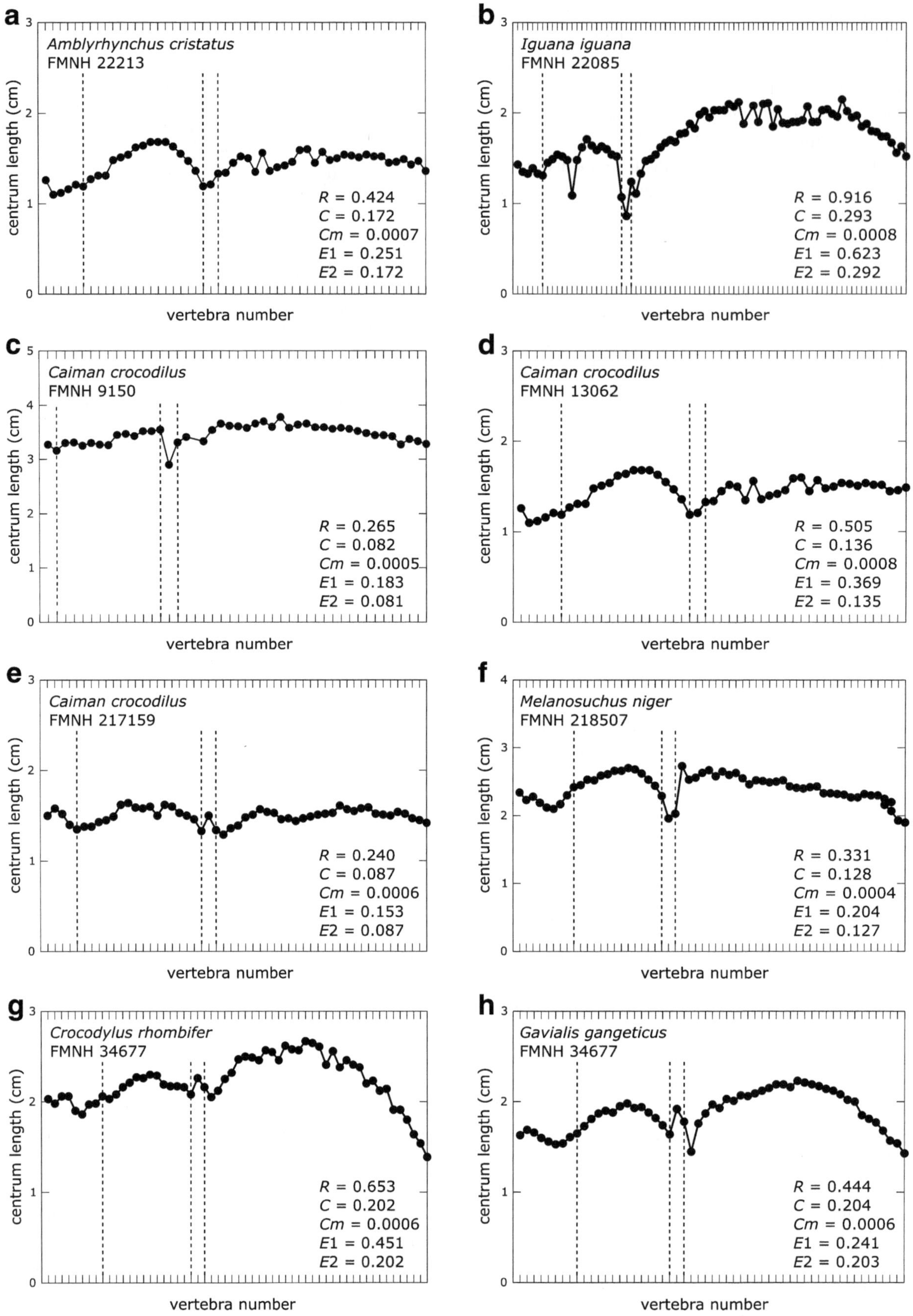

Fig. 3.3 Plots showing centrum length profiles for *Amblyrhynchus*, *Iguana*, *Caiman*, *Melanosuchus*, *Crocodylus*, and *Gavialis*. *Dashed lines* mark the last cervical, dorsal, and sacral vertebrae in each specimen

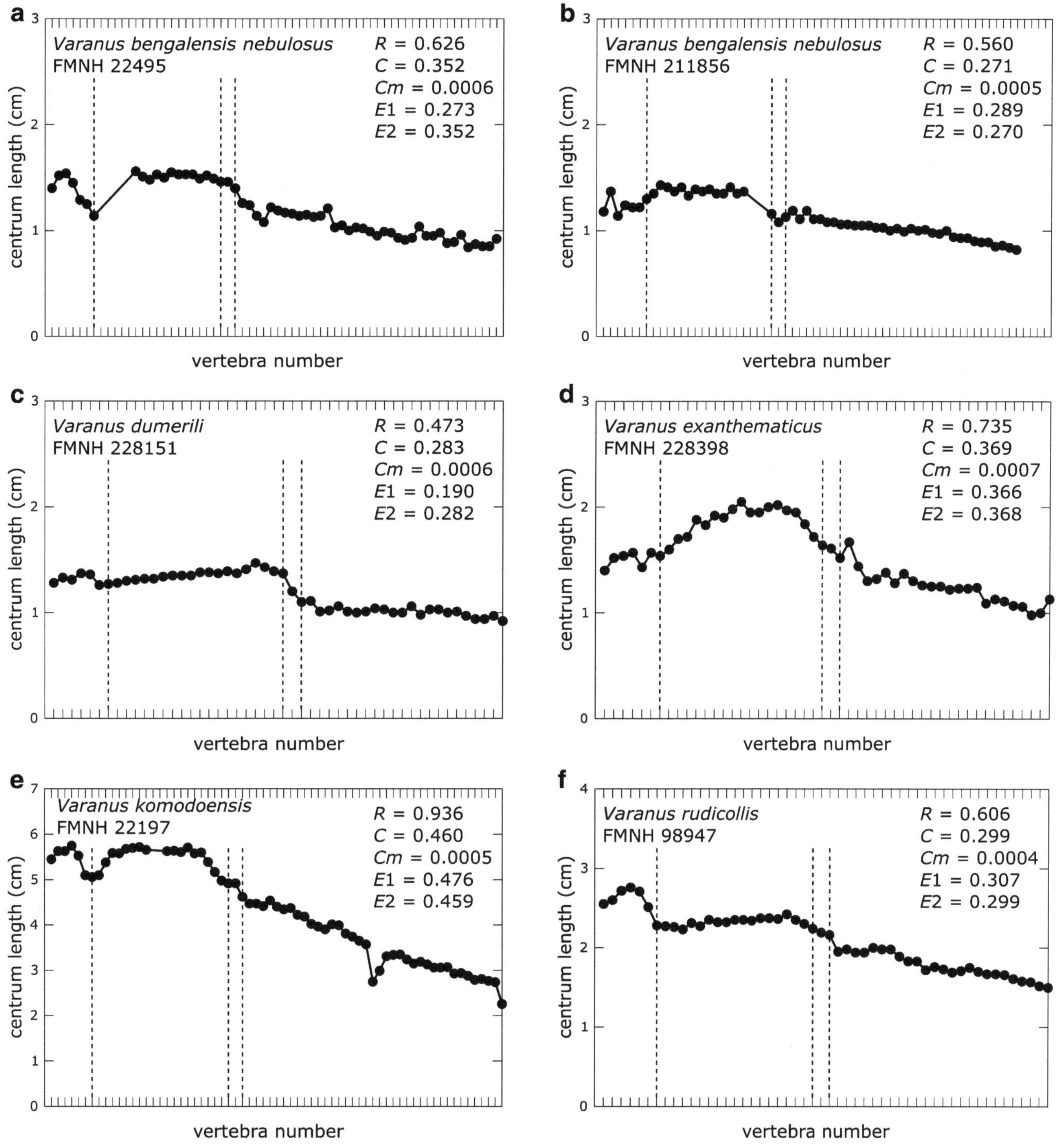

Fig. 3.4 Plots showing centrum length profiles for the terrestrial varanid specimens. *Dashed lines* mark the last cervical, dorsal, and sacral vertebrae in each specimen

Following McShea (1993), we removed size effects by log transforming all centrum measurements before computing the metrics, and then omitting the log operation from the above equations.

To visualize patterns captured by these metrics, we performed two principal components analyses (PCA) on variance-covariance matrices of the metrics. The first PCA used only the five metrics calculated for our centrum length

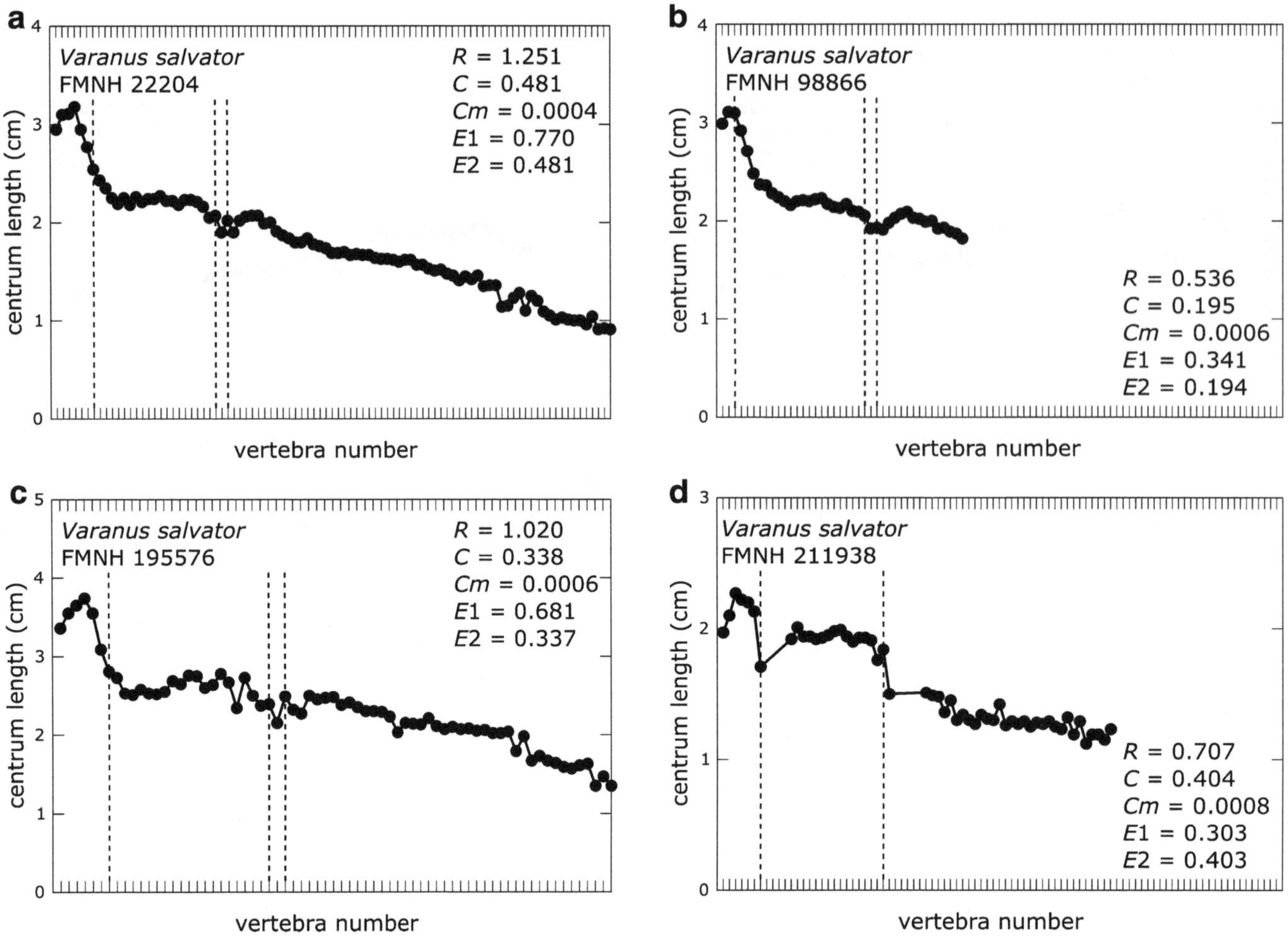

Fig. 3.5 Plots showing centrum length profiles for *Varanus salvator* specimens. *Dashed lines* mark the last cervical, dorsal, and sacral vertebrae in each specimen. The last sacral is missing in FMNH 211938

measurements because we were able to measure this dimension on nearly all specimens. To avoid spurious results, we excluded seven fossil specimens (FMNH P 12841, FMNH UC 112, FMNH UC 690, FMNH UC 709, FMNH UR 607/UR 616, FMNH UR 827, UCMP V2841/40096) because they preserve only limited sections of the vertebral column and therefore likely do not capture the full range of variation in centrum lengths originally displayed by the specimens. The second PCA was based on the metric values for centrum, length, width, and height (for a total of 15 variables). Although this analysis has the potential to provide a more nuanced view of variation in vertebral proportions, it has the drawback of including a smaller number of specimens because all three dimensions could not be accurately measured for all of the fossil specimens (Table 3.1). All specimens but one (FMNH UR607/UR616) that had length, height, and width metrics were included in this analysis. Note that two specimens, FMNH UC 112 (*Dimetrodon giganhomogenes*) and FMNH P 12841 (*Varanops* sp.), which we excluded from the length-only PCA were included in this analysis despite the relatively limited number of vertebrae available for each to ensure that these taxa had at least one representative in the data set.

As noted above, Romer and Price (1940) and Kemp (1982) also cited the disparity in limb lengths (hind limbs longer than fore limbs) observed in *Ophiacodon* as evidence for a semi-aquatic lifestyle. To determine whether such a difference should be expected, we measured fore and hind limb lengths (measured as the summed length of the humerus and radius or femur and tibia) in all of the examined specimens in which these elements were preserved. We excluded the manus and pes from these measurements because they were infrequently preserved in the fossil specimens. We then carried out two Kolmogorov-Smirnov two-sample tests on the ratios of fore limb length to hind limb length. In the first test, we included only the extant specimens, whereas we included the extant and fossil specimens (excluding *Ophiacodon*) in the second test. To test whether *Ophiacodon* is indeed more similar to semi-aquatic taxa in this regard, we calculated whether the

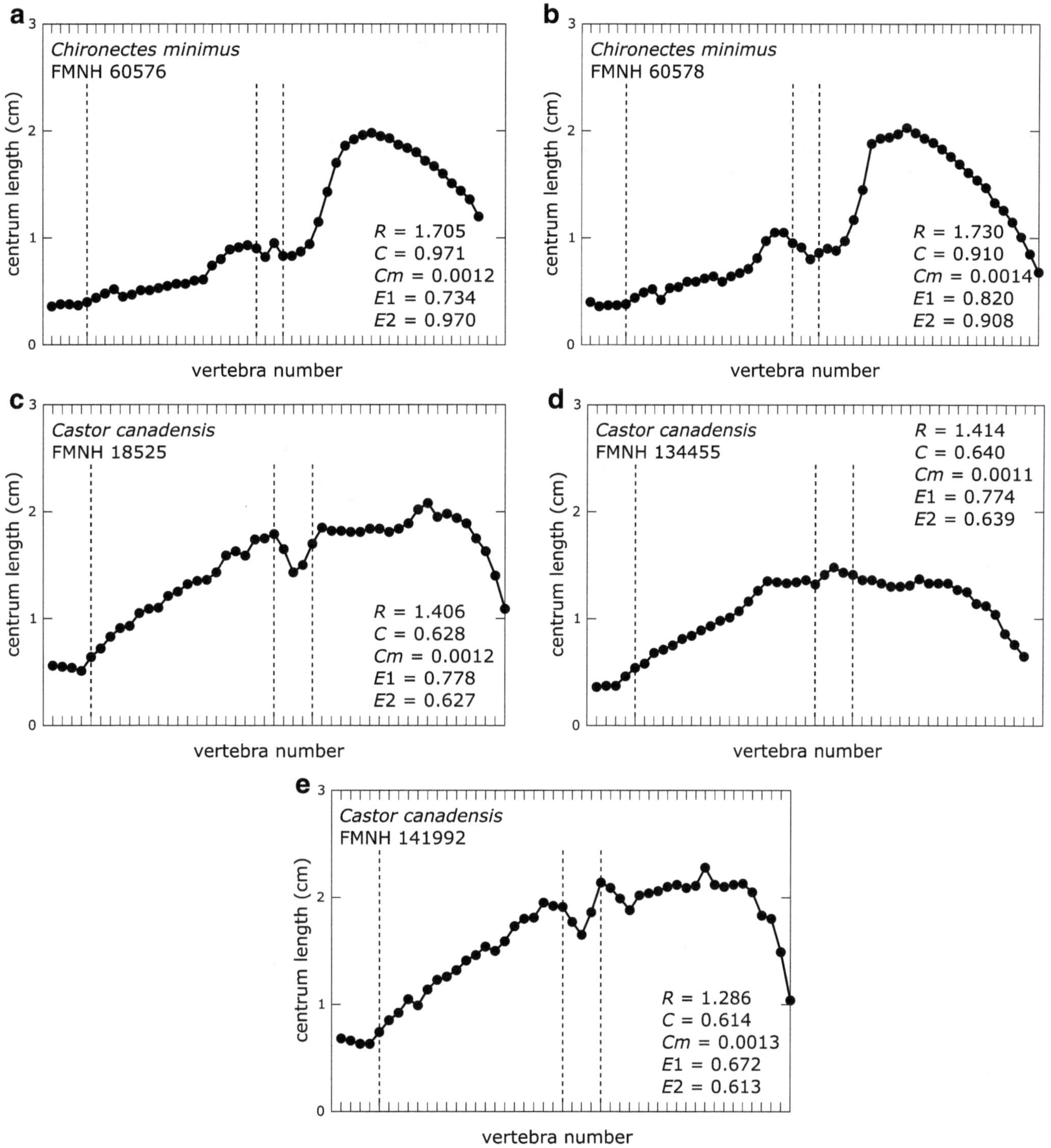

Fig. 3.6 Plots showing centrum length profiles for *Chironectes* and *Castor*. *Dashed lines* mark the last cervical, lumbar, and sacral vertebrae in each specimen

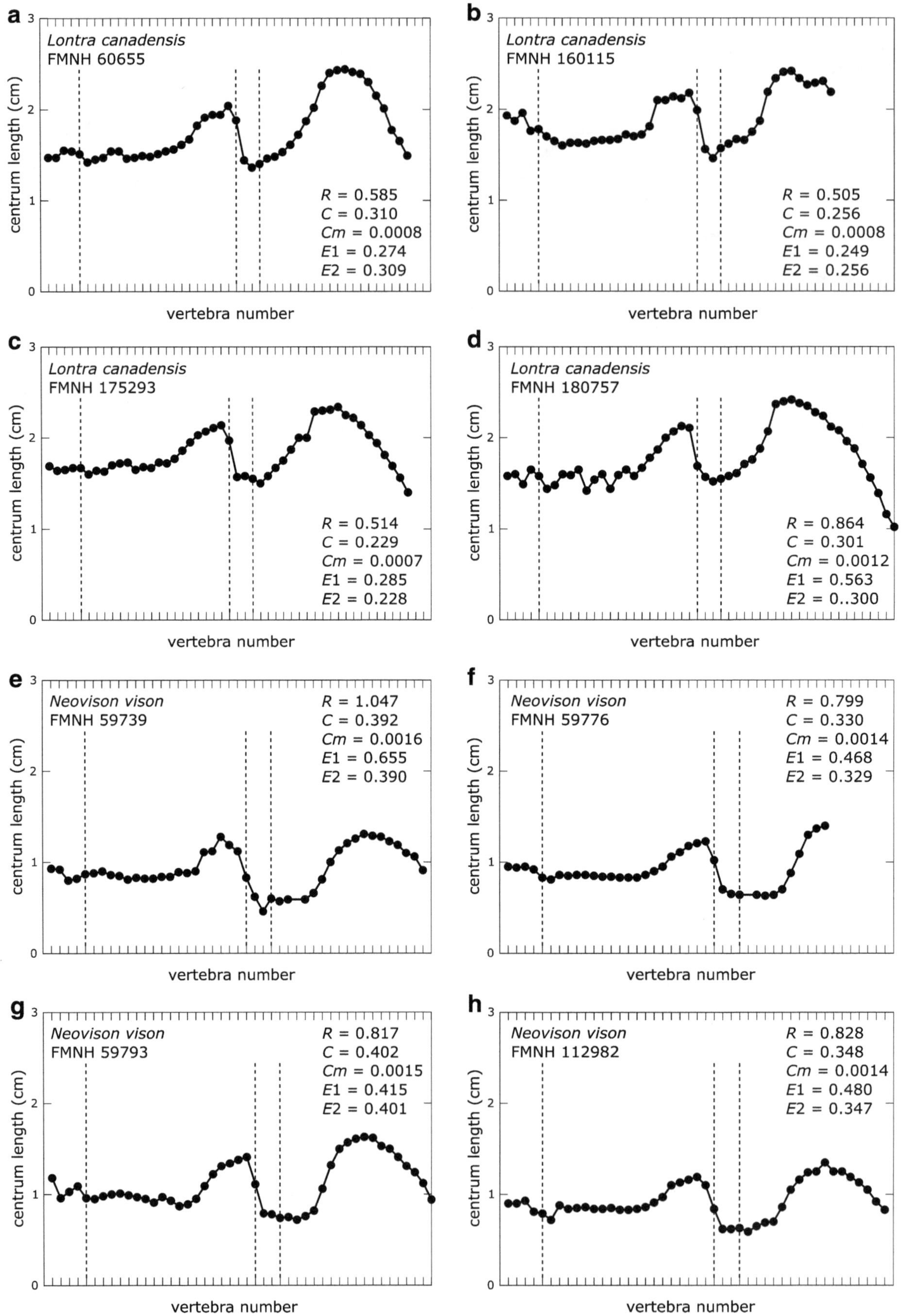

Fig. 3.7 Plots showing centrum length profiles for *Lontra* and *Neovison*. *Dashed lines* mark the last cervical, lumbar, and sacral vertebrae in each specimen

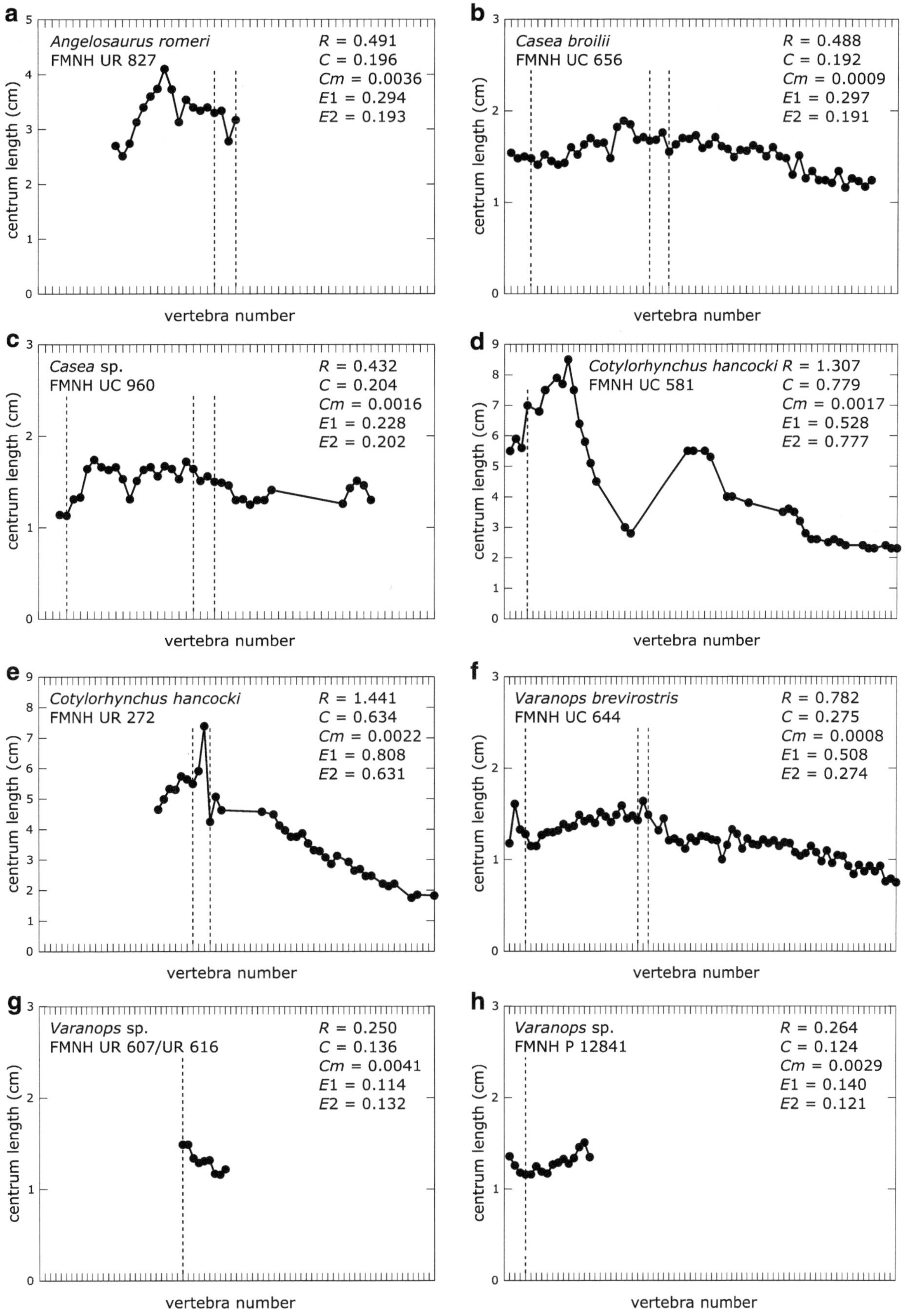

Fig. 3.8 Plots showing centrum length profiles for the caseid and varanopid specimens. *Dashed lines* mark the last cervical, dorsal, and sacral vertebrae in each specimen. The last cervical is not preserved in FMNH UR 827 and FMNH UR 607/UR 616. The last sacral is not preserved in FMNH UR 607/UR 616, and no sacrals are preserved in FMNH UC 581 and FMNH P12841

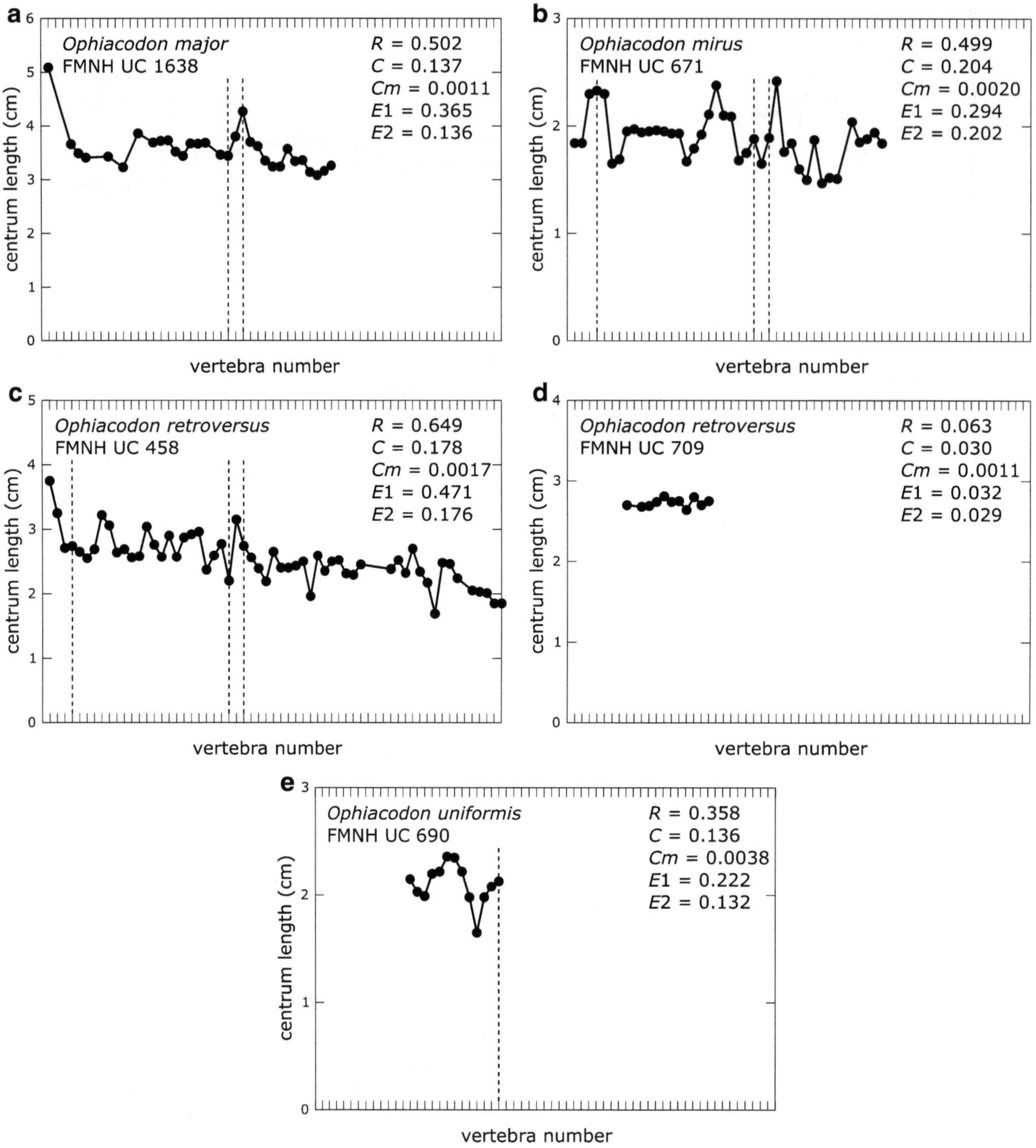

Fig. 3.9 Plots showing centrum length profiles for *Ophiacodon* specimens. *Dashed lines* mark the last cervical, dorsal, and sacral vertebrae in each specimen. The last cervical is not preserved in FMNH UC 1683, FMNH UC 709, and FMNH UC 690. The last sacral is not preserved in FMNH UC 690, and the last dorsal and the last sacral are not preserved in FMNH UC 709

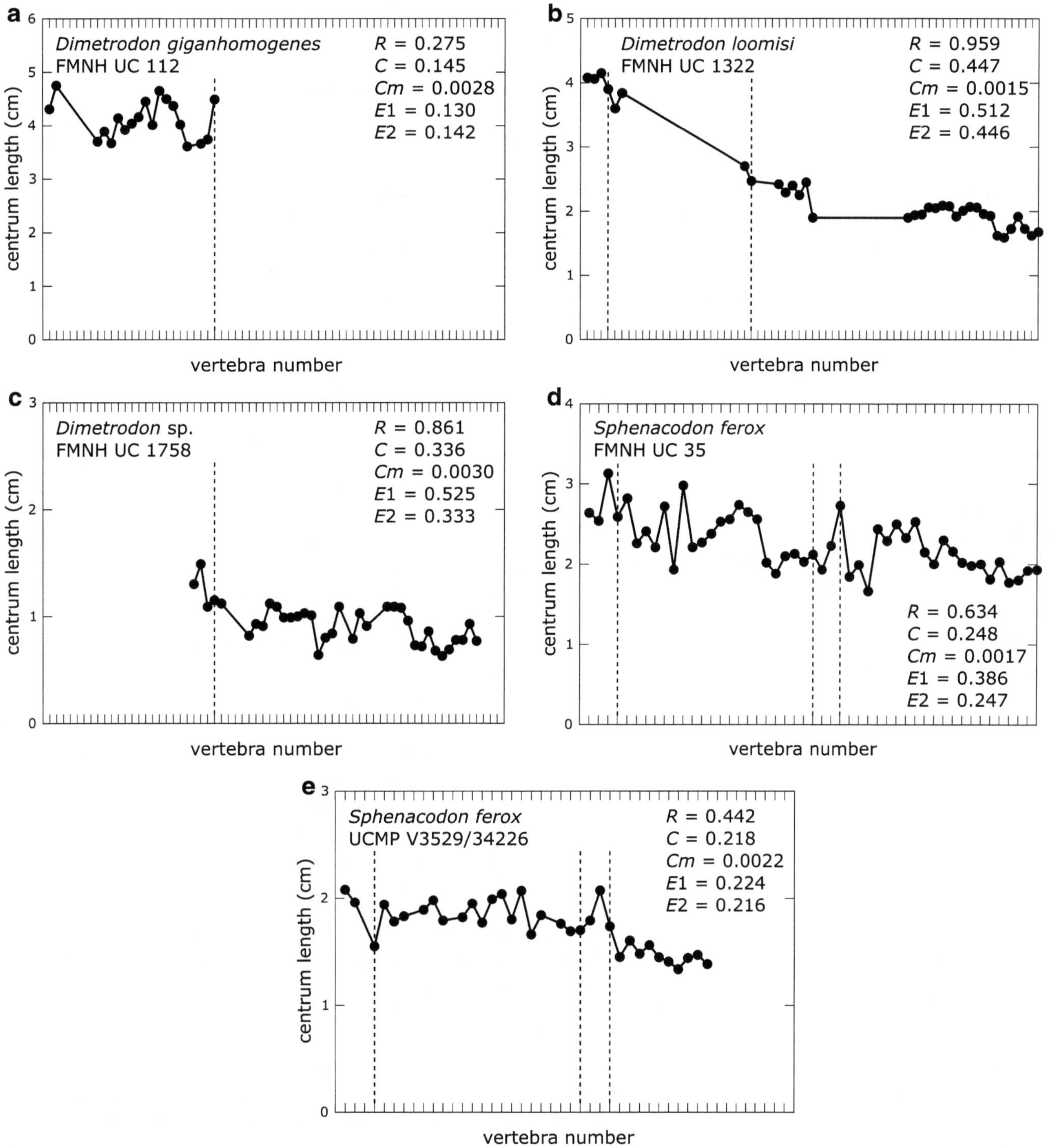

Fig. 3.10 Plots showing centrum length profiles for the sphenacodontid specimens. *Dashed lines* mark the last cervical, dorsal, and sacral vertebrae. The last cervical and last sacral are not preserved in FMNH UC 112 and FMNH UC 1758. The last sacral is not preserved in FMNH UC 1322

Ophiacodon specimen we measured (FMNH UC 671) was closer to the mean value for the terrestrial or aquatic groups.

Results

The extant taxa show several distinct patterns of vertebral morphology. Although many of these are congruent with previous observations (e.g., there is a great deal of similarity between our plots and those of Buchholtz (1998) for those taxa sampled in both studies), some unexpected patterns emerged, especially among the reptiles. Because of the importance centrum length has played in previous works on swimming style, we focus on these results.

The *Amblyrhynchus*, *Iguana*, *Caiman*, *Crocodylus*, and *Gavialis* specimens we measured all show a peak in centrum length between the pectoral and pelvic girdles, corresponding to the area of lateral undulation during terrestrial locomotion (Buchholtz 1998), although the shape of the peak (sharp vs. more plateau-like) varied from taxon to taxon (Fig. 3.3). These taxa also all show a peak or plateau in vertebral length in the caudal vertebrae, which Buchholtz (1998) stated represents the region where traveling undulatory waves pass along the tail during swimming (in semiaquatic taxa) or terrestrial locomotion (in taxa with reduced limbs). In general, these taxa tended to have relatively low to moderate values for the metrics we calculated (Table 3.1) for length, indicating that they have comparatively uniform vertebral columns. Interestingly, among these taxa, the terrestrial *Iguana iguana* had the highest scores for all metrics.

The six *Varanus* species we measured show several vertebral length profiles, all of which are unlike those of the other reptiles (Figs. 3.3, 3.4). *Varanus exanthematicus*, *V. bengalensis nebulosus,* and *V. komodoensis* possess a peak or plateau in centrum length between the pectoral and pelvic girdles, much like the other reptiles, but then show a steady decline that starts near the pelvic girdle and continues throughout the tail. A distinct peak between the pectoral girdles is absent in *V. dumerilii* and *V. rudicollis*. Instead, after a decrease in length near the pectoral girdle, there is a gradual increase in centrum length to just before the sacrum. Centrum length then declines and either continues a gradual decrease (*V. rudicollis*) or is relatively constant for the remainder of the tail (*V. dumerilii*). Finally, *V. salvator* shows a sharp drop in length in the first few dorsal vertebrae followed by very slowly decreasing centrum lengths for the rest of the vertebral column. In general, the varanids were characterized by moderate ranges of centrum length (*R*), moderate to high polarization (*C*), low to moderate irregularity (*Cm*), a wide range of *E*1 values, and moderate to high *E*2 values (Table 3.1). These patterns reflect the fact that the varanids tend to be characterized by small changes in length between successive vertebrae (i.e., low amounts of noise), but often have relatively polarized columns, with longer presacrals and shorter caudals.

The plots of vertebral length in extant mammals (Figs. 3.6, 3.7) also fall into several categories. The two mustelids measured, *Lontra canadensis* and *Neovison vison*, display distinctly bimodal patterns of centrum length. Both show uniform centrum length in the thorax, a peak in centrum length in the lumbar region, and a second peak in centrum length in the distal caudal region. *Chironectes minimus* also displays a bimodal centrum length profile, but the lumbar peak in centrum length is less well defined than in the mustelid genera. Unlike the other extant mammal taxa examined, *Castor canadensis* does not display a bimodal pattern in centrum length. Rather, it shows a uniform increase in centrum length, width, and height through the trunk, followed by a plateau in the dimensions of the caudal vertebrae. The mammals tended to be characterized by moderate to high ranges of centrum length (*R*) and polarization (*C*), low to moderate irregularity (*Cm*), and moderate to high *E*1 and *E*2 values (Table 3.1). This reflects the fact that most of the mammals have vertebral columns characterized by relatively great peaks of centrum length, as well as relatively close correspondence in centrum length in successive vertebrae (i.e., low amounts of noise).

The synapsid specimens measured also display several centrum length profiles, but many do not show an exact correspondence with the extant taxa (Figs. 3.8, 3.9, 3.10). The caseid specimens (Fig. 3.8) are the most basal, and perhaps among the most terrestrial, synapsids we measured. Most show evidence of a plateau or peak in centrum length between the pectoral and pelvic girdles, although its exact placement along the column seems to vary. In contrast, FMNH UC 656 (*Casea broilii*) and perhaps FMNH UC 272 (*Cotylorhynchus hancocki*) show a weak pattern of increase in centrum length to the pelvic girdle. However, FMNH UC 656 includes vertebrae from three individuals (Williston 1911), and FMNH UC 272 preserved a limited number of presacrals, so they may provide a biased picture of the vertebral proportions of the presacral column. All caseids with well-preserved tails show a gradual decline in centrum length in the caudal series. Taken together, the centrum length profile that emerges for caseids includes a weak plateau in length between the girdles, followed by gradually decreasing centrum lengths in the tail, a pattern resembling some of the *Varanus* species we measured (e.g., Fig. 3.4). Although it occupies a more crownward position in synapsid phylogeny than the caseids, the varanopid *Varanops* also displays a centrum length profile resembling that of the extant varanids, particularly *V. dumerilii* (Fig. 3.4). Most of the caseid and *Varanops* specimens show low to moderate ranges of centrum length (*R*), and polarization (*C*) (Table 3.1), although

Cotylorhynchus hancocki is characterized by higher *R* and *C* values, likely reflecting its large size (McShea 1993). Like all of the fossil specimens, the caseids and *Varanops* have relatively high irregularity values (*Cm*), perhaps because of minor amounts of deformation (McShea 1993). Finally, these specimens show a wide range of *E*1 and *E*2 values. Although this likely reflects genuine differences in some cases (e.g., centrum lengths for *Casea broilii* cluster more closely around their mean value than is the case for *Cotylorhynchus hancocki*), some of the values may be artifacts reflecting the fact that certain of the measured columns represent only small sections of the original vertebral series of those specimens.

A single, clear pattern of region variation is not apparent in the *Ophiacodon* specimens that we measured (Fig. 3.9). FMNH UC 671 shows a weak plateau in centrum length in the anterior part of the thoracic column, but such a plateau is not apparent in FMNH UC 1638 or FMNH UC 709, and this section of the column is very "noisy" in FMNH UC 458. A narrow peak in centrum length is present five to seven vertebrae anterior to the sacrum in most of the *Ophiacodon* specimens in which this region is preserved, although the noisiness of FMNH UC 458 makes the pattern less distinct than in the other specimens. None of the specimens preserves a complete caudal series, but although some variation in centrum length among the caudals of the two most complete specimens exists (FMNH UC 458 and FMNH UC 1638), there is no evidence of a distinct peak or plateau comparable to that observed in the extant crocodilians, iguanas, or mammals. Taken together, these observations suggest that while some undulation may have occurred between the girdles of *Ophiacodon*, there was no obvious specialization in the tail for the propagation of undulatory waves used in swimming. The *Ophiacodon* specimens typically displayed low to moderate ranges of centrum length (*R*) and polarization (*C*), high irregularity (*Cm*), a wide range of *E*1 values and low *E*2 values (Table 3.1). These patterns reflect the fact that the specimens displayed relatively few distinct peaks or plateaus in centrum length, but had comparatively noisy vertebral columns.

A consistent pattern of regional variation also was not apparent for the sphenacodontids we measured (Fig. 3.10). The two *Sphenacodon* specimens present very different centrum length profiles: the presacral, sacral and anterior caudal centra of UCMP V3529/34226 are very uniform in length, with no apparent peaks or plateaus. The centra of FMNH UC 35 are much more variable, with an extremely noisy region in the cervicals and anterior dorsals, a peak in the mid-dorsal region, and a weakly developed peak in the anterior caudals. None of the *Dimetrodon* specimens available for this study preserve a complete vertebral column, but a relatively complete composite picture can be assembled from FMNH UC 112, FMNH UC 1322, and FMNH UC 1758. A peak in centrum length is present in the mid-to-posterior dorsal series of FMNH UC 112, as well as an increase in length towards the sacrum. The caudal series of FMNH UC 1758 is relatively noisy, but shows some evidence of a plateau or multiple peaks. The caudals of FMNH UC 1322 are much more uniform, but show a weak plateau in length near the distal end of the tail. However, the caudal peaks and plateaus in these specimens differ from those observed in the iguanas, crocodilians, and mammals because the centrum lengths in the peak/plateau are less than those in the presacral series, whereas they are as long or longer in the extant taxa. The sphenacodontids spanned a wide range of centrum length (*R*), polarization (*C*), *E*1 and *E*2 values, and possessed high irregularity (*Cm*) values (Table 3.1). These patterns likely reflect several causes, both real (e.g., wide range of lengths in FMNH UC 1322) and artifactual (e.g., only portions of the total column preserved in FMNH UC 112; noise in the measurements from minor deformation).

In the analysis based only on length metrics, the first two PC axes captured approximately 94.9 and 5.1 % of the variance, respectively, so we focus our interpretation on these two axes. The metric *R* had a high positive loading on axis 1, and *C*, *E*1, and *E*2 had moderate positive loading on this axis. The metrics *R* and *E*2 had moderate and high negative loadings on PC 2, respectively, whereas *C* and *E*2 had moderate positive loading. The metric *Cm* had a loading very close to zero on both axes. On a plot of axes 1 and 2, most of the extant semi-aquatic species fall together in an area near the origin, although *Iguana iguana* also falls within this region (Fig. 3.11). The terrestrial varanid species also are located near the origin on axis 1, but extend higher on axis 2 than the semi-aquatic reptiles (except one outlier in *V. salvator*, FMNH 211938, which is a captive specimen). *Castor canadensis* and *Chironectes minimus* have higher scores on axis 1 than the reptiles, but much like them *Castor* falls mostly on the negative side of axis 2. *Chironectes* has positive scores on axis 2. All of the basal synapsids were included in the PCA but two fall in the region occupied primarily by the extant semi-aquatic reptiles, with the three *Ophiacodon* specimens falling closest to the range of the extant crocodilians. The basal synapsid specimens that do not fall among the extant semi-aquatic reptiles are two specimens of *Cotylorhynchus*, one of which falls near *Castor canadensis* whereas the other is located in a unique position in quadrant 1. The greater noisiness of the basal synapsid vertebral columns is reflected by the fact the variable *Cm* has high positive loadings on PC axis 3 and the extant specimens (aside from *Neovison vison*) and fossil specimens segregate relatively well this axis. However, this axis accounts for a trivial proportion (approximately 0.02 %) of the variance in the data set.

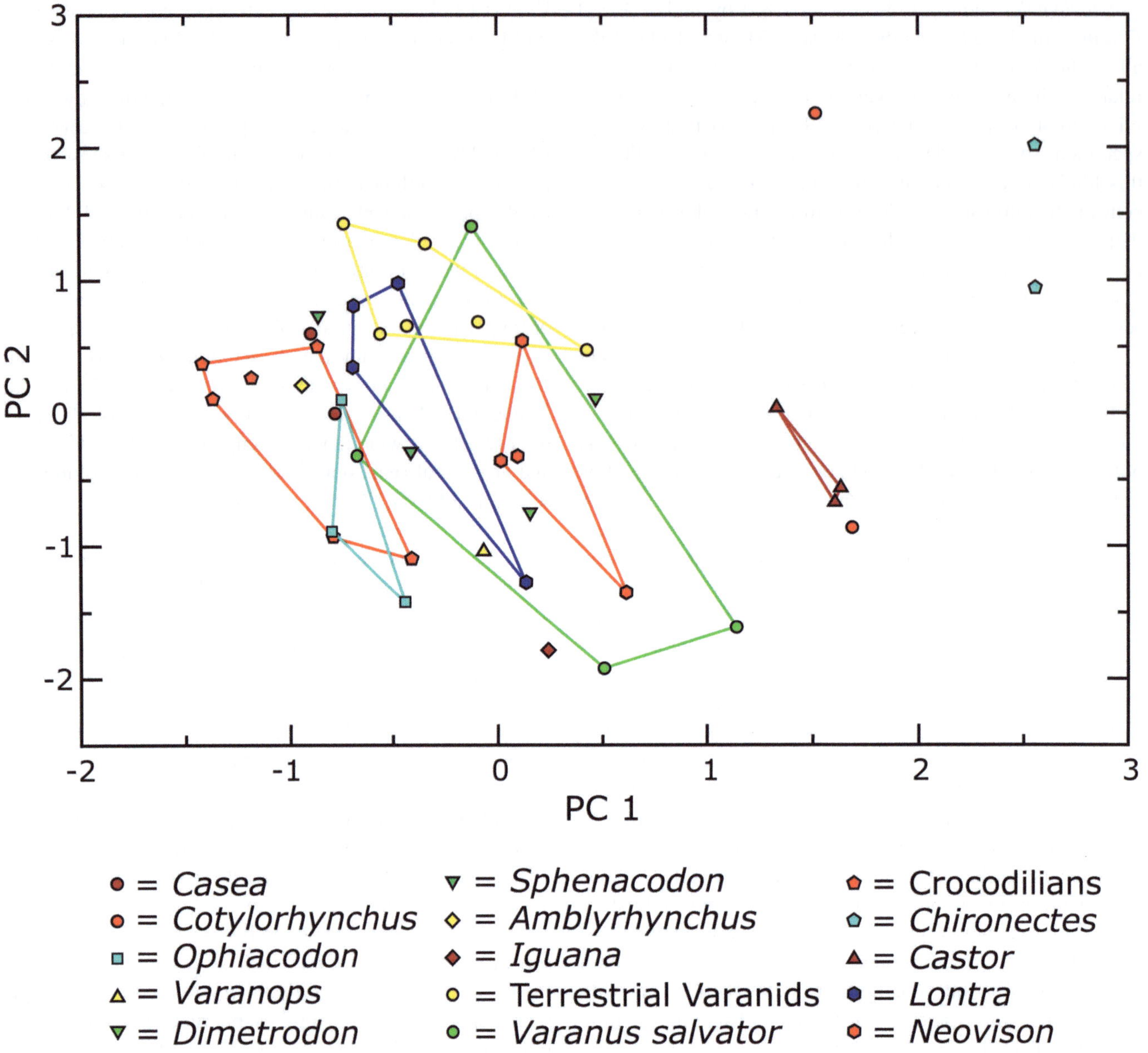

Fig. 3.11 Plot showing the distribution of specimens on principal components one and two of the PC analysis of centrum length metrics. Convex hulls have been drawn for taxa represented by three or more specimens

In the analysis based on all of the length, width, and height metrics, the first three PC axes captured approximately 66.9, 22.2, and 6.3 % of the variance respectively. The remaining axes captured 2.5 % of the variance or less, so our interpretation focuses on the first three axes. All of the length, width, and height metrics except the *Cm* metrics had positive loadings on PC 1. Most of these loadings were low to moderate, but R_{width} was noteworthy for having the highest positive loading of all of the metrics. The metrics *R*, *C*, *E*1, and *E*2 for length all had moderate to high positive loadings on PC 2. *R*, *C*, and *E*2 for height had low positive loading on PC 2, but *E1* had a low negative loading. Loadings for *R*, *C*, *E*1, and *E*2 for width were weakly to moderately negative on PC2. The *Cm* metrics had slightly positive loadings on PC 2 in all cases. The metrics *R*, *C*, *E*1, and *E*2 for length had low to moderate negative loadings on PC 3. *R* and *E1* for width also had moderate negative loadings on PC 3, but *C* and *E*2 for width had slightly positive loadings. All four of these metrics for height had moderate to strong positive loadings on PC 3. The *Cm* metrics had slightly negative loadings on PC 3 in all cases. The basal synapsids were differentiated from most of the extant specimens along PC 1 in this analysis (Fig. 3.12), and the area occupied by the synapsids was closest on

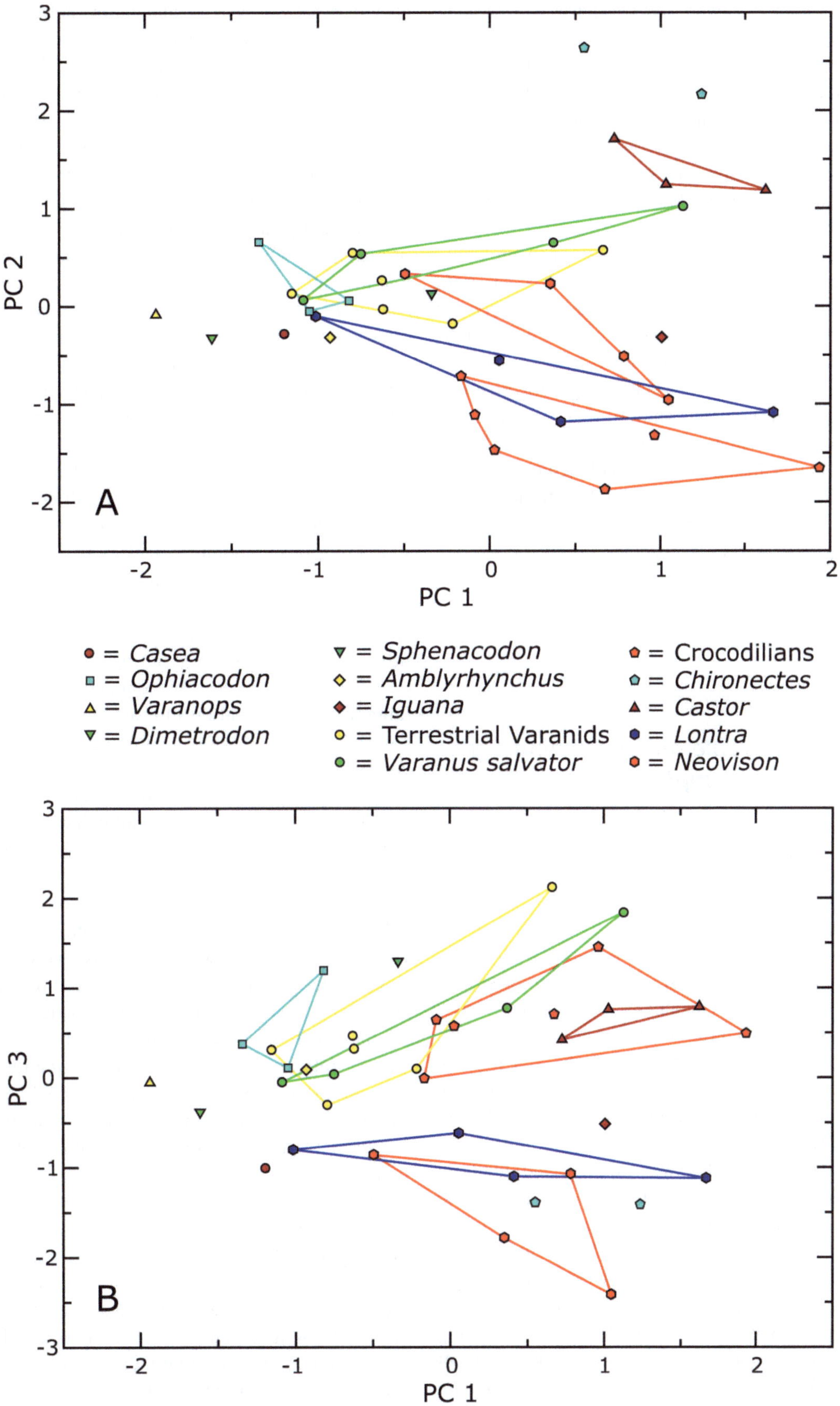

Fig. 3.12 Plots showing the distribution of specimens on principal components one, two, and three of the PC analysis of length, width, and height metrics. **a** PCs 1 and 2. **b** PCs 1 and 3. Convex hulls have been drawn for taxa represented by three or more specimens

average to the area occupied by the terrestrial varanid specimens. Among the basal synapsids, the three *Ophiacodon* specimens and one *Sphenacodon* specimen were located especially close to or within the area occupied by the terrestrial varanids. Surprisingly, the specimen of *Amblyrhynchus cristatus* fell close to the basal synapsids and terrestrial varanids, especially on PCs 1 and 3, whereas the *I. iguana* specimen was closer to the crocodiles and semi-aquatic mammals.

Finally, only two of the specimens we measured that preserved fore and hind limb elements (FMNH UC 656, *Casea broilii*; FMNH UR 272, *Cotylorhynchus hancocki*) possessed fore limbs that were longer than hind limbs (Table 3.1). The remaining specimens all possessed longer hind limbs, regardless of ecology and phylogenetic position. The Kolmogorov-Smirnov tests showed that terrestrial and aquatic taxa did not differ significantly in limb length regardless of whether fossil taxa were included ($p = 0.295$) or not ($p = 0.809$). The fore limb to hind limb ratio of the measured *Ophiacodon* specimen was closer to the mean for terrestrial taxa regardless of whether fossil taxa were included or excluded.

Discussion

Centrum Length Profile and Swimming Style in the Extant Taxa Sampled

The centrum length profiles of several extant and extinct taxa have received attention in the literature (e.g., Buchholtz 1998, 2001a, b; Gingerich 1998; Buchholtz and Shur 2004; Buchholtz et al. 2005, 2007; Pierce et al. 2011), but most of this work has focused on mammals, particularly whales, and highly pelagic reptiles (i.e., ichthyosaurs). Our results for the extant taxa we examined suggest that there is likely more diversity in centrum length profile than has been appreciated previously, even intraspecifically, and that the correlation between centrum length profile and aquatic locomotion in semi-aquatic taxa may be less exact than in highly aquatic tetrapods.

Among the extant reptiles we sampled, the majority possess a peak in centrum length between the pelvic and pectoral girdles, corresponding to the area of lateral undulation during terrestrial locomotion (Buchholtz 1998). However, three of the varanid species we measured (*Varanus dumerilii*, *V. rudicollis*, and *V. salvator*) lack this peak, despite having sprawling limbs and laterally undulating during terrestrial locomotion, as do some specimens of *Caiman crocodilus*. The semi-aquatic reptiles *Caiman*, *Crocodylus*, *Gavialis*, *Melanosuchus*, and *Amblyrhynchus* also all possess a peak or plateau in centrum length in their tails, corresponding to the region where traveling waves pass along the vertebral column during swimming (Buchholtz 1998). The presence of this peak or plateau does not seem to be a foolproof indicator of a semi-aquatic lifestyle, though, because a similar plateau is present in the vertebral profile of the terrestrial *Iguana iguana* but is lacking in *Varanus salvator*, which is an adept swimmer.

A similarly inexact relationship between centrum length profile and swimming style is apparent when the extant mammalian results are considered. *Lontra canadensis*, *Neovison vison*, and *Chironectes minimus* all display a bimodal centrum length profile, with peaks in the lumbar and distal caudal regions of the vertebral column. Buchholtz (1998) documented a comparable pattern in *Enhydra lutra*, which she noted was consistent with the fact that dorsoventral undulation occurs in the lumbar and caudal regions of the body in swimming sea otters (also see Kenyon 1981). Therefore, similarity in centrum length profile to *E. lutra* is expected for *L. canadensis*, given that similar undulatory swimming is a component of its aquatic locomotor regime (Fish 1994), but it is surprising for *N. vison* considering that quadrupedal paddling is the main swimming style of the species (Williams 1983). Likewise, this profile likely has little to do with undulation during swimming in *C. minimus* because this species uses bipedal paddling of the hind limbs (alternate pelvic paddling) for aquatic propulsion and keeps the back horizontal while swimming (Fish 1993). Instead of corresponding only to swimming style, these bimodal centrum length profiles may represent retention or slight modification of an ancestral mammalian organization of the vertebral column, considering that a comparable bimodal pattern can also be found in fully terrestrial, cursorial taxa such as *Pachyaena* and *Canis* (Buchholtz 1998; Gingerich 1998). *Castor canadensis* differs from the other mammals we measured in showing a steady increase in centrum length in the presacral column, followed by a plateau in the proximal caudals. The unique pattern observed in *C. canadensis* likely reflects its derived caudal morphology, but the relationship between this morphology and swimming style is somewhat obscure. Much like *Chironectes*, alternate pelvic paddling is the primary source of propulsion in *Castor* (Fish 1996, 2001), with the tail playing at best a supporting role (Fish 2001).

Despite these qualitative differences, the length metric scores of the extant semi-aquatic taxa do differ on average from those of the terrestrial taxa. The extant semi-aquatic taxa have higher range (R), higher polarization (C), higher irregularity (Cm), higher $E1$, and higher $E2$ values on average for length than the terrestrial taxa, and a MANOVA on the metrics indicates that there is a significant difference between the two groups (Wilk's $\lambda = 0.0941$, d.f. = 5, 25, $F = 48.14$, $p \ll 0.001$).

Does Centrum Length Profile Support a Semi-aquatic Lifestyle in Ophiacodon?

Our observations of extant semi-aquatic taxa suggest that there is an inexact correlation between centrum length profile and swimming style, but that the length metrics can discriminate terrestrial and aquatic taxa effectively. Although the former observation adds complexity to the interpretation of our results for *Ophiacodon*, the available data largely refute the semi-aquatic *Ophiacodon* hypothesis.

The centrum length profiles of the *Ophiacodon* specimens do not show strong qualitative similarities to those of the extant semi-aquatic mammals or reptiles. In particular, *Ophiacodon* lacks the peak or plateau in caudal centrum length observed in *Amblyrhynchus* and the crocodilians that may be associated with undulatory swimming. The three *Ophiacodon* specimens included in the PC analysis of length metrics do fall within the area of morphospace occupied by most of the extant semi-aquatic taxa. Specifically, they largely overlap the area occupied by extant crocodiles. These results for *Ophiacodon* might suggest a semi-aquatic lifestyle, but they would be more convincing as evidence if the other basal synapsids fell in a different area of morphospace. However, nearly all of the basal synapsids fall in the area of morphospace occupied by most of the extant semi-aquatic taxa. Even the two *Cotylorhynchus* specimens, which fall outside of this area, are still closer in morphospace to the *Castor* and *Chironectes* specimens than to the terrestrial varanids or *Iguana*. Although this might imply that all of the basal synapsid taxa, including *Ophiacodon*, were semi-aquatic, such an interpretation would contradict almost all previous interpretations of their ecologies (e.g., Romer and Price 1940). This is particularly true of the caseids and *Varanops*, since caseids and varanopids often have been considered to be components of highly terrestrial, upland faunas (e.g., Olson 1952, 1962, 1968; Reisz 2005; Maddin et al. 2006; Evans et al. 2009; Berman et al. 2013). Instead, it seems more likely that the basal synapsid vertebral columns were organized in a different manner than those of the extant taxa, given that all of the basal synapsids also tend to cluster away from the extant taxa in the PC analysis based on the length, width, and height metrics.

The length metric results are somewhat more equivocal. The length metrics for *Ophiacodon* tend to be notably lower than either the extant terrestrial or semi-aquatic taxa (regardless of whether the three relatively incomplete specimens are included), although they are (slightly) closer to the values of the extant terrestrial taxa. In contrast, the length metrics for the other basal synapsids are closer on average to those of the extant semi-aquatic and terrestrial taxa, and their average values are intermediate between the two extant groups if only relatively complete specimens are considered. These differences imply that in terms of length, the vertebral column of *Ophiacodon* tends to be less polarized on average than those of the other basal synapsids and the extant tetrapods, as well as somewhat noisier. However, the relationship between this variation and lifestyle for basal synapsids is uncertain: *Ophiacodon* is different from the other basal synapsids, but not in the way we would predict if it was aquatic based on the extant specimens. Inclusion of a greater sample of terrestrial taxa, particularly extant terrestrial reptiles, in analyses such as those we present here may be useful for identifying a suitable modern analogue for *Ophiacodon* and the other basal synapsids. Nevertheless, when combined with the other observations on centrum length profile, we do not consider the length metrics for *Ophiacodon* to provide strong support for its hypothesized semi-aquatic lifestyle.

Is the Disparity in Limb Lengths in Ophiacodon Indicative of a Semi-aquatic Lifestyle?

Romer and Price (1940) and Kemp (1982) cited the fact that the hind limbs of *Ophiacodon* are longer than its fore limbs as evidence in support of a semi-aquatic lifestyle, but our results suggest that there is little importance to this observation. There was no significant difference in the ratio of limb lengths between aquatic and terrestrial taxa we sampled, with only two of the specimens we measured, FMNH UC 656 (*Casea broilii*) and FMNH UR 272 (*Cotylorhynchus hancocki*) possessing longer fore limbs than hind limbs. Although this is unlikely to be an artifact because the limb material in FMNH UC 656 represents only one individual (in contrast to its vertebral column; Williston 1911), it does emphasize that *Ophiacodon* is not unusual in its relative limb lengths. Furthermore, inclusion of manus and pes lengths most likely would not alter these conclusions. All of the basal synapsid taxa we examined possess relatively conservative manus and pes morphologies, and if anything show trends towards the development of short, broad feet rather than the long, triangular feet of paddling swimmers (Reisz 1986). Likewise, extant semi-aquatic reptiles do not show modification of the feet for swimming (Romer 1956), and only three extant lizards have been documented to use the feet for propulsion when swimming (Russell and Bels 2001). Among the semi-aquatic mammals we sampled, the hind limbs play a major role in aquatic locomotion in three of the four species, and they possess larger feet than hands. Taken together, these data indicate that limb length disparity should be eliminated from the evidence cited in support of the semi-aquatic *Ophiacodon* hypothesis.

Were Basal Synapsid Vertebral Columns Less Regulated than Those of Extant Taxa?

On average, the irregularity (*Cm*) of the basal synapsids are higher than those of the extant taxa, and their *E*2 (the metric quantifying the tendency of adjacent measurements to have similar values; McShea 1993) values are lower for height, length, and width. These statistics imply that a given vertebra in a basal synapsid column is less constrained to resemble the preceding vertebra than is the case for the extant taxa. McShea (1993) noted that the fossil taxa included in his analysis also tended to have more irregular columns than extant taxa, and attributed this to taphonomic artifacts such as deformation and rough bone preservation. Similar processes may be responsible for the greater noise in the basal synapsids, given that some show evidence of deformation (e.g., FMNH UC 112, FMNH UR 272), but others seem to be close to pristinely preserved (e.g., FMNH UC 1638). Alternatively, the greater noisiness of the basal synapsid columns may reflect a genuine difference because the centra of these taxa essentially are hour glass-shaped tubes that surrounded the notochord, which in turn widened in the intervertebral spaces (Romer and Price 1940). Given this mode of connection, centrum dimensions may have been less of a constraint on the function of the column, and therefore may have been less tightly regulated. This is a potentially important consideration because if the vertebral columns of the basal synapsids were operating under a different set of constraints than the extant taxa, then extant taxa may make poor predictors of expected morphologies for particular locomotor patterns and ecologies. A full test of this hypothesis is beyond the scope of this paper, but would be possible given a more systematic exploration of vertebral dimensions across a wider range of non-mammalian synapsids. It would also be interesting to determine whether an increase in regularity of vertebral dimensions accompanied the evolution of more regionalized vertebral columns in advanced non-mammalian synapsids.

Paleoecological Implications

The paleoecology of Early Permian terrestrial communities has been of interest for some time, especially in the context of the evolution of terrestrial communities with diverse, abundant tetrapod herbivores (e.g., Olson 1952, 1961, 1966, 1977, 1983, 1985a, b; Sullivan and Reisz 1999; Eberth et al. 2000; Berman et al. 2000, 2004, 2013; Reisz 2005; Evans et al. 2009). *Ophiacodon* has not figured prominently in these discussions, but when included it has been hypothesized to have fed primarily on aquatic food sources and perhaps to have been a food source for larger terrestrial carnivores like *Dimetrodon* (Olson 1977, 1983, 1985a, b). If *Ophiacodon* had a more terrestrial lifestyle, these hypotheses will need to be revised to include more terrestrial animals as potential prey items. The degree to which these communities were dependent on aquatic producers and primary consumers may also need to be reassessed. Nevertheless, the fact that *Ophiacodon* is not found in the most upland communities of the time, unlike some carnivorous basal synapsids (Berman et al. 2001, 2013), and became extinct as environments became drier (e.g., Olson 1983), implies that it did have some important connection to wetter, lowland environments. Further exploration of the functional morphology and paleoecology of *Ophiacodon* will be needed to determine whether this connection was related to its diet or other aspects of its physiology.

Conclusions

The hypothesis that *Ophiacodon* was semi-aquatic has a long history in the paleontological literature, and many lines of evidence have been advanced to support it. However, much of this evidence is inconclusive or has been reinterpreted. The unusually slow skeletal ossification of *Ophiacodon* and its bone histology may represent the data most suggestive of a semi-aquatic lifestyle that has been presented to date, but more extensive investigation of the bone histology of pelycosaur-grade synapsids is necessary before its true significance can be understood. Unfortunately, centrum length patterns across the vertebral column do not provide an unambiguous answer to the question of whether *Ophiacodon* was semi-aquatic. Although *Ophiacodon* does show centrum length patterns that are similar to some extant semi-aquatic tetrapods, these properties are also present in most of the other basal synapsids we examined, including taxa such as caseids and varanopids that are widely regarded as highly terrestrial. Therefore, we propose that the vertebral organization of *Ophiacodon* is more a function of its phylogenetic position than its ecology. The limb proportions of *Ophiacodon* also do not provide support for a semi-aquatic lifestyle. Given that there is almost no unequivocal evidence for an aquatic lifestyle in *Ophiacodon*, we do not think that such an ecology should be assumed. Instead, a more suitable null hypothesis is that it was terrestrial, and future studies should focus on falsifying that conjecture.

Acknowledgments Data collection for this project was supported by a Field Museum Internship awarded to RNF. We thank P. Holroyd, A. Resetar, W. Simpson, and W. Stanley for assistance with specimens. K. Melstrom measured the terrestrial varanid specimens. J. Caruso, D. Heins, and R. Parsley provided feedback on a previous version of the manuscript. S. Pierce, S. Sumida, and an anonymous reviewer also provided helpful suggestions. Presentation of results from this study at the 2008 Annual Meeting of the Society of Vertebrate Paleontology was made possible by a grant from The Jackson School of Geosciences, University of Texas, Austin.

References

Berman, D. S, Reisz, R. R., Bolt, J. R., & Scott, D. (1995). The cranial anatomy and relationships of the synapsid *Varanosaurus* (Eupelycosauria: Ophiacodontidae) from the early Permian of Texas and Oklahoma. *Annals of Carnegie Museum, 64*, 99–133.

Berman, D. S, Henrici, A. C., Sumida, S. S., & Martens, T. (2000). Redescription of *Seymouria sanjuanensis* (Seymouriamorpha) from the Lower Permian of Germany based on complete, mature specimens with a discussion of the paleoecology of the Bromacker Locality Assemblage. *Journal of Vertebrate Paleontology, 20*, 253–268.

Berman, D. S, Reisz, R. R., Martens, T., & Henrici, A. C. (2001). A new species of *Dimetrodon* (Synapsida: Sphenacodontidae) from the Lower Permian of Germany records first occurrence of genus outside of North America. *Canadian Journal of Earth Sciences, 38*, 80–812.

Berman, D. S, Henrici, A. C., Kissel, R. A., Sumida, S. S., & Martens, T. (2004). A new diadectid (Diadectomorpha), *Orobates pabsti*, from the Early Permian of central Germany. *Bulletin of Carnegie Museum of Natural History, 35*, 1–36.

Berman, D. S, Henrici, A. C., Sumida, S. S., Martens, T., & Pelletier, V. (2013). First European record of a varanodontine (Synapsida: Varanopidae): Member of a unique Early Permian upland ecosystem, Tambach Basin, central Germany. In C. F. Kammerer, K. D. Angielczyk, & J. Fröbisch (Eds.), *Early evolutionary history of the Synapsida* (pp. 69–86). Dordrecht: Springer.

Braziatis, P. (1973). The identification of living crocodilians. *Zoologica, 58*, 59–101.

Brinkman, D. (1988). Size-independent criteria for estimating relative age in *Ophiacodon* and *Dimetrodon* (Reptilia, Pelycosauria) from the Admiral and Lower Belle Plains Formations of West-Central Texas. *Journal of Vertebrate Paleontology, 8*, 172–180.

Buchholtz, E. A. (1998). Implications of vertebral morphology for locomotor evolution in early Cetacea. In J. G. M. Thewissen (Ed.), *The emergence of whales: The evolutionary patterns in the origin of Cetacea* (pp. 325–352). New York: Plenum Press.

Buchholtz, E. A. (2001a). Swimming styles in Jurassic ichthyosaurs. *Journal of Vertebrate Paleontology, 21*, 61–73.

Buchholtz, E. A. (2001b). Vertebral osteology and swimming style in living and fossil whales (Order: Cetacea). *Journal of Zoology, 253*, 175–190.

Buchholtz, E. A., & Schur, S. A. (2004). Vertebral osteology in Delphinidae (Cetacea). *Zoological Journal of the Linnean Society, 140*, 383–401.

Buchholtz, E. A., Wolkovitch, E. M., & Cleary, R. J. (2005). Vertebral osteology and complexity in *Lagenorhynchus acutus* (Delphinidae) with comparison to other delphinoid genera. *Marine Mammal Science, 21*, 411–428.

Buchholtz, E. A., Booth, A. C., & Webbink, K. E. (2007). Vertebral anatomy in the Florida Manatee, *Trichechus manatus latirostris*: A developmental and evolutionary analysis. *The Anatomical Record, 290*, 624–637.

Canoville, A., & Laurin, M. (2010). Evolution of humeral microanatomy and lifestyle in amniotes, an some comments on palaeobiological inferences. *Biological Journal of the Linnean Society, 100*, 384–406.

Carpenter, K. (2009). Role of lateral body bending in crocodylian track making. *Ichnos, 16*, 202–207.

Carroll, R. L. (1986). The skeletal anatomy and some aspects of the physiology of primitive reptiles. In N. Hotton, P. D. MacLean, J. J. Roth, & E. C. Roth (Eds.), *The ecology and biology of mammal-like reptiles* (pp. 25–45). Washington, DC: Smithsonian Institution Press.

Case, E. C. (1907). *Revision of the Pelycosauria of North America*. Washington, DC: Carnegie Institution of Washington.

Case, E. C. (1915). *The Permo-Carboniferous red beds of North America and their vertebrate fauna*. Washington, DC: Carnegie Institution of Washington.

Cope, E. D. (1878). Descriptions of extinct Batrachia and Reptilia from the Permian formation of Texas. *Proceedings of the American Philosophical Society, 17*, 505–530.

de Ricqlès, A. (1974). Recherches paléohistologiques sur les os longs des tétrapodes. IV—éothériodontes et pélycosaures. *Annales de Paléontologie (Vertébratés), 60*, 3–39.

de Ricqlès, A. (1989). Les méchanismes hétérochroniques dans le retour des tétrapodes au Milieu aquatique. *Geobios Mémoir Spécial, 12*, 337–348.

de Ricqlès, A., & de Buffrénil, V. (2001). Bone histology, heterochronies and the return of tetrapods to life in water. In J. M. Mazin & V. de Buffrénil (Eds.), *Secondary adaptation of tetrapods to life in water* (pp. 289–310). München: Verlag Dr. Friedrich Pfeil.

Eberth, D. A., & Miall, A. D. (1991). Stratigraphy, sedimentology and evolution of a vertebrate-bearing, braided to anastomosed fluvial system, Cutler Formation (Permian-Pennsylvanian), north-central New Mexico. *Sedimentary Geology, 72*, 225–252.

Eberth, D. A., & Berman, D. S (1993). Stratigraphy, sedimentology, and vertebrate paleoecology of the Cutler Formation redbeds (Pennsylvanian-Permian) of north central New Mexico. In S. G. Lucas & J. Zidek (Eds.) *Vertebrate Paleontology in New Mexico* (pp. 33–48). New Mexico Museum of Natural History and Science 2.

Eberth, D. A., Berman, D. S, Sumida, S. S., & Hopf, H. (2000). Lower Permian terrestrial paleoenvironments and vertebrate paleoecology of the Tambach Basin (Thuringia, central Germany): The upland Holy Grail. *Palaios, 15*, 293–313.

Enlow, D. H., & Brown, S. O. (1957). A comparative histological study of fossil and recent bone tissues. Part II. *Texas Journal of Science, 9*, 186–214.

Evans, D. C., Maddin, H. C., & Reisz, R. R. (2009). A re-evaluation of sphenacodontid synapsid material from the Lower Permian fissure fills near Richards Spur, Oklahoma. *Palaeontology, 52*, 219–227.

Fish, F. E. (1984). Kinematics of undulatory swimming in the American Alligator. *Copeia, 4*, 839–843.

Fish, F. E. (1993). Comparison of swimming kinematics in terrestrial and semiaquatic opossums. *Journal of Mammalogy, 74*, 275–284.

Fish, F. E. (1994). Association of propulsive swimming mode with behavior in river otters (*Lutra canadensis*). *Journal of Mammalogy, 75*, 989–997.

Fish, F. E. (1996). Transitions from drag-based to lift-based propulsion in mammalian aquatic swimming. *American Zoologist, 36*, 628–641.

Fish, F. E. (2001). A mechanism for evolutionary transition in swimming mode by mammals. In J. M. Mazin & V. de Buffrénil (Eds.), *Secondary adaptation of tetrapods to life in water* (pp. 261–287). München: Verlag Dr. Friedrich Pfeil.

Germain, D., & Laurin, M. (2005). Microanatomy of the radius and lifestyle in amniotes (Vertebrata, Tetrapoda). *Zoologica Scripta, 34*, 335–350.

Gingerich, P. D. (1998). Paleobiological perspectives on Mesonychia, Archaeoceti, and the origin of whales. In J. G. M. Thewissen (Ed.), *The emergence of whales: The evolutionary patterns in the origin of Cetacea* (pp. 423–449). New York: Plenum Press.

Gingerich, P. D. (2003). Land-to-sea transition in early whales: Evolution of Eocene Archaeoceti (Cetacea) in relation to skeletal proportions and locomotion of living semiaquatic mammals. *Paleobiology, 29*, 429–454.

Girondot, M., & Laurin, M. (2003). Bone profiler: A tool to quantify, model, and statistically compare bone-section compactness. *Journal of Vertebrate Paleontology, 23*, 458–461.

Gould, S. J. (1965). Evolutionary patterns in pelycosaurian reptiles: A factor-analytic study. *Evolution, 21*, 385–401.

Hentz, T. F. (1988). Lithostratigraphy and paleoenvironments of upper Paleozoic continental red beds, north-central Texas: Bowie (new) and Wichita (revised) groups. *University of Texas at Austin Bureau of Economic Geology Report of Investigations, 170*, 1–55.

Hentz, T. F. (1989). Depositional environments of the Early Permian coastal plain, north-cental Texas: A synopsis. In R. W. Hook (Ed.), *Permo-Carboniferous vetebrate paleontology, lithostratigraphy, and depositional environments of North-Central Texas* (pp. 22–39). 49th Annual Meeting of the Society of Vertebrate Paleontology, Field Trip Guidebook 2.

Houssaye, A. (2009). "Pachyostosis" in aquatic amniotes: A review. *Integrative Zoology, 4*, 325–340.

Hunt, A. P., & Lucas, S. G. (1998). Vertebrate tracks and the myth of the belly-dragging, tail-dragging tetrapods of the late Paleozoic. In S. G. Lucas, J. W. Estep, & J. M. Hoffer (Eds.), *Permian stratigraphy and paleontology of the Robledo Mountains, New Mexico* (pp. 67–69). *New Mexico Museum of Natural History and Science Bulletin, 12*.

Jasinoski, S. C., & Chinsamy-Turan, A. (2012). Biological inferences of the cranial microstructure of the dicynodonts *Oudenodon* and *Lystrosaurus*. In A. Chinsamy-Turan (Ed.), *The Forerunners of mammals: Radiation, histology, biology* (pp. 149–176). Bloomington: Indiana University Press.

Jasinoski, S. C., Rayfield, E. J., & Chinsamy, A. (2009). Comparative feeding biomechanics of *Lystrosaurus* and the generalized dicynodont *Oudenodon. Anatomical Record, 292*, 862–874.

Jasinoski, S. C., Rayfield, E. J., & Chinsamy, A. (2010a). Functional implications of dicynodont cranial suture morphology. *Journal of Morphology, 271*, 705–728.

Jasinoski, S. C., Rayfield, E. J., & Chinsamy, A. (2010b). Mechanics of the scarf premaxilla-nasal suture in the snout of *Lystrosaurus. Journal of Vertebrate Paleontology, 30*, 1283–1288.

Jasinoski, S. C., Cluver, M. J., Chinsamy, A., & Reddy, B. D. (2013). Anatomical plasticity in the snout of *Lystrosaurus*. In C. F. Kammerer, K. D. Angielczyk, & J. Fröbisch (Eds.), *Early evolutionary history of the Synapsida* (pp. 139–149). Dordrecht: Springer.

Jenkins, I., Thomason, J. J., & Norman, D. B. (2002). Primates and engineering principles: Applications to craniodental mechanisms in ancient terrestrial predators. *Senckenbergiana Lethaea, 82*, 223–240.

Kemp, T. S. (1982). *Mammal-like reptiles and the origin of mammals*. London: Academic Press.

Kemp, T. S. (2005). *The origin and evolution of mammals*. Oxford: Oxford University Press.

Kenyon, K. W. (1981). Sea Otter, *Enhydra lutra* (Linnaeus, 1758). In S. H. Ridgway & R. J. Harrison (Eds.), *Handbook of marine mammals* (Vol. 1, pp. 209–223). London: Academic Press.

Kriloff, A., Germain, D., Canoville, A., Vincent, P., Sache, M., & Laurin, M. (2008). Evolution of bone microanatomy of the tetrapod tibia and its use in palaeobiological inference. *Journal of Evolutionary Biology, 21*, 807–826.

Laurin, M., Girondot, M., & Loth, M.-M. (2004). The evolution of long bone microstructure and lifestyle in lissamphibians. *Paleobiology, 30*, 589–613.

Madar, S. I. (1998). Structural adaptations of early archaeocete long bones. In J. G. M. Thewissen (Ed.), *The emergence of whales: The evolutionary patterns in the origin of Cetacea* (pp. 353–377). New York: Plenum Press.

Maddin, H. C., & Reisz, R. R. (2007). The morphology of the terminal phalanges in Permo-Carboniferous synapids: An evolutionary perspective. *Canadian Journal of Earth Sciences, 44*, 267–274.

Maddin, H. C., Evans, D. C., & Reisz, R. R. (2006). An Early Permian varanodontine varanopid (Synapsida: Eupelycosauria) from the Richards Spur locality of Oklahoma. *Journal of Vertebrate Paleontology, 26*, 957–966.

Marsh, O. C. (1878). Notice of new fossil reptiles. *American Journal of Science, 15*, 409–411.

Martens, T., Berman, D. S, Henrici, A. C., & Sumida, S. S. (2005). The Bromacker Quarry—the most important locality of Lower Permian terrestrial vertebrate fossils outside of North America. In S. G. Lucas & K. E. Zeigler (Eds.), *The nonmarine Permian* (pp. 67–69). *New Mexico Museum of Natural History and Science Bulletin, 30*.

Massare, J. A. (1987). Tooth morphology and prey preference of Mesozoic marine reptiles. *Journal of Vertebrate Paleontology, 7*, 121–137.

McHenry, C. R., Clausen, P. D., Daniel, W. J. T., Meers, M. B., & Pendharkar, A. (2006). Biomechanics of the rostrum in crocodilians: A comparative analysis using finite element modeling. *The Anatomical Record Part A, 288A*, 827–849.

McShea, D. W. (1992). A metric for the study of evolutionary trends in the complexity of serial structures. *Biological Journal of the Linnean Society, 45*, 39–55.

McShea, D. W. (1993). Evolutionary change in the morphological complexity of the mammalian vertebral column. *Evolution, 47*, 730–740.

Nakajima, Y. (2010). Evaluating the utility of limb bone internal structure as an indicator for aquatic adaptation of Testudines. *Abstract from the 2010 Annual Symposium of Vertebrate Palaeontology and Comparative Anatomy, Cambridge*. http://www.svpca.org/general/pages/abstractPage.php?i=1540&r=talksAndPosters.php&y=2010.

Northover, J., Rybczynski, N., & Schroder-Adams, C. (2010). Evidence for correlated evolution between long bone compactness, swimming behavior and body mass in Arctoidea (Mammalia: Carnivora). *Program and Abstracts, Society of Vertebrate Paleontology Annual Meeting, Pittsburgh, Pennsylvania*, 149A.

Olson, E. C. (1941). New specimens of Permian vertebrates in Walker Museum. *Journal of Geology, 49*, 753–763.

Olson, E. C. (1952). The evolution of a Permian vertebrate chronofauna. *Evolution, 6*, 181–196.

Olson, E. C. (1961). The food chain and the origin of mammals. *Koninklijke Vlaamse Academie voor Wetenschappen, Letteren en Schone Kunsten van Belgie: Klasse der Wetenschappen, 1961*, 97–116.

Olson, E. C. (1962). Late Permian terrestrial vertebrates, U.S.A. and U.S.S.R. *Transactions of the American Philosophical Society, New Series, 52*, 1–224.

Olson, E. C. (1966). Community evolution and the origin of mammals. *Ecology, 47*, 291–302.

Olson, E. C. (1968). The family Caseidae. *Fieldiana: Geology, 17*, 225–349.

Olson, E. C. (1977). Permian lake faunas: A study in coevolution. *Journal of the Palaeontological Society of India, 20*, 146–163.

Olson, E. C. (1983). Coevolution or coadaptation? Permo-Carboniferous vertebrate chronofauna. In M. H. Nitecki (Ed.), *Coevolution* (pp. 307–338). Chicago: University of Chicago Press.

Olson, E. C. (1985a). Permo-Carboniferous vertebrate communities. In J. T. Dutro & H. W. Pfefferkorn (Eds.), *Neuvième Congrès International de Stratigraphie et de Géologie du Carbonifère. Compte Rendu 5: Paleontology, Paleoecology, Paleogeography* (pp. 331–345). Carbondale: Southern Illinois University Press.

Olson, E. C. (1985b). Nonmarine vertebrates and late Paleozoic climates. In J. T. Dutro & H. W. Pfefferkorn (Eds.), *Neuvième Congrès International de Stratigraphie et de Géologie du Carbonifère. Compte Rendu 5: Paleontology, Paleoecology, Paleogeography* (pp. 403–414). Carbondale: Southern Illinois University Press.

Panko, L. J. (2001). *Evolution and functional morphology of the axial skeleton in the Synapsida*. Unpublished Ph.D. dissertation, University of Chicago.

Paton, R. L. (1974). Lower Permian pelycosaurs from the English midlands. *Palaeontology, 17*, 541–552.

Pierce, S. E., Clack, J. A., & Hutchinson, J. R. (2011). Comparative axial morphology in pinnipeds and its correlation with aquatic locomotory behaviour. *Journal of Anatomy, 219*, 502–514.

Rayfield, E. J., & Milner, A. C. (2008). Establishing a framework for archosaur cranial mechanics. *Paleobiology, 34*, 494–515.

Rayfield, E. J., Milner, A. C., Xuan, V. B., & Young, P. G. (2007). Functional morphology of spinosaur 'crocodile mimic' dinosaurs. *Journal of Vertebrate Paleontology, 27*, 892–901.

Reisz, R. R. (1986). Pelycosauria. In P. Wellnhofer (Ed.), *Handbuch der Paläoherpetologie* (Vol. 17A). Stuttgart: Gustav Fischer Verlag.

Reisz, R. R. (2005). *Oromycter*, a new caseid from the Lower Permian of Oklahoma. *Journal of Vertebrate Paleontology, 25*, 905–910.

Ritter, D. (1992). Lateral bending during lizard locomotion. *Journal of Experimental Biology, 173*, 1–9.

Ritter, D. (1996). Axial muscle function during lizard locomotion. *Journal of Experimental Biology, 199*, 2499–2510.

Romer, A. S. (1925). An ophiacodont reptile from the Permian of Kansas. *Journal of Geology, 33*, 173–182.

Romer, A. S. (1948). Ichthyosaur ancestors. *American Journal of Science, 246*, 109–121.

Romer, A. S. (1956). *Osteology of the reptiles*. Chicago: University of Chicago Press.

Romer, A. S. (1957). Origin of the amniote egg. *The Scientific Monthly, 85*, 57–63.

Romer, A. S. (1958). Tetrapod limbs and early tetrapod life. *Evolution, 12*, 365–369.

Romer, A. S., & Price, L. I. (1940). Review of the Pelycosauria. *Geological Society of America Special Papers, 28*, 1–538.

Russell, A. P., & Bels, V. (2001). Biomechanics and kinematics of limb-based locomotion in lizards: Review, synthesis and prospectus. *Comparative Biochemistry and Physiology Part A, 131*, 89–112.

Samuels, J. X., & Van Valkenburgh, B. (2008). Skeletal indicators of locomotor adaptations in living and extinct rodents. *Journal of Morphology, 269*, 1387–1411.

Sullivan, C. S., & Reisz, R. R. (1999). First record of *Seymouria* (Vertebrata: Seymouriamorpha) from Early Permian fissure fills at Richards Spur, Oklahoma. *Canadian Journal of Earth Sciences, 36*, 1257–1266.

Sumida, S. S., & Modesto, S. (2001). A phylogenetic perspective on locomotory strategies in early amniotes. *American Zoologist, 41*, 586–597.

Taylor, M. A. (1994). Stone, bone, or blubber? Buoyancy control strategies in aquatic tetrapods. In L. Maddock, Q. Bone, & J. M. V. Rayner (Eds.), *Mechanics and physiology of animal swimming* (pp. 151–161). Cambridge: Cambridge University Press.

Thewissen, J. G. M., & Fish, F. E. (1997). Locomotor evolution in the earliest cetaceans: Functional model, modern analogues, and paleontological evidence. *Paleobiology, 23*, 482–490.

Thomason, J. J., & Russell, A. P. (1986). Mechanical factors in the evolution of the mammalian secondary palate: A theoretical analysis. *Journal of Morphology, 189*, 199–213.

Voight, S., Berman, D. S, & Henrici, A. C. (2007). First well-established trackmaker association of Paleozoic tetrapods based on *Ichniotherium* trackways and diadectid skeletons from the Lower Permian of Germany. *Journal of Vertebrate Paleontology, 27*, 553–570.

Webb, P. W. (1988). Simple physical principles and vertebrate aquatic locomotion. *American Zoologist, 28*, 709–725.

Williams, T. M. (1983). Locomotion in the North American mink, a semi-aquatic mammal. I. Swimming energetics and body drag. *Journal of Experimental Biology, 103*, 155–168.

Williston, S. W. (1911). *American Permian vertebrates*. Chicago: University of Chicago Press.

Williston, S. W. (1914). *Water reptiles of the past and present*. Chicago: University of Chicago Press.

Williston, S. W., & Case, E. C. (1913). Description of a nearly complete skeleton of *Ophiacodon* Marsh. *Carnegie Institute of Washington Publication, 181*, 37–59.

Chapter 4
Postcranial Description and Reconstruction of the Varanodontine Varanopid *Aerosaurus wellesi* (Synapsida: Eupelycosauria)

Valerie Pelletier

Abstract The postcranial skeleton of the varanopid synapsid *Aerosaurus wellesi* is reconstructed based on several specimens, including partly articulated material of an apparently fully adult specimen. Comparisons are made with other known varanopid taxa, reaffirming its position as a basal varanodontine. *Aerosaurus* was an obligatory sprawling-gaited animal with an extremely long tail and a presacral column that descended posteriorly at an angle of 25°. *Aerosaurus* shares many synapomorphies with other varanodontines: tall neural spines, double headed ribs, presence of a supraglenoid foramen, broadly expanded proximal and distal heads of humerus, high degree of twist or torque of the humeral heads about the shaft, radius shorter than humerus, expanded heads of femur, femur lacking a sigmoid curvature, and humerus and femur roughly subequal in length. The high degree of torque in the humerus, the extremely long tail, and the subequal lengths of the humerus and femur are considered primitive features of basal tetrapods, whereas the length of the centra being 23 % greater than the width and the expanded clavicular plate are specializations also seen in some later, derived varanodontines.

Keywords Arroyo del Agua • Cutler Formation • Lower Permian • New Mexico • Pelycosaur

Introduction

The diversification of amniotes and their close relatives during the Late Paleozoic marked a key transition in terrestrial vertebrate evolution. From the Middle Carboniferous through the Late Permian the two major groups of amniotes, Synapsida (ultimately including mammals), and Reptilia (ultimately including extant reptiles and birds) quickly became the dominant constituents of terrestrial vertebrate ecosystems. The early synapsid record is arguably more complete than that of basal Reptilia; particularly in the Middle to Late Carboniferous (Wideman et al. 2005). Though some have a large body size which can influence preservation potential, a clear understanding of basal synapsids may offer the best insight into the early evolution of Amniota (Hopson 1991).

Phylogenetic and Taxonomic Context

The phylogeny of the primitive basal Synapsida, frequently referred to as "pelycosaurs" or pelycosaurian-grade synapsids, is well studied (Reisz 1986; Reisz et al. 1992; Berman et al. 1995, 2013; Reisz et al. 1998; Maddin et al. 2008; Sumida et al. 2013). Phylogenetic analyses are based primarily on cranial characters [e.g., 82, 83, and 89 % for Maddin et al. (2006), Campione and Reisz (2010) and Anderson and Reisz (2004), respectively], but postcranial features can also be useful although they are often overlooked. Further, an understanding of postcranial anatomy can offer significant insight into functional and behavioral biology (Fröbisch and Reisz 2009).

Reisz et al. (1998) and others (Reisz 1986; Reisz et al. 1992; Berman et al. 1995; Reisz and Dilkes 2003; Maddin et al. 2008) divide the pelycosaurian-grade synapsids into two clades, Caseasauria (including Eothyrididae and Caseidae) and Eupelycosauria (including Varanopidae, Ophiacodontidae, Edaphosauridae, and Sphenacodontia). Among caseasaurians, the postcranial skeleton of the basalmost recognized taxa (*Oedaleops* and *Eothyris*) are represented by few remains (Reisz et al. 2009; Sumida et al. 2009, 2013). In contrast, the postcrania of caseids are well known. However, they demonstrate significant specializations

V. Pelletier (✉)
Department of Biology, California State University San Bernardino, 5500 University Parkway, San Bernardino, CA 92407, USA
e-mail: uraprimate@msn.com

Christian F. Kammerer, Kenneth D. Angielczyk and Jörg Fröbisch (eds.), *Early Evolutionary History of the Synapsida*, Vertebrate Paleobiology and Paleoanthropology, DOI: 10.1007/978-94-007-6841-3_4,

associated with high-fiber herbivory and therefore may not be representative of the basal synapsid condition (Hotton et al. 1997; Botha-Brink and Modesto 2007). Among eupelycosaurs, the family Varanopidae is considered the sister taxon to all other eupelycosaurs, and therefore has the potential of offering considerable insight into the evolution of the later-diverging clades (Reisz 1986; Reisz and Berman 2001; Maddin et al. 2008). Varanopids have attracted considerable attention recently because they have one of the longest temporal ranges of any Paleozoic synapsid group, from the latest Carboniferous to the late Middle Permian (Anderson and Reisz 2004), a cosmopolitan distribution (Dilkes and Reisz 1996; Reisz et al. 1998; Reisz and Dilkes 2003), and because they provide one of the earliest examples of potential parental care in the amniote fossil record (Botha-Brink and Modesto 2007). As noted above, the phylogeny of varanopids is well-studied but this work is based overwhelmingly on cranial data (Maddin et al. 2006; Botha-Brink and Modesto 2009; Campione and Reisz 2010). This is somewhat surprising because Varanopidae is the basalmost family of eupelycosaurs for which postcranial information is abundant.

Twelve varanopid genera are currently recognized: *Aerosaurus* (Romer and Price 1940; Langston 1953; Langston and Reisz 1981); *Apsisaurus* (Laurin 1991; Reisz et al. 2010); *Archaeovenator* (Reisz and Dilkes 2003); *Elliotsmithia* (Dilkes and Reisz 1996; Reisz et al. 1998); *Heleosaurus* (Carroll 1976; Reisz and Modesto 2007; Botha-Brink and Modesto 2009); *Mesenosaurus* (Reisz and Berman 2001); *Mycterosaurus* (Berman and Reisz 1982; Reisz et al. 1997); *Pyozia* (Anderson and Reisz 2004), which may not belong in the Varanopidae (see Maddin et al. 2006); *Tambacarnifex* (Berman et al. 2013); *Varanodon* (Olson 1965); *Varanops* (Williston 1911, 1914; Campione and Reisz 2010); and *Watongia* (Reisz and Laurin 2004). With the exception of *Watongia* and *Tambacarnifex*, most of these varanopid taxa are known from reasonably complete cranial materials. Several taxa also have articulated or associated postcranial materials, few of which have been described (Williston 1911; Carroll 1976; Botha-Brink and Modesto 2009; Campione and Reisz 2010).

There are two species in the genus *Aerosaurus*: *A. greenleeorum* from the Upper Pennsylvanian of El Cobre Canyon, New Mexico, represented by partial skull and postcranial remains (Romer and Price 1940) and *A. wellesi*. *Aerosaurus wellesi* is represented by some of the best preserved and most complete postcrania of any varanopid. Although these materials were available to Langston and Reisz (1981), with the exception of the pes, little else was illustrated in that publication and the postcranial description was minimal. Preparation and study of material assignable to *Aerosaurus* at the University of California Museum of Paleontology (UCMP) indicates that it preserves one of the most complete postcranial skeletons of any pelycosaurian-grade synapsid, and therefore provides a basis for a reasonably confident reconstruction of the entire postcranium. Postcranial features known to be characteristic of the Varanopidae are: (1) midventral margin of dorsal centra ridged but without a distinct keel; (2) lateral excavation at base of dorsal neural spines; (3) a plate-like head of the interclavicles; (4) two to three subequal sacral ribs are present (Reisz and Dilkes 2003); and (5) a long, slender femur with a length to distal width greater than 3:1 (Reisz and Modesto 2007).

Geological and Geographic Context

The specimens described here as attributable to *Aerosaurus wellesi* originated at a locality that was discovered by Charles L. Camp in 1928. The Camp Quarry (UCMP Locality V-2814) is located in the Lower Permian (Wolfcampian) redbeds of the Cutler Formation in south central Rio Arriba County, New Mexico, near Arroyo del Agua (Lucas et al. 2005). The Cutler Formation of central New Mexico was recently elevated to group status by Lucas et al. (2005) and includes the El Cobre Canyon Formation, which spans the Pennsylvanian-Permian boundary, and the Arroyo del Agua Formation. The Camp Quarry is located in the upper El Cobre Canyon Formation and therefore remains Wolfcampian in age (Lucas et al. 2005). All of the quarries of the El Cobre Canyon Formation are considered part of a single biostratigraphic assemblage by Lucas et al. (2005); therefore the only other species in the genus *Aerosaurus*, *A. greenleeorum*, may be similar in age to *A. wellesi*.

In 1935, Samuel P. Welles collected six siltstone blocks totaling about 9.15 m^2. These blocks contain two almost complete skeletons of *Aerosaurus wellesi*, one juvenile and one adult, plus scattered bones then attributed to *Limnosceloides* (now considered a *nomen dubium*, Wideman et al. 2005), the basal eothyridid caseasaurian synapsid *Oedaleops* (see Sumida et al. 2013), and the larger sphenacodontid synapsid *Sphenacodon* (Sumida et al. 2009). Langston (1953) suggested that the character of the sediments in this area seemed to depict a flood plain deposit. However, Eberth and Miall (1991) demonstrated that the Cutler Formation in Arroyo del Agua represented an arid climate with ephemeral anastomosing streams running south-southwest from the San Luis-Uncompahgre Uplift. Crevasse channels that spilled into sheet splays may have formed ephemeral ponds for vertebrates. The channels formed U-shaped, mixed-fill units that were eventually filled by flooding and aggradation. The disarticulated vertebrate remains were probably washed here as lag deposits during a flooding event, whereas the better articulated remains may be the result of attritional accumulation (Eberth and Miall 1991). Regardless of the depositional environment, the well-articulated condition of the *Aerosaurus* remains suggests rapid burial; yet Langston (1953) suggested that the partial disarticulation of the larger specimen meant that death and burial were not coeval events. On the basis of the excellent condition of the *Aerosaurus* skeletons relative to the few scattered remains of the other taxa, Langston (1953, p. 357) wrote that *Aerosaurus* may have been dragging dead animals here to a "nesting" site. Notably, the earliest record for nesting behavior in a basal amniote has been demonstrated for another varanopid, *Heleosaurus* (Botha-Brink and Modesto 2007, 2009), lending some credence to Langston's hypothesis.

Varanopids have a wide geographic range, including North America, Europe, Russia, and South Africa (Reisz and Berman 2001; Reisz and Laurin 2004; Maddin et al. 2006; Reisz and Modesto 2007; Berman et al. 2013). Varanopidae is a highly conservative lineage (without the specializations for herbivory of Caseidae and Edaphosauridae or the greatly elongated neural spines of Sphenacodontidae) that was able to survive the climatic changes of the Permian and coexist with therapsids in both Laurasia and Gondwana while other groups of eupelycosaurs were replaced (Reisz et al. 1998; Reisz and Berman 2001). Unfortunately, fossils of varanopids are rare in lowland aquatic ecosystems (Reisz and Modesto 2007), which are by far the most common depositional environments among Late Carboniferous and Permian tetrapod-bearing fossiliferous sites (Sumida et al. 2004). Thus, it has been suggested that they were more abundant members of Early Permian upland terrestrial ecosystems, where they may have been apex predators, but where fossils were less likely to be preserved (Maddin et al. 2006; Berman et al. 2013). Only two such sites are currently known and documented, the Lower Permian Bromacker locality of central Germany (Eberth et al. 2000) and the Richards Spur (Dolese Brothers Limestone Quarry, also known as Fort Sill) locality of central Oklahoma (Reisz et al. 1997; Maddin et al. 2006; Evans et al. 2009). Published and illustrated accounts from these localities have so far presented mostly fragmentary varanopid remains (although see Berman 2013), one species in the former site and three in the latter. Other sites are mainly mixed assemblage aquatic to semi-terrestrial with occasional terrestrial components dominated by sphenacodontids, ophiacodontids, and edaphosaurids (Evans et al. 2009).

Our understanding of the postcrania of other varanopids is usually based on one, often poorly preserved, specimen per locality (Reisz et al. 1998; Reisz and Berman 2001; Reisz and Dilkes 2003). Thus, the excellent preservation and complete representation of virtually the entire skeleton of *Aerosaurus wellesi* from north-central New Mexico is fortuitous, offering insight into the postcranial construction and biology of early amniotes. A study of the postcranium can allow us to interpret agility as a predator compared to other tetrapods of the time and any terrestrial specializations that may have been present. As a basal varanodontine, *Aerosaurus* also can give insight into the body plan and mode of life from which other varanodontines evolved.

This study provides: (1) the first complete documentation of the postcranial skeleton of the varanopid *Aerosaurus*, (2) a basis for comparison with other varanopid and eupelycosaurian postcrania, and (3) allows a full body reconstruction of *Aerosaurus*.

Materials

UCMP 40093, humerus, femur, vertebrae, and ribs; UCMP 40094, ribs; UCMP 40097, a partly articulated skeleton consisting of a string of dorsal vertebrae, some with associated intercentra and ribs, three strings of caudal vertebrae, disarticulated ribs including one sacral, clavicle, scapulocoracoid, two ulnae, radius, disarticulated manus elements, ilium, ischium, pubis, femur, two tibiae, two fibulae, two partially articulated pedes; UCMP 40098, clavicle, ribs, vertebrae.

Description

Axial Skeleton

The vertebral column of *Aerosaurus* consists of at least 110 vertebrae, including 27 presacral vertebrae, the standard number for eupelycosaurs (Reisz and Dilkes 2003). Previous sacral vertebrae counts for varanopids suggest two vertebrae (Romer and Price 1940), but the exact number cannot be determined in the specimens of *Aerosaurus* described here. At least 80 caudal vertebrae can be confidently accounted for. This includes three strings plus a few unassociated caudals, and is a much higher count than previously reported for eupelycosaurs, which on average is 60 caudals (Romer and Price 1940). The primitive varanopid *Archaeovenator* has approximately 66 caudals, 22 articulated and approximately 2/3 of the length missing (Reisz and Dilkes 2003), and *Varanops* has ~53, 47 of which are in articulation and approximately six missing (Williston 1911). Given the disparity in the sizes of the fore and hindlimbs, the presacral column would have descended posteriorly at an angle of about 25° from a horizontal plane as reconstructed by Romer and Price (1940).

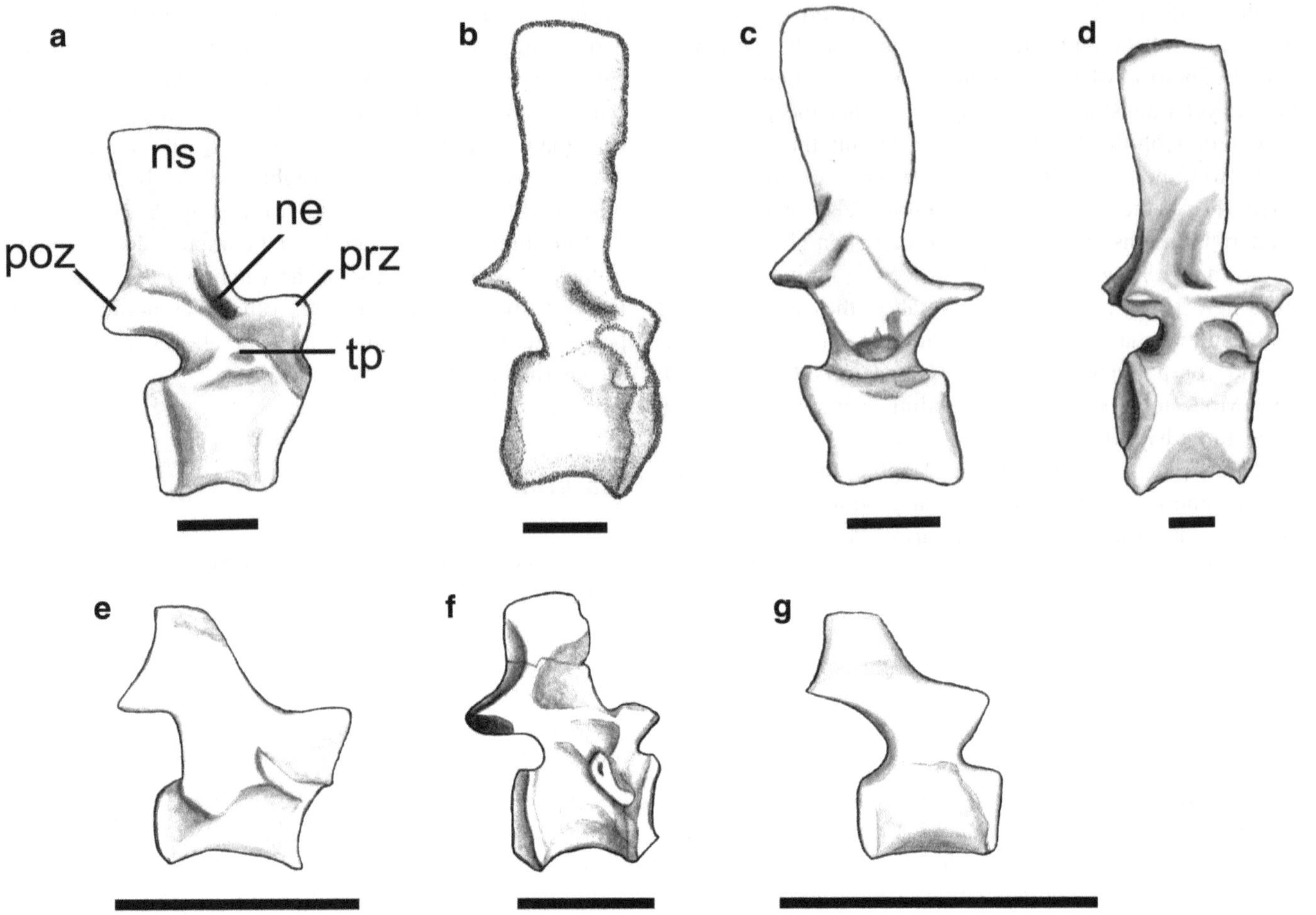

Fig. 4.1 Dorsal vertebrae in right lateral view of varanodontines and mycterosaurines. **a** *Aerosaurus wellesi* (UCMP 40097). **b** *Varanops brevirostris* (after Campione and Reisz 2010). **c** *Varanodon agilis* (after Olson 1965). **d** *Watongia meieri* (after Reisz and Laurin 2004). **e** *Heleosaurus scholtzi* (after Carroll 1976). **f** *Mycterosaurus longiceps* (after Reisz et al. 1997). **g** *Archaeovenator hamiltonensis* (after Reisz and Dilkes 2003). *ne* neural excavation, *ns* neural spine, *poz* postzygapophysis, *prz* prezygapophysis, *tp* transverse process. Scale bars equal 1 cm

The average length of the middorsal centra (approximately 13 mm) is 23 % greater than the average width (approximately 10 mm; Fig. 4.1). This is in contrast to the condition in *Varanops*, in which the centra are 8 % longer than wide (Campione and Reisz 2010) and *Mycterosaurus*, in which the centra are as wide as they are long (Berman and Reisz 1982).

The entire length of the reconstructed column in *Aerosaurus* is approximately 1170–1180 mm. The length of the caudal centra decrease posteriorly from 12 mm in the first caudal vertebrae to 9 mm in the 38th, and 6 mm in the posteriormost caudals. By way of contrast, in *Archaeovenator* there is a slight increase in the length of the caudal centra anteriorly (Reisz and Dilkes 2003). The dorsal centra are amphicoelous, as in other "pelycosaurs" (Romer and Price 1940). A midventral ridge is present, similar to that in *Watongia* (Reisz and Laurin 2004), *Pyozia* (Anderson and Reisz 2004), and *Varanops* (Maddin et al. 2006). The ridge is rounded compared to the sharp keel in sphenacodontids and forms a ventral lip on both the anterior and posterior centrum rims (Fig. 4.2) similar to that in *Aerosaurus greenleeorum* (Romer and Price 1940). In marked contrast, a shallow midventral groove is formed by parallel ridges in *Heleosaurus* (Reisz and Modesto 2007). The presence of a midventral ridge is uncertain in the sacral and caudal vertebrae of *Aerosaurus*. The anterior edges of the centra are concave in lateral view with a slight bevel for the intercentrum, as in *Varanops* (Maddin et al. 2006), whereas the posterior edge is convex. As in *Elliotsmithia* (Reisz et al. 1998), there are no longitudinal ridges present on the lateral surfaces of the centra below the transverse process.

Intercentra are unfused and present throughout most of the dorsal column. Chevrons, or haemal arches, are present in the caudal series after the fourth caudal vertebra and can be traced to the 22nd. The chevrons increase in length posteriorly from 22 to 26 mm for the fourth chevron, and then decrease gradually in length. The last one is on caudal five and is a small spike-like form. The chevron spine is

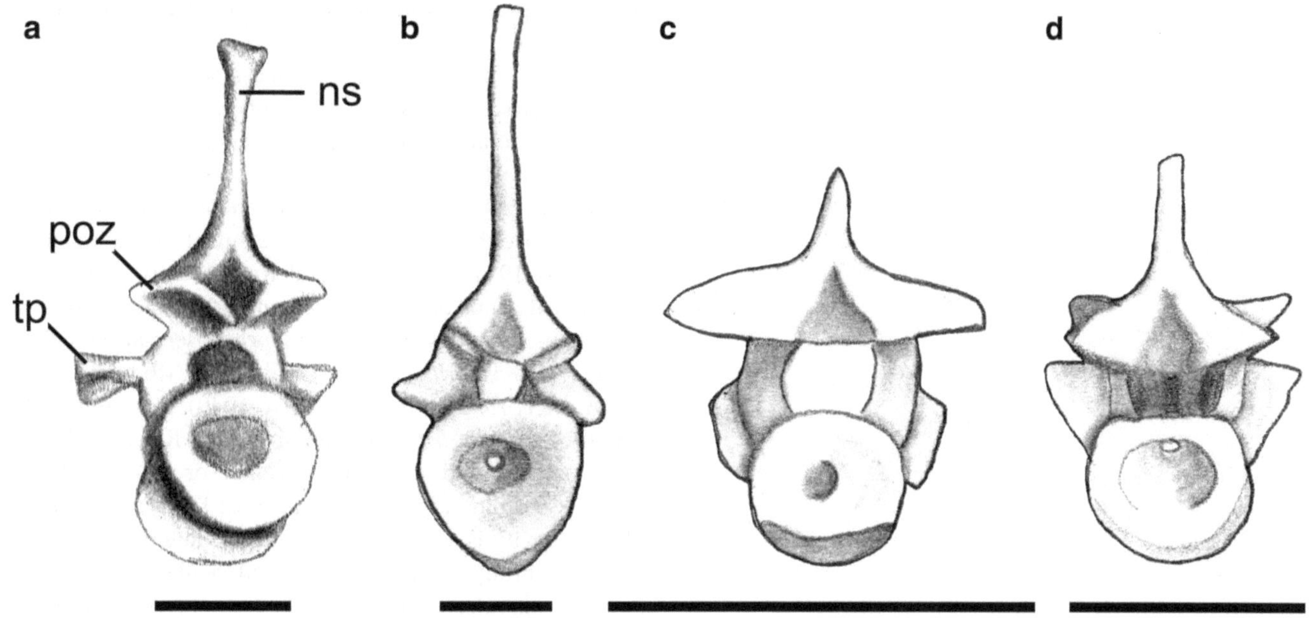

Fig. 4.2 Dorsal vertebrae in posterior view. **a** *Aerosaurus wellesi* (UCMP 40098) as preserved, showing partial distortion. **b** *Varanops brevirostris* (after Campione and Reisz 2010). **c** *Heleosaurus scholtzi* (after Carroll 1976). **d** *Mycterosaurus longiceps* (after Reisz et al. 1997). *ns* neural spine, *poz* postzygapophysis, *tp* transverse process. Scale bars equal 1 cm

laterally flattened and slightly sigmoidal in lateral view, ending in a blunt tip.

The transverse processes of the dorsal vertebrae, which extend laterally and slightly dorsally 5–7 mm from the centrum, are located just anterior to the centrum midlength. The zygapophyses are positioned approximately 2.5 mm above the centrum and do not extend beyond its lateral margins. The prezygapophyses and postzygapophyses face dorsomedially and ventrolaterally, respectively. In *Mycterosaurus* (Berman and Reisz 1982), the zygapophyses extend slightly beyond the lateral surfaces of the centra and in *Heleosaurus* they extend even further (Botha-Brink and Modesto 2009) (Fig. 4.2).

The surface between the neural spine and the postzygapophyses in *Aerosaurus* is concave, whereas in cotylosaurs it is convex, and in ophiacodonts and *Varanops* it is flat (Romer and Price 1940). The dorsal neural spines are subrectangular in shape and extend as much as 20 mm above the centrum, a height almost twice that of the centra (Fig. 4.1). This is similar to the condition in *Mycterosaurus* (Berman and Reisz 1982), but considerably shorter than *Varanops*, where the spines are three times the height of the centra (Williston 1911; Campione and Reisz 2010). The spine widens from 7 mm at its base to 9 mm at its dorsal end. At the base of the lateral surface of the neural spine is a deep elongate lateral excavation, similar to that in *Watongia* (Reisz and Laurin 2004) and *Varanops* (Maddin et al. 2006; Campione and Reisz 2010). Lateral excavations also are present in *Mycterosaurus* (Berman and Reisz 1982) and *Heleosaurus*, but are shallower and in *Heleosaurus* occur only in the cervical region (Reisz and Modesto 2007). Although the number of sacral vertebrae in the referred specimen, UCMP 40097, cannot be determined, Langston and Reisz (1981) reported three in the holotype, UCMP 40096. This is unusual in that most varanopids, including the closely related *Varanops*, have only two sacral vertebrae (Williston 1911; Olson 1965; Reisz and Dilkes 2003), which is the plesiomorphic condition for synapsids.

The ribs are double-headed with distinct tubercular and capitular articulations, similar to those in *Varanodon* (Olson 1965) and *Varanops* (Williston 1911; Campione and Reisz 2010). In contrast, the ribs in *Heleosaurus* (Botha-Brink and Modesto 2009), *Mycterosaurus* (Berman and Reisz 1982), and *Archaeovenator* (Reisz and Dilkes 2003) are holocephalous with a single, expanded, triangular head. The thin sheet of bone extending between the two heads is notched for the segmental artery. The longest rib in *Aerosaurus* is 92.5 mm and along its proximodorsal surface is a ridge for attachment of the iliocostalis muscle (Olson 1936; Romer and Price 1940). Posteriorly the ribs curve downward, but are less curved distally (Fig. 4.3). This suggests that the trunk was rather deep and narrow. The sacral ribs are Y-shaped with a short, cylindrical proximal neck and a plate-like distal portion that descends along the inner surface of the ilium. The articular face is straight and the one preserved rib is not fused to its centrum. As in *Varanodon* (Olson 1965), the first four caudal ribs curve directly posteriorly. The first caudal rib is the longest, extending beyond

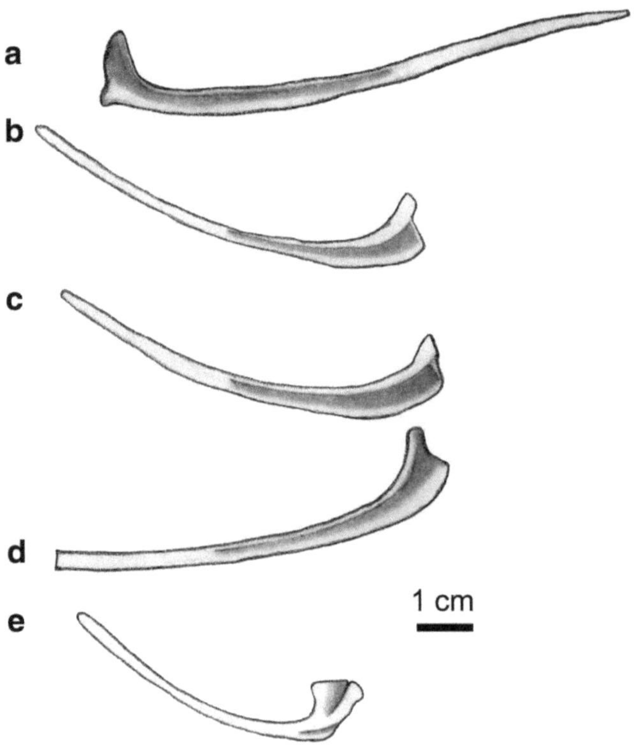

Fig. 4.3 Ribs of *Aerosaurus* (UCMP 40097). **a** Left rib in dorsal view. **b–d** Left ribs in ventral view. **e** Left caudal rib in ventral view

the succeeding four caudals, thus paralleling the subsequent three ribs, which become progressively shorter, with the fourth just reaching the posterior margin of the centrum of its origin.

Although technically dermal elements and not ribs, mention of the gastralia is made here for the sake of completeness. The gastralia are slender and rod-shaped and in life were probably aligned in a chevron pattern (Romer and Price 1940), not unlike those in *Varanodon* (Olson 1965) and *Varanops* (Williston 1911).

Appendicular Skeleton

Neither the cleithrum nor the interclavicle has been found in any of the studied specimens (Langston and Reisz 1981). The clavicle (UCMP 40097) consists of a narrow, rod-like dorsal shaft that curves posteriorly and an expanded ventral portion that curves medially and anteriorly (Fig. 4.4). The plate has a greater expansion than that in *Watongia* (Reisz and Laurin 2004), *Archaeovenator* (Reisz and Dilkes 2003) or *Heleosaurus* (Reisz and Modesto 2007).

The clavicle measures approximately 59 mm in vertical length, whereas the transverse width of the plate is 27 mm. Striations extend from the base of the shaft across to the ventral plate.

The upper portion of the scapular blade in UCMP 40097 has been broken off, leaving only about 38 mm remaining. A suture separating the scapula and anterior coracoid is not visible in most varanopids (Berman and Reisz 1982; Reisz and Dilkes 2003) with the exception of *Varanops*, where sutures are either present or the elements have separated (Williston 1911; Campione and Reisz 2010). A posterior coracoid has not been identified in any of the *Aerosaurus* specimens, whereas in *Mycterosaurus* (Berman and Reisz 1982), *Heleosaurus* (Botha-Brink and Modesto 2009), *A. greenleeorum* (Romer and Price 1940), *Varanodon* (Olson 1965), and *Varanops* (Campione and Reisz 2010) it is fully ossified and suturally attached or fused with the scapulocoracoid (Fig. 4.5). As in most basal amniotes, the glenoid cavity is screw-shaped, with the anterior end facing posteroventrally and the posterior end facing anterodorsally.

The triangular supraglenoid buttress faces posterolaterally and is triangular in shape. As in *A. greenleeorum* (Romer and Price 1940), *Varanops* (Williston 1911), and *Pyozia* (Anderson and Reisz 2004), a supraglenoid foramen is located just anterior to the midwidth of the scapular blade. A supraglenoid foramen has not been identified in *Heleosaurus* (Reisz and Modesto 2007), *Mycterosaurus* (Berman and Reisz 1982), or *Archaeovenator* (Reisz and Dilkes 2003). A well-developed anteroventrally oriented ridge is present anterior to the glenoid, ventral to which is a deep depression containing the coracoid foramen. The anteroventral margin of the anterior coracoid is broadly convex. The basal portion of the scapular blade is angled posteriorly about 80° from the coracoid plate. A notch is present at the presumed division of the scapula and anterior coracoid.

The humerus (UCMP 40093) is typically tetrahedral in shape, with a well-defined shaft and expanded proximal and distal heads (Fig. 4.6). It measures approximately 78 mm in length, and whereas the width of the incomplete proximal head cannot be determined, the width of the complete distal head is 29 mm, which is 39 % of the humeral length. In *Varanops* (Williston 1911; Maddin et al. 2006; Campione and Reisz 2010) and *Watongia* (Reisz and Laurin 2004), the proximal and distal heads are also broadly expanded. This is not the case in *Archaeovenator* (Reisz and Dilkes 2003), *Mycterosaurus* (Berman and Reisz 1982), and *Heleosaurus* (Botha-Brink and Modesto 2009), in which the proximal and distal heads are 36 and 33 % the length of the humerus, respectively. The planes of the proximal and distal heads meet at an angle of approximately 90°. This angle, termed the torque, is very high for a varanopid. The torque is 70°, 60°, and 40° in *Varanops* (Williston 1911), *Watongia* (Reisz and Laurin 2004), and *Heleosaurus* (Botha-Brink and Modesto 2009)

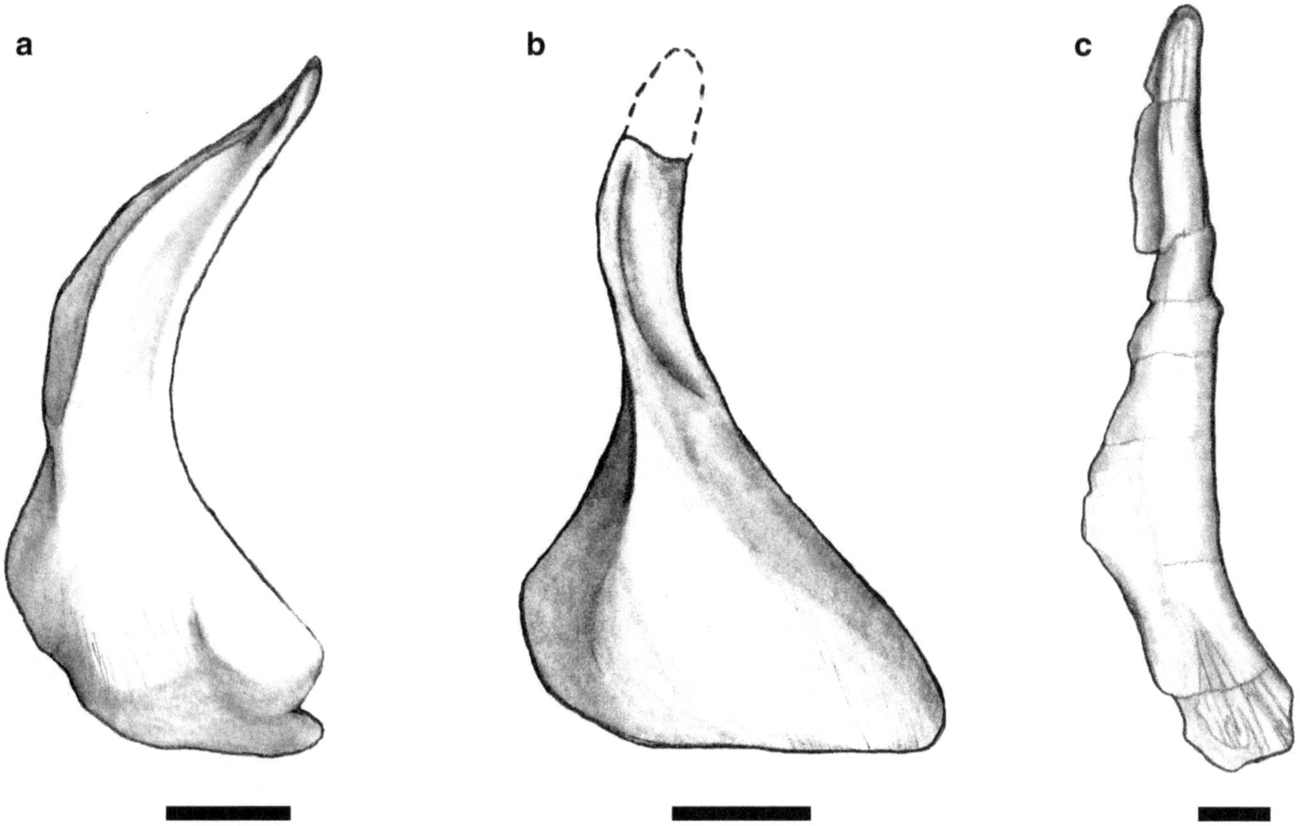

Fig. 4.4 Left clavicles of varanopids in lateral view. **a** *Aerosaurus wellesi* (UCMP 40097). **b** *Varanops brevirostris* (after Williston 1911). **c** *Watongia meieri* (after Reisz and Laurin 2004). Scale bars equal 1 cm

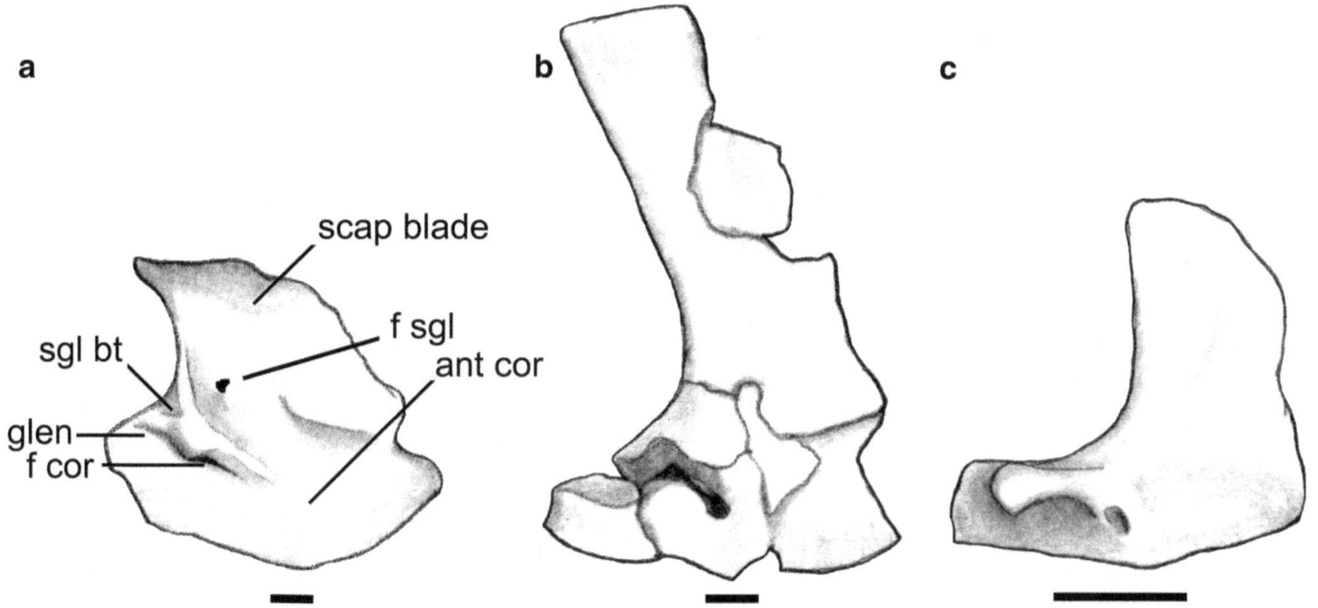

Fig. 4.5 Right scapulocoracoids of varanopids in lateral view. **a** *Aerosaurus wellesi* (UCMP 40097). **b** *Varanops brevirostris* (after Campione and Reisz 2010). **c** *Mycterosaurus longiceps* (after Reisz et al. 1997). *ant cor* anterior coracoid, *f cor* coracoid foramen, *glen* glenoid cavity, *scap blade* scapular blade, *sgl bt* supraglenoid buttress. Scale bars equal 1 cm, *f sgl* supraglenoid foramen

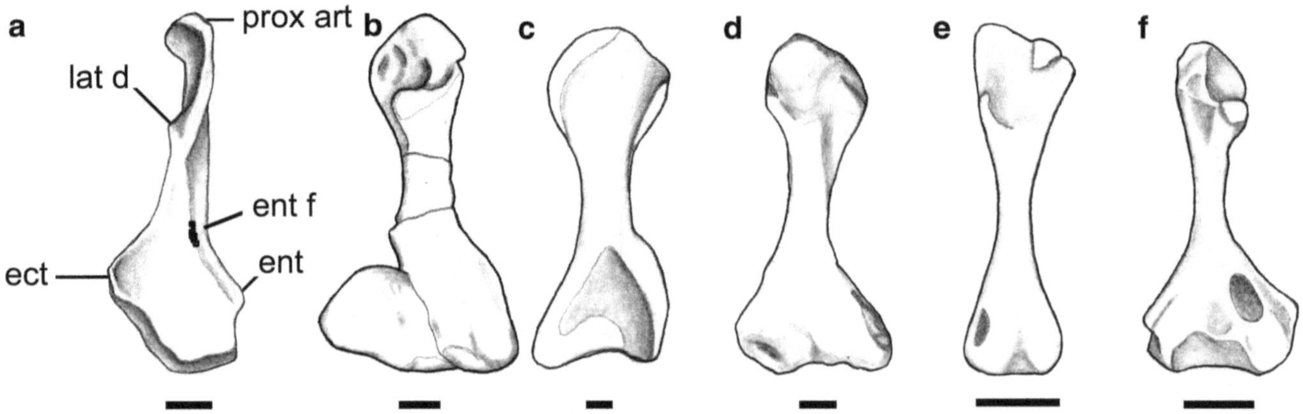

Fig. 4.6 Left humeri of varanopids. **a** *Aerosaurus wellesi* (UCMP 40093) in dorsal view. **b** *Varanops brevirostris* in dorsal view (after and reversed from Campione and Reisz 2010). **c** *Varanodon agilis* in dorsal view (after Olson 1965). **d** *Watongia meieri* in ventral view (after Reisz and Laurin 2004). **e** *Heleosaurus scholtzi* in dorsal view (after Botha-Brink and Modesto 2009). **f** *Mycterosaurus longiceps* in ventral view (after Reisz et al. 1997). *ect* ectepicondyle, *ent* entepicondyle, *lat d* tubercle for latissimus dorsi, *prox art* proximal articular surface. Scale bars equal 1 cm, *ent f* entepicondylar foramen

respectively, although in *Watongia* it has been flattened slightly (Reisz and Laurin 2004). In *Aerosaurus* a distinct tubercle for the latissimus dorsi is present on the posterior edge of the proximal dorsal surface. The entepicondylar foramen is an oval opening on a ridge that extends from the proximal head to the entepicondyle, where it ends in a somewhat truncated margin. The supinator process is not strongly differentiated and is separated from the ectepicondyle by what appears to be a shallow groove for the radial nerve and accessory blood vessels. An ectepicondylar foramen is not present, as in most other varanopids (Berman and Reisz 1982; Reisz and Laurin 2004; Maddin et al. 2006), with the exception of *Heleosaurus* (Botha-Brink and Modesto 2009).

The radius (Fig. 4.7) is short compared to the humerus, with an approximate length of 54 mm, 69 % of the humeral length. Similarly, in *Watongia* and *Varanops* the radius is 75 % the length of the humerus (Williston 1911; Reisz and Laurin 2004), whereas in *Heleosaurus* the same metric for the radius is much greater at 83 % (Botha-Brink and Modesto 2009). The radius in *Aerosaurus* is very similar in shape to those of other varanopids in having slightly expanded proximal and distal heads and a ridge that extends the length of its dorsal surface. The radius is somewhat concave medially, and the humeral facet is concave and faces slightly dorsomedially.

As in most varanopids, the ulna (Fig. 4.8a) is dorsoventrally flattened and expanded at both ends, more so proximally, and is similar in shape to that of *Varanops* (Campione and Reisz 2010). It is thicker and longer than the radius, though not as bowed. The well-developed olecranon process accounts for 15 % of its 60 mm length; similarly, the olecranon process of *Varanodon* is well ossified (Olson 1965), though in *Watongia* it is thin and modest (Reisz and Laurin 2004). The proximal portion of its dorsal face is slightly concave.

The description and reconstruction of the manus is based on the disarticulated left manus of UCMP 40097 (Fig. 4.8b, c). The radiale is broad and short, similar to that in *Watongia* (Reisz and Laurin 2004), with the facet for the medial centrale being much larger than that for the lateral centrale. There are two raised areas on the radiale, probably for insertion of the extensor carpi radialis superficialis and origin of the extensor digitorum communis brevis (Holmes 1977). The pisiform is small, in contrast to the condition in other varanodontines (Reisz and Laurin 2004). The medial centrale has a wide facet for the radiale, whereas its distal surface has two facets for distal carpals 1 and 2. There is also a facet for the lateral centrale, which meets the medial centrale in a slightly offset angulation. The lateral centrale is triangular with slightly convex lateral and medial margins and a distally pointed apex. Of the five distal carpals, the fourth is longer anteroposteriorly, with the third longer mediolaterally. These are followed in size by the first and second, with the fifth being greatly reduced. The dorsal surfaces of the distal carpals are flat with raised edges on the proximal and distal margins. The phalangeal formula as reconstructed conforms to the standard 2-3-4-5-3. The phalanges decrease serially in length from the first to the penultimate, with the ungual being slightly longer than the penultimate phalanx. The phalanges are dorsoventrally flattened and expanded at both ends, with the proximal expansion being greater. The unguals are flat dorsally and taper to a distal point. Proximally, on the ventral surface is a well-developed flexor tubercle, probably for the attachment of flexor tendons. On the lateral surface is a longitudinal blood groove that nourished the keratinous sheath of the claw (Maddin et al. 2007).

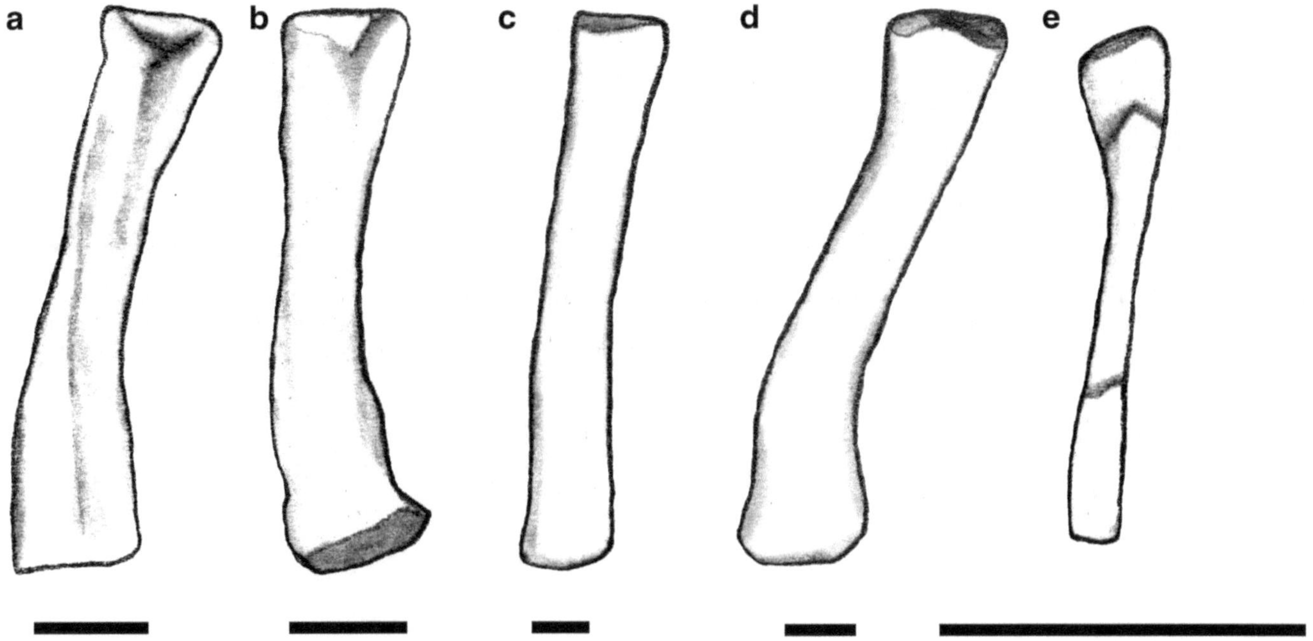

Fig. 4.7 Right radii of varanopids in lateral view. **a** *Aerosaurus wellesi* (UCMP 40097). **b** *Varanops brevirostris* (after Campione and Reisz 2010). **c** *Varanodon agilis* (after and reversed from Olson 1965). **d** *Watongia meieri* (after and reversed from Reisz and Laurin 2004). **e** *Heleosaurus scholtzi* (after Botha-Brink and Modesto 2009). Scale bars equal 1 cm

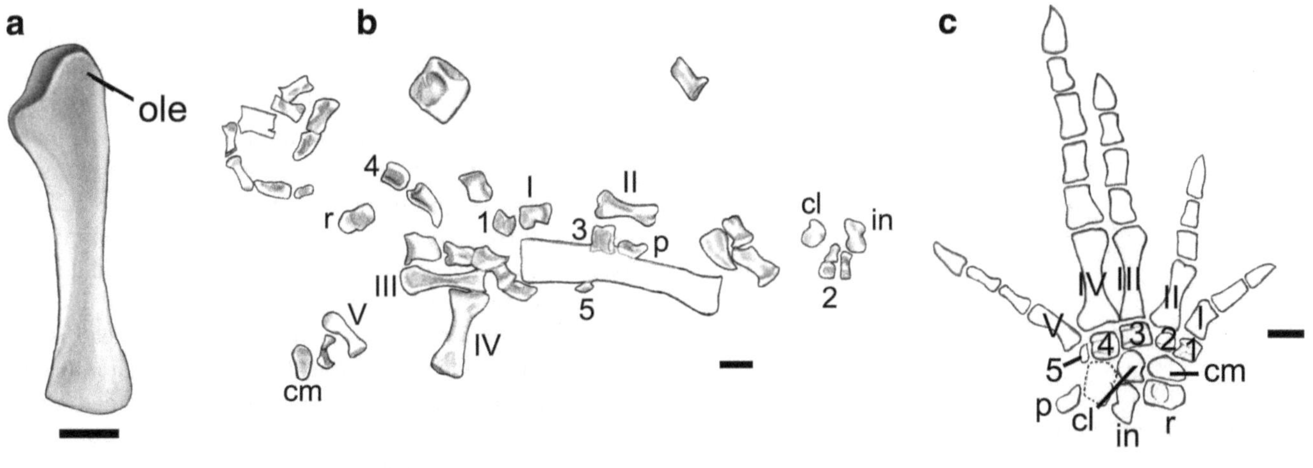

Fig. 4.8 Elements of forelimb of *Aerosaurus* (UCMP 40097). **a** Right ulna in ventral view. **b** Left manus as preserved in matrix and **c** reconstructed in dorsal view. *1–5* distal carpals, *I–V* metacarpals, *cl* lateral centrale, *cm* medial centrale, *in* intermedium, *ole* olecranon, *p* pisiform, *r* radiale. Scale bars equal 1 cm

In the pelvic girdle of UCMP 40097 (Figs. 4.9, 4.10, 4.11), the sutures of the ilium, ischium, and pubis are not fused, which is typical of varanopids with the exception of *Archaeovenator* (Reisz and Dilkes 2003). The iliac blade of UCMP 40097 (Fig. 4.9) extends posteriorly and slightly dorsally for 37.5 mm, with a greatest width at its base of 10 mm. A ridge extends anteroposteriorly and adjacent to the dorsal margin of the ilium above the acetabulum, presumably for attachment of the of the iliofemoralis muscle (Romer and Price 1940). The iliac blade tapers distally, similar to the condition in other varanopids with the exception of *Varanops* in which the distal end of the iliac blade is broad (Campione and Reisz 2010). The articular facets of the ilium for the ischium and pubis meet at an angle of approximately 70°.

The contact for the pubis is shorter than that for the ischium, similar to the condition in *Varanops* (Maddin et al. 2006). The ischium (Fig. 4.10) is hatchet shaped, as in

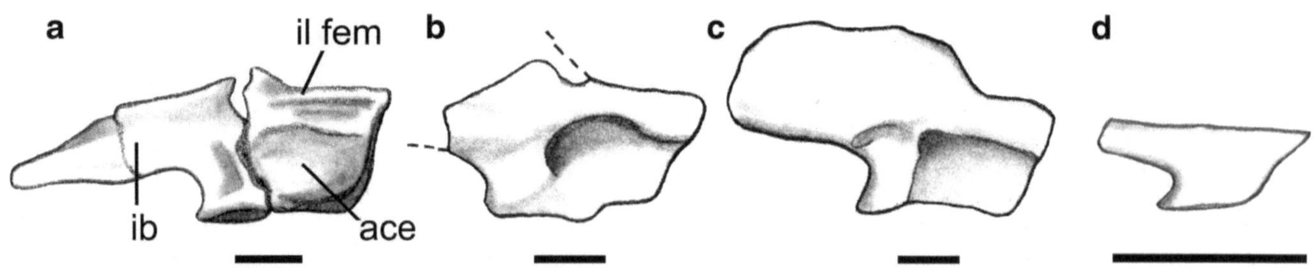

Fig. 4.9 Right ilia of varanopids in lateral view. **a** *Aerosaurus wellesi* (UCMP 40097). **b** *Aerosaurus greenleeorum* (after and reversed from Romer and Price 1940). **c** *Varanops brevirostris* (after Campione and Reisz 2010). **d** *Archaeovenator hamiltonensis* (after and reversed from Reisz and Dilkes 2003). *ace* acetabulum, *ib* iliac blade, *il fem* ridge for attachment of iliofemoralis muscle. Scale bars equal 1 cm

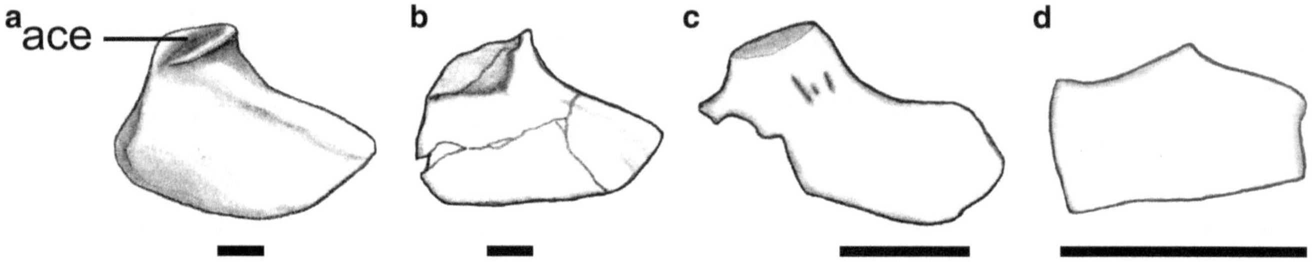

Fig. 4.10 Left ischia of varanopids in lateral view. **a** *Aerosaurus wellesi* (UCMP 40097). **b** *Varanops brevirostris* (after Campione and Reisz 2010). **c** *Heleosaurus scholtzi* (after Carroll 1976). **d** *Archaeovenator hamiltonensis* (after Reisz and Dilkes 2003). *ace* acetabulum. Scale bars equal 1 cm

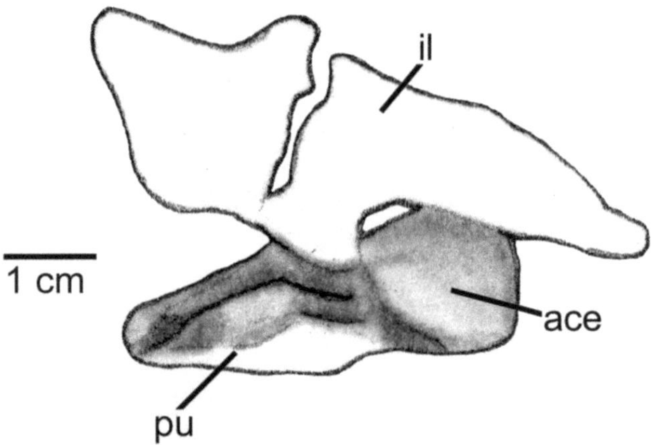

Fig. 4.11 Right pubis in ventrolateral view and ilium of *Aerosaurus* (UCMP 40097). *ace* acetabulum, *il* ilium, *pu* pubis

Varanops (Williston 1911; Campione and Reisz 2010), with a blade-like distal portion as in *Heleosaurus* (Reisz and Modesto 2007).

A thickened ridge along the ventral margin of the acetabulum is also present in other varanopids (Williston 1911; Berman and Reisz 1982; Reisz and Dilkes 2003). The ischium (UCMP 40097) is about 60 mm long, which is 15 % greater than its height (Fig. 4.10). It is not possible to identify the obturator foramen or whether a pubic tubercle is present in UCMP 40097 (Fig. 4.11). Similar to other varanopids, the pubis has a thickened dorsal ridge with a flattened plate-like area dorsally.

The femur of *Aerosaurus* UCMP 40093 is approximately 84 mm long with expanded heads and a slender shaft, and is similar in shape to that of *Varanops* (Williston 1911; Campione and Reisz 2010). The femur is longer than the humerus by about 7 %, which indicates that *Aerosaurus* has the longest humerus relative to the femur of any varanopid genus in which the comparison is available; the same measurement is approximately 23 % in *Varanops* (Campione and Reisz 2010). The expansions of the proximal and distal heads of the femur of *Aerosaurus* are 31 and 26 % of its length, respectively (Fig. 4.12). An almost identical value of 32 % for the width of the proximal head is calculated in *Varanops* (Williston 1911), but is greater than the same measurement for *Archaeovenator* (Reisz and Dilkes 2003), *Heleosaurus* (Botha-Brink and Modesto 2009), and *Mycterosaurus* (Berman and Reisz 1982).

The relative width of the distal end in *Aerosaurus* is less than in any other varanopid. The articular surface for the acetabulum is located terminally on the proximal end. The femur does not exhibit a substantial sigmoidal curvature, similar to the condition in *Varanops* (Campione and Reisz 2010) and in contrast to *Archaeovenator* (Reisz and Dilkes 2003), *Heleosaurus* (Botha-Brink and Modesto 2009), and *Mycterosaurus* (Berman and Reisz 1982). In *Aerosaurus* the anterior and posterior margins of the femur are concave, the anterior margin more so than the posterior, which among varanopids is most similar to the condition in *Varanops* (Williston 1911). As in other early amniotes, the posterior condyle of the distal end of the femur extends a short

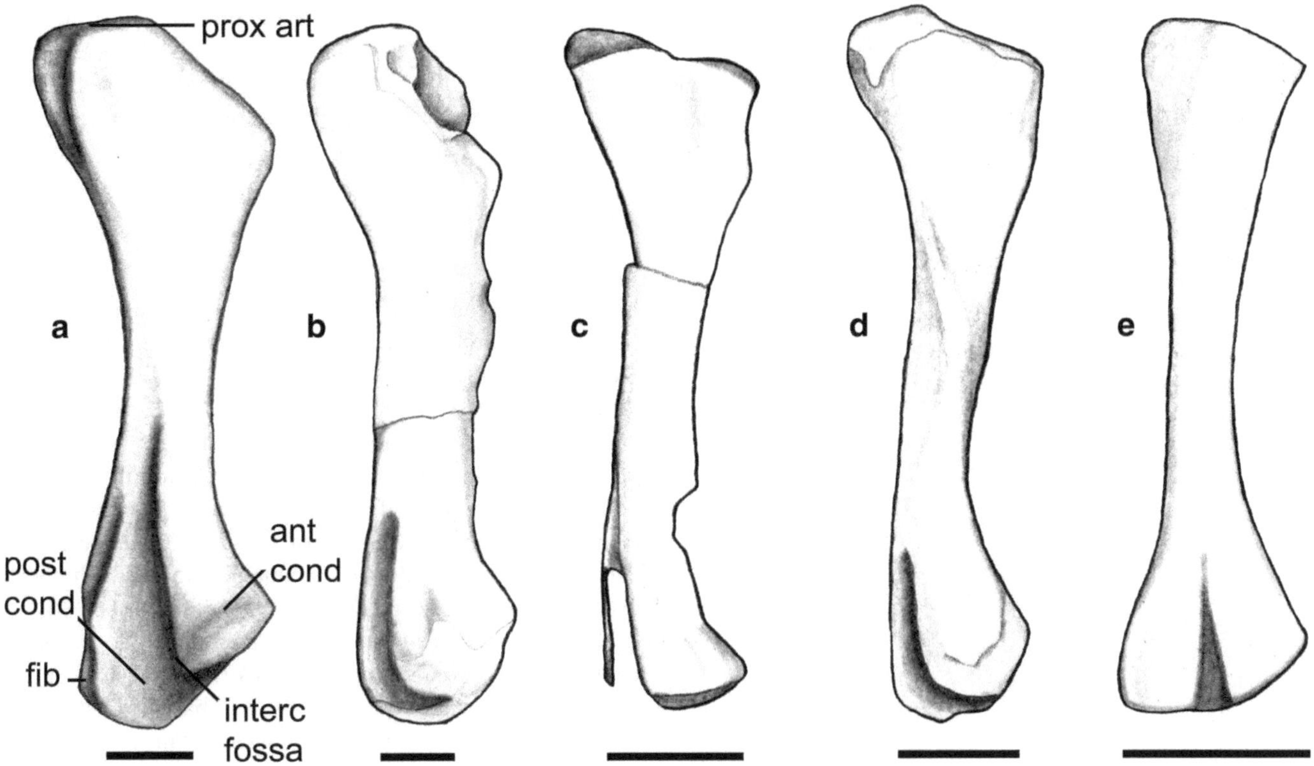

Fig. 4.12 Right femora of varanopids in dorsal view. **a** *Aerosaurus wellesi* (UCMP 40097). **b** *Varanops brevirostris* (after and reversed from Campione and Reisz 2010). **c** *Heleosaurus scholtzi* (after Carroll 1976). **d** *Mycterosaurus longiceps* (after and reversed from Reisz et al. 1997). **e** *Archaeovenator hamiltonensis* (after Reisz and Dilkes 2003). *ant cond* anterior condyle, *fib* articular surface for fibula, *interc fossa* intercondylar fossa, *post cond* posterior condyle, *prox art* proximal articular surface. Scale bars equal 1 cm

distance beyond the anterior condyle. The intercondylar fossa is deep and extends 31.3 mm onto the shaft. As in *Varanops*, this is approximately 37 % of the femoral length (Williston 1911), whereas in *Mycterosaurus*, the same measurement is only 21 % (Berman and Reisz 1982). The fibular facet is also long, extending 23.3 mm onto the shaft or approximately 28 % of the femur length.

The tibia in *Aerosaurus* has a narrow shaft and greatly expanded proximal and distal heads (Fig. 4.13). The lateral margin is concave, giving the entire bone a slightly bowed appearance as in *Archaeovenator* (Reisz and Dilkes 2003) and *Varanops* (Campione and Reisz 2010). The tibia in *Aerosaurus* UCMP 40097 is approximately 52 mm long, 62 % the length of the femur. This is much less than the values for *Archaeovenator* (Reisz and Dilkes 2003), *Mycterosaurus* (Berman and Reisz 1982), or *Heleosaurus* (Botha-Brink and Modesto 2009), which are 75, 78, and 95 % respectively, and also contrasts with the condition in *Varanops*, in which the tibia is 15 % longer than the femur (Campione and Reisz 2010). The dorsal surface of the proximal head bears a deep groove that divides the proximal surface into distinct medial and lateral facets, whereas the distal end of the tibia is dorsoventrally flattened with a nearly flat facet for the astragalus.

The proximal end of the fibula is expanded, with a pronounced tubercle on its dorsal surface. The shaft is slender and twisted so that distally the dorsal surface faces posteriorly, similar to the condition in *Varanops* (Williston 1911). The medial surface of the fibula is concave and the lateral surface nearly straight. The length of the fibula (UCMP 40097) is approximately 66 mm, making it longer than the tibia, as in *Varanops* (Williston 1911), whereas in *Mycterosaurus* they are subequal in length (Berman and Reisz 1982).

The pedes of UCMP 40097 (Fig. 4.14) are partially preserved and articulated. Together they provide a firm basis for the reconstruction of the entire pes except for distal tarsal 5 and the medial and lateral centrales, which may have been coosified (Fig. 4.14c). The articular surface of the astragalus for the fibula is convex. The dorsal surface is slightly concave with raised margins adjacent to the facets as in *Varanops* (Williston 1911; Campione and Reisz 2010), *Mycterosaurus* (Berman and Reisz 1982), and *Heleosaurus* (Botha-Brink and Modesto 2009). The notch for the perforating artery cannot be observed in UCMP 40097 because it is covered by the calcaneum, which cannot be removed without risk of damage (Fig. 4.14a). The outer margin of the calcaneum is convex as in other varanopids except

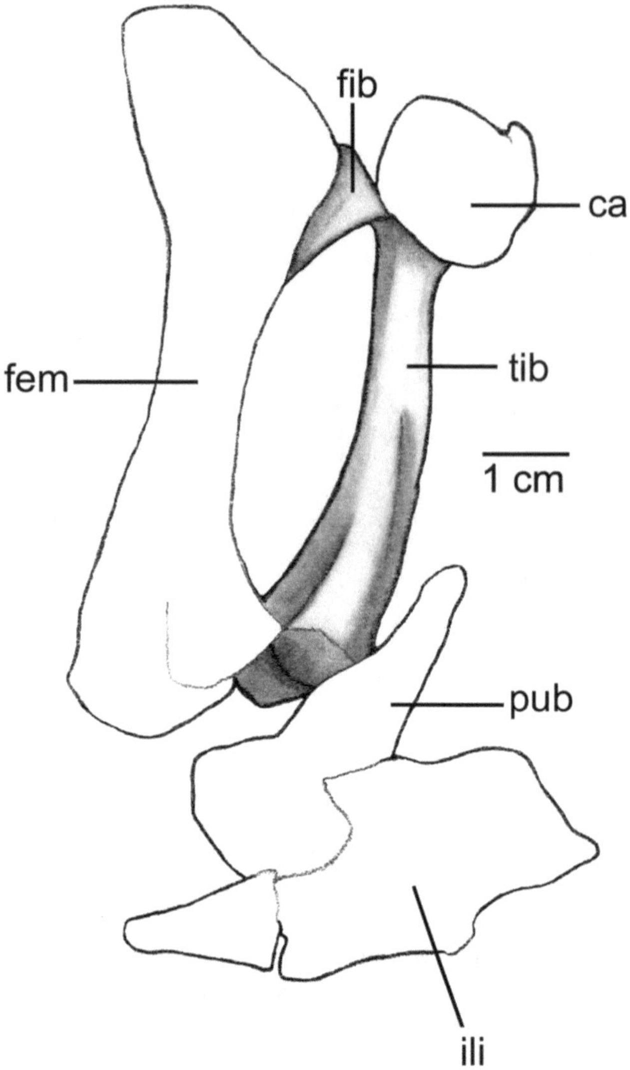

Fig. 4.13 Right tibia of *Aerosaurus* (UCMP 40097) in lateral view and associated elements of pelvis and limbs. *ca* calcaneum, *fem* femur, *fib* fibula, *ili* ilium, *pub* pubis, *tib* tibia

Archaeovenator, where the lateral margin is nearly straight (Reisz and Dilkes 2003). The medial margin is essentially straight and thickened, with a distal notch for the perforating artery. As in *Varanops*, the fourth distal tarsal is the largest, followed in decreasing size by the first, third, and second (Williston 1911). The metatarsals are strongly expanded at either end and dorsoventrally flattened. The phalangeal formula of the pes is the standard 2-3-4-5-4. There is a serial increase in the length of the digits from the first to the fourth with the fifth digit being intermediate in length between the second and third. The phalanges increase in width serially from the first digit to the fourth, with those of the fifth being more slender than those of the other digits. Within each digit the phalanges decrease serially in length distally except for the ungual, which is longer than all others except for the proximalmost phalanx. As in the manus, the unguals taper to a distal point. At the proximal end of the ventral surface there is a well-developed flexor tubercle for attachment of the flexor tendons.

Discussion

Aerosaurus is an undisputed varanodontine and the sister taxon to the larger, younger taxa, *Varanops* and *Varanodon* (Anderson et al. 2004; Maddin et al. 2006; Botha-Brink and Modesto 2009; Campione and Reisz 2010; Reisz et al. 2010). Therefore, a thorough knowledge of its anatomy provides insight into the postcranial evolution of varanodontines, and its differentiation with mycterosaurines (Fig. 4.15). The high degree of twist in the humerus, the extremely long tail, and the subequal lengths of the humerus and femur are primitive features (Romer and Price 1940) of *Aerosaurus* not seen in other varanopids. A torque of approximately 90° is reported in *Cotylorhynchus* (Stovall et al. 1966) and *Limnoscelis*, a basal diactomorph (Wideman et al. 2005). An elongated tail is present in Ophiacodontidae and *Varanosaurus* (Romer and Price 1940), and subequal lengths of humerus and femur are present in *Casea broilii* (Olson 1954) and *Cotylorhynchus* (Stovall et al. 1966). On the other hand, the greater length of the centra compared to the width and the expanded clavicular plate are specializations seen in more derived varanopids such as *Varanops* (Williston 1911; Campione and Reisz 2010). The inclusion of *Aerosaurus* as a varanodontine is supported by several synapomorphies that it shares with *Varanops, Watongia*, and *Varanodon*: a deep and elongate excavation at the base of the neural spine, in contrast to the shallow and anteroposteriorly restricted excavations of mycterosaurines (Maddin et al. 2006, Campione and Reisz 2010; Berman et al. 2013, Sumida et al. 2013), distortion precludes exact measurement of depth of excavation; dicephalic ribs; presence of a supraglenoid foramen; expanded proximal and distal ends of the femur; femur lacking a strong sigmoid curvature; and length of the tibia less than 70 % that of the femur. These results are supported by phylogenetic analyses such as Berman et al. (2013) and Sumida et al. (2013). Many of these features, including the broadly expanded proximal and distal ends of the humerus and femur may be due to an overall larger body size in varanodontines relative to mycterosaurines. The proximal articular surface encompassing the terminal end of the femur and the screw-shaped glenoid of the scapulocoracoid show that *Aerosaurus* was an obligatory sprawling-gaited animal (Holmes 1977, 2003). A similar articular surface is present in ophiacodonts,

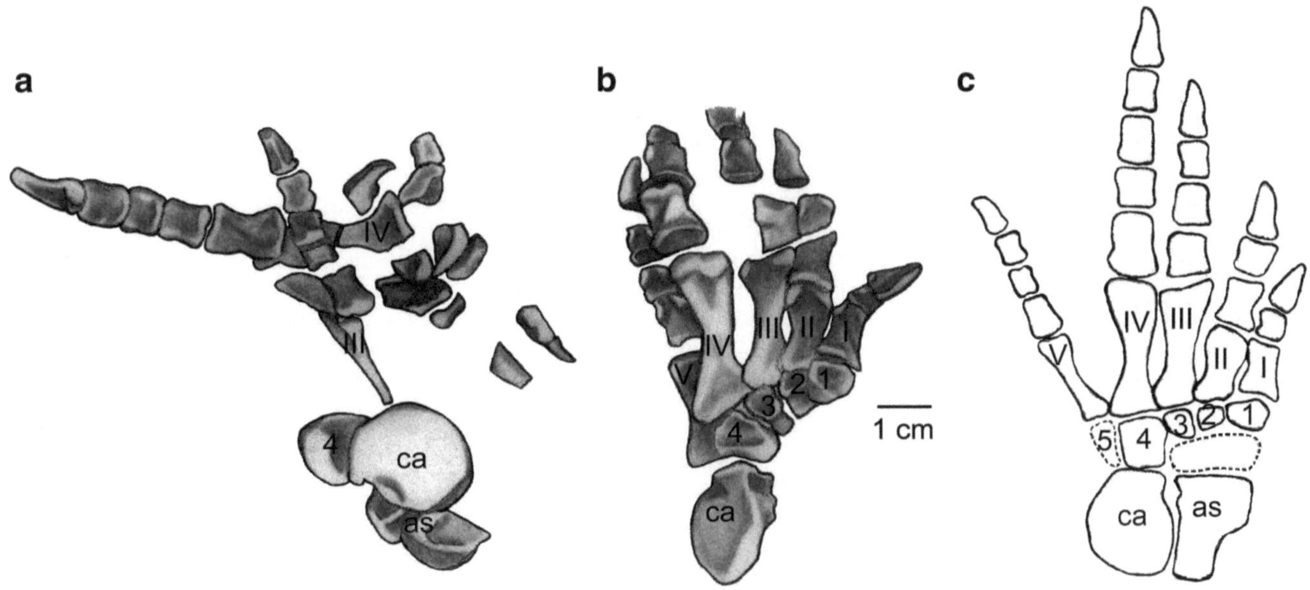

Fig. 4.14 Pedes of *Aerosaurus wellesi* (UCMP 40097). **a** Right and **b** left as preserved. **c** Left reconstructed in dorsal view. *1–5* distal tarsals, *I–V* metatarsals, *as* astragalus, *ca* calcaneum

caseids, and edaphosaurids. However, in sphenacodontids the articular surface only occupies two-thirds of the proximal end (Romer and Price 1940), with the expansion of the proximal and distal ends of the humerus and femur allowing more area for muscle attachment in a relatively larger animal.

The subequal lengths of the humerus and femur and the epipodials being shorter than the propodials might suggest that *Aerosaurus* was still a relatively slower, less agile, sprawling animal than the smaller mycterosaurines, as leg length can often be an indicator of stride length (Romer and Price 1940). However, *Aerosaurus* was probably not a confrontational threat to the larger sphenacodontids also found in the Arroyo de Agua deposits, which were larger and longer legged (Romer and Price 1940).

Aerosaurus is larger than the mycterosaurines but smaller than other varanodontines, depending on the phylogenetic placement of *Elliotsmithia* (Olson 1965; Berman and Reisz 1982; Reisz and Laurin 2004; Botha-Brink and Modesto 2009; Campione and Reisz 2010). Therefore, it appears that the evolution of varanodontines is marked by an increase in size, and it is possible that they became dominant predators in the upland ecosystems of Laurasia where other large synapsid predators such as large sphenacodontids appear to be rare or absent (Langston and Reisz 1981; Maddin et al. 2006; Berman et al. 2013). Such hypotheses have been suggested by Olson (1991), Sullivan and Reisz (1999), and Maddin et al. (2006) for the Richards Spur locality, as well as by Eberth et al. (2000), Reisz and Modesto (2007), and Berman et al. (2013) for the Bromacker Quarry in Germany. The smaller size of *Aerosaurus* compared to other varanodontines may be related to the presence of larger synapsid carnivores such as sphenacodontids in the Arroyo del Agua deposits. Mycterosaurines, however, remained small, agile predators, and are more widespread, including examples from both Laurasia and Gondwana.

Acknowledgments I wish to thank Drs. Kevin Padian and Patricia Holroyd of the University of California Museum of Paleontology for loan of the specimens. I am greatly indebted to Dr. Stuart Sumida for suggesting this project and providing advice. Elaine Bayer provided assistance in transporting the specimens and many helpful discussions. I would also like to thank David Pelletier for all his support and understanding, and Dr. Elizabeth Rega and Darwin and Owen Sumida for allowing Stuart the time to help with this project.

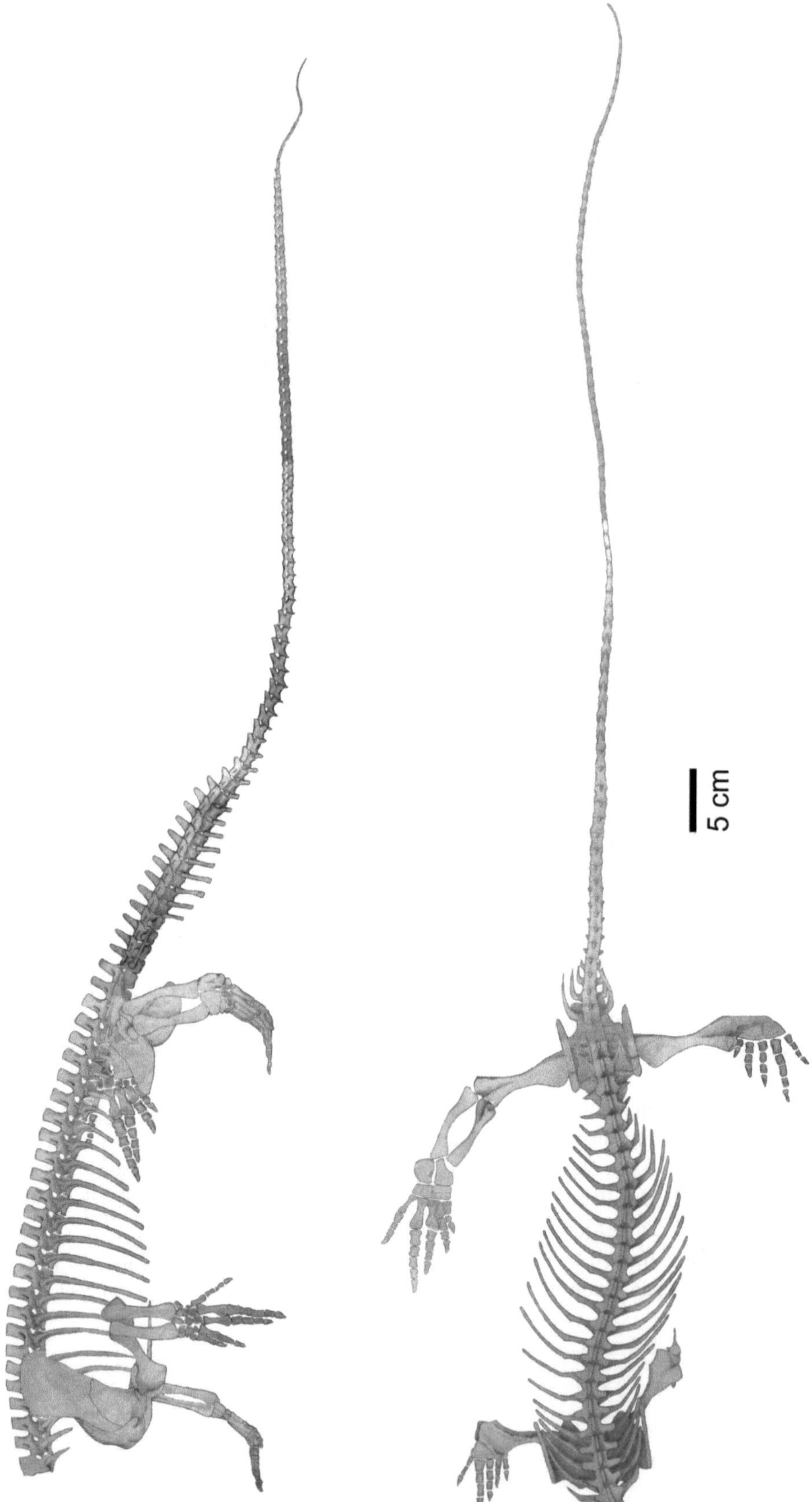

Fig. 4.15 Postcranial reconstruction of *Aerosaurus wellesi*, based on UCMP 40097, 40093, 40098, and 40094. **a** Lateral and **b** dorsal views

References

Anderson, J. S., & Reisz, R. R. (2004). *Pyozia mesenensis*, a new, small varanopid (Synapsida, Eupelycosauria) from Russia: "Pelycosaur" diversity in the Middle Permian. *Journal of Vertebrate Paleontology, 24*, 173–179.

Berman, D. S, & Reisz, R. R. (1982). Restudy of *Mycterosaurus longiceps* (Reptilia, Pelycosauria) from the Lower Permian of Texas. *Annals of Carnegie Museum, 51*, 423–453.

Berman, D. S, Reisz, R. R., Bolt, J. R., & Scott, D. (1995). The cranial anatomy and relationships of the synapsid *Varanosaurus* (Eupelycosauria: Ophiacodontidae) from the Early Permian of Texas and Oklahoma. *Annals of Carnegie Museum, 64*, 100–133.

Berman, D. S, Henrici, A. C., Sumida, S. S., Martens, T., & Pelletier, V. (2013). First European record of a varanodontine (Synapsida: Varanopidae): Member of a unique Early Permian upland paleoecosystem, Tambach Basin, central Germany. In C. F. Kammerer, K. D. Angielczyk, & J. Fröbisch (Eds.), *Early evolutionary history of the Synapsida* (pp. 69–86). Dordrecht: Springer.

Botha-Brink, J., & Modesto, S. P. (2007). A mixed-age classed "pelycosaur" aggregation from South Africa: Earliest evidence of parental care in amniotes? *Proceedings of the Royal Society B, 274*, 2829–2834.

Botha-Brink, J., & Modesto, S. P. (2009). Anatomy and relationships of the Middle Permian varanopid *Heleosaurus scholtzi* based on a social aggregation from the Karoo Basin of South Africa. *Journal of Vertebrate Paleontology, 29*, 389–400.

Campione, N., & Reisz, R. R. (2010). *Varanops brevirostris* (Eupelycosauria:Varanopidae) from the Lower Permian of Texas, with discussion of varanopid morphology and interrelationships. *Journal of Vertebrate Paleontology, 30*, 724–746.

Carroll, R. L. (1976). Eosuchians and the origin of archosaurs. In C. S. Churcher (Ed.), *Athlon: Essays on Palaeontology in Honour of Loris Shano Russell* (pp. 58–79). Toronto: The Royal Ontario Museum.

Dilkes, D. W., & Reisz, R. R. (1996). First record of a basal synapsid ("mammal-like reptile") in Gondwana. *Proceedings of the Royal Society B, 263*, 1165–1170.

Eberth, D. A., & Miall, A. D. (1991). Stratigraphy, sedimentology and evolution of a vertebrate bearing, braided to anastomosed fluvial system, Cutler Formation (Permian-Pennsylvanian), north-central New Mexico. *Sedimentary Geology, 72*, 225–252.

Eberth, D. A., Berman, D. S, Sumida, S. S., & Hopf, H. (2000). Lower Permian terrestrial paleoenvironments and vertebrate paleoecology of the Tambach Basin (Thuringia, central Germany): The upland holy grail. *Palaios, 15*, 293–313.

Evans, D. C., Maddin, H. C., & Reisz, R. R. (2009). A re-evaluation of sphenacodontid synapsid material from the Lower Permian fissure fills near Richards Spur, Oklahoma. *Palaeontology, 52*, 219–227.

Fröbisch, J., & Reisz, R. R. (2009). The Late Permain herbivore *Suminia* and the early evolution of arboreality in terrestrial vertebrate ecosystems. *Proceedings of the Royal Society B, 276*, 3611–3618.

Holmes, R. (1977). The osteology and musculature of the pectoral limb of small captorhinids. *Journal of Morphology, 152*, 101–140.

Holmes, R. B. (2003). The hind limb of *Captorhinus aguti* and the step cycle of basal amniotes. *Canadian Journal of Earth Sciences, 40*, 515–526.

Hopson, J. A. (1991). Systematics of the non-mammalian Synapsida and implications for patterns of evolution in synapsids. In H.-P. Schultze & L. Trueb (Eds.), *The origin of higher groups of tetrapods: Controversy and consensus* (pp. 635–693). Ithaca: Cornell University Press.

Hotton, N., III, Olson, E. C., & Beerbower, R. (1997). The amniote transition and the discovery of herbivory. In S. S. Sumida & K. L. M. Martin (Eds.), *Amniote origins: Completing the transition to land* (pp. 207–264). San Diego: Academic Press.

Langston, W., Jr. (1953). Permian amphibians from New Mexico. *University of California Publications in Geological Sciences, 29*, 349–416.

Langston, W., Jr, & Reisz, R. R. (1981). *Aerosaurus wellesi*, new species, a varanopseid mammal-like reptile (Synapsida: Pelycosauria) from the Lower Permian of New Mexico. *Journal of Vertebrate Paleontology, 1*, 73–96.

Laurin, M. (1991). The osteology of a Lower Permian eosuchian from Texas and a review of diapsid phylogeny. *Zoological Journal of the Linnean Society, 101*, 59–95.

Lucas, S. G., Harris, S. K., Spielman, J. A., Berman, D. S, Henrici, A. C., Heckert, A. B., et al. (2005). Early Permian biostratigraphy at Arroyo del Agua, Rio Arriba County, New Mexico. In S. G. Lucas, K. E. Zeigler, & J. A. Spielman (Eds.), *The Permian of Central New Mexico. New Mexico Museum of Natural History and Science Bulletin, 31*, 163–169.

Maddin, H. C., Evans, D. C., & Reisz, R. R. (2006). An Early Permian varanodontine varanopid (Synapsida: Eupelycosauria) from the Richards Spur locality, Oklahoma. *Journal of Vertebrate Paleontology, 26*, 957–966.

Maddin, H. C., Musat-Marcu, S., & Reisz, R. R. (2007). Histological microstructure of the claws of the African clawed frog, *Xenopus laevis* (Anura: Pipidae): Implications for the evolution of claws in tetrapods. *Journal of Experimental Zoology*, Part B: *Molecular and Developmental, 308*, 259–268.

Maddin, H. C., Sidor, C. A., & Reisz, R. R. (2008). Cranial anatomy of *Ennatosaurus tecton* (Synapsida: Caseidae) from the Middle Permian of Russia and the evolutionary relationships of Caseidae. *Journal of Vertebrate Paleontology, 28*, 160–180.

Olson, E. C. (1936). The dorsal axial musculature of certain primitive Permian tetrapods. *Journal of Morphology, 59*, 265–311.

Olson, E. C. (1954). Fauna of the Vale and Choza: 7. Pelycosauria: family caseidae. *Fieldiana: Geology, 10*, 193–204.

Olson, E. C. (1965). Chickasha vertebrates. *Oklahoma Geological Survey Circular, 70*, 1–70.

Olson, E. C. (1991). An eryopid (Amphibia: Labyrinthodontia) from the Fort Sill fissures, Lower Permian, Oklahoma. *Journal of Vertebrate Paleontology, 11*, 130–132.

Reisz, R. R. (1986). Pelycosauria. In P. Wellnhofer (Ed.), *Handbuch der Paläoherpetologie* (Vol. 17A). Stuttgart: Gustav Fischer Verlag.

Reisz, R. R., & Berman, D. S (2001). The skull of *Mesenosaurus romeri*, a small varanopseid (Synapsida: Eupelycosauria) from the Upper Permian of the Mezen River basin, northern Russia. *Annals of Carnegie Museum, 70*, 113–132.

Reisz, R. R., & Dilkes, D. W. (2003). *Archaeovenator hamiltonensis*, a new varanopid (Synapsida: Eupelycosauria) from the Upper Carboniferous of Kansas. *Canadian Journal of Earth Sciences, 40*, 667–678.

Reisz, R. R., & Laurin, M. (2004). A reevaluation of the enigmatic Permian synapsid *Watongia* and its stratigraphic significance. *Canadian Journal of Earth Sciences, 41*, 377–386.

Reisz, R. R., & Modesto, S. P. (2007). *Heleosaurus scholtzi* from the Permian of South Africa: A varanopid synapsid, not a diapsid reptile. *Journal of Vertebrate Paleontology, 27*, 734–739.

Reisz, R. R., Berman, D. S, & Scott, D. (1992). The cranial anatomy of *Secodontosaurus obtusidens*, an unusual mammal-like reptile (Synapsida: Sphenacodontidae) from the Lower Permian of Texas. *Zoological Journal of the Linnean Society, 104*, 127–184.

Reisz, R. R., Wilson, H., & Scott, D. (1997). Varanopseid synapsid skeletal elements from Richards Spur, a Lower Permian fissure fill near Fort Sill, Oklahoma. *Oklahoma Geology Notes, 57*, 160–170.

Reisz, R. R., Dilkes, D. W., & Berman, D. S (1998). Anatomy and relationships of *Elliotsmithia longiceps* Broom, a small synapsid (Eupelycosauria: Varanopseidae) from the Late Permian of South Africa. *Journal of Vertebrate Paleontology, 18*, 602–611.

Reisz, R. R., Godfrey, S. J., & Scott, D. (2009). *Eothyris* and *Oedaleops*: Do these Early Permian synapsids from Texas and New Mexico form a clade? *Journal of Vertebrate Paleontology, 29*, 39–47.

Reisz, R. R., Laurin, M., & Marjanovic, D. (2010). *Apsisaurus witteri* from the Lower Permian of Texas: yet another small varanopid synapsid, not a diapsid. *Journal of Vertebrate Paleontology, 30*, 1628–1631.

Romer, A. S., & Price, L. I. (1940). Review of the Pelycosauria. *Geological Society of America Special Paper, 28*, 1–538.

Stovall, J. W., Price, L. I., & Romer, A. S. (1966). The postcranial skeleton of the giant Permian pelycosaur *Cotylorhynchus romeri. Bulletin of the Museum of Comparative Zoology, 135*, 1–30.

Sullivan, C., & Reisz, R. R. (1999). First record of *Seymouria* (Vertebrata: Seymouriamorpha) from early Permian fissure fills at Richards Spur, Oklahoma. *Canadian Journal of Earth Sciences, 36*, 1257–1266.

Sumida, S. S., Berman, D. S, Eberth, D. A., & Henrici, A. C. (2004). A terrestrial vertebrate assemblage from the Late Paleozoic of Central Germany, and its bearing on Lower Permian paleoenvironments. In G. C. Young (Ed.), *Lower vertebrates from the Paleozoic. First International Paleontological Congress, Sydney. Australia, July 2002* (pp. 113–123). Oslo: Taylor and Francis.

Sumida, S. S., Pelletier, V., Berman, D. S, & English, L. (2009). New information on the basal pelycosaurian-grade synapsid *Oedaleops. Journal of Vertebrate Paleontology, 29*, 188A.

Sumida, S. S., Pelletier, V., & Berman, D. S (2013). New information on the basal pelycosaurian-grade synapsid *Oedaleops.* In C. F. Kammerer, K. D. Angielczyk, & J. Fröbisch (Eds.), *Early evolutionary history of the Synapsida* (pp. 7–23). Dordrecht: Springer.

Wideman, N. K., Sumida, S. S., & O'Neil, M. (2005). A reassessment of the taxonomic status of the materials assigned to the Early Permian tetrapod genera *Limnosceloides* and *Limnoscelops*. In S. G. Lucas & K. E. Zeigler (Eds.), *The nonmarine Permian. New Mexico Museum of Natural History and Science Bulletin 30*, 358–362.

Williston, S. W. (1911). *American Permian vertebrates*. Chicago: University of Chicago Press.

Williston, S. W. (1914). The osteology of some American Permian vertebrates. *Contributions of the Walker Museum, 1*, 107–162.

Chapter 5
First European Record of a Varanodontine (Synapsida: Varanopidae): Member of a Unique Early Permian Upland Paleoecosystem, Tambach Basin, Central Germany

David S Berman, Amy C. Henrici, Stuart S. Sumida, Thomas Martens, and Valerie Pelletier

Abstract A new genus and species of varanodontine varanopid, *Tambacarnifex unguifalcatus*, is described on the basis of the greater portion of the postcranium and a closely associated partial left dentary from the Lower Permian (Wolfcampian) Tambach Formation, the lowermost unit of the Upper Rotliegend, of the Bromacker locality in the midregion of the Thuringian Forest near Gotha, central Germany. *Tambacarnifex unguifalcatus* can be distinguished from all other varanopids on the basis of unique features of its vertebrae and unguals. A cladistic analysis of Varanopidae resolves *T. unguifalcatus* as nested within the varanodontines as the sister taxon of *Varanops* in a terminal dichotomy, which in turn forms the sister clade of the terminal dichotomy *Varanodon+Watongia*. The position of *Aerosaurus* is unaltered from previous analyses as the basal taxon of Varanodontinae. *Elliotsmithia*, which has been assigned alternately to both the varanodontines and the mycterosaurines, is resolved as a member of the latter. *Tambacarnifex unguifalcatus* is, therefore, the only varanodontine known from outside of North America. Within the Mycterosaurinae clade *Mycterosaurus* and *Mesenosaurus* resolve as a terminal dichotomy with *Elliotsmithia* and *Heleosaurus* related as successive sister taxa. As in previous analyses, *Archaeovenator* retains its position as the basal taxon of Varanopidae. *Tambacarnifex unguifalcatus* was an apex predator in a unique, heretofore undocumented Early Permian paleoecosystem in which the vertebrates were highly terrestrial inhabitants of an upland terrestrial setting, and constituted an early stage in the evolution of the modern terrestrial vertebrate trophic system, with herbivores greatly outnumbering apex predators in diversity, abundance, and biomass.

Keywords Bromacker locality • Tambach Formation • Paleoenvironment • Paleobiology • Trophic system • Eupelycosauria

This chapter includes one or more new nomenclatural-taxonomic actions, registered in Zoobank, and for such purposes the official publication date is Sep 2013.

D. S Berman (✉) · A. C. Henrici
Section of Vertebrate Paleontology, Carnegie Museum of Natural History, 4400 Forbes Avenue, Pittsburgh, PA 15213, USA
e-mail: bermand@carnegiemnh.org

A. C. Henrici
e-mail: henricia@carnegiemnh.org

S. S. Sumida · V. Pelletier
Department of Biology, California State University San Bernardino, 5500 University Parkway, San Bernardino, CA 92407, USA
e-mail: ssumida@csusb.edu

V. Pelletier
e-mail: uraprimate@msn.com

T. Martens
Abteilung Paleontology, Museum der Natur, PSF 217, 99853 Gotha, Germany
e-mail: martens@stiftung-friedenstein.de

Introduction

The varanodontine varanopid described herein is a member of an extensive terrestrial vertebrate assemblage collected from the Lower Permian Tambach Formation of the Bromacker locality, an area of small, abandoned sandstone quarries scattered over an area of less than 0.5 km^2 in the midregion of the Thuringian Forest, Thuringia, central Germany. The Tambach Formation, which forms the lowermost formational unit of the Upper Rotliegend Group or Series in this area, comprises a 200 to 400-m-thick unit of conglomerates, sandstones, and mudstones. It has been interpreted generally as divided into three informal units: a lower and upper conglomerate separated by a middle interval of sandstone and minor mudstones referred to as the Tambach Sandstone,

Christian F. Kammerer, Kenneth D. Angielczyk and Jörg Fröbisch (eds.), *Early Evolutionary History of the Synapsida*, Vertebrate Paleobiology and Paleoanthropology, DOI: 10.1007/978-94-007-6841-3_5,

which has long been known for its exceptionally well-preserved vertebrate trackways (Voigt et al. 2007). Outcrops of the Tambach Formation are restricted to an area of about 50 km^2 and were deposited in a small, internally drained paleograben, termed the Tambach Basin, whose original aerial extent was approximately 200–300 km^2.

Vertebrate remains were first discovered in the 'Tambach Sandstone' at the Bromacker locality in 1974 (Martens 1980, 1982), and a program of intensive, systematic excavation was initiated in 1993 (by the authors DSB, ACH, SSS, and TM) and has continued to the present. To date, three closely associated quarries have been excavated, collectively regarded as the Bromacker locality, which are located near the center of the Tambach Basin and cover a total area of about 700 m^2. In terms of abundance of specimens, diversity of taxa, and quality of preservation the Bromacker locality has become the most productive locality for Lower Permian terrestrial vertebrates in Europe. To date, 12 taxa of tetrapods have been identified from the Bromacker locality. Anamniotes ("amphibians") include: the ostodolepidid microsaur *Tambaroter carrolli* Henrici, Martens, Berman, and Sumida, 2011; the amphibamid *Georgenthalia clavinasica* Anderson, Henrici, Sumida, Martens, and Berman, 2008; the trematopids *Tambachia trogallas* Sumida, Berman, and Martens, 1998 and *Rotaryus gothae* Berman, Henrici, Martens, Sumida, and Anderson, 2011; the seymouriamorph *Seymouria sanjuanensis* Vaughn 1966, (Berman and Martens 1993; Berman et al. 2000a; Klembara et al. 2005, 2006, 2007); and the diadectomorphs *Diadectes absitus* Berman, Sumida, and Martens, 1998, and *Orobates pabsti* Berman, Henrici, Kissel, Sumida, and Martens, 2004. The amniotes include: the eureptile (rather than the defunct taxa 'Captorhinidae' and 'Protorothyrididae') *Thuringothyris mahlendorffae* Boy and Martens, 1991 (Müller et al. 2006); the bolosaurid parareptile *Eudibamus cursoris* Berman, Reisz, Scott, Henrici, Sumida, and Martens, 2000; the sphenacodontid synapsid *Dimetrodon teutonis* Berman, Reisz, Martens, and Henrici, 2001 (Berman et al. 2004b); a new, undescribed caseid synapsid (Berman et al. 2009; Reisz et al. 2010); and a new varanopid synapsid first reported in 2009 (Berman et al. 2009) and described herein as *Tambacarnifex unguifalcatus*. Of these, all but one was recovered from the Tambach Sandstone of the Bromacker locality proper. The skull of *Tambaroter carrolli* (Henrici et al. 2011) and a closely associated articulated skeleton of the diadectomorph *Diadectes absitus*, an ubiquitous member of the Bromacker locality assemblage, were recovered from the upper conglomerate member in the village of Tambach-Dietharz. All of the vertebrates from the Tambach Formation are otherwise unreported from Europe, yet share a strong commonality with those of the mixed terrestrial-aquatic assemblages from the well-documented, lowland terrestrial environments almost exclusively found in the Lower Permian sediments of the United States (Berman and Martens 1993; Berman et al. 1997, 1998, 2000a, 2001, 2009; Eberth et al. 2000). As such, the Bromacker vertebrates have provided the first irrefutable, biological evidence of a predrift continent of Euramerica and the first substantial use of terrestrial vertebrates to correlate a Lower Permian horizon in Europe, the Tambach Formation, with the standard terrestrial Lower Permian section of north-central Texas, indicating an earliest Permian Wolfcampian age (Sumida et al. 1996; Berman and Martens 1993; Berman et al. 1997, 1998, 2001).

The Bromacker locality and its vertebrate assemblage comprise a unique paleoecosystem, with highly terrestrial vertebrates inhabiting an upland terrestrial setting, that is otherwise undocumented in the Early Permian. As such, it constitutes an initial stage in the evolution of the modern terrestrial vertebrate trophic system or food chain, in which the herbivores greatly outnumber the apex predators in diversity, abundance, and biomass.

Systematic Paleontology

Synapsida Osborn, 1903
Eupelycosauria Kemp, 1982
Varanopidae Romer, 1936
Varanodontinae Reisz and Berman, 2001
Tambacarnifex gen. nov.

Type species: *Tambacarnifex unguifalcatus* sp. nov.

Diagnosis: Varanodontine varanopid eupelycosaur characterized by the following autapomorphies of the neural spines (observed in the partially preserved articulated series of posterior dorsals of serial position 13–23) and in the unguals of the manus and pes: neural spines inclined anteriorly, gradually increasing in serial height from either end of column to serial position 18, and alternating in width; unguals of digits I, III, and IV of the manus and IV of the pes (the only ones preserved) strongly recurved and greatly elongated, with that of manus digit I being subequal to the combined lengths of its penultimate phalanx and metacarpal; unguals of manus digits III and IV and pes digit V exceed the combined lengths of their penultimate and antepenultimate phalanges by 59, 50, and 29 %, respectively; extremely slender (distal to basal) retractor tubercle with a depth only slightly exceeding the width, and with the medial and lateral surfaces converging dorsally to a narrow, rounded ridge.

Etymology: *Tamba*, referring to the Tambach Basin, which the holotypic individual inhabited, and from the Latin *carnifex*, meaning executioner, referring to its role as an apex predator.

Tambacarnifex unguifalcatus sp. nov.

Holotype: MNG (collections of the Museum der Natur Gotha, Germany) 10596, greater part of postcranial skeleton preserved both in bone and as impressions on counterpart blocks.

Paratype: MNG 15037, partial left dentary exposed in lateral view.

Horizon and Locality: Tambach Formation of the Bromacker locality, an area of small abandoned and intermittently active sandstone quarries scattered over an area of less than 0.5 km^2 in the midregion of the Thuringian Forest approximately 1.5 km north of the village of Tambach-Dietharz and 20 km south of the town of Gotha, central Germany. Two superimposed stratigraphic successions at the Bromacker locality, characterized by their facies associations, are referred to as the Lower and Upper beds (Eberth et al. 2000). Both the holotype and paratype were recovered from the lower of two massive, red-brown, very fine-grained sandstone and siltstone sheetflood deposits separated by a 50 cm stratigraphic interval within the massive siltstones to very-fine-grained-sandstones of the 50–100 m thick Tambach Sandstone. This level is near the base of the Upper Beds, as defined by Eberth et al. (2000). On the basis of its vertebrate assemblage the Tambach Sandstone of the Bromacker locality is considered Early Permian (Wolfcampian) (Sumida et al. 1996).

Note: Lucas (2006) subdivided the North American Wolfcampian and Leonardian into a series of five faunachrons, which in ascending order are the Coyotean, Seymouran, Mitchellcreekian, Redtankian, and Littlecrotonian. He placed the vertebrate assemblage from the Bromacker locality of the Tambach Formation in the Seymouran faunachron, which spans the Wolfcampian-Leonardian boundary. The Seymouran was defined by the first occurrence of *Seymouria*, which is currently best known by *S. sanjuanensis* Vaughn, 1966, and *S. baylorensis* Broili, 1904, and is restricted to the Seymouran and Redtankian. Since *S. sanjuanensis* has a widespread occurrence that includes Utah, New Mexico, and Germany (Bromacker locality), Lucas (2006) suggested that it is probably the best index taxon for the Seymouran faunachron. Even though the Seymouran faunachron spans the Wolfcampian-Leonardian boundary, North American occurrences of *S. sanjuanensis* are restricted to the Wolfcampian (Berman et al. 1987; Sumida et al. 2004), indicating that the Bromacker locality is best regarded as Wolfcampian on this basis.

Diagnosis: As for genus.

Etymology: From the Latin *unguis*, nail, claw, or talon, and *falcatus*, sickle-shaped, referring to the long, strongly recurved unguals of the holotype.

Fig. 5.1 *Tambacarnifex unguifalcatus*, paratype, MNG 15037. Partial left dentary in lateral view

Description and Comparisons

General

Assignment of the isolated, partial paratypic dentary MNG 15037 (Fig. 5.1) to *Tambacarnifex unguifalcatus* is based on its dentition being characteristic of the varanodontines and quite distinct from those of the other members of the Bromacker assemblage, and its size being appropriate to belong to the holotype. The holotypic postcranial skeleton MNG 10596 is preserved on two large counterpart blocks (Figs. 5.2, 5.3), and at the time of excavation many elements were partially or entirely lost, but in most instances are represented as impressions. However, because the impressions often provide useful information, they have been whitened for better visibility, and in the description that follows those elements represented as impressions are indicated as such. Because there is no consistency among authors, nor do they typically provide an explanation of how the various limb elements are oriented relative to one another or to the axial column when describing them in various views, it seems best to clarify the scheme used here to prevent confusion. It essentially follows that of Romer and Price (1940, p. 137): the propodials are viewed as extending directly laterally from the girdles and the epipodials, manus, and pes as extending directly anteriorly with the entire limb in the same horizontal plane. Therefore, the four major descriptive aspects of the propodials are dorsal, ventral, anterior, and posterior and for the epipodials dorsal, ventral, medial, and lateral, whereas in the descriptions of the manus and pes the standard dorsal and ventral orientations are used.

Based on the cladistic analysis of Maddin et al. (2006), with alterations suggested by Campione and Reisz (2010), several diagnostic features clearly support the assignment of

Fig. 5.2 *Tambacarnifex unguifalcatus*, holotype, MNG 10596. Partial postcranial skeleton exposed mainly in dorsal aspect on opposite side of counterpart block in Fig. 5.3

Fig. 5.3 *Tambacarnifex unguifalcatus*, holotype, MNG 10596. Partial postcranial skeleton exposed mainly in ventral aspect on opposite side of counterpart block in Fig. 5.2. *dv* dorsal vertebra, *il* ilium, *pu* pubis

Tambacarnifex unguifalcatus to Varanodontinae: marginal teeth strongly recurved and mediolaterally flattened; deep, elongate excavations at base of the neural spine; and heads of the humerus greatly expanded and connected by a short, narrow, rounded shaft. Comparisons of *T. unguifalcatus* with other varanodontines are limited to a single species from each of five genera and rely on a short list of pertinent descriptions. These include: Dilkes and Reisz (1996) and Reisz et al. (1998) for *Elliotsmithia longiceps* Broom, 1937; Langston and Reisz (1981; although also see Pelletier 2013) for *Aerosaurus wellesi* Langston and Reisz, 1981; Maddin et al. (2006) and Campione and Reisz (2010) for *Varanops brevirostris* (Williston, 1911) (reassigned from *Varanosaurus* by Williston 1914); Olson (1965) for *Varanodon agilis* Olson, 1965; and Olson (1974) and Reisz and Laurin (2004) for *Watongia meieri* Olson, 1974. Repetitive citation of these references, therefore, has been substantially eliminated below as unnecessary.

Isolated Dentary

The isolated dentigerous jaw element (MNG 10595, paratype) found closely associated with the holotype is tentatively identified as a left dentary in lateral view (Fig. 5.1). Narrow, sutural scars along the ventral margin may indicate the contacts of splenial and angular. Furthermore, there are no signs of a canine swelling or a dorsal process to suggest that the jaw element is a maxilla. Twelve teeth of varying completeness are preserved. As in varanodontines, they are recurved, although possibly less so than in some varanopids, mediolaterally flattened, sharply pointed, and non-serrated. The series is incomplete anteriorly and posteriorly, and the teeth increase in height and width anteriorly except for the anteriormost two being narrower. Surface sculpturing consists of densely packed, short, minute longitudinal ridges.

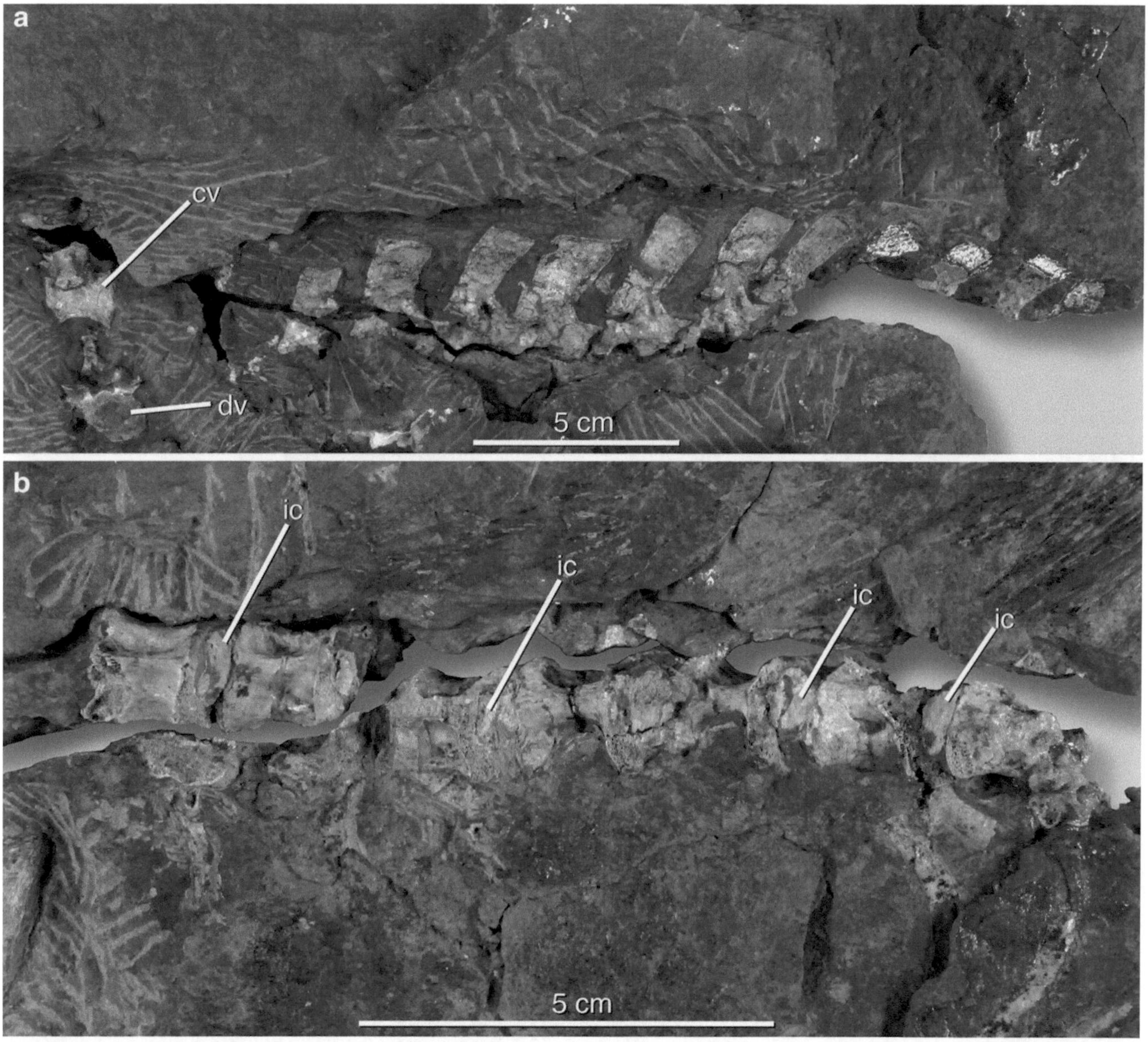

Fig. 5.4 *Tambacarnifex unguifalcatus*, holotype, MNG 10596. Partially preserved series of 11 articulated posterior dorsal vertebrae believed to represent serial positions 11–23 (anterior to *right*). **a** Lateral view of series as exposed on opposite side of counterpart block in Fig. 5.3, and **b** ventral view of centra 6–11 as exposed on counterpart block in Fig. 5.3. *cv* cervical vertebra, *dv* dorsal vertebra, *in* intercentrum

Axial Skeleton

The axial skeleton is represented by a partially preserved series of 11 articulated presacral vertebrae (Figs. 5.3, 5.4) and several scattered, isolated vertebrae. The first four anterior vertebrae of the articulated series are represented only by the distal ends of the neural spines and the fifth only by the neural arch and spine. The last six are mostly complete, primarily missing fragments along a break in the block that extends through the ventral margins of the centra. The series is believed to approximate serial positions 13–23 and, therefore, would include only dorsal vertebrae. This estimate is based on two features: (1) the gap between the posterior end of the series and the approximate position of the missing sacrum, which could have accommodated four vertebrae (the position of the sacrum is based on the assumption that the partially preserved left pelvis and proximal end of the left femur are preserved in their correct relationship to the axial column; Fig. 5.3); and (2) according to Langston and Reisz (1981) the dorsal vertebrae in *Aerosaurus wellesi* reach their greatest height at serial position 18, which in the holotype

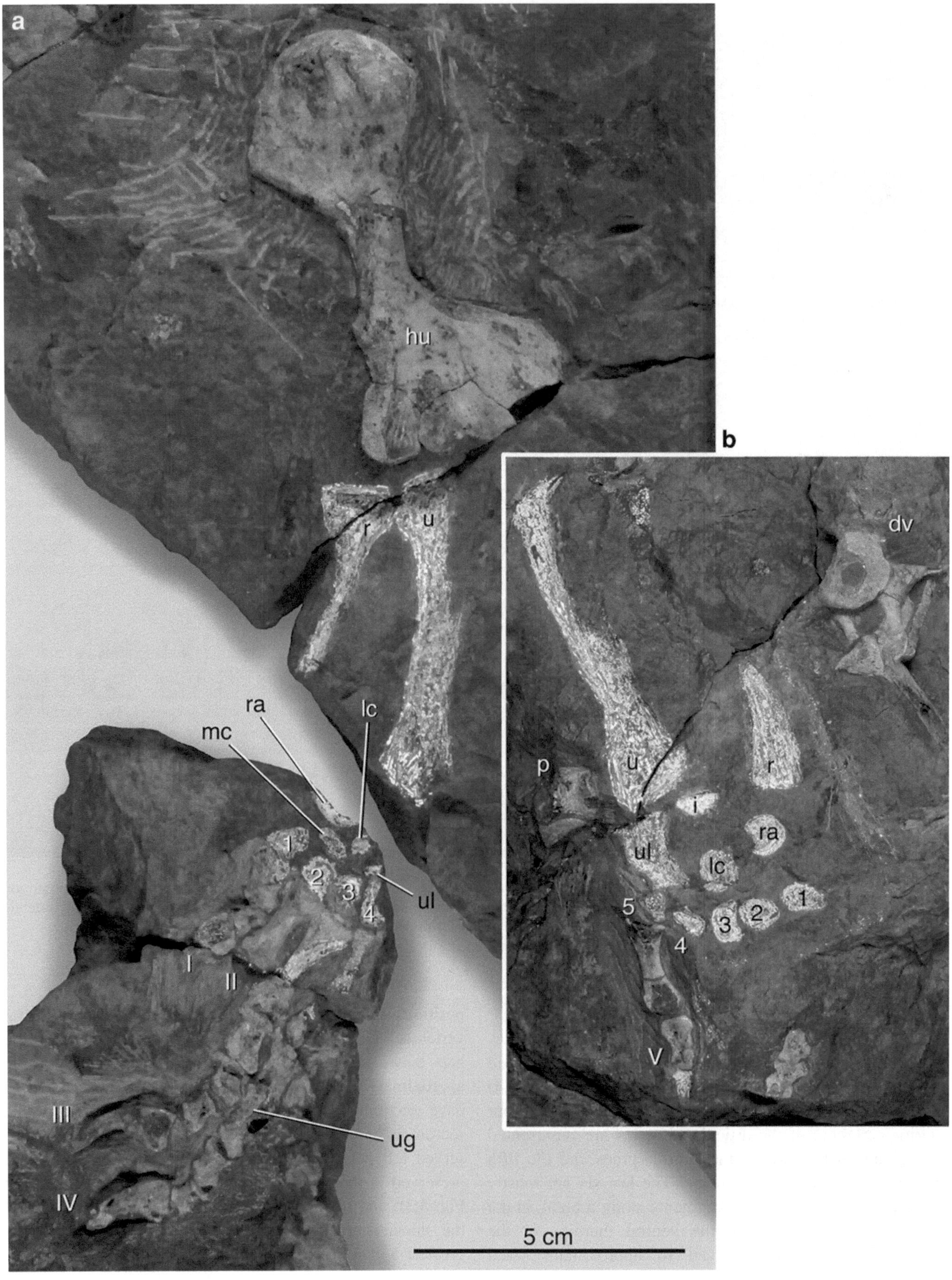
a
hu
r
u
ra
lc
mc
1
2
3
4
ul
I
II
III
IV
ug
5 cm
b
dv
p
u
r
i
ra
ul
lc
5
4
3
2
1
V

◀ **Fig. 5.5** Left forelimb and manus of *Tambacarnifex unguifalcatus*, holotype, MNG 10596. **a**, **b** As exposed on counterpart blocks in Figs. 5.2 and 5.3, respectively. *I–V* digits, *1–5* distal carpals, *dv* dorsal vertebra, *hu* humerus, *i* intermedium, *lc* lateral centrale, *mc* medial centrale, *p* pisiform, *r* radius, ra radiale, *u* ulna, *ug* ungual, *ul* ulnare

of *Tambacarnifex* would be the sixth vertebra of the series. The presacral count, therefore, is estimated at 27, which is the typical number of presacrals in basal synapsids and recorded in *Varanops brevirostris* (Romer and Price 1940) and *Aerosaurus wellesi* (Langston and Reisz 1981; Pelletier 2013). Only the centra of the fifth through eleventh vertebrae of the series are preserved, and, although they are slightly incomplete and distorted, rough measurements of their greatest centrum length and height yield a rather consistent 1.4 and 1.6 cm, respectively. As in other varanodontines, the neural spines are anteroposteriorly broad in lateral view, moderately tall, and subrectangular with a slight expansion distally and a flat dorsal margin. Furthermore, as in *Varanops* and *Aerosaurus*, there are deep, narrow, elongate, lateral excavations at the base of the neural spines (Fig. 5.4a) that are slightly inclined posteroventrally from the vertical. In *Varanodon* and *Watongia* the excavations differ in being much shallower and far less elongated.

Although the articulated series of the 11 dorsal vertebrae in *Tambacarnifex* exhibit an overall morphology very similar to that of other varanodontines, the neural spines can be distinguished by three unique features: (1) a pronounced anterior inclination, which is made especially evident by the anteroventral sloping of the dorsal margins; (2) a serial increase in height from either end of the series to the sixth vertebra, which is directly measurable in the last six vertebrae from 2.8 to about 1.7 cm; and (3) an alternation in spine width in the last seven vertebrae. The vertebrae lack a suture between the centrum and neural arch, indicating a mature stage of development. The midventral keel of the series has, as expected, the form of a low, but distinct, rounded ridge. The dorsal series also exhibits well-preserved intercentra, but there is no evident beveling of the centra to accommodate them, which may be not be obvious due to distortion and imperfect preservation. Three additional, incomplete, partially exposed, isolated vertebrae are preserved on the counterpart blocks containing the holotype. A short distance from the posterior end of the articulated dorsal series in Fig. 5.4a is a centrum and partial neural arch of a cervical exposed in left lateral view (centrum length 1.8 cm) and an anterior dorsal vertebra missing most of its neural spine exposed in anterior view (centrum width and height 1.5 and 1.9 cm, respectively). In Figs. 5.3 and 5.4b an anterior dorsal is visible in posterior view (overall height 2.0 cm and centrum width and height 1.6 and 1.9 cm, respectively) lying beside the impression of the partial left forelimb.

Ribs from the anterior to the midregion of the dorsal series of vertebrae are represented as narrow, posterolaterally arching rods that reach a maximum length in the anterior part of the dorsal series, then become much reduced posteriorly. The rib heads are preserved only as impressions, but appear characteristic of varanodontines. The capitulum expands slightly proximally, whereas the tuberculum is greatly reduced to a short, rectangular, dorsally directed protuberance on the upper surface of the rib. There is no triangular web of bone connecting the capitulum and tuberculum and the ribs are, thus, double-headed and can be considered dichocephalous.

Appendicular Skeleton

The pectoral girdle is not represented and only the left forelimb and manus are partially preserved as bone and impression (Figs. 5.2, 5.3, 5.5) (lengths of propodials and epipodials are given in Table 5.1). The humerus is nearly complete and exposed in dorsal aspect, but obviously has been greatly flattened, so as to expose both heads in the same plane. The heads are moderately damaged, but show characters that are closely comparable to those of other varanodontines. They are greatly expanded and connected by a short, relatively narrow, rounded shaft. As in varanopids, the smooth, strongly convex articular margin of the proximal head is set off from a well-developed deltopectoral crest. Only the base of the supinator process remains, but appears to indicate that the process was directed strongly distomedially. Despite the loss of the lateral margin of the ectepicondyle, the distal head is noticeably more expanded than the proximal head, due mainly to the greatly flared entepicondyle. An elongate entepicondylar foramen is well exposed. The left radius and ulna are represented only as very incomplete dorsal and ventral impressions. All that the impressions safely reveal are modestly expanded ends.

With the exception of the pisiform, all of the carpals are represented entirely or almost entirely as partial ventral (Figs. 5.2, 5.5) or dorsal (Fig. 5.3) impressions and, therefore, do not allow reliable description. Although only the impression of the distal margin of the radiale is preserved, it suggests that the element had a mediolateral width equal to or greater than its proximodistal length. The nearly complete pisiform is exposed in ventral view and presumably in its correct orientation near the junction of the ulna and the ulnare (Fig. 5.3). It is trapezoidal in outline, with the

Table 5.1 Length measurements (in mm) of limb bones of *Tambacarnifex unguifalcatus*

Left humerus	84
Left radius	57
Left ulna	63
Left femur	85
Left tibia	82
Right tibia	81
Right fibula	85

Fig. 5.6 *Tambacarnifex unguifalcatus*, holotype, MNG 10596. Right ilium in lateral view as exposed in Fig. 5.3

proximal and distal margins being parallel to one another and the proximal margin being longer. The smoothly finished ventral surface is concave in transverse section with thickened medial and lateral margins. Metacarpals I–V are essentially complete, with I–IV exposed in dorsal view (Figs. 5.2, 5.5) and V in ventral view (Fig. 5.3). They exhibit the standard serial increase in length of I–IV, with the length of V being about 70 % of that of IV. The phalanges of the digits I, III, and IV are well preserved and exhibit the expected counts of 2, 4, and 5, respectively. In several features, however, the unguals are unique among varanodontines in being: (1) much more strongly recurved; (2) considerably longer, with the first ungual being subequal to the combined lengths of the preceding phalanx and metacarpal and the third and fourth unguals being approximately 59 and 50 % longer than the combined lengths of their penultimate and antepenultimate phalanges, respectively (where the same measurements are available in the manus digits of *Aerosaurus*, *Varanops*, *Varanodon*, and *Watongia*, the unguals are either far shorter or subequal); and (3) extremely slender dorsoventrally and mediolaterally distal to the flexor tubercle, with the third and fourth unguals having depths of about 4 and 3 mm and widths of about 4 and 5 mm, respectively, and the medial and lateral surfaces converging dorsally to a narrow, rounded ridge. In *Aerosaurus*, *Varanops*, *Varanodon*, and *Watongia* the unguals are described as flattened dorsoventrally, with a width that is considerably greater than the depth. Maddin and Reisz (2007, p. 270) described the unguals in *Varanops* as having "a strikingly flat, broad morphology in cross-section that is reminiscent of the condition in diadectids." Yet, Campione and Reisz (2010) described the first ungual of the manus in a large specimen of *Varanops* as dorsoventrally thick, but that of the pes as dorsoventrally flattened.

All that remains of the pelvic girdles are the right ilium (Figs. 5.3, 5.6) and the basal portion of the left in lateral views (Fig. 5.2), and what appears to be the anteroventral portion of the left pubis (Figs. 5.3, 5.7a), which we tentatively regard to be exposed in lateral view with its convex ventral margin directed laterally from the vertebral column.

Only the left femur is represented, mainly as impressions of its anterior (Fig. 5.2) and posterior (Figs. 5.3, 5.7a) aspects except in the latter, where the distal end is preserved in anterior aspect, exposing the anterior head and the distal portion of the posterior head. The entire length and profile of the femur is represented by the impression of its anterior aspect, which clearly exhibits the typical varanopid features of being slender and having a slight sigmoid curvature, with the proximal head bent slightly upward and the distal head bent slightly downward.

The left tibia (Figs. 5.3, 5.7a) is complete, whereas the right (Figs. 5.3, 5.8a) is missing two short sections of the shaft, and both elements are exposed in medial aspect (the right also as impression in Figs. 5.3, 5.8b). As is typical in varanopids, the tibia is gracile and strongly bowed dorsally. The cnemial crest of the right tibia dominates the medial aspect of the proximal head, giving it a broadly triangular outline. The crest diminishes quickly distally, occupying the proximal third of the bone. Just medial and parallel to the distal end of the crest is a well-developed, highly rugose ridge that extends nearly to the distal end of the bone (Fig. 5.7a). The cnemial crest is separated from the lateral margin of the proximal head by a channel, termed the cnemial trough by Pawley and Warren (2006). Mediolateral crushing, however, has greatly narrowed the channel to a narrow slit.

Both fibulae are preserved (Figs. 5.3, 5.8), but only the left fibula is complete and exposed in medial aspect, whereas the right is exposed in ventral aspect and represented in part as bone and impression. The left tibia obscures all but the distal half of the left fibula, which appears as an anteroventrally narrow strut that is bowed slightly dorsally. Preservation of the right fibula includes the proximal head (partially hidden by the tibia) and the greater medial portion of the distal head, which is joined proximally by a short, narrow strip of the shaft. Preservation as impression is limited to the medial portion of the distal

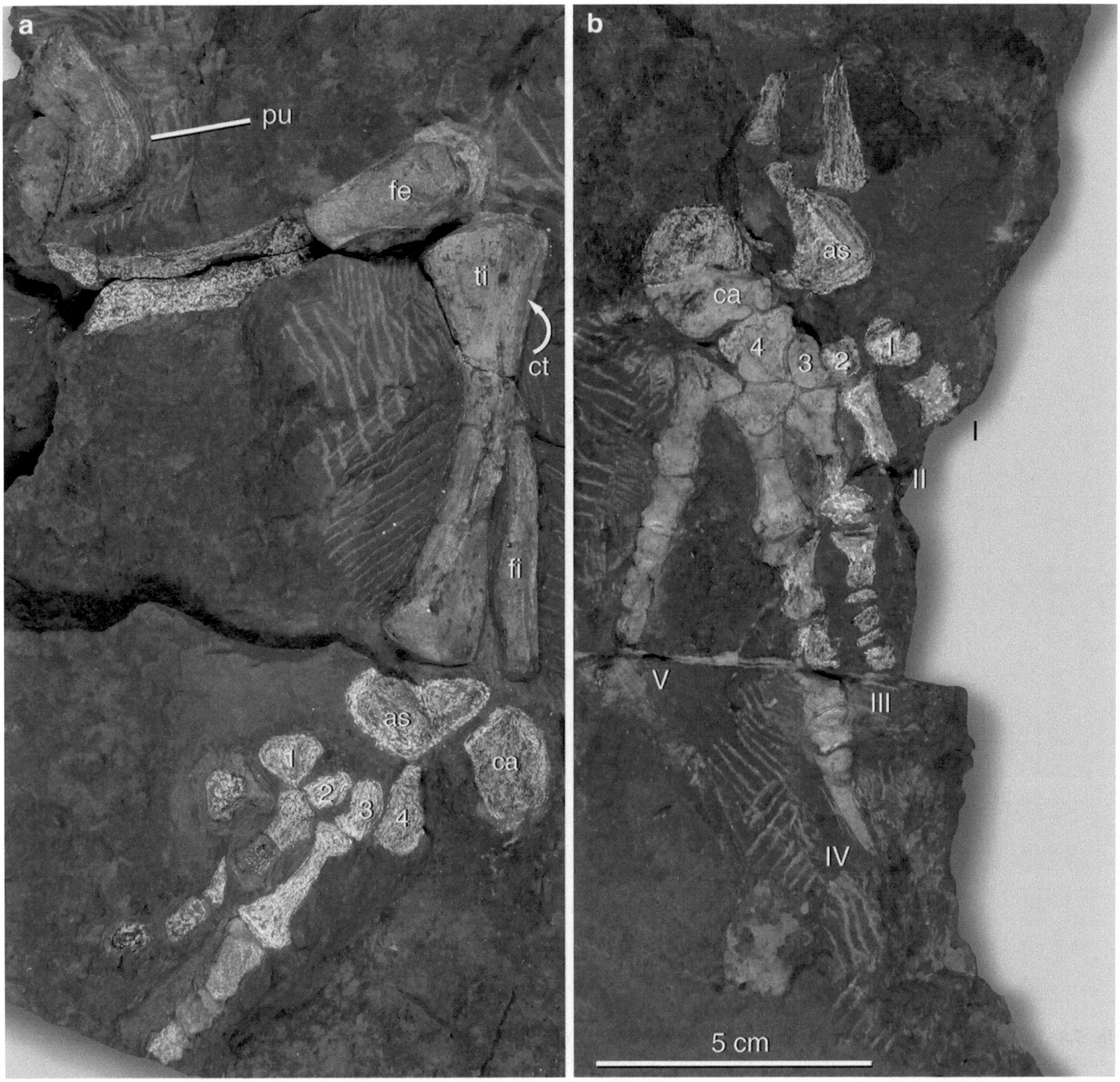

Fig. 5.7 *Tambacarnifex unguifalcatus*, holotype, MNG 10596. **a** Partial left hindlimb and pes, and **b** partial left pes as exposed in Figs. 5.3 and 5.2, respectively. *as* astragalus, *ca* calcaneum, *ct* cnemial trough, *I–V* digits, *1–4* distal tarsals, *fe* femur, *fi* fibula, *pu* pubis, *ti* tibia

head and a continuing narrow strip of the medial margin of the shaft. Both heads flare medially, giving the fibula a moderately concave medial margin and a slight lateral bowing. The impression of the ventral aspect of the distal end of the right fibula (Fig. 5.8b) indicates a broad, flattened, triangular surface. The fibula is slightly longer than the tibia.

The astragalus and calcaneum are best exemplified by their dorsal exposure in the partial right pes (Figs. 5.2, 5.8b), where they are essentially complete, articulated, and well preserved except for some erosion along the distal margin of the calcaneum. Partial impressions of the dorsal and ventral surfaces of the left astragalus and the dorsal surface of the right (Figs. 5.2, 5.3, 5.7, 5.8a) provide no additional information. The stoutly constructed astragalus has the standard sharply angled L-shape outline, with the length of the short vertical arm (0.6 cm) being about 24 % of the total 2.5 cm of the total proximodistal length of the element. The astragalus-calcaneum contact is slightly bowed medially for most of its proximal length, with the remainder of the contact being straight except for a notch-like interruption near its distal end for the perforating artery.

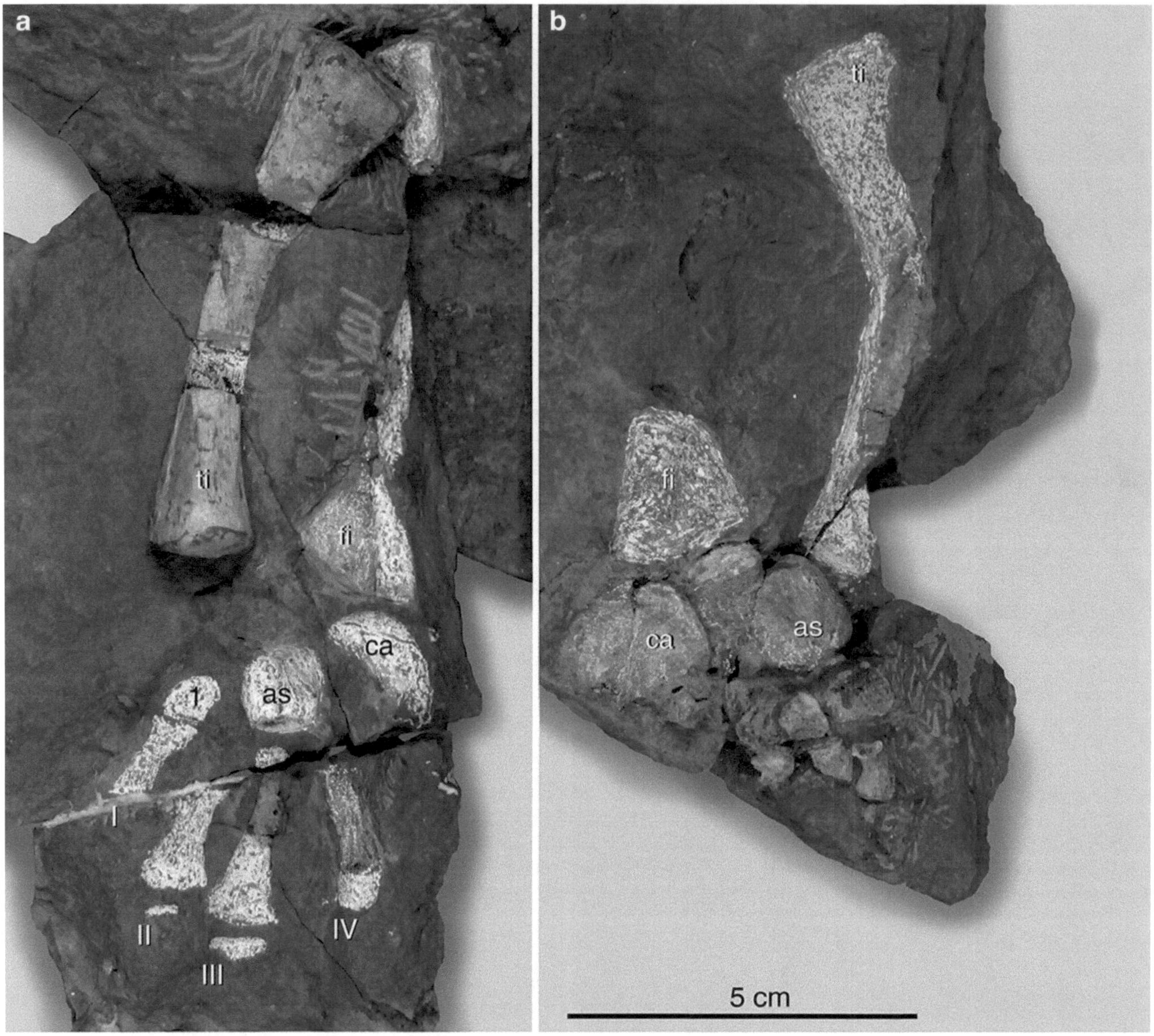

Fig. 5.8 *Tambacarnifex unguifalcatus*, holotype, MNG 10596. **a**, **b** Counterparts of partial right hindlimb and pes as exposed in Figs. 5.2 and 5.3, respectively. *as* astragalus, *ca* calcaneum, *I–IV* digits, *1* distal tarsal, *fi* fibula, *ti* tibia

The dorsal surface of the astragalus is smoothly finished and relatively flat except for where it rises along the contact margins with adjacent elements. The proximal articular facet of the vertical arm for the fibula is well developed, slightly convex, and transversely oval except for a slight lateral narrowing, whereas the horizontal arm ends in a prominent, well-developed, transversely convex, condyle-like articular facet for the tibia. The proximodistal length of the astragalus is subequal to that of the calcaneum, with the proximal and distal margins of both elements ending at approximately the same transverse planes. The dorsally exposed calcaneum of the right pes is nearly complete, whereas in the left pes (Figs. 5.2, 5.8a) only the distal half is well preserved in ventral view, but the outline of the dorsal surface impression of the proximal half is accurately depicted. The calcaneum has in general a proximodistally elongate, oval outline with a thinning in thickness toward the lateral margin. It thickens toward the facets on the convex proximomedial and medial margins, and the straight distal marginal contacts with the fibula, astragalus, and fourth distal tarsal.

The medial and lateral centrales are not represented, but the space they occupied between the astragalus and distal tarsals 1–3 is well defined in the ventral view of the left pes (Figs. 5.2, 5.7b). Distal tarsal 5 is not represented, but 1–4 are in the left pes (Fig. 5.7b), with the first and second preserved as impressions of their dorsal surfaces and the third and fourth exposed in ventral view. Distal tarsal 1 is roughly kidney-shaped in outline, with a notch-like indentation on its medial margin. The outline of the second distal tarsal, smallest of the four, describes a proximodistally elongated rectangle with a slightly convex lateral margin. The first and second distal tarsals have slightly convex distal articular margins, which match the concave proximal margins of their respective metatarsals. The third distal tarsal has roughly the outline of a proximodistally elongated ellipse with a slightly concave medial margin that contacts the greater proximal portion of the medial margin of the second distal tarsal, and a convex lateral margin that fits snugly into a complementary concave medial margin of the fourth distal tarsal. The strongly convex distal margin of the third distal tarsal is opposed by a slightly convex proximal margin of the metatarsal. The fourth distal tarsal, the largest of the series, is pentagonal in outline with a broad, flat contact with the calcaneum, a small projection of its proximomedial margin that likely contacted the distolateral corner of the astragalus, a concave medial margin that received the lateral margin of the third, and broad, mutually flat contacts with metatarsals IV and V that meet at an oblique angle.

Metatarsals I–IV increase serially in length, with that of the fifth being 79 % of the fourth. Metatarsal I, the shortest, is about 52 % of the length of the fourth. Only the fourth and fifth digits are preserved well enough to exhibit the standard phalangeal counts of five and four, respectively. The fourth and only preserved ungual of the pes is complete and exposed in ventral view. In contrast to the unguals in *Aerosaurus*, *Varanops*, *Varanodon*, and *Watongia* it has a morphology identical to those of the manus in *Tambacarnifex* in being: (1) much more strongly recurved than the unguals of the other taxa; (2) longer than the combined lengths of its respective penultimate and antepenultimate phalanges by 29 %; (3) extremely slender dorsoventrally and mediolaterally distal to the flexor tubercle, with a depth of about 4 mm and a width of about 5 mm, respectively; (4) and the medial and lateral surfaces converge dorsally to a narrow, rounded ridge.

Gastralia

Remarkably, the abdominal ribs or gastralia are preserved nearly intact and complete, on both counterpart blocks (Figs. 5.2, 5.3), but they have shifted to the right side of the trunk region. Their shape and arrangement are best exemplified at the anterior end of the structure (Fig. 5.9), where they are arranged in tightly packed, chevron-shaped rows with the sharply angled apices directed anteriorly. Individual elements are slender, forming an approximately one centimeter long rod that is sharply pointed at both ends. The anterior end of each element anteriorly overlaps the distal end of the medially adjacent element.

Fig. 5.9 *Tambacarnifex unguifalcatus*, holotype, MNG 10596. Gastralia superimposed on ribs as exposed in Fig. 5.3. Anterior is towards the *top* of the figure

Phylogenetic Analysis

In a redescription of the varanopid *Elliotsmithia longiceps*, Reisz et al. (1998) were the first to recognize that the varanopids consist of two distinct clades, one consisting of *Mesenosaurus* Efremov, 1938 (Reisz and Berman 2001) and *Mycterosaurus* Williston, 1915 (Berman and Reisz 1982) and a second consisting of *Elliotsmithia*, *Aerosaurus*, *Varanodon*, and *Varanops*. However, they did not assign them formal systematic designations. Subsequently, in a

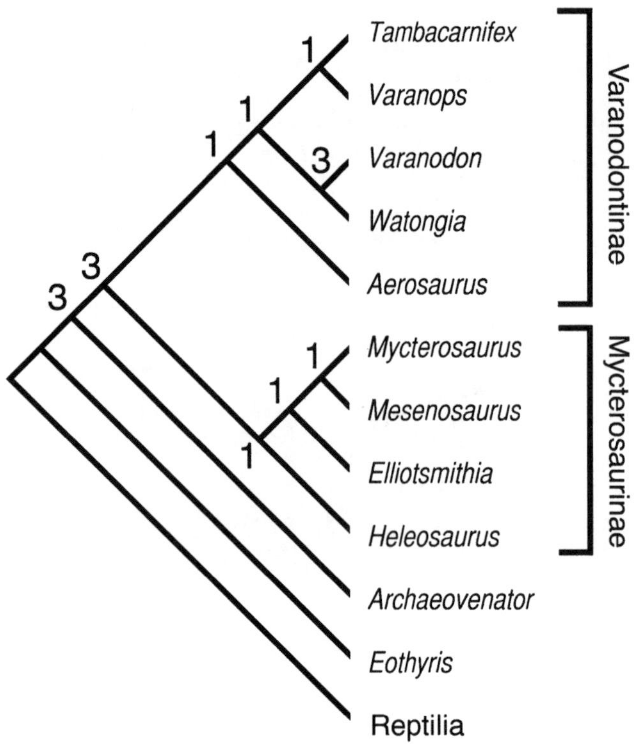

Fig. 5.10 Hypothesized relationships of varanopids based on a single most parsimonious tree (Tree Length = 87). Bremer decay values are given next to nodes

redescription of the varanopid *Mesenosaurus romeri*, Reisz and Berman (2001) proposed the formal designations Mycterosaurinae and Varanodontinae, respectively, for the two varanopid clades, which have been accepted by later authors with some alterations (although see Kammerer and Angielczyk 2009). A fifth species of varanodontine, *Watongia meieri*, based originally on a single specimen described by Olson (1974) as a gorgonopsian therapsid, was reassigned to Varanodontinae in a redescription by Reisz and Laurin (2004). Modesto et al. (2001) assigned a second specimen, BP/1/5678 (Bernard Price Institute for Palaeontological Research), to *E. longiceps* and, using a composite character coding based on it and the holotype, reassigned BP/1/5678 to Mycterosaurinae. Reisz and Dilkes (2003), however, suspected that the new specimen may pertain to a different species based on their differing interpretations from those of Modesto et al. (2001) on certain characters inferred to be present in the holotype. In a subsequent reconsideration of the affinities of BP/1/5678, however, Botha-Brink and Modesto (2009) cautioned that none of the autapomorphies of the mycterosaurines are determinable in BP/1/5678 and suggested that it may be referable to *Heleosaurus scholtzi* Broom, 1907. *H. scholtzi*, described originally as a diapsid by Broom (1907), was recently reclassified by Reisz and Modesto (2007) to Mycterosaurinae. With the discovery of an aggregation of five specimens of *H. scholtzi*, Botha-Brink and Modesto (2007, 2009) were able to provide significant new anatomical information that not only provided a greatly expanded diagnosis of the species, but also established its mycterosaurinae affinities in a revised phylogenetic analysis of the varanopids. It should be noted also that their analysis did not use BP/1/5678 to code either *E. longiceps* or *H. scholtzi*. Modesto et al. (2011) reported an additional fragmentary mycterosaurine specimen from the rocks assigned to the *Pristerognathus* Assemblage Zone in the Karoo Basin, but did not name it. Lastly, two new species of varanopids, *Archaeovenator hamiltonensis* and *Pyozia mesenensis*, were recently described by Reisz and Dilkes (2003) and Anderson and Reisz (2004), respectively. Neither species, however, was assigned to either the Varanodontinae or Mycterosaurinae clades, but rather, in a cladistic analysis by Anderson and Reisz (2004), *Pyozia* and *Archaeovenator* were resolved as successive sister taxa.

In order to evaluate the phylogenetic relationships of *Tambacarnifex unguifalcatus* within Varanopidae, three recently published cladistic analyses were utilized. Maddin et al. (2006) were the first to offer a comprehensive data matrix based on 60 characters in their evaluation of the phylogenetic position of a varanopid from the Lower Permian fissure-fill deposits of the Richards Spur locality near Fort Sill, Oklahoma. Their analysis of isolated cranial and postcranial elements of three specimens, which were described by them as virtually indistinguishable from *Varanops brevirostris* of the Lower Permian of Texas, resolved the unnamed Richards Spur varanopid as the sister taxon of *V. brevirostris*.

In a more recent analysis of the varanopids, Botha-Brink and Modesto (2009) used a slightly altered version of the character data matrix of Maddin et al. (2006) in a phylogenetic assessment of the mycterosaurine *Heleosaurus scholtzi* Broom, 1907. The analysis, the first to include *H. scholtzi*, was prompted by the discovery of five additional, closely associated, articulated specimens from the Middle Permian of South Africa (Botha-Brink and Modesto 2007). Changes made to the Maddin et al. (2006) data matrix by Botha-Brink and Modesto (2009) were limited to the addition of two characters, 61 (character 59 here, absence or presence of squamosal tubercle) and 62 (character 60 here, absence or presence of osteoderms). Interestingly, their analysis resolved *Elliotsmithia* from the Middle Permian of South Africa within the Mycterosaurinae as the sister taxon of *Heleosaurus*. Most recently, Campione and Reisz (2010) provided a comprehensive revision of *Varanops brevirostris* based on a large, well-preserved, nearly complete, articulated skeleton from the Lower Permian of Texas. Their cladistic analysis also adopted in great part the characters used by Maddin et al. (2006), with the following alterations: (1) inclusion of *Heleosaurus*; (2)

exclusion of *Pyozia* from the Middle Permian of Russia, as it is based on a specimen they believed does not exhibit the diagnostic characters of Varanopidae; (3) removal of the Richards Spur varanopid as separate taxon from *V. brevirostris*, but still utilizing it to score the amended data matrix (characters 36 and 39, the morphology of the parasphenoid); (4) removal of characters 45 (presence of a lateral temporal fenestra) and 60 (length to distal width ratio of femur) as uninformative; and (5) the addition of new information provided by the specimen described by Campione and Reisz (2010) that they used to recode *V. brevirostris* (characters 3, presence or absence of tooth serrations, and 22, morphology of postorbital). Furthermore, in the data matrix of Campione and Reisz (2010) three of the five outgroups used by Maddin et al. (2006) (Reptilia, the eothyridid synapsid *Eothyris*, and the ophiacodontid synapsid *Archaeothyris*) were eliminated, but the caseid *Cotylorhynchus* and ophiacodontid *Ophiacodon* were retained.

In evaluating the phylogenetic relationships of *Tambacarnifex unguifalcatus* within Varanopidae, the data matrix of Maddin et al. (2006) was used with the alterations listed above by Botha-Brink and Modesto (2009) and Campione and Reisz (2010). The data matrix of the latter was not used, because of numerous transcription errors. In addition, Reptilia and *Eothyris* are the only outgroups of Maddin et al. (2006) retained in the data matrix used here. Unfortunately, *Tambacarnifex* could be coded only for eight of the 60 characters of Botha-Brink and Modesto (2009) and Campione and Reisz (2010), as all but 11 are cranial. The applicable characters include 1, 3, 49–51, and 55–57, which were coded 1, 0, 1, 2, 1, 0, 0, and 1, respectively. MacClade 2.0 (Maddison and Maddison 1992) was used to construct the data matrix, consisting of 12 taxa (Fig. 5.10), and a cladistic analysis utilizing the branch-and-bound algorithm of PAUP (Swofford 1993) was conducted to determine the most parsimonious tree(s). All the characters were equally weighted, and multiple, derived states were unordered. Delayed transformation (DELTRAN) was used to optimize characters. The analysis produced only one most parsimonious tree with a length of 87 steps, a Consistency Index of 0.782, a Retention Index of 0.810, and a Rescaled Consistency Index of 0.633. The robustness of each varanopid clade was determined using the Bremer (1994) decay analysis.

The resultant cladogram (Fig. 5.10) resolves *Tambacarnifex* as nested within the varanodontines as the sister taxon to *Varanops* in a terminal dichotomy, which in turn forms a sister clade to the terminal dichotomy of *Varanodon+Watongia* from the Middle Permian of Oklahoma. The position of *Aerosaurus* from the Lower Permian of New Mexico as the basal taxon of Varanodontinae is unaltered from previous analyses. Within Mycterosaurinae, *Mycterosaurus* from Lower Permian of Texas and *Mesenosaurus* from the Middle Permian of Russia form a terminal dichotomy, with *Elliotsmithia* and *Heleosaurus* as successive basal sister taxa. This relationship, therefore, supports Modesto et al.'s (2001) contention that *Elliotsmithia* is a mycterosaurine. In this regard, it is worth noting that in the cladistic analysis of Campione and Reisz (2010) *Elliotsmithia* is poorly supported as a varanodontine, yet it is also poorly supported as a mycterosaurine in their stratocladistic analysis. The position of *Archaeovenator* from the Upper Pennsylvanian of Kansas as a basal taxon to the clade Varanodontinae+Mycterosaurinae is unaltered here from previous analyses. In the present analysis, however, relatively high Bremer support values of three are obtained only for those nodes that include *Varanodon+Watongia*, Varanodontinae+Mycterosaurinae, and *Archaeovenator*+all other varanopids, whereas the remaining varanopid nodes are supported by a low value of one. Importantly, the analysis implies that *Tambacarnifex* is the only varanodontine known outside of North America. This is not totally unexpected, however, considering the strong taxonomic commonality between the vertebrates of the Bromacker locality and those of the Early Permian of North America.

Environmental and Biological Uniqueness of the Bromacker Locality

General

The Bromacker locality is distinguishable from all other well-known Early Permian terrestrial vertebrate-fossil sites in its paleoenvironmental and paleobiological features (Eberth et al. 2000; Berman et al. 2000a, 2001), which for the first time fully document a paleoecosystem in the Early Permian that represents an initial stage in the evolution of the modern terrestrial vertebrate trophic system or food chain. The small, internally-drained Tambach Basin and its Bromacker locality currently document the best example of an Early Permian assemblage from a rarely encountered upland terrestrial setting, and the strictly terrestrial Bromacker vertebrate assemblage is dominated by large herbivores that greatly outnumbered the apex predators in diversity and abundances. Because of their diversity and abundance, the large herbivorous tetrapods played a vital role in sustaining the ecosystem by providing a major link between primary producers and a greatly reduced number and variety of apex predators, a situation that contrasts that of the well-known, lowland, paralic, mixed terrestrial-aquatic assemblages of comparable age. It is not until nearly 50 million years later in the Early Triassic that modern terrestrial vertebrate herbivore-based ecosystems appear, but then as fully developed and typifying most terrestrial vertebrate communities. Together, both aspects of the

Bromacker locality have provided the unique opportunity to assess the mutual influences of the environmental and biological components of community assembly and structure in an Early Permian upland terrestrial paleoecosystem.

Paleoenvironmental Features

Deposition of the Bromacker sediments and their associated vertebrate fossil assemblage in the Tambach Basin was interpreted as representative of a rarely encountered paleoenvironment of a 'truly upland' terrestrial setting (Eberth et al. 2000; Berman et al. 2000a, 2001). This was defined as far removed and up-dip from regional-scale drainage systems of extensive coastal or alluvial plains bordering non-coal-forming wetlands, which constitute the overwhelming source of Late Pennsylvanian-Early Permian terrestrial vertebrates. All of the hundreds of vertebrate specimens collected from the Bromacker locality, ranging from isolated elements to partial and complete articulated skeletons, were recovered from two massive, red-brown, very fine-grained sandstone and siltstone sheetflood (non-channelized) deposits separated by 50 cm that occur in a stratigraphic interval of 1.2 m within the massive siltstones to very-fined-grained-sandstones of the 50–100 m thick 'Tambach Sandstone' (the informal name given to the middle unit of the tripartite division that separates a lower and upper conglomerate unit of the Tambach Formation). Collectively, the Tambach Sandstone consists of alluvial paleochannel and sheetflood facies and lacustrine suspension deposits. The beds of the Bromacker locality are interpreted as having been deposited on an upland alluvial plain with minor stream channels under seasonal-to-subseasonal cycles of flooding in an ephemeral setting of a savanna-type climate. The sheetflood deposits containing the vertebrates were interpreted as representing meteoric flooding events, and they were undoubtedly the agents of death of the vertebrates.

Ample sedimentological and taphonological evidence was provided (Eberth et al. 2000) to support the conclusion that the Bromacker assemblage is representative of a single community living within the Tambach Basin. The sheetflood deposits containing the vertebrates occur as two, discrete, massive units, representing two, closely spaced, catastrophic flooding events of very short duration over a very wide area. They probably originated at the basin margin and when of sufficient magnitude spread across the low relief at the center of the basin and the Bromacker locality. The thickness of the fossiliferous sheet-flood deposits indicate a magnitude of deposition sufficient to have had the potential of preserving individuals much larger than those recovered to date or possessing a cursorial ability to escape entrapment. In many instances death and burial were apparently coeval events, sometimes including small groups of partial-to-complete articulated specimens, with some being preserved in natural poses, with the limbs extended away from the body and the elbows pointed posteriorly and the knees pointed anteriorly. Size ranges of the different taxa do not differ depending on whether they are represented by isolated elements or partial-to-complete skeletons. The excellent state of preservation of the fossil vertebrates indicates that subaerial exposures were either lacking or of short duration, and reworking and transport were limited or in some instances absent. The location of the Bromacker locality near the center of a lowland topography of the Tambach Basin was undoubtedly a critical factor in the superb preservation of the many articulated specimens.

Paleobiological Features

All the taxa from the Tambach Formation are highly terrestrial, but most remarkably, all are otherwise unique to Europe yet share a strong commonality with those of the mixed terrestrial-aquatic assemblages of the well-documented, lowland terrestrial environments of the Early Permian of the southwestern United States (Sumida et al. 1996; Eberth et al. 2000). The uniqueness of the Bromacker vertebrate assemblage is most profoundly expressed in the composition of its constituents and the relative abundances of its members. Palaeoniscoid fish, xenacanth sharks, and dipnoan lungfish are completely absent, as are aquatic and semi-terrestrial amphibians. Instead, amphibians are limited to the highly terrestrial trematopids *Tambachia trogallas* (Sumida et al. 1998) and *Rotaryus gothae* (Berman et al. 2011), the amphibamid *Georgenthalia clavinasica* (Anderson et al. 2008), the ostodolepidid microsaur *Tambaroter carrolli* (Henrici et al. 2011), *Seymouria sanjuanensis* (Berman et al. 2000a), and the diadectomorphs *Diadectes absitus* (Berman et al. 1998) and *Orobates pabsti* (Berman et al. 2004b). The abundance and variety of herbivores is unusually high and includes *Diadectes absitus* and *Orobates pabsti* (Berman et al. 1998, 2004b), an as yet undescribed, primitive basal caseid synapsid (Berman et al. 2009; Reisz et al. 2010), and the bolosaurid parareptile *Eudibamus cursoris* (Berman et al. 2000a). In contrast, the abundance and variety of apex predators is unusually low, including only two primitive basal synapsids: several specimens of the diminutive-sized *Dimetrodon teutonis*, weighing about 24 kg (compared to 37–259 kg for previously described species) (Berman et al. 2001, 2004b) and the relatively large-sized varanodontine varanopid *Tambacarnifex unguifalcatus*, first reported but undescribed by Berman et al.

(2009). On the basis of these paleobiological features it was hypothesized (Eberth et al. 2000) that the Bromacker assemblage documents an initial stage in the evolution of the modern terrestrial trophic ecosystem based on herbivory, in which herbivorous tetrapods represented a significant source for the direct introduction of plant food into the animal food chain.

The relative abundances of the Bromacker locality taxa were originally estimated by preliminary, albeit rough, specimen counts (Eberth et al. 2000). In the time since that report, relative abundances were continually tracked using minimum number of individuals counts during continued excavation of the Bromacker locality. These data clearly indicate that collectively the four herbivores outnumbered the apex predators by a ratio of at least 8 to 1. Using the same census, the combined numbers of individuals counts of the herbivores and carnivores for the entire assemblage is about 50 and 6 %, respectively. In addition, documentation of the great abundance of herbivores is also revealed by the superb vertebrate trackways preserved in paleochannel sandstones subadjacent to the Bromacker locality. Of the five recognized ichnospecies, a track-trackmaker association was firmly established between the ichnospecies *Ichniotherium cottae* (Pohlig 1885) and *I. sphaerodactylum* (Pabst 1895) and the skeletal fossils of *Diadectes absitus* and *Orobates pabsti*, respectively (Voigt et al. 2007). Collectively, the *Ichniotherium* species comprise about 95 % of the more than 600 collected tracks or trackways (Voigt et al. 2007).

An equally compelling measure of whether the herbivores constituted a major standing crop capable of sustaining the apex predators would be to calculate the difference between the relative total biomasses of the two groups. Unfortunately, there is no reliable technique for accurately calculating body mass of vertebrate fossils (Laurin 2004), which is especially exaggerated by the vagaries of preservation. An alternate, although rough, method of indexing relative body masses is maximum snout-vent length (Laurin 2004). This convention, of course, ignores the fact that the large herbivores had much more rotund torsos and stockier limbs that the apex predators and snout-vent lengths would, therefore, actually tend to underestimate the herbivore biomass. Nevertheless, the snout-vent lengths of the three large herbivores (*Orobates*, *Diadectes*, and the caseid) and the two large apex predators (*Dimetrodon* and *Tambacarnifex*) fall into a narrow range of 50–60 cm. Again, preliminary minimum number of individuals counts indicates a great numerical disparity in the abundance ratio of the two groups of 22 herbivores to three apex predators, suggesting a comparable disparity between their total biomasses. Thus, as in a modern terrestrial trophic ecosystems the herbivores would appear to have substantially supported the apex predators. Undoubtedly, the apex predators did not prey solely on the large herbivores, and the food web must have also included alternative pathways in which the apex predators preyed on the smaller carnivores and insectivores.

Until recently, Olson (1952, 1966, 1971, 1983) was the primary researcher to theorize the existence of an Early Permian upland terrestrial vertebrate community or ecosystem. Olson's (1966, 1971, 1983) evidence of an Early Permian upland terrestrial vertebrate community was based of on the rare occurrences of faunal elements of what he referred to as the Caseid Chronofauna, which was characterized mainly by herbivorous species of the primitive basal synapsid Caseidae and eureptiles, and the carnivorous Varanopidae. The unexpected occurrences of these taxa were explained as "erratics," forms that originally inhabited an upland community, but were transported or introduced into a lowland, water-based vertebrate community, Olson's Permo-Carboniferous Chronofauna. Interestingly, both a caseid and a varanopid are members of the Bromacker assemblage.

Other than the Bromacker locality, our understanding of Early Permian vertebrates from an upland setting is dependent mainly on a single site, the Richards Spur locality near Fort Sill, Oklahoma, which has been widely regarded as Leonardian in age and therefore younger than the Bromacker locality. Woodhead et al. (2010), however, utilizing radiometric analysis to determine the age of a stalagmite from the cave deposits of the same locality, assessed an age of Wolfcampian, making it roughly coeval with the Bromacker locality. The vertebrates of the Richards Spur locality are preserved in clay-rich fissure fills in the Ordovician limestone of the Arbuckle Formation and represent possibly the richest late Paleozoic vertebrate locality in the world, comprising a very diverse assemblage of 25 or more taxa (Sullivan and Reisz 1999; Sullivan et al. 2000; Anderson and Reisz 2004; Kissel et al. 2002; Reisz 2005; Maddin et al. 2006; Fröbisch and Reisz 2008; Reisz et al. 2009). Carroll (1968) theorized the existence of an upland ecosystem based on a diapsid parietal he described from the Richards Spur locality, believing that it might have been transported from an upland fauna. Unfortunately, the allochthonous origin of the Richards Spur vertebrates prevents any description of the physical and climatic aspects of the environment(s) that influenced and supported this assemblage in its original paleoenvironment (Olson 1991; Sullivan and Reisz 1999). Furthermore, the vertebrates are randomly concentrated and typically disarticulated to single elements, and therefore may have be subjected to sorting biases during transport, making calculations of the relative abundances of the taxa and the determination of the trophic organization of the assemblage highly speculative at best. In dramatic contrast, the Bromacker locality provides a unique opportunity to assess the interplay between physical, climatic, and biological ecosystem components in an upland setting, and to test previously proposed hypotheses about upland paleoecosystem composition and dynamics. In this context, the Bromacker locality offers a critical understanding of late Paleozoic tetrapod evolution and paleobiogeography, and ecosystem evolution (e.g., Vaughn 1966, 1969, 1970; Olson 1975, 1976, 1979; Eberth and Berman 1993; Sullivan and Reisz 1999).

Acknowledgments Research for this project was supported in part by a grants from the Deutsche Forschungsgemeinschaft (DFG) (to TM), and the National Geographic Society (to DSB, ACH, & SSS). Sincere thanks are also due the Rotary Club of Gotha, Germany, for continued financial support for the Bromacker project. We are greatly indebted to Mr. Mark Klinger of the Carnegie Museum of Natural History for his expertise in producing the figures. We acknowledge the invaluable contributions of the numerous, dedicated volunteer field assistants, whose tedious backbreaking labors since 1993 have been responsible for the discovery and recovery of many of the specimens from the Bromacker quarry. Special thanks are extended to the reviewers S. P. Modesto and R. A. Kissel, whose constructive comments and criticisms had considerable impact on the improvement of this paper. Finally, we are especially grateful to David A. Eberth, who initially brought our attention the uniqueness of the Bromacker locality paleoecosystem, which was first described in his seminal publication Eberth et al. (2000).

References

Anderson, J. S., & Reisz, R. R. (2004). *Pyozia mesenensis*, a new small varanopid (Synapsida, Eupelycosauria) from Russia: "Pelycosaur" diversity in the Middle Permian. *Journal of Vertebrate Paleontology, 24*, 173–179.

Anderson, J., Henrici, A. C., Sumida, S. S., Martens, T., & Berman, D. S (2008). *Georgenthalia clavinasica*, a new genus and species of dissorophoid temnospondyl from the Early Permian of Germany, and the relationships of the Family Amphibamidae. *Journal of Vertebrate Paleontology, 28*, 61–75.

Berman, D. S, Reisz, R. R., & Eberth, D. A. (1987). Seymouria sanjuanensis (Amphibia, Batrachosauria) from the Lower Permian Cutler Formation of north-central New Mexico and the occurrence of sexual dimorphism in that genus questioned. *Canadian Journal of Earth Sciences, 24*, 1769–1784.

Berman, D. S, & Martens, T. (1993). First occurrence of *Seymouria* (Amphibia: Batrachosauria) in the Lower Permian Rotliegend of central Germany. *Annals of Carnegie Museum, 62*, 63–79.

Berman, D. S, & Reisz, R. R. (1982). Restudy of *Mycterosaurus longiceps* (Reptilia, Pelycosauria) from the Lower Permian of Texas. *Annals of Carnegie Museum, 51*, 423–453.

Berman, D. S, Sumida, S. S., & Lombard, R. E. (1997). Biogeography of primitive amniotes. In S. S. Sumida & K. L. M. Martin (Eds.), *Amniote origins: Completing the transition to land* (pp. 85–139). San Diego: Academic Press.

Berman, D. S, Sumida, S. S., & Martens, T. (1998). *Diadectes* (Diadectomorpha: Diadectidae) from the Early Permian of central Germany, with description of a new species. *Annals of Carnegie Museum, 67*, 53–93.

Berman, D. S, Reisz, R. R., Scott, D., Henrici, A. C., Sumida, S. S., & Martens, T. (2000). Early Permian bipedal reptile. *Science, 290*, 969–972.

Berman, D. S, Henrici, A. C., Sumida, S. S., & Martens, T. (2000a). Redescription of *Seymouria sanjuanensis* (Seymouriamorpha) from the Lower Permian of Germany based on complete, mature specimens with a discussion of paleoecology of the Bromacker locality assemblage. *Journal of Vertebrate Paleontology, 20*, 253–268.

Berman, D. S, Reisz, R. R., Scott, D., Henrici, A. C., Sumida, S. S., & Martens, T. (2000b). Early Permian bipedal reptile. *Science, 290*, 969–972.

Berman, D. S, Reisz, R. R., Martens, T., & Henrici, A. C. (2001). A new species of *Dimetrodon* (Synapsida: Sphenacodontidae) from the Lower Permian of Germany records first occurrence of genus outside of North America. *Canadian Journal of Earth Sciences, 38*, 803–812.

Berman, D. S, Henrici, A. C., Kissel, R. A., Sumida, S. S., & Martens, T. (2004a). A new diadectid (Diadectomorpha), *Orobates pabsti*, from the Early Permian of central Germany. *Bulletin of Carnegie Museum of Natural History, 35*, 1–36.

Berman, D. S, Henrici, A. C., Sumida, S. S., & Martens, T. (2004b). New materials of *Dimetrodon teutonis* (Synapsida: Sphenacodontidae) from the Lower Permian of Germany. *Annals of Carnegie Museum, 73*, 48–56.

Berman, D. S, Henrici, A. C., & Sumida, S. S. (2009). Pelycosaurian-grade synapsids from the Lower Permian Bromacker locality, central Germany. *Journal of Vertebrate Paleontology, 29*, 62A.

Berman, D. S, Henrici, A. C., Martens, T., Sumida, S. S., & Anderson, J. S. (2011). *Rotaryus gothae*, a new trematopid (Temnospondyli: Dissorophoidea) from the Lower Permian of central Germany. *Annals of Carnegie Museum, 80*, 49–65.

Botha-Brink, J., & Modesto, S. P. (2007). A mixed-aged classed 'pelycosaur' aggregation from south Africa: Earliest evidence of parental care in amniotes? *Proceedings of the Royal Society B, 274*, 2829–2834.

Botha-Brink, J., & Modesto, S. P. (2009). Anatomy and relationships of the Middle Permian varanopid *Heleosaurus scholtzi* based on a social aggregation from the Karoo Basin of South Africa. *Journal of Vertebrate Paleontology, 29*, 389–400.

Boy, J. A., & Martens, T. (1991). A new captorhinomorph reptile from the Rotliegend of Thuringia (Lower Permian; eastern Germany). *Paläontologische Zeitschrift, 65*, 363–389.

Bremer, K. (1994). Branch support and tree stability. *Cladistics, 10*, 295–304.

Broili, F. (1904). Permische Stegocephalen und Reptilien aus Texas. *Palaeontographica, 51*, 80–84.

Broom, R. (1907). On some new fossil reptiles from the Karroo beds of Victoria West, South Africa. *Transactions of the South African Philosophical Society, 18*, 31–42.

Broom, R. (1937). A further contribution to our knowledge of the fossil reptiles of the Karroo. *Proceedings of the Zoological Society Series B, 1937*, 299–318.

Campione, N. E., & Reisz, R. R. (2010). *Varanops brevirostris* (Eupelycosauria: Varanopidae) from the Lower Permian of Texas, with discussion of varanopid morphology and interrelationships. *Journal of Paleontology, 30*, 724–746.

Carroll, R. L. (1968). A diapsid (Reptilia) parietal from the Lower Permian of Oklahoma. *Postilla, 117*, 1–7.

Dilkes, D., & Reisz, R. R. (1996). The first record of a basal synapsid ("mammal-like reptile") in Gondwana. *Proceedings of the Royal Society Series B, 264*, 1165–1170.

Eberth, D. A., & Berman, D. S (1993). Stratigraphy, sedimentology, and vertebrate paleoecology of the Cutler Formation redbeds (Pennsylvanian-Permian) of north-central New Mexico. In S. G. Lucas & J. Zidek (Eds.), *Vertebrate Paleontology in New Mexico* (pp. 33–48). *New Mexico Museum of Natural History and Science Bulletin, 2*.

Eberth, D. A., Berman, D. S, & Hopf, H. (2000). Lower Permian terrestrial paleoenvironments and vertebrate paleoecology of the Tambach Basin (Thuringia, central Germany): The upland Holy Grail. *Palaios, 15*, 293–313.

Efremov, J. A. (1938). Some new Permian reptiles of the USSR. *Academy of Sciences URSS C.R., 19*, 121–126.

Fröbisch, N. B., & Reisz, R. R. (2008). A new Lower Permian amphibamid (Dissorophoidea, Temnospondyli) from the fissure fill deposits near Richards Spur, Oklahoma. *Journal of Vertebrate Paleontology, 28*, 1015–1030.

Henrici, A. C., Martens, T., Berman, D. S, & Sumida, S. S. (2011). An ostodolepid "microsaur" (Lepospondyli) from the Lower Permian Tambach Formation of Central Germany. *Journal of Vertebrate Paleontology, 31*, 997–1004.

Kammerer, C. F., & Angielczyk, K. D. (2009). A proposed higher taxonomy of anomodont therapsids. *Zootaxa, 2018*, 1–24.

Kemp, T. S. (1982). *Mammal-like reptiles and the origin of mammals*. London: Academic Press.

Kissel, R. A., Dilkes, D. W., & Reisz, R. R. (2002). *Captorhinus magnus*, a new captorhinid (Amniota: Eureptilia) from the Lower Permian of Oklahoma, with new evidence on the homology of the astragalus. *Canadian Journal of Earth Sciences, 39*, 1363–1372.

Klembara, J., Berman, D. S, Henrici, A. C., & Čerčanský, A. (2005). New structures and reconstructions of the skull of the seymouriamorph *Seymouria sanjuanensis*, Vaughn. *Annals of Carnegie Museum, 74*, 1–8.

Klembara, J., Berman, D. S, Henrici, A. C., Čerčanský, A., & Werneburg, R. (2006). Comparison of cranial anatomy and proportions of similarly sized *Seymouria sanjuanensis* and *Discosauriscus austriacus*. *Annals of Carnegie Museum, 75*, 37–49.

Klembara, J., Berman, D. S, Henrici, A. C., Čerčanský, A., Werneburg, R., & Martens, T. (2007). First description of skull of Lower Permian *Seymouria sanjuanensis* (Seymouriamorpha: Seymouridae) at an early juvenile growth stage. *Annals of Carnegie Museum, 76*, 53–72.

Langston, W., Jr., & Reisz, R. R. (1981). *Aerosaurus wellesi*, new species, a varanopseid mammal-like reptile (Synapsida: Pelycosauria) from the Lower Permian of New Mexico. *Journal of Vertebrate Paleontology, 1*, 73–96.

Laurin, M. (2004). The evolution of body size, Cope's Rule and the origin of the amniotes. *Systematic Biology, 53*, 594–622.

Lucas, S. G. (2006). Global Permian tetrapod biostratigraphy and biochronology. In S. G. Lucas, G. Cassinis, & J. W. Schneider (Eds.), *Non-marine Permian biostratigraphy and biochronology* (Vol. 265, pp. 65–93). London: Geological Society, Special Publications.

Maddin, H. C., & Reisz, R. R. (2007). The morphology of the terminal phalanges in Permo-Carboniferous synapsids: An evolutionary perspective. *Canadian Journal of Earth Sciences, 44*, 267–274.

Maddin, H. C., Evans, D. C., & Reisz, R. R. (2006). An Early Permian varanodontine varanopid (Synapsida: Eupelycosauria) from the Richards Spurs locality, Oklahoma. *Journal of Vertebrate Paleontology, 26*, 957–966.

Maddison, D. R., & Maddison, W. R. (1992). *MacClade: Analysis of phylogeny and character evolution*. Sunderland: Sinauer Associates.

Martens, T. (1980). Beitrag zur Taxonomie und Ökologie des Oberrotliegenden im Elgersburger Becken in Thüringen. *Abhandlungen und Berichte des Museums der Natur Gotha, 10*, 21–32.

Martens, T. (1982). Zur Stratigraphie, Taxonomie, Ökologie und Klimaentgwicklung des Oberrotliegenden (Unteres Perm) im Thüringer Wald (DDR). *Abhandlungen und Berichte des Museums der Natur Gotha, 11*, 33–57.

Modesto, S., Sidor, C. A., Rubidge, B. S., & Welman, J. (2001). A second varanopseid skull from the Upper Permian of South Africa: Implications for Late Permian 'pelycosaur' evolution. *Lethaia, 34*, 249–259.

Modesto, S. P., Smith, R. M. H., Campione, N. C., & Reisz, R. R. (2011). The last "pelycosaur": A varanopid synapsid from the *Pristerognathus* Assemblage Zone, Middle Permian of South Africa. *Naturwissenschaften, 98*, 1027–1034.

Müller, J., Berman, D. S, Henrici, A. C., Martens, T., & Sumida, S. S. (2006). The basal reptile *Thuringothyris mahlendorffae* (Amniota: Eureptilia) from the Lower Permian of Germany. *Journal of Paleontology, 80*, 726–739.

Olson, E. C. (1952). The evolution of the Permian vertebrate chronofauna. *Evolution, 6*, 181–196.

Olson, E. C. (1965). New Permian vertebrates from the Chickasha Formation in Oklahoma. *Oklahoma Geological Survey Circular, 70*, 1–69.

Olson, E. C. (1966). Community evolution and the origin of mammals. *Evolution, 47*, 291–301.

Olson, E. C. (1971). *Vertebrate paleozoology*. New York: Wiley Interscience.

Olson, E. C. (1974). On the source of therapsids. *Annals of the South African Museum, 64*, 27–46.

Olson, E. C. (1975). Permo-Carboniferous paleoecology and morphotypic series. *American Zoologist, 15*, 371–389.

Olson, E. C. (1976). The exploitation of land by early tetrapods. In A. D. Bellairs & C. B. Cox (Eds.), *Morphology and biology of Reptiles. Linnean Society Symposium Series, 3*, 207–264.

Olson, E. C. (1979). Biological and physical factors in the dispersal of Permo-Carboniferous terrestrial vertebrates. In J. Gray & A. J. Boucot (Eds.), *Historical biogeography, plate tectonics and the changing environment* (pp. 227–238). Corvallis: Oregon State University Press.

Olson, E. C. (1983). Coevolution or coadaptation: Permo-Carboniferous vertebrate Chronofauna. In M. Nitecki (Ed.), *Coevolution* (pp. 301–338). Chicago: University of Chicago Press.

Olson, E. C. (1991). An eryopid (Amphibian: Labyrinthodontia) from the Fort Sill fissures, Lower Permian, Oklahoma. *Journal of Vertebrate Paleontology, 11*, 130–132.

Osborn, H. F. (1903). On the primary division of the Reptilia into two sub-classes, Synapsida and Diapsida. *Science, 17*, 275–276.

Pabst, W. (1895). Thierfährten aus dem Rothliegenden von Friedrichroda, Tambach and Kabarz in Thüringen. *Zeitschrift der deutschen geologischen Gesellschaft, 47*, 570–576.

Pawley, K., & Warren, A. (2006). The appendicular skeleton of *Eryops megacephalus* Cope, 1877 (Temnospondyli: Eryopoidae) from the Lower Permian of North Americas. *Journal of Paleontology, 80*, 561–580.

Pelletier, V. (2013). Postcranial description and reconstruction of the varanodontine varanopid *Aerosaurus wellesi* (Synapsida: Eupelycosauria). In C. F. Kammerer, K. D. Angielczyk, & J. Fröbisch (Eds.), *Early evolutionary history of the Synapsida* (pp. 53–68). Dordrecht: Springer.

Pohlig, H. (1885). Saurierfährten in dem Unteren Rotliegenden von Friedrichroda. *Verhandlungen des naturhistorischen Vereins der preussischen Rheinlande und Westfalens, 42*, 285–286.

Reisz, R. R. (2005). *Oromycter*, a new caseid from the Lower Permian of Oklahoma. *Journal of Vertebrate Paleontology, 25*, 905–910.

Reisz, R. R., & Berman, D. S (2001). The skull of *Mesenosaurus romeri*, a small varanopseid (Synapsida: Eupelycosauria) from the Upper Permian of the Mezen River Basin, Northern Russia. *Annals of Carnegie Museum, 70*, 113–132.

Reisz, R. R., & Dilkes, D. W. (2003). *Archaeovenator hamiltonensis*, a new varanopid (Synapsida: Eupelycosauria) from the Upper Carboniferous of Kansas. *Canadian Journal of Earth Sciences, 40*, 667–678.

Reisz, R. R., & Laurin, M. (2004). A reevaluation of the enigmatic Permian synapsid *Watongia* and of its stratigraphic significance. *Canadian Journal of Earth Sciences, 41*, 377–386.

Reisz, R. R., & Modesto, S. P. (2007). *Heleosaurus scholtzi* from the Permian of South Africa: A varanopid synapsid, not a diapsid reptile. *Journal of Vertebrate Paleontology, 27*, 234–239.

Reisz, R. R., Dilkes, D. W., & Berman, D. S (1998). Anatomy and relationships of *Elliotsmithia longiceps* Broom, a small synapsid (Eupelycosauria: Varanopseidae) from the Late Permian of South Africa. *Journal of Vertebrate Paleontology, 18*, 602–611.

Reisz, R. R., Schoch, R. R., & Anderson, J. S. (2009). The armoured dissorophid *Cacops* from the Early Permian of Oklahoma and the exploitation of the terrestrial realm by amphibians. *Naturwissenschaften, 96*, 789–796.

Reisz, R. R., Fröbisch, J., Berman, D. S, & Henrici, A. C. (2010). New Permo-Carboniferous caseid synapsids from North America and Europe and their evolutionary significance. *Program and Abstracts, Society of Vertebrate Paleontology Annual Meeting*, Pittsburgh, Pennsylvania, 150A.

Romer, A. S. (1936). Studies on American Permo-Carboniferous tetrapods. *Problems of Paleontology, 1*, 85–93.

Romer, A. S., & Price, L. I. (1940). Review of the Pelycosauria. *Geological Society of America Special Papers, 28*, 1–538.

Sullivan, C., & Reisz, R. R. (1999). First record of *Seymouria* (Vertebrata: Seymouriamorpha) from Early Permian fissure fills at Richards Spur, Oklahoma. *Journal of Earth Sciences, 36*, 1257–1266.

Sullivan, C., Reisz, R. R., & May, W. J. (2000). Large dissorophoid skeletal elements from the Lower Permian Richards Spur fissure, Oklahoma, and their paleoecological implications. *Journal of Vertebrate Paleontology, 20*, 456–461.

Sumida, S. S., Berman, D. S, & Martens, T. (1996). Biostratigraphic correlation between the Lower Permian of North America and central Europe using the first record of an assemblage of terrestrial tetrapods from Germany. In C. J. Bell & S. S. Sumida (Eds.), *The uses of vertebrates fossils in biostratigraphic correlation. PaleoBios, 17*, 1–12.

Sumida, S. S., Berman, D. S, & Martens, T. (1998). A new trematopid amphibian from the Lower Permian of central Germany. *Palaeontology, 41*, 605–629.

Sumida, S. S., Berman, D. S, Eberth, D. A., & Henrici, A. C. (2004). A terrestrial vertebrate assemblage from the late Paleozoic of central Germany, and its bearing on Lower Permian paleoenvironments. In G. C. Young (Ed.), *Lower vertebrates from the Paleozoic. First International Palaeontological Congress, Sidney, Australia, July 2002, Proceedings of Symposium 6. Fossils and Strata, 50*, 113–123.

Swofford, D. L. (1993). *Phylogenetic analysis using parsimony*. Champaign: Illinois Natural History Survey.

Vaughn, P. P. (1966). *Seymouria* from the Lower Permian of southeastern Utah, and possible sexual dimorphism in that genus. *Journal of Paleontology, 40*, 603–612.

Vaughn, P. P. (1969). Comparisons of the Early Permian vertebrate faunas of the Four Corners region and north-central Texas. *Los Angeles County Museum of Natural History, Contributions in Science, 105*, 1–13.

Vaughn, P. P. (1970). Lower Permian vertebrates of the Four Corners and the Midcontinent as indices of climate differences. *Proceedings of the 1969 North American Paleontological Convention*, 388–408.

Voigt, S., Berman, D. S, & Henrici, A. C. (2007). First well-established track-trackmaker association of Paleozoic tetrapods based on *Ichniotherium* trackways and diadectid skeletons from the Lower Permian of Germany. *Journal of Vertebrate Paleontology, 27*, 553–570.

Williston, S. W. (1911). *American Permian vertebrates*. Chicago: University of Chicago Press.

Williston, S. W. (1914). The osteology of some American Permian vertebrates. *Journal of Geology, 22*, 364–419.

Williston, S. W. (1915). A new genus and species of American Theromorpha, *Mycterosaurus longiceps*. *Journal of Geology, 23*, 554–559.

Woodhead, J. R., Reisz, R. R., Fox, D., Drysdale, R., Hellstrom, J., Maas, R., et al. (2010). Speleothem climate records from deep time? Exploring the potential with an example from the Permian. *Geology, 38*, 445–458.

Part II
Anomodontia

Chapter 6
Anomodontia: Introduction

Jörg Fröbisch

Among non-mammalian synapsids, Anomodontia represents the taxonomically and morphologically most diverse of all major subclades. Owen described the first anomodont material in 1845, based on several skulls from the South African Karoo Basin that he assigned to three separate species of the genus *Dicynodon*. Anomodont research experienced an initial burst in the early twentieth century, particularly through Robert Broom's work, although he named a large number of new taxa of doubtful validity. For most of the past century, this oversplitting remained a serious problem with respect to macroevolutionary studies on anomodonts. However, several recent alpha taxonomic revisions of particularly species-rich anomodont genera and subclades (Keyser 1973, 1975, 1993; King 1993; King and Rubidge 1993; Grine et al. 2006; Fröbisch and Reisz 2008; Angielczyk et al. 2009) resulted in a much better understanding of anomodont diversity as well as intra- and interspecific variation. These efforts culminated in the recent monograph by Kammerer et al. (2011), which tackled *Dicynodon*, one of the last remaining wastebasket genera among fossil tetrapods from the early days of vertebrate paleontology. This study reduced the number of valid species of *Dicynodon* to two (South African *D. lacerticeps* and east African *D. huenei*), synonymizing others or transferring them to separate genera.

A recent conservative richness estimate of anomodonts listed 128 species in 68 genera (Fröbisch 2009). When considering the above-mentioned revisions, these numbers change to a decreased number of 123 species and an increased count of 75 genera. The majority of anomodonts, about 90 % of the valid species, fall with the subclade Dicynodontia, which achieved a global distribution and comprised the dominant herbivores of their time. In contrast, the basal grade of non-dicynodontian anomodonts mainly includes species from South Africa and Russia (Brinkman 1981; Ivakhnenko 1994, 1996; Modesto et al. 1999) with recent additions from China and Brazil (Liu et al. 2010; Cisneros et al. 2011). Interestingly, the description of more and more basal anomodonts in the past two decades, including *Patranomodon*, *Suminia*, *Anomocephalus*, *Biseridens*, and *Tiarajudens*, has changed our picture of the most basal representatives of the clade significantly. Anomodontia historically also included dinocephalians (King 1988; see Kammerer and Angielczyk 2009 for a review of the complex history of this taxon), but Grine (1997) argued that the latter represent a distinct clade outside of anomodonts, and this has been borne out by subsequent cladistic analyses (e.g., Sidor and Hopson 1998; Liu et al. 2010). In addition to the taxonomic diversity of the group, anomodonts further display a surprisingly high morphological diversity, considering that the cranial anatomy of dicynodontian anomodonts is often described as extremely conservative. However, anomodonts include a wide range of body sizes and an extensive ecomorphological diversity, including small fossorial, slender arboreal, potentially semi-aquatic, and large ground-dwelling animals. Moreover, early anomodonts, including basal toothed dicynodonts, document an enormous dental diversity, ranging from simple, peg-like to cingulated, serrated cheek teeth to transversely expanded palatal teeth. Enlarged caniniform teeth also evolved at least twice within Anomodontia, once as sabre teeth in *Tiarajudens* and once in the form of the continuously-growing tusks of dicynodonts (Fröbisch 2011). Besides the important alpha-taxonomic work that forms the foundation of our discipline, recent studies of anomodonts have investigated various aspects of their paleobiology, including their phylogenetic relationships, biogeography, biostratigraphy, biodiversity, life history (bone histology), dental pathology, and functional morphology (Angielczyk and Kurkin 2003; Ray and Chinsamy 2003, 2004; Ray et al. 2005, 2009, 2012; Botha

J. Fröbisch (✉)
Museum für Naturkunde, Leibniz-Institut für Evolutions- und Biodiversitätsforschung an der Humboldt-Universität zu Berlin, Invalidenstraße 43, 10115 Berlin, Germany
e-mail: joerg.froebisch@mfn-berlin.de

Christian F. Kammerer, Kenneth D. Angielczyk and Jörg Fröbisch (eds.), *Early Evolutionary History of the Synapsida*, Vertebrate Paleobiology and Paleoanthropology, DOI: 10.1007/978-94-007-6841-3_6,

and Smith 2007; Fröbisch 2008, 2009; Fröbisch and Reisz 2008, 2009; Jasinoski et al. 2009; Botha-Brink and Angielczyk 2010; Jasinoski and Chinsamy-Turan 2012).

This second section of *Early Evolutionary History of the Synapsida* comprises three chapters that focus on anomodonts and capture the diversity of recent research in the field very well.

The chapter by Jasinoski et al. (2013) investigates cranial variability in the skull of the Late Permian to Early Triassic dicynodont genus *Lystrosaurus* using a number of different approaches, including gross osteological investigations, histology and CT data. The study takes a very thorough and broad approach, considering all four currently recognized species of *Lystrosaurus* from South Africa as well as the Late Permian genera *Dicynodon* and *Daptocephalus*. The investigations reveal anatomical plasticity with respect to the occurrence of supernumerary bones in the snout of *Lystrosaurus*, and the developmental and functional implications of these bones are discussed. Interestingly, the supernumerary bones occur in the nasofrontal region, which bears a pronounced depression in *Dicynodon*, raising the possibility that this variability could already have been present in closely related but more basal dicynodonts than *Lystrosaurus*. Despite the plastic occurrence of these wedge-shaped supernumerary elements, their function seems to have been related to stabilizing the dorsal part of the snout. However, at the moment this extra element could only be identified in immature individuals of *Lystrosaurus*, which leaves much freedom for interpretation. Future studies will be needed to test whether these elements are also present in adult individuals or whether they ultimately fuse to the frontals or nasals in later ontogenetic stages.

The chapter by Vega and Maisch (2013) identifies and discusses a variety of pathologies preserved in an Upper Permian dicynodont from Tanzania (*Geikia locusticeps*) and in a Middle Triassic form from Brazil (*Stahleckeria potens*). In the *Geikia* specimen, the ventral surface of the skull displays a perfectly circular and deeply concave pit with a prominent rim at the anterior edge of the right suborbital bar between maxilla and jugal. The authors identify this lesion as a slowly growing benign process that led to a pressure erosion of the bone. This pathological feature was most likely caused by a post-traumatic epidermal inclusion cyst, or alternatively by an infection caused by the parasitic tapeworm genus *Echinococcus*, resulting in hydatid disease. In contrast to *Geikia*, the *Stahleckeria* skeleton, which represents a composite mount consisting of several individuals, preserves a variety of pathologies in elements of the appendicular skeleton only, particularly on articular surfaces of the limb bones. The skeletal elements with lesions include the right scapula, right humerus, left femur, left tibia and left fibula of the mounted skeleton, displaying a variety of pathological features such as holes, rugosities, concavities of various sizes, irregular prominences, and well-defined, smooth and irregular protuberances. Not all the lesions could be narrowed down to a single cause, but the discussed pathological phenomena include epidermal inclusion cysts, muscular avulsion, fungal disease, hydatid disease, and osteomyelitis. Although at least some of the pathologies present in *Stahleckeria* were previously noted by von Huene (1935), they have never been discussed in such detail. The study of pathologies in the fossil record is an exciting and promising but often underrepresented branch in our field. Beyond helping to prevent the misinterpretation of pathologies as original morphologies, and thus reducing the risk of incorrect taxonomic, phylogenetic, and functional interpretations, an understanding of these features can provide fascinating insights into the early evolution of certain illnesses as well.

The chapter by Angielczyk et al. (2013) represents an extensive review of the dicynodont faunas from Zambia. The study summarizes previous work in the Permian and Triassic of Zambia, including a thorough discussion of every dicynodont occurrence noted in the past, and furthermore greatly expands our knowledge of the fauna by presenting new material that has been collected by the authors in recent years. The authors recognize 14 dicynodont species from the Upper Permian Upper Madumabisa Mudstone, including *Pristerodon*, *Endothiodon*, *Diictodon*, *Compsodon*, *Emydops*, *Dicynodontoides*, *Katumbia*, *Kitchinganomodon*, *Oudenodon*, *Odontocyclops*, *Dicynodon huenei*, *Syops*, a new unnamed cistecephalid, and a new unnamed lystrosaurid. The new data indicate the presence of a single faunal assemblage in the Upper Permian rather than a geographically and stratigraphically subdivided fauna as it has previously been suggested. Most importantly, this contribution provides a solid framework for biostratigraphic correlation of the various Karoo basins in southern and eastern Africa. Specifically, the Madumabisa fauna seems to be best correlated with the *Cistecephalus* Assemblage Zone of South Africa and shared taxa between the Luangwa Basin and the Ruhuhu Basin suggests the same age for the Usili Formation in Tanzania as well. Four dicynodont species are present in the Middle Triassic Ntawere Formation in Zambia, including *Kannemeyeria lophorhinus*, "*Kannemeyeria*" *latirostris*, *Zambiasaurus submersus*, and *Sangusaurus edentatus*, occuring at two stratigraphic levels that correspond to the subzones B and C of the *Cynognathus* Assemblage Zone in South Africa, respectively. While the occurrence of the Triassic dicynodonts, their stratigraphic subdivision and correlation were already well established prior to the study by Angielczyk and colleagues (see Fröbisch 2009), the Permian data certainly represents a breakthrough in Karoo biostratigraphy, providing an increasingly resolved understanding of faunal similarity in southern and eastern Africa throughout the Permian and Triassic.

Altogether these three chapters reflect the vibrant research on anomodont therapsids, and when combined with the other sections, they document the diversity that characterizes current research efforts on synapsids.

References

Angielczyk, K. D., & Kurkin, A. A. (2003). Phylogenetic analysis of Russian Permian dicynodonts (Therapsida: Anomodontia): Implications for Permian biostratigraphy and Pangaean biogeography. *Zoological Journal of the Linnean Society, 139*, 157–212.

Angielczyk, K. D., Sidor, C. A., Nesbitt, S. J., Smith, R. M. H., & Tsuji, L. A. (2009). Taxonomic revision and new observations on the postcranial skeleton, biogeography, and biostratigraphy of the dicynodont genus *Dicynodontoides*, the senior subjective synonym of *Kingoria* (Therapsida, Anomodontia). *Journal of Vertebrate Paleontology, 29*, 1174–1187.

Angielczyk, K. D., Steyer, J.-S., Sidor, C. A., Smith, R. M. H., Whatley, R. L., & Tolan, S. (2013). Permian and Triassic dicynodont (Therapsida: Anomodontia) faunas of the Luangwa Basin, Zambia: Taxonomic update and implications for dicynodont biogeography and biostratigraphy. In C. F. Kammerer, K. D. Angielczyk, & J. Fröbisch (Eds.), *Early evolutionary history of the Synapsida* (pp. 93–138). Dordrecht: Springer.

Botha, J., & Smith, R. M. H. (2007). *Lystrosaurus* species composition across the Permo-Triassic boundary in the Karoo Basin of South Africa. *Lethaia, 40*, 125–137.

Botha-Brink, J., & Angielczyk, K. D. (2010). Do extraordinarily high growth rates in Permo-Triassic dicynodonts (Therapsida, Anomodontia) explain their success before and after the end-Permian extinction? *Zoological Journal of the Linnean Society, 160*, 341–365.

Brinkman, D. (1981). The structure and relationships of the dromasaurs (Reptilia: Therapsida). *Breviora, 465*, 1–34.

Cisneros, J. C., Abdala, F., Rubidge, B. S., Dentzien-Dias, P. C., & Bueno, A. (2011). Dental occlusion in a 260-million-year-old therapsid with saber canines from the Permian of Brazil. *Science, 331*, 1603–1605.

Fröbisch, J. (2008). Global taxonomic diversity of anomodonts (Tetrapoda, Therapsida) and the terrestrial rock record across the Permian-Triassic boundary. *PLoS One, 3*(11), e3733. doi: 10.1371/journal.pone.0003733.

Fröbisch, J. (2009). Composition and similarity of global anomodont-bearing tetrapod faunas. *Earth-Science Reviews, 95*, 119–157.

Fröbisch, J. (2011). On dental occlusion and saber teeth. *Science, 331*, 1525–1528.

Fröbisch, J., & Reisz, R. R. (2008). A new species of *Emydops* (Synapsida, Anomodontia) and a discussion of dental variability and pathology in dicynodonts. *Journal of Vertebrate Paleontology, 28*, 770–787.

Fröbisch, J., & Reisz, R. R. (2009). The Late Permian herbivore *Suminia* and the early evolution of arboreality in terrestrial vertebrate ecosystems. *Proceedings of the Royal Society B-Biological Sciences, 276*, 3611–3618.

Grine, F. E. (1997). Dinocephalians are not anomodonts. *Journal of Vertebrate Paleontology, 17*, 177–183.

Grine, F. E., Forster, C. A., Cluver, M. A., & Georgi, J. A. (2006). Cranial variability, ontogeny, and taxonomy of *Lystrosaurus* from the Karoo Basin of South Africa. In M. T. Carrano, T. J. Gaudin, R. W. Blob, & J. R. Wible (Eds.), *Amniote paleobiology: Perspectives on the evolution of mammals, birds, and reptiles* (pp. 432–503). Chicago: The University of Chicago Press.

Huene, F. von. (1935). *Die fossilen Reptilien des südamerikanischen Gondwanalandes an der Zeitenwende (Denwa—Molteno—Unterkeuper = Ober-Karnisch). Ergebnisse der Sauriergrabungen in Südbrasilien 1928/29. Lieferung I.* Tübingen: Heine.

Ivakhnenko, M. F. (1994). A new Late Permian dromasaurian (Anomodontia) from Eastern Europe. *Paleontological Journal, 28*, 96–103.

Ivakhnenko, M. F. (1996). Primitive anomodonts, venyukoviids, from the Late Permian of Eastern Europe. *Paleontological Journal, 30*, 575–582.

Jasinoski, S. C., Rayfield, E. J., & Chinsamy, A. (2009). Comparative feeding biomechanics of *Lystrosaurus* and the generalized dicynodont *Oudenodon*. *The Anatomical Record: Advances in Integrative Anatomy and Evolutionary Biology, 292*, 862–874.

Jasinoski, S. C., & Chinsamy-Turan, A. (2012). Biological inferences of the cranial microstructure of the dicynodonts *Oudenodon* and *Lystrosaurus*. In A. Chinsamy-Turan (Ed.), *Forerunners of mammals: Radiation histology biology* (pp. 149–176). Bloomington: Indiana University Press.

Jasinoski, S. C., Cluver, M. A., Chinsamy, A., & Reddy, B. D. (2013). Anatomical plasticity in the snout of *Lystrosaurus*. In C. F. Kammerer, K. D. Angielczyk, & J. Fröbisch (Eds.), *Early evolutionary history of the Synapsida* (pp. 139–149). Dordrecht: Springer.

Kammerer, C. F., & Angielczyk, K. D. (2009). A proposed higher taxonomy of anomodont therapsids. *Zootaxa, 2018*, 1–24.

Kammerer, C. F., Angielczyk, K. D., & Fröbisch, J. (2011). A comprehensive taxonomic revision of *Dicynodon* (Therapsida, Anomodontia) and its implications for dicynodont phylogeny, biogeography, and biostratigraphy. *Society of Vertebrate Paleontology Memoir, 11*, 1–158.

Keyser, A. W. (1973). A re-evaluation of the genus *Tropidostoma* Seeley. *Palaeontologia Africana, 16*, 25–35.

Keyser, A. W. (1975). A re-evaluation of the cranial morphology and systematics of some tuskless Anomodontia. *Memoirs of the Geological Survey of South Africa, 67*, 1–110.

Keyser, A. W. (1993). A re-evaluation of the smaller Endothiodontidae. *Memoirs of the Geological Survey of South Africa, 82*, 1–53.

King, G. M. (1988). Anomodontia. In P. Wellnhofer (Ed.), *Handbuch der Paläoherpetologie* (Vol. 17C). Stuttgart: Gustav Fischer Verlag.

King, G. M. (1993). How many species of *Diictodon* were there? *Annals of the South African Museum, 102*, 303–325.

King, G. M., & Rubidge, B. S. (1993). A taxonomic revision of small dicynodonts with postcanine teeth. *Zoological Journal of the Linnean Society, 107*, 131–154.

Liu, J., Rubidge, B., & Li, J. (2010). A new specimen of *Biseridens qilianicus* indicates its phylogenetic position as the most basal anomodont. *Proceedings of the Royal Society B: Biological Sciences, 277*, 285–292.

Modesto, S. P., Rubidge, B. S., & Welman, J. (1999). The most basal anomodont therapsid and the primacy of Gondwana in the evolution of the anomodonts. *Proceedings of the Royal Society of London B, 266*, 331–337.

Owen, R. (1845). Description of certain fossil crania, discovered by A. G. Bain, Esq., in sandstone rocks at the south-eastern extremity of Africa, referable to different species of extinct genus of Reptilia (*Dicynodon*), and indicative of a new tribe or sub-order of Sauria. *Quarterly Journal of the Geological Society of London, 1*, 318–322.

Ray, S., & Chinsamy, A. (2003). Functional aspects of the postcranial anatomy of the Permian dicynodont *Diictodon* and their ecological implications. *Palaeontology, 46*, 151–183.

Ray, S., & Chinsamy, A. (2004). *Diictodon feliceps* (Therapsida, Dicynodontia): Bone histology, growth, and biomechanics. *Journal of Vertebrate Paleontology, 24*, 180–194.

Ray, S., Chinsamy, A., & Bandyopadhyay, S. (2005). *Lystrosaurus murrayi* (Therapsida, Dicynodontia): Bone histology, growth and lifestyle adaptations. *Palaeontology, 48*, 1169–1185.

Ray, S., Bandyopadhyay, S., & Bhawal, D. (2009). Growth patterns as deduced from bone microstructure of some selected neotherapsids with special emphasis on dicynodonts: Phylogenetic implications. *Palaeoworld, 18*, 53–66.

Ray, S., Botha-Brink, J., & Chinsamy-Turan, A. (2012). Dicynodont growth dynamics and lifestyle adaptations. In A. Chinsamy-Turan (Ed.), *Forerunners of mammals: Radiation histology biology* (pp. 120–146). Bloomington: Indiana University Press.

Sidor, C. A., & Hopson, J. A. (1998). Ghost lineages and "mammalness": Assessing the temporal pattern of character acquisition in the Synapsida. *Paleobiology, 24*, 254–273.

Vega, C. S., & Maisch, M. W. (2013). Pathological features in Upper Permian and Middle Triassic dicynodonts (Synapsida, Therapsida). In C. F. Kammerer, K. D. Angielczyk, & J. Fröbisch (Eds.), *Early evolutionary history of the Synapsida* (pp. 151–161). Dordrecht: Springer.

Chapter 7
Permian and Triassic Dicynodont (Therapsida: Anomodontia) Faunas of the Luangwa Basin, Zambia: Taxonomic Update and Implications for Dicynodont Biogeography and Biostratigraphy

Kenneth D. Angielczyk, Jean-Sébastien Steyer, Christian A. Sidor, Roger M. H. Smith, Robin L. Whatley, and Stephen Tolan

Abstract Dicynodont fossils were first collected in the Luangwa Basin, Zambia, in the 1920s, but limited detailed study and taxonomic uncertainty have obscured their biostratigraphic utility and their implications for topics such as dicynodont biogeography and the effects of the end-Permian extinction. Here we present a comprehensive taxonomic revision of the dicynodonts of the Luangwa Basin, taking into account specimens in all major museum collections and new material collected by our team in 2009. We recognize 14 dicynodont species from the Upper Permian Upper Madumabisa Mudstone: *Pristerodon mackayi*, *Endothiodon* sp., *Diictodon feliceps*, *Compsodon helmoedi*, *Emydops* sp., *Dicynodontoides* cf. *D. nowacki*, a new tusked cistecephalid, cf. *Katumbia parringtoni*, *Kitchinganomodon crassus*, *Oudenodon bainii*, *Odontocyclops whaitsi*, *Dicynodon huenei*, *Syops vanhoepeni*, and a new lystrosaurid. Previous reports of *Lystrosaurus* in the basin appear to be in error. In addition, we found no significant partitioning of dicynodont taxa in the northern and southern parts of the basin, despite substantial differences in preservation, indicating the presence of a single faunal assemblage in the Upper Permian. The Madumabisa dicynodont assemblage is best correlated with the *Cistecephalus* Assemblage Zone of South Africa. The shared presence of *Dicynodon huenei* and possibly *Katumbia* in the Luangwa Basin and the Ruhuhu Basin of Tanzania suggests that the Tanzanian Usili Formation also can be correlated with the *Cistecephalus* zone. Interestingly, the Madumabisa assemblage from Zambia is more similar to the coeval assemblage from South Africa, despite its closer geographic proximity to Tanzania. The Karoo and Ruhuhu basins also include more endemic species in the Permian than the Luangwa Basin. The Middle Triassic Ntawere Formation preserves four dicynodont species (*Kannemeyeria lophorhinus*, "*Kannemeyeria*" *latirostris*, *Zambiasaurus submersus*, *Sangusaurus edentatus*), which occur at two stratigraphic levels. The lower Ntawere assemblage resembles that of the Omingonde Formation of Namibia in the presence of *Kannemeyeria lophorhinus* and potentially *Dolichuranus* (if "*K.*" *latirostris* represents this taxon). The upper Ntawere assemblage shares the genus *Sangusaurus* with that of the Manda beds of Tanzania and includes the endemic *Zambiasaurus*. Comparisons of these assemblages to the Omingonde and Manda suggest that both are best correlated with the *Cynognathus* C subzone. When combined with data on other tetrapod taxa, our revised dicynodont assemblages contribute to an emerging picture of broad faunal similarity in southern and eastern Africa during the Late Permian, and increasing differentiation between the South African and other Karoo basins following the end-Permian extinction.

K. D. Angielczyk (✉)
Department of Geology, Field Museum of Natural History, 1400 South Lake Shore Drive, Chicago, IL 60605, USA
e-mail: kangielczyk@fieldmuseum.org

J.-S. Steyer
Department of Earth History, CNRS-Museum d'Histoire Naturelle, 8 rue Buffon, CP 38, 75005 Paris, France
e-mail: steyer@mnhn.fr

C. A. Sidor
Burke Museum and Department of Biology, University of Washington, Seattle, WA 98195, USA
e-mail: casidor@u.washington.edu

R. M. H. Smith
Department of Karoo Palaeontology, Iziko: South African Museum, Cape Town 8000, South Africa
e-mail: rsmith@iziko.org.za

R. L. Whatley
Department of Science and Mathematics, Columbia College Chicago, 600 South Michigan Avenue, Chicago, IL 60605, USA
e-mail: rwhatley@colum.edu

S. Tolan
Chipembele Wildlife Education Centre, Chowo Site, Malama Road, P. O. Box 67Mfuwe, Zambia
e-mail: info@chipembele.org

Christian F. Kammerer, Kenneth D. Angielczyk and Jörg Fröbisch (eds.), *Early Evolutionary History of the Synapsida*, Vertebrate Paleobiology and Paleoanthropology, DOI: 10.1007/978-94-007-6841-3_7,

Keywords *Cistecephalus* Assemblage Zone • East Africa • Karoo Basins • Ntawere Formation • Upper Madumabisa Mudstone

Introduction

Dicynodonts number among the most successful Permian and Triassic nonmammalian synapsids in terms of their species richness, abundance, and stratigraphic distribution. The clade was also cosmopolitan, with dicynodont fossils having been discovered on every continent (King 1992; Rubidge 2005; Fröbisch 2009). However, the detail with which the dicynodont faunas from particular areas are known varies greatly depending on factors such as available outcrop area and the cumulative collecting effort expended by paleontologists. Thus, the Karoo Basin of South Africa, which has extensive fossiliferous exposures and a history of continuous paleontological research spanning over 150 years, has produced more than 15,000 cataloged dicynodont specimens (Nicolas and Rubidge 2009, 2010), making it by far the best window into the evolutionary history of the clade. However, even other well-studied areas, such as the fore-Ural region of Russia, have produced far fewer specimens (e.g., Ivakhnenko et al. 1997), and some geographically significant dicynodont records consist of very small samples (e.g., 10 specimens from Scotland: Cruickshank et al. 2005; five published specimens from Laos, one of which is lost: Battail 2009a, b; Steyer 2009). This unevenness in sampling obscures biogeographic and biostratigraphic patterns, and it makes it difficult to determine whether apparent absences of particular dicynodont taxa in a given area are real or artifactual. An extreme example can be found in the geographic distribution of *Diictodon feliceps*, which is known from South Africa, Zambia, and China, but not fossiliferous basins in between (Angielczyk and Sullivan 2008). Obviously there was a route that *D. feliceps* used to disperse between these widely separated areas, but was that route outside of areas where Permian tetrapod fossils were preserved, or would more intensive collecting in intermediate areas reveal novel geographic occurrences?

A related issue is the complex taxonomic history of dicynodonts. Over the course of the past four decades, much of dicynodont taxonomy, particularly for Permian taxa, has been extensively revised (e.g., Cox 1964; Keyser 1973a, b, 1975, 1993; Tollman et al. 1980; Cluver and Hotton 1981; King 1988; King and Rubidge 1993; Sullivan and Reisz 2005; Grine et al. 2006; Botha and Angielczyk 2007; Fröbisch and Reisz 2008; Angielczyk et al. 2009; Kammerer et al. 2011), greatly improving our knowledge of the clade's diversity. However, because the majority of named dicynodont species are based on material from the Karoo Basin, most revisions have focused on South African taxa. Even when non-South African taxa have been included, it is often difficult to trace a particular valid name or synonym through the literature. If a particular name or reported occurrence has not been dealt with explicitly in a revision, it can be a daunting task to attempt to identify a specimen short of personally examining it. Many of the original reports of material from outside of South Africa consist of very brief descriptions of fragmentary specimens accompanied by figures that are little more than sketches (e.g., Haughton 1926, 1932; Boonstra 1938). The dicynodont faunas of the Luangwa Basin of Zambia exemplify many of these issues.

The first tetrapod fossils discovered in the Luangwa Basin were two fragmentary pieces of dicynodont postcrania (SAM-PK-7424, SAM-PK-7425) collected in 1925 by G. Prentice (Dixey 1937). The geologist F. Dixey made the first significant collections of fossils, mostly from the northern part of the basin, in 1928 and 1935 (Dixey 1937; Boonstra 1938) (Fig. 7.1a). Additional collecting in the northern part of the basin was carried out in 1960 and 1961 by the Geological Survey of Northern Rhodesia and the Bernard Price Institute for Palaeontological Research (Brink 1963; Drysdall and Kitching 1963; Kitching 1963), and by the British Museum (Natural History)—University of London Joint Palaeontological Expedition in 1963 (Attridge et al. 1964). A decade later, in 1972, members of the Geological Survey of Northern Rhodesia discovered additional localities in the central Luangwa Basin. As a direct result, fossil collections were made in 1974 in collaboration with the Oxford University Museum of Natural History (Kerr 1974; Kemp 1975). A short subsequent reconnaissance to the same area was made in 2000 by T. S. Kemp (Oxford University Museum of Natural History), J. G. Theime (former director of the Geological Survey of Zambia), and associates (T. S. Kemp, personal communication, 2009). Most recently, our team spent three weeks in July, 2009 working in both the northern and central parts of the basin.

To date, collecting efforts have resulted in several hundred specimens that are distributed among the Iziko: South African Museum (Prentice's and Dixey's collections), the Bernard Price Institute for Palaeontological Research (specimens from the 1960 and 1961 expeditions), The Livingstone Museum (a small number of specimens from the 1963 expedition), and The Natural History Museum (most specimens from the 1963 expedition); material from the 1974 expedition is currently housed at Oxford University but it and specimens from the 2009 expedition will be returned to the National Heritage Conservation Commission of Zambia. Much of this material is all but unstudied, and only a handful of papers focusing on dicynodonts from the

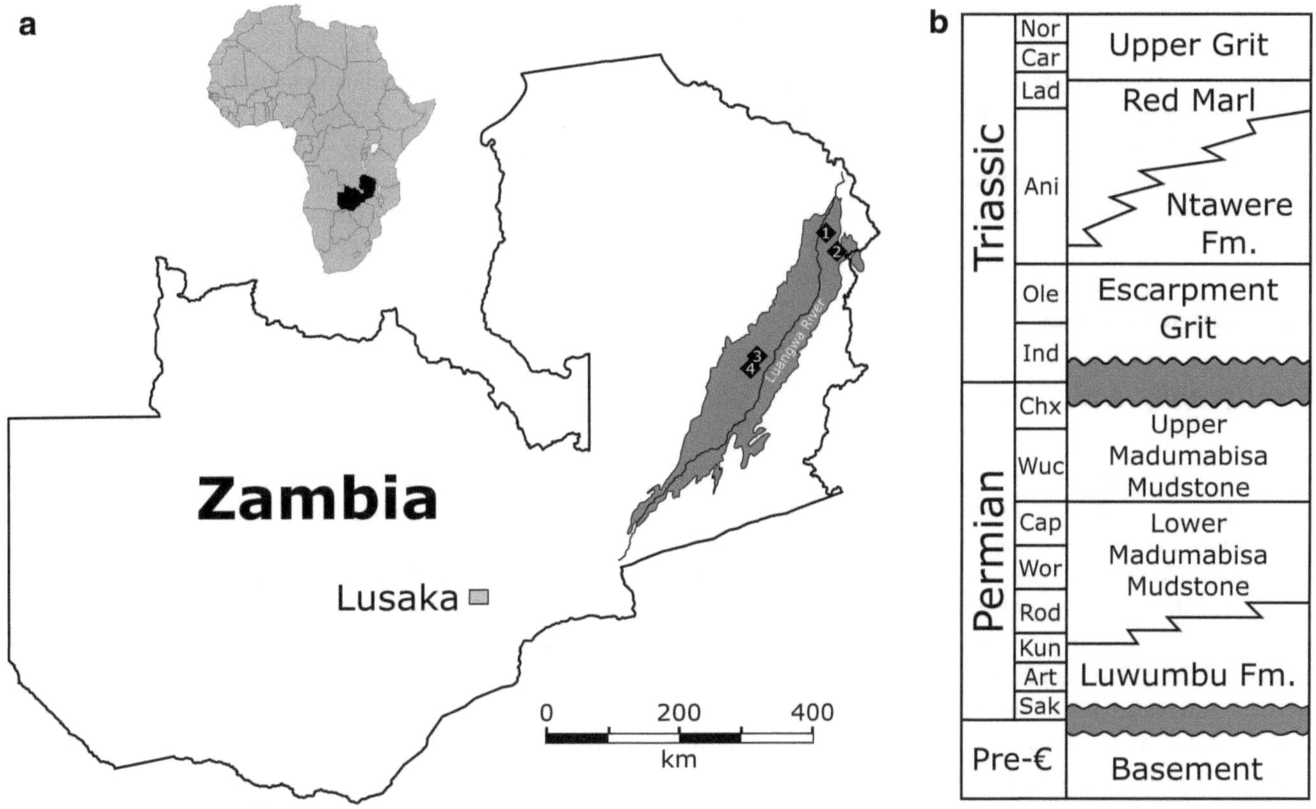

Fig. 7.1 a Map showing location of the Luangwa Basin and the approximate locations of the main areas in which vertebrate fossils have been collected; *inset* map shows the location of Zambia in Africa. *Area 1* corresponds to the northern Permian localities of Dixey (1937; Boonstra 1938), Drysdall and Kitching (1962, 1963), Kitching (1963), and Attridge et al. (1964). *Area 2* corresponds to the Triassic localities of Drysdall and Kitching (1962, 1963), Kitching (1963), and Attridge et al. (1964). *Area 3* corresponds to the localities of Kerr (1974) and Kemp (1975) in North Luangwa National Park. *Area 4* corresponds to the localities of Kerr (1974) and Kemp (1975) in the Munyamadzi Game Management Area. Specimens were collected in all four of these areas by the 2009 expedition. **b** Generalized stratigraphy of the Luangwa Basin. Lithostratigraphy based on Banks et al. (1995). Correlations between lithostratigraphy and marine stages approximate and based on Nyambe and Utting (1997), Nyambe (1999), Cairncross (2001), and Catuneanu et al. (2005). Note that relative thicknesses of the marine stages are not scaled to their relative temporal durations

Luangwa Basin have been published (Boonstra 1938; Cox 1969; Crozier 1970; Keyser 1979; Keyser and Cruickshank 1979; King 1981; Gale 1988; King and Jenkins 1997; Angielczyk 2002), although some described particular specimens in great detail. Of the material that has been published, a considerable proportion has been included in taxonomic revisions, and in some cases multiple revisions (Keyser 1973c, 1975; Keyser and Cruickshank 1979; Angielczyk 2002; Renaut et al. 2003; Botha and Angielczyk 2007; Kammerer et al. 2011). There is also an unpublished report of at least one additional new taxon that has never been formally described (Freeman 1993). These facts make it difficult to compile accurate faunal lists for the Luangwa Basin, but such data are necessary if broader studies of dicynodont biogeography and biostratigraphy, and the end-Permian mass extinction, are to produce meaningful results.

Here we review the Permian and Triassic dicynodont faunas of the Luangwa Basin, based on our personal observations of most of the Zambian dicynodont specimens in museum collections and supplemented with observations of dicynodont fossils in the field. Although the review uses up-to-date dicynodont taxonomy, we provide links between modern and older names, and provide justifications for our identifications and images of voucher specimens for each taxon. Finally, we discuss the biogeographic and biostratigraphic implications of the revised faunal lists.

Anatomical Abbreviations: Al, alveolus; Ect, ectepicondylar foramen; Dpc, deltopectoral crest; Nb, nasal boss; Pct, "postcanine" tooth; Pds, posterior dentary sulcus; Sq, squamosal.

Institutional Abbreviations: AMNH, American Museum of Natural History, New York City, NY, USA; BP, Bernard Price Institute for Palaeontological Research, Johannesburg, South Africa; CAMZM, University Museum of Zoology, Cambridge, UK; CGP, Council for Geosciences, Pretoria, South Africa; GPIT, Institut für Geowissenschaften, Tübingen, Germany; IVPP, Institute of Vertebrate Paleontology and Paleoanthropology, Beijing, China; LM, Livingstone

Mueum, Livingstone, Zambia; NHCC, National Heritage Conservation Commission, Lusaka, Zambia; NHMUK, Natural History Museum, London, UK; NMQR, National Museum, Bloemfontein, South Africa; NMT, National Museum of Tanzania, Dar es Salaam, Tanzania; RC, Rubidge Collection, Graaff-Reinet, South Africa; TSK, T. S. Kemp Collection, Oxford University, Oxford, UK.

Geological Context

Stratigraphy and Sedimentology

The Karoo basins of south-central Africa formed during the assembly and breakup of Pangaea under two distinct tectonic regimes sourced from the southern and northern margins of Gondwana. The southern tectonic regime, generated by subduction and orogenesis along the Panthalassan (paleo-Pacific) margin of Gondwana resulted in the formation of the Gondwanide mountain belt with a series of retroarc foreland basins. Subsidence and sedimentation in these basins was primarily controlled by flexural and dynamic loading of the crust (Catuneanu et al. 2005). The main Karoo Basin in South Africa is the best exposed of these foreland basins and contains the litho- and biostratigrapic reference sections for the Upper Carboniferous–Middle Jurassic Karoo Supergroup.

North of the main Karoo Basin, tectonic regimes were dominated by extensional or transtensional stresses that propagated southwards into the supercontinent from the Tethyan margin of Gondwana. The sedimentary fills of these rift basins show a pronounced similarity due to their similar structural history; the older Karoo deposits were laid down and preserved within the oldest graben structures, most of which occupy the deepest parts of the basins today. As the rifts expanded, younger sedimentary sequences progressively overstepped onto domino-style tilted horsts and younger grabens. Thus, almost continuous sedimentation took place within the deep parts of the rifts whereas the successions on the rift shoulders were interrupted by hiatuses and erosion, evidenced by unconformities and reduced sections (Tankard et al. 2009).

Climatic fluctuations also left a mark on the stratigraphic record, providing a common trend that can be identified in the sedimentary fill of most of the Karoo-aged basins formed under different tectonic regimes. The climate changed from cold to humid temperate and semi-arid during the Late Carboniferous–earliest Permian interval, to warmer and eventually hot with fluctuating precipitation from the Early Triassic through to Early Jurassic (Smith et al. 1993).

In Zambia, Karoo-aged sequences occur in the Luangwa, Luano and Zambezi rift basins. To date, tetrapod fossils have been collected in large numbers only from the Luangwa Basin (Fig. 7.1a), although there is a report of fragmentary dinocephalian material from the Zambezi Basin (Gair 1959). The Luangwa Basin itself comprises two non-overlapping opposing half-graben separated by a transfer zone or accommodation zone that forms a structural high (Banks et al. 1995). The sub-basins are structurally and depositionally similar, and preserve essentially the same stratigraphic sequence (Fig. 7.1b). Tetrapod fossils occur in the Upper Permian Madumabisa Mudstone Formation and the Middle Triassic Ntawere Formation in the Luangwa Basin. Numerous Permian and Triassic fossils have been collected in the northern sub-basin, but nearly all of the material from the transfer zone is of Permian age. The specimen of *Luangwa drysdalli* described by Kemp (1980a; also see Kerr 1974; Kemp 1975) is the only Triassic tetrapod collected in the transfer zone to date. The vertebrate paleontology of the southern sub-basin remains almost entirely unexplored.

The Madumabisa Mudstone sediments accumulated on the floor of a wide flat-bottomed rift valley with a gentle regional slope towards the south-southwest. The sequence is interpreted as having initially been an alluvial plain dominated by low sinuosity river channels (Lower Member of Drysdall and Kitching 1963). As the graben widened and the rates of sedimentation increased, the rivers became progressively higher in sinuosity with more and more ponds and lakes until eventually the valley floor became predominantly sub-aqueous as evidenced by the extensive, thick beds of massive grey and green mudstones in the Upper Member of the Madumabisa Mudstone Formation (Yemane and Kelts 1990; Banks et al. 1995). Most of the Late Permian vertebrate fossils are found in greenish grey and pale brown massive siltstone beds, associated with and often partly enclosed within smooth surfaced calcareous nodules. The fossil rich beds are interpreted as having accumulated by episodic, possibly catastrophic, flooding of a vegetated floodplain.

The contact with the overlying pebbly sandstone of the Escarpment Grit Formation is regarded as erosional throughout the Luangwa Basin. However, the same contact in the nearby Zambezi basin is an abrupt change of depositional style, disconformable rather than unconformable (Bond 1967). In the main Karoo Basin, time-equivalent strata show a transition through the argillaceous Palingkloof member of the Balfour Formation into the arenaceous Katberg Formation, with no disconformity, and vertebrate fossils that record the End-Permian mass extinction event (Smith 1995). The synchroneity of this relatively rapid switch from cool-wet lacustrine to warm-dry fluvial depositional environments across all the Karoo basins in southern Gondwana has been attributed to CO_2 degassing from basaltic floods in northern Pangaea causing rapid global warming, a shift in precipitation belts, and aridification of inland regions (Ward et al. 2005).

In the Luangwa sequence the Lower Triassic Escarpment Grit grades upwards into semi-arid fluvio-lacustrine redbed strata of the Middle Triassic Ntawere Formation, where the mudrocks are predominantly dark reddish brown with horizons of calcareous rhizocretions and nodules. Trough cross bedded coarse-grained gritstones with intraformational conglomerates interbedded with structureless light-red siltstone beds are interpreted as ephemeral stream channel fills incised into wind deposited loess. Stromatolitic limestone drapes over lenses of reworked brecciated mudstone with bone clasts are part of the playa lake shoreline facies in which many of the vertebrate fossils are found.

Vertebrate Taphonomy

The preservation style of dicynodonts in the fossil rich localities within the Upper Madumabisa Mudstone is most commonly isolated skulls, mostly without articulated lower jaws, and isolated limb and girdle elements, although a few complete articulated skeletons (some curled- up) and semi-associated skeletons also can be found. Most of the fully articulated specimens are of *Diictodon* and *Pristerodon* and most of the disarticulated, semi-associated skeletons are of the medium and large dicynodonts *Oudenodon* and *Odontocyclops*. The close association of the articulated skeletons with mictitic nodules suggests these bones were buried while skin and connective tissue were still present. The organic matter subsequently decomposed to release hydrogen sulphide into the surrounding silt, creating reduction halos that later induced calcium carbonate to precipitate from the groundwater. Preservation of specimens in hematite-rich nodules is more common in the northern sub-basin than in the transfer zone, although a thin hematite rind sometimes is present on the bone surface of specimens from the latter area.

The dicynodonts of the Ntawere Formation have a similar taphonomic range to those of the Upper Madumabisa Formation, but with far fewer complete articulated specimens and none found in curled-up pose. This may be due to the disappearance of small-bodied dicynodonts such as *Diictodon, Pristerodon*, and *Emydops*, which may have inhabited underground burrows (Smith 1987). The larger kannemeyeriiforms are commonly found as patches of scattered (i.e., disarticulated but still associated) postcrania within an area of three to five square meters, suggesting that scavenging was more prevalent in the mid-Triassic. This interpretation is reinforced by the occurrence of bone bearing coprolites in and around the bone scatters (this study and Drysdall and Kitching 1963).

Note on Treatment of Fossil Localities

The various individuals and groups who collected fossils in the Luangwa Basin used different systems for identifying localities. We use three sets of locality numbers in the Systematic Paleontology section. For specimens collected in the northern part of the basin by Dixey and the 1960, 1961, and 1963 expeditions, we use the Drysdall and Kitching (1963; also see Kitching 1963) locality numbering system, which incorporates and standardizes all localities up to that time. For specimens collected by the 1974 expedition, we use the numbering system of Kerr (1974), which was used in most of the papers describing material from this collection (Kemp 1979, 1980b; Davies 1981; King 1981) and can be directly related to information provided in other publications that do not refer to localities by number (Kemp 1975; King and Jenkins 1997). Finally, we use our locality numbers for specimens collected during the 2009 expedition (i.e., NHCC specimens with locality numbers starting with "L"). Detailed locality information is available to qualified researchers from the respective museums or from KDA in the case of specimens collected by our team.

Permian Dicynodont Fauna

We use the higher-level taxonomy of Kammerer and Angielczyk (2009) for Permian dicynodonts, with minor changes reflecting the results of Kammerer et al. (2011). Our taxonomic results for Permian dicynodonts are summarized in Table 7.1.

Systematic Paleontology

Therapsida Broom, 1905
Anomodontia Owen, 1860a
Chainosauria Nopcsa, 1923
Dicynodontia Owen, 1860a
Endothiodontia Owen, 1876
Endothiodon sp.
Figure 7.2a, c

Material: BP/1/3574, NHCC LB11, NHCC LB12.

Localities: Locality 3 of Drysdall and Kitching (1963) (BP/1/3574), locality L32 (NHCC LB12), locality L49 (NHCC LB11).

Identifying Characteristics: Cox (1964), Cluver and King (1983), King (1988), and Ray (2000) provided diagnoses for *Endothiodon*, and Cox's paper is noteworthy for its consideration of intra- and interspecific variation in the genus. The available Zambian *Endothiodon* specimens are

Table 7.1 Dicynodont taxa present in the Upper Permian Upper Madumabisa Mudstone, Luangwa Basin, Zambia, and synonyms used in the literature on the Luangwa Basin

Taxon	Synonyms in Luangwa Basin literature
Endothiodon sp.	*Endothiodon uniseries*
Pristerodon mackayi	*Parringtoniella, Emydops, Emydopsis*
Diictodon feliceps	*Dicynodon grimbeeki, Dicynodon sollasi, Dicynodon clarencei*
Compsodon helmoedi	None
Emydops sp.	*Emydops* sp. indet.
Dicynodontoides cf. *D. nowacki*	None
Cistecephalidae n. g. & sp.	*Cistecephalus, Cistecephalus microrhinus, Cistecephalus planiceps*
cf. *Katumbia parringtoni*	None
Odontocyclops whaitsi	*Rhachiocephalus dubius, Odontocyclops dubius, Dicynodon* cf. *breviceps, Dicynodon, Rhachiocephalus magnus*
Oudenodon bainii	*Dicynodon lutriceps, Dicynodon* cf. *breviceps, Dicynodon corstorphinei, Dicynodon* cf. *corstorphinei, Dicynodon* cf. *milletti, Dicynodon latirostris, Dicynodon luangwanensis, Dicynodon helenae, Dicynodon euryceps, Dicynodon parabreviceps, Oudenodon luangwanensis, Oudenodon luangwaensis, Oudenodon luangwensis*
Kitchinganomodon crassus	None
Dicynodon huenei	*Dicynodon lacerticeps, Dicynodon trigonocephalus, "Dicynodon" trigonocephalus*
Syops vanhoepeni	*Dicynodon vanhoepeni, Dicynodon roberti, "Dicynodon" vanhoepeni, "Dicynodon" roberti*
Lystrosauridae n. g. & sp.	*Lystrosaurus* cf. *curvatus*

See text for details

all very fragmentary, and appear to represent parts of the palate and/or jaws. The most diagnostic features of these fragments are their relatively large sizes and the presence of long, medially-placed tooth rows, both of which give them a close resemblance to *Endothiodon* specimens from areas such as South Africa and Mozambique (Fig. 7.2a–d). For example, comparison to more complete specimens suggests that NHCC LB12 was part of a skull with a basal length of approximately 380 mm and NHCC LB11 originated in a mandible with a length of approximately 300 mm, values that are comparable to large *Endothiodon* specimens such as AMNH 5565 (basal length of skull 430 mm; length of jaw ramus 400 mm) or BP/1/1659 (basal length of skull 345 mm). However, the specimens are too fragmentary to allow a species-level identification.

Synonyms in Luangwa Basin Literature: Fröbisch (2009) stated that Drysdall and Kitching (1963) and Kitching (1963) recorded *Endothiodon uniseries* from the Luangwa Basin. However, we can find no mention of that species in those publications. Cox (1964) reported a specimen of *E. uniseries* collected by John Attridge from the "Madumabisa shales," but of Zimbabwe, not Zambia.

Previous Reports: Kitching (1963) and Drysdall and Kitching (1963) reported collecting two fragmentary specimens of *Endothiodon* at their Locality 3, which they included in their lower fossiliferous horizon. We relocated one of those specimens (BP/1/3574) and confirm its identification. Anderson and Cruickshank (1978), King (1988, 1992), Rubidge (2005), and Fröbisch (2009) all included *Endothiodon* in the faunal lists they compiled. Cooper (1982) and Angielczyk (2002) both mentioned *Endothiodon* in their discussion of the biostratigraphic correlation of the Madumabisa Mudstone.

Eumantelliidae Broom, 1935
Pristerodon mackayi Huxley, 1868
Figure 7.2e–g

Material: BP/1/3386, BP/1/3399, BP/1/3410, BP/1/3601, NHCC LB4, NHCC LB5, NHCC LB8, NHCC LB9, NHCC LB10, SAM-PK-K7933.

Localities: Locality 4 of Drysdall and Kitching (1963) (BP/1/3386, BP/1/3399, BP/1/3410, BP/1/3601, SAM-PK-K7933), locality L6 (NHCC LB4), locality L7 (NHCC LB5), locality L50 (NHCC LB8), locality L52 (NHCC LB9, NHCC LB10).

Identifying Characteristics: Keyser (1993) and King and Rubidge (1993) provided recent diagnoses of *Pristerodon*. The Zambian *Pristerodon* specimens we identified vary greatly in quality of completeness and the degree to which they have been prepared. The most informative specimens (e.g., BP/1/3410; Fig. 7.2e, f), show numerous characters diagnostic of *Pristerodon*, including a broad temporal region in which the parietals are exposed between the postorbitals, leaf-shaped palatine pads, and the presence of maxillary "postcanine" teeth arranged in a row that is oblique to the sagittal plane of the skull. Identifications for more fragmentary and/or unprepared specimens are based on a combination of size and the presence of one or more diagnostic characters. For example, the material from locality L52 (NHCC LB9 and NHCC LB10) consists of the remains of at least three individuals, and diagnostic material includes a palate with an exposed tusk and an oblique row of "postcanine" teeth, and two toothed dentaries, one of which also possesses the remains of a relatively large, rounded lateral dentary shelf.

Synonyms in Luangwa Basin Literature: *Parringtoniella* (Drysdall and Kitching 1963; King 1988, 1992). Keyser (1993) and King and Rubidge (1993) discussed the synonymy of *Pristerodon* and *Parringtoniella.* Boonstra (1938) referred one specimen (SAM-PK-K7933) to *Emydops* or *Emydopsis*. Although the former taxon is valid, and the latter appears to be a junior synonym of it (King 1988), SAM-PK-K7933 most plausibly represents a poorly preserved specimen of *Pristerodon.*

Previous Reports: If our identification of SAM-PK-K7933 is correct, then Boonstra's (1938) report of this specimen is the first time a Zambian specimen of *Pristerodon* was mentioned in the literature, although it was not identified as such at the time. Drysdall and Kitching (1963) noted the occurrence of "*Parringtoniella*" at their Locality 3, but it is unclear whether they collected any of these specimens because all of the *Pristerodon* material we identified at the BP originated at their Locality 4. Anderson and Cruickshank (1978), Rubidge (2005), and Fröbisch (2009) all included *Pristerodon* in their compilations. King (1988) included "*Parringtoniella*" in the faunal list for Zambia, but later King (1992) suggested that this most likely was a synonym of *Pristerodon.*

Therochelonia Seeley, 1894
Pylaecephalidae (van Hoepen, 1934)
Diictodon feliceps (Owen, 1876)
Figure 7.2h–m

Material: BP/1/3598, NHCC LB1, NHCC LB2, NHCC LB3, NHCC LB6, NHCC LB7, NHCC LB27, TSK 77, TSK 98.

Localities: Locality 4 of Drysdall and Kitching (1963) (BP/1/3598). Locality L31 (NHCC LB1), locality L38 (NHCC LB2, NHCC LB6, NHCC LB7, NHCC LB27), locality L48 (NHCC LB3). Kerr's (1974) Locality 11 (TSK 98), Kerr's (1974) Locality 13 (TSK 77).

Identifying Characteristics: Sullivan and Reisz (2005) and Angielczyk and Sullivan (2008) recently discussed diagnostic characters and ranges of discrete and morphometric variability for *Diictodon feliceps*. The specimens we refer to *D. feliceps* all are relatively small with square-cut caniniform processes set off from the palatal rim by a notch (Fig. 7.2i, j). In the specimens where the intertemporal region is preserved and exposed, it is relatively narrow and the postorbitals extensively overlap the parietals. A mandible from locality L38, NHCC LB6 (Fig. 7.2h), lacks postcanine teeth, has a dentary table that grades into a short broad posterior dentary sulcus (see Angielczyk and Rubidge 2013 for information on the interpretation of the homologies of these characters), and the remains of a tall, convex cutting blade on the medial side of the dorsal surface of the dentary. The shape of the well-preserved deltopectoral crest of NHCC LB7 (Fig. 7.2k) closely resembles those of South African *D. feliceps* specimens, and other humeral fragments from locality L38 (e.g. NHCC LB27; Fig. 7.2l) also show the presence of an ectepicondylar foramen.

Synonyms in Luangwa Basin Literature: *Dicynodon grimbeeki*, *Dicynodon sollasi*, *Dicynodon clarencei* (Drysdall and Kitching 1963; Gale 1989). See King (1993) and Sullivan and Reisz (2005) for discussion of the synonymies of the first two species with *D. feliceps*. *Dicynodon clarencei* was recognized as a synonym of *Dicynodontoides recurvidens* by Angielczyk et al. (2009).

Previous Reports: Based on the species identifications given in their subsequent publications, at least some of the specimens identified as *Dicynodon* by Drysdall and Kitching (1962) likely represent *Diictodon*. Drysdall and Kitching (1962, 1963) reported specimens in what they considered the lower (Locality 3) and upper (Locality 4) fossiliferous beds of the Madumabisa Mudstone. However, they did not state whether these occurrences were based on collected specimens or field reports, making confirmation of the identifications difficult. Gale (1988) referred an assemblage of juvenile dicynodont specimens from Zambia to *Diictodon*, and King (1993) followed this identification in her discussion of *Diictodon* taxonomy. However, this identification is questionable because their sizes are larger than would be expected for *Diictodon* given their presumed early ontogenetic stage, and they lack the distinctive notched caniniform process that is typical of *Diictodon*. Gale (1989) referred the same specimens to *Dicynodon clarencei* (a synonym of *Dicynodontoides recurvidens*; see Angielczyk et al. 2009), but they also lack diagnostic characters for that species. Anderson and Cruickshank (1978), King (1988, 1992), King and Jenkins (1997), Rubidge (2005), and Fröbisch (2009) included *Diictodon* in the faunal lists they compiled for Zambia. Angielczyk and Sullivan (2008) figured a largely unprepared but diagnostic Zambian *Diictodon* specimen (BP/1/3598).

Emydopoidea (van Hoepen, 1934)
Compsodon helmoedi van Hoepen, 1934
Figure 7.3a–h

Material: NHCC LB13, NHCC LB14.

Localities: Locality L26 (NHCC LB13), locality L45 (NHCC LB14).

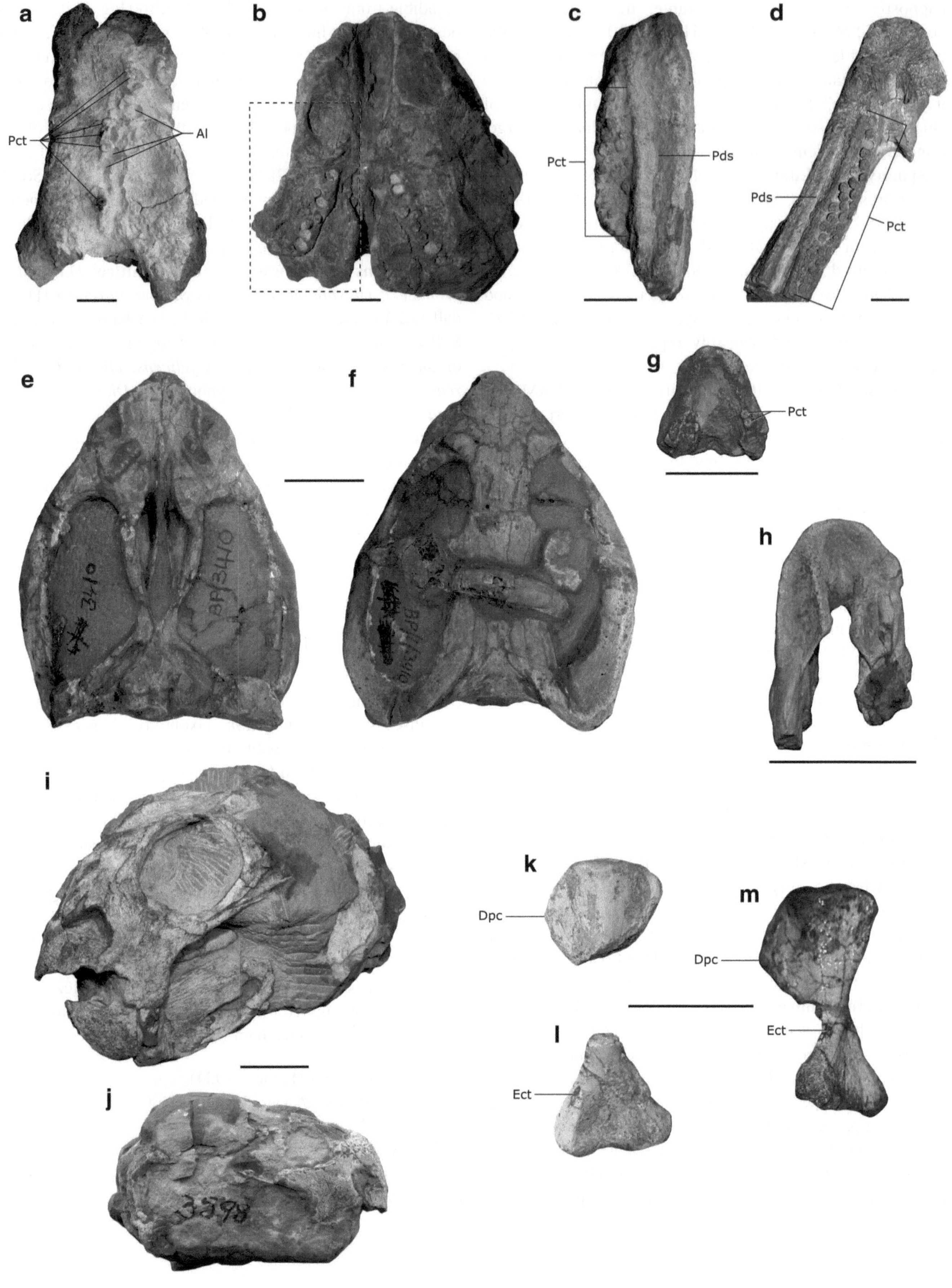
a
b
c
d
Pct
Al
Pct
Pds
Pds
Pct
e
f
g
Pct
h
i
k
Dpc
m
Dpc
l
Ect
Ect
j

◀ **Fig. 7.2** Zambian specimens of *Endothiodon*, *Pristerodon*, and *Diictodon*, and comparative material. **a** Palate fragment of *Endothiodon* sp. from Zambia (NHCC LB12) in ventral view. The fragment includes a partial premaxilla, partial palatine, and several "postcanine" teeth and alveoli. **b** Comparative snout of *Endothiodon uniseries* (NHMUK R4042) from South Africa in ventral view. The *dashed box* shows the section of the palate preserved in NHCC LB12. **c** Partial right dentary of *Endothiodon* sp. from Zambia in dorsal view. **d** Partial mandible of *Endothiodon* sp. from Mozambique (BP/1/5489) in dorsal view. Note the similarity of the location and morphology of the "postcanine" teeth and posterior dentary sulcus to those of NHCC LB12. **e** Skull of *Pristerodon mackayi* from Zambia (BP/1/3410) in ventral view. **f** Skull of *Pristerodon mackayi* from Zambia (BP/1/3410) in dorsal view. **g** Partial mandible of *Pristerodon mackayi* from Zambia (NHCC LB9) in dorsal view. **h** Partial mandible of *Diictodon feliceps* fom Zambia (NHCC LB6) in dorsal view. **i** Partially-prepared skull and mandible of *Diictodon feliceps* from Zambia (NHCC LB3) in anterolateral view. **j** Partially-prepared skull and mandible of *Diictodon feliceps* from Zambia (BP/1/3598) in right lateral view. **k** Proximal end of a left humerus of *Diictodon feliceps* from Zambia (NHCC LB7) in dorsal view. **l** Distal end of a right humerus of *Diictodon feliceps* from Zambia (NHCC LB27) in ventral view. **m** Left humerus of *Diictodon feliceps* from South Africa (CGP STH 36). Note the similarity in shape of the deltopectoral crest to that of NHCC LB7 and the presence of an ectepicondylar foramen. Scale bars are 20 mm

Identifying Characteristics: *Compsodon helmoedi* was described by van Hoepen (1934) based on a small skull (basal length approximately 100 mm) collected in the Karoo Basin of South Africa. Toerien (1954) provided additional information on the holotype and suggested that *C. helmoedi* may be related to *Emydops*, but Cluver and King (1983) and King (1988) treated it as a valid species of uncertain affinities. Brink and Keyser (1983) considered *C. helmoedi* to be a synonym of *Tropidostoma microtrema*, but did not provide justification for this synonymy.

The holotype of *C. helmoedi* (NMQR 1460) (Fig. 7.3a, d, g) is a small, laterally compressed skull with tusks and "postcanine" teeth, relatively large but smooth palatine pads that are pierced by a palatine foramen, and long, straight anterior pterygoid rami that bear prominent, triangular ventral keels. Van Hoepen (1934) stated that a fragment of tooth associated with the specimen was serrated, and speculated that *C. helmoedi* may have had serrated tusks. However, this would be unprecedented among dicynodonts, particularly because serrations are structures associated with enamel whereas dicynodont tusks are composed only of dentine (Camp and Welles 1956; Poole 1956), and the portions of the tusks preserved in situ in NMQR 1460 show no evidence of serrations. The "postcanine" preserved on the left side of the specimen also shows no sign of serrations. An embayment of the palatal rim is present anterior to the caniniform process, and a postcaniniform keel is present. Anterior palatal ridges are absent, but a posterior median ridge is present that is flanked laterally by longitudinal depressions (although the depressions are poorly preserved due to lateral compression). A nasal boss with a continuous posterior border is present on the snout, and the lateral surface of the maxilla bears a distinctive, pocket-like depression posterior to the external naris between the anterior orbital margin and the caniniform process. On the skull roof, the midfrontal suture is slightly raised, and the edges of the orbit are slightly raised above the surface of the frontals. Posterior to the orbital rim, the postfrontals and postorbitals are raised above the posterior portion of the frontals and the preparietals, giving the skull roof in this area a slightly depressed appearance. The preparietal itself is elongate, and roughly triangular in shape, with its apex reaching the parietal foramen. The edges of the preparietal are slightly raised, forming weak ridges that extend to the parietal foramen, which itself is surrounded by a slightly raised lip. The parietals are exposed between the postorbitals on the skull roof, although this exposure appears to be narrower than is the case in *Emydops* or *Pristerodon*. Based on this combination of character states, including the distinctive depression on the lateral surface of the premaxilla and the raised postfrontals and postorbitals, we consider *C. helmoedi* a valid taxon that is likely part of Emydopoidea. A full redescription and investigation of its phylogenetic relationships will be the subject of a subsequent publication.

NHCC LB13 (Fig. 7.3b, e, h) and NHCC LB14 (Fig. 7.3c, f, i) bear a striking resemblance to the type of *C. helmoedi*. NHCC LB13 is almost exactly the same size as NMQR 1460 (basal length approximately 103 mm) and is tusked. "Postcanines" are not preserved, but two empty alveoli are present posterior to the tusk on each side of the skull. Median anterior ridges are absent on the secondary palate, although lateral ridges similar to those found in *Diictodon* or *Emydops* are present. A posterior median ridge is also present, and is flanked by rounded grooves. Interestingly, the anterior portion of the median ridge forms a flattened, Y-shaped expanded area that is reminiscent of that seen in *Eosimops newtoni* (Angielczyk and Rubidge 2013). An embayment on the palatal rim anterior to the caniniform process is present, as is a postcaniniform keel. The palatine pads are relatively large, smooth, and pierced by a palatal foramen, and the left anterior pterygoid ramus preserves a ventral keel nearly identical to that of NMQR 1460 (the right side is damaged). The snout of NHCC LB13 is damaged, but the remains of a median nasal boss appear to be present, and a pocket-like depression is present on the lateral surface of the maxilla. The midfrontal suture is slightly raised, as are the orbital margins, and the postorbitals and postfrontals are raised above the level of the posterior portion of the frontals and the preparietal. The preparietal is of similar shape as that of NMQR 1460, and its edges form distinct ridges that extend to the parietal foramen. The

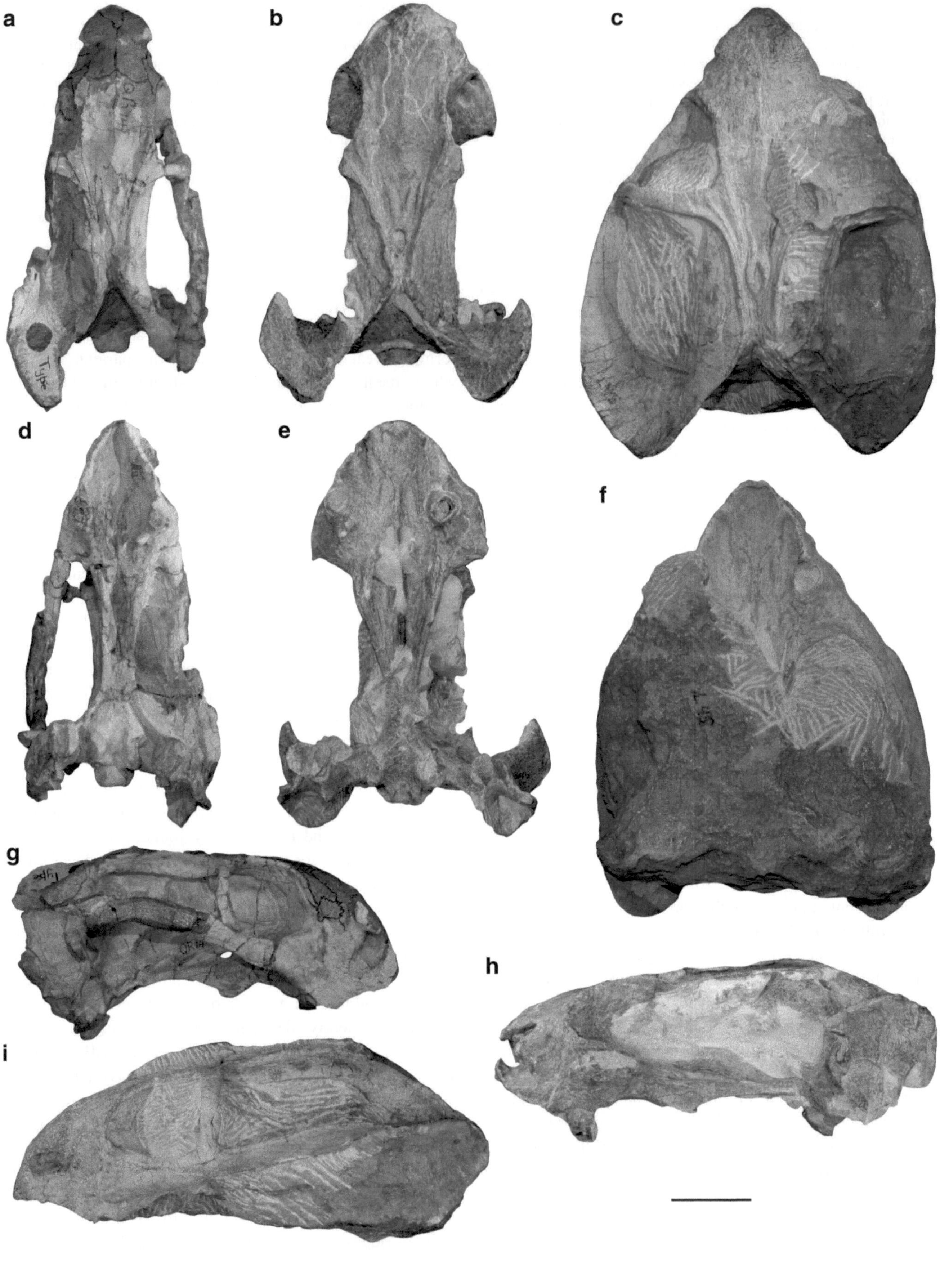
a
b
c
d
e
f
g
h
i

◀ **Fig. 7.3** Zambian specimens of *Compsodon helmoedi* and comparative material. **a** Holotype skull of *Compsodon helmoedi* from South Africa (NMQR 1460) in dorsal view. Note that the specimen has been laterally compressed. **b** Skull of *Compsodon hemoedi* from Zambia (NHCC LB13) in dorsal view. **c** Partially prepared skull of *Compsodon hemoedi* from Zambia (NHCC LB14) in dorsal view. **d** Holotype skull of *Compsodon helmoedi* from South Africa (NMQR 1460) in ventral view. **e** Skull of *Compsodon hemoedi* from Zambia (NHCC LB13) in ventral view. **f** Partially prepared skull of *Compsodon hemoedi* from Zambia (NHCC LB14) in ventral view. **g** Holotype skull of *Compsodon helmoedi* from South Africa (NMQR 1460) in right lateral view. **h** Skull of *Compsodon hemoedi* from Zambia (NHCC LB13) in left lateral view. **i** Partially prepared skull of *Compsodon hemoedi* from Zambia (NHCC LB14) in left lateral view. Scale bar is 20 mm

parietal foramen is surrounded by a slightly raised lip. The postorbitals of NHCC LB13 have a larger exposure on the skull roof than those of NMQR 1460 and overlap the parietals more extensively, but given the otherwise great degree of similarity between the specimens, we interpret this as likely individual variation or preservation differences.

NHCC LB14 is larger than NMQR 1460 (basal length approximately 113 mm), and at the time of writing has only been partially prepared. Part of a tusk is exposed on the right side of the specimen, and empty alveoli for a tusk and at least one "postcanine" are present on the left side. Median anterior palatal ridges are absent, but the lateral anterior ridges are well developed. The posterior median palatal ridge also is present and bears a flattened expanded anterior section similar to NHCC LB13. The posterior median ridge is also flanked by rounded depressions. An embayment of the palatal rim anterior to the caniniform process is present, as is a postcaniniform keel. Only the left palatine pad is exposed, but it has the same shape as that of NHCC LB13 and is pierced by a palatal foramen. A median nasal boss with a continuous posterior border is present on the snout, and the pocket-like depression is well developed on the lateral surface of the maxilla. The midfrontal suture and the orbital margins are raised, and the postfrontals and postorbitals are raised above the level of the frontals and preparietal. The preparietal is triangular with raised edges that continue posteriorly to meet with the raised lip that surrounds the parietal foramen. The parietals are slightly more exposed on the skull roof than in NHCC LB13, giving this region an appearance more similar to that of NMQR 1460.

Synonyms in Luangwa Basin Literature: None.

Previous Reports: *Compsodon* has not been reported previously from the Luangwa Basin.

Emydopidae (van Hoepen, 1934)
Emydops sp.
Figure 7.4a, b, d, f

Material: BP/1/3347, NHCC LB15.

Localities: Locality 4 of Drysdall and Kitching (1963) (BP/1/3347), locality L52 (NHCC LB15).

Identifying Characteristics: Angielczyk et al. (2005) and Fröbisch and Reisz (2008) provided recent reviews of diagnostic characters of *Emydops*. Both specimens are small, possess an intertemporal region in which the parietals are broadly exposed between the postorbitals, and display a squared-off profile of the occiput in posterior view (better exposed in BP/1/3347). The mandible is preserved in NHCC LB15, and although unprepared, it is suggestive of the presence of a prominent lateral dentary shelf and shovel-like symphyseal region. Unfortunately, neither BP/1/3347 or NHCC LB15 is prepared enough at this time to determine whether it represents *Emydops arctatus*, *E. oweni*, or a new species.

Synonyms in Luangwa Basin Literature: *Emydops* sp. indet. (Fröbisch, 2009).

Previous Reports: Boonstra (1938) reported a fragmentary specimen with tusks and "postcanines" that he tentatively identified as *Emydops* or *Emydopsis*. However, as noted above, this specimen (SAM-PK-K7933) is most likely *Pristerodon*. Drysdall and Kitching (1962, 1963) and Kitching (1963) noted field observations of *Emydops* from their Localities 3 and 17, which they considered part of the lower and middle fossiliferous beds of the Madumabisa Mudstone, respectively. Strangely, they did not mention *Emydops* at Locality 4, despite the fact that the only identifiable *Emydops* specimen collected during that fieldwork of which we are aware (BP/1/3347) is from Locality 4. Anderson and Cruickshank (1978), King (1988, 1992), Rubidge (2005), and Fröbisch (2009) included *Emydops* in the faunal lists they presented, undoubtedly based on previous reports.

Kistecephalia Seeley, 1894
Kingoriidae King, 1988
Dicynodontoides cf. *D. nowacki* (von Huene, 1942)
Figure 7.4h, j, l, m

Material: NHMUK R15944, NHCC LB16. NHCC LB17 also may be *Dicynodontoides*, but the specimen is unprepared.

Localities: Locality 4 of Drysdall and Kitching (1963) (NHMUK R15944), locality L29 (NHCC LB16). NHCC LB17 was collected at locality L64.

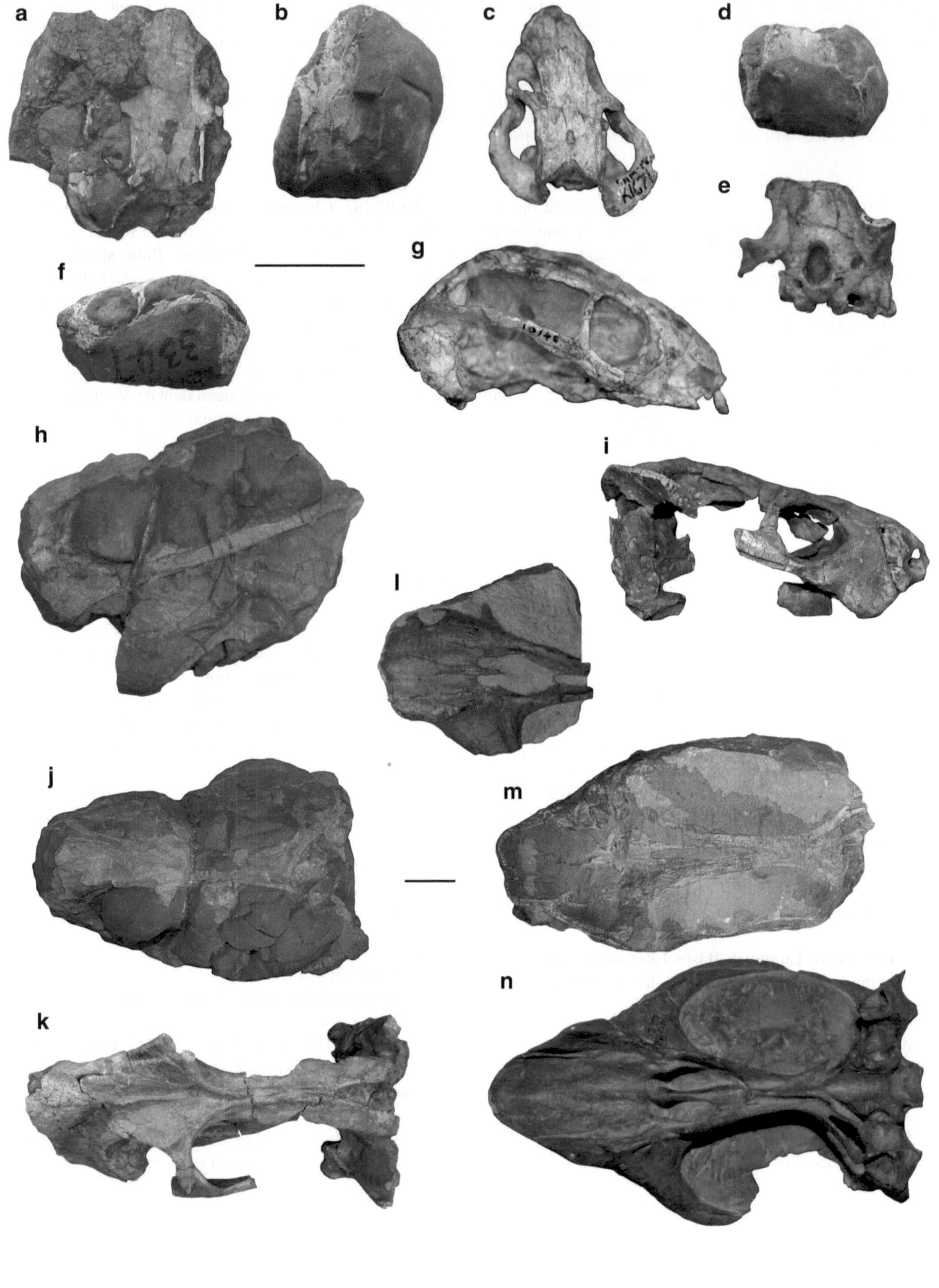
a
b
c
d
e
f
g
h
i
l
j
m
k
n

◀ **Fig. 7.4** Zambian specimens of *Emydops* sp. and *Dicynodontoides* cf. *D. nowacki*, and comparative material. **a** Partially-prepared skull of *Emydops* sp. from Zambia (NHCC LB15) in dorsal view. **b** Partially-prepared skull of *Emydops* sp. from Zambia (BP/1/3347) in dorsal view. **c** Skull of *Emydops arctatus* from South Africa (SAM-PK-K1671) in dorsal view. **d** Partially-prepared skull of *Emydops* sp. from Zambia (BP/1/3347) in posterior view. **e** Skull of *Emydops arctatus* from South Africa (SAM-PK-11060) in posterior view. **f** Partially-prepared skull of *Emydops* sp. from Zambia (BP/1/3347) in left lateral view. **g** Skull of *Emydops arctatus* from South Africa (SAM-PK-10148) in right lateral view. **h** Partially-prepared skull and mandible of *Dicynodontoides* cf. *D. nowacki* from Zambia (NHMUK R15944) in left lateral view. **i** Skull of *Dicynodontoides nowacki* from Tanzania (CAMZM T747) in right lateral view. **j** Partially-prepared skull of *Dicynodontoides* cf. *D. nowacki* from Zambia (NHMUK R15944) in dorsal view. **k** Partial skull of *Dicynodontoides nowacki* from Tanzania (NMT RB2) in dorsal view. **l** Snout of *Dicynodontoides* cf. *D. nowacki* from Zambia (NHCC LB16) in ventral view. **m** Coronal section through the skull of *Dicynodontoides* cf. *D. nowacki* from Zambia (NHCC LB17). Note the long, straight anterior pterygoid rami. **n** Skull of *Dicynodontoides nowacki* from Tanzania (GPIT K12) in ventral view. Note the long, straight anterior pterygoid rami and the similarity of the anterior palate to that of NHCC LB16. *Upper left* scale bar applies to *panels* **a–g**; *lower* scale bar applies to *panels* **h–n**. Scale bars are 20 mm

Identifying Characteristics: Angielczyk et al. (2009) provided a revised diagnosis for the two valid species of *Dicynodontoides*. NHMUK R15944 (Fig. 7.4h, j) is a poorly preserved specimen that is preserved in a hematitic nodule, and it has undergone only rudimentary preparation. We refer the specimen to *Dicynodontoides* primarily based on the absence of a postfrontal, the apparent occlusion of the mandibular fenestra by a lamina of the dentary, and the preserved anterior portion of the dentary seeming to be consistent with the original presence of an elongate, shovel-shaped symphysis. Our identification of NHCC LB16 (Fig. 7.4l) as *Dicynodontoides* is based on the absence of median anterior ridges on the palate, the presence of lateral anterior palatal ridges, the presence of a posterior median ridge flanked by elongate depressions, the presence of an embayment of the palatal rim anterior to the caniniform process, the presence of a postcaniniform keel, the absence of "postcanines," the presence of very small, smooth palatine pads, and the absence of postfrontals on the skull roof. NHCC LB17 (Fig. 7.4m) is a skull and lower jaw that are preserved in a nodule. The nodule was recently burned when collected, and appears to have split open during the burning process, exposing a coronal section through the palate. NHCC LB17 is tusked, and possesses the very long, straight anterior pterygoid rami that are typical of *Dicynodontoides* (e.g., Fig. 7.4n). However, this identification must remain tentative until the specimen is more fully prepared. It is difficult to assign any of the specimens to one of the two species of *Dicynodontoides* with certainty. However, given that two of the three specimens are tuskless and all are relatively large (particularly NHCC LB16) they may be part of *D. nowacki* since that species seems to have attained large sizes and was more frequently tuskless than *D. recurvidens* (Angielczyk et al. 2009).

Synonyms in Luangwa Basin Literature: None.

Previous Reports: Angielczyk et al. (2009) were the first to report the presence of *Dicynodontoides* in the Luangwa Basin, based on NHMUK R15944. Gale (1989) referred an assemblage of juvenile dicynodonts to *Dicynodon clarencei*, a synonym of *Dicynodontoides recurvidens* (Angielczyk et al. 2009). These specimens do not appear to represent *Dicynodontoides*, but their exact identification is uncertain (see above).

Cistecephalidae Broom, 1903
New Taxon
Figure 7.5a–j

Material: BP/1/3337, BP/1/3591, BP/1/3603, NHCC LB18, NHCC LB19. According to a handwritten note in the BP collections by J.W. Kitching, dated October 6, 1992, BP/1/3437 may represent a sixth specimen. However, we have been unable to locate this specimen and assume that it is lost.

Localities: Locality 4 of Drysdall and Kitching (1963) (BP/1/3337, BP/1/3591, BP/1/3603), locality L53 (NHCC LB18), locality L55 (NHCC LB19). BP/1/3437 was collected at Locality 5 of Drysdall and Kitching (1963).

Identifying Characteristics: Following Freeman (1993), we consider these specimens to represent a new cistecephalid taxon. Their cistecephalid affinities are indicated by their anteroposteriorly short skulls with broad intertemporal regions, the presence of a stapedial foramen, the absence of a preparietal, the absence of an interpterygoid vacuity, and large olecranon process of the ulna (see Kammerer and Angielczyk 2009 for a compilation of cistecephalid apomorphies). Freeman (1993) hypothesized that the specimens represented a new species of *Cistecephalus*, but we are hesitant to endorse this conclusion until the specimens are formally described and included in a phylogenetic analysis. Nevertheless, they do appear to differ in several ways from the three currently recognized cistecephalids, *Cistecephalus microrhinus*, *Cistecephaloides boonstrai*, and *Kawingasaurus fossilis*.

The most obvious diagnostic character of the Zambian cistecephalid is the presence of tusks, whereas all other cistecephalid species are tuskless (e.g., Cox 1972; Keyser 1973b; Cluver 1974a). Three of the specimens (BP/1/3337, BP/1/3591, and BP/1/3603) (Fig. 7.5f) possess tusks, one specimen NHCC LB18 (Fig. 7.5e) possesses empty tusk alveoli, and one specimen is tuskless (NHCC LB19) (Fig. 7.5g). We suspect that the presence of empty alveoli in NHCC LB18 likely represents a taphonomic artifact instead

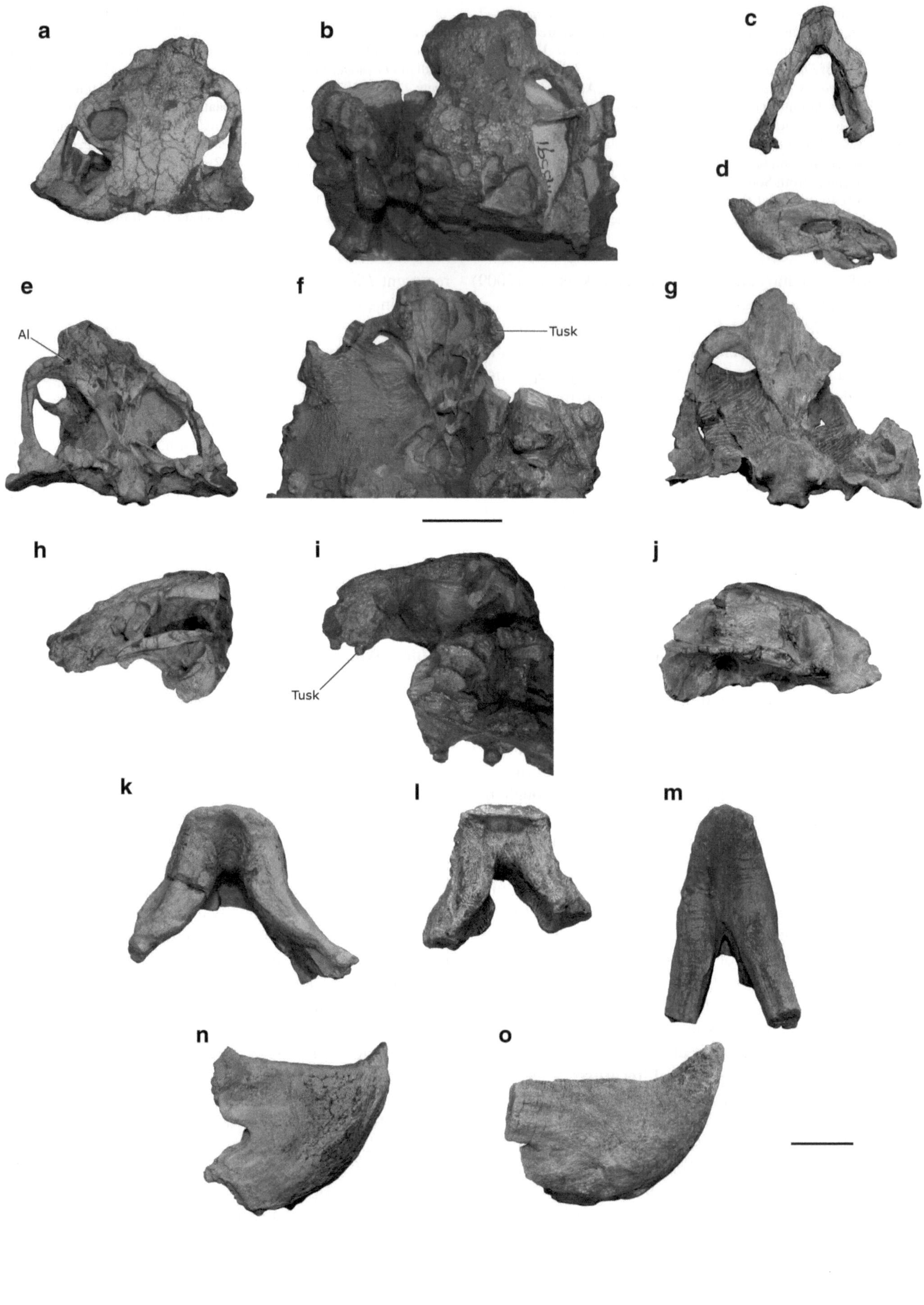
a
b
c
d
e
f
g
Al
Tusk
h
i
j
Tusk
k
l
m
n
o

◀ **Fig. 7.5** Zambia specimens of Cistecephalidae n. g. & sp. and cf. *Katumbia parringtoni*, and comparative material. **a** Skull of Cistecephalidae n. g. & sp. from Zambia (NHCC LB18) in dorsal view. **b** Partially-prepared skull of Cistecephalidae n. g. & sp. from Zambia (BP/1/3591) in dorsal view. **c** Mandible of Cistecephalidae n. g. & sp. from Zambia (NHCC LB18) in dorsal view. **d** Mandible of Cistecephalidae n. g. & sp. from Zambia (NHCC LB18) in left lateral view. **e** Skull of Cistecephalidae n. g. & sp. from Zambia (NHCC LB18) in ventral view. Note the presence of an empty tusk alveolus on the right maxilla. **f** Partially-prepared skull of Cistecephalidae n. g. & sp. from Zambia (BP/1/3591) in ventral view. Note the tusk in the left maxilla. **g** Partial skull of Cistecephalidae n. g. & sp. from Zambia (NHCC LB19) in ventral view. Note that this specimen is tuskless. **h** Skull of Cistecephalidae n. g. & sp. from Zambia (NHCC LB18) in left lateral view. **i** Partially-prepared skull of Cistecephalidae n. g. & sp. from Zambia (BP/1/3591) in left lateral view. **j** Skull of Cistecephalidae n. g. & sp. from Zambia (NHCC LB19) in right lateral view. **k** Partial mandible of cf. *Katumbia parringtoni* from Zambia (NHCC LB20) in dorsal view. **l** Partial mandible of *Katumbia parringtoni* from Tanzania (CAMZM T791) in dorsal view. **m** Partial mandible of *Oudenodon bainii* (NMT RB37) in dorsal view. Note that the symphyseal region is longer in NMT RB37 than in NHCC LB20 and CAMZM T791, and that the dentary rami diverge at a shallower angle in NMT RB37. **n** Partial mandible of cf. *Katumbia parringtoni* from Zambia (NHCC LB20) in right lateral view. **o** Partial mandible of *Oudenodon bainii* (NMT RB37) in right lateral view. Note the shorter, more sharply-upturned symphysis in NHCC LB20). *Upper* scale bar applies to *panels* **a–j**; *lower right* scale bar applies to *panels* **k–o**. Scale bars are 20 mm

of tooth replacement or another biological process, since both specimens of *Compsodon* we collected feature empty tusk or "postcanine" alveoli, and NHCC LB14 in particular shows a preservation style that is extremely similar to NHCC LB18. The absence of tusks in NHCC LB19 may represent sexual dimorphism or another form of polymorphism. Such variability is not surprising given that several other dicynodonts recently have been shown to be sexually dimorphic or polymorphic for tusks (e.g., Angielczyk 2002; Sullivan et al. 2003; Botha and Angielczyk 2007; Fröbisch and Reisz 2008; Angielczyk et al. 2009), but a larger sample of specimens will be needed to determine whether sexual dimorphism is a likely cause.

The new Zambian cistecephalid can be further distinguished from *Cistecephalus microrhinus* by the absence of a depression or notch on the ventral surface of the maxilla lateral to the caniniform process (see Cluver 1974b), the presence of a small, triangular, ventrally-directed flange on the anterior pterygoid ramus, a mid-ventral vomerine plate that is wide and trough-like anteriorly, and a more robust, block-like crista oesophagea on the median pterygoid plate. It can be distinguished from *Cistecephaloides boonstrai* by the presence of a single embayment anterior to the caniniform process, the presence of a small, triangular, ventrally-directed flange on the anterior pterygoid ramus, a mid-ventral vomerine plate that is wide and trough-like anteriorly, a robust, block-like crista oesophagea on the median pterygoid plate, a larger lateral dentary shelf, and the absence of a tall cutting blade on the dorsal surface of the dentary near the level of the lateral dentary shelf. Finally, it can be distinguished from *Kawingasaurus fossilis* by larger size, the presence of a small, triangular, ventrally-directed flange on the anterior pterygoid ramus, a mid-ventral vomerine plate that is wide and trough-like anteriorly, and a robust, block-like crista oesophagea on the median pterygoid plate. The only mandible of *K. fossilis* (GPIT K55f) is poorly preserved, but the Zambian cistecephalid may additionally differ from this species in the presence of a posterior dentary sulcus and a larger lateral dentary shelf.

Synonyms in Luangwa Basin Literature: *Cistecephalus*, *Cistecephalus microrhinus*, *Cistecephalus planiceps* (Drysdall and Kitching 1962, 1963; Kitching 1963; Anderson and Cruickshank 1978; Cooper 1982; King 1988, 1992; Smith and Keyser 1995; Lucas 2002, 2005, 2006; Angielczyk 2002; Rubidge 2005; Fröbisch 2009).

Previous Reports: Drysdall and Kitching (1962, 1963) and Kitching (1963) were the first to report *Cistecephalus* from the Luangwa Basin. Specifically, they reported at least 13 specimens from localities in their middle and upper fossiliferous beds, but most of these occurrences appear to represent field identifications because they provided no specimen numbers or photographs. The only Zambian specimens in the BP collection that could be mistaken for *Cistecephalus* are BP/1/3337, BP/1/3591, and BP/1/3603, so we consider Drysdall and Kitching's (1962, 1963) and Kitching's (1963) reports to instead represent this new taxon. Numerous authors have cited Drysdall and Kitching's papers as a basis for including *Cistecephalus* in the Zambian dicynodont fauna (Anderson and Cruickshank 1978; Cooper 1982; King 1988, 1992; Smith and Keyser 1995; Lucas 2002, 2005, 2006; Angielczyk 2002; Rubidge 2005; Fröbisch 2009), but only one author (Freeman 1993) recognized that the specimens represented a new taxon.

Bidentalia Owen, 1876
cf. *Katumbia parringtoni* (von Huene, 1942)
Figure 7.5k, n

Material: NHCC LB20.

Localities: Locality L59 (NHCC LB20).

Identifying Characteristics: NHCC LB20 consists of the symphyseal region of a dicynodont jaw that was collected as float. It is noteworthy in possessing extremely short dentary tables, dentary rami that strongly diverge posteriorly, and an upturned anterior margin of the symphysis that forms a relatively thin edge (Fig. 7.5k, n). In these characters, the specimen is very similar to the jaw of *Katumbia parringtoni* (Fig. 7.5l; also see Angielczyk 2007), but differs from jaws of most dicynodonts, which have

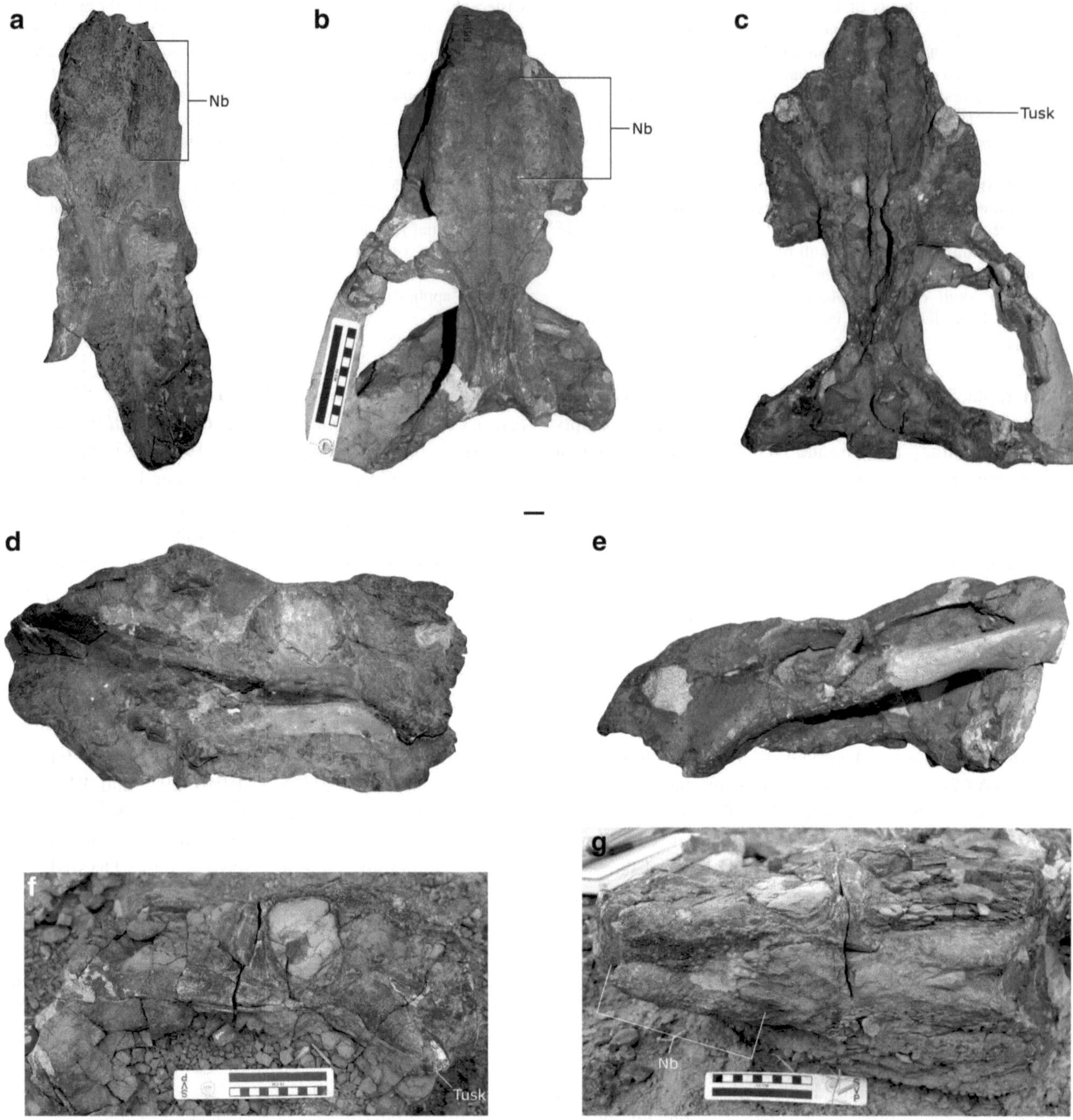

Fig. 7.6 Zambian specimens of *Odontocyclops whaitsi*. **a** Partial skull of *Odontocyclops whaitsi* (holotype of *Rhachiocephalus dubius*) (SAM-PK-11313) in dorsal view. **b** Partial skull of *Odontocyclops whaitsi* (BP/1/3419) in dorsal view. Note the elongate nasal bosses in this specimen and in SAM-PK-11313. **c** Partial skull of *Odontocyclops whaitsi* in ventral view. Note the presence of tusks. **d** Partial skull and mandible of *Odontocyclops whaitsi* (holotype of *Rhachiocephalus dubius*) (SAM-PK-11313) in right lateral view. **e** Partial skull of *Odontocyclops whaitsi* (BP/1/3419) in left lateral view. **f** Field photograph of partial skull of *Odontocyclops whaitsi* (NHCC LB24) in right lateral view. Note the presence of a tusk. **g** Field photograph of partial skull of *Odontocyclops whaitsi* (NHCC LB24) in dorsal view. Note the presence of elongate nasal bosses similar to those of BP/1/3419 and SAM-PK-11313. *Upper* scale bar applies to *panels* **a–e** and is 20 mm. Scale bar in field photographs is 100 mm

proportionally longer dentary tables (e.g., *Oudenodon bainii*; Fig. 7.5m, o). Therefore, we tentatively refer the specimen to *K. parringtoni*, although confirmation of the presence of this taxon in the fauna of the Upper Madumabisa Mudstone must await more complete and diagnostic material.

Synonyms in Luangwa Basin Literature: None.

Previous Reports: There are no previous reports of *K. parringtoni* in the Luangwa Basin.

Cryptodontia Owen, 1860a
Odontocyclops whaitsi (Broom, 1913)
Figure 7.6a–g

Material: BP/1/3244, BP/1/3419, BP/1/3585, BP/1/3586, BP/1/3587, BP/1/3589, NHCC LB24, SAM-PK-11313. SAM-PK-K7936 also likely represents *O. whaitsi*, although poor preservation of the specimen makes this identification tentative.

Localities: Locality 4 of Drysdall and Kitching (1963) (BP/1/3244, BP/1/3419, BP/1/3585, BP/1/3586, BP/1/3587, BP/1/3589, SAM-PK-11313), locality L39 (NHCC LB24). SAM-PK-K7936 was collected at Locality 3 of Drysdall and Kitching (1963).

Identifying Characteristics: Boonstra (1938), Keyser (1979), Keyser and Cruickshank (1979), and Angielczyk (2002) provided diagnoses of *O. whaitsi* based primarily on Zambian material, although Broom's (1913) initial description of the species was based on a South African specimen. The most distinctive autapomorphies of *O. whaitsi* are the elongate nasal bosses that extend from the posterodorsal corner of the external nares to contact the prefrontal bosses, and a concave dorsal surface of the snout between the nasal bosses (Fig. 7.6a, b, g). All of the specimens listed above except SAM-PK-K7936 show this character. Additional characters diagnostic of *Odontocyclops* include large size, variable presence of tusks, presence of a postcaniniform crest, absence of a labial fossa, and a relatively narrow temporal bar in which the parietals are well exposed between the postorbitals. SAM-PK-K7936 displays these characters, and the presence of tusks in that specimen allow it to be differentiated from similarly-sized *Oudenodon* specimens.

Synonyms in Luangwa Basin Literature: *Rhachiocephalus dubius*, *Odontocyclops dubius* (Boonstra 1938; Drysdall and Kitching 1963; Kitching 1963; Keyser 1979; Keyser and Cruickshank 1979). Boonstra (1938) identified SAM-PK-K7936 as *Dicynodon* cf. *D. breviceps*, and if this specimen is indeed *O. whaitsi*, then this would be an additional synonym. Cluver and King (1983) suggested that *Odontocyclops* was likely a synonym of *Dicynodon*, and King (1988) listed the genus *Odontocyclops* as synonym of *Dicynodon*. However, she included the species *Rhachiocephalus dubius* Boonstra, 1938 as a synonym of *Rhachiocephalus magnus*, despite the fact that Keyser (1979) and Keyser and Cruickshank (1979) used the former as the type species of *Odontocyclops*. As noted below, at least some of Kitching's (1962, 1963) and Drysdall and Kitching's (1963) field reports of *Aulacephalodon* likely represent *Odontocyclops*.

Previous Reports: Boonstra's (1938) description of "*Rhachiocephalus*" *dubius* is the first report of *O. whaitsi* from the Luangwa Basin. Drysdall and Kitching (1963), Kitching (1963), Cooper (1982), Anderson and Cruickshank (1978), Keyser (1979), Keyser and Cruickshank (1979), Angielczyk (2002), and Fröbisch (2009) all discuss Luangwa Basin specimens of *O. whaitsi* using various names.

Oudenodontidae Cope, 1871
Oudenodon bainii Owen, 1860b
Figure 7.7j–n

Material: BP/1/3420, NHCC LB21, NHCC LB22, SAM-PK-11310, SAM-PK-11312, SAM-PK-11316, SAM-PK-11319, SAM-PK-K7940, SAM-PK-K7941, SAM-PK-K7944, TSK 67, TSK 69, TSK 70, TSK 95, TSK 101, TSK 107. NHCC LB23, SAM-PK-K7934, SAM-PK-K7943, SAM-PK-K7947, TSK 103, TSK 112 also are likely specimens of *O. bainii*, but incompleteness and/or lack of preparation make these identifications tentative.

Localities: Locality 4 of Drysdall and Kitching (1963) (BP/1/3420, SAM-PK-11310, SAM-PK-11312, SAM-PK-11316, SAM-PK-11319, SAM-PK-K7940, SAM-PK-K7941, SAM-PK-K7944), locality L30 (NHCC LB21), locality L37 (NHCC LB22), Kerr's (1974) Locality 11 (TSK 67, TSK 69, TSK 70, TSK 95, TSK 101, TSK 107). NHCC LB23 was collected at locality L59. TSK 103 was collected at Kerr's (1974) Locality 11, and TSK 112 was collected at Kerr's (1974) Locality 13. The locality information for SAM-PK-K7943 and SAM-PK-K7947 is: "Probably from horizon in Upper Green Marls. About 1 ¾ miles South of Mpundu. Horizon 4" (Boonstra 1938, p. 37), which would correspond to Drysdall and Kitching's (1963) Locality 4. Specific locality information is not available for SAM-PK-K7943; the SAM collections database only states that it is from the Luangwa Valley (S. Kaal, personal communication, 2010).

Identifying Characteristics: Keyser (1975), Cluver and Hotton (1981), and Botha and Angielczyk (2007) are the most detailed recent papers to address the ways in which *Oudenodon bainii* can be differentiated from other dicynodonts, and we follow the latter's hypothesis that Zambian *Oudenodon* specimens most likely represent *O. bainii* and not a distinct species. The Zambian specimens we refer to *O. bainii* vary widely in the quality of their preservation and the degree to which they have been prepared. The following characters are ones we focused on for identifying *O. bainii* specimens, although not all are preserved or visible in all specimens: medium size; thin ridge present on anterior surface of premaxilla; paired nasal bosses that are rounded and overhang the external nares; pineal boss absent; temporal bar in which the parietals are exposed between the postorbitals and are often slightly depressed relative to the postorbitals; tusks and "postcanines" absent;

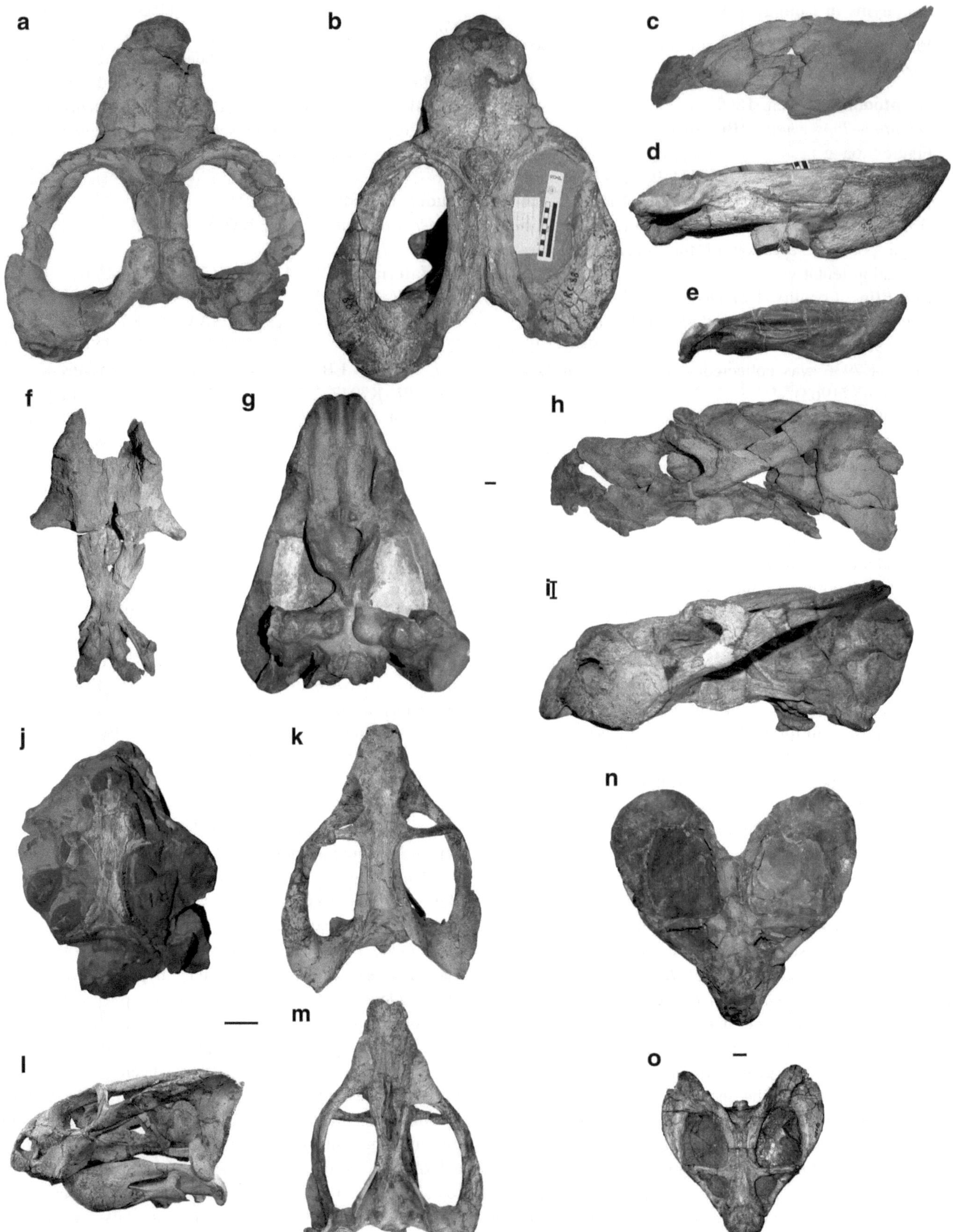
a
b
c
d
e
f
g
h
i
j
k
l
m
n
o

◀ **Fig. 7.7** Zambian specimens of *Kitchinganomodon crassus* and *Oudenodon bainii*, and comparative material. **a** Skull of *Kitchinganomodon crassus* from Zambia (TSK 23) in dorsal view. **b** Holotype skull of *Kitchinganomodon crassus* from South Africa (RC 88) in dorsal view. **c** Mandible of *Kitchinganomodon crassus* from Zambia (TSK 23) in right lateral view. **d** Holotype mandible of *Kitchinganomodon crassus* from South Africa (RC 88) in right lateral view. **e** Mandible of *Rhachiocephalus magnus* from Tanzania [GPIT K30 g(uk)] in right lateral view. Note the steeper angulation of the posteroventral corner of the dentary in the *Kitchinganomodon* specimens. **f** Skull of *Kitchinganomodon crassus* from Zambia (TSK 23) in ventral view. Note that this photograph was taken before the specimen was completely reassembled. **g** Skull of *Kitchinganomodon crassus* from South Africa (BP/1/819) in ventral view. Note the robust anterior pterygoid rami in this specimen and in TSK 23. **h** Skull of *Kitchinganomodon crassus* from Zambia (TSK 23) in left lateral view. **i** Skull of *Kitchinganomodon crassus* from South Africa (RC 88) in left lateral view. **j** Unprepared partial skull of *Oudenodon bainii* (holotype of *Dicynodon luangwanensis*) from Zambia (SAM-PK-11310) in dorsal view. **k** Skull of *Oudenodon bainii* from Zambia (TSK 67) in dorsal view. **l** Skull and mandible of *Oudenodon bainii* from Zambia (TSK 67) in left lateral view. **m** Skull of *Oudenodon bainii* from Zambia (TSK 67) in ventral view. **n** Skull of *Oudenodon bainii* (holotype of *Dicynodon helenae*) from Zambia (SAM-PK-11312) in dorsal view, showing the heart shape sometimes used as a character to differentiate Zambian *Oudenodon* specimens from South African specimens. **o** Skull of *Oudenodon bainii* from South Africa (CGP MIF 133) in dorsal view. Note that this specimen shows a similar heart shape in dorsal view as SAM-PK-11312. *Upper* scale bar applies to *panels* **a–i**, *lower left* scale bar applies to *panels* **j–m**, and *lower right* scale bar applies to *panels* **n–o**. Scale bars are 20 mm. Photographs in *panels* **a** and **h** courtesy of C. Kammerer; photographs in *panels* **k–m** courtesy of S. Jasinoski

postcaniniform crest present; labial fossa surrounded by the palatine, maxilla, and jugal absent; palatal surface of the palatine possesses raised rugose posterior section and a smoother anterior section that is flush with the secondary palate; interpterygoid vacuity relatively long, reaching the level of the palatal surface of the palatines; mid-ventral plate of vomer narrow and blade-like in ventral view; intertuberal ridge between basioccipital tubera absent.

Keyser (1975) provided a detailed justification for why the holotypes of four species described by Boonstra (1938) (SAM-PK-11301, type of *Dicynodon luangwanensis* (Fig. 7.7j, n); SAM-PK-11312, type of *Dicynodon helenae*; SAM-PK-11316, type of *Dicynodon euryceps*; SAM-PK-11319, type of *Dicynodon parabreviceps)* are best regarded as part of *Oudenodon*. He favored retaining them as a species (*O. luangwanensis*) distinct from South African *O. bainii* on account of the wide zygomatic arches that give the skull a heart shape in dorsal view. We do not consider this to be a valid diagnostic character because a similar morphology can be found among South African specimens (Fig. 7.7o), and the morphometric results of Botha and Angielczyk (2007) are consistent with the presence of only one species.

The remaining specimens show varying numbers of the characters listed above depending on the quality of their preservation and the degree to which they have been prepared (e.g., compare Fig. 7.7j–m). For example, TSK 67 is extremely well preserved and completely prepared, and it shows all of the diagnostic characters listed above (Fig. 7.7k, l, m). TSK 101 is a nearly complete skull, but it is mostly unprepared. Nevertheless, it possesses a thin snout ridge, rounded nasal bosses, and a temporal bar in which the parietals are exposed but depressed below the level of the postorbitals; tusks are absent, but a postcaniniform crest is present. NHCC LB21 also is relatively complete, but even though it is entirely unprepared, the snout ridge, nasal bosses, temporal bar, and caniniform processes are sufficiently visible to confirm that it is *O. bainii*. TSK 69 consists only of a snout, but it is well-prepared, and paired, rounded nasal bosses, a median snout ridge, palatines with rugose posterior surfaces and smoother, flush anterior surfaces, postcaniniform crest, and absence of tusks can all be easily observed. Finally, TSK 103 and NHCC LB23 are examples of specimens that can only be tentatively identified as *O. bainii*. TSK 103 is a well preserved and prepared occiput that includes part of the temporal bar and part of the zygomatic portion of the squamosal. The size of the specimen and the morphology of the preserved portion of the temporal bar are consistent with it representing *O. bainii*, but it presents too few diagnostic features to confirm its identity. Likewise, NHCC LB23 is a fragmentary, unprepared snout. The specimen is tuskless, and the general shapes of the snout and caniniform process are consistent with *O. bainii*, but additional preparation would be necessary to confirm this identification.

Synonyms in Luangwa Basin Literature: *Dicynodon lutriceps*, *Dicynodon* cf. *breviceps*, *Dicynodon corstorphinei*, *Dicynodon* cf. *corstorphinei*, *Dicynodon* cf. *milletti*, *Dicynodon latirostris*, *Dicynodon luangwanensis*, *Dicynodon helenae*, *Dicynodon euryceps*, *Dicynodon parabreviceps*, *Oudenodon luangwanensis* (Boonstra 1938; Drysdall and Kitching 1963; Kitching 1963; Keyser 1975; Fröbisch 2008, 2009). The majority of these species are based on types from the Karoo Basin of South Africa, and justifications of these synonymies can be found in Keyser (1975). Two misspellings of the species name *luangwanensis* Boonstra, 1938 are commonly used in the literature with the genus name *Oudenodon*: *O. luangwaensis* (Drysdall and Kitching 1963; Kitching 1963; Keyser 1972, 1975; Cluver and Hotton 1981; Jacobs et al. 2005) and *O. luangwensis* (King 1988; Botha and Angielczyk 2007).

Previous Reports: Boonstra (1938) made the first report of material from the Luangwa Basin that was eventually referred to *O. bainii*. Drysdall and Kitching (1962, 1963),

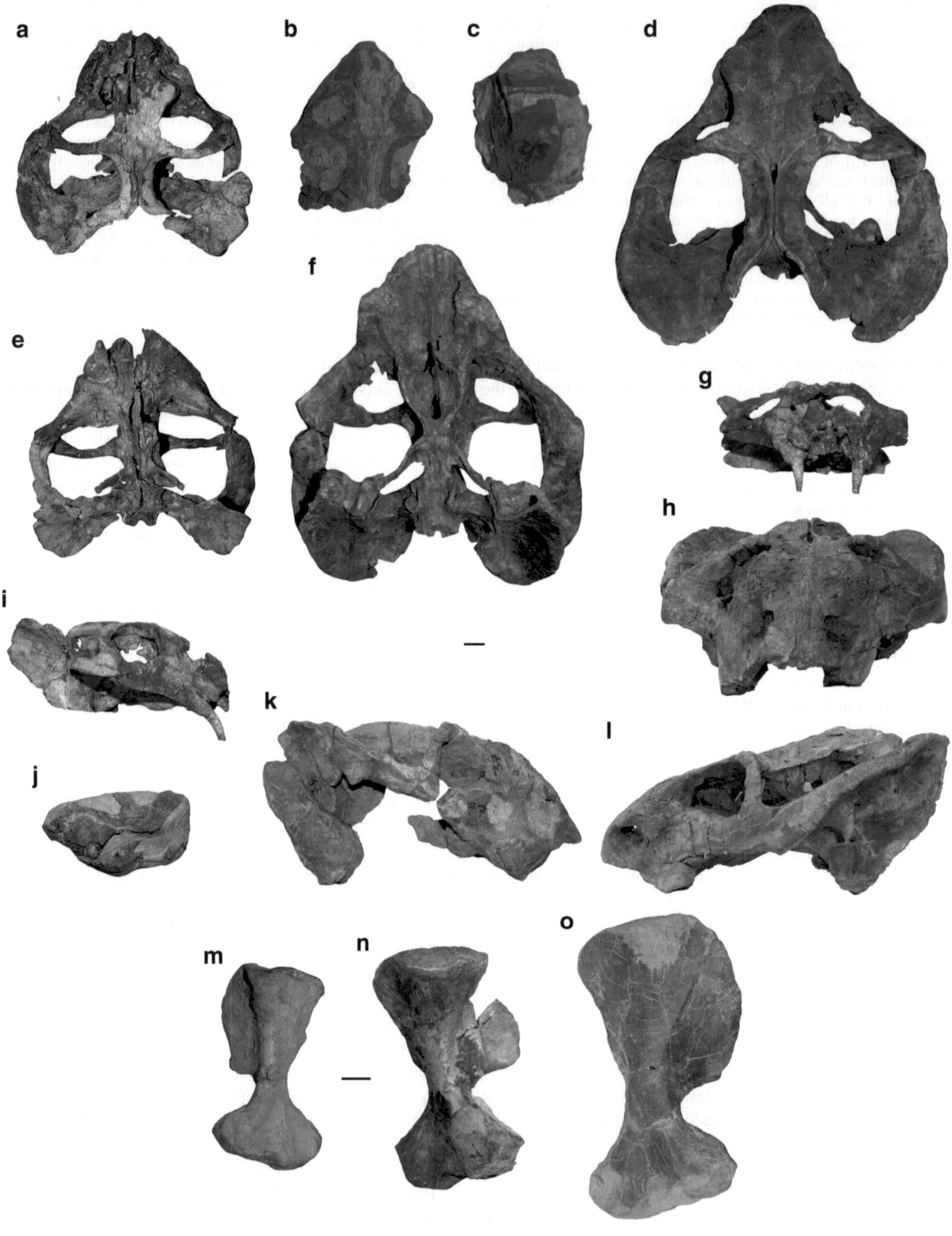
a
b
c
d
e
f
g
h
i
j
k
l
m
n
o

◀ **Fig. 7.8** Zambian specimens of *Dicynodon huenei* and comparative material. **a** Skull of *Dicynodon huenei* from Zambia (TSK 14) in dorsal view. **b** Unprepared skull likely of *Dicynodon huenei* from Zambia (NHMUK field number 5-10) in dorsal view. **c** Unprepared partial skull likely of *Dicynodon huenei* from Zambia (NHMUK field number 5-2) in left dorsolateral view. NHMUK field number 5-2 and NHMUK field number 5-10 are the primary evidence for the presence of *Dicynodon huenei* in the northern sub-basin of the Luangwa Basin. **d** Skull of *Dicynodon huenei* from Tanzania (CAMZM T1089) in dorsal view. **e** Skull of *Dicynodon huenei* from Zambia (TSK 14) in ventral view. **f** Skull of *Dicynodon huenei* from Tanzania (CAMZM T1089) in ventral view. **g** Skull of *Dicynodon huenei* from Zambia (TSK 14) in anterior view. **h** Skull of *Dicynodon huenei* from Tanzania (CAMZM T1089) in anterior view. Note the expanded suborbital bar and plate-like distal end of the postorbital bar in both specimens. **i** Skull of *Dicynodon huenei* from Zambia (TSK 14) in right lateral view. **j** Unprepared skull of *Dicynodon huenei* from Zambia (TSK 27) in anterolateral view. **k** Skull of *Dicynodon huenei* from Zambia (TSK 37) in right lateral view. **l** Skull of *Dicynodon huenei* from Tanzania (CAMZM T1089) in left lateral view. Note the thickened anterior portion of the zygomatic arch, especially in TSK 14, TSK 27, and CAMZM T1089. **m** Left humerus of *Dicynodon huenei* from Zambia (TSK 14) in dorsal view. **n** Right humerus of *Dicynodon huenei* from Zambia (TSK 37) in dorsal view. **o** Right humerus of *Dicynodon huenei* from Tanzania (NMT RB44) in dorsal view. *Upper* scale bar applies to *panels* **a–l**; *lower* scale bar applies to *panels* **m–o**. Scale bars are 20 mm

Kitching (1963), Keyser (1972, 1975), Anderson and Cruickshank (1978), Cluver and Hotton (1981), Cooper (1982), King (1988, 1992), King and Jenkins (1997), Angielczyk (2002), Rubidge (2005), and Fröbisch (2009) all mentioned the presence of *Oudenodon* in the Luangwa Basin, although Drysdall and Kitching (1963) and Kitching (1963) primarily did so using synonyms.

Rhachiocephalidae Maisch, 2000
Kitchinganomodon crassus Maisch, 2002a
Figure 7.7a, c, f, h

Material: TSK 23.

Localities: Kerr's (1974) Locality 2 (TSK 23).

Identifying Characteristics: Maisch (2002a; also see Maisch 1999) erected the genus *Kitchinganomodon* and identified characters that can be used to distinguish *K. crassus* from *Rhachiocephalus magnus*. Among Maisch's (2002a) characters, TSK 23 possesses broadened anterior pterygoid rami, a wide, robust snout, and large nasal bosses that are located directly above the external nares. A rod of bone that seems to be formed by the vomer also extends along the midline of the interpterygoid vacuity, much like the condition in *K. crassus* that Maisch (2002a) described as closure of the interpterygoid vacuity by the vomer. Maisch's (2002a) remaining diagnostic characters are difficult to assess because of preservation, but TSK 23 is suggestive of the presence of at least some of these (e.g., the extensive ossification of the lateral wall of the braincase). In addition, the overall shape of the skull of TSK 23 (Fig. 7.7a), especially in dorsal view, shows a strong resemblance to the type of *K. crassus* (Fig. 7.7b). Finally, the shape of the mandible in lateral view in TSK 23 (Fig. 7.7c) is much more similar to *K. crassus* (Fig. 7.7d) than that of *R. magnus* (Fig. 7.7e). The symphyseal region of the dentary is much deeper than the postdentary bones in TSK 23 and RC 88 (the holotype of *K. crassus*), with the ventral and posterior edges of the symphyseal region meeting in a sharp corner. In contrast, although symphyseal region of *R. magnus* also is deeper than the postdentary bones, the disparity is not as marked, and the posterior and ventral edges form a much more obtuse angle.

Synonyms in Luangwa Basin Literature: Strictly speaking, there are no synonyms of *Kitchinganomodon crassus* in the Luangwa Basin literature. However, Angielczyk and Kurkin (2003), Angielczyk (2007), and Angielczyk and Rubidge (2010) included TSK 23 among the *Rhachiocephalus* specimens they consulted for character state codings in their phylogenetic analyses.

Previous Reports: There are no previous reports of *Kitchinganomodon crassus* from Zambia.

Dicynodontoidea (Owen, 1860a)
Dicynodon huenei Haughton, 1932
Figure 7.8a–c, e, g, i–k, m, n

Material: TSK 14, TSK 27, TSK 37. TSK 40 may represent *D. huenei*, but is mostly unprepared. TSK 83 includes several juvenile specimens that were described by Gale (1988) and referred to *Diictodon*, but this identification has been questioned (Angielczyk and Sullivan 2008; also see above). *Dicynodon huenei* may be a better identification but is somewhat tentative because independent data on the earliest ontogenetic stages of this taxon are unavailable. Three specimens in the NHMUK that were collected by the 1963 expedition (field numbers 5-2, 5-4, 5-10) may represent *D. huenei*, but these specimens are unprepared.

Localities: Kerr's (1974) Locality 1 (TSK 14), Kerr's (1974) Locality 3 (TSK 27, TSK 37). TSK 40 was collected at Kerr's (1974) Locality 7. TSK 83 was collected at "Locality 14." This may correspond to Kerr's (1974) Locality 6 (because the locality is described as producing several small dicynodont skulls in that reference), but this is uncertain. The NHMUK specimens with field numbers were collected at Locality 5 of Drysdall and Kitching (1963).

Identifying Characteristics: Kammerer et al. (2011) discussed the basis for recognizing *Dicynodon huenei* as a distinct and valid species. TSK 14 was described in detail by King (1981), and she referred the specimen to *Dicynodon trigonocephalus*. Kammerer et al. (2011) concluded that the holotype of *D. trigonocephalus* (RC 38) is a somewhat distorted juvenile of *Dicynodon lacerticeps*, and that *D. lacerticeps* and *D. huenei* are closely related. Two of the primary distinguishing features of *D. huenei* are an autapomorphic thickening of the zygomatic arch, such that the structure appears somewhat flattened in lateral view, and twisting and widening of the postorbital bar, such that its distal end forms a mediolaterally-oriented, flattened plate on the zygomatic arch. Together, these characteristics give the suborbital portion of the face a wide, flattened appearance in anterior view (e.g., Fig. 7.8g). TSK 14 displays all of these characters, and also shows a strong resemblance to other specimens of *D. huenei* in other respects (e.g., CAMZM T1089; Fig. 7.8d, f, h, l). TSK 27 (Fig. 7.8j) is mostly unprepared, but the left zygomatic arch is well exposed, showing the autapomorphic thickening of the zygomatic arch typical of *D. huenei*. TSK 37 (Fig. 7.8k) consists of a relatively well preserved but somewhat unprepared skull, jaw, and portions of the postcranial skeleton. The skull resembles that of other *D. huenei* specimens, including in the presence of a thickened zygoma, and the postcranial elements are also comparable to other *D. huenei* material (e.g., Fig. 7.8m, n, o). TSK 40 is a nearly unprepared skull preserved in a broken nodule, but a thickened zygoma is apparent. Specimen 5-2 (Fig. 7.8c) and 5-4 are incomplete and unprepared, but both have portions of their temporal bars exposed. These show extensive overlap of the postorbitals by the parietals, a morphology most consistent with *D. huenei* among Zambian dicynodonts. Specimen 5-10 (Fig. 7.8b) is somewhat more complete, and appears to have originally possessed a plate-like zygomatic arch and postorbital bar, although these areas are currently not well preserved. Although their poor preservation makes their identification somewhat tentative, these three specimens are important data points for establishing *D. huenei* in the northern part of the Luangwa Basin.

Synonyms in Luangwa Basin Literature: *Dicynodon lacerticeps*, *Dicynodon trigonocephalus*, "*Dicynodon*" *trigonocephalus* (Boonstra 1938; Drysdall and Kitching 1963; Kitching 1963; King 1981, 1988; King and Jenkins 1997; Fröbisch 2008, 2009).

Previous Reports: Boonstra (1938) reported a fragmentary specimen consisting of an occiput and an intertemporal bar in which the postorbitals strongly overlapped the parietals. He referred this specimen to *Dicynodon lacerticeps*, and referred to it by the field number R.40. None of the catalogued Zambian material at the SAM has this field number associated with it (S. Kaal, personal communication, 2011), so we were unable to examine the specimen. Although Boonstra's description is not detailed enough to definitively state whether the specimen is *D. lacerticeps* or *D. huenei*, we regard the most parsimonious interpretation of this report as the latter species. Drysdall and Kitching (1962, 1963) and Kitching (1963) also reported *Dicynodon* from the Luangwa Basin, but most of these reports represent *Diictodon* or *Oudenodon* instead (see above). Drysdall and Kitching's (1963) and Kitching's (1963) specific mentions of *Dicynodon lacerticeps* are only repetitions of Boonstra's (1938) original report. Anderson and Cruickshank (1978), Cooper (1982), King (1988, 1992), and Rubidge (2005) all noted the presence of *Dicynodon* in Zambia, but did not refer to a particular species. King's (1981) paper focused on the skeletal morphology and function of TSK 14, but she did note the Zambian origin of the specimen and referred it to *D. trigonocephalus*. King and Jenkins (1997) also mentioned the presence of *D. trigonocephalus* in the Luangwa Basin as part of the biostratigraphic context for their putative specimen of *Lystrosaurus*. In the taxonomic framework used here, both of these reports should be considered to represent *D. huenei*. Lucas (1997, 1998a, 2002, 2005, 2006, 2009) used the occurrence of *Dicynodon* in the Luangwa Basin as part of his tetrapod biochronology for the Permian, but in most cases did not explicitly discuss any particular species of the genus. He does mention *D. trigonocephalus* in Lucas (1997, 1998a) following King (1981), and his later citations of King and Jenkins (1997) would imply that his usage focuses on this species as well (here recognized as *D. huenei*; see above and Kammerer et al. 2011). However, the locality he gives [e.g., "'Horizon 5' of Boonstra in the Luangwa Valley, 4.8–6.4 km north of Nt'awere, Zambia" (Lucas 2006, p. 83; also see Lucas 1997, 1998a, 2001)] corresponds to the type locality of *Dicynodon roberti*, a junior synonym of *Syops vanhoepeni* (see Kammerer et al. 2011, and below). Fröbisch (2009) regarded four species of *Dicynodon* sensu lato as potentially valid and occurring in the Luangwa Basin: *D. lacerticeps*, "*D.*" *trigonocephalus*, "*D.*" *roberti*, and "*D.*" *vanhoepeni*. The first two of these correspond to material we assign to *D. huenei*.

Syops vanhoepeni (Boonstra, 1938)
Figure 7.9a–i

Material: NHCC LB25, SAM-PK-11311, SAM-PK-11325a, SAM-PK-11325b.

Localities: Locality 4 of Drysdall and Kitching (1963) (SAM-PK-11311), Locality 5 of Drysdall and Kitching (1963) (SAM-PK-11325a, SAM-PK-11325b), locality L61 (NHCC LB25).

Identifying Characteristics: Kammerer et al. (2011) discuss the basis for recognizing *Syops vanhoepeni* as a valid species and the rationale for considering *D. roberti* to be its junior synonym. The four specimens of *S. vanhoepeni* are unprepared, but enough morphology is exposed in each to allow them to be grouped together confidently. All four possess a similar long, low snout profile, large external nares, caniniform processes with a postcaniniform crest, and large, robust tusks. SAM-PK-11325a and NHCC LB25 also share a unique pattern of ornamentation on the dorsal surface of the snout: a thick, rounded median ridge that is flanked laterally by shallow depressions, which in turn are bounded laterally by elongate, ridge-like nasal bosses

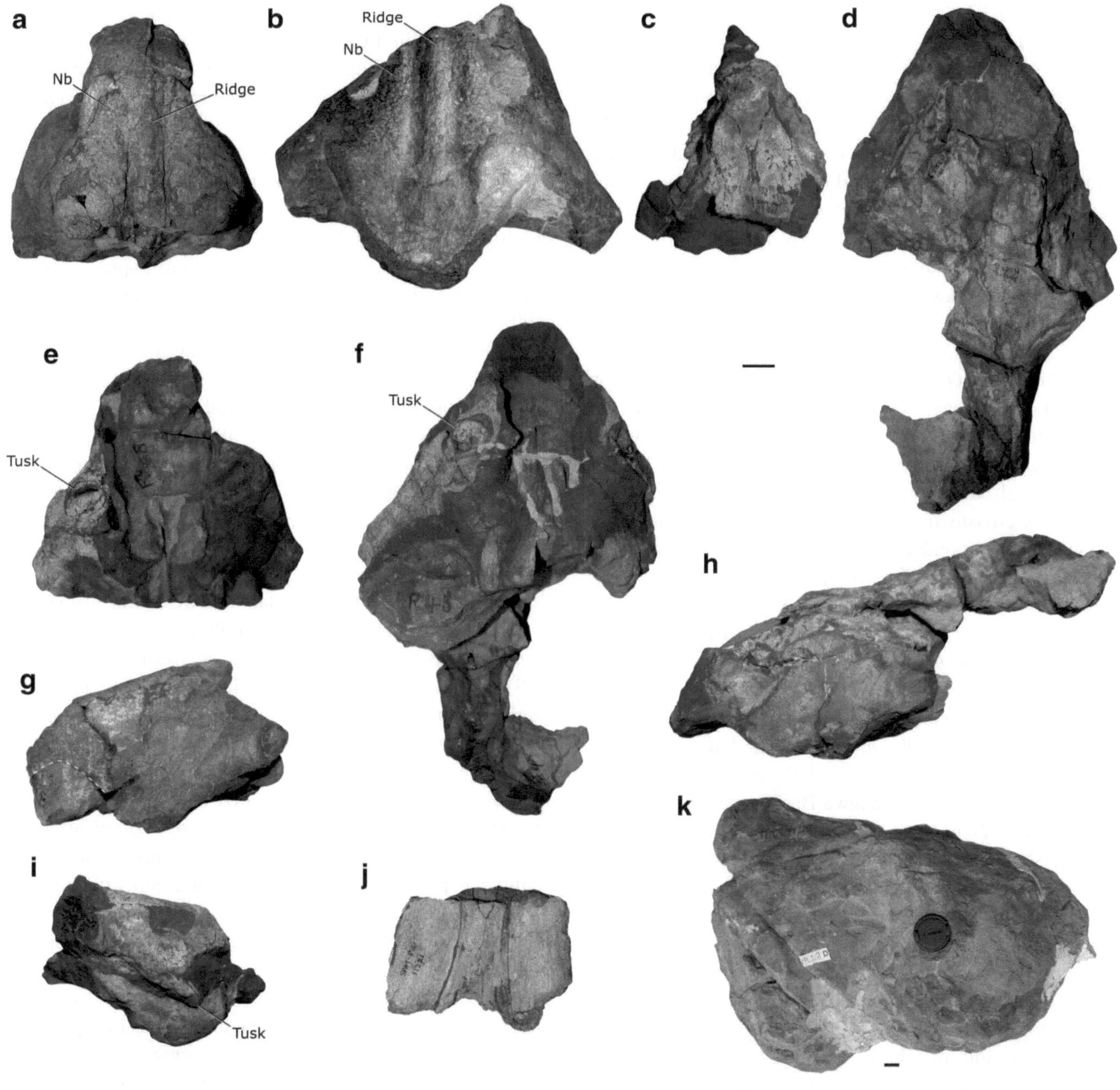

Fig. 7.9 Zambian specimens of *Syops vanhoepeni* and *Haughtoniana magna*, and supposed Zambian specimen of *Aulacephalodon bainii*. **a** Snout of *Syops vanhoepeni* [SAM-PK-11325a (paratype of *Dicynodon roberti*)] in dorsal view. **b** Snout of *Syops vanhoepeni* (NHCC LB25) in dorsal view. Note the similar configuration of paired nasal bosses separated by a strong median ridge in NHCC LB25 and SAM-PK-11325a. **c** Snout of *Syops vanhoepeni* (holotype of *Dicynodon vanhoepeni*) (SAM-PK-11311) in dorsal view. **d** Partial skull of *Syops vanhoepeni* (holotype of *Dicynodon roberti*) (SAM-PK-11325b) in dorsal view. **e** Snout of *Syops vanhoepeni* (paratype of *Dicynodon roberti*) (SAM-PK-11325a) in ventral view. **f** Partial skull of *Syops vanhoepeni* (holotype of *Dicynodon roberti*) (SAM-PK-11325b) in ventral view. **g** Snout of *Syops vanhoepeni* (paratype of *Dicynodon roberti*) (SAM-PK-11325a) in left lateral view. *Dashed line* highlights the alveolar margin (*below* the *line* is matrix). **h** Partial skull of *Syops vanhoepeni* (holotype of *Dicynodon roberti*) (SAM-PK-11325b) in left lateral view. **i** Snout and partial mandible of *Syops vanhoepeni* (holotype of *Dicynodon vanhoepeni*) (SAM-PK-11311) in right lateral view. **j** Posterior portion of temporal bar of the holotype of *Haughtoniana magna* (SAM-PK-11321) in dorsal view. **k** Probable left lateral view of an unprepared large dicynodont skull (BP/1/3242) identified in the BP collections catalogue as *Aulacephalodon bainii*. The exact orientation of this specimen is uncertain because no diagnostic structures are exposed on its surface. *Upper* scale bar applies to *panels* **a–j**; *lower* scale bar applies to *panel* **k**. Scales bars are 20 mm. Photograph in *panel* **j** courtesy of C. Kammerer; photograph in *panel* **k** courtesy of B. Rubidge

(Fig. 7.9a, b). The snout region is more poorly preserved in SAM-PK-11311 and SAM-PK-11325b, making it impossible to confirm whether these specimens had the same pattern of ornamentation, but part of a median ridge is preserved on the premaxilla of SAM-PK-11325b. NHCC LB25 and SAM-PK-11325b (e.g., Fig. 7.9d) also share a relatively narrow intertemporal bar in which the postorbitals extensively overlap the parietals, but this portion of the skull is not preserved in SAM-PK-11325a or SAM-PK-11311. *Syops vanhoepeni* shows a number of superficial similarities to *Odontocyclops* and *Kitchinganomodon*, but can be differentiated from both even with the fragmentary material currently available. *Syops vanhoepeni* can be distinguished from *Odontocyclops* based on the latter taxon's much more concave dorsal snout surface, absence of a strong median ridge on the dorsal surface of the snout, and wider exposure of the parietals between the postorbitals on the temporal bar. *Syops vanhoepeni* differs from *Kitchinganomodon crassus* in possessing tusks, more elongate nasal bosses, a wide rounded median ridge on the dorsal surface of the snout, and the absence of a pineal boss.

Synonyms in Luangwa Basin Literature: *Dicynodon vanhoepeni*, *Dicynodon roberti* (Boonstra 1938; King 1988); "*Dicynodon*" *vanhoepeni*, "*Dicynodon*" *roberti* (Fröbisch 2008, 2009).

Previous Reports: Boonstra (1938) described *Dicynodon vanhoepeni* and *D. roberti*. Drysdall and Kitching (1963) included both *D. vanhoepeni* and *D. roberti* in their list of taxa reported from the Luangwa Basin, but Kitching (1963) only included *D. vanhoepeni*. King (1988) listed both species in the systematic section of her monograph, but only included the genus *Dicynodon* without reference to particular species in her faunal list for Zambia. Anderson and Cruickshank (1978), King (1992), and Rubidge (2005) also only included the genus *Dicynodon* without reference to particular species. As noted above, at least some of the material referred to the genus *Dicynodon* by Lucas (1997, 1998a, 2001, 2002, 2005, 2006, 2009) and used in his tetrapod biochronology corresponds to specimens of *S. vanhoepeni*. Fröbisch (2009; also see Fröbisch 2008) included both *D. vanhoepeni* and *D. roberti* in his compilation, but noted that their validity had not been reassessed since their description.

Permian Dicynodonts Whose Presence in Zambia Cannot be Confirmed

Pachytegos stockleyi Haughton, 1932

Previous Reports: King (1988) included *Pachytegos* in her faunal list for Zambia. However, we are aware of no other reports of *Pachytegos* from Zambia and suspect this is an error since *Pachytegos* is not included in her faunal list for Tanzania, despite the fact that the only published material attributed to the taxon originated in Tanzania (Haughton 1932; Cox 1964; Gay and Cruickshank 1999).

Cistecephalus microrhinus Owen, 1876

Synonyms in Luangwa Basin Literature: *Cistecephalus planiceps* (Drysdall and Kitching 1963; Kitching 1963). Keyser (1973b) discusses the synonymy of *C. planiceps* and *C. microrhinus*.

Previous Reports: Drysdall and Kitching (1962, 1963) and Kitching (1963) were the first authors to report specimens of *Cistecephalus* in the Luangwa Basin. Several additional authors subsequently cited these records, primarily in biogeographic and biostratigraphic contexts (Anderson and Cruickshank 1978; Cooper 1982; King 1988, 1992; Smith and Keyser 1995; Lucas 2002, 2005, 2006; Angielczyk 2002; Rubidge 2005; Fröbisch 2009). However, as noted above, Drysdall and Kitching did not provide photographs of or specimen numbers for any of their *Cistecephalus* records. Because all Zambian cistecephalid material available in collections appears to be referable to a new taxon (see above), we consider there to be no reliable evidence of *Cistecephalus* in the Luangwa Basin at this time.

Tropidostoma microtrema (Seeley, 1889)

Previous Reports: Drysdall and Kitching (1963) and Kitching (1963) stated that they collected specimens of *Dicynodon acutirostris* in Zambia, a species that Keyser (1973a) and Botha and Angielczyk (2007) considered to be a junior synonym of *Tropidostoma microtrema*. Based on this taxonomic change, Fröbisch (2009) included *Tropidostoma* in the dicynodont fauna of the Madumabisa Mudstone in Zambia. Keyser (1981) also stated that *Tropidostoma* was present in the Luangwa Basin. The type of *D. acutirostris* was collected in South Africa, and we have not identified any Zambian specimens in our search of museum collections or our fieldwork that can be referred to *T. microtrema*. Drysdall and Kitching (1963) and Kitching (1963) did not provide numbers for any of the specimens they identified as *D. acutirostris*, so the accuracy of their identification cannot be checked. Therefore, because there are no voucher specimens documenting the presence of *T. microtrema* (=*T. dubium*; see Kammerer et al. 2011), it should not be included in the Permian dicynodont fauna of the Luangwa Basin.

Rhachiocephalus magnus (Owen, 1876)

Previous Reports: Boonstra (1938) was the first to refer a Zambian dicynodont specimen to *Rhachiocephalus* (the holotype of *Rhachiocephalus dubius*). Keyser (1975) included this species in his taxonomic review but hinted that the specimens in question may represent a distinct taxon, an observation that was confirmed with the erection of *Odontocyclops* (Keyser 1979; Keyser and Cruickshank 1979; Angielczyk 2002). Drysdall and Kitching (1963) and Kitching (1963) reported *Neomegacyclops* and *Platycyclops*, now recognized as junior synonyms of *Rhachiocephalus* (Keyser 1975; Cluver and King 1983; also see

Maisch 2002a). Anderson and Cruickshank (1978), King (1988, 1992), and Fröbisch (2009) included *Rhachiocephalus* in their faunal tabulations for the Luangwa Basin following Drysdall and Kitching (1963). However, Drysdall and Kitching's (1963) and Kitching's (1963) reports of *Rhachiocephalus* appear to be based on field observations, and we know of no voucher specimens that can confirm the presence of this taxon in Zambia. The best potential candidate (TSK 23) instead represents *Kitchinganomodon*. Therefore we cannot include *Rhachiocephalus* in the fauna of the Madumabisa Mudstone at this time.

Haughtoniana magna Boonstra, 1938
Figure 7.9j

Material: SAM-PK-11321.

Localities: Locality 4 of Drysdall and Kitching (1963) (SAM-PK-11321).

Identifying Characteristics: The holotype of *Haughtoniana magna* consists of fragmentary cranial and postcranial material of a large dicynodont. Boonstra (1938) erected the species primarily on the basis of the construction of the intertemporal bar. In particular, he noted that the postorbitals were wide, nearly horizontal, and in the same plane as the parietals; the parietals were relatively narrow; and the interparietal extended onto the dorsal surface of the skull roof (Fig. 7.9j). Keyser (1975) and Cluver and King (1983) considered it to be a likely nomen dubium, and King (1988) also cast doubt on its validity, although she suggested it might be referable to *Aulacephalodon*. We agree that the type of *H. magna* is too fragmentary to allow a definitive identification, either as a valid species or as a synonym of another species. Moreover, the temporal bar fragment appears to have lost some of its original bone surface, either through weathering or preparation in which the hematitic matrix was not removed cleanly from the bone. We suggest that this may account for the fact that the parietals and postorbitals are in the same plane and potentially explains the oddly-shaped exposure of the interparietal on the skull roof. The general size and appearance of the type is similar to the temporal bar of *Odontocyclops* (e.g., BP/1/3419; Fig. 7.6b), although uncertainty about whether the exposure of the interparietal on the skull roof is real or an artifact prevents the definitive identification of *H. magna* as a synonym of *O. whaitsi* (the interparietal is not exposed on the skull roof in *O. whaitsi*; Angielczyk 2002).

Synonyms in Luangwa Basin Literature: None.

Previous Reports: Boonstra (1938) described *Haughtoniana magna*, and Drysdall and Kitching (1963) and Kitching (1963) noted its occurrence in the Luangwa Basin. Keyser (1975), Cluver and King (1983), and King (1988) considered it in their systematic treatments, and King (1988) also included it in her faunal list for Zambia. Fröbisch (2009) also mentioned *H. magna*, but noted that its taxonomic status was uncertain and did not include it in his final faunal list for Zambia.

Aulacephalodon bainii (Owen, 1845)
Figure 7.9k

Previous Reports: Drysdall and Kitching (1962, 1963) and Kitching (1963) were the first to report *Aulacephalodon*, including the species *A. laticeps* (a synonym of *A. bainii*; see Cluver and King 1983) from the Luangwa Basin. The records appear to represent field identifications of specimens from at least their Localities 1 and 4 because no specimen numbers for collected material were cited. Kitching (1963) also used the name *Aulacocephalodon*, which is an often-repeated misspelling of *Aulacephalodon* (Tollman et al. 1980). Several subsequent authors included *Aulacephalodon* in the Zambian dicynodont fauna based on these reports (Anderson and Cruickshank 1978; Cooper 1982; King 1988, 1992; Angielczyk 2002; Fröbisch 2009). However, in our examination of material in collections, we found no specimens that can be positively identified as *Aulacephalodon*. BP/1/3242, a specimen originating at Drysdall and Kitching's Locality 1 that is identified in the BP catalogue as *Aulacephalodon*, is completely unprepared (Fig. 7.9k) and displays no characters allowing it to be identified. King's (1988) suggestion that *Haughtoniana magna* might represent *Aulacephalodon* also is likely incorrect (see above). We did not observe any specimens that could be positively identified as *Aulacephalodon* during our fieldwork, and we suspect that previous reports likely represent *Odontocyclops* specimens, *Syops vanhoepeni* specimens, or large *Oudenodon* specimens that were misidentified in the field. Therefore, *Aulacephalodon* should not be included in faunal lists for the Madumabisa Mudstone.

Dicynodon lissops Broom, 1913

Previous Reports: Drysdall and Kitching (1963) stated that *Dicynodon lissops* was among a collection of four small anomodonts collected at their Locality 21, but provided no figures of or specimen numbers for this material. The holotype of *Dicynodon lissops* (AMNH 5508) is from the *Dicynodon* Assemblage Zone of South Africa, and was considered a junior synonym of *Daptocephalus leoniceps* by Kammerer et al. (2011). Given that this would be the only known occurrence of *Daptocephalus* from the Luangwa Basin of Zambia, we are hesitant to consider the report valid due to the lack of voucher specimens. Therefore we recommend that *Dicynodon lissops* and its senior synonym *Daptocephalus leoniceps* be excluded from the dicynodont fauna of the Luangwa Basin until positively identifiable material comes to light.

Dicynodon rhodesiensis

Previous Reports: Boonstra (1938, p. 384) included the name *D. rhodesiensis* in a list of *Dicynodon* species from Zambia. However, he does not mention or describe the species elsewhere in the paper, and we are unaware of any

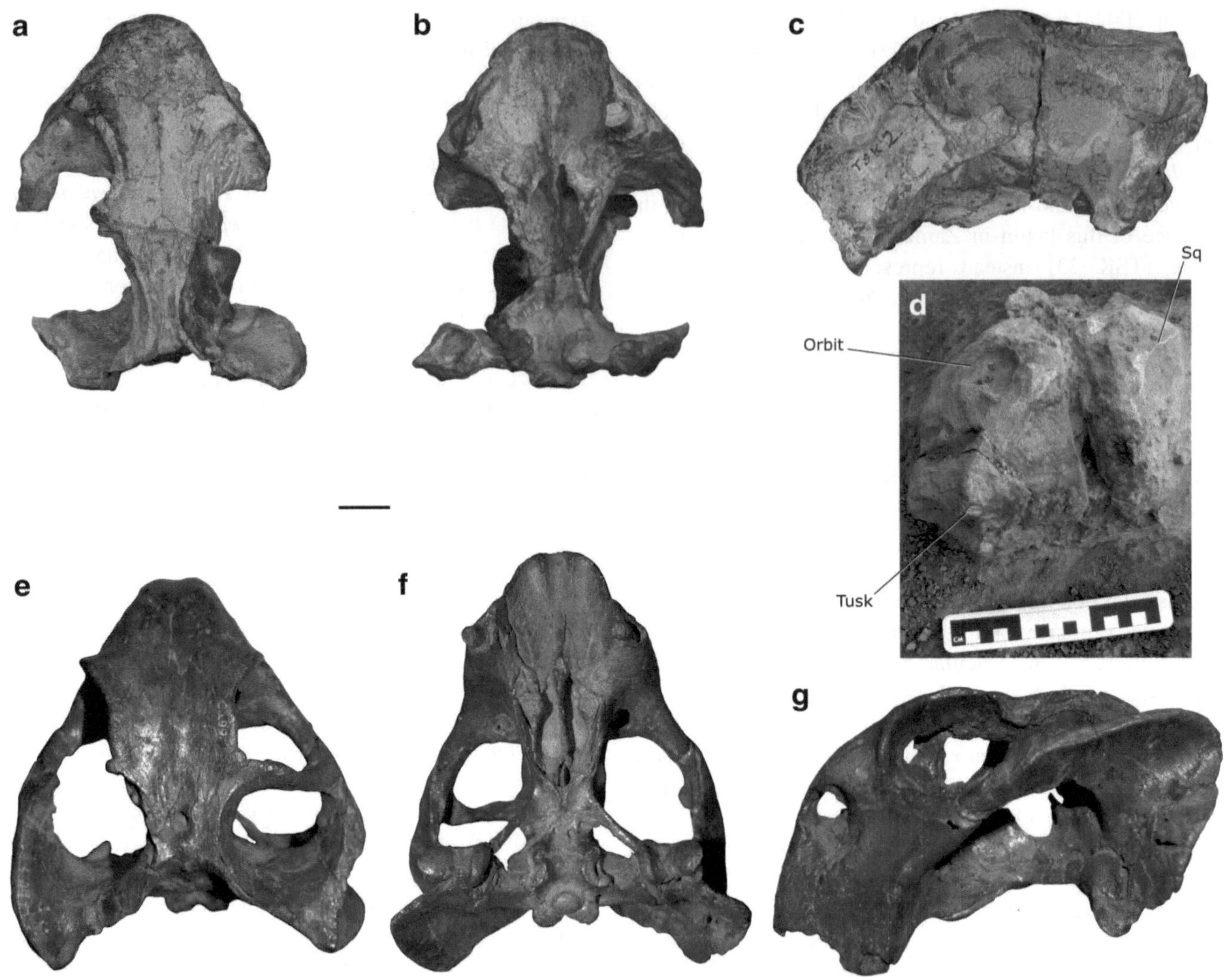

Fig. 7.10 Zambian specimens of Lystrosauridae n. g. & sp. and comparative specimen of *Lystrosaurus curvatus*. **a** Skull of Lystrosauridae n. g. & sp. from Zambia (TSK 2) in dorsal view. **b** Skull of Lystrosauridae n. g. & sp. from Zambia (TSK 2) in ventral view. **c** Skull of Lystrosauridae n. g. & sp. from Zambia (TSK 2) in left lateral view. **d** Field photograph of a probable specimen of the same species represented by TSK 2. **e** Skull of *Lystrosaurus curvatus* from South Africa (NMQR 3595, formerly NMQR C299) in dorsal view. **f** Skull of *Lystrosaurus curvatus* from South Africa (NMQR 3595) in ventral view. **g** Skull of *Lystrosaurus curvatus* from South Africa (NMQR 3595) in left lateral view. *Central* scale bar applies to *panels* **b–c** and **e–g**, and is 20 mm. Scale bar in field photograph is 150 mm

mention of the species in the literature before or after this aside from Drysdall and Kitching's (1963) and Kitching's (1963) inclusion of the species in their lists of dicynodonts reported previously from the basin. Therefore, we conclude that the inclusion of the name must have been a mistake or oversight by Boonstra, and no species of this name was ever described from Zambia or elsewhere.

Lystrosaurus cf. *L. curvatus* (Owen, 1876)
Figure 7.10a–d
Material: TSK 2.

Localities: "East side of hunter's track from Luangwa River, along north side of Munyamadzi River, Luangwa Valley, Zambia; Madumabisa Mudstones, Upper Permian" (King and Jenkins 1997, p. 152). This corresponds to Kerr's (1974) Locality 1 (also see Davies, 1981).

Identifying Characteristics: In their discussion of TSK 2, King and Jenkins (1997) listed a shortened basicranial axis, exposure of the parietals between the postorbitals on the skull roof, the deepened, ventrally-angled snout, the smooth premaxilla-maxilla suture, the extension of the premaxilla to the level of the prefrontals, and the pear shaped external naris bounded posteroventrally by a rugose ridge as characters that were typical of *Lystrosaurus*. They also suggested that the smoothly curving snout profile, the absence of a nasofrontal ridge and ornamentation on the frontals, the absence of strong prefrontal bosses, and a laterally flared squamosal implied the specimen most closely resembled *L. curvatus* since these characters were included in diagnoses of *L. curvatus* available at the time (e.g., Cluver 1971; Cosgriff et al. 1982). They also have been included in diagnoses of *L. curvatus* in subsequent works dealing with the species composition of *Lystrosaurus* (e.g., Ray 2005; Grine et al. 2006; Botha and Smith 2007).

Although we agree that the TSK 2 shows some features similar to *Lystrosaurus*, and *L. curvatus* in particular, there are other characters that do not fit well with this identification and seem to fall outside of the ranges of intraspecific and intrageneric variation identified by authors such as Ray (2005) or Grine et al. (2006). For example, TSK 2 possesses an ectopterygoid, whereas the ectopterygoid is absent in *Lystrosaurus* (Cluver 1971) and recent authors have not identified this as a variable character within the taxon. Similarly, although the parietals are exposed between the postorbitals on the skull roof, the exposure is narrower than typical in *Lystrosaurus*, and the temporal bar is relatively longer anteroposteriorly in TSK 2 (e.g., compare Fig. 7.10a–e). The latter character is especially interesting because both Ray (2005) and Grine et al. (2006) noted that the temporal region displays negative allometry in *Lystrosaurus*. NMQR 3595 (*L. curvatus*; Fig. 7.10e) is consistent with this pattern, with the temporal bar being approximately 9 % of the basal length of the skull. In contrast, the temporal bar is approximately 21 % of the basal skull length in TSK 2 (Fig. 7.10a), despite the two specimens having nearly identical basal skull lengths. The frontal region, although slightly damaged in TSK 2, appears to have been narrower than typical in *L. curvatus*. This also is inconsistent with TSK 2 being part of *Lystrosuarus* because Ray (2005) and Grine et al. (2006) found that this measurement was isometric to positively allometric. The snout is angled ventrally in TSK 2, but the angle of deflection is less than in *L. curvatus* and it does not extend as far downwards (e.g., compare Fig. 7.10c–g). TSK 2 also lacks most of the conspicuous skull ornamentation in *Lystrosaurus*, such as a sagittal ridge on the premaxilla or a prefrontal nasal crest. Although these characters tend to be weakly developed in *L. curvatus* (Grine et al. 2006) and show evidence of sexual dimorphism (Ray 2005), their complete absence in a relatively large specimen specimen such as TSK 2 (basal length approximately 144 mm) is surprising. These characters usually manifest in specimens with basal lengths in the range of 80–100 mm and are present in at least some *L. curvatus* specimens with sizes comparable to TSK 2 (Grine et al. 2006). Taken together, the differences between TSK 2 and definite specimens of *L. curvatus* (and other *Lystrosaurus*) species do not appear consistent with patterns of ontogenetic variation or sexual dimorphism identified by previous authors. Because of this, as well as the fact that a ventrally-extended snout is present in other dicynodonts such as *Kwazulusaurus shakai*, *Euptychognathus bathyrhynchus*, and *Basilodon woodwardi* (Maisch 2002b; Kammerer et al. 2011), we do not think that TSK 2 can be identified unequivocally as *Lystrosaurus curvatus* or even *Lystrosaurus*. It may instead represent a new taxon (likely a lystrosaurid; see Kammerer et al. 2011), and it should be possible to collect additional material to characterize this taxon more fully. For example, Fig. 7.10d shows a specimen that we observed in 2009 but did not collect that has a relatively long, downturned snout and a narrow temporal region.

Synonyms in Luangwa Basin Literature: None.

Previous Reports: King and Jenkins (1997) were the first to report *Lystrosaurus* from the Luangwa Basin, and the occurrence was noted in other compilations examining therapsid biogeography (Angielczyk and Kurkin 2003; Rubidge 2005; Fröbisch 2009). It also was cited in a number of papers considering biostratigraphic correlations between the Luangwa and other basins (e.g., Lucas 1998b, 2006; Gay and Cruickshank 1999; Ray 1999; Catuneanu et al. 2005), as well as in studies of the end-Permian extinction and the origin and survivorship of *Lystrosaurus* during that event (e.g., Rubidge and Sidor 2001; Maisch 2002b; Botha and Smith 2006, 2007; Fröbisch 2007, 2008; Lucas 2009).

Triassic Dicynodont Fauna

We use the higher-level taxonomy of Maisch (2001) for Triassic dicynodonts, with minor changes reflecting the results of Kammerer et al. (2011). Our taxonomic results for Triassic dicynodonts are summarized in Table 7.2.

Systematic Paleontology

Dicynodontoidea (Owen, 1860a)
Kannemeyeriiformes Maisch, 2001
Kannemeyeriidae von Huene, 1948
Kannemeyeria lophorhinus Renaut et al., 2003
Figure 7.11a–d

Material: BP/1/3638.

Localities: Locality 16 of Drysdall and Kitching (1963) (BP/1/3636). This locality is in Drysdall and Kitching's (1963; also see Kitching 1963) lower fossiliferous horizon of the Ntawere Formation.

Identifying Characteristics: Renaut et al. (2003) provided the first diagnosis of *K. lophorhinus*, which was based on the detailed morphological study of Renaut (2000). They implied that it possessed all of the diagnostic characters of *Kannemeyeria* identified by Renaut (2000), such as a prominent median ridge on the snout, prominent caniniform

Table 7.2 Dicynodont taxa present in the Middle Triassic Ntawere Formation, Luangwa Basin, Zambia, and synonyms used in the literature on the Luangwa Basin

Taxon	Synonyms in Luangwa Basin literature
Kannemeyeria lophorhinus	*Rechnisaurus cristarhynchus*, *Rechnisaurus*, *Kannemeyeria cristarhynchus*, *Kannemeyeria*
"*Kanneyemeria*" *latirostris*	*Kannemeyeria*, *Kannemeyeria latirostris*, "*Kannemeyeria*" *latirostris*, *Dolichuranus latirostris*, *Dolichuranus*, *Shansiodon*
Zambiasaurus submersus	*Zambiosaurus*, *Zambiasaurus submerses*
Sangusaurus edantatus	*Sanguasaurus*, *Sangausaurus*
Kannemeyeriiformes *incertae sedis*	None

See text for details

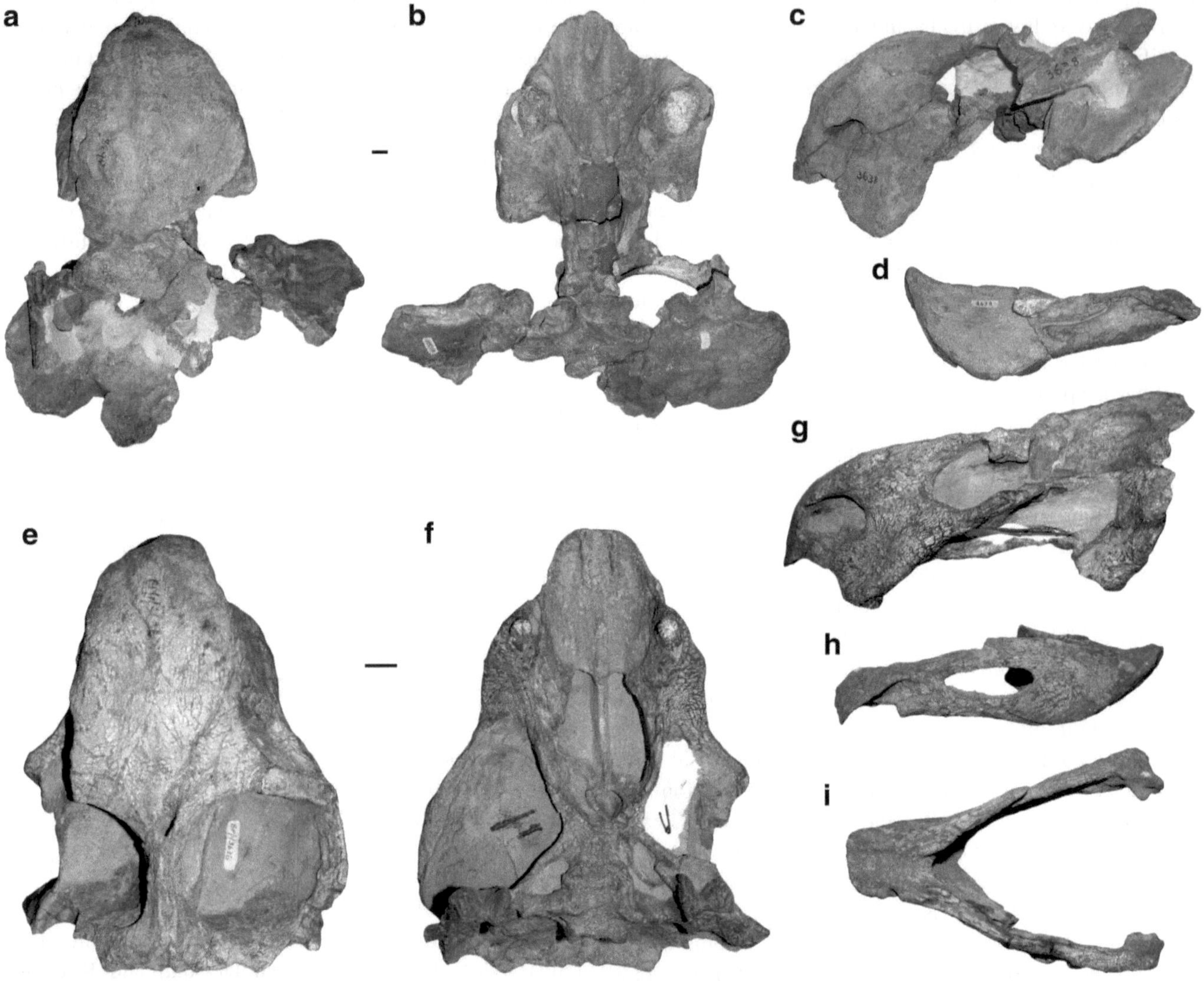

Fig. 7.11 Zambian specimens of *Kannemeyeria lophorhinus* and "*Kannemeyeria*" *latirostris*. **a** Holotype skull of *Kannemeyeria lophorhinus* (BP/1/3638) in dorsal view. **b** Holotype skull of *Kannemeyeria lophorhinus* (BP/1/3638) in ventral view. **c** Holotype skull of *Kannemeyeria lophorhinus* (BP/1/3638) in left lateral view. **d** Holotype mandible of *Kannemeyeria lophorhinus* (BP/1/3638) in left lateral view. **e** Skull of "*Kannemeyeria*" *latirostris* (holotype of *Kannemeyeria latirostris*) (BP/1/3636) in dorsal view. **f** Skull of "*Kannemeyeria*" *latirostris* (holotype of *Kannemeyeria latirostris*) (BP/1/3636) in ventral view. **g** Skull of "*Kannemeyeria*" *latirostris* (holotype of *Kannemeyeria latirostris*) (BP/1/3636) in left lateral view. **h** Mandible of "*Kannemeyeria*" *latirostris* (holotype of *Kannemeyeria latirostris*) (BP/1/3636) in right lateral view. **i** Mandible of "*Kannemeyeria*" *latirostris* (holotype of *Kannemeyeria latirostris*) (BP/1/3636) in dorsal view. *Upper* scale bar applies to *panels* **b–d**; *lower* scale bar applies to *panels* **e–i**. Scale bars are 20 mm. Photographs in *panels* **a–d** courtesy of C. Kammerer

processes, tusks, a crest-like temporal bar that forms a sharp angle with the interorbital region, an anteriorly-sloping occipital plate, limited facial exposure of the lacrimal, the presence of a labial fossa, and the absence of an ectopterygoid. In addition, they differentiated *K. lophorhinus* from *K. simocephalus* on the basis of a more robust skull, a stronger median ridge on the snout that is flanked by depressions, a deeper, wider snout, larger caniniform processes, a broader intertemporal region, shorter temporal openings and secondary palate, and the absence of a fossa on the ventral surface of the median pterygoid plate, among other characters. Although BP/1/3638 is not completely preserved, most of Renaut et al.'s (2003) diagnostic characters are visible on the specimen, and the robust snout, strong median snout ridge, and large caniniform processes are especially apparent (Fig. 7.11a–c).

Synonyms in Luangwa Basin Literature: *Rechnisaurus cristarhynchus*, *Rechnisaurus*, *Kannemeyeria cristarhynchus*, *Kannemeyeria*. Renaut et al. (2003; also see Renaut 2000) provided an excellent review of the complex taxonomic history of BP/1/3638. In her initial description, Crozier (1970) referred the specimen to *Rechnisaurus cristarhynchus* Roy Chowdhury, 1970, with Keyser (1973c), Battail (1978, 1993), and Ochev and Shishkin (1989) following this identification. Keyser (1974) expressed uncertainty about whether *R. cristarhynchus* (including BP/1/3638) was distinct from *Kannemeyeria* and Keyser and Cruickshank (1979) elaborated this argument, concluding that it should be treated as a species of *Kannemeyeria* (*K. cristarhynchus*). A number of authors followed this taxonomy (Anderson and Cruickshank 1978; Cooper 1980, 1982; Cox and Li 1983; Cruickshank 1986; Bandyopadhyay 1988). However, Bandyopadhyay (1985, 1989) argued that the Indian holotype of *Rechnisaurus cristarhynchus* Roy Chowdhury, 1970 could not be assigned to *Kannemeyeria* and was distinct from the Namibian and Zambian specimens that Keyser and Cruickshank (1979) considered. She retained *R. cristarhynchus* Roy Chowdhury, 1970 for the Indian specimen, and used the name "*Kannemeyeria cristarhynchus* (Crozier, 1970; Keyser and Cruickshank, 1979)" for the Namibian and Zambian specimens. King (1988) followed this taxonomy, although she referred to the Zambian and Namibian specimens as "*Kannemeyeria cristarhynchus* Keyser and Cruickshank, 1979" and erroneously stated that Keyser and Cruickshank used the emended spelling "*cristarhyncha.*" King (1990), Cox (1991), Renaut (2000), and Renaut and Hancox (2001) used the name *K. cristarhynchus* for the Zambian and Namibian specimens and accepted that they were distinct from *Rechnisaurus*, although Lucas (1993b, 1996, 1998b, 1999, 2001, 2010; also see Lucas and Wild 1995) argued repeatedly for their synonymy. Renaut et al. (2003) argued that the name *Kannemeyeria cristarhynchus* violated Article 49 of the International Code of Zoological Nomenclature and coined the new species name *Kannemeyeria lophorhinus* as its replacement. Abdala et al. (2005) and Fröbisch (2008, 2009) used the name *Kannemeyeria lophorhinus*.

Previous Reports: Crozier (1970) was the first to formally describe and figure BP/1/3638, but the specimen is one of two "*Kannemeyeria*-like" dicynodonts that Drysdall and Kitching (1963) and Kitching (1963) mentioned collecting at Locality 16 (also see Brink 1963; Cox 1969; note that Chernin 1974 mistakenly reported these specimens as originating in the upper fossiliferous horizon at Drysdall and Kitching's (1963) Locality 15). Keyser (1974), Keyser and Cruickshank (1979), Cox and Li (1983), Bandyopadhyay (1985, 1988, 1989), King (1988), Renaut (2000), Renaut and Hancox (2001), and Renaut et al. (2003) discussed various aspects of the taxonomy and phylogenetic relationships of BP/1/3638 using various names. Many authors noted the presence of *Kannemeyeria "cristarhynchus," K. lophorhinus*, or more generally *Kannemeyeria* in Zambia in a biogeographic or biostratigraphic framework (Keyser 1973c, 1981; Anderson and Cruickshank 1978; Cooper 1980, 1982; Cruickshank 1986; King 1988, 1990; Cox 1991; Lucas 1993b, 1996, 1998b, 1999, 2001, 2010; Lucas and Wild 1995; Abdala et al. 2005; Fröbisch 2009; although note that Keyser 1981 erroneously reported the occurrence in the Upper Madumabisa Mudstone). Battail (1978, 1993) and Ochev and Shishkin (1989) also considered the biostratigraphic implications of BP/1/3638, but used the name *Rechnisaurus*. Battail (1993) also mistakenly reported that it occurred in the upper horizon of the Ntawere Formation. The studies of DeFauw (1989), and Fröbisch (2008) are more evolutionary in focus, but they do mention *Kannemeyeria* from Zambia. Finally, it is important to note that even though BP/1/3638 was initially referred to *Rechnisaurus* by Crozier (1970) and BP/1/3636 (see below) was referred to *Kannemeyeria* in the same paper, the rapid reassignment of these specimens to *Kannemeyeria* and *Dolichuranus* (respectively) means that nearly all subsequent literature reports of *Kannemeyeria* from Zambia refer to BP/1/3638, not BP/1/3636.

"*Kannemeyeria*" *latirostris* Crozier, 1970
Figure 7.11e–i

Material: BP/1/3636.

Localities: Locality 16 of Drysdall and Kitching (1963) (BP/1/3636). This locality is in Drysdall and Kitching's (1963; also see Kitching 1963) lower fossiliferous horizon of the Ntawere Formation.

Identifying Characteristics: In her diagnosis of "*K.*" *latirostris*, Crozier (1970) emphasized the broad snout, absence of any ridges on the snout, vertical orientation of the tusks, antero-posteriorly short interpterygoid vacuity, and short dentary symphysis as distinguishing features of the species. There have been three main suggestions for the generic affinities of BP/1/3636: (1) it is a specimen of *Kannemeyeria*, perhaps representing a distinct species (e.g., Crozier 1970); (2) it is a specimen of *Dolichuranus*, perhaps representing a distinct species (e.g., Keyser 1973c; Keyser and Cruickshank 1979; King 1988); (3) it is a specimen of *Shansiodon*, although its species-level taxonomy in this scenario has not been discussed (e.g., Cooper 1980; Lucas 1993a, b, 1996, 2001).

A full consideration of the taxonomic and phylogenetic status of "*K.*" *latirostris* is beyond the scope of this study, but some discussion of the taxonomic problem is warranted. It is unlikely to be a part of *Shansiodon* sensu stricto for several reasons. For example, BP/1/3636 (basal skull length 241 mm) is notably larger than most adult *Shansiodon* specimens (e.g., IVPP V2416 has a basal length of 150 mm; IVPP V2417 has a basal length of 165 mm). It also has proportionally much smaller tusks that are positioned farther anteriorly relative to the anterior orbital margin, a longer, wider preorbital region, and anteroposteriorly shorter temporal openings. Finally, it is worth noting that the reason BP/1/3636 was referred to *Shansiodon* was the hypothesis that *Dolichuranus* was its junior synonym. However, recent phylogenetic analyses that included both *Dolichuranus* and *Shansiodon* did not recover a close relationship between the two taxa (Damiani et al. 2007; Govender and Yates 2009; Kammerer et al. 2011).

BP/1/3636 also does not fit perfectly within *Dolichuranus* or *Kannemeyeria*. Crozier (1970) did not provide a detailed justification for her referral of BP/1/3636 to *Kannemeyeria*, only noting that it was of "*Kannemeyeria* type." In general appearance, the specimen does resemble *Kannemeyeria* in features such as its relatively large snout and narrow, crest-like temporal bar (e.g., Fig. 7.11d, g). However, comparison with the Renaut's (2000) diagnosis of *Kannemeyeria* highlights several differences. For example, there is no midline ridge on the snout, the occipital plate is relatively vertical, and the temporal bar is not strongly angled dorsally, although it is somewhat offset from the interorbital region of the skull. Furthermore, although Renaut (2000) did not explicitly state what he thought was the correct identity for BP/1/3636, he did not include it in his list of referred specimens for either of the species of *Kannemeyeria* that he recognized (*K. simocephalus* and *K. lophorhinus*).

Keyser (1973c) referred the specimen to *Dolichuranus* based on overall similarities in shape with the type material from Namibia, although he noted that the secondary palate of BP/1/3636 was somewhat shorter. BP/1/3636 does possess similar proportions of the snout and temporal openings to *Dolichuranus* specimens such as CGP/1/711 (the holotype of *D. primaevus*), as well as a similarly long, straight midventral vomerine plate and small interpterygoid vacuity (Fig. 7.11e). However, it also departs from Damiani et al.'s (2007) diagnosis of *Dolichuranus* in characters such as its more curved alveolar margin in lateral view, its lack of strongly differentiated nasal bosses and a trough-like furrow on the anterior surface of the snout, and the absence of an ectopterygoid.

An additional complication is the possible juvenile status of BP/1/3636. Crozier (1970) suggested the specimen might be a sub-adult based on its relatively small size compared to other *Kannemeyeria* specimens, and the specimen shows other potential juvenile features as well. For example, the orbits are large relative to the overall size of the skull. Orbit length shows negative allometry in *Kannemeyeria* (Renaut 2000) and other dicynodonts (Tollman et al. 1980; Ray 2005; Angielczyk 2007; although see Grine et al. 2006), and the proportions of BP/1/3636 are comparable to similarly-sized juvenile specimens of *K. simocephalus* (e.g., BP/1/2092; Renaut 2000). The tusks are also relatively small, and although different dicynodont taxa show different patterns of allometry for this character (compare Renaut 2000; Ray 2005; Grine et al. 2006), tusk diameter is positively allometric in *Kannemeyeria*. At the same time, these comparisons assume that the taxon represented by BP/1/3636 underwent a *Kannemeyeria*-like ontogeny, which would be logical if BP/1/3636 is eventually shown to be part of *K. lophorhinus* but potentially incorrect if it represents a distinct taxon.

Given these uncertainties, and the fact that BP/1/3636 cannot be easily accommodated within other roughly coeval taxa such as *Tetragonias*, *Vinceria*, *Dinodontosaurus*, or *Angonisaurus*, it is clear that the identity of BP/1/3636 requires further investigation. We follow Fröbisch (2009) in referring to the specimen as "*Kannemeyeria*" *latirostris* until its affinities can be resolved. However, we add that use of this specimen for making biogeographic or biostratigraphic inferences is questionable because of its uncertain identity.

Synonyms in Luangwa Basin Literature: *Kannemeyeria*, *Kannemeyeria latirostris*, "*Kannemeyeria*" *latirostris*, *Dolichuranus latirostris*, *Dolichuranus*, *Shansiodon*. Much like BP/1/3638, BP/1/3636 has had a complex taxonomic history. Crozier (1970) named "*K.*" *latirostris*. Keyser (1973c; also see Keyser 1973d) considered the specimen to be referable to his newly-created genus *Dolichuranus*, but retained *latirostris* Crozier, 1970 as a valid species. Many subsequent authors followed this synonymy (Keyser 1974; Battail 1978, 1993; Anderson and Cruickshank 1978; Keyser and Cruickshank 1979; Cooper 1982; King 1988, 1990; Surkov 2000; Renaut 2000; Rubidge 2005). Given that Ochev and Shishkin (1989) also report *Rechnisaurus* from Zambia, it appears that their record of *Kannemeyeria* from the Ntawere Formation refers to BP/1/3636. Cox (1991) stated that only *Kannemeyeria* was present in the lower fossiliferous horizon of the Ntawere Formation, although he did not discuss the species *latirostris* Crozier, 1970 specifically. Other workers, often arguing from a biostratigraphic perspective, considered *Dolichuranus*, including *D. latirostris* from Zambia, to be a junior synonym of *Shansiodon* (Cooper 1980; Lucas 1993a, b, 1996, 2001; Lucas and Wild 1995). In their redescription of *Dolichuranus*, Damiani et al. (2007) stated that they did not consider the species *latirostris* Crozier, 1970 to be referable to this genus, but they did not elaborate on why they concluded this or their preferred placement for the species. Based on Damiani et al. (2007), Fröbisch (2008, 2009) referred to the species as "*Kannemeyeria*" *latirostris*, reflecting its uncertain taxonomic status.

Previous Reports: Although Crozier (1970) described BP/1/3636, the specimen is one of two "*Kannemeyeria*-like" dicynodonts that Drysdall and Kitching (1963) and Kitching (1963) mentioned collecting at Locality 16 [also see Brink 1963; Cox 1969; note that Chernin 1974 mistakenly reported these specimens as originating in the upper fossiliferous horizon at Drysdall and Kitching's (1963) Locality 15]. Kitching (1977) included a photograph of BP/1/3636 and referenced it in a discussion of Karoo taphonomy. Anderson and Cruickshank (1978), King (1988), Surkov (2000), Rubidge (2005), and Fröbisch (2009) included "*K.*" *latirostris* in their biogeographic compilations under various names. Battail (1978, 1993) and Cooper (1982) used the occurrence of "*K.*" *latirostris* (called *Dolichuranus* in those papers) as a datum for correlating the lower Ntawere Formation with units in other basins, as did Ochev and Shishkin (1989), but using the name *Kannemeyeria*. Lucas (1993a, b, 1996, 2001; also see Lucas and Wild 1995), following Cooper (1980), regarded "*K.*" *latirostris* as a synonym of *Shansiodon*, and discussed the biostratigraphic implications of this synonymy. In other works (e.g., Lucas 1998b, 2010), however, he reported only *Kannemeyeria* from the lower Ntawere Formation. Keyser (1973c, d, 1974), Keyser and Cruickshank (1979), Cooper (1980), King (1990), Renaut (2000), and Damiani et al. (2007) discussed "*K.*" *latirostris* (often under the name *Dolichuranus*) in taxonomic or phylogenetic contexts. Finally, it is important to note that even though BP/1/3636 was initially referred to *Kannemeyeria* by Crozier (1970) and BP/1/3638 (see above) was referred to *Rechnisaurus* in the same paper, the rapid reassignment of these specimens to *Dolichuranus* and *Kannemeyeria* (respectively) means that nearly all subsequent literature reports of *Kannemeyeria* from Zambia refer to BP/1/3638, not BP/1/3636.

Stahleckeriidae (Lehman, 1961)
Zambiasaurus submersus Cox, 1969
Figure 7.12a–j

Material: Cox (1969) provided a list of 499 identifiable elements or fragments of elements, but provides specimen numbers for 174 specimens in two series: LM/NH 2 to LM/NH 35 and NHMUK R9001 to NHMUK R9140. The collection includes at least 18 juvenile individuals (based on the number of distal right humeri) and at least one large adult.

Localities: All specimens originated at Locality 15 of Drysdall and Kitching (1963). The type locality of *Z. submersus* is in Drysdall and Kitching's (1963; also see Kitching 1963) upper fossiliferous horizon of the Ntawere Formation (Cox 1969).

Identifying Characteristics: Cox (1969) provided a diagnosis for *Zambiasaurus submersus*. Diagnostic characters he listed include edentulous skull and jaws; short median suture between nasals; preparietal absent; interparietal that does not extend far forwards on skull roof; sharp transition between the skull roof and occipital plate; paired anterior ridges present on the palatal surface of the premaxilla; at least four sacral vertebrae; tall, narrow scapular blade with a ridge on its lateral surface; coracoid foramen entirely within the procoracoid.

Synonyms in Luangwa Basin Literature: *Zambiosaurus*, *Zambiasaurus submerses* (Surkov 2000; Fröbisch 2009). These appear to be misspellings.

Previous Reports: Although Cox (1969) provided the first description of *Zambiasaurus submersus*, Attridge et al. (1964) made a passing reference to the specimens that eventually were assigned to this species. King (1988, 1990), Fröbisch (2009), and Sues and Fraser (2010) included *Zambiasaurus* in their faunal lists for the Ntawere Formation. Battail (1978, 1993), Cox (1991), and DeFauw (1993) included *Zambiasaurus* in their discussions of Triassic biostratigraphy, and Surkov (2000) mentioned it in his biogeographic study. Various authors considered *Zambiasaurus* in taxonomic or phylogenetic contexts (e.g., Roy Chowdhury 1970; Keyser and Cruickshank 1979, 1980; Cooper 1980; Cox and Li 1983; Bandyopadhyay 1988, 1989; Cox 1998; Maisch 2001; Irmis 2005; Kemp 2005, Surkov et al. 2005).

Sangusaurus edentatus Cox, 1969
Figure 7.12k–l

Material: LM/NH 9/1.

Localities: Locality 15 of Drysdall and Kitching (1963) (LM/NH 9/1). The type locality of *S. edentatus* is in Drysdall and Kitching's (1963; also see Kitching 1963) upper fossiliferous horizon of the Ntawere Formation (Cox, 1969).

Identifying Characteristics: Cox (1969) and Cruickshank (1986) provided diagnoses of *Sangusaurus edentatus* and information on how it could be distinguished from the Tanzanian species *S. parringtonii*. The primary diagnostic

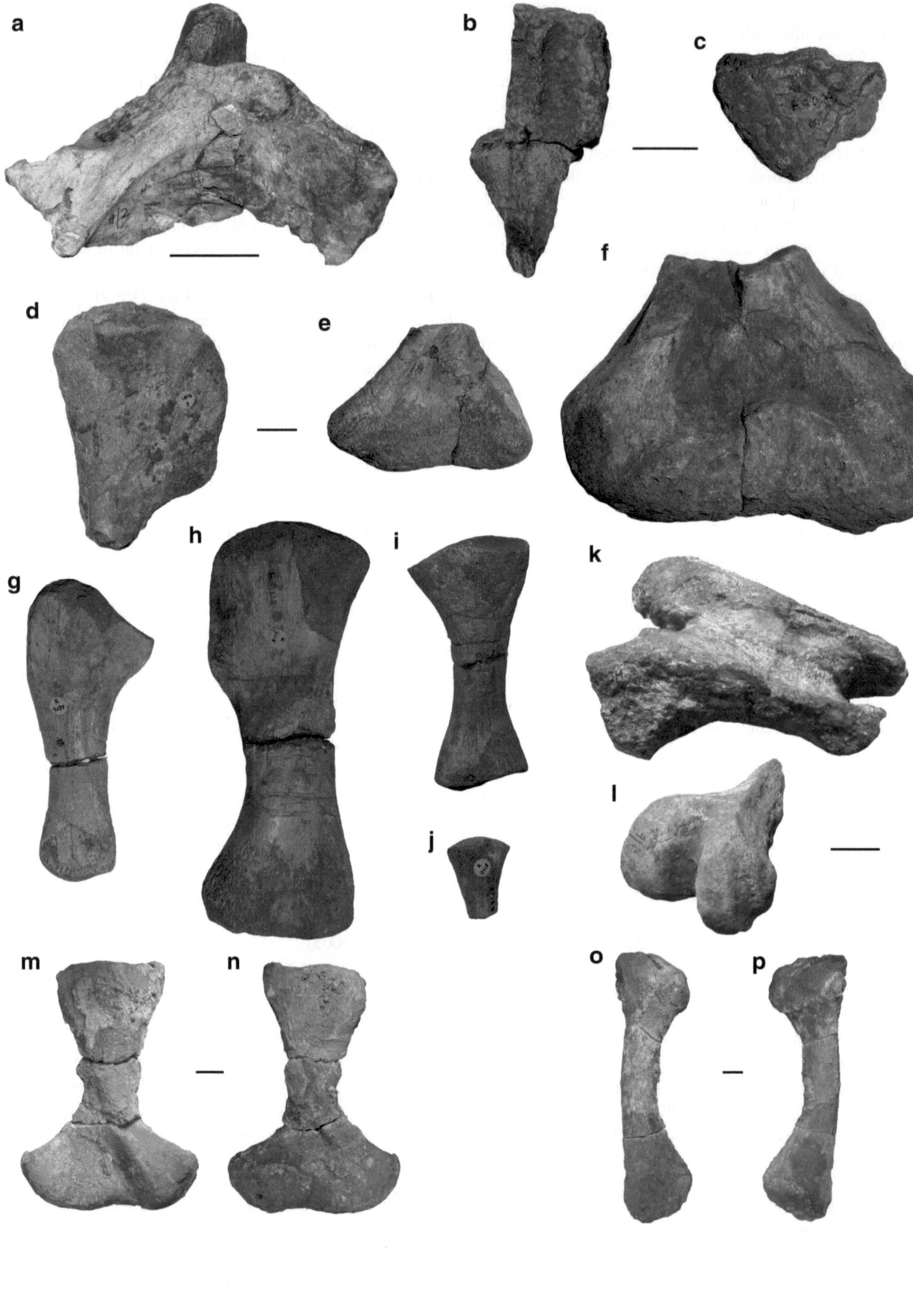
a
b
c
d
e
f
g
h
i
j
k
l
m
n
o
p

◀ **Fig. 7.12** Zambian specimens of *Zambiasaurus submersus*, *Sangusaurus edentatus*, and Kannemeyeriiformes *incertae sedis*. **a** Holotype temporal bar of *Zambiasaurus submersus* (LM 9/2) in left anterolateral view. **b** Partial premaxilla of *Zambiasaurus submersus* (NHMUK R9002) in ventral view. **c** Partial mandible of *Zambiasaurus submersus* (NHMUK R9039) in left lateral view. **d** Proximal portion of a juvenile right humerus of *Zambiasaurus submersus* (NHMUK R9091) in dorsal view. **e** Distal portion of a juvenile left humerus of *Zambiasaurus submersus* (NHMUK R9089) in ventral view. **f** Distal portion of an adult left humerus of *Zambiasaurus submersus* (NHMUK R9140) in ventral view. **g** Juvenile right ulna of *Zambiasaurus submersus* (NHMUK R9098) in anterior view. **h** Juvenile right femur of *Zambiasaurus submersus* (NHMUK R9118) in anterior view. **i** Juvenile right tibia of *Zambiasaurus submersus* (NHMUK R9123) in posterior view. **j** Proximal end of juvenile right fibula of *Zambiasaurus submersus* (NHMUK R9128) in lateral view. **k** Holotype temporal bar of *Sangusaurus edentatus* (LM 9/1) in right dorsolateral view. **l** Holotype right quadrate of *Sangusaurus edentatus* (LM 9/1) in anterodorsal view. **m** Right humerus of Kannemeyeriiformes *incertae sedis* (NHCC LB26) in dorsal view. **n** Right humerus of Kannemeyeriiformes *incertae sedis* (NHCC LB26) in ventral view. **o** Right fibula of Kannemeyeriiformes *incertae sedis* (NHCC LB26) in anterior view. **p** Right fibula of Kannemeyeriiformes *incertae sedis* (NHCC LB26) in posterior view. *Upper* left scale bar applies to *panel* **a**, *upper right* scale bar applies to *panels* **b**, **c**, *middle left* scale bar applies to *panels* **d–j**, *middle right* scale bar applies to *panels* **k**, **l**, *lower left* scale bar applies to *panels* **m**, **n**, and *lower right* scale bar applies to *panels* **o**, **p**. Scale bars are 20 mm. Photographs in *panels* **a**, **k**, and **l** courtesy of C. Mateke; photographs in *panels* b–j courtesy of C. Kammerer

characters for *Sangusaurus* are absence of tusks, temporal bar with a midline groove, presence of a boss posterior to the pineal foramen, exposure of the interparietal on the dorsal surface of the temporal bar. Cruickshank (1986) differentiated *S. edentatus* from *S. parringtonii* on the basis of the shape and size of the caniniform process.

Synonyms in Luangwa Basin Literature: *Sanguasaurus*, *Sangausaurus* (Keyser and Cruickshank 1979). Both of these names appear to be misspellings of *Sangusaurus*.

Previous Reports: Cox (1969) was the first to describe *Sangusaurus* from the Luangwa Basin. Anderson and Cruickshank (1978), King (1988, 1990), and Fröbisch (2009) included *Sangusaurus* in their biogeographic compilations. Battail (1978, 1993), Jain and Roy Chowdhury (1987), Cox (1991), and DeFauw (1993) discussed *S. edentatus* in a primarily biostratigraphic context, and various authors have considered it in a taxonomic or phylogenetic context (e.g., Roy Chowdhury 1970; Keyser and Cruickshank 1979, 1980; Cooper 1980; Cox and Li 1983; Cruickshank 1986; Bandyopadhyay 1988, 1989; Cox 1998; Maisch 2001). Sues and Fraser (2010) included *Sangusaurus* in their faunal list for the Ntawere Formation.

Kannemeyeriiformes *incertae sedis*

Figure 7.12m–p

Material: NHCC LB26.

Localities: Locality L12 (NHCC LB26). This locality is within the outcrops designated as Locality 16 by Drysdall and Kitching (1963), which is part of their lower fossiliferous horizon of the Ntawere Formation.

Identifying Characteristics: The material in this collection consists of a medium-sized dicynodont humerus (length approximately 183 mm) (Fig. 7.12m, n), a large dicynodont fibula (length approximately 279 mm) (Fig. 7.12o, p), and a rib fragment that likely belonged to an animal of similar size as that which produced the fibula. Of the dicynodonts known from the Triassic of Zambia, only *Zambiasaurus* includes definite postcranial material (Cox 1969), although Govender and Yates (2009) described specimens from Namibia that they assigned to *Dolichuranus*, cf. *Dolichuranus*, and cf. *Kannemeyeria lophorhinus* that make for relevant comparisons. The humerus of NHCC LB26 is comparable in size to the juvenile humerus (consisting of specimens NHMUK R9088 and NHMUK R9089; Fig. 7.12d, e) figured by Cox (1969), but it differs in having much more fully ossified joint surfaces (particularly on the distal end) and more strongly flaring ect- and entepicondyles (also compare to NHMUK R9091; Fig. 7.12d). The adult distal humerus of *Zambiasaurus* has comparably well-ossified joint surfaces, although it is much larger and has less flared ect- and entepicondyles. The humerus shows greater similarity to the humeri of cf. *Dolichuranus* and cf. *K. lophorhinus* described by Govender and Yates (2009). Their specimens of cf. *K. lophorhinus* (both from CGP R316) possess similarly flared ect- and entepicondyles, but the entepicondyles of their specimens of cf. *Dolichuranus* (CGP/1/191A and CGP/1/412) do not seem as strongly flared as in NHCC LB26. Both the specimens of cf. *K. lophorhinus* and cf. *Dolichuranus* figured by Govender and Yates (2009) also are noteworthy in having a tab-like projection on the posterior surface of the proximal end, near the insertion of M. subcoracoscapularis. This area is somewhat damaged in NHCC LB26, but the preserved morphology is suggestive of a similar tab having been present originally.

In addition to *Sangusaurus*, *Zambiasaurus*, *K. lophorhinus*, and *Dolichuranus*, several other dicynodont taxa are documented from coeval beds in southern and eastern Africa: *Kannemeyeria simocephalus*, *Tetragonias njalilus*, *Rechnisaurus cristarhynchus*, *Angonisaurus cruickshanki*, and *Shansiodon* sp. (e.g., von Huene 1942; Cruickshank 1967; Cox and Li 1983; King 1988; Hancox and Rubidge 1997, 2001; Hancox 2000; Rubidge 2005, Fröbisch 2009; Hancox et al. 2013). There are also reports of an additional taxon from Tanzania that was informally named "*Ruhuhuungulasaurus croucheri*" in an unpublished thesis (Larkin 1994; this specimen, NHMUK R12710, was listed as *Shansiodon* in Surkov and Benton 2004). Although humeri are not available for *Rechnisaurus* or

"*Ruhuhuungulasaurus,*" comparisons are possible with the other taxa. NHCC LB26 differs from the humeri of *K. simocephalus* (e.g., Govender et al. 2008) in having more flared distal ect- and entepicondyles. NHCC LB26 may differ from *K. simocephalus* in having a tab on the posterior surface of the proximal end if such a structure was originally present. Cox and Li (1983) did not discuss the humerus of *Angonisaurus* that is part of the holotype (NHMUK R9723), but it is relatively complete with minor damage to the proximal end, deltopectoral crest, and distal end. As preserved this specimen appears to have a less flared distal end than NHCC LB26 and no tab-like projection on the proximal end, and these observations are confirmed by a more fragmentary specimen (NMT RB155) that likely represents *Angonisaurus*, which our team collected in Tanzania in 2007. The humeri are somewhat better preserved in the holotype of *Tetragonias* (GPIT 292) than the referred specimen (CAMZM T753) described by Cruickshank (1967). GPIT 292 possesses a comparably flared distal end and a well-developed tab-like projection on the posterior surface of the proximal end. The humeral head is also prominent and well-ossified in GPIT 292, and a similar situation seems to be the case in NHCC LB26, although this area is also somewhat damaged. Unfortunately, the deltopectoral crest of NHCC LB26 is not preserved, so we cannot determine whether it possessed the hook-like anterodistal corner present in *Tetragonias* (Cruickshank 1967). A humerus for the South African *Shansiodon* specimen is not available for comparison. However, comparison to Chinese material of *Shansiodon* (e.g., IVPP V.2415; see Yeh 1959) shows that NHCC LB26 may show some similarities (e.g., well-defined humeral head, tab-like projection on the posterior surface of proximal end). At the same time, NHCC LB26 is larger and also appears to have a more flared distal end.

The fibula of *Zambiasaurus* is known only from juvenile material (NHMUK R9128, R9129) (e.g., Fig. 7.12j). Those elements show a similar degree of curvature to NHCC LB26, but the joint surfaces are nearly entirely unossified, precluding detailed comparisons. The fibula of cf. *K. lophorhinus* (CGP R316) figured by Govender and Yates (2009) has a much straighter shaft than that of NHCC LB26. Although not prepared, NHCC LB26 appears to lack the groove on the posterior surface seen in CGP R316. The proximal and distal ends of CGP R316 also seem less expanded than those of NHCC LB26, although Govender and Yates' (2009) photographs make it seem like these areas may be somewhat weathered. NHCC LB26 shows similarity to the fibula of *Dolichuranus* (BP/1/4578) described by Govender and Yates (2009), particularly in the curvature of the shaft. However, the curvature of the shaft is greater in NHCC LB26 (and this does not appear to be a taphonomic artifact given that the specimen is well preserved and does not show signs of crushing or plastic deformation) and there is no evidence of a groove on the posterior surface of the shaft comparable to that in BP/1/4578.

Fibulae also are available for comparison for *Tetragonias*, *K. simocephalus*, and "*Ruhuhuungulasaurus.*" The fibula of the holotype of *Tetragonias* (GPIT 292) is strongly curved, but the profile of this curvature is different than in NHCC LB26. In the latter specimen the shaft smoothly curves, whereas in GPIT K292 the offset between the proximal and distal ends of the fibula is more of a distinct kink. The fibula of the referred specimen (CAMZM T754) described by Cruickshank (1967) is more smoothly curved, giving it a profile more comparable to that of NHCC LB26. The proximal and distal ends of NHCC LB26 are more expanded and more strongly ossified than those of either *Tetragonias* specimen, though, and NHCC LB26 is considerably larger. The fibula of *K. simocephalus* (e.g., Govender et al. 2008) also is smoothly curved, but again is typically somewhat smaller than NHCC LB26. The latter specimen also differs from *K. simocephalus* in having a well developed ridge-like edge that forms the posterior margin of the distal articular surface, although some of this difference may stem from the fact that NHCC LB26 represents an animal that is larger than most known *Kannemeyeria* specimens. The longitudinal groove on the posterior surface of the shaft of the fibula in *K. simocephalus* described by Govender et al. (2008) appears to be absent in NHCC LB26. Finally, the fibula of "*Ruhuhuungulasaurus*" (NHMUK R12710) shows a similar degree of curvature as NHCC LB26, and its proximal and distal ends are relatively expanded. The articular surfaces are well-defined, but the distal surface is somewhat damaged, making it uncertain whether the posterior edge of distal surface had the form of a distinct ridge. NHCC LB26 also represents a considerably larger animal.

Only two major clades of dicynodonts are known to occur in the Middle Triassic, the emydopoids (represented only by *Kombuisia frerensis* in the Karoo Basin), and the much more diverse Kannemeyeriiformes. The elements included in NHCC LB26 are much larger than any known emydopoid specimens (Permian or Triassic), so we are confident in their referral to Kannemeyeriiformes. However, the limited amount of material available, and the fact that the specimens show a mixture of similarities to and differences from coeval dicynodonts known from southern and eastern Africa, prevent us from unequivocally assigning them to a specific dicynodont taxon. Their similarities to likely *Kannemeyeria lophorhinus* specimens from Namibia and their occurrence in outcrops assigned to Locality 16 by Drysdall and Kitching (1963) make *K. lophorhinus* a potential identification that deserves more scrutiny as additional postcranial material referrable to that species becomes available. Likewise, as more comparative material of *Z. submersus* and *S. edentatus* become available, and the taxonomic uncertainty regarding BP/1/3636 ("*K.*" *latirostris*) is resolved, it will be important to determine whether NHCC LB26 falls within the ranges of variation for any of these taxa. Finally, the fact that the humerus of NHCC LB26 shows similarities to shansiodontids such as *Tetragonias* and *Shansiodon* is intriguing because this specimen could represent the first occurrence of this clade in Zambia. However, additional specimens, particularly cranial material, will be necessary to confirm this possibility.

Triassic Dicynodonts Whose Presence in Zambia Cannot be Confirmed

Shansiodon Yeh, 1959

Previous Reports: Cooper (1980) and Lucas (1993a, b, 1996, 2001) suggested that *Shansiodon* was present in the Ntawere Formation. This occurrence was based on a two-step reasoning process that accepted Keyser's (1973c) referral of BP/1/3636 to *Dolichuranus* and then posited that *Dolichuranus* was a junior synonym of *Shansiodon*. Neither Cooper (1980) nor Lucas (1993a, b, 1996, 2001) specified whether the putative Zambian occurrence of *Shansiodon* represented the type species *Shansiodon wangi* Yeh, 1959 or a different species. As noted above, both steps in this reasoning are questionable. BP/1/3636 diverges from the diagnosis of *Dolichuranus* and is not clearly referable to the genus (Damiani et al. 2007). Even if BP/1/3636 is eventually shown to be part of *Dolichuranus*, the synonymy between it and *Shansiodon* is unlikely because recent phylogenetic analyses suggest that *Shansiodon* and *Dolichuranus* are not closely related (Damiani et al. 2007; Govender and Yates 2009; Kammerer et al. 2011). We are unaware of any Zambian specimens that can be referred unequivocally to *Shansiodon*. Therefore, we do not consider it part of the Zambian Triassic dicynodont fauna.

Angonisaurus cruickshanki Cox and Li, 1983

Previous Reports: Sues and Fraser (2010) included *Angonisaurus* in their list of dicynodont taxa known from the Ntawere Formation in the Luangwa Basin, and cited Cox (1969, 1991) as sources. However, *Angonisaurus* was not reported from Zambia in either of these papers, and we are unaware of any unpublished specimens that would support this record. Therefore, we do not consider *Angonisaurus* to be part of the Zambian Triassic dicynodont fauna.

Rechnisaurus cristarhynchus Roy Chowdhury, 1970

Previous Reports: Crozier (1970) initially referred BP/1/3638 to *Rechnisaurus cristarhynchus*. Keyser (1973c; also see Battail 1978, 1993 and Ochev and Shishkin 1989) followed this taxonomy, but by 1974 was expressing doubts that later culminated in the transfer of the specimen to *Kannemeyeria* (Keyser and Cruickshank 1979). Several subsequent authors further highlighted the distinctions between BP/1/3638 (and the specimen of *K. lophorhinus* from Namibia) and *Rechnisaurus* (Bandyopadhyay 1985, 1989; King 1988; Renaut 2000; Renaut et al. 2003), and if these observations are accepted, then *Rechnisaurus* is not present in the Ntawere Formation of Zambia. Lucas argued repeatedly that *Rechnisaurus* and *K. lophorhinus* (usually *K. cristarhynchus* in his papers) are synonyms (1993b, 1996, 1998, 1999, 2001, 2010; also see Lucas and Wild, 1995). If this synonymy is correct, it still would not imply the presence of *Rechnisaurus* in Zambia, since that taxon would be a junior synonym of *Kannemeyeria* (i.e., it would imply the presence of *Kannemeyeria* in India instead). Given these observations, and the fact that there are currently no other specimens from the Ntawere Formation that could represent *Rechnisaurus* (if it is a valid taxon), we conclude that *Rechnisaurus* is not part of the Zambian Triassic dicynodont fauna.

Dolichuranus Keyser, 1973c

Previous Reports: Keyser (1973c) was the first author to suggest that *Dolichuranus* occurred in the Ntawere Formation when he referred BP/1/3636, the holotype of *Kannemeyeria latirostris* Crozier, 1970, to the genus. Many subsequent authors followed this synonymy (Keyser 1974; Anderson and Cruickshank 1978; Battail 1978, 1993; Keyser and Cruickshank 1979; Cooper 1982; King 1988, 1990; Surkov 2000; Renaut 2000; Rubidge 2005). Most only referred to the genus *Dolichuranus* in Zambia, but Keyser and Cruickshank (1979) and King (1988) listed *Dolichuranus latirostris* as a valid species. As noted above however, BP/1/3636 differs from Damiani et al.'s (2007) diagnosis of *Dolichuranus*, making its assignment to the genus questionable. Likewise, although NHCC LB26 shows some similarities to *Dolichuranus* postcrania collected in Namibia, we do not consider it complete enough to provide a definitive identification. Until the identity of BP/1/3636 is resolved and/or new specimens are discovered that can be unequivocally identified as *Dolichuranus*, we do not consider it to be a part of the Zambian Triassic dicynodont fauna.

Discussion

How Many Faunal Assemblages are Preserved in the Fossiliferous Beds of the Upper Madumabisa Mudstone?

The idea that the Upper Madumabisa Mudstone preserves multiple Permian assemblages can be traced back to the earliest works on the paleontology of the Luangwa Basin. Dixey (1937) reported fossils in five horizons that are now considered part of the Madumabisa Mudstone (Drysdall and Kitching 1963) in the northern part of the basin. Based on comparisons of the fossils Dixey collected with specimens in South Africa, Boonstra (1938) concluded that some of these horizons might be coeval, but that assemblages corresponding to the *Endothiodon* and *Cistecephalus* zones of the South African Karoo (equivalent to the *Tropidostoma*, *Cistecephalus*, and *Dicynodon* assemblage zones of Rubidge et al. 1995) were present. Drysdall and Kitching (1962, 1963) and Kitching (1963) considered the problem in more detail, recognized that sets of Dixey's horizons were parts of single layers offset by faulting, and added several additional fossil localities to Dixey's list. Based on perceived faunal differences, they recognized lower, middle, and upper fossiliferous horizons in their Upper Madumabisa Mudstone. They considered the lower horizon to be equivalent to rocks of the *Endothiodon* zone of South Africa (now the *Tropidostoma* Assemblage Zone), and the middle and upper horizons to be equivalent to rocks of the *Cistecephalus* zone (equivalent to the current *Cistecephalus* and/or *Dicynodon* assemblage zones). The discovery of Permian fossils in the central Luangwa Basin (Kemp 1975) generally has been regarded as adding a fourth fossiliferous horizon that is equivalent to rocks of the *Dicynodon* Assemblage Zone of South Africa, a conclusion reinforced by the suggested co-occurrence of *Dicynodon* and *Lystrosaurus* in one of these localities (King and Jenkins 1997). Most recent biostratigraphic works included two assemblages in the Upper Madumabisa Mudstone equivalent to those of the *Cistecephalus* and *Dicynodon* assemblage zones of South Africa (e.g., Lucas 1998a, 2002, 2005, 2006; Rubidge 2005), although Fröbisch (2009) included four assemblages (consisting of Drysdall and Kitching's three horizons from the north of the basin and Kemp's horizon from the central basin) in his biogeographic study.

Our taxonomic revision of the dicynodonts of the Upper Madumabisa Mudstone provides an opportunity to reassess whether there is strong evidence of multiple assemblages in the formation or if the various localities throughout the basin are better regarded as sampling a single assemblage. To test the hypotheses of single versus multiple assemblages, we compiled faunal lists for each of Drysdall and Kitching's (1963) three fossiliferous horizons and Kemp's (1975) horizon based on voucher specimens identified from historical localities and our new localities in their immediate proximity. Because none of our new fossil localities fall within Drysdall and Kitching's (1963) lower and middle horizons, and we identified few or no voucher specimens from these horizons in museum collections, we supplemented the faunal lists with our reidentifications of Drysdall and Kitching's (1963) field identifications.

A clear pattern emerges from these results (Table 7.3). Kemp's (1975) central Luangwa Basin localities have the greatest taxonomic richness, with 14 species represented. Drysdall and Kitching's (1963) upper horizon is a close second (10 species), followed by the lower horizon (six species) and the middle horizon (at least two species). More importantly, the assemblages of all three of Drysdall and Kitching's (1963) horizons consist of subsamples of the assemblage present in Kemp's (1975) localities; no taxa are confined only to one or more of the lower horizons. Importantly, the subsamples all include taxa that have been hypothesized to have biostratigraphic utility, such as *Endothiodon*, *Odontocyclops*, *Dicynodon huenei*, and the new cistecephalid, not just stratigraphically long-ranging taxa such as *Pristerodon* or *Diictodon*. Based on these observations, we consider it most conservative to posit only a single assemblage in the Upper Madumabisa Mudstone, similar to the situation recognized for the Usili Formation of Tanzania (Sidor et al. 2010). We hypothesize that the assemblage from the central Luangwa Basin localities is more completely sampled primarily because of taphonomic issues. In particular, specimens from these localities tend to be more complete and easier to prepare than those from the northern localities, which are often encased in highly resistant hematite nodules. This nodular preservation style makes field identifications of specimens and collecting decisions difficult, in addition to slowing preparation. Nevertheless, Drysdall and Kitching's (1963) upper horizon, particularly their Locality 4, shows that it is possible to gain a relatively complete picture of the assemblage even in areas characterized by preservation in hematite nodules.

Permian Biostratigraphy and Biogeography

As noted above, most previous workers who considered the biostratigraphy of the Upper Madumabisa Mudstone correlated it with rocks belonging to one or more of the biozones of the South African Karoo Basin (e.g., Boonstra 1938; Drysdall and Kitching 1963; Kemp 1975; King and Jenkins 1997; Lucas 1998a, 2002, 2005, 2006; Angielczyk 2002; Rubidge 2005; Fröbisch 2009). Stratigraphic ranges for taxa that occur in South Africa and Zambia, as well as *Dicynodon lacerticeps* (which is closely related to *D. huenei*; Kammerer et al. 2011) and *Cistecephalus* (the likely closest South African relative of the Zambian tusked cistecephalid) are shown in Fig. 7.13. Given the long ranges of several of the taxa, it is easy to see why previous authors suggested the presence of multiple faunal assemblages corresponding to two to three South African assemblage zones.

The assumption of a single faunal assemblage in the Upper Madumabisa Mudstone greatly simplifies the problem. The taxa found in both the Karoo Basin and the

Table 7.3 Occurrences of dicynodont taxa in previously-recognized horizons of the Upper Madumabisa Mudstone

Taxon	Drysdall and Kitching Lower Horizon	Drysdall and Kitching Middle Horizon	Drysdall and Kitching Upper Horizon	Kemp Horizon
Endothiodon sp.	X			X
Pristerodon mackayi	X[a]		X	X
Diictodon feliceps	X[b]		X	X
Compsodon helmoedi				X
Emydops sp.	X[c]	X[f]	X	X
Dicynodontoides cf. *D. nowacki*			X	X
Cistecephalidae n. g. & sp.		X[g, h]	X	X
cf. *Katumbia parringtoni*				X
Odontocyclops whaitsi	X[d]		X	X
Oudenodon bainii	X[e]		X	X
Kitchinganomodon crassus			X[i]	X
Dicynodon huenei			X[j]	X
Syops vanhoepeni			X	X
Lystrosauridae n. g. & sp.				X

Drysdall and Kitching's (1963) horizons crop out in the northern part of the Luangwa Basin (Area 1 in Fig. 7.1); Kemp's horizon crops out in the central part of the basin (Area 3 and Area 4 in Fig. 7.1). The Lower Horizon includes Drysdall and Kitching's (1963) localities 3, 18, 19, 20, and 22. The Middle Horizon includes Drysdall and Kitching's (1963) localities 2, 11, and 17. The Upper Horizon includes Drysdall and Kitching's (1963) localities 1, 4, 5, 7, 8, 9, 10, 12, 13, and 21, as well as our localities L6 and L7. Kemp's (1975) Horizon includes Kerr's (1974) localities 1, 2, 3, 4, 5, 6, 7, 8, 9, 11, and 14 (although this might correspond to locality 4; see text), as well as our localities L26, L29, L30, L31, L32, L37, L38, L45, L48, L49, L50, L52, L53, L55, L59, and L61

[a] Record based on Drysdall and Kitching's (1963) field observations of *Parringtoniella*

[b] Record based on Drysdall and Kitching's (1963) field observations of *Dicynodon sollasi* and *Dicynodon grimbeeki*

[c] Record based on Drysdall and Kitching's (1963) field observations of *Emydops*

[d] Tentative record based on SAM-PK-K7936

[e] Record based on Drysdall and Kitching's (1963) field observations of *Dicynodon breviceps* and *Dicynodon corstorphinei*

[f] Record based on Drysdall and Kitching's (1963) field observations of *Emydops*

[g] Tentative record based on Drysdall and Kitching's (1963) field observations of *Cistecephalus*

[h] Drysdall and Kitching (1963) note that medium and large anomodonts occur at their Middle Horizon localities, but do not provide identifications or specimen numbers for this material

[i] Tentative record based on Drysdall and Kitching's (1963) field observations of *Platycyclops and Neomegacyclops*

[j] Tentative record based NHMUK 5-2, NHMUK 5-4, and NHMUK 5-10

Luangwa Basin all only overlap stratigraphically in the *Cistecephalus* Assemblage Zone of Rubidge et al. (1995; roughly equivalent to Faunachron I of Lucas 2002 and the Steilkransian of Lucas 2005, 2006) (Fig. 7.13). Even if some diachroneity in stratigraphic ranges is allowed between the two basins, it still seems likely that the Upper Madumabisa Mudstone primarily represents *Cistecephalus* Assemblage Zone time, with only limited overlap with the *Tropidostoma* and/or *Dicynodon* assemblage zones. This conclusion is significant because if true, it implies that the Upper Madumabisa Mudstone cannot provide direct insight into faunal turnover at the Permo-Triassic boundary.

Our faunal revision also demonstrates that Zambia is an important biostratigraphic link between the South African Karoo Basin and the Ruhuhu Basin of Tanzania. A number of authors noted that the Ruhuhu Basin includes a mixture of widespread and endemic taxa, as well as taxa that do not overlap stratigraphically in the Karoo Basin, complicating correlations between the two areas (e.g., Gay and Cruickshank 1999; Maisch 2002c; Abdala and Allinson 2005; Angielczyk 2007; Weide et al. 2009; Sidor et al. 2010). However, the presence of *Dicynodon huenei* in both Tanzania and Zambia allows a direct correlation between the Usili Formation and the Upper Madumabisa Mudstone. A second direct correlation between the two formations may be possible with *Katumbia parringtoni* if more definitive specimens than the jaw we collected in 2009 come to light. In turn, the well-supported correlation between the Upper Madumabisa Mudstone and the rocks of the *Cistecephalus* Assemblage Zone in the Karoo Basin implies that the Usili Formation also may best be regarded as primarily representing *Cistecephalus* Assemblage Zone time.

Finally, our results also have implications for biogeographic patterns in the Late Permian of southern and eastern Africa. As noted above, there has been much discussion of endemism in the Usili Formation assemblage (e.g., Gay and Cruickshank 1999; Maisch 2002c; Abdala and Allinson 2005; Angielczyk 2007; Weide et al. 2009; Sidor et al.

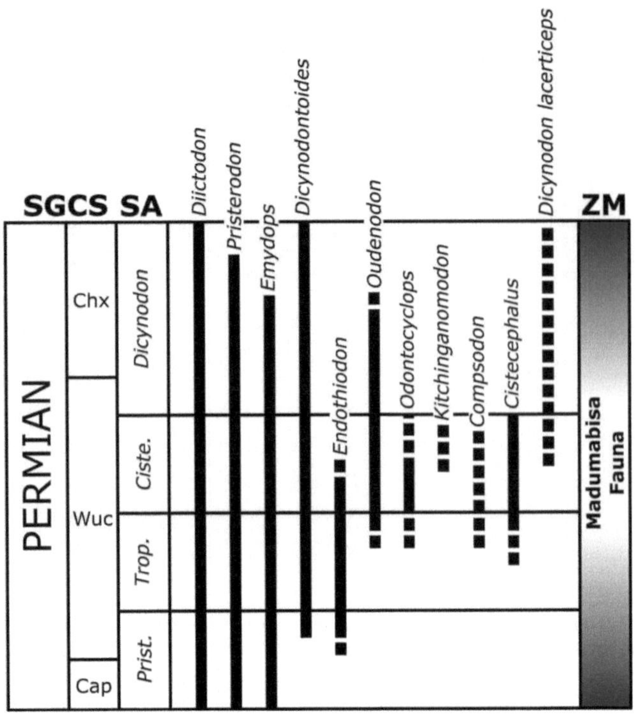

Fig. 7.13 Biostratigraphic comparison of the South African Beaufort Group (Karoo Basin) and the Zambian Upper Madumabisa Mudstone (Luangwa Basin). Stratigraphic ranges of nine dicynodont genera in both basins are plotted; *Cistecephalus* and *Dicynodon lacerticeps* occur only in the Karoo Basin but are closely related to the tusked Zambian cistecephalid and *Dicynodon huenei*, respectively. Ranges for the Karoo Basin and are based primarily on Rubidge et al. (1995), with modifications from Angielczyk (2002), Angielczyk et al. (2005, 2009), Botha and Smith (2006), Botha and Angielczyk (2007), and Botha-Brink et al. (2013). The range for *Compsodon* is uncertain because it is known from only a single specimen in the Karoo Basin. This specimen was discovered at a locality that also yielded *Rhachiocephalus* and *Oudenodon* (Kitching, 1977), and the stratigraphic range shown for *Compsodon* represents the overlap of the ranges of these two taxa. The stratigraphic range of *Dicynodon lacerticeps* is uncertain because the taxonomic revision of Kammerer et al. (2011) has greatly modified the content of that species compared to other recent usages The upper limit of its range is based on Botha-Brink et al.'s (2013) report of *D. lacerticeps* specimen BP/1/4026 close to the Permo-Triassic boundary. *SA* South African vertebrate assemblage zones, *SGCS* standard global chronostratigraphic scale, *ZM* biostratigraphic correlation of the Upper Madumabisa Mudstone (Luangwa Basin, Zambia) vertebrate fauna based on dicynodonts

2010), but most of this work focuses on comparisons with the South African Karoo Basin. Despite its intermediate geographic location, the Upper Madumabisa Mudstone assemblage received little attention on its own or in relation to the Tanzanian assemblage. Table 7.4 shows dicynodont species present in the *Cistecephalus* Assemblage Zone of South Africa, the Upper Madumabisa Mudstone, and the Usili Formation. Our focus on the species level introduces some uncertainty because species-level identifications for some of the taxa occurring in Zambia are difficult (e.g., *Endothiodon*, *Emydops*, *Dicynodontoides*). Nevertheless, we think it is important to use as detailed information as possible because previous research on this question showed that the results can be influenced by choice of taxonomic level (Abdala and Allinson 2005).

Three important points emerge from this comparison. First, in terms of their dicynodont assemblages, the Karoo, Luangwa, and Ruhuhu basins form something of a faunal gradient: eight to nine of the 14 (57–64 %) species present in Zambia also are found in the Karoo Basin, whereas only five of the 14 (36 %) species present in Tanzania also occur in South Africa. However, this pattern is not perfect because each basin shares different taxa with the others. Thus, the Luangwa Basin and the Karoo basin show roughly the same degree of similarity to the Ruhuhu Basin (only four to six, or 29–43 %, of the species in the Luangwa Basin also are found in the Ruhuhu Basin; five of 20 or 25 % of the species in the Karoo Basin are present in the Luangwa Basin). This fact is surprising given the closer proximity of the Ruhuhu and Luangwa basins now and during the Permian. Second, both the Luangwa Basin and the Ruhuhu Basin include endemic dicynodont species, but Tanzania is characterized by slightly more endemism (three of 14, or 21 %, endemic in Zambia; five to seven of 14, or 36–50 %, endemic in Tanzania). Interestingly, the Karoo Basin itself shows a previously unappreciated degree of endemism during *Cistecephalus* Assemblage Zone times (eight to 11 species, or 40–55 %), so the Luangwa Basin may be unusual in having fewer endemic species than its neighbors. Third, there does not seem to be an obvious relationship between ecology and dispersal abilities among the dicynodonts in the three basins. For example, it might be expected that larger species were more likely to be widespread given their greater resource needs. However, more than half of the species that are definitely restricted to a single basin (*Aulacephalodon baini*, *Dicynodon lacerticeps*, *Syops vanhoepeni*, *Dinanomodon gilli*, *Oudenodon grandis*, *Pachytegos stockleyi*, *Rhachiocephalus behemoth*) also are of large body size (maximum skull lengths in excess of 300 mm; estimated for *Pachytegos* based on comparisons with *Endothiodon*), whereas several small-bodied species are found in two or more basins (e.g., *Compsodon helmoedi*, *Diictodon feliceps*, *Pristerodon mackayi*, either *Emydops arctatus* or *Emydops oweni*). Ecological specialization also seems to have an inconsistent effect. *Cistecephaloides boonstrai*, *Kawingasaurus fossilis*, and the Zambian tusked cistecephalid each are restricted to a single basin, as might be expected for animals characterized by a specialized fossorial lifestyle (Cox 1972; Cluver 1974a), but *Cistecephalus microrhinus* shares a comparable lifestyle (Cluver 1978) and also is found in India (Kutty 1972; Ray 1997, 1999, 2000, 2001; Ray and Bandyopadhyay 2003). Taken together, these observations emphasize Angielczyk's (2007) and Angielczyk and Sullivan's (2008) suggestions that the

Table 7.4 Comparison of the dicynodont faunas of the *Cistecephalus* Assemblage Zone of the Karoo Basin, South Africa, the Upper Madumabisa Mudstone, Luangwa Basin, Zambia, and the Usili Formation, Ruhuhu Basin, Tanzania

Taxon	*Cistecephalus* Assemblage Zone (South Africa)	Upper Madumabisa Mudstone (Zambia)	Usili Formation (Tanzania)
Endothiodon n. sp.			X[a]
Endothiodon uniseries	X	X[b]	X
Pachytegos stockleyi			X
Pristerodon mackayi	X	X	X
Diictodon feliceps	X	X	
Compsodon helmoedi	X	X	
Emydops arctatus	X	X[c]	
Emydops oweni	X	X[c]	
Dicynodontoides recurvidens	X	X[d]	
Dicynodontoides nowacki		X[d]	X
Myosauroides minaari	X		
Cistecephalus microrhinus	X		
Cistecephaloides boonstrai	X		
Kawingasaurus fossilis			X
Cistecephalidae n. g. & sp.		X	
Katumbia parringtoni		X[e]	X
Cryptodontia n. g. & sp.			X
Keyseria benjamini	X[f]		
Oudenodon bainii	X	X	X
Oudenodon grandis	X[g]		
Odontocyclops whaitsi	X	X	
Rhachiocephalus magnus	X		X
Rhachiocephalus behemoth			X
Kitchinganomodon crassus	X	X	
Aulacephalodon bainii	X		
Geikia locusticeps			X
Dicynodon lacerticeps	X		
Dicynodon huenei		X	X
Dinanomodon gilli	X		
Syops vanhoepeni		X	
Euptychognathus bathyrhynchus	X[h]		X
Lystrosauridae n. g. & sp.		X	
Basilodon woodwardi	X		

Faunal list for the *Cistecephalus* Assemblage Zone based on Rubidge et al. (1995), with additional data from Angielczyk (2002), Botha and Angielczyk (2007), Angielczyk et al. (2009), Fröbisch (2009), and Kammerer et al. (2011). Faunal list for the Upper Madumabisa Mudstone based on this paper. Faunal list for the Ruhuhu Basin based on Sidor et al. (2010)

[a] Attridge et al. (1964) and Cox (1964, 1991) reported this material was preserved in the Ruhuhu Formation of Tanzania, but Sidor et al. (2010) suggested that it may have come from the basal portion of the Usili Formation

[b] For the purposes of this comparison, the Zambian specimens of *Endothiodon* are assumed to be *E. uniseries*, although they are too fragmentary to identify to the species level with certainty

[c] It is uncertain which species of *Emydops* is present in Zambia

[d] It is uncertain which species of *Dicynodontoides* is present in Zambia

[e] For the purposes of this comparison, the jaw assigned to cf. *Katumbia parringtoni* is considered to definitely represent that species

[f] Because it has only been recently been confirmed as a valid species (Kammerer et al. 2011), the stratigraphic range of *Keyseria benjamini* in the Karoo Basin is not well-constrained

[g] It is uncertain whether *Oudenodon grandis* is a valid species (e.g., Botha and Angielczyk 2007)

[h] Kammerer et al. (2011) were the first to report *Euptychognathus bathyrhynchus* from the Karoo Basin of South Africa, and the range of this species in that basin is poorly constrained because it is known from only three specimens

factors controlling Permian dicynodont distributions were complex, and a combination of quantitative approaches such as those of Fröbisch (2009) and techniques that incorporate phylogeny (see Angielczyk and Kurkin 2003 for a simple example) will be important in resolving their biogeographic history.

Triassic Biostratigraphy and Biogeography

Although our review of the Zambian Triassic dicynodonts does not result in changes in the number or identities of the currently-recognized species, it does underscore the fact that a good deal of uncertainty surrounds these taxa, making their use in biogeographic and biostratigraphic studies difficult. The dicynodonts of the lower Ntawere Formation show the greatest similarity to the Triassic dicynodonts of Namibia. Indeed, the presence of *Kannemeyeria lophorhinus* in the lower Ntawere Formation provides a direct correlation between these rocks and those of the upper Omingonde Formation of Namibia (e.g., Keyser 1973c; Keyser and Cruickshank 1979; Cooper 1982; Cox 1991; Lucas 1998a, b; Rubidge 2005). However, considering that this species is known from only two specimens (i.e., one specimen each from Zambia and Namibia; Renaut 2000; Renaut et al. 2003), its stratigraphic ranges in Namibia and Zambia are not well-constrained and correlations based on it will necessarily be imprecise until additional material can be recovered. If "*Kannemeyeria*" *latirostris* eventually proves to pertain to *Dolichuranus*, it would provide a second direct link between the lower Ntawere and the upper Omingonde formations (e.g., Keyser 1973c; Keyser and Cruickshank 1979; Cooper 1982; Battail 1993; Abdala et al. 2005; Rubidge 2005), and potentially to the *Cynognathus* C subzone of South Africa if a temporal bar tentatively referred to *Dolichuranus* by Abdala et al. (2005) truly represents this taxon. Alternatively, if it is not *Dolichuranus*, "*K.*" *latirostris* could represent an endemic species or provide biogeographic and/or biostratigraphic links with other basins both in Africa and elsewhere. This issue will not be resolved until the morphology, taxonomy and phylogenetic relationships of the specimen are reexamined in detail. The problem of "*K.*" *latirostris* also highlights the fact that alphataxonomic work on Triassic dicynodonts has lagged behind corresponding Permian work, and that this discrepancy has implications extending beyond simple measures of dicynodont diversity.

The dicynodonts of the upper Ntawere Formation are known from much more fragmentary material than those of the lower Ntawere, even though they are represented by a much greater number of individual specimens. *Sangusaurus* has received more biostratigraphic attention than *Zambiasaurus* because it also occurs in the Lifua Member of the Manda beds of Tanzania, providing a direct link between the Luangwa and Ruhuhu basins (Cruickshank 1986; Jain and Roy Chowdhury 1987; Cox 1991; Battail 1993; DeFauw 1993). However, the Zambian specimen is very fragmentary, and nearly all of the descriptive and phylogenetic work carried out on *Sangusaurus* focuses on Tanzanian material (Cruickshank 1986; Bandyopadhyay 1989; Maisch 2001; Surkov and Benton 2004; Kammerer et al. 2011). The question of whether *S. edentatus* and *S. parringtonii* are distinct species will be particularly important to address with future material from Zambia. *Zambiasaurus* is of little biostratigraphic utility because it is endemic to the Luangwa Basin. It may be of biogeographic significance if it is closely related to *Stahleckeria* (Cox 1969; Cox and Li 1983; Bandyopadhyay 1988; King 1988; Maisch 2001) since this would suggest an African origin for the lineage, but the only phylogenetic analysis to include *Zambiasaurus* did not recover such a relationship (Surkov et al. 2005). Again, both a detailed reassessment of the currently available *Zambiasaurus* specimens and the discovery of more complete specimens, particularly cranial material, are needed to improve our understanding of this taxon.

Conclusions

(1) The Upper Permian Upper Madumabisa Mudstone in the Luangwa Basin of Zambia preserves a single assemblage of dicynodonts consisting of 14 taxa: *Pristerodon mackayi*, *Endothiodon* sp., *Diictodon feliceps*, *Compsodon helmoedi*, *Emydops* sp., *Dicynodontoides* cf. *D. nowacki*, a new tusked cistecephalid represented by five specimens (BP/1/3337, BP/1/3591, BP/1/3603, NHCC LB18, NHCC LB19), cf. *Katumbia parringtoni*, *Kitchinganomodon crassus*, *Oudenodon bainii*, *Odontocyclops whaitsi*, *Dicynodon huenei*, S*yops vanhoepeni*, and a new lystrosaurid taxon represented by one specimen (TSK 2). Importantly, we find no evidence of *Lystrosaurus* sensu stricto in the Upper Madumabisa Mudstone. Previous reports of a number of additional taxa are duplications of one of the above taxa, mistakes, or based on non-diagnostic material.

(2) The Middle Triassic Ntawere Formation preserves two dicynodont assemblages. The lower Ntawere assemblage consists of *Kannemeyeria lophorhinus* and "*Kannemeyeria*" *latirostris*. The upper Ntawere assemblage includes *Zambiasaurus submersus* and *Sangusaurus edentatus*. Previous reports of additional dicynodont taxa primarily reflect the complex taxonomic histories of *K. lophorhinus* and "*K.*" *latirostris*.

(3) The Upper Madumabisa Mudstone dicynodont assemblage is best correlated with the *Cistecephalus* Assemblage Zone of the Karoo Basin of South Africa. In turn, the presence of *Dicynodon huenei* in the Luangwa and Ruhuhu basins, as well as the probable occurrence of *Katumbia parringtoni* in both basins, suggests that the dicynodont assemblage of the Tanzanian Usili Formation also can be correlated with the *Cistecephalus* Assemblage Zone.

(4) The Upper Madumabisa Mudstone dicynodont assemblage shows greater similarity to the dicynodont fauna of the South African *Cistecephalus* Assemblage Zone than to the assemblage preserved in the Tanzanian Usili Formation, despite the closer proximity of the Ruhuhu and Luangwa Basins. Both the Usili Formation and the *Cistecephalus* Assemblage Zone include more endemic species than the Madumabisa Mudstone, but the distribution of species in these and other basins suggests that the factors controlling the geographic ranges of Permian dicynodonts were complex.

(5) The lower Ntawere Formation dicynodont assemblage resembles the upper Omingonde Formation of Namibia in the presence of *Kannemeyeria lophorhinus*. However, the stratigraphic range of this species is poorly constrained because it is represented by a single specimen in each place. If "*Kannemeyeria*" *latirostris* is part of *Dolichuranus*, it would provide an additional tie between these formations, but resolution of this issue must await a reconsideration of the taxonomy and phylogenetic relationships of "*Kannemeyeria*" *latirostris*.

(6) *Sangusaurus* provides a link between the upper Ntawere Formation and the Lifua Member of the Manda beds, and an important question to address in the future is whether *S. edentatus* and *S. parringtonii* are distinct species. *Zambiasaurus submersus* is endemic to the Luangwa Basin and therefore is of little biostratigraphic utility. It may have biogeographic significance if it is a close relative of *Stahleckeria*, but additional work is needed to test this hypothesis.

Acknowledgments Fieldwork in Zambia in 2009 was made possible by a grant from the National Geographic Society (CRE 8571-08 to J.S.S.) and additional support from the Field Museum of Natural History. R.L.W.'s participation in the 2009 expedition was made possible by a grant from The Field Museum/IDP Foundation, Inc. African Training Fund (to K.D.A.). K. Mwamulowe and C. Chipote (NHCC) provided much helpful assistance in planning and carrying out the 2009 expedition, and in obtaining our temporary export permits. J.S.S. thanks the UMR 7207 of the CNRS (Paris) for additional support. A. Goulding and J. Menke made valuable contributions to the fieldwork and provided additional logistical support in Zambia. B. Rubidge and B. Zipfel (BP) arranged the loan of BP/1/3337, BP/1/3591, and BP/1/3603 to K.D.A. A. Shinya, L. Herzog, and C. Van Beek (Field Museum of Natural History) prepared many of the specimens figured in this paper, including all of the figured material from the 2009 expedition. We thank the anonymous reviewers for their helpful comments.

References

Abdala, F., & Allinson, M. (2005). The taxonomic status of *Parathrinaxodon proops* (Therapsida: Cynodontia), with comments on the morphology of the palate in basal cynodonts. *Palaeontologia Africana, 41*, 45–52.

Abdala, F., Hancox, P. J., & Neveling, J. (2005). Cynodonts from the uppermost Burgersdorp Formation, South Africa, and their bearing on the biostratigraphy and correlation of the Triassic *Cynognathus* Assemblage Zone. *Journal of Vertebrate Paleontology, 25*, 192–199.

Anderson, J. M., & Cruickshank, A. R. I. (1978). The biostratigraphy of the Permian and the Triassic. Part 5. A review of the classification and distribution of Permo-Triassic tetrapods. *Palaeontologia Africana, 21*, 15–44.

Angielczyk, K. D. (2002). Redescription, phylogenetic position, and stratigraphic significance of the dicynodont genus *Odontocyclops* (Synapsida: Anomodontia). *Journal of Paleontology, 76*, 1047–1059.

Angielczyk, K. D. (2007). New specimens of the Tanzanian dicynodont "*Cryptocynodon*" *parringtoni* von Huene, 1942 (Therapsida, Anomodontia), with an expanded analysis of Permian dicynodont phylogeny. *Journal of Vertebrate Paleontology, 27*, 116–131.

Angielczyk, K. D., & Kurkin, A. A. (2003). Phylogenetic analysis of Russian Permian dicynodonts (Therapsida: Anomodontia): Implications for Permian biostratigraphy and Pangaean biogeography. *Zoological Journal of the Linnean Society, 139*, 157–212.

Angielczyk, K. D., & Rubidge, B. S. (2010). A new pylaecephalid dicynodont (Therapsida, Anomodontia) from the *Tapinocephalus* Assemblage Zone, Karoo Basin, Middle Permian of South Africa. *Journal of Vertebrate Paleontology, 30*, 1396–1409.

Angielczyk, K. D., & Sullivan, C. (2008). *Diictodon feliceps* (Owen, 1876), a dicynodont (Therapsida, Anomodontia) species with a Pangaean distribution. *Journal of Vertebrate Paleontology, 28*, 788–802.

Angielczyk, K. D., Fröbisch, J., & Smith, R. M. H. (2005). On the stratigraphic range of the dicynodont taxon *Emydops* (Therapsida: Anomodontia) in the Karoo Basin, South Africa. *Palaeontologia Africana, 41*, 23–33.

Angielczyk, K. D., Sidor, C. A., Nesbitt, S. J., Smith, R. M. H., & Tsuji, L. A. (2009). Taxonomic revision and new observations on the postcranial skeleton, biogeography, and biostratigraphy of the dicynodont genus *Dicynodontoides*, the senior subjective synonym of *Kingoria* (Therapsida, Anomodontia). *Journal of Vertebrate Paleontology, 29*, 1174–1187.

Angielczyk, K. D., & Rubidge, B. S. (2013). Skeletal morphology, phylogenetic relationships, and stratigraphic range of *Eosimops newtoni* Broom, 1921, a pylaecephalid dicynodont (Therapsida, Anomodontia) from the Middle Permian of South Africa. *Journal of Systematic Palaeontology, 11*, 191–231.

Attridge, J., Ball, H. W., Charig, A. J., & Cox, C. B. (1964). The British Museum (Natural History)—University of London joint palaeontological expedition to Northern Rhodesia and Tanganyika. *Nature, 201*, 445–449.

Bandyopadhyay, S. (1985). *Dicynodont reptiles from the Triassic Yerrapalli Formation and their importance in stratigraphy and palaeontology*. Unpublished Ph.D. thesis, Calcutta University, Calcutta.

Bandyopadhyay, S. (1988). A kannemeyeriid dicynodont from the Middle Triassic Yerrapalli Formation. *Philosophical Transactions of the Royal Society Series B, 320*, 185–233.

Bandyopadhyay, S. (1989). The mammal-like reptile *Rechnisaurus* from the Triassic of India. *Palaeontology, 32*, 305–312.

Banks, N. L., Bardwell, K. A., & Musiwa, S. (1995). Karoo rift basins of the Luangwa Valley, Zambia. In J. J. Lambiase (Ed.), *Hydrocarbon habitat in rift basins* (Vol. 80, pp. 285–295). London: Geological Society, Special Publications

Battail, B. (1978). Les Reptiles Thérapsides dans la stratigraphie du Trias continental: les faunes d'âge Stormberg en Afrique et leurs équivalents dans le monde. *Société Géologique du Nord Annales, 97*, 343–350.

Battail, B. (1993). On the biostratigraphy of Triassic therapsid-bearing formations. In S. G. Lucas & M. Morales (Eds.), *The nonmarine Triassic* (pp. 31–35). *New Mexico Museum of Natural History and Science Bulletin, 3*.

Battail, B. (2009a). Late Permian dicynodont fauna from Laos. In E. Buffetaut, G. Cuny, J. Le Loeuff, & V. Suteethorn (Eds.), *Late Paleozoic and Mesozoic ecosystems in Southeast Asia* (Vol. 315, pp. 33–40). London: Geological Society, Special Publications.

Battail, B. (2009b). *Dicynodon* from South Africa, '*Dicynodon*' from Russia, and *Dicynodon*(?) from southeast Asia. *Palaeontologia Africana, 44*, 143–145.

Bond, G. (1967). A review of Karroo sedimentation and lithology in southern Rhodesia. In *Reviews prepared for the First Symposium on Gondwana Stratigraphy* (pp. 173–195). Mar del Plata: IUGS.

Boonstra, L. D. (1938). A report on some Karroo reptiles from the Luangwa Valley, Northern Rhodesia. *Quarterly Journal of the Geological Society of London, 94*, 371–384.

Botha, J., & Angielczyk, K. D. (2007). An integrative approach to distinguishing the Late Permian dicynodont species *Oudenodon bainii* and *Tropidostoma microtrema* (Therapsida, Anomodontia). *Palaeontology, 50*, 1175–1209.

Botha, J., & Smith, R. M. H. (2006). Rapid vertebrate recuperation in the Karoo Basin of South Africa following the end-Permian extinction. *Journal of African Earth Sciences, 45*, 502–514.

Botha, J., & Smith, R. M. H. (2007). *Lystrosaurus* species composition across the Permo-Triassic boundary in the Karoo Basin of South Africa. *Lethaia, 40*, 125–137.

Botha-Brink, J., Huttenlocker, A. K., & Modesto, S. P. (2013). Vertebrate paleontology of Nooitgedacht 68: A *Lystrosaurus maccaigi*-rich Permo-Triassic boundary locality in South Africa. In C. F. Kammerer, K. D. Angielczyk, & J. Fröbisch (Eds.), *Early evolutionary history of the Synapsida* (pp. 289–304). Dordrecht: Springer.

Brink, A. S. (1963). Two cynodonts from the N'tawere Formation in the Luangwa Valley of Northern Rhodesia. *Palaeontologia Africana, 8*, 77–96.

Brink, A. S., & Keyser, A. W. (1983). Illustrated bibliographic catalogue of the Synapsida. *Geological Survey of South Africa Handbook, 10*, J212A231B1.

Broom, R. (1903). On the classification of the theriodonts and their allies. *Report of the South African Association for the Advancement of Science, 1*, 286–294.

Broom, R. (1905). On the use of the term Anomodontia. *Albany Museum Records, 1*, 266–269.

Broom, R. (1913). On some new genera and species of dicynodont reptiles, with notes on a few others. *American Museum of Natural History Bulletin, 32*, 441–457.

Broom, R. (1935). A new type of anomodont reptile. *Nature, 135*, 583–584.

Cairncross, B. (2001). An overview of the Permian (Karoo) coal deposits of southern Africa. *Journal of African Earth Sciences, 33*, 529–562.

Camp, C. A., & Welles, S. P. (1956). Triassic dicynodont reptiles. Part I. The North American genus *Placerias*. *Memoirs of the Univeristy of California, 13*, 255–304.

Catuneanu, O., Wopfner, H., Eriksson, P. G., Cairncross, B., Rubidge, B. S., Smith, R. M. H., et al. (2005). The Karoo basins of south-central Africa. *Journal of African Earth Sciences, 43*, 211–253.

Chernin, S. (1974). Capitosaurid amphibians from the Upper Luangwa Valley, Zambia. *Palaeontologia Africana, 17*, 33–55.

Cluver, M. A. (1971). The cranial morphology of the dicynodont genus *Lystrosaurus*. *Annals of the South African Museum, 56*, 155–274.

Cluver, M. A. (1974a). The skull and mandible of a new cistecephalid dicynodont. *Annals of the South African Museum, 64*, 137–155.

Cluver, M. A. (1974b). The cranial morphology of the Lower Triassic dicynodont *Myosaurus gracilis*. *Annals of the South African Museum, 66*, 35–54.

Cluver, M. A. (1978). The skeleton of the mammal-like reptile *Cistecephalus* with evidence for a fossorial mode of life. *Annals of the South African Museum, 76*, 213–246.

Cluver, M. A., & Hotton, N. (1981). The genera *Dicynodon* and *Diictodon* and their bearing on the classification of the Dicynodontia (Reptilia, Therapsida). *Annals of the South African Museum, 83*, 99–146.

Cluver, M. A., & King, G. M. (1983). A reassessment of the relationships of Permian Dicynodontia (Reptilia, Therapsida) and a new classification of dicynodonts. *Annals of the South African Museum, 91*, 195–273.

Cooper, M. R. (1980). "The origins and classification of Triassic Dicynodonts" by A. W. Keyser and A. R. I. Cruickshank. Discussion. *Transactions of the Geological Society of South Africa, 83*, 107–110.

Cooper, M. R. (1982). A mid-Permian to earliest Jurassic tetrapod biostratigraphy and its significance. *Arnoldia Zimbabwe, 9*, 77–104.

Cope, E. D. (1871). On the homologies of some of the cranial bones of the Reptilia, and on the systematic arrangement of the class. *Proceedings of the American Association for the Advancement of Science, 19*, 194–247.

Cosgriff, J. W., Hammer, W. R., & Ryan, W. J. (1982). The Pangaean reptile *Lystrosaurus maccaigi* in the Lower Triassic of Antarctica. *Journal of Paleontology, 56*, 371–385.

Cox, C. B. (1964). On the palate, dentition, and classification of the fossil reptile *Endothiodon* and related genera. *American Museum Novitates, 2171*, 1–25.

Cox, C. B. (1969). Two new dicynodonts from the Triassic Ntawere Formation, Zambia. *Bulletin of the British Museum (Natural History), 17*, 257–294.

Cox, C. B. (1972). A new digging dicynodont from the Upper Permian of Tanzania. In K. A. Joysey & T. S. Kemp (Eds.), *Studies in vertebrate evolution* (pp. 173–189). Edinburgh: Oliver & Boyd.

Cox, C. B. (1991). The Pangaea dicynodont *Rechnisaurus* and the comparative biostratigraphy of Triassic dicynodont faunas. *Palaeontology, 34*, 767–784.

Cox, C. B. (1998). The jaw function and adaptive radiation of the dicynodont mammal-like reptiles of the Karoo Basin of South Africa. *Zoological Journal of the Linnean Society, 122*, 349–384.

Cox, C. B., & Li, J.-L. (1983). A new genus of Triassic dicynodont from East Africa and its classification. *Palaeontology, 26*, 389–406.

Crozier, E. A. (1970). Preliminary report on two Triassic dicynodonts form Zambia. *Palaeontologia Africana, 13*, 39–45.

Cruickshank, A. R. I. (1967). A new dicynodont genus from the Manda Formation of Tanzania (Tanganyika). *Journal of Zoology, 153*, 163–208.

Cruickshank, A. R. I. (1986). Biostratigraphy and classification of a new Triassic dicynodont from East Africa. *Modern Geology, 10*, 121–131.

Cruickshank, A. R. I., Clark, N. D. L., & Adams, C. (2005). A new specimen of *Dicynodon traquairi* (Newton) (Synapsida:

Anomodontia) from Elgin, Scotland. *Palaeontologia Africana, 41*, 35–43.

Damiani, R., Vasconcelos, C., Renaut, A., Hancox, J., & Yates, A. (2007). *Dolichuranus primaevus* (Therapsida: Anomodontia) from the Middle Triassic of Namibia and its phylogenetic relationships. *Palaeontology, 50*, 1531–1546.

Davies, K. C. (1981). Heulandite deposition in fossil reptile bones from the Karroo of Zambia. *Journal of Paleontology, 55*, 47–61.

DeFauw, S. L. (1989). Patterns of evolution in the Dicynodontia, with special reference to austral taxa. In J. A. Crame (Ed.), *Origins and evolution of the Antarctic Biota* (Vol. 47, pp. 63–84). London: Geological Society, Special Publications.

DeFauw, S. L. (1993). The Pangaean dicynodont *Rechnisaurus* from the Triassic of Argentina. In S. G. Lucas & M. Morales (Eds.), *The nonmarine Triassic* (pp. 101–105). *New Mexico Museum of Natural History and Science Bulletin, 3*.

Dixey, F. (1937). The geology of part of the upper Luangwa Valley, north-eastern Rhodesia. *Quarterly Journal of the Geological Society of London, 93*, 52–74.

Drysdall, A. R., & Kitching, J. W. (1962). The Karoo succession of the Upper Luangwa Valley, Northern Rhodesia. *Transactions and Proceedings of the Geological Society of South Africa, 65*, 75–90.

Drysdall, A. R., & Kitching, J. W. (1963). A re-examination of the Karroo succession and fossil localities of part of the upper Luangwa Valley. *Geological Survey of Northern Rhodesia Memoir, 1*, 1–62.

Freeman, L. (1993). *The cranial morphology of a new tusked species of the genus* Cistecephalus (*Therapsida, Dicynodontia*). Unpublished B.Sc. Honours thesis, University of the Witwatersrand, Johannesburg.

Fröbisch, J. (2007). The cranial anatomy of *Kombuisia frerensis* Hotton (Synapsida, Dicynodontia) and a new phylogeny of anomodont therapsids. *Zoological Journal of the Linnean Society, 150*, 117–144.

Fröbisch, J. (2008). Global taxonomic diversity of anomodonts (Tetrapoda, Therapsida) and the terrestrial rock record across the Permian-Triassic boundary. *PLoS ONE, 3*, e3733. doi: 10.1371/journal.pone.0003733.

Fröbisch, J. (2009). Composition and similarity of global anomodont-bearing tetrapod faunas. *Earth-Science Reviews, 95*, 119–175.

Fröbisch, J., & Reisz, R. R. (2008). A new species of *Emydops* (Synapsida, Anomodontia) and a discussion of dental variability and pathology in dicynodonts. *Journal of Vertebrate Paleontology, 28*, 770–787.

Gair, S. H. (1959). The Karroo System and coal resources of the Gwembe District, north-east section. *Geological Survey of Northern Rhodesia Bulletin, 1*, 1–88.

Gale, T. M. (1988). Comments on a "nest" of juvenile dicynodont reptiles. *Modern Geology, 13*, 119–124.

Gale, T. M. (1989). *Function, ecology, and relationships of fossil herbivores*. Unpublished Ph.D. thesis, University of Oxford, Oxford.

Gay, S. A., & Cruickshank, A. R. I. (1999). Biostratigraphy of the Permian tetrapod faunas from the Ruhuhu Valley, Tanzania. *Journal of African Earth Sciences, 29*, 195–210.

Govender, R., & Yates, A. (2009). Dicynodont postcrania from the Triassic of Namibia and their implications for the systematics of Kannemeyeriiforme dicynodonts. *Palaeontologia Africana, 44*, 41–57.

Govender, R., Hancox, P. J., & Yates, A. M. (2008). Re-evaluation of the postcranial skeleton of the Triassic dicynodont *Kannemeyeria simocephalus* form the *Cynognathus* Assemblage Zone (Subzone B) of South Africa. *Palaeontologia Africana, 43*, 19–37.

Grine, F. E., Foster, C. A., Cluver, M. A., & Georgi, J. A. (2006). Cranial variability, ontogeny, and taxonomy of *Lystrosaurus* from the Karoo Basin of South Africa. In M. T. Carrano, T. J. Gaudin, R. W. Blob, & J. R. Wible (Eds.), *Amniote paleobiology* (pp. 432–503). Chicago: University of Chicago Press.

Hancox, P. J. (2000). The continental Triassic of South Africa. *Zentralblatt für Geologie und Paläontologie Teil I, 1998*, 1285–1324.

Hancox, P. J., & Rubidge, B. S. (1997). The role of fossils in interpreting the development of the Karoo Basin. *Palaeontologia Africana, 33*, 41–54.

Hancox, P. J., & Rubidge, B. S. (2001). Breakthroughs in the biodiversity, biogeography, biostratigraphy, and basin analysis of the Beaufort Group. *Journal of African Earth Sciences, 33*, 563–577.

Hancox, P. J., Angielczyk, K. D., & Rubidge B. S. (2013). *Angonisaurus* and *Shansiodon*, dicynodonts (Therapsida, Anomodontia) from subzone C of the *Cynognathus* Assemblage Zone (Middle Triassic) of South Africa. *Journal of Vertebrate Paleontology, 33*, 655–676.

Haughton, S. H. (1926). On Karroo vertebrates from Nyasaland. *Transactions of the Geological Society of South Africa, 27*, 69–83.

Haughton, S. H. (1932). On a collection of Karroo vertebrates from Tanganyika Territory. *Quarterly Journal of the Geological Society of London, 88*, 634–668.

Huene, F. von. (1942). Die Anomodontier des Ruhuhu-Gebietes in der Tübinger Sammlung. *Palaeontographica Abteilung A, 44*, 154–184.

Huene, F. von. (1948). Short review of the lower tetrapods. In A. Du Toit (Ed.), *Robert Broom Commemorative Volume* (pp. 65–106). Cape Town: Royal Society of South Africa.

Huxley, T. H. (1868). On *Saurosternon Bainii,* and *Pristerodon McKayi*, two new fossil lacertillian reptiles from South Africa. *Geological Magazine, 5*, 201–205.

Irmis, R. B. (2005). The vertebrate fauna of the Upper Triassic Chinle Formation in northern Arizona. In S. J. Nesbitt, W. G. Parker, & R. B. Irmis (Eds.), *Guidebook to the Triassic Formations of the Colorado Plateau in Northern Arizona* (pp. 63–88). *Mesa Southwest Museum Bulletin, 9*.

Ivakhnenko, M. F., Golubev, V. K., Gubin, Y. M., Kalandadze, N. N., Novikov, I. V., Sennikov, A. G., et al. (1997). *Permian and Triassic Tetrapods of Eastern Europe*. Geos: Moscow.

Jacobs, L. L., Winkler, D. A., Newman, K. D., Gomani, E. M., & Deino, A. (2005). Therapsids from the Late Permian Chiweta Beds and the age of the Karoo Supergroup in Malawi. *Palaeontologia Electronica, 8*, 1–23.

Jain, S. L., & Roy Chowdhury, T. (1987). Fossil vertebrates from the Pranhita-Godavari Valley (India) and their stratigraphic correlation. In J. W. Collinson, D. H. Elliot, S. M. Haban, & G. D. McKenzie (Eds.), *Gondwana six: Stratigraphy, sedimentology, and paleontology* (pp. 219–228). *Geophysical Monograph, 41*.

Kammerer, C. F., & Angielczyk, K. D. (2009). A proposed higher taxonomy of anomodont therapsids. *Zootaxa, 2018*, 1–24.

Kammerer, C. F., Angielczyk, K. D., & Fröbisch, J. (2011). A comprehensive taxonomic revision of *Dicynodon* (Therapsida, Anomodontia), and its implications for dicynodont phylogeny, biogeography, and biostratigraphy. *Society of Vertebrate Paleontology Memoir, 11*, 1–158.

Kemp, T. S. (1975). Vertebrate localities in the Karroo System of the Luangwa Valley, Zambia. *Nature, 254*, 415–416.

Kemp, T. S. (1979). The primitive cynodont *Procynosuchus*: Functional anatomy of the skull and relationships. *Philosophical Transactions of the Royal Society of London. Series B, 285*, 73–122.

Kemp, T. S. (1980a). Aspects of the structure and functional anatomy of the Middle Triassic cynodont *Luangwa*. *Journal of Zoology, 191*, 193–239.

Kemp, T. S. (1980b). The primitive cynodont *Procynosuchus*: Structure, function and evolution of the postcranial skeleton. *Philosophical Transactions of the Royal Society of London. Series B, 288*, 217–258.
Kemp, T. S. (2005). *The origin and evolution of mammals*. Oxford: Oxford University Press.
Kerr, C. D. (1974). *Preliminary report on the 1974 Geological Survey of Zambia & Oxford University Museum Joint Palaeontological Expedition to Luangwa Valley*. Unpublished manuscript, Geological Survey of Zambia, Lusaka.
Keyser, A. W. (1972). A re-evaluation of the systematics and morphology of certain anomodont Therapsida. *Palaeontologia Africana, 14*, 15–16.
Keyser, A. W. (1973a). A re-evaluation of the genus *Tropidostoma* Seeley. *Palaeontologia Africana, 16*, 25–35.
Keyser, A. W. (1973b). A preliminary study of the type area of the *Cistecephalus* Zone of the Beaufort Series, and a revision of the anomodont family Cistecephalidae. *Geological Survey of South Africa Memoir, 62*, 1–71.
Keyser, A. W. (1973c). A new Triassic vertebrate fauna from South West Africa. *Palaeontologia Africana, 16*, 1–15.
Keyser, A. W. (1973d). New Triassic vertebrate fauna from South West Africa. *South African Journal of Science, 69*, 113–115.
Keyser, A. W. (1974). Evolutionary trends in Triassic Dicynodontia. *Palaeontologia Africana, 17*, 57–68.
Keyser, A. W. (1975). A reevaluation of the cranial morphology and systematics of some tuskless Anomodontia. *Geological Survey of South Africa Memoir, 67*, 1–110.
Keyser, A. W. (1979). A new dicynodont genus and its bearing on the origin of the Gondwana Triassic Dicynodontia. In B. Laskar & C. S. Raja Rao (Eds.), *Proceedings and papers of the 4th IUGS Gondwana symposium* (pp. 184–198). Delhi: Hindustan Publishing Corporation.
Keyser, A. W. (1981). The stratigraphic distribution of the Dicynodontia of Africa reviewed in a Gondwana context. In M. M. Creswell & P. Vella (Eds.), *Gondwana five* (pp. 61–63). Rotterdam: A.A. Balkema.
Keyser, A. W. (1993). A re-evaluation of the smaller Endothiodontidae. *Geological Survey of South Africa Memoir, 82*, 1–53.
Keyser, A. W., & Cruickshank, A. R. I. (1979). The origins and classification of Triassic dicynodonts. *Transactions of the Geological Society of South Africa, 82*, 81–108.
Keyser, A. W., & Cruickshank, A. R. I. (1980). "The origins and classification of Triassic Dicynodonts" by A. W. Keyser and A. R. I. Cruickshank. Author's reply to discussion. *Transactions of the Geological Society of South Africa, 83*, 110–111.
King, G. M. (1981). The functional anatomy of a Permian dicynodont. *Philosophical Transactions of the Royal Society of London. Series B, 291*, 243–322.
King, G. M. (1988). Anomodontia. In P. Wellnhofer (Ed.), *Handbuch der Paläoherpetologie* (Vol. 17C). Stuttgart: Gustav Fischer Verlag.
King, G. M. (1990). *The dicynodonts: A study in palaeobiology*. London: Chapman and Hall.
King, G. M. (1992). The paleobiogeography of Permian anomodonts. *Terra Nova, 4*, 633–640.
King, G. M. (1993). How many species of *Diictodon* were there? *Annals of the South African Museum, 102*, 303–325.
King, G. M., & Jenkins, I. (1997). The dicynodont *Lystrosaurus* from the Upper Permian of Zambia: Evolutionary and stratigraphical implications. *Palaeontology, 40*, 149–156.
King, G. M., & Rubidge, B. S. (1993). A taxonomic revision of small dicynodonts with postcanine teeth. *Zoological Journal of the Linnean Society, 107*, 131–154.
Kitching, J. W. (1963). The fossil localities and mammal-like reptiles of the upper Luangwa Valley, Northern Rhodesia. *South African Journal of Science, 59*, 259–264.
Kitching, J. W. (1977). The distribution of the Karroo vertebrate fauna. *Bernard Price Institute for Palaeontological Research Memoir, 1*, 1–131.
Kutty, T. S. (1972). Permian reptilian fauna from India. *Nature, 237*, 462–463.
Larkin, N. (1994). *Description of a new Triassic dicynodont from the Manda Formation of Tanzania*. Unpublished M.Sc. thesis, University College London, London.
Lehman J.-P. (1961). Dicynodontia. In Piveteau, J. (Ed.), *Traité de Paléontologie. VI(i) Mammifères. Origine Reptilienne. Evolution* (pp. 287–351). Paris: Masson et Cie.
Lucas, S. G. (1993a). The *Shansiodon* Biochron, nonmarine Middle Triassic of Pangaea. *Albertiana, 11*, 40–42.
Lucas, S. G. (1993b). Vertebrate biochronology of the Triassic of China. In S. G. Lucas & M. Morales (Eds.), *The nonmarine Triassic* (pp. 301–306). *New Mexico Museum of Natural History and Science Bulletin, 3*.
Lucas, S. G. (1996). Vertebrate biochronology of the Mesozoic of China. *Memoirs of the Beijing Natural History Museum, 55*, 110–148.
Lucas, S. G. (1997). *Dicynodon* and Late Permian Pangaea. In W. Naiwen & J. Remane (Eds.), *Proceedings of the 30th International Geological Congress* (Vol. 11, pp. 133–141). Utrecht: VSP International Science Publishers.
Lucas, S. G. (1998a). Toward a tetrapod biochronology of the Permian. In S. G. Lucas, J. W. Estep, & J. M. Hoffer (Eds.), *Permian stratigraphy and paleontology of the Robledo Mountains, New Mexico* (pp. 71–91). *New Mexico Museum of Natural History and Science Bulletin, 12*.
Lucas, S. G. (1998b). Global Triassic tetrapod biostratigraphy and biochronology. *Palaeogeography, Palaeoclimatology, Palaeoecology, 143*, 347–384.
Lucas, S. G. (1999). A tetrapod-based Triassic timescale. *Albertiana, 22*, 31–40.
Lucas, S. G. (2001). *Chinese fossil vertebrates*. New York: Columbia University Press.
Lucas, S. G. (2002). Tetrapods and the subdivision of Permian time. In L. V. Hillis, C. M. Henderson, & E. W. Bamber (Eds.), *Carboniferous and Permian of the World* (pp. 479–491). *Canadian Society of Petroleum Geologists Memoir, 19*.
Lucas, S. G. (2005) Permian tetrapod faunachrons. In S. G. Lucas & K. E. Zeigler (Eds.), *The nonmarine Permian* (pp. 197–201). *New Mexico Museum of Natural History and Science Bulletin, 30*.
Lucas, S. G. (2006). Global Permian tetrapod biostratigraphy and biochronology. In S. G. Lucas, G. Cassinis, & J. W. Schneider (Eds.), *Non-marine Permian biostratigraphy and biochronology* (Vol. 265, pp. 65–93). London: Geological Society, Special Publications.
Lucas, S. G. (2009). Timing and magnitude of tetrapod extinctions across the Permo-Triassic bounary. *Journal of Asian Earth Sciences, 36*, 491–502.
Lucas, S. G. (2010). The Triassic timescale based on nonmarine tetrapod biostratigraphy and biogeography. In S. G. Lucas (ed.), *The Triassic timescale* (Vol. 334, pp. 447–500). London: Geological Society, Special Publications.
Lucas, S. G., & Wild, R. (1995). A Middle Triassic dicynodont from Germany and the biochronology of Triassic dicynodonts. *Stuttgarter Beiträge zur Naturkunde Serie B (Geologie und Paläontologie), 220*, 1–16.
Maisch., M. W. (1999). *The tetrapods from the Late Permian of Tanzania in the collections of the Institut und Museum für Geologie und Paläontologie der Universität Tübingen, with special reference to the pristerodontian dicynodonts* Rhachiocephalus *and* Pelanomodon. Unpublished Ph.D. thesis, Eberhard Karls Universität Tübingen, Tübingen.

Maisch, M. W. (2000). Observations on Karoo vertebrates. Part 1. The taxonomic status of *Rhachiocephalus usiliensis* (von Huene, 1942) (Therapsida, Dicynodontia) from the Upper Permian Kawinga Formation of Tanzania. *Neues Jahrbuch für Geologie und Paläontologie Monatshefte, 2000*, 15–28.

Maisch, M. W. (2001). Observations on Karoo and Gondwana vertebrates. Part 2: A new skull-reconstruction of *Stahleckeria potens* von Huene, 1935 (Dicynodontia, Middle Triassic) and a reconsideration of kannemeyeriiform phylogeny. *Neues Jahrbuch für Geologie und Paläontologie Abhandlungen, 220*, 127–152.

Maisch, M. W. (2002a). Observations on Karoo and Gondwana vertebrates. Part 4: The taxonomic status of the Late Permian rhachiocephalid *Platycyclops crassus* Broom, 1948 (Therapsida: Dicynodontia) from the South African Karoo. *Neues Jahrbuch für Geologie und Paläontologie Monatshefte, 2002*, 362–372.

Maisch, M. W. (2002b). A new basal lystrosaurid dicynodont from the Upper Permian of South Africa. *Palaeontology, 45*, 343–359.

Maisch, M. W. (2002c). Observation on Karoo and Gondwana vertebrates. Part 3: Notes on the gorgonopsians from the Upper Permian of Tanzania. *Neues Jahrbuch für Geologie und Paläontologie Monatshefte, 2002*, 237–251.

Nicolas, M., & Rubidge, B. S. (2009). Assessing content and bias in South African Permo-Triassic Karoo tetrapod fossil collections. *Palaeontologia Africana, 44*, 13–20.

Nicolas, M., & Rubidge, B. S. (2010). Changes in Permo-Triassic ecological representation in the Beaufort Group (Karoo Supergroup) of South Africa. *Lethaia, 43*, 45–59.

Nopcsa, F. (1923). *Die Familien der Reptilien*. Berlin: Verlag von Gebrüder Borntraeger.

Nyambe, I. A. (1999). Tectonic and climatic controls on sedimentation during deposition of the Sinakumbe Group and Karoo Supergroup, in the mid-Zambezi Valley Basin, southern Zambia. *Journal of African Earth Sciences, 28*, 443–463.

Nyambe, I. A., & Utting, J. (1997). Stratigraphy and palynostratigraphy, Karoo Supergroup (Permian and Triassic), mid-Zambezi Valley, southern Zambia. *Journal of African Earth Sciences, 24*, 563–583.

Ochev, V. G., & Shishkin, M. A. (1989). On the principles of global correlation of the continental Triassic on the tetrapods. *Acta Palaeontologica Polonica, 34*, 149–173.

Owen, R. (1845). Report on the reptilian fossils of South Africa. Part I.—Description of certain fossil crania, discovered by A. G. Bain, Esq., in sandstone rocks at the south-eastern extremity of Africa, referable to different species of an extinct genus of Reptilia (*Dicynodon*), and indicative of a new tribe or sub-order of Sauria. *Transactions of the Geological Society of London, Second Series, 7*, 59–84.

Owen, R. (1860a). On the orders of fossil and recent Reptilia and their distribution in time. *Report of the British Association for the Advancement of Science, 1859*, 153–166.

Owen, R. (1860b). On some reptilian fossils from South Africa. *Quarterly Journal of the Geological Society of London, 16*, 49–63.

Owen, R. (1876). *Descriptive and illustrated catalogue of the fossil reptilia of South Africa in the collections of the British Museum*. London: Taylor and Francis.

Poole, D. F. G. (1956). The structure of the teeth of some mammal-like reptiles. *Quarterly Journal of Microscopical Science, 97*, 303–312.

Ray, S. (1997). Some contributions to the lower Gondwanan stratigraphy of the Pranhita-Godavari Valley, Deccan, India. *Journal of the Geological Society of India, 50*, 633–640.

Ray, S. (1999). Permian reptilian fauna from the Kundaram Formation, Pranhita-Godvari Valley, India. *Journal of African Earth Sciences, 29*, 211–218.

Ray, S. (2000). Endothiodont dicynodonts from the Late Permian Kundaram Formation, India. *Palaeontology, 43*, 375–404.

Ray, S. (2001). Small Permian dicynodonts from India. *Paleontological Research, 5*, 177–191.

Ray, S. (2005). *Lystrosaurus* (Therapsida, Dicynodontia) from India: Taxonomy, relative growth and cranial dimorphism. *Journal of Systematic Palaeontology, 3*, 203–221.

Ray, S., & Bandyopadhyay, S. (2003). Late Permian vertebrate community of the Pranhita-Godavari Valley, India. *Journal of Asian Earth Sciences, 21*, 643–654.

Renaut, A. J. (2000). *A re-evaluaiton of the cranial morphology and taxonomy of the Triassic dicynodont genus* Kannemeyeria. Unpublished Ph.D. thesis, University of the Witwatersrand, Johannesburg.

Renaut, A. J., & Hancox, P. J. (2001). Cranial description and taxonomic re-evaluation of *Kannemeyeria argentinensis* (Therapsida: Dicynodontia). *Palaeontologia Africana, 37*, 81–91.

Renaut, A. J., Damiani, R. J., Yates, A. M., & Hancox, P. J. (2003). A taxonomic note concerning a dicynodont (Synapsida: Anomodontia) from the Middle Triassic of East Africa. *Palaeontologia Africana, 39*, 93–94.

Roy Chowdhury, T. (1970). Two new dicynodonts from the Triassic Yerrapalli Formation of Central India. *Palaeontology, 13*, 132–144.

Rubidge, B. S. (2005). Reuniting lost continents—fossil reptiles from the ancient Karoo and their wanderlust. *South African Journal of Geology, 108*, 135–172.

Rubidge, B. S., & Sidor, C. A. (2001). Evolutionary patterns among Permo-Triassic therapsids. *Annual Review of Ecology and Systematics, 32*, 449–480.

Rubidge, B. S., Johnson, M. R., Kitching, J. W., Smith, R. M. H., Keyser, A. W., & Groenewald, G. H. (1995). An introduction to the biozonation of the Beaufort Group. In B. S. Rubidge (Ed.), *Biostratigraphy of the Beaufort Group (Karoo Supergroup)* (pp. 1–2). South African Committee for Stratigraphy Biostratigraphic Series, 1.

Seeley, H. G. (1889). Researches on the structure, organisation, and classification of the fossil Reptilia.—VI. On the anomodont Reptilia and their allies. *Philosophical Transactions of the Royal Society of London B, 180*, 215–296.

Seeley, H. G. (1894). Researches on the structure, organisation, and classification of the fossil Reptilia.—Part IX., Section 1. On the Therosuchia. *Philosophical Transactions of the Royal Society of London. Series B, Biological sciences, 185*, 987–1018.

Sidor, C. A., Angielczyk, K. D., Weide, D. M., Smith, R. M. H., Nesbitt, S. J., & Tsuji, L. A. (2010). Tetrapod fauna of the lowermost Usili Formation (Songea Group, Ruhuhu Basin) of southern Tanzania, with a new burnetiid record. *Journal of Vertebrate Paleontology, 30*, 696–703.

Smith, R. M. H. (1987). Helical burrow casts of therapsid origin from the Beaufort Group (Permian) of South Africa. *Palaeogeography, Palaeoclimatology, Palaeoecology, 60*, 155–170.

Smith, R. M. H. (1995). Changing fluvial environments across the Permian-Triassic boundary in the Karoo Basin, South Africa and possible causes of tetrapod extinctions. *Palaeogeography, Palaeoclimatology, Palaeoecology, 117*, 81–104.

Smith, R. M. H., & Keyser, A. W. (1995). Biostratigraphy of the *Cistecephalus* Assemblage Zone. In B. S. Rubidge (Ed.), *Biostratigraphy of the Beaufort Group (Karoo Supergroup)* (pp. 23–28). South African Committee for Stratigraphy Biostratigraphic Series, 1.

Smith, R. M. H., Eriksson, P. G., & Botha, W. J. (1993). A review of the stratigraphy and sedimentary environments of the Karoo-aged basins of Southern Africa. *Journal of African Earth Sciences, 16*, 143–169.

Steyer, J. S. (2009). The geological and palaeontological exploration of Laos—following in the footsteps of J.B.H. Counillon. *Journal of the Geological Society, London, 315*, 25–32.

Sues, H.-D., & Fraser, N. C. (2010). *Triassic life on land*. New York: Columbia University Press.

Sullivan, C., & Reisz, R. R. (2005). Cranial anatomy and taxonomy of the Late Permian dicynodont *Diictodon*. *Annals of Carnegie Museum, 74*, 45–75.

Sullivan, C., Reisz, R. R., & Smith, R. M. H. (2003). The Permian mammal-like herbivore *Diictodon*, the oldest known example of sexually dimorphic armament. *Proceedings of the Royal Society of London B, 270*, 173–178.

Surkov, M. V. (2000). On the historical biogeography of Middle Triassic anomodonts. *Paleontological Journal, 34*, 84–88.

Surkov, M. V., & Benton, M. J. (2004). The basicranium of dicynodonts (Synapsida) and its use in phylogenetic analysis. *Palaeontology, 47*, 619–638.

Surkov, M. V., Kalandadze, N. N., & Benton, M. J. (2005). *Lystrosaurus georgi*, a dicynodont from the Lower Triassic of Russia. *Journal of Vertebrate Paleontology, 25*, 402–413.

Tankard, A., Welsink, H., Aukes, P., Newton, R., & Stettler, E. (2009). Tectonic evolution of the Cape and Karoo basins of South Africa. *Marine and Petroleum Geology, 26*, 1379–1412.

Toerien, M. J. (1954). Note on the systematic position of *Compsodon*, van H. *Navorsinge van die Nasionale Museum, 1*, 131–132.

Tollman, S. M., Grine, F. E., & Hahn, B. D. (1980). Ontogeny and sexual dimorphism in *Aulacephalodon* (Reptilia, Anomodontia). *Annals of the South African Museum, 81*, 159–186.

van Hoepen, E. C. N. (1934). Oor die indeling van die Dicynodontidae na aanleiding van nuew vorme. *Paleontologiese Navorsing van die Nasionale Museum, 2*, 67–101.

Ward, P. D., Botha, J., Buick, R., De Kock, M. O., Erwin, D. H., Garrison, G. H., et al. (2005). Abrupt and gradual extinction among Late Permian land vertebrates in the Karoo Basin, South Africa. *Science, 307*, 709–714.

Weide, D. M., Sidor, C. A., Angielczyk, K. D., & Smith, R. M. H. (2009). A new record of *Procynosuchus delaharpeae* (Therapsida: Cynodontia) from the Upper Permian Usili Formation, Tanzania. *Palaeontologia Africana, 44*, 21–26.

Yeh, H.-K. (1959). New dicynodont from *Sinokannemeyeria*-fauna from Shansi. *Vertebrata PalAsiatica, 3*, 187–204.

Yemane, K., & Kelts, K. (1990). A short review of palaeoenvironments for Lower Beaufort (Upper Permian) Karoo sequences from southern to central Africa: A major Gondwana lacustrine episode. *Journal of African Earth Sciences, 10*, 169–185.

Chapter 8
Anatomical Plasticity in the Snout of *Lystrosaurus*

Sandra C. Jasinoski, Michael A. Cluver, Anusuya Chinsamy, and B. Daya Reddy

Abstract The skull of *Lystrosaurus*, characterized by an elongated snout and a scarf premaxilla-nasal suture, differs from the generalized Permian dicynodont form. The sutural relationships of the bones of the *Lystrosaurus* snout are further investigated here using several anatomical lines of evidence: gross osteology, histological and serial sections, and micro-computed tomography scans. Novel evidence was found for supernumerary bone(s) in the dorsal region of the snout in a few specimens of *Lystrosaurus*. The developmental and functional implications of this anatomical plasticity are discussed. It is hypothesized that the supernumerary bones may have formed from separate ossification centers of the frontal bone.

Keywords Dicyndontia • Feeding • Skull • Supernumerary bone • Suture morphology

Introduction

Dicynodonts were a diverse group of herbivorous non-mammalian therapsids that lived mainly during the Permian and Triassic, although contentious remains were recently found in the Cretaceous of Australia (Thulborn and Turner 2003). This highly successful group of herbivores had a worldwide distribution and occupied a large range of ecological niches (King 1988, 1990).

Lystrosaurus, one of the best-known and most abundant dicynodont genera, occurs in Late Permian and Early Triassic strata. It is the only dicynodont genus known to have crossed the Permo-Triassic extinction boundary (Smith et al. 2012) and had a cosmopolitan distribution, with remains found in South Africa (e.g., Smith and Botha 2005), Antarctica (Colbert 1974), India (Tripathi and Satsangi 1963), China (Yuan and Young 1934), and Russia (Battail and Surkov 2000). Its abundance in South Africa led to its designation as a biostratigraphic marker in the earliest Triassic (e.g., Kitching 1977), and may allow for morphological variation among individuals to be more easily detected.

The first specimen of *Lystrosaurus* was described by Huxley in 1859 as *Dicynodon murrayi*, but was later recognized to be a species of the genus *Lystrosaurus* described by Cope in 1870 (Cluver 1971, and see review by King 1988). Since then, the functional and taxonomic significance of the cranial morphology of *Lystrosaurus* has been investigated in several studies (e.g., Cluver 1971; Hotton 1986; King and Cluver 1991; Thackeray et al. 1998; Ray 2005; Grine et al. 2006; Jasinoski et al. 2009, 2010a, b; Jasinoski and Chinsamy-Turan 2012). Despite the numerous studies of its cranial and postcranial anatomy, the ecology of *Lystrosaurus* continues to be widely debated (e.g., Cox 1991; King 1991; Germain and Laurin 2005; Ray et al. 2005; Botha-Brink and Angielczyk 2010). The cranial morphology of *Lystrosaurus*, with its ventrally elongated snout and anteroposteriorly shortened skull (Cluver 1971; King and Cluver 1991; Fig. 8.1a), is distinct from that of other contemporary dicynodonts. The combination of a ventrally elongated snout, which presumably lowered its feeding level, along with contentious evidence for a 'narial flap', were traditionally interpreted as a suite of adaptations for a semi-aquatic lifestyle (see King 1991 for a review).

S. C. Jasinoski (✉) · A. Chinsamy
Department of Biological Sciences, University of Cape Town, Private Bag X3, Rondebosch, Cape Town, 7701, South Africa
e-mail: sandra_jas@hotmail.com

M. A. Cluver
Iziko: South African Museum, PO Box 61, Cape Town, 8000, South Africa

B. D. Reddy
Centre for Research in Computational and Applied Mechanics, University of Cape Town, Private Bag X3, Rondebosch, Cape Town, 7701, South Africa

Christian F. Kammerer, Kenneth D. Angielczyk and Jörg Fröbisch (eds.), *Early Evolutionary History of the Synapsida*, Vertebrate Paleobiology and Paleoanthropology, DOI: 10.1007/978-94-007-6841-3_8,

However, the elongated snout was not necessarily a semi-aquatic adaptation (King 1991), and may have been useful for grubbing activities (King and Cluver 1991). Based on observations of gross osteological specimens, King and Cluver (1991) surmised that the premaxilla-nasal suture in the snout of *Lystrosaurus* permitted small sliding movements. Sections through the *Lystrosaurus* snout confirmed that constrained sliding at the premaxilla-nasal sutural contact was possible because of the extensive and relatively straight scarf suture separating these two bones (Jasinoski et al. 2010a). The scarf premaxilla-nasal suture decreased stress and strain in the anterior surface of the snout during a snapping bite (Jasinoski et al. 2009, 2010b), and may have also reduced the forces associated with grubbing activities (King and Cluver 1991).

In this study we investigate the anatomical and sutural relationships of the bones in the snout of *Lystrosaurus* using several lines of evidence: gross osteology, histological and serial sections, and micro-computed tomography (micro-CT) scans. The results of the survey provide novel evidence for the existence of a supernumerary bone in the dorsal snout region of *Lystrosaurus*. We discuss the functional and developmental implications of this anatomical plasticity.

Institutional Abbreviations: BP, Bernard Price Institute for Palaeontological Research, University of the Witwatersrand, Johannesburg, South Africa; BRSUG, Bristol University Geology Museum, Bristol, UK; CGP, Council for Geosciences, Pretoria, South Africa; NHMUK, The Natural History Museum, London, UK; SAM, Iziko: South African Museum, Cape Town, South Africa; TM, Ditsong: National Museum of Natural History (formerly Transvaal Museum), Pretoria, South Africa; UCL, University College London, London, UK.

Materials and Methods

Several gross osteological specimens of *Lystrosaurus* crania, mainly housed in the Karoo Palaeontological Collections at the SAM (413 cranial specimens listed in the database), were examined (Table 8.1; Figs. 8.1, 8.2). This survey was supplemented by high resolution digital photographs of *Lystrosaurus* from the BP, NHMUK and TM. The four South African species of *Lystrosaurus* (*L. declivis, L. murrayi, L. curvatus L. maccaigi*) recognized by Grine et al. (2006) were investigated. Also examined were a few well-preserved crania of the basal dicynodontoids *Dicynodon lacerticeps* and *Daptocephalus leoniceps* (Kammerer and Angielczyk 2009; Kammerer et al. 2011). These specimens are housed in the SAM, and digital photographs of a few specimens from the BP, CGP, and Graaff-Reinet Museum (A. T. Bremner Collection, Graaff-Reinet) were also examined. In addition, high resolution digital images of *Kwazulusaurus* (BP/1/2792), closely related to *Lystrosaurus* either as its sister taxon or within the genus (Maisch 2002; Kammerer et al. 2011), were also scrutinized.

In addition to examination of the gross osteology, we utilized histological sections, serial sections of skulls, and micro-CT scan slices to establish the internal morphology of the snout and anterior skull roof. A longitudinal histological section was taken through the midline of the snout of three *Lystrosaurus* specimens (Section 2, Table 8.1; Jasinoski et al. 2010a). In addition, an offcut block of specimen SAM-PK-K1342 (block 5, anterior skull roof; Table 8.1) was cut longitudinally. Serial sections of *Lystrosaurus*, prepared by M. A. Cluver and housed in the SAM (Cluver 1971), were also examined (Table 8.1). These included serial transverse sections of SAM-PK-K90 (*L. curvatus*) and oblique sections from SAM-PK-K1284 (*L. declivis*).

Micro-CT scans of a skull of *Lystrosaurus declivis* (BRSUG 22211; Table 8.1) were produced at three different resolutions using an HMX microfocus X-ray system (Metris, Tring, UK). The isotropic voxel size for two micro-CT scans of the entire cranium was 73.3 and 90 μm, and was 35.4 μm for a partial scan of the dorsal snout/anterior skull roof. A three-dimensional model was reconstructed from the slice data using Mimics software (version 13.1, Materialise, Belgium).

Specimens SAM-PK-K1250 (*L. declivis*) and SAM-PK-8988 (*Lystrosaurus* sp.) were also micro-CT scanned (X-Sight, Stellenbosch, ZA), and SAM-PK-K4800 (*L. declivis*) and UCL R329 (*L. murrayi*) were previously scanned using a conventional medical CT scanner (Jasinoski et al. 2010a). However, the contrast and resolution of these scans were insufficient to distinguish suture morphology in the dorsal snout region.

Results

Gross Osteological Evidence

Although there were many well-preserved cranial specimens of *Lystrosaurus* in the examined collections, the area of the nasal-frontal suture had a tendency of not being well-preserved or, in some cases, not fully prepared. In addition, the sutural region in several specimens of *L. murrayi* and *L. declivis* (e.g., Fig. 8.2a) and in all specimens of *L. maccaigi* was obscured by frontal tuberosities, an interfrontal ridge, and/or a prominent frontonasal ridge (=prefrontal-nasal crest of Grine et al. 2006), which separates the snout from the anterior skull roof. This greatly reduced the

Table 8.1 Partial list of *Lystrosaurus* specimens described in this study

Taxon	Specimen number	Type of sample	Basal skull length (mm)	Possible ontogenetic stage[a]	Supernumerary bone(s) present?	Supernumerary bone(s) width (mm)
L. declivis	BRSUG 22211	Gross osteological, micro-CT	107	Late juvenile	Equivocal (paired)	N/A
L. sp.	SAM-PK-8988	Gross osteological	N/A	Late juvenile	Yes (paired)	3.9
L. murrayi	SAM-PK-11186	Gross osteological	92	Late juvenile	Yes (paired)	4.1 (approx)
L. declivis	SAM-PK-K1250	Gross osteological	N/A	Subadult	Yes (single)	7.4
L. declivis	SAM-PK-K10616	Gross osteological	96	Late juvenile	Yes (paired)	2.3 (approx)
L. murrayi	SAM-PK-1195	Gross osteological	N/A	Late juvenile	Equivocal (paired)	N/A
L. murrayi	SAM-PK-4326	Gross osteological	113	Subadult	Equivocal (paired)	N/A
L. murrayi	SAM-PK-8985	Gross osteological	120	Subadult	Equivocal (paired)	N/A
L. murrayi	SAM-PK-K10687	Gross osteological	138	Subadult	Equivocal (single)	N/A
L. curvatus	SAM-PK-K1269-2	Histological	125[b]	Subadult	Post-nasal[c] (paired)	N/A
L. murrayi	SAM-PK-K1342-2, K1342-5 (block)	Histological, offcut section	135	Subadult	No	N/A
L. curvatus	SAM-PK-K1422-2	Histological	N/A	Subadult	N/A	N/A
L. curvatus	SAM-PK-K90	Serial section	N/A	Subadult	Post-nasal (paired)	N/A
L. declivis	SAM-PK-K1284	Serial section	N/A	Subadult	Post-nasal (paired)	N/A

[a] Based upon gross osteological features, Ray et al.'s (2005) ontogenetic categories, and/or cranial histological features (Jasinoski and Chinsamy-Turan 2012)
[b] Cranium is compressed anteroposteriorly
[c] Designated as a 'post-nasal' bone because it may either represent the anterior process of the frontal bone or a supernumerary bone

number of specimens that could provide unequivocal anatomical information in the region of the nasal-frontal suture. In addition, these complicating factors may have led to misinterpretation of the nature of the nasal-frontal suture, which in some cases was previously described as a straight transverse suture (see summary in Grine et al. 2006). As the nasal-frontal suture was not discernible in any specimens of *L. maccaigi*, the following description applies only to the other three South African species of *Lystrosaurus*.

The paired anterior processes of the frontal bones extend anteriorly along the midline of the skull to insert between the paired nasal bones (Figs. 8.1b, 8.2b; Cluver 1971: Fig. 27). These anterior frontal processes are usually triangular, with the ends tapering anteroventrally. The shape and size of the paired frontal processes can vary across specimens, and in some cases, may differ on either side of the sagittal midline within the same specimen (e.g., Fig. 8.2b). The anteroventral termination of the frontal processes occurs at or near the frontonasal ridge (if present) (Figs. 8.1b, 8.2b), although some specimens (especially *L. curvatus*) have shorter frontal processes.

In specimen SAM-PK-K1397 (*L. murrayi*), the right anterior process of the frontal is taphonomically displaced outwards, indicating that the frontal process overlaps the underlying nasal bone. The widely patent nasal-frontal suture of specimen NHMUK R7889 (*Lystrosaurus* sp.) indicates that the margins of the frontal processes and the adjacent nasal bones are relatively straight, whereas the sutural contacts lateral to this are tightly sinuous. This indicates that the sutural overlap occurs only in the region of the anterior processes of the frontal.

Paired anterior frontal processes were rarely observed in *L. declivis* (Fig. 8.1b) mainly because of the presence of prominent frontal rugosities, an interfrontal ridge, and a frontonasal ridge (Fig. 8.2a). An alternate type of variation observed in *L. declivis* is described in the subsequent section 'Micro-CT scan evidence.'

A single median bone or paired bones, separated by sutures from the surrounding nasal and frontal bones, could be unequivocally identified in four gross osteological specimens of *Lystrosaurus* (Table 8.1; Fig. 8.1c, d). Because these bone(s) could not be attributed to either the nasal or frontal bones, they are designated as supernumerary or neomorphic bones. These supernumerary bone(s) were observed in specimens of both *L. declivis* and *L. murrayi*, but not in *L. curvatus* or *L. maccaigi* (Table 8.1). The bone(s) extend anteroventrally to a point at or near the frontonasal ridge. Interestingly, an extra bone with a similar position between the nasal and frontal bones was described in the Late Triassic dicynodont *Jachaleria candelariensis* (Vega-Dias and Schultz 2004), but has not been described in any other nonmammalian synapsid.

Paired supernumerary bones appear to be present in at least three specimens (Table 8.1; Fig. 8.1c); however, in

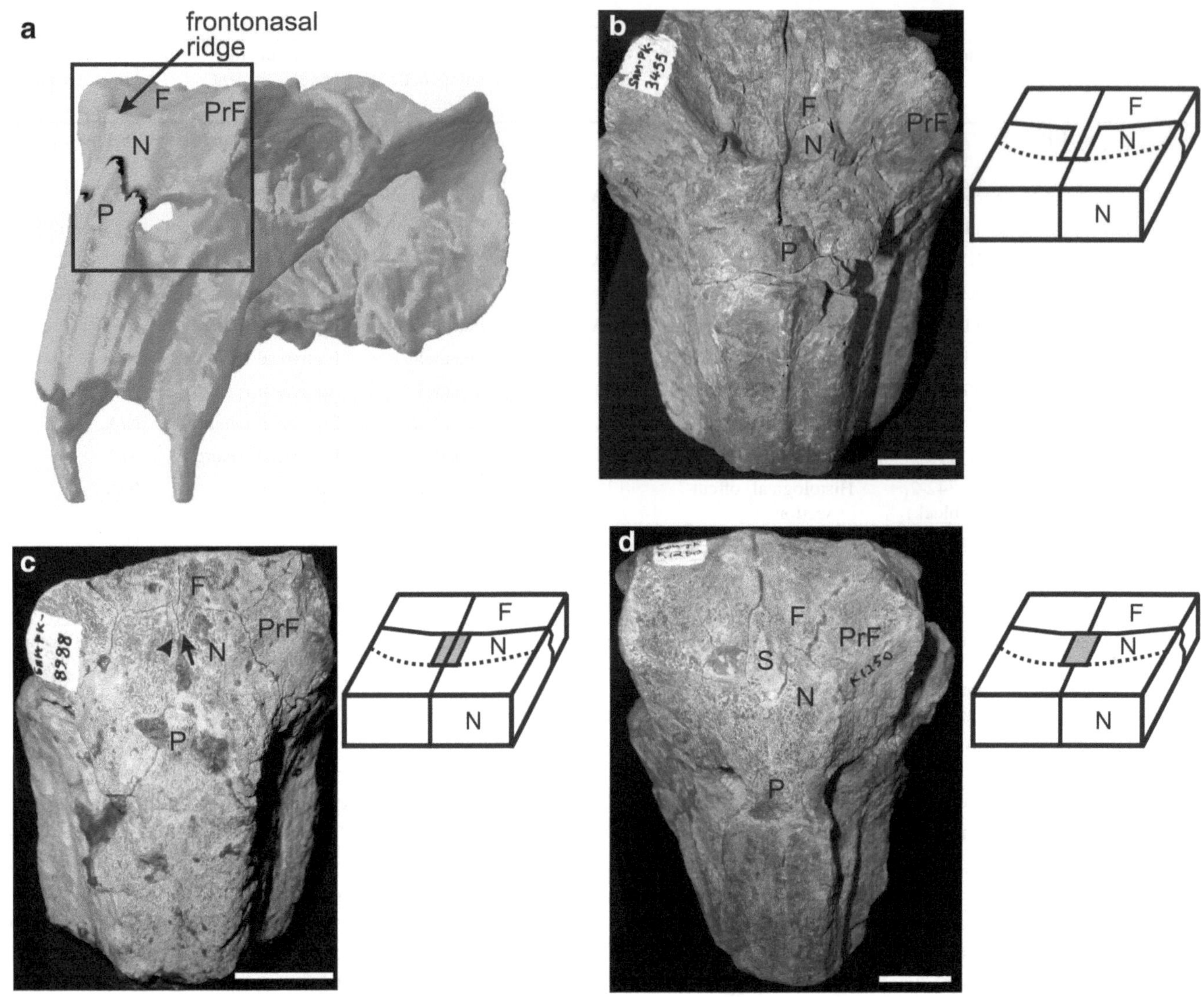

Fig. 8.1 **a** Oblique view of the digitally reconstructed cranium of *Lystrosaurus* (*L. declivis*, SAM-PK-K4800), showing the location of the premaxilla-nasal suture. *Boxed area* demarcates the dorsal region of the snout and anterior region of the skull roof. Figure modified from Jasinoski et al. (2010b). © Copyright 2010 The Society of Vertebrate Paleontology. Reprinted and distributed with permission of the Society of Vertebrate Paleontology. **b–d** Specimens and corresponding schematic diagrams of the dorsal snout and anterior skull roof regions pertaining to each type of bone configuration (see text for details). *Dotted line* indicates frontonasal ridge. **b** Anterodorsal view of the snout and anterior skull roof of SAM-PK-3455 (*L. declivis*) showing paired anterior processes of the frontal. **c** Anterodorsal view of the snout and anterior skull roof of SAM-PK-8988 (*Lystrosaurus* sp.), showing variation within the paired supernumerary bone configuration: one supernumerary bone is isolated (*arrowhead*) and one is fused posterodorsally with the frontal bone (*arrow*). **d** Anterodorsal view of the snout and anterior skull roof of SAM-PK-K1250 (*L. declivis*) showing a single median supernumerary bone. Abbreviations for Figs. 8.1, 8.2, 8.3 and 8.4: *F* frontal, *N* nasal, *P* premaxilla, *pn* 'postnasal' bone (designated as such because we could not determine unequivocally if this represents the anterior process of the frontal bone or a supernumerary bone), *PrF* prefrontal, *S* supernumerary bone. All scale bars are 2 cm

each specimen, one of the paired supernumerary bones is either fused posteriorly with the frontal (Fig. 8.1c) or the suture surrounding the supernumerary bone is not fully discernible. The supernumerary bones are longer than wide and taper ventrally. The total width across the narrow paired bones ranges from approximately 2.3 to 4.1 mm (Table 8.1). The presence of paired supernumerary bones is equivocal in an additional four specimens (Table 8.1) because the sutural relationships of the supposed supernumerary bones with the surrounding bones are not completely clear due to poor preservation.

The supernumerary bone manifests as a single median element in two specimens (Table 8.1). It is somewhat diamond-shaped in specimen SAM-PK-K1250 (*L. declivis*; Fig. 8.1d), but it is narrower in SAM-PK-K10687 (*L. murrayi*). In both specimens, the median supernumerary bone is

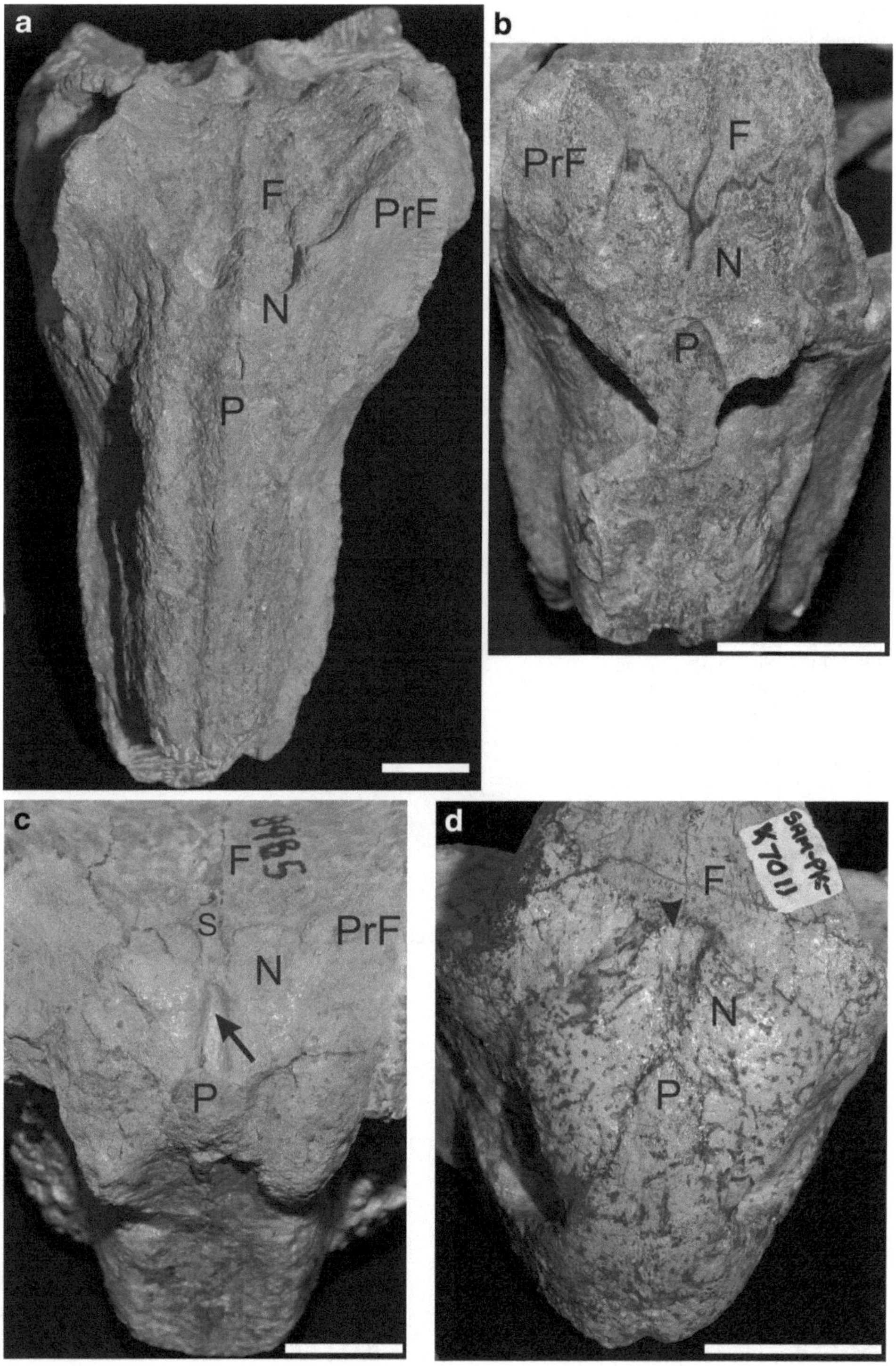

Fig. 8.2 **a** *Lystrosaurus declivis* (SAM-PK-K10373) in anterodorsal view showing frontal tuberosities and a prominent frontonasal and interfrontal ridge, which prevent the nasal-frontal suture from being unequivocally documented. **b** An example of variation between the left and right anterior frontal processes within the same individual (*L. murrayi*, SAM-PK-K10686). **c** Specimen SAM-PK-8985 (*L. murrayi*) showing the widely separated nasal bones (*arrow*), which may have accommodated a nasal tunnel similar to that observed in BRSUG 22211 (Fig. 8.4b). The internasal gap does not appear to be an artefact because mechanical preparation was carefully done without artificially creating the gap. Note that an equivocal supernumerary bone is present only on the right side of the skull. **d** *Dicynodon lacerticeps* (SAM-PK-K7011) showing a depression on the posterodorsal part of the nasal bones (*arrowhead*), which roughly corresponds to the area where the supernumerary bones occur in *Lystrosaurus* (e.g., Fig. 8.1c, d). All scale bars are 2 cm

wider than the combined width of the paired supernumerary bones found in the other specimens (Table 8.1).

A shallow midline depression on the surface of the dorsal snout, occurring on either the nasal or frontal bones, was observed in two specimens of *Dicynodon lacerticeps* (SAM-PK-K7011, -K7806; Fig. 8.2d) and one specimen of *Daptocephalus leoniceps* (BP/1/2188). This nasofrontal depression roughly corresponds to the area where the supernumerary bones occur in *Lystrosaurus*. In *Kwazulusaurus*, the midline region between the ascending process (dorsal tip) of the premaxilla and the anterior paired frontal processes was covered in matrix, thus the presence of a nasofrontal depression could not be ascertained.

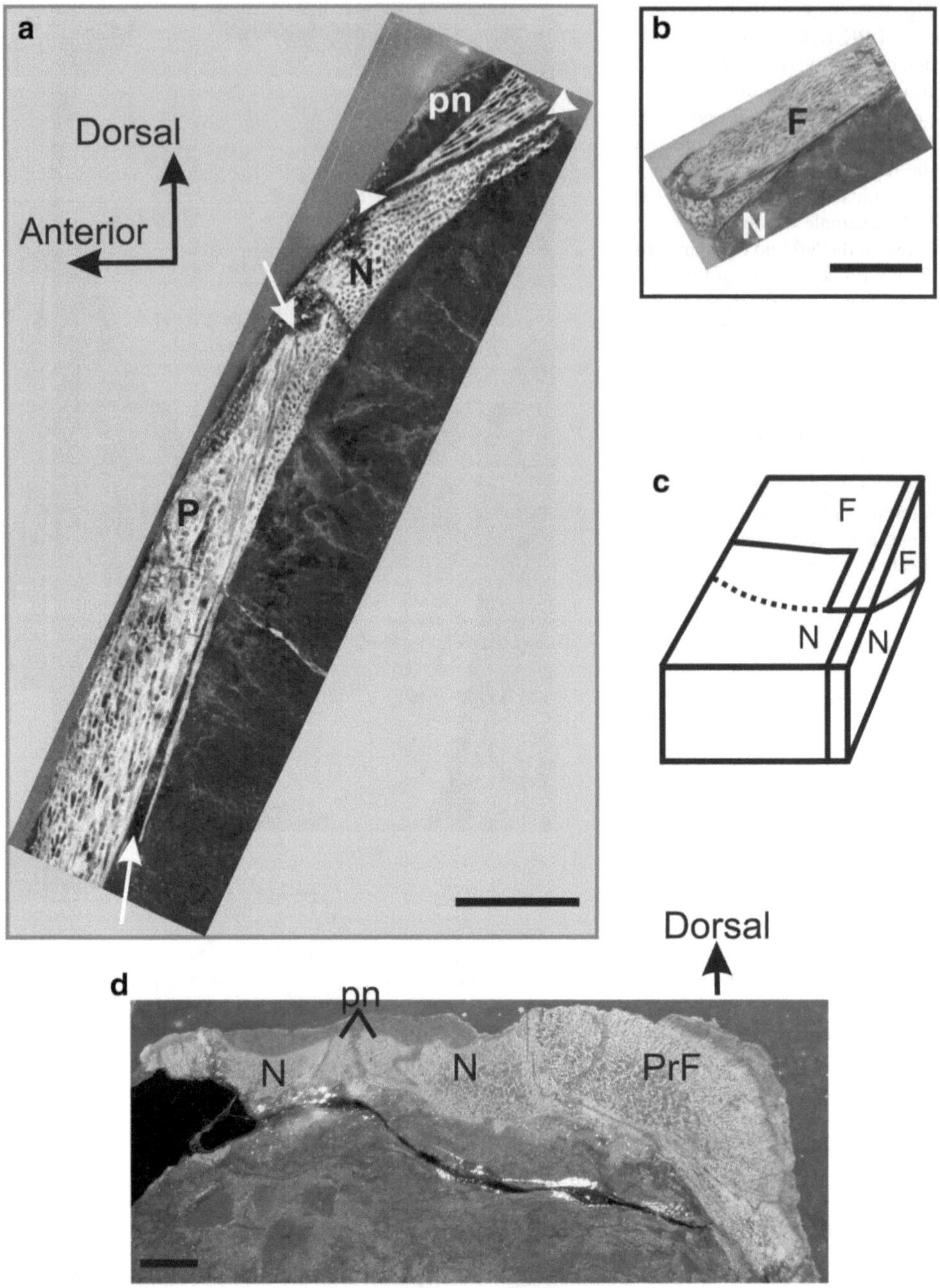

Fig. 8.3 **a** Longitudinal section through the dorsal snout region of SAM-PK-K1269-2 (*Lystrosaurus curvatus*; Table 8.1), showing the overlapping suture between the nasal and the 'post-nasal' bone (*arrowheads*), which is dorsal to the scarf premaxilla-nasal suture (*arrows*). **b** Longitudinal section through the anterior skull roof of SAM-PK-K1342 (block 5; *L. murrayi*), showing hook-like process of the nasal bone that cupped the frontal bone anteriorly. This section includes only the dorsal-most part of the snout, and so the anteroventral part of the nasal has been cut off (unlike **a**). The orientation arrows for the section are the same as in **a**. **c** Schematic of the nasal-frontal suture (or nasal-'postnasal' suture) in longitudinal and surficial views. **d** Transverse serial section through the dorsal part of the snout region of SAM-PK-K90 (*L. curvatus*) showing the 'post-nasal' bones wedged between the nasal bones. Abbreviations are the same as in Fig. 8.1. All scale bars are 5 mm

Evidence from Sections

Of the three histological specimens, only section SAM-PK-K1269-2 (*L. curvatus*; Table 8.1; Fig. 8.3a) sampled the dorsal part of the snout. Dorsal to the premaxilla-nasal scarf suture, the section through specimen SAM-PK-K1269-2 intersected what appears to be a straight scarf suture between the nasal and a bone dorsal to the nasal (Fig. 8.3a). No suture is apparent between the frontal and the bone dorsal to the nasal, but the section was not taken far enough posterodorsally to rule out the presence of such a suture, and the skull surface was not visible before thin-sectioning. Thus, the bone dorsal to the nasal is herein termed 'post-nasal' because it could either represent an anterior process of the frontal bone, or a supernumerary bone. The 'post-nasal' bone is, however, clearly distinguishable from the nasal bone by histological features: the nasal bone consisted of transversely oriented channels that appear circular in longitudinal section; whereas the 'post-nasal' bone has longitudinally oriented channels that appear laminar or elongated in longitudinal section (Fig. 8.3a).

A transverse offcut through the bone block of specimen SAM-PK-K1269-2 revealed that the 'post-nasal' bone is in fact paired, and that the two relatively narrow elements are wedged between and dorsal to the nasal bones (not figured). This arrangement was also observed in transverse serial sections of SAM-PK-K90 (*L. curvatus*; Fig. 8.3d) and SAM-PK-K1284 (*L. declivis*; Table 8.1). Specimen SAM-PK-K90 also revealed an interlocking lateral contact between the nasal and 'post-nasal' bones (Fig. 8.3d).

The longitudinal cut through specimen SAM-PK-K1342 (block 5) revealed an overlapping suture between the right anterior frontal process and the underlying nasal bone. However, the nasal has a hook-like process that cups the frontal bone anteriorly (Fig. 8.3b).

The combination of data observed in both longitudinal and transverse planes indicate that although an overlapping suture exists between the 'post-nasal' bones and the underlying nasal bones (Fig. 8.3c), no sliding adjustive movement would have been possible along the sutural contact because the 'post-nasal' bones are wedged between the nasal bones.

Micro-CT Scan Evidence

In the longitudinal plane, the anterior surface of the snout of specimen BRSUG 22211 (*L. declivis*) appeared to incorporate two scarf joints (Fig. 8.4a), similar to those observed in the histological specimen SAM-PK-K1269-2 (*L. curvatus*; Fig. 8.3a). In the former specimen, the nasal bone is overlapped by the ascending process of the premaxilla, and by the anteroventral tip of paired 'post-nasal' bones that form the anteroventralmost part of the frontonasal ridge (Fig. 8.4a). These 'post-nasal' bones more than likely represent supernumerary bones because a shallow interdigitated suture separates them from the posteriorly-situated frontal bones (Table 8.1; Fig. 8.4a).

When slices in all three planes were reconstructed into a three-dimensional skull model (Fig. 8.4b), it became apparent that the 'scarf suture' between the nasal bone and the supernumerary bone is actually part of an inclined and elongated tunnel (Fig. 8.4a, c). This structure was not discernible on the surface of the gross osteological specimen because it was infilled by matrix (darker grey color on the micro-CT slice in Fig. 8.4a). Once this matrix was digitally removed, the tunnel appeared at the surface just anteroventral to the frontonasal ridge (Fig. 8.4b). In the transverse slices, it was difficult to continuously trace the suture between the nasal bones and the wedge-shaped paired supernumerary bones (especially on the right side of the skull; Fig. 8.4c), suggesting that incipient fusion had occurred between these bones. This lateral fusion with the nasal bones was unlike the fused condition observed in other gross osteological specimens, in which one of the supernumerary bones may fuse with the posteriorly-situated frontal bone.

A nasal tunnel similar to that revealed in BRSUG 22211 was not observed in any other gross osteological specimens. However, in a few specimens (SAM-PK-1195, -8985, -8988; Fig. 8.2c), the nasal bones are widely separated along the internasal suture. Within this gap, a nasal tunnel may have occurred posteriorly under the frontonasal ridge, but only high contrast, high resolution micro-CT scans can unequivocally determine its presence.

Discussion

All lines of anatomical evidence (gross osteological, histological and serial sections, micro-CT scans) indicate that there was variation in the number of bones and sutural contacts in the dorsal snout region of *Lystrosaurus*. This variation includes the presence of either: (1) paired anterior processes of the frontal bone (Fig. 8.1b); (2) paired supernumerary bones separated by a midline suture (Fig. 8.1c); or (3) a median supernumerary bone with no midline suture (Fig. 8.1d). The majority of specimens observed have condition 1, whereas supernumerary bone(s) occur in at least four specimens (Table 8.1) [note that some specimens housed in collections in the United States may also have supernumerary bones (J. Camp, personal communication, 2010), although the authors could not confirm this by personal observation]. In all observed cases, the supernumerary bone(s)/anterior processes of the frontal form an anteroventral extension between the nasal bones. The paired supernumerary bones are similar in relative position, shape, and number to the paired anterior processes of the frontal bones (although sometimes only a single supernumerary bone is present). These similar features suggest that the supernumerary bone(s) are homologous to the anterior processes of the frontals, which further implies that the frontal bone had at least two separate centers of ossification.

Despite the differences in the manifestation of these bones, all anatomical evidence indicated that the supernumerary bone(s)/anterior frontal processes were narrow and wedged between the nasals. The variations in morphology may be attributed to several factors, such as (1) preservational bias; (2) ontogenetic differences; (3) taxonomic differences; and/or (4) anatomical/developmental plasticity. Each of these explanations is considered below.

The limited number of well-preserved specimens available for this study may have contributed to the small number of specimens in which the isolated supernumerary bone(s) were identified. As expected, specimens that were well-prepared allowed easy identification of sutures, and in some cases, further preparation of several specimens housed at the SAM enabled identification of the supernumerary bones. Also, specimens that were slightly weathered or

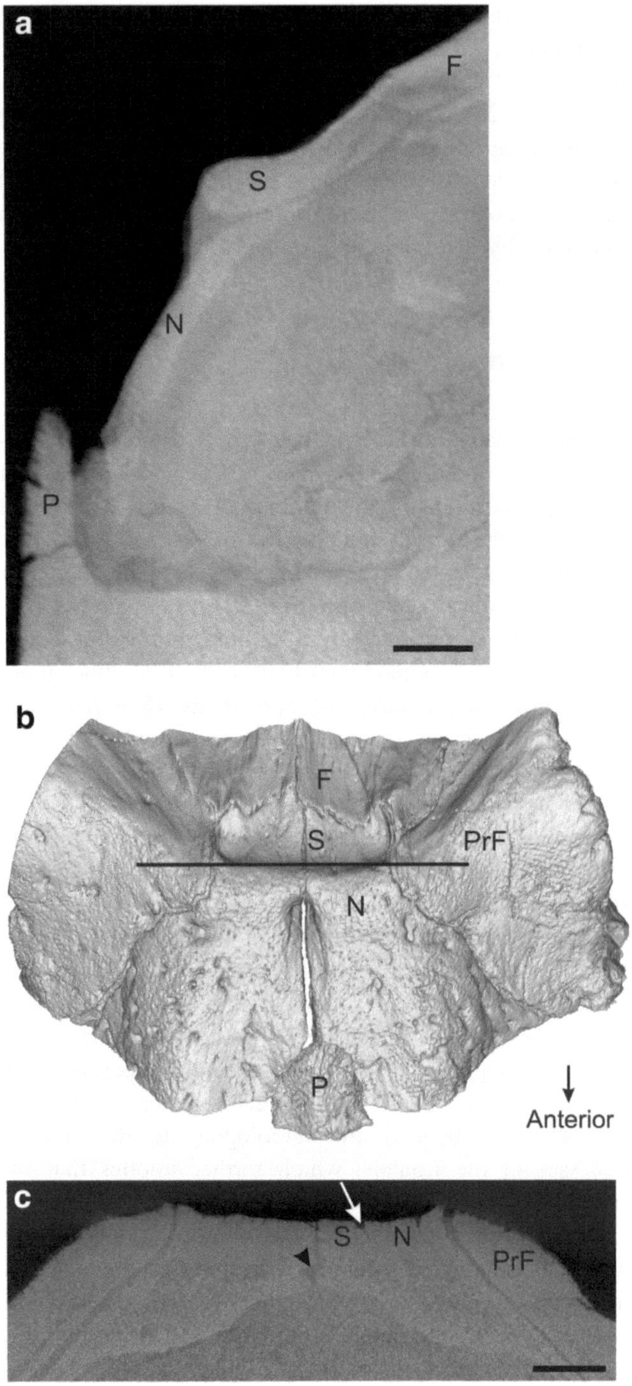

◀ **Fig. 8.4** **a** Longitudinal micro-CT slice through the dorsal snout and anterior skull roof region of BRSUG 22211 (*L. declivis*) near the mid-sagittal plane of the cranium. The apparent scarf suture between the nasal and the supernumerary bone is actually an inclined, elongated tunnel (see **b**). The premaxilla has pivoted outwards away from the nasal bone, a taphonomic feature that was observed in several other specimens of *Lystrosaurus* (e.g., King and Cluver 1991: Fig. 5b). The suture between the supernumerary bone and the frontal bone was crushed in this particular slice, but in other slices it is shallowly interdigitated. **b** Anterodorsal view of a three-dimensional cranium reconstructed from the partial micro-CT scan of BRSUG 22211. The matrix was digitally removed to show the elongated nasal tunnel that continues into the snout and separates the supernumerary bones from the underlying nasal bones. **c** Transverse micro-CT slice through the anterior skull roof of BRSUG 22211 (approximate location on cranium indicated by a *black line* in **b**) showing a distinctive suture (*white arrow*) between the left wedge-shaped supernumerary bone and the nasal bone. The suture between the right supernumerary bone and the nasal was not apparent on this slice, which may indicate that partial fusion has occurred between these bones. The elongated tunnel is indicated with an *arrowhead*. Abbreviations are the same as in Fig. 8.1. All scale bars are 5 mm

eroded were usually the most informative, as the sutures on the snout and skull roof were more clearly visible. It was difficult to determine if large, adult specimens have this isolated bone(s) due to the increased bony ornamentation, such as rugosities and ridges, on the anterior skull roof.

Variation in the presence of the supernumerary bones(s) may also be related to ontogeny. From this survey, it appears that the isolated bone(s) tend to be visible in smaller specimens that represent late juvenile or subadult growth stages (Table 8.1). However, it is difficult to determine whether the presence of this supernumerary bone is related to ontogeny because of several confounding factors. Firstly, adult size has not been firmly established for any of the known species of *Lystrosaurus*, and the size ranges for the skull of *Lystrosaurus* given in Grine et al. (2006) may be artificially narrow due to the relatively small number of specimens examined. Early juveniles (basal skull length of less than 52 mm) in the current sample, such as SAM-PK-3531, SAM-PK-K1396, and BP/1/3904, unfortunately do not have the frontonasal region well-preserved, so it is unknown whether supernumerary bones were present at this early ontogenetic stage. However, paired supernumerary bones were present in at least three larger specimens that were late juveniles in age, and a single median supernumerary bone was present in two subadult specimens (Table 8.1). The supernumerary bone(s) were not detected in any large (adult) specimens, although it was more difficult to clearly trace sutures in the dorsal snout/anterior skull roof region of such specimens, and there is also a paucity of adult specimens in collections (see Ray et al. 2012). Further investigation of large specimens of *Lystrosaurus* is required to determine if the paired supernumerary bones observed in some late juveniles later fuse to form a single median bone, or perhaps fuse with the frontal bones.

The different manifestation of the supernumerary bone(s) may also be related to taxonomic differences. From this preliminary survey, it appears that specimens of *L. curvatus* and *L. maccaigi* do not have supernumerary bones. This presumed difference may be due to the relatively small number of *L. curvatus* specimens included in the survey, as well as the lack of immature specimens of *L. maccaigi* at the time of this study (although see Botha-Brink et al.

2013). The very steep, straight snout of adult *L. maccaigi*, which ends at the strongly developed frontonasal ridge, together with what appear to be tightly-knit contacts among the premaxilla, nasals, and frontals, may imply a divergent pattern of cranial function in this species, but further investigation is required.

Regardless of the potential effects of the above factors, the different manifestations of the supernumerary bone(s) indicates the presence of anatomical or developmental plasticity in the dorsal snout region of *Lystrosaurus*. Plasticity in this region of the skull is supported by the observation that paired anterior frontal processes were not always the same size or length within the same individual or among individuals. A similar type of variation was also observed in some of the *Lystrosaurus* specimens that possess the supernumerary bones. Variation in the fusion of the supernumerary bones with either the frontal or nasal bones was also apparent in a few specimens of *Lystrosaurus*. Interestingly, there may have been some variation in the dorsal snout region that already existed among other basal dicynodontoids. Some specimens of *Dicynodon lacerticeps* had a depression near the nasal-frontal suture, which was situated close to where the supernumerary bone(s) in *Lystrosaurus* occurred.

Besides the supernumerary elements described here, evidence of variation in other bones of the skull roof has been noted by previous authors. For example, variation in the shape of the preparietal bone, a neomorphic median element in dicynodonts and certain other nonmammalian therapsids (Sidor 2001), and its surrounding sutural contacts in the posterior skull roof has long been known (e.g., Toerien 1953: Figs. 43–45; Keyser 1975: Fig. 22). In certain dicynodont taxa, the postfrontal bone in the dorsal orbital rim also may have a variable occurrence within individuals of the same taxon (Angielczyk 2007: supplementary information; Angielczyk et al. 2009). In addition, an extra median bone in the snout, which is likely homologous to the supernumerary bone(s) described here in *Lystrosaurus*, was described in *Jachaleria candelariensis* (Vega-Dias and Schultz 2004). Further investigation of the dorsal snout/anterior skull roof region of other dicynodont taxa is required in order to determine if variation within this region is common. If individual variation in the presence of supernumerary bones in the dicynodont snout is widespread, it is unlikely to be a useful character for species diagnosis. However, intraspecific variation in the vertebrate skull is generally not well known, and not many studies have investigated "its nature, range, developmental/genetic basis, or ecological/evolutionary causes" (Hanken and Hall 1993, p. 11).

Cranial supernumerary bones may form from separate ossification centers of a single bone that have not fused (Sidor 2001; Moore and Dalley 2006). Thus the supernumerary bones in *Lystrosaurus* may represent separate ossification centers of the frontal bone that did not fuse with the other centers during ontogeny. The bones of the snout are dermal in origin, and thus there were no cartilaginous precursors of these bones. It is possible that the nasal tunnel observed in BRSUG 22211 was bridged by fibrous tissue, and eventually would have been infilled with advancing bone from the growing nasal bone.

Functional Implications

Despite the anatomical differences in the manifestation of the supernumerary bone(s) in *Lystrosaurus* (i.e., fusion or non-fusion with frontals, or partial fusion with the nasals [BRSUG 22211]), their functional significance remained essentially the same. The supernumerary bone(s)/frontal processes formed a wedge between and dorsal to the paired nasals in the midline of the snout (Figs. 8.1b–d, 8.3d, 8.4c). During biting, the premaxilla could slide over the nasal bones (King and Cluver 1991; Jasinoski et al. 2010a), which may have caused instability in the dorsal part of the snout. This small adjustive movement of the single (fused) premaxilla bone over the paired nasals (which are separated by a butt-ended internasal suture) may have destabilized the nasal bones. The supernumerary bone(s) (or, if absent, the frontal processes) may have acted as a wedge, preventing displacement or torsion of the paired nasal bones. Even though the supernumerary bone(s) are not fused with the frontals, their ability to act as a wedge was not diminished because they are strongly abutted against the frontals via a shallow interdigitated suture (Fig. 8.4a, b), which would have stabilized their position.

The simulation of the scarf premaxilla-nasal suture in a finite element model of *Lystrosaurus* (Jasinoski et al. 2010b) caused low to moderate tension in the lateral part of the nasals and prefrontals. Near the internasal suture, the principal tensile strains were oriented dorsally on the surface of the snout, whereas they were oriented dorsally and slightly laterally near the lateral edges of the nasals and the prefrontal bones. These tensile strain vectors suggested that the flexibility at the premaxilla-nasal suture may have forced the nasal bones dorsally. However, resistance to this displacement may have been accomplished by the overlapping anterior frontal processes/supernumerary bone(s), which were firmly wedged between the nasals.

Conclusions

Lystrosaurus represents a radical departure from the typical generalized Permian dicynodont skull architecture, which suggests that its feeding activities, such as a snapping bite or grubbing activities, also may have differed from earlier dicynodonts (e.g., King and Cluver 1991; Jasinoski et al. 2009, 2010a). Thus, differences in its skull, its associated jaw musculature, and possibly its feeding habits may be reflected in its cranial bone structure. In a few specimens of *Lystrosaurus*, supernumerary bone(s) in the snout were documented, indicating the occurrence of intraspecific cranial variability within this genus. It is significant that the supernumerary bones occurred in the same region as the nasofrontal depression observed in *Dicynodon lacerticeps*, since this may reflect variability already present in dicynodonts that are more basal, but closely related to *Lystrosaurus*.

Despite the anatomical plasticity in the snout of *Lystrosaurus*, the function of the wedge-shaped supernumerary bones/anterior frontal processes may have been to stabilize the dorsal part of the snout during adjustive movement at the scarf premaxilla-nasal suture.

Our analysis found evidence of supernumerary bone(s) only in immature individuals of *Lystrosaurus*. Further investigation of additional specimens, including the implementation of micro-CT scanning and histological sampling, is required to determine whether the supernumerary bone(s) are present across all ontogenetic stages, or if they fuse with the frontals or nasals in older individuals.

Acknowledgments We thank R. Smith (SAM) for permission to histologically sample several cranial specimens of *Lystrosaurus*, and for allowing further preparation of other specimens. We also thank the SAM technicians N. Mtalana and Z. Erasmus for their help with mechanical preparation of specimens. The collection managers and curators at the BP, BRSUG, CGP, NHMUK, SAM, TM, and UCL are also acknowledged for access to specimens. We would especially like to thank A. Ramsey, G. Dermody, B. Morgan and M. Robinson of Metris UK (Tring, UK), and P. Keanly (X-Sight, ZA) for micro-CT scanning specimens of *Lystrosaurus*. We also thank K. Angielczyk and the three reviewers J. Camp, C. Sullivan, and C. Vega for their useful comments and critiques. Heidi Fourie and J. Camp are thanked for providing additional photographs of dicynodonts. Funding for this project was provided by the Claude Leon Foundation (SCJ) and NRF through SA Research Chair in Computational Mechanics (SCJ, BDR).

References

Angielczyk, K. D. (2007). New specimens of the Tanzanian dicynodont "*Cryptocynodon*" *parringtoni* von Huene, 1942 (Therapsida, Anomodontia), with an expanded analysis of Permian dicynodont phylogeny. *Journal of Vertebrate Paleontology, 27*, 116–131.

Angielczyk, K. D., Sidor, C. A., Nesbitt, S. J., Smith, R. M. H., & Tsuji, L. A. (2009). Taxonomic revision and new observations on the postcranial skeleton, biogeography, and biostratigraphy of the dicynodont genus *Dicynodontoides*, the senior subjective synonym of *Kingoria* (Therapsida, Anomodontia). *Journal of Vertebrate Paleontology, 29*, 1174–1187.

Battail, B., & Surkov, M. V. (2000). Mammal-like reptiles from Russia. In M. J. Benton, M. A. Shishkin, D. M. Unwin, & E. N. Kurochkin (Eds.), *The age of dinosaurs in Russia and Mongolia* (pp. 86–119). Cambridge: Cambridge University Press.

Botha-Brink, J., & Angielczyk, K. D. (2010). Do extraordinarily high growth rates in Permo-Triassic dicynodonts (Therapsida, Anomodontia) explain their success before and after the end-Permian extinction? *Zoological Journal of the Linnean Society, 160*, 341–365.

Botha-Brink, J., Huttenlocker, A. K., & Modesto, S. P. (2013). Vertebrate paleontology of Nooigedacht 68: A *Lystrosaurus maccaigi*-rich Permo-Triassic boundary locality in South Africa. In C. F. Kammerer, K. D. Angielczyk, & J. Fröbisch (Eds.), *Early evolutionary history of the Synapsida* (pp. 289–304). Dordrecht: Springer.

Cluver, M. A. (1971). The cranial morphology of the dicynodont genus *Lystrosaurus*. *Annals of the South African Museum, 56*, 155–274.

Colbert, E. H. (1974). *Lystrosaurus* from Antarctica. *American Museum Novitates, 2535*, 1–44.

Cox, C. B. (1991). The Pangaea dicynodont *Rechnisaurus* and the comparative biostratigraphy of Triassic dicynodont faunas. *Palaeontology, 34*, 767–784.

Cope, E. D. (1870). Remarks. *Proceedings of the American Philosophical Society, 11*, 419.

Germain, D., & Laurin, M. (2005). Microanatomy of the radius and lifestyle in amniotes (Vertebrata, Tetrapoda). *Zoologica Scripta, 34*, 335–350.

Grine, F. E., Forster, C. A., Cluver, M. A., & Georgi, J. A. (2006). Cranial variability, ontogeny, and taxonomy of *Lystrosaurus* from the Karoo Basin of South Africa. In M. T. Carrano, T. J. Gaudin, R. W. Blob, & J. R. Wible (Eds.), *Amniote paleobiology: Perspectives on the evolution of mammals, birds, and reptiles* (pp. 432–503). Chicago: University of Chicago Press.

Hanken, J., & Hall, B. K. (1993). Mechanisms of skull diversity and evolution. In J. Hanken & B. K. Hall (Eds.), *The skull, Volume 3: Functional and evolutionary mechanisms* (pp. 1–36). Chicago: University of Chicago Press.

Hotton, N. (1986). Dicynodonts and their role as primary consumers. In N. Hotton, P. D. MacLean, J. J. Roth, & E. C. Roth (Eds.), *The ecology and biology of mammal-like reptiles* (pp. 71–82). Washington: Smithsonian Institutional Press.

Huxley, T. (1859). On a new species of *Dicynodon* (*D. murrayi*), from near Colesberg, South Africa; and on the structure of the skull in the dicynodonts. *Quarterly Journal of the Geological Society, 15*, 649–659.

Jasinoski, S. C., & Chinsamy-Turan, A. (2012). Biological inferences of the cranial microstructure of the Dicynodonts *Oudenodon* and *Lystrosaurus*. In A. Chinsamy-Turan (Ed.), *Forerunners of mammals: Radiation, biology, histology* (pp. 149–176). Bloomington: Indiana University Press.

Jasinoski, S. C., Rayfield, E. J., & Chinsamy, A. (2009). Comparative feeding biomechanics of *Lystrosaurus* and the generalized dicynodont *Oudenodon*. *Anatomical Record, 292*, 862–874.

Jasinoski, S. C., Rayfield, E. J., & Chinsamy, A. (2010a). Functional implications of dicynodont cranial suture morphology. *Journal of Morphology, 271*, 705–728.

Jasinoski, S. C., Rayfield, E. J., & Chinsamy, A. (2010b). Mechanics of the patent premaxilla-nasal suture in the snout of *Lystrosaurus*. *Journal of Vertebrate Paleontology, 30*, 1283–1288.

Kammerer, C. F., & Angielczyk, K. D. (2009). A proposed higher taxonomy of anomodont therapsids. *Zootaxa, 2018*, 1–24.

Kammerer, C. F., Angielczyk, K. D., & Fröbisch, J. (2011). A comprehensive taxonomic revision of *Dicynodon* (Therapsida, Anomodontia), and its implications for dicynodont phylogeny, biogeography, and biostratigraphy. *Society of Vertebrate Paleontology Memoir, 11*, 1–158.

Keyser, A. W. (1975). A re-evaluation of the cranial morphology and systematics of some tuskless Anomodontia. *Memoirs of the Geological Survey of South Africa, 67*, 1–110.

King, G. M. (1988). Anomodontia. In P. Wellnhofer (Ed.), *Handbuch der Paläoherpetologie* (Vol. 17C). Stuttgart: Gustav Fischer Verlag.

King, G. M. (1990). *The dicynodonts: A study in palaeobiology*. London: Chapman and Hall.

King, G. M. (1991). The aquatic *Lystrosaurus*: A palaeontological myth. *Historical Biology, 4*, 285–321.

King, G. M., & Cluver, M. A. (1991). The aquatic *Lystrosaurus*: An alternative lifestyle. *Historical Biology, 4*, 323–341.

Kitching, J. W. (1977). The distribution of the Karroo vertebrate fauna. *Memoirs of the Bernard Price Institute for Palaeontological Research, 1*, 1–131.

Maisch, M. W. (2002). A new basal lystrosaurid dicynodont from the Upper Permian of South Africa. *Palaeontology, 45*, 343–359.

Moore, K. L., & Dalley, A. F. (2006). *Clinically oriented anatomy* (5th ed.). Baltimore: Lippincott Williams & Wilkins.

Ray, S. (2005). *Lystrosaurus* (Therapsida, Dicynodontia) from India: Taxonomy, relative growth and cranial dimorphism. *Journal of Systematic Palaeontology, 3*, 203–221.

Ray, S., Chinsamy, A., & Bandyopadhyay, S. (2005). *Lystrosaurus murrayi* (Therapsida, Dicynodontia): Bone histology, growth and lifestyle adaptations. *Palaeontology, 48*, 1169–1185.

Ray, S., Botha-Brink, J., & Chinsamy-Turan, A. (2012). Dicynodont growth dynamics and lifestyle adaptations. In A. Chinsamy-Turan (Ed.), *Forerunners of mammals: Radiation, biology, histology* (pp. 121–146). Bloomington: Indiana University Press.

Sidor, C. A. (2001). Simplification as a trend in synapsid cranial evolution. *Evolution, 55*, 1419–1442.

Smith, R., & Botha, J. (2005). The recovery of terrestrial vertebrate diversity in the South African Karoo Basin after the end-Permian extinction. *Comptes Rendus Palevol, 4*, 623–636.

Smith, R., Rubidge, B., & van der Walt, M. (2012). Therapsid biodiversity patterns and paleoenvironments of the Karoo Basin, South Africa. In A. Chinsamy-Turan (Ed.), *Forerunners of mammals: Radiation, histology, biology* (pp. 31–62). Bloomington: Indiana University Press.

Thackeray, J. F., Durand, J. F., & Meyer, L. (1998). Morphometric analysis of South African dicynodonts attributed to *Lystrosaurus murrayi* (Huxley, 1859) and *L. declivis* (Owen, 1860): Probabilities of conspecificity. *Annals of the Transvaal Museum, 36*, 413–420.

Thulborn, T., & Turner, A. (2003). The last dicynodont: An Australian Cretaceous relict. *Proceedings of the Royal Society of London. Series B, 270*, 985–993.

Toerien, M. (1953). The evolution of the palate in South African Anomodontia and its classificatory significance. *Palaeontologia Africana, 1*, 49–118.

Tripathi, C., & Satsangi, P. P. (1963). *Lystrosaurus* fauna of the Panchet series of the Raniganj coalfield. *Palaeontologia Indica, 37*, 1–66.

Vega-Dias, C., & Schultz, C. L. (2004). Postcranial material of *Jachaleria candelariensis* Araújo and Gonzaga 1980 (Therapsida, Dicynodontia), Upper Triassic of Rio Grande do Sul, Brazil. *PaleoBios, 24*, 7–31.

Yuan, P. L., & Young, C. C. (1934). On the occurrence of *Lystrosaurus* in Sinkiang. *Bulletin of the Geological Society of China, 13*, 575–580.

Chapter 9
Pathological Features in Upper Permian and Middle Triassic Dicynodonts (Synapsida, Therapsida)

Cristina Silveira Vega and Michael W. Maisch

Abstract Morphological peculiarities that are attributable to pathologies are described in an Upper Permian dicynodont from Tanzania (*Geikia locusticeps*) and in a Middle Triassic species from Brazil (*Stahleckeria potens*). In ventral view, the *Geikia* skull (GPIT/RE/7187) presents an unusual feature between the right maxilla and jugal, a perfectly concave circular pit with a very prominent rim. This lesion could correspond to a slowly growing benign process that produced a pressure erosion of the bone, such as an epidermal inclusion cyst, probably post-traumatic, or a parasite infection caused by *Echinococcus*, also known as hydatid disease. In *Stahleckeria* (GPIT/RE/8001, mounted skeleton), various pathologies appear in different bones of distinct individuals, particularly on articular surfaces of the limb bones. The anterodorsal portion of the scapula bears a lesion that is interpreted as caused by an epidermal inclusion cyst or a muscular avulsion. The deltopectoral crest of the right humerus bears lesions that are attributed to fungal disease or to muscular avulsion. The entepicondyle of the same humerus bears lesions that are attributed to a fungal disease or a parasite infection caused by *Echinococcus*. The great trochanter of the left femur also shows signs of fungal disease. The distal extremity of the same femur, particularly the popliteal fossa and lateral condyle, show features that could correspond to healed osteomyelitis, fungal disease or a parasitic infection caused by *Echinococcus*. The proximal end of the left tibia and its cnemial crest show evidence of an infection caused by osteomyelitis or a fungal disease. The proximal end of the left fibula also may have been affected by fungal disease. The identification of these features as pathologies prevents their misinterpretation as original morphologies, reducing the risk of incorrect taxonomic, phylogenetic, and functional interpretations.

Keywords *Echinococcus* • Muscular avulsion • Fungal disease • Epidermal inclusion cyst • Dicynodontia

Introduction

Skeletal pathologies are common in extant and fossil tetrapods, but they are often neglected in the fossil record. Nevertheless, pathological conditions of the skeleton have long been recorded in fossil tetrapods as old as the Carboniferous (Moodie 1918). Since then, a variety of pathologies have been described in a wide range of fossil amniotes. These include the possible occurrence of cysts in Brazilian rhynchosaurs (Schultz 1999), avascular necrosis in mosasaurs (Rothschild and Martin 1987), osteoarthritis in plesiosaurs (Wells 1964), osteomyelitis in crocodilians (Ferigolo 1993a), bone infections in aetosaurs (Lucas 2000), fractures and infections in dinosaurs (Rothschild 1988; Hanna 2002; Peterson et al. 2009), and luxations, exostoses and arthrosis in mammals (Moodie 1930; Choquette et al. 1975; Ferigolo 1985, 1993b; Lucas and Schoch 1987; Wang and Rothschild 1992; Gillette and Madsen 1993; Henriques et al. 1998; Scott and Rooney 2001). Recently, there has been increasing interest in pathological features of basal synapsids (Huttenlocker et al. 2010; Huttenlocker and Rega 2012) and other Paleozoic tetrapods (Reisz et al. 2011). Among dicynodonts, pathologies have been reported on several occasions. von Huene (1935) noted some usual features of the scapula, tibia, and fibula of *Stahleckeria potens*, but did not speculate on the disease or diseases that caused them. An unusual structure attributed to osteomyelitis was noted by Vega-Dias and Schultz (2003) on a scapula of *Jachaleria candelariensis*, and it was mentioned in the

C. S. Vega (✉)
Departamento de Geologia, Universidade Federal do Paraná, Centro Politécnico, Caixa Postal 19.001, Curitiba-PR 81531-990, Brazil
e-mail: cvega@ufpr.br

M. W. Maisch
Staatliches Museum für Naturkunde, Rosenstein 1, 70191 Stuttgart, Germany
e-mail: michael.maisch@smns-bw.de

Christian F. Kammerer, Kenneth D. Angielczyk and Jörg Fröbisch (eds.), *Early Evolutionary History of the Synapsida*, Vertebrate Paleobiology and Paleoanthropology, DOI: 10.1007/978-94-007-6841-3_9,

description of the postcranial skeleton of this species (Vega-Dias and Schultz 2004). A number of dental pathologies and potential abnormalities also have been noted in dicynodonts. Occurrence of double-tusked dicynodonts, an unusual feature likely produced by mutation, is known to occur in at least three dicynodont species, *Eodicynodon oosthuizeni*, *Emydops oweni*, and *Kannemeyeria simocephalus* (Camp and Welles 1956; Jinnah and Rubidge 2007; Fröbisch and Reisz 2008). Fröbisch and Reisz (2008) also reported evidence of a likely tooth abscess in the double-tusked specimen of *K. simocephalus*, and speculated that it might have caused a splitting of the tooth germ. Some individuals of the dicynodont species *Odontocyclops whaitsi* and *Tropidostoma dubium* only possess an erupted tusk on one side of the skull (Angielczyk 2002; Botha and Angielczyk 2007; see Kammerer et al. 2011 for information of the taxonomy of *Tropidostoma*). Angielczyk (2002) also figured a specimen of *O. whaitsi* that possessed a large unerupted tusk in the caniniform process. This tusk is particularly noteworthy because instead of being a smooth cylinder like those of other dicynodonts, it has a lobed appearance.

In this work, we describe and interpret the presence of unusual pathologies in two dicynodonts in the collections of the University of Tübingen. The first of these pathological specimens is a skull of *Geikia locusticeps* from the Late Permian of Tanzania, which presents a single lesion on its ventral surface. The second specimen is a mounted skeleton of the Middle Triassic *Stahleckeria potens* from Brazil, which shows multiple lesions in the pectoral girdle, fore and hind limbs. This is the same pathological specimen described by von Huene (1935), and we expand upon his work by noting additional pathologies and investigating their possible origins.

Materials

The material discussed in this article comprises:

- GPIT/RE/7187 (=K114): *Geikia locusticeps* (von Huene, 1942), holotype of *Pelanomodon tuberosus* von Huene, 1942. This specimen consists of a nearly complete skull and lower jaw. The specimen was collected at the locality of Kingori in the Upper Permian Usili Formation (formerly Kawinga Formation) of the Ruhuhu Basin, southwest Tanzania (von Huene 1942; Maisch and Gebauer 2005)
- GPIT/RE/8001: *Stahleckeria potens* von Huene, 1935. This specimen is a mounted skeleton, and the material used in the mount was collected in the Middle Triassic Santa Maria Formation of Rio Grande do Sul in southeastern Brazil (von Huene 1935). von Huene (1935) described three skulls and numerous postcranial elements of the species originating at a single locality, much of which was used for the mounted skeleton. Because the cranial remains of several individuals were present at the locality, von Huene (1935) considered it highly unlikely that all the postcranial elements are from a single individual. Here we focus on the right scapula, right humerus, left femur, left tibia and left fibula of the mounted skeleton.

Description of Pathological Features

Geikia locusticeps

The skull of *Geikia locusticeps* (GPIT/RE/7187) is almost complete (Fig. 9.1a), lacking just the anterior portion of the premaxilla, which is restored in plaster. In ventral view, the specimen presents an unusual feature between the right maxilla and jugal, a perfectly concave circular pit (2.5 cm in diameter and 0.5 cm deep) with a very prominent rim. The rim has a completely smooth margin with intact cortical bone, indicating an advanced state of healing. Different textures are not observed in the margin (Fig. 9.1b), and there are no rugosities of the bone surface associated with this feature; instead the bone is quite smooth. On the left side of skull, there is no indication of a similar structure, and there appear to be no other pathological features present on the skull (although certain bones such as the nasals and squamosals display the bosses and ornamentation that is typical of geikiid dicynodonts).

Stahleckeria potens

Von Huene (1935) figured some unusual features of the right scapula (von Huene 1935: plate 7, Fig. 1a, b), tibia and fibula (von Huene 1935: plate 9, Figs. 6a, 7a, b) and mentioned that they may be due to pathologies, particularly with respect to the tibia, but he did not indicate the disease to which they could be attributed. During a re-study of the postcranial skeleton of *Stahleckeria*, we also found some distinct pathological features in the right humerus and left femur. All these pathologies are described below.

The right scapula (Fig. 9.2a) of *S. potens* has a length of 54 cm and its dorsal margin is about 32 cm wide. A well-developed acromion process is present, and there is a marked spine on the lateral surface of the bone. In lateral view there are two continuous concavities in the dorsal half of its anterior border. In the most anterodorsal part of the scapula (Fig. 9.2b), there is a hole of almost rectangular shape, 6 cm in length and 2 cm in depth. The margins are

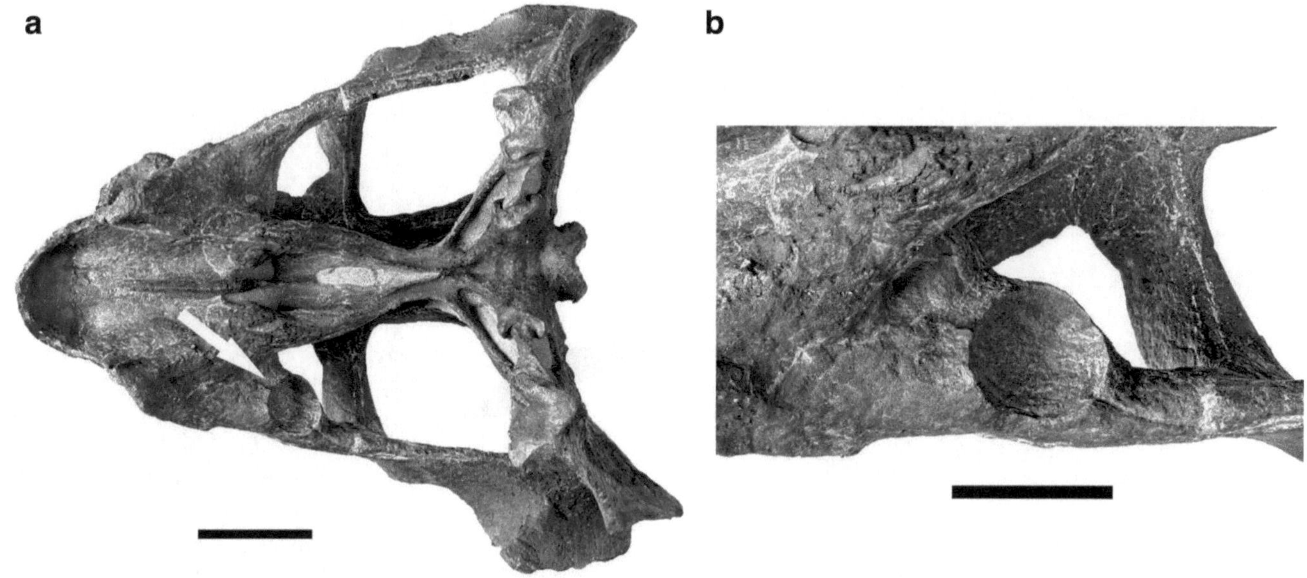

Fig. 9.1 Skull of *Geikia locusticeps* (GPIT/RE/7187). **a** Skull in ventral view with *arrow* indicating the pathology; scale bar equals 5 cm. **b** Detail of the pathology located between the right squamosal and jugal; scale bar equals 3 cm

sharp, without any rugosities, and there is no evidence of reactive bone. In the ventral portion of the scapula, near the contact with the procoracoid, there are two concavities that are not very pronounced and may be best attributed to areas of the attachment of the supracoracoideus muscle. In anterior view, above the acromion process and mediolaterally accompanying the scapular spine that extends from the acromion process, there is a portion of the bone surface that bears three rugosities that are around 0.5 cm in length, roughened, and almost rounded in shape.

The right humerus of *S. potens* has a length of about 49 cm (Fig. 9.3a). In anterior view, the deltopectoral crest of the humerus displays 11 rugosities that begin on its lateral surface and extend medially up the end of the deltopectoral crest (Fig. 9.3b). In dorsal view, the distal end of the humerus bears 12 rugosities that are concentrated on the anterior portion of the ectepicondyle (on the articular surface) and in the concavity that corresponds to the trochlea, where it is possible to observe seven rugosities and a deep concavity (Fig. 9.3a). This concavity of the trochlea is deeper than expected in dicynodonts. The margin of the trochlea is flanked by somewhat pronounced, strap-like rugose patches, a morphology completely distinct from that observed in the left humerus of the mounted skeleton.

The left femur measures 48 cm in length (Fig. 9.4a) and shows significant pathologies in comparison with the right femur of the mounted skeleton. There is a deep concavity in its proximal region (Fig. 9.4b) on the right side of the head of the bone. This feature makes the bone appear much less thick anteroposteriorly than its counterpart. The concavity is surrounded by many protuberances. The region of the great trochanter also has many irregular prominences that cannot be attributed to the attachment of the femorotibialis and iliofemoralis muscles. In dicynodonts, the great trochanter can be aligned with or oblique to the shaft of the bone, or even present as "S"-shaped morphology. However, the attachment sites of these muscles are not highly rugose, but instead almost smooth bosses with no major irregularities on its contour. Comparable irregularities are also not observed on the great trochanter of the right femur of the mounted *Stahleckeria* skeleton, which is in stark contrast to the completely irregular structures in this area on the pathological left femur. In the distal part of the diaphysis (Fig. 9.4c) there is an accentuated concavity that extends mediolaterally. Distally, this concavity bears many protuberances, and they continue into the region corresponding to the popliteal fossa posteriorly, and on the lateral condyle. On the medial condyle, a deep furrow (Fig. 9.4c, d) extends along the condylar articular surface in an anteroposterior direction.

The left tibia of *S. potens* has a length of 34 cm. In its proximal region, the cnemial crest has been nearly destroyed by pathologies. There is an accentuated concavity containing at least three smaller concavities inside, the largest almost 4 cm in depth, surrounded by a series of three well-defined, smooth but irregularly-shaped protuberances (Fig. 9.5a). In the posterior portion of the cnemial crest, corresponding to the articular area for the lateral condyle of the femur, there are two prominent protuberances surrounded by many small prominences, which may be

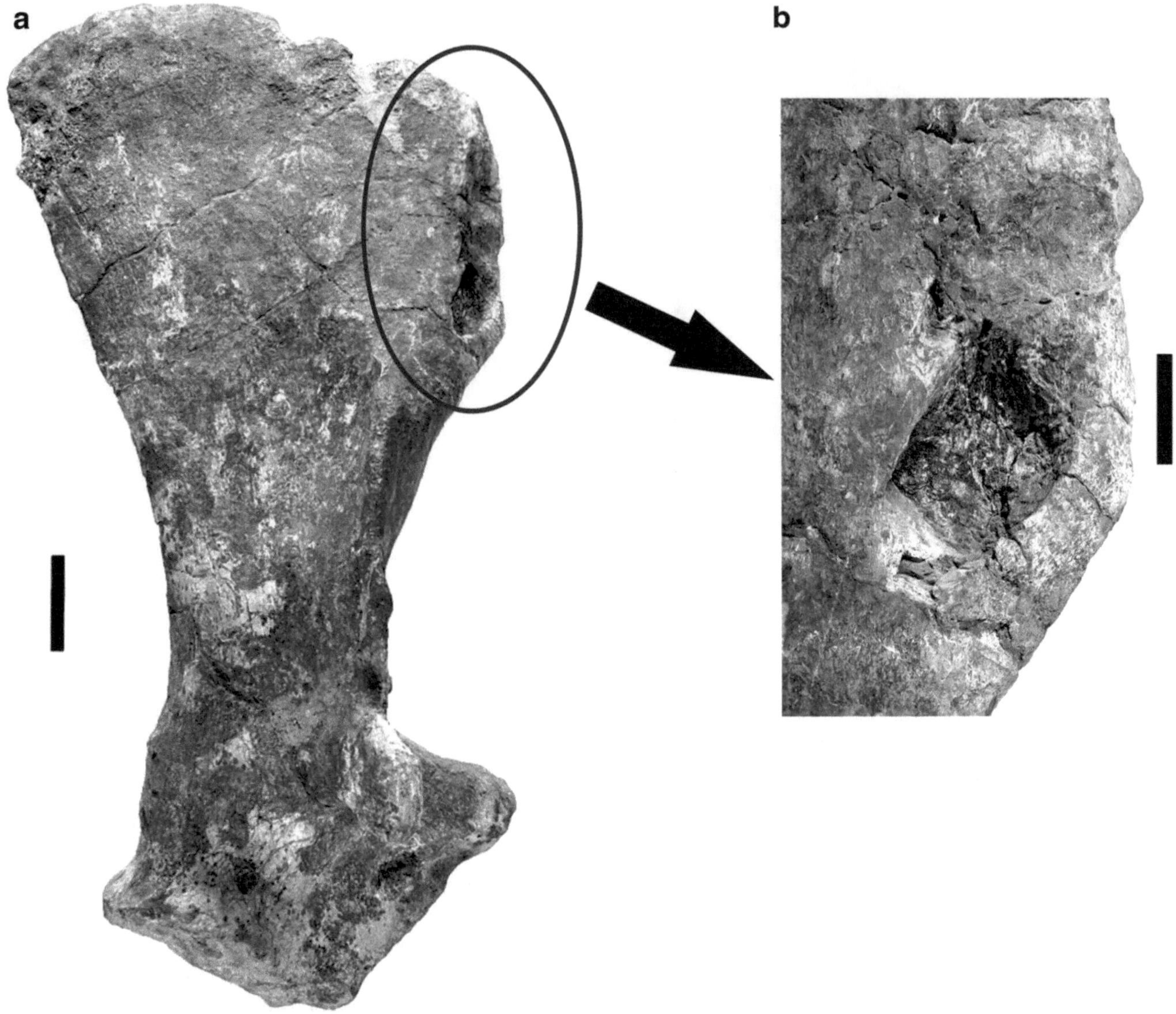

Fig. 9.2 Scapula of *Stahleckeria potens* (GPIT/RE/8001). **a** Right scapula in anterolateral view; scale bar equals 5 cm. **b** Detail of the pathology at the anterodorsal border of the scapula, in lateral view; scale bar equals 3 cm

evidence of reactive bone. In proximal view, the tibia shows a deep concavity in the center of the bone, between the cnemial crest and the region for articulation with the medial and lateral condyles of femur (Fig. 9.5b).

The left fibula of *S. potens* has a length of 32 cm (Fig. 9.6a). In the medial portion of its proximal end (Fig. 9.6b), an extensive protuberance is present anteromedially, almost with the appearance of a trochanter. The whole proximal region also displays a series of rugosities that are very similar to those seen in the left tibia.

Discussion

Pathological features of the skeleton are common phenomena in vertebrates today and in the past, but they are often neglected and generally understudied in the fossil record. However, an understanding of paleopathologies is important for the correct interpretation of bone morphology. For example, *S. potens* is known from relatively little other postcranial material aside from that incorporated into the

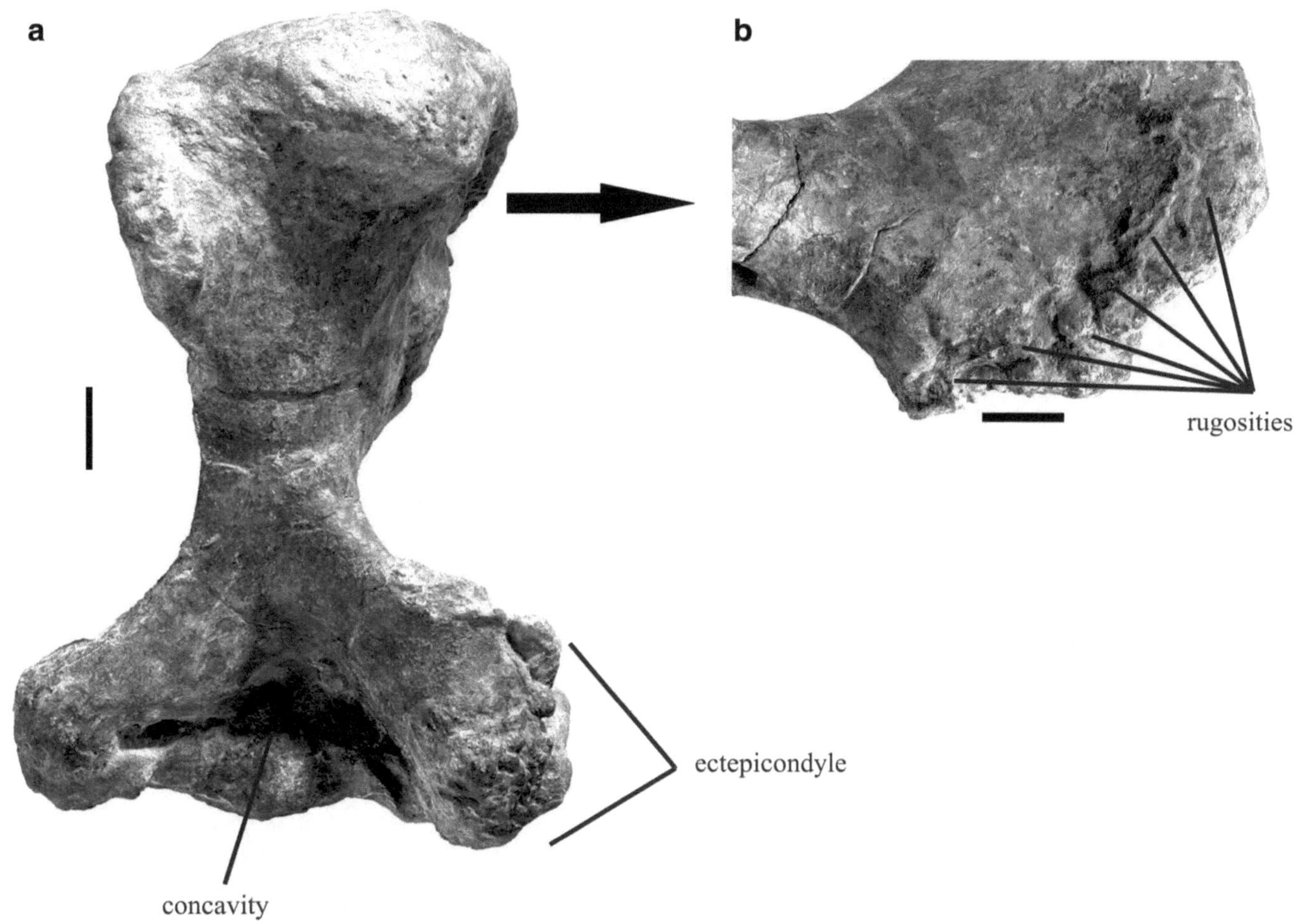

Fig. 9.3 Right humerus of *Stahleckeria potens* (GPIT/RE/8001). **a** Humerus in dorsal view, showing the concavity on the trochlea and rugosities on the ectepicondyle in the distal portion. **b** Pathologies on the deltopectoral crest in anterior view. Scale bars equal 5 cm

mounted skeleton (Vega-Dias et al. 2005). Therefore, some of the unusual features of the skeleton could be misinterpreted as autapomorphies of the species or evidence of significant intraspecific variation if pathology was not explicitly considered. In the following section we discuss the pathologies observed in the two dicynodonts (*Geikia* and *Stahleckeria*) in the context of general pathological classification, and consider alternative hypotheses for their causes.

Cysts

Resnick et al. (1995) divided tumors into tumors and tumor-like lesions of miscellaneous or unknown origin, and benign tumors.

According to Ortner (2003c), a benign tumor, or cyst, is a lesion characterized by a fluid-filled cavity enclosed by a lining usually composed of connective tissue. There are many types of cysts that occur in bones: simple (solitary or unicameral) bone cysts, epidermoid cysts, aneurysmal bone cysts, or intraosseous ganglion cysts (Resnick et al. 1995). A traumatic epidermal inclusion cyst can be represented by "a central cavity, usually not more than 1 cm in diameter, with or without evidence of healed fracture" (Ortner 2003c, p. 505). A congenital epidermal inclusion cyst occurs only in the calvarium, and commonly occurs as a solitary lesion. Occasionally, these cysts could measure 10 cm or more in diameter (Ortner 2003c). Other kinds of cysts (as unicameral bone cyst and aneurysmal bone cyst) are an intraosseous phenomenon. According to Rothschild and Martin (2006, p. 149), an "aneurysmal bone cyst is expansile ovoid, lytic, thin walled, blood-filled lesions with thin eggshell-like bony margins." Cortical erosion is rare.

Considering that the lesion present in the skull of *G. locusticeps* is not intra-osseous but rather represented by a perfectly concave circular pit bounded by a rim with completely smooth margins, its probable cause likely was a post-traumatic epidermal inclusion cyst. The same type of feature is observed on the dorsal extremity of the scapula of

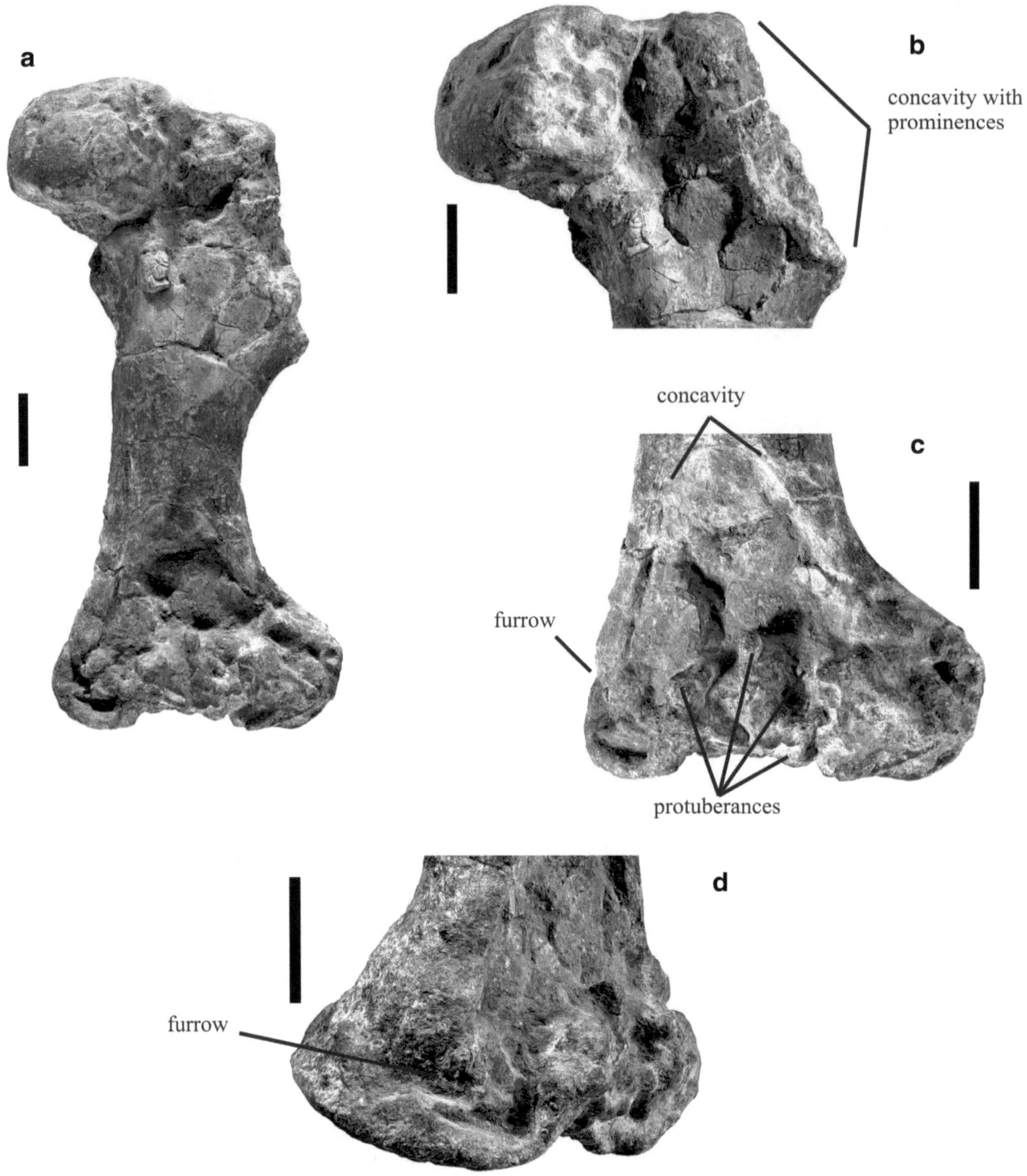

Fig. 9.4 Left femur of *Stahleckeria potens* (GPIT/RE/8001). **a** Femur in anterior view. **b** Proximal portion of femur in anterior view, showing the concavity with prominences. **c** Distal portion of femur in anterior view. The concavity that extends mediolaterally and the protuberances on the popliteal fossa and lateral condyle are indicated. **d** Detail of the distal portion of the femur in medial view, showing the furrow in the medial condyle. Scale bars equal 5 cm

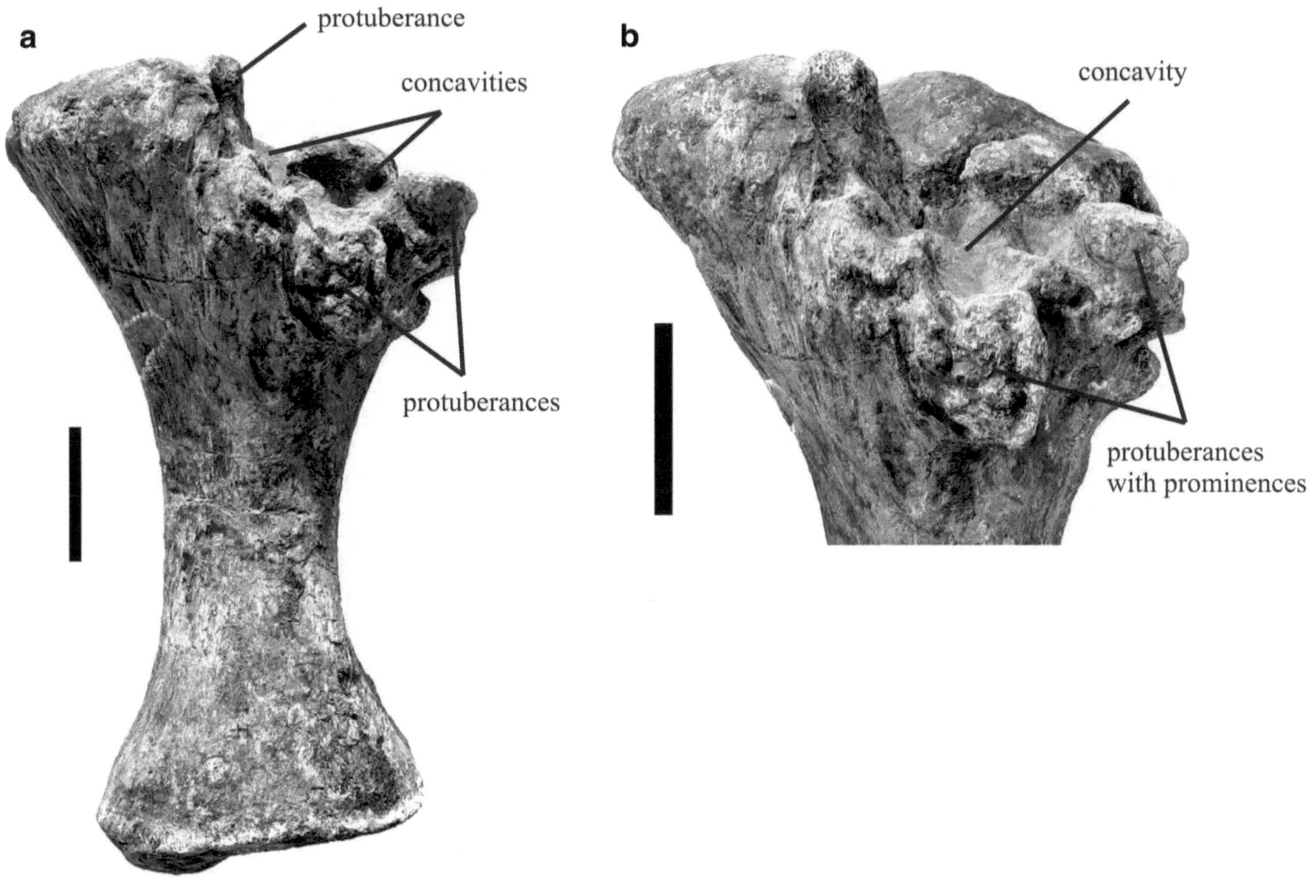

Fig. 9.5 Left tibia of *Stahleckeria potens* (GPIT/RE/8001). **a** Tibia in anterior view, showing the protuberances and the concavities. **b** Pathologies on the cnemial crest and on the articular facet for the lateral condyle of the femur in lateral view. Scale bars equal 5 cm

Stahleckeria (Fig. 9.2b); the morphology of the lesion, without any exostosis, could also correspond to a traumatic epidermal inclusion cyst. This region includes the insertions of the trapezius and deltoideus muscles, and if this feature is not taphonomic, the lesion observed could also correspond to an avulsion of these muscles.

Hydatid Disease

Some parasitic infections can also affect bones. One of them is caused by the cestode *Echinococcus* and is called hydatid disease (Rothschild and Martin 2006; Resnick and Niwayama 1995b). The lesions caused by hydatid disease are erosive and expansile, and the "progression occurs under the periosteum with small, variably sized vesicles. One or more round or oval, central or lateral lacunae form a 'bunch of grapes'" (Rothschild and Martin 2006, p. 92). The cyst is formed by a pressure effect, producing periosteal expansion.

The lesion observed in the skull of *Geikia locusticeps* does not present any vesicles. On the contrary, the lesion is rounded and seems to be produced by pressure erosion of the bone. Despite the absence of vesicles, this pathology resembles features produced by *Echinococcus*, and thus could be an alternative cause of the lesion seen in the specimen.

With respect to the pathologies observed in *Stahleckeria potens*, hydatid disease could have affected the anterior portion of the articular surface of the ectepicondyle of the right humerus (Fig. 9.3a), and possibly caused the irregular prominences observed on the distal extremity of the left femur (Fig. 9.4c).

Bacterial Infection and Osteomyelitis

A third type of bone paleopathology is the result of bacterial infection. According to Resnick and Niwayama (1995a), osteomyelitis implies an infection of bone and marrow. Stages of osteomyelitis are designated acute, subacute and chronic, and each of them leaves different features on bone, beginning with vascular changes and edema of soft tissues, to infectious penetration of the periosteum and formation of

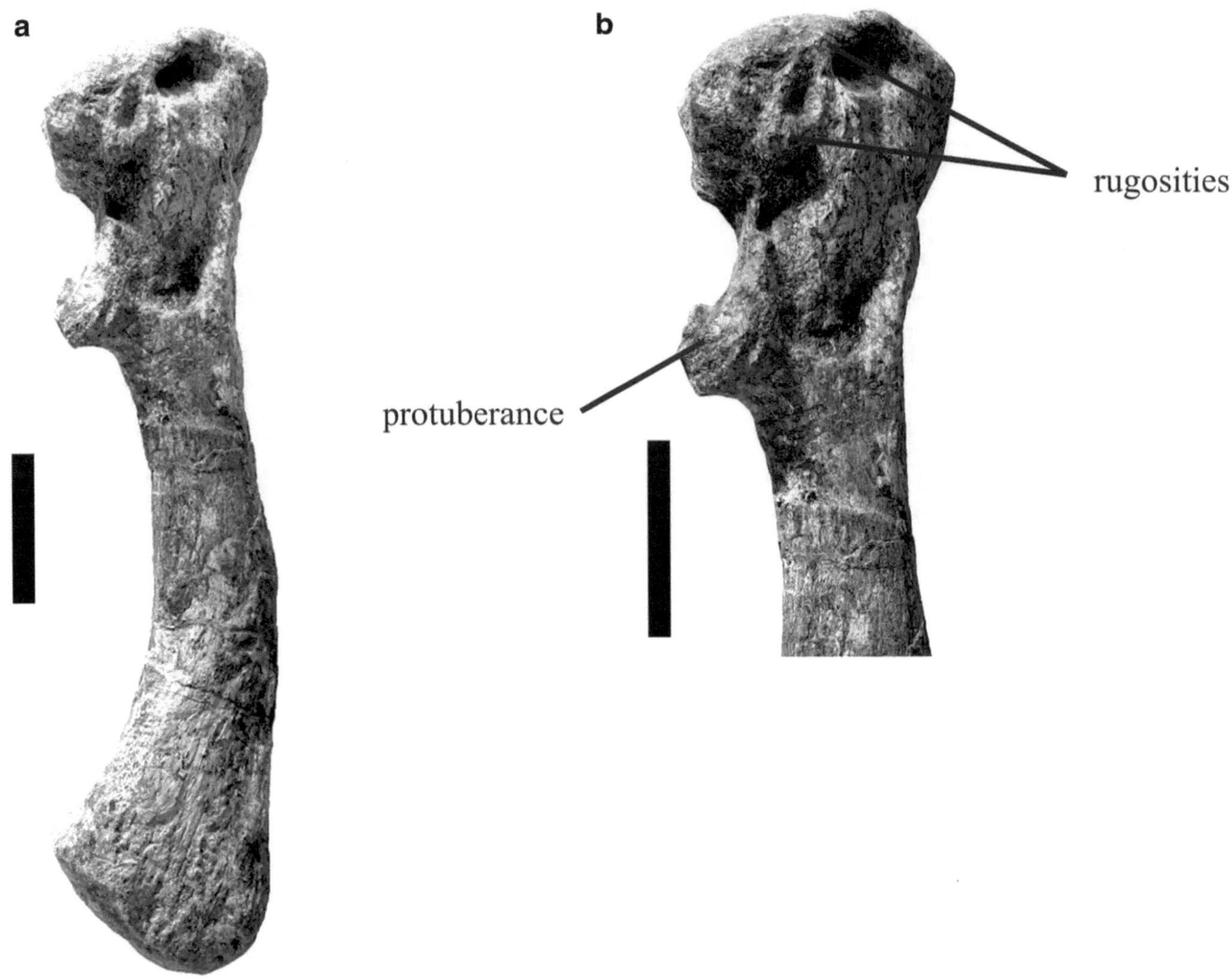

Fig. 9.6 Left fibula of *Stahleckeria potens* (GPIT/RE/8001). **a** Fibula in anterior view. **b** Proximal portion of the fibula, showing the pathological features. Scale bars equal 5 cm

abscesses, and finally cortical necrosis. According to Halstead (1990), the surface of bone becomes necrotic and develops a pitted texture; when osteomyelitis is established, "bacteria proliferate and pus accumulates within the cavity. Sinuses develop to drain away the pus and large areas of bone become necrotic" (Halstead 1990, p. 383). The bone is replaced by new tissue, and the mixture of necrotic and new tissue yields a characteristically roughened surface on the bone. If the bacteria gain entry into deep tissues an abscess is formed, and when the abscess occurs adjacent to hard tissue, a cavity is eroded. According to Ortner (2003a, p. 181), osteomyelitis is most often the result of the introduction of pyrogenic bacteria into bone. However, other agents, such as viruses, fungi, and multicelled parasites can also infect bone marrow (Ortner 2003a; Resnick and Niwayama 1995a). Infection may result in the full-blown manifestation of acute and chronic osteomyelitis, but is more often limited and localized. According to Ortner (2003a), the result would be focal periosteal bone deposition around a partial cortical defect, with or without a small sequestrum, and with some sclerotic response in the vicinity. Such local infections may heal with sclerotic scarring around a depression that may be effaced subsequently by remodelling. The aspect of the bone is rugose, eventually presenting one or more cloacae.

The pathology observed in *Geikia locusticeps* presents a pronounced concave pit (an abscess) surrounded by a lip of bone. The margins are completely smooth and no exostosis is observed on them (Fig. 9.1b). Therefore, the absence of scarring shows that the action of osteomyelitis is improbable in *Geikia*.

On the other hand, the morphology of the pathologies observed in *Stahleckeria potens*, on the distal extremity of the left femur (Fig. 9.4c) and on the proximal portion of tibia (Fig. 9.5b) could correspond to osteomyelitis, although no cloacae are observed. If these pathologies stem from osteomyelitis, the infection attacked the periosteum and the bones became a mixture of necrotic and healthy tissue, creating the typically roughened surfaces. The rugose surfaces, especially on the tibia where a concave surface is surrounded by scars, would correspond to remodelling in this scenario.

Fungal Disease

Fungal diseases rarely affect human skeletal remains (Ortner 2003b). According to Ortner (2003b), blastomycosis is represented by lesions with a lytic, well-defined border, and periosteal reactive bone may form at the margins of the lytic focus. Blastomycosis is not well represented in the literature, but according to Hershkovitz et al. (1998) this disease has a distinct osseous impact that can be distinguished from other pathologies. The explanations of fungal diseases presented below are adopted from Ortner (2003b). Paracoccidiomycosis is represented by single or multiple rounded and lytic lesions. This disease is common in humans on the clavicles, vertebrae and extremities. Cyptococcosis is common on bony prominences, cranial bones and vertebrae, although any bone may be affected. The lesions caused by cryptococcosis appear lytic and well circumscribed. Coccidioidomycosis is represented by solitary or multiple bone foci with symmetrical involvement. The bone lesions tend to be lytic and may be associated with periosteal reactive bone formation. The most characteristic aspect is that this disease prefers areas usually affected by other infections, especially in the axial region. Histoplasmosis is represented by multiple rounded lytic lesions, usually on the cranial vault and in long and small bones of the extremities. Sporotrichosis is very rare on developing bone lesions. In the more severe disseminated form on bone, the lesions may be mild and heal without residual alterations, although localized bone abscesses and ossifying focal or extensive periostitis have been observed. The bones affected (in decreasing order of frequency) are: tibia, bones of the manus and pes, ulna and radius, skull and facial bones, ribs, clavicles and vertebrae. In aspergillosis, bone involvement is rare, but may occur as a lesion in the skin or hematogenous dissemation most commonly involving the ribs, sternum or vertebrae. Destruction and necrosis of bone in the paranasal sinuses and orbit, as well as anterior cranial fossa, can be observed. Mucormycosis (phycomycosis) is normally limited to skull and facial bones, resulting in a perforation of the hard palate.

In the case of the *Stahleckeria* skeleton, blastomycosis, paracoccidiomycosis, coccidioidomycosis and sporotrichosis are the most plausible fungal diseases observed on the deltopectoral crest of the humerus, on the anterior articular portion of ectepicondyle of the humerus, the great trochanter and distal extremity of the left femur, and the tibia and fibula. The features of these lesions are very similar to the irregular new bone formation seen on a fibula in Fig. 66–74 (B) of Resnick and Niwayama (1995b), and also on the periosteal reaction associated with a serpentine rugosity on a tibia in Fig. 5 of Hershkovitz et al. (1998). In these pathologies, the lesions have a lytic well-defined border, and may be associated with periosteal reactive bone formation, with some aspects similar to osteomyelitis. The eburnation observed on the deltopectoral crest of the humerus represents the degeneration of cartilage likely associated with fungal disease. In the same position, small and regular rugosities are expected that are associated with the muscular insertion of the deltoideus muscle. Given this muscle attachment in the area, if the lesions are not the result of fungal disease, they could also stem from muscular avulsion. On the ectepicondyle, the small rugosities could be associated with periosteal reactive bone, although further examination via X-ray is needed to confirm this interpretation.

On the great trochanter of the left femur, the observed rugosities are very similar to those observed on the deltopectoral crest of the humerus. Again, rugosities are expected to be present at this position related to the insertion of the iliofemoralis muscle. However, the lesions of the femur are not restricted to the trochanter, but extend onto the body of the bone. Given this observation, it is unlikely that the lesion is related to muscular avulsion. On the distal extremity of this femur, the same eburnation observed on the deltopectoral crest of the humerus also represents degradation of cartilage. In anterior view, a concave morphology without rugosities is expected between the condyles. However, the observed protuberances are not restricted to the condyles, but also extend onto the body of the bone.

It is possible that the tibia and fibula of *Stahleckeria* were also affected by fungal disease. The rugosities observed on the proximal portion of tibia and fibula could represent bone abscesses, resulting from healing processes.

Conclusions

As discussed above, it is most likely that the lesion observed in the skull of *Geikia locusticeps* corresponded to an epidermal inclusion cyst, probably post-traumatic, or hydatid disease. In the composite skeleton of *Stahleckeria potens*, the observed lesion on the right scapula could have been caused by an epidermal inclusion cyst or a muscular avulsion. On the right humerus, fungal disease, hydatid disease or even muscular avulsion could be responsible for the distinct features observed. Also in *Stahleckeria*, the pathologies observed on the left femur could have been caused by fungal disease, healed osteomyelitis, or hydatid disease. On the tibia, a fungal disease or osteomyelitis is most plausible. Finally, on the fibula of the composite specimen of *Stahleckeria*, fungal disease is also most likely.

Acknowledgments The first author thanks the Conselho Nacional de Desenvolvimento Científico e Tecnológico (CNPq) of Brazil for the financial support of this work. We would also like to thank Philipe Havlik from the Eberhard Karls Universität Tübingen for assistance and access to the collection. Thanks also to the photographer Wolfgang Gerber for the photos. Additional thanks to the three reviewers for their suggestions.

References

Angielczyk, K. D. (2002). Redescripton, phylogenetic position, and stratigraphic significance of the dicynodont genus *Odontocyclops* (Synapsida: Anomodontia). *Journal of Paleontology, 76*, 1047–1049.

Botha, J., & Angielczyk, K. D. (2007). An integrative approach to distinguishing the Late Permian dicynodont species *Oudenodon bainii* and *Tropidostoma microtrema* (Therapsida: Anomodontia). *Palaeontology, 50*, 1175–1209.

Camp, C. L., & Welles, S. P. (1956). Triassic dicynodont reptiles. Part I. The North American genus *Placerias*. *Memoirs of the University of California, 13*, 255–304.

Choquette, L. P. E., Harington, C. R., & Archibald, J. (1975). Paleopathology: Exostoses of the third metacarpal in Pleistocene horses from the Yukon Territory. *Canadian Journal of Earth Science, 12*, 1053–1058.

Ferigolo, J. (1985). Paleopatologia em preguiças terrícolas – artrose. *Coletânea de Trabalhos Paleontológicos. Brasília: DNPM, sér. Geologia, 27, seção Paleontologia e Estratigrafia, 2*, 35–41.

Ferigolo, J. (1993a). Fratura com osteomielite em fêmur de crocodiliano do Neógeno do Estado do Amazonas, Brasil. *IX Jornadas Argentinas de Paleontologia de Vertebrados. Ameghiniana, 30*, 105–106.

Ferigolo, J. (1993b). Lesões pós-traumáticas em mamíferos do Pleistoceno Superior do Brasil. *IX Jornadas Argentinas de Paleontologia de Vertebrados. Ameghiniana, 30*, 106.

Fröbisch, J., & Reisz, R. R. (2008). A new species of *Emydops* (Synapsida, Anomodontia) and a discussion of dental variability and pathology in dicynodonts. *Journal of Vertebrate Paleontology, 28*, 770–787.

Gillette, D. D., & Madsen, D. B. (1993). The Columbian mammoth, *Mammuthus columbi*, from the Wasatch Mountains of Central Utah. *Journal of Paleontology, 67*, 669–680.

Halstead, L. B. (1990). 4.15. Palaeopathology. In D. E. G. Briggs & P. R. Crowther (Eds.), *Palaeobiology: A synthesis* (pp. 381–385). Oxford: Blackwell.

Hanna, R. R. (2002). Multiple injury and infection in a sub-adult theropod dinosaur *Allosaurus fragilis* with comparisons to allosaur pathology in the Cleveland-Lloyd dinosaur quarry collection. *Journal of Vertebrate Paleontology, 22*, 76–90.

Henriques, D. D. R., Soares, A. A., & Mello, M. G. S. (1998). Registros de reação óssea em *Panochthus* Burmeister, 1866 do Pleistoceno do Estado da Paraíba. *Acta Geologica Leopoldensia, 21*, 149–155.

Hershkovitz, I., Rothschild, B. M., Dutour, O., & Greenwald, C. (1998). Clues to recognition of fungal origin of lytic skeletal lesions. *American Journal of Physical Anthropology, 106*, 47–60.

Huene, F. von. (1935). *Die fossilen Reptilien des südamerikanischen Gondwanalandes an der Zeitenwende. Ergebnisse der Sauriergrabungen in Südbrasilien 1928/29. Lieferung 1: Anomodontia.* München: C.H. Beck'sche Verlagsbuchhandlung.

Huene, F. von. (1942). Die Anomodontier des ostafrikanischen Ruhuhu-Gebietes in der Tübinger Sammlung. *Palaeontographica, Abteilung A, 94*, 154–184.

Huttenlocker, A. K., & Rega, E. (2012). The paleobiology and bone microstructure of pelycosaurian-grade synapsids. In A. Chinsamy-Turan (Ed.), *Forerunners of mammals: Radiation, histology, biology* (pp. 90–119). Bloomington: Indiana University Press.

Huttenlocker, A. K., Rega, E., & Sumida, S. S. (2010). Comparative anatomy and osteohistology of hyperelongate neural spines in the sphenacodontids *Sphenacodon* and *Dimetrodon* (Amniota: Synapsida). *Journal of Morphology, 271*, 1407–1421.

Jinnah, Z. A., & Rubidge, B. S. (2007). A double-tusked dicynodont and its biostratigraphic significance. *South African Journal of Science, 103*, 51–53.

Kammerer, C. F., Angielczyk, K. D., & Fröbisch, J. (2011). A comprehensive taxonomic revision of *Dicynodon* (Therapsida, Anomodontia) and its implications for dicynodont phylogeny, biogeography, and biostratigraphy. *Society of Vertebrate Paleontology Memoir, 11*, 1–158.

Lucas, S. G. (2000). Pathological aetosaur armor from the Upper Triassic of Germany. *Stuttgarter Beiträge zur Naturkunde, 281*, 1–6.

Lucas, S. G., & Schoch, R. M. (1987). Paleopathology of Early Cenozoic *Coryphodon* (Mammalia: Pantodonta). *Journal of Vertebrate Paleontology, 7*, 145–154.

Maisch, M. W., & Gebauer, E. V. I. (2005). Reappraisal of *Geikia locusticeps* (Therapsida: Dicynodontia) from the Upper Permian of Tanzania. *Palaeontology, 48*, 309–324.

Moodie, R. L. (1918). Paleontological evidences of the antiquity of disease. *The Scientific Monthly, 1918*, 265–281.

Moodie, R. L. (1930). Studies in Paleopathology, XXVI. Pleistocene luxations. *The American Journal of Surgery, 9*, 348–362.

Ortner, D. J. (2003a). Chapter 9. Infectious diseases: Introduction, Biology, Osteomyelitis, Periostitis, Brucellosis, Glanders, and Septic Arthritis. In D. J. Ortner (Ed.), *Identification of pathological conditions in human skeletal remains* (pp. 179–226). Amsterdam: Academic Press.

Ortner, D. J. (2003b). Chapter 12. Infectious diseases: Mycotic, viral and multicelled parasitic diseases of the human skeleton. In D. J. Ortner (Ed.), *Identification of pathological conditions in human skeletal remains* (pp. 325–341). Amsterdam: Academic Press.

Ortner, D. J. (2003c). Chapter 20. Tumors and tumor-like lesions of bones. In D. J. Ortner (Ed.), *Identification of pathological conditions in human skeletal remains* (pp. 503–544). Amsterdam: Academic Press.

Peterson, J. E., Henderson, M. D., Scherer, R. P., & Vittore, C. P. (2009). Face biting on a juvenile tyrannosaurid and behavioral implications. *Palaios, 24*, 780–784.

Reisz, R. R., Scott, D. M., Pynn, B. R., & Modesto, S. P. (2011). Osteomyelitis in a Paleozoic reptile: Ancient evidence for bacterial infection and its evolutionary significance. *Naturwissenschaften, 98*, 551–555.

Resnick, D., & Niwayama, G. (1995a). Osteomyelitis, septic arthritis and soft tissue infection: Mechanisms and situations. In D. Resnick (Ed.), *Diagnosis of bone and joint disorders* (3rd ed., Vol. 4, pp. 2325–2418). Philadelphia: W.B. Saunders Company.

Resnick, D., & Niwayama, G. (1995b). Osteomyelitis, septic arthritis and soft tissue infection: Organisms. In D. Resnick (Ed.), *Diagnosis of bone and joint disorders* (3rd ed., Vol. 4, pp. 2448–2558). Philadelphia: W.B. Saunders Company.

Resnick, D., Kyriakos, M., & Greenway, G. D. (1995). Tumors and tumor-like lesions of bone: Imaging and pathology of specific lesions. In D. Resnick (Ed.), *Diagnosis of bone and joint disorders*

(3rd ed., Vol. 4, pp. 3628–3938). Philadelphia: W.B. Saunders Company.

Rothschild, B. M. (1988). Stress fracture in a ceratopsian phalanx. *Journal of Paleontology, 62*, 302–303.

Rothschild, B., & Martin, L. D. (1987). Avascular necrosis: Occurrence in diving Cretaceous mosasaurs. *Science, 236*, 75–77.

Rothschild, B. M., & Martin, L. D. (2006). Skeletal impact of disease. *New Mexico Museum of Natural History and Science Bulletin, 33*, 1–226.

Schultz, C. L. (1999). An example of paleopathology in rhynchosaurs from the Triassic of Southern Brazil. *Ameghiniana, 36*, 20R.

Scott, E., & Rooney, J. R. (2001). Non-articular periostosis of a proximal phalanx of *Equus conversidens*. *Paleobios, 21*, 12–14.

Vega-Dias, C., & Schultz, C. L. (2003). A paleopathology in *Jachaleria candelariensis* Araújo & Gonzaga 1980 (Synapsida, Dicynodontia) from the Upper Triassic of Southern Brazil. *Ameghiniana, 40*, 74R.

Vega-Dias, C., & Schultz, C. L. (2004). Postcranial material of *Jachaleria candelariensis* Araújo and Gonzaga 1980 (Therapsida, Dicynodontia), Upper Triassic of Rio Grande do Sul, Brazil. *PaleoBios, 24*, 7–31.

Vega-Dias, C., Maisch, M. W., & Schwanke, C. (2005). The taxonomic status of *Stahleckeria impotens* (Therapsida, Dicynodontia): Redescription and discussion of its phylogenetic position. *Revista Brasileira de Paleontologia, 8*, 221–228.

Wang, X., & Rothschild, B. M. (1992). Multiple hereditary osteochondroma in Oligocene *Hesperocyon* (Carnivora: Canidae). *Journal of Vertebrate Paleontology, 12*, 387–394.

Wells, C. (1964). Pathological epipodials and tarsus in *Stretosaurus macromerus* from the Kimmeridge Clay, Stretham, Cambridgeshire. *Quarterly Journal of the Geological Society of London, 120*, 299–304.

Part III
Theriodontia

Chapter 10
Theriodontia: Introduction

Christian F. Kammerer

With their prominent canines, semi-sprawling gait, vaguely dog-like aspect, and probable hairy integument in at least some species, theriodonts represent the archetypal pre-mammalian therapsids. The fossil record of theriodonts preserves in exquisite detail the morphological transition between 'reptilian' and 'mammalian'-grade synapsids (Hopson 1991), and as such they have historically received the lion's share of research effort into synapsid evolution. But though the most thoroughly-studied theriodonts (e.g., *Thrinaxodon*, *Pachygenelus*) hew close to the mammalian stem, the group as a whole includes far more than just our own Permo-Triassic forebears. Permo-Triassic theriodonts are known from every continent and exhibit vast ecological diversity, ranging from shrew- to bear-sized and including an array of carnivores, insectivores, and herbivores (Rubidge and Sidor 2001). Even without including Mammalia they are the longest-ranging synapsid group, with the last non-mammalian cynodonts surviving until the Early Cretaceous (Tatarinov and Matchenko 1999).

Theriodontia was initially established by Owen (1860, 1876) to include various carnivorous therapsids from the Karoo Basin of South Africa (e.g., *Galesaurus* and *Gorgonops*). The modern composition of the group was solidified by the highly influential therapsid classification of Watson and Romer (1956), which divided Therapsida into the primarily herbivorous Anomodontia (including Dinocephalia, Venyukovioidea, and Dicynodontia) and the primarily carnivorous Theriodontia (including Gorgonopsia, Therocephalia, and Cynodontia). Watson and Romer's (1956) "Anomodontia" has since been overturned by phylogenetic analysis (Sidor 2000; Liu et al. 2009), but Theriodontia is still generally recognized (Hopson and Barghusen 1986; Rubidge and Sidor 2001), although its monophyly is uncertain. A close relationship [either as sister taxa (Hopson and Barghusen 1986; Huttenlocker 2009) or deriving cynodonts from within Therocephalia (Botha-Brink et al. 2007; Abdala 2007)] between therocephalians and cynodonts (Eutheriodontia) is strongly supported, but cladistic analyses of higher level therapsid phylogeny have produced conflicting results as to whether gorgonopsians or anomodonts form the sister-group of Eutheriodontia (Rowe 1986; Kemp 1988; Sidor 2000; Modesto et al. 1999). In addition to general paucity of data, two particular issues underlie this conflict: the long branch of Anomodontia and the proclivity for Biarmosuchia and Gorgonopsia to clade together based on postcranial characters (Sidor 2000; Liu et al. 2009). However, the majority of higher-level analyses of therapsid phylogeny predate the discovery of early anomodonts (e.g., Liu et al. 2010; Cisneros et al. 2011) that help break up this clade's long branch. Hopefully future higher-level analyses incorporating these taxa will help to elucidate theriodont relationships.

Gorgonopsia is made up exclusively of sabre-toothed predators, although there is extensive variation in their (presumed) adult body size [basal skull length in the group ranges from ~15 cm in *Cyonosaurus* to ~60 cm in *Inostrancevia* (Kammerer, personal observation)]. Middle Permian gorgonopsians are rare and thus far only known from the Karoo (Kammerer 2013), but the group was abundant and geographically widespread in the Late Permian before going extinct at the Permo-Triassic boundary (Sigogneau-Russell 1989). Unfortunately, they remain one of the most problematic synapsid clades from a taxonomic standpoint, despite two comprehensive, monographic reviews (Sigogneau 1970; Gebauer 2007). Gorgonopsian cranial morphology is highly conservative and the majority of taxa are distinguished by proportional characters (Sigogneau-Russell 1989). A few

C. F. Kammerer (✉)
Division of Paleontology and Richard Gilder Graduate School, American Museum of Natural History, Central Park West at 79th Street, New York, NY 10024, USA

and

Museum für Naturkunde, Leibniz-Institut für Evolutions- und Biodiversitätsforschung, Humboldt-Universität zu Berlin, Invalidenstra. 43, 10115 Berlin, Germany
e-mail: christian.kammerer@mfn-berlin.de

Christian F. Kammerer, Kenneth D. Angielczyk and Jörg Fröbisch (eds.), *Early Evolutionary History of the Synapsida*,
Vertebrate Paleobiology and Paleoanthropology, DOI: 10.1007/978-94-007-6841-3_10,

distinct gorgonopsian subclades can be recognized, particularly the gigantic rubidgeines and inostranceviines, but beyond the general recognition of these groups there is no consensus as to their composition. For example, Sigogneau-Russell (1989) recognized seven rubidgeine genera with 18 species, whereas Gebauer (2007) recognized four rubidgeine genera with 16 species.

A particularly acute problem for gorgonopsian taxonomy is the referral of complete, well-preserved skulls to species based on extremely poor holotypes, as well as the erection of new species in a genus whose type species is based on a poor holotype. In these cases, a nominal taxon (e.g., *Arctops*, *Gorgonops*) becomes well known in the literature based on the referred material/species, which is highly problematic if the holotype is indeterminate or, worse, actually represents a different taxon. In groups where detailed taxonomic study has established discrete diagnostic characters for a species, these problems can be readily addressed, either by confirmation of a fragmentary holotype as conspecific with well-preserved referred specimens [as was the case for *Dicynodon lacerticeps* in anomodonts and *Diademodon tetragonus* in cynodonts (Hopson and Kitching 1972; Bradu and Grine 1979; Kammerer et al. 2011)] or by the erection of a new taxon to accommodate specimens previously referred to a taxon based on indeterminate material [as in the case of the biarmosuchian *Rubidgina*, referred material of which was transferred to the new genus *Herpetoskylax* (Sidor and Rubidge 2006)]. However, because the currently recognized gorgonopsian taxa are distinguished mainly by proportional differences, it is very difficult to establish conspecificity between fragmentary holotypes and referred specimens; at the same time, it would be premature to consider many of these gorgonopsian taxa *nomina dubia* in the absence of discrete diagnoses for even the well-preserved material. Most of the nominal gorgonopsian species differ only in characters such as orbit size, snout length, and postcanine number that are known to be ontogenetically variable in other theriodonts (Anderson 1968; Hopson 1991). It is probable that the majority of nominal gorgonopsian species represent ontogenetic or taphonomic variants of relatively few real morphospecies. Repreparation and redescription of numerous gorgonopsian holotypes will be necessary before a stable species-level taxonomy for this clade can be established, with two pertinent contributions presented in this volume.

Kammerer (2013) redescribes the stratigraphically earliest-known gorgonopsian, the South African *Eriphostoma microdon*, based on new computed tomographic images of the holotype. This taxon has generally been considered a *nomen dubium* (Theriodontia indet.) (Boonstra 1935, 1969; Sigogneau 1970; Sigogneau-Russell 1989), but Kammerer demonstrates that it is indeed a gorgonopsian and represents a diagnosable species. Middle Permian gorgonopsians are extremely rare, and although incomplete, this specimen provides important new information on the palatal anatomy of the group and how it changes during gorgonopsian evolution.

Gebauer (2013) redescribes one of the most complete and best-preserved gorgonopsian specimens, the holotype of *Scymnognathus parringtoni*. This specimen has been referred to multiple genera over the years; in this latest iteration Gebauer determines that it represents the first African representative of the genus *Sauroctonus*, previously known only from the Russian species *Sauroctonus progressus*. This conclusion remains to be tested by phylogenetic analysis, but if correct would be the first evidence of a transcontinental sister-group relationship within Gorgonopsia, echoing the distribution of the cynodont *Procynosuchus* with significant implications for dispersal in the Late Permian. Modern research on gorgonopsians is still in its infancy, but several recent papers suggest a complex biogeographic pattern for the group: Smiley et al. (2008) described a probable gorgonopsian fragment as the only therapsid fossil from the equatorial Moradi Formation of Niger and Botha-Brink et al. (2013) figure a South African gorgonopsian specimen from near the PTB that appears very similar to the Russian endemic taxon *Inostrancevia*.

Therocephalians are remarkably morphologically and ecologically diverse compared to gorgonopsians. The earliest known therocephalians (Middle Permian Lycosuchidae and Scylacosauridae) were large, sabre-toothed predators of gorgonopsian aspect, but Late Permian taxa (Eutherocephalia) include tiny insectivores, pug-nosed carnivores, taxa with boomerang-shaped jaws completely lacking postcanine teeth, and the earliest venomous amniote in the fossil record (*Euchambersia mirabilis* from the South African *Cistecephalus* Assemblage Zone) (Mendrez 1975). Few phylogenetic analyses of Therocephalia exist and species-level relationships are poorly understood (Hopson and Barghusen 1986; Abdala 2007; Huttenlocker 2009), but there exists a general consensus on higher-level phylogeny within the group, with a tripartite breakdown of eutherocephalians into whaitsioids (e.g., *Theriognathus*), akidnognathids (e.g., *Euchambersia*), and baurioids (e.g., *Regisaurus*). Although they suffered heavy losses in the end-Permian extinction, therocephalians are relatively abundant in the *Lystrosaurus* Assemblage Zone recovery fauna compared with other Permian clades (Damiani et al. 2004; Huttenlocker et al. 2011) and the group had at least one major Triassic radiation (Bauriamorpha). Bauriamorphs were extremely mammal-like herbivores with complete secondary palates and complexly occluding postcanine dentition, but many aspects of their taxonomy and paleobiology remain poorly understood.

Abdala et al. (2013) present a detailed description of the most complete and best-preserved specimen of the bauriamorph therocephalian *Microgomphodon oligocynus*. This provides the foundation both for a much-needed review of South African Middle Triassic bauriamorphs (reducing the nominal taxa to *M. oligocynus* and *Bauria cynops*) and a functional interpretation of bauriamorph cranial anatomy. Additionally, Abdala et al. present a three-dimensional morphometric analysis of differences in cranial morphology between bauriid morphotypes that should serve as a useful example for future comparative studies between closely related synapsid taxa.

Because they include mammals, cynodonts have been more heavily studied than any other therapsid group. Although subject to the same historical oversplitting as other therapsids, cynodont alpha taxonomy has experienced extensive scrutiny, and all major cynodont groups have been revised within the past 40 years (Hopson and Kitching 1972; Bradu and Grine 1979; Gow 1980; Abdala and Giannini 2000, 2002; Sidor and Smith 2004; Watabe et al. 2007; Liu 2007; Liu et al. 2008). Cynodonts are the only major therapsid group not to appear in the Middle Permian and the origins of the clade remain poorly known. The recently-described purported Middle Permian cynodont *Novocynodon kutorgai* (Ivakhnenko 2012) is known only from a partial dentary without any cynodont apomorphies: it lacks a masseteric fossa, has a cuspulated postcanine cingulum but not discrete cusps, and exhibits a tooth replacement style typical of basal therapsids. *Novocynodon* is not a cynodont; it may represent a juvenile dinocephalian or anomodont (Kammerer, personal observation). The earliest known definitive cynodonts are found in the *Tropidostoma* Assemblage Zone (Botha et al. 2007), but the cynodont radiation is primarily a Middle Triassic phenomenon (Abdala and Ribeiro 2010; Liu and Olsen 2010; Botha-Brink et al. 2011). By the Middle Triassic, cynodonts include an array of large and small predators as well as herbivores in two major clades, Cynognathia and Probainognathia (Hopson and Kitching 2001). Cynodonts remain species-rich and locally abundant through the Late Triassic, and three cynodont subclades survive across the Triassic-Jurassic boundary: Mammalia, Tritheledontidae (Early Jurassic), and Tritylodontidae (which survive through the Early Cretaceous).

Although most cynodont research has been undertaken with a view towards mammal origins, the contributions in this volume focus on an important branch of cynodont evolution far removed from mammalian ancestry. Traversodontidae is one of the most diverse and abundant groups of tetrapod herbivores in the Triassic and were major components of terrestrial communities from the Anisian to the Norian (Abdala and Ribeiro 2010). Traversodontids are particularly common in Gondwana, where they are the numerically dominant herbivores in many Triassic terrestrial assemblages (e.g., the Chañares Formation of Argentina and the basal 'Isalo II' beds of Madagascar) (Abdala and Giannini 2002; Ranivoharimanana et al. 2011). New discoveries continue to expand the geographic range and morphological diversity of the group: Hopson and Sues (2006) described the first definitive traversodontid from Europe (*Nanogomphodon wildi* from Germany) and Reichel et al. (2009) described a bizarre traversodontid with bony nodes lining the ribs [*Protuberum cabralense* from Brazil (necessarily emended from the original *P. cabralensis* because *Protuberum* is a neuter generic name)]. Although traversodontid morphology and systematics have had extensive recent study (Abdala and Ribeiro 2003; Kammerer et al. 2008; Liu 2008; Sues and Hopson 2010), traversodontid phylogeny remains problematic. A derived group of mostly Late Triassic traversodontids (Gomphodontosuchinae) is recovered in most analyses, but relationships between the basal members of the clade are essentially unresolved. In this volume, we include two studies directly addressing this portion of the tree.

Hopson (2013) provides a long overdue reevaluation of the Tanzanian traversodontid cynodont "*Scalenodon*" *hirschsoni*. Although named as a distinct species by Crompton (1972) on the basis of its unusual dentition, the holotype and referred mandibular material have never been described in full. Hopson remedies this in what is hopefully the first of several revisions of the Manda beds cynodonts, driven in part by new discoveries from the Middle Triassic of Tanzania (Sidor et al. 2009).

Finally, Liu and Abdala (2013) review the Traversodontidae in its entirety and present a new phylogeny for the group, incorporating new information on various basal taxa. Their phylogeny represents the first in which the basal portion of the tree is well resolved (albeit still not robustly supported), representing a springboard for future studies on this segment of Triassic cynodont evolution. In particular, establishing character polarity at the base of Traversodontidae represents a crucial step in understanding the convergent evolution of mammal-like features in later members of the clade, a recurring feature in therapsid evolution.

The chapters in this section illustrate the varying stages in systematic study of the major theriodont groups. Gorgonopsians and therocephalians still require extensive alpha taxonomic revision, whereas cynodont species are relatively well-resolved but higher level phylogeny requires further analysis. Resolution of these issues is particularly important for understanding the divergent fortunes of these three major clades across the Permo-Triassic boundary, and it is hoped that the contributions in this volume will help spur renewed interest in Permian theriodont research.

References

Abdala, F. (2007). Redescription of *Platycraniellus elegans* (Therapsida, Cynodontia) from the Lower Triassic of South Africa, and the cladistic relationships of eutheriodonts. *Palaeontology, 50*, 591–618.

Abdala, F., & Giannini, N. P. (2000). Gomphodont cynodonts of the Chañares Formation: The analysis of an ontogenetic sequence. *Journal of Vertebrate Paleontology, 20*, 501–506.

Abdala, F., & Giannini, N. P. (2002). Chiniquodontid cynodonts: Systematic and morphometric considerations. *Palaeontology, 45*, 1151–1170.

Abdala, F., & Ribeiro, A. M. (2003). A new traversodontid cynodont from the Santa Maria Formation (Ladinian-Carnian) of southern Brazil, with a phylogenetic analysis of Gondwanan traversodontids. *Zoological Journal of the Linnean Society, 139*, 529–545.

Abdala, F., & Ribeiro, A. M. (2010). Distribution and diversity patterns of Triassic cynodonts (Therapsida, Cynodontia) in Gondwana. *Palaeogeography, Palaeoclimatology, Palaeoecology, 286*, 202–217.

Abdala, F., Jashashvili, T., Rubidge, B. S., & van den Heever, J. (2013). New material of *Microgomphodon oligocynus* (Eutherapsida, Therocephalia) and the taxonomy of southern African Bauriidae. In C. F. Kammerer, K. D. Angielczyk, & J. Fröbisch (Eds.), *Early evolutionary history of the Synapsida* (pp. 209–231). Dordrecht: Springer.

Anderson, J. M. (1968). The confused state of classification of the family Procynosuchidae. *Palaeontologia Africana, 11*, 77–84.

Boonstra, L. D. (1935). On the South African gorgonopsian reptiles preserved in the American Museum of Natural History. *American Museum Novitates, 772*, 1–14.

Boonstra, L. D. (1969). The fauna of the *Tapinocephalus* Zone (Beaufort beds of the Karoo). *Annals of the South African Museum, 56*, 1–73.

Botha, J., Abdala, F., & Smith, R. (2007). The oldest cynodont: New clues on the origin and early diversification of the Cynodontia. *Zoological Journal of the Linnean Society, 149*, 477–492.

Botha-Brink, J., Abdala, F., & Chinsamy-Turan, A. (2011). The radiation and osteohistology of nonmammaliaform cynodonts. In A. Chinsamy-Turan (Ed.), *Forerunners of mammals: Radiation histology biology* (pp. 222–246). Bloomington: Indiana University Press.

Botha-Brink, J., Huttenlocker, A. K., & Modesto, S. P. (2013). Vertebrate paleontology of Nooitgedacht 68: A *Lystrosaurus maccaigi* rich Permo-Triassic boundary locality in South Africa. In C. F. Kammerer, K. D. Angielczyk, & J. Fröbisch (Eds.), *The early evolutionary history of the Synapsida* (pp. 289–304). Dordrecht: Springer.

Bradu, D., & Grine, F. E. (1979). Multivariate analysis of diademodontine crania from South Africa and Zambia. *South African Journal of Science, 75*, 441–448.

Cisneros, J. C., Abdala, F., Rubidge, B. S., Dentzien-Dias, P. C., & Bueno, A. (2011). Dental occlusion in a 260-million-year-old therapsid with saber canines from the Permian of Brazil. *Science, 331*, 1603–1605.

Crompton, A. W. (1972). Postcanine occlusion in cynodonts and tritylodontids. *Bulletin of the British Museum (Natural History) Geology, 21*, 29–71.

Damiani, R., Neveling, J., Modesto, S., & Yates, A. (2004). Barendskraal, a diverse amniote locality from the *Lystrosaurus* Assemblage Zone, Early Triassic of South Africa. *Palaeontologia Africana, 39*, 53–62 (for 2003).

Gebauer, E. V. I. (2007). *Phylogeny and evolution of the Gorgonopsia with a special reference to the skull and skeleton of GPIT/RE/7113* ('Aelurognathus' parringtoni). Unpublished Ph.D. thesis, Eberhard-Karls Universität Tübingen.

Gebauer, E. V. I. (2013). Re-assessment of the taxonomic position of the specimen GPIT/RE/7113 (*Sauroctonus parringtoni* comb. nov., Gorgonopsia). In C. F. Kammerer, K. D. Angielczyk, & J. Fröbisch (Eds.), *Early evolutionary history of the Synapsida* (pp. 185–207). Dordrecht: Springer.

Gow, C. E. (1980). The dentitions of the Tritheledontidae (Therapsida: Cynodontia). *Proceedings of the Royal Society of London Series B, 208*, 461–481.

Hopson, J. A. (1991). Systematics of the nonmammalian Synapsida and implications for patterns of evolution in synapsids. In H.-P. Schultze & L. Trueb (Eds.), *Origin of the higher groups of tetrapods, controversy and consensus* (pp. 635–693). Ithaca: Cornell University Press.

Hopson, J. A. (2013). The traversodontid cynodont *Mandagomphodon hirschsoni* from the Middle Triassic of the Ruhuhu Valley, Tanzania. In C. F. Kammerer, K. D. Angielczyk, & J. Fröbisch (Eds.), *Early evolutionary history of the Synapsida* (pp. 233–253). Dordrecht: Springer.

Hopson, J. A., & Barghusen, H. R. (1986). An analysis of therapsid relationships. In N. Hotton, III, P. D. MacLean, J. J. Roth, & E. C. Roth (Eds.), *The ecology and biology of the mammal-like reptiles* (pp. 83–106). Washington, D. C.: Smithsonian Institution Press.

Hopson, J. A., & Kitching, J. W. (1972). A revised classification of cynodonts (Reptilia; Therapsida). *Palaeontologia Africana, 14*, 71–85.

Hopson, J. A., & Kitching, J. W. (2001). A probainognathian cynodont from South Africa and the phylogeny of nonmammalian cynodonts. *Bulletin of the Museum of Comparative Zoology, 156*, 5–35.

Hopson, J. A., & Sues, H.-D. (2006). A traversodont cynodont from the Middle Triassic (Ladinian) of Baden-Württemberg (Germany). *Paläontologische Zeitschrift, 80*, 124–129.

Huttenlocker, A. K. (2009). An investigation into the cladistic relationships and monophyly of therocephalian therapsids (Amniota: Synapsida). *Zoological Journal of the Linnean Society, 157*, 865–891.

Huttenlocker, A. K., Sidor, C. A., & Smith, R. M. H. (2011). A new specimen of *Promoschorhynchus* (Therapsida: Therocephalia: Akidnognathidae) from the lowermost Triassic of South Africa and its implications for therocephalian survival across the Permo-Triassic boundary. *Journal of Vertebrate Paleontology, 31*, 405–421.

Ivakhnenko, M. F. (2012). Permian Cynodontia (Theromorpha) of Eastern Europe. *Paleontological Journal, 46*, 199–207.

Kammerer, C. F. (2013). A redescription of *Eriphostoma microdon* Broom, 1911 (Therapsida, Gorgonopsia) from the *Tapinocephalus* Assemblage Zone of South Africa and a review of Middle Permian gorgonopsians. In C. F. Kammerer, K. D. Angielczyk, & J. Fröbisch (Eds.), *The early evolutionary history of the Synapsida* (pp. 171–184). Dordrecht: Springer.

Kammerer, C. F., Flynn, J. J., Ranivoharimanana, L., & Wyss, A. R. (2008). New material of *Menadon besairiei* (Cynodontia: Traversodontidae) from the Triassic of Madagascar. *Journal of Vertebrate Paleontology, 28*, 445–462.

Kammerer, C. F., Angielczyk, K. D., & Fröbisch, J. (2011). A comprehensive taxonomic revision of *Dicynodon* (Therapsida, Anomodontia) and its implications for dicynodont phylogeny, biogeography, and biostratigraphy. *Society of Vertebrate Paleontology Memoir, 11*, 1–158.

Kemp, T. S. (1988). Interrelationships of the Synapsida. In M. J. Benton (Ed.), *The phylogeny and classification of the tetrapods Volume 2: Mammals. Systematics Association Special Volume No. 35*. Oxford: Clarendon University Press.

Liu, J. (2007). The taxonomy of the traversodontid cynodonts *Exaeretodon* and *Ischignathus*. *Revista Brasileira de Paleontologia, 10*, 133–136.

Liu, J., & Abdala, F. (2013). Phylogeny and taxonomy of the Traversodontidae. In C. F. Kammerer, K. D. Angielczyk, & J. Fröbisch (Eds.), *Early evolutionary history of the Synapsida* (pp. 255–279). Dordrecht: Springer.

Liu, J., & Olsen, P. (2010). The phylogenetic relationships of Eucynodontia (Amniota: Synapsida). *Journal of Mammalian Evolution, 17*, 151–176.

Liu, J., Soares, M. B., & Reichel, M. (2008). *Massetognathus* (Cynodontia, Traversodontidae) from the Santa Maria Formation of Brazil. *Revista Brasileira de Paleontologia, 11*, 27–36.

Liu, J., Rubidge, B., & Li, J. (2009). New basal synapsid supports Laurasian origin for therapsids. *Acta Palaeontologica Polonica, 54*, 393–400.

Liu, J., Rubidge, B., & Li, J. (2010). A new specimen of *Biseridens qilianicus* indicates its phylogenetic position as the most basal anomodont. *Proceedings of the Royal Society of London B, 277*, 285–292.

Mendrez, C. (1975). Principales variations du palais chez les Thérocéphales sud-africains (Pristerosauria et Scaloposauria) au cours du Permien supérieur et du Trias inférieur. *Problémes Actuels de Paléontologie—Évolution des Vertébrés. Coloques Internationaux de Centre National de la Recherche Scientifique, 218*, 379–408.

Modesto, S. P., Rubidge, B. S., & Welman, J. (1999). The most basal anomodont therapsid and the primacy of Gondwana in the evolution of the anomodonts. *Proceedings of the Royal Society of London Series B, 266*, 331–337.

Owen, R. (1860). On the orders of fossil and recent Reptilia, and their distribution in time. *Report of the Twenty-Ninth Meeting of the British Association for the Advancement of Science, 1859*, 153–166.

Owen, R. (1876). *Descriptive and illustrated catalogue of the fossil Reptilia of South Africa in the collection of the British Museum*. London: Taylor and Francis.

Ranivoharimanana, L., Kammerer, C. F., Flynn, J. J., & Wyss, A. R. (2011). New material of *Dadadon isaloi* (Cynodontia, Traversodontidae) from the Triassic of Madagascar. *Journal of Vertebrate Paleontology, 31*, 1292–1302.

Reichel, M., Schultz, C., & Soares, M. B. (2009). A new traversodontid cynodont (Therapsida, Eucynodontia) from the Middle Triassic Santa Maria Formation of Rio Grande do Sul, Brazil. *Palaeontology, 52*, 229–250.

Rowe, T. (1986). *Osteological diagnosis of Mammalia, L. 1758, and its relationship to extinct Synapsida*. Unpublished Ph.D. thesis, University of California, Berkeley, California.

Rubidge, B. S., & Sidor, C. A. (2001). Evolutionary patterns among Permo-Triassic therapsids. *Annual Reviews of Ecology and Systematics, 32*, 449–480.

Sidor, C. A. (2000). *Evolutionary trends and relationships within the Synapsida*. Unpublished Ph.D. thesis, University of Chicago.

Sidor, C. A., & Rubidge, B. S. (2006). *Herpetoskylax hopsoni*, a new biarmosuchian (Therapsida: Biarmosuchia) from the Beaufort Group of South Africa. In M. T. Carrano, T. J. Gaudin, R. W. Blob, & J. R. Wible (Eds.), *Amniote paleobiology: Perspectives on the evolution of mammals, birds, and reptiles* (pp. 76–113). Chicago: The University of Chicago Press.

Sidor, C. A., & Smith, R. M. H. (2004). A new galesaurid (Therapsida: Cynodontia) from the Lower Triassic of South Africa. *Palaeontology, 47*, 535–556.

Sidor, C. A., Angielczyk, K. D., Hopson, J. A., Kammerer, C. F., & Smith, R. M. H. (2009). New information about cynodonts from the Middle Triassic Manda beds (Ruhuhu Basin) of Tanzania. *Journal of Vertebrate Paleontology, 29*, 181A.

Sigogneau, D. (1970). *Révision systématique des Gorgonopsiens sud-africains. Cahiers de Paléontologie*. Paris: Centre National de la Recherche Scientifique.

Sigogneau-Russell, D. (1989). Theriodontia I. In P. Wellnhofer (Ed.), *Handbuch der Paläoherpetologie* (Vol. 17 B/I). Stuttgart: Gustav Fischer Verlag.

Smiley, T. M., Sidor, C. A., Maga, A., & Ide, O. (2008). The vertebrate fauna of the Upper Permian of Niger. VI. First evidence of a gorgonopsian therapsid. *Journal of Vertebrate Paleontology, 28*, 543–547.

Sues, H.-D., & Hopson, J. A. (2010). Anatomy and phylogenetic relationships of *Boreogomphodon jeffersoni* (Cynodontia: Gomphodontia) from the Upper Triassic of Virginia. *Journal of Vertebrate Paleontology, 30*, 1202–1220.

Tatarinov, L. P., & Matchenko, E. N. (1999). A find of an aberrant tritylodont (Reptilia, Cynodontia) in the Lower Cretaceous of the Kemerovo Region. *Paleontological Journal, 33*, 422–428.

Watabe, M., Tsubamoto, T., & Tsogtbataar, K. (2007). A new tritylodontid synapsid from Mongolia. *Acta Palaeontologica Polonica, 52*, 263–274.

Watson, D. M. S., & Romer, A. S. (1956). A classification of therapsid reptiles. *Bulletin of the Museum of Comparative Zoology, 114*, 35–89.

Chapter 11
A Redescription of *Eriphostoma microdon* Broom, 1911 (Therapsida, Gorgonopsia) from the *Tapinocephalus* Assemblage Zone of South Africa and a Review of Middle Permian Gorgonopsians

Christian F. Kammerer

Abstract The problematic *Tapinocephalus* Assemblage Zone (AZ) theriodont *Eriphostoma microdon* Broom, 1911 is redescribed based on computed tomographic images of the type and only known specimen. *Eriphostoma* is identified as a gorgonopsian, one of the few representatives of this clade known from the Middle Permian. *Eriphostoma microdon* represents a valid taxon diagnosed by the combination of elongate delta-shaped palatine bosses with numerous small teeth and a short, downward-sloping snout. Among gorgonopsians, *Eriphostoma* is most similar to *Gorgonops torvus*, but can be distinguished from that taxon by a trough separating the palatal bosses, relatively larger fossa for the lower canine, relatively shorter, taller snout, and mediolaterally thinner premaxilla at its posterior alveolar margin. Other nominal gorgonopsian taxa from the *Tapinocephalus* AZ exhibit the same general cranial morphology as *Eriphostoma*, suggesting that most are synonymous. Additional preparation and study of these other taxa will be necessary before their synonymy can be confirmed, however.

Keywords Karoo Basin • Theriodontia • CT scan • *Gorgonops* • *Eoarctops*

C. F. Kammerer (✉)
Division of Paleontology and Richard Gilder Graduate School, American Museum of Natural History, Central Park West at 79th Street, New York, NY 10024, USA

and

Museum für Naturkunde, Leibniz-Institut für Evolutions- und Biodiversitätsforschung, Humboldt-Universität, Invalidenstra. 43, 10115 Berlin, Germany
e-mail: christian.kammerer@mfn-berlin.de

Introduction

Gorgonopsians were the dominant predatory therapsids of the Late Permian (Sigogneau 1970; Gebauer 2007). They were abundant and at least nominally diverse in the Late Permian Karoo Basin of South Africa (Sigogneau 1970; Kitching 1977; Rubidge 1995) and are also common in coeval deposits in eastern Africa, Russia, and India (von Huene 1950; Sigogneau-Russell 1989; Ray and Bandyopadhyay 2003; Ivakhnenko 2008). Additionally, they are the only therapsid group known from the otherwise sauropsid-dominated Moradi Formation in Niger (Smiley et al. 2008). Gorgonopsian cranial anatomy has been thoroughly studied (Owen 1881; Broom 1930; Boonstra 1953; Laurin 1998) and the functional morphology of the skull is well understood (Kemp 1969). Gorgonopsian postcranial anatomy has received comparatively little study, but nearly complete skeletons have been described for a few taxa (Pravoslavlev 1927; Broili and Schröder 1935; Colbert 1948; von Huene 1950; Gebauer 2013). Over a hundred species of Late Permian gorgonopsians have been named, although it is likely that this number has been exaggerated by the taxonomic oversplitting typical for Karoo therapsids (Wyllie 2003).

Despite their Late Permian dominance, the early evolutionary history of the Gorgonopsia is poorly understood. The earliest known gorgonopsians (represented by fragmentary material indeterminate to genus) are found in the same Middle Permian *Eodicynodon* Assemblage Zone (AZ) deposits yielding the earliest known members of the other major therapsid groups (with the exception of the late-appearing cynodonts) (Rubidge 1995). However, whereas biarmosuchians, dinocephalians, anomodonts, and therocephalians are all reasonably abundant and diverse by the *Tapinocephalus* AZ, contemporaneous gorgonopsians are extremely rare. Only a handful of specimens are known, nearly each of which has been named as a separate species.

Christian F. Kammerer, Kenneth D. Angielczyk and Jörg Fröbisch (eds.), *Early Evolutionary History of the Synapsida*, Vertebrate Paleobiology and Paleoanthropology, DOI: 10.1007/978-94-007-6841-3_11,

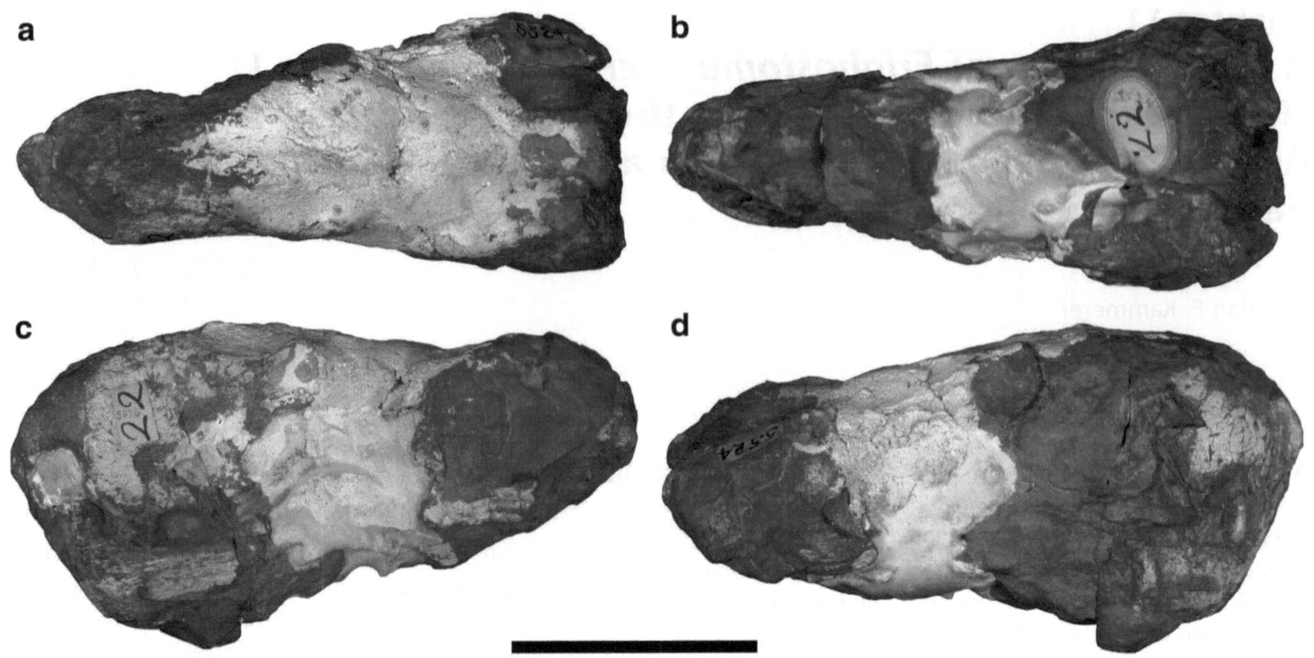

Fig. 11.1 AMNH FARB 5524, the holotype of *Eriphostoma microdon*, in **a** dorsal (anterior is left), **b** ventral (anterior is left), **c** left lateral, and **d** right lateral views. Scale bar equals 5 cm

Broom (1911) described *Eriphostoma microdon* as the first gorgonopsian recorded from the *Tapinocephalus* AZ (at the time the lowest known section of the therapsid-bearing Beaufort Group). He considered *E. microdon* to be closely related to *Ictidosaurus* and *Lycosaurus*, but distinguished it from those taxa by lower incisor and postcanine counts. *Ictidosaurus* has since been recognized as a scylacosaurid therocephalian (Haughton and Brink 1954; van den Heever 1987; Abdala et al. 2008), and *Lycosaurus* is generally considered a *nomen dubium* (Sigogneau 1970; Sigogneau-Russell 1989), rendering comparisons with these taxa of little importance for gorgonopsian taxonomy. Since Broom's description, *Eriphostoma* has been addressed rarely and briefly, which is understandable given the exceedingly poor state of the *E. microdon* holotype. AMNH FARB 5524, the holotype of *E. microdon*, consists of two largely unprepared skull fragments (a snout and occipital portion) held together (and partially obscured) by a mass of plaster occupying what would be the orbital region (Fig. 11.1). Most of the bone surface on the snout portion has been lost, with exposed preserved bone limited to fragments of the maxillae and the anterior portion of the dentary. The dentition is poorly preserved and all teeth are broken to varying degrees. Most of the right side of the snout is missing, although the position of the orbit is evident. A large hole is present in the left subnarial region and has been filled in with plaster, obscuring the incisors. Distinct bone grain is visible on the surface of the occipital portion, but no external morphology can be discerned other than the position of the foramen magnum and the presence of a jugal/squamosal fragment on the left side.

The poor quality of the holotype led all subsequent authors to consider *Eriphostoma microdon* a *nomen dubium*. Indeed, many authors have even questioned the gorgonopsian status of *Eriphostoma*. Boonstra (1935, p. 2) stated that, "it is not possible to state with certainty whether *Eriphostoma* is a gorgonopsian or a therocephalian", and considered this taxon to be an indeterminate gorgonopsian in his review of the *Tapinocephalus* AZ (Boonstra 1969). Watson and Romer (1956) placed *Eriphostoma* among indeterminate gorgonopsians. Sigogneau (1970) considered *E. microdon* to be identifiable only as Theriodontia *incertae sedis*, and later (Sigogneau-Russell 1989, p. 115) wrote that it "could as well be a therocephalian."

Although the holotype of *Eriphostoma microdon* is very poor, given the extreme rarity of gorgonopsians in the *Tapinocephalus* AZ and the fact that *Eriphostoma* was potentially the first gorgonopsian taxon named from this zone, an accurate determination of this specimen as, at the very least, gorgonopsian or therocephalian is necessary. In order to address this problem, computed tomographic (CT) scanning of the holotype was undertaken to provide new information on this specimen.

Institutional abbreviations: AMG, Albany Museum, Grahamstown, South Africa; AMNH FARB, American Museum of Natural History (Fossil Amphibians, Reptiles, and Birds Collection), New York City, NY, USA; BP, Bernard Price Institute, University of the Witwatersrand,

Johannesburg, South Africa; BSPG, Bayerische Staatssammlung für Paläontologie und Geologie, Munich, Germany; GPIT, Institut und Museum für Geologie und Paläontologie der Universität Tübingen, Tübingen, Germany; NHMUK, the Natural History Museum, London, UK; PIN, Paleontological Institute of the Russian Academy of Sciences, Moscow, Russia; SAM, Iziko, the South African Museum, Cape Town, South Africa; TM, Ditsong, the National Museum of Natural History, Pretoria, South Africa.

Materials and Methods

AMNH FARB 5524, the holotype of *Eriphostoma microdon*, was CT-scanned at the AMNH Imaging Facility along the coronal axis. Segmentation of the CT image data and reconstruction of the preserved portions of the skull was done using the volumetric rendering software VGStudio-Max© v.2.0. Comparisons with other gorgonopsian taxa were made based on first-hand observation of 360 specimens (including the types of most nominal species) and reference to the monographic reviews of Sigogneau (1970) and Sigogneau-Russell (1989).

Description

The scan has exposed a number of previously hidden anatomical features and clarified aspects of AMNH FARB 5524 that have traditionally confounded interpretation of this specimen (Figs. 11.2, 11.3, 11.4). The rear portion of the skull was displaced relative to the snout portion during Broom's plaster assembly of the 'complete' skull. From an occipital view, the rear portion has been rotated slightly clockwise relative to the snout portion. Additionally, judging by the unnatural amount of separation between the quadrate rami of the pterygoid and the palatine bosses of the pterygoid on the scan (Fig. 11.3), too much plaster was used to reconstruct the skull; the actual skull would have been anteroposteriorly shorter (Fig. 11.5).

Cranium

Little bone is preserved on the dorsal surface of AMNH FARB 5524, with only fragments of the nasals and parietals present. The anterior border of the orbit is preserved on the right side. Presumably this area is formed by the prefrontal, lacrimal, and jugal as in other gorgonopsians, but no sutures can be observed; all that is preserved is a thin semicircle of bone around the orbital margin. A fragment of bone ventrolateral to the palate and below the orbit may also represent a fragment of jugal given its position, but it is too poorly preserved to be certain. Laterally, the premaxillae and maxillae are evident, although neither is complete. The palatal surface of the maxilla is eroded off the right side of the skull, but is reasonably intact on the left side. Conversely, the premaxilla is intact and well preserved on the right side of the skull but almost completely lost on the left side. The upper right incisors are intact, but no clear teeth can be made out on the thin fragment of upper left premaxilla. The crown of the left upper canine is broken at midlength (estimated by comparison with other gorgonopsians), and parts of the root of the right upper canine can be seen in the scan. Only root fragments of the upper postcanines are present.

Palatally, the vomer, palatines, and pterygoids are preserved and well resolved on the scan. Only the palatal bosses of the pterygoids are preserved in the snout portion of the skull; the left transverse process and both quadrate rami of the pterygoids are preserved in the rear portion. Ventrally, the rear portion of the skull also preserves a well-resolved parasphenoid and basisphenoid, and poorly-preserved, largely indistinct basioccipital and opisthotic. Fragments of the jugals and squamosals that made up the temporal arches are present on both sides of the rear portion of the skull (with the left fragment larger and better preserved). Additionally, parts of the squamosals, prootics, and epipterygoids making up the lateral wall of the braincase are preserved. Although the foramen magnum can be made out on the actual specimen, the occipital plate itself is not visible on the scan and appears to have eroded off.

The snout of AMNH FARB 5524 is relatively short for a gorgonopsian (Fig. 11.2), similar to the holotypes of *Aelurosaurus breviceps* (AMNH FARB 5514), *Arctognathus curvimola* (NHMUK 47339), and *Eoarctops vanderbyli* (SAM-PK-5598). Also like *Eoarctops*, the dorsal surface of the snout curves downward anteriorly. The dorsal margins of the maxillae are lost, but it appears that they were tallest above the upper canine. Neither upper canine is well preserved but they appear typical for gorgonopsians: large and bladelike with serrated edges. The right canine is in the posterior canine alveolus and was in the process of replacement at the time of death; an anterior replacement canine is visible beneath the broken surface of the maxilla. Four clear incisors are present on the right premaxilla. This is an unusual number for a gorgonopsian, which usually have five upper incisors. Because of the poor preservation of AMNH FARB 5524, it is likely that the fifth incisor was present in life but lost in this specimen. The midline area between the two premaxillae is damaged and it is possible that the missing incisor (which would be I1) was lost there.

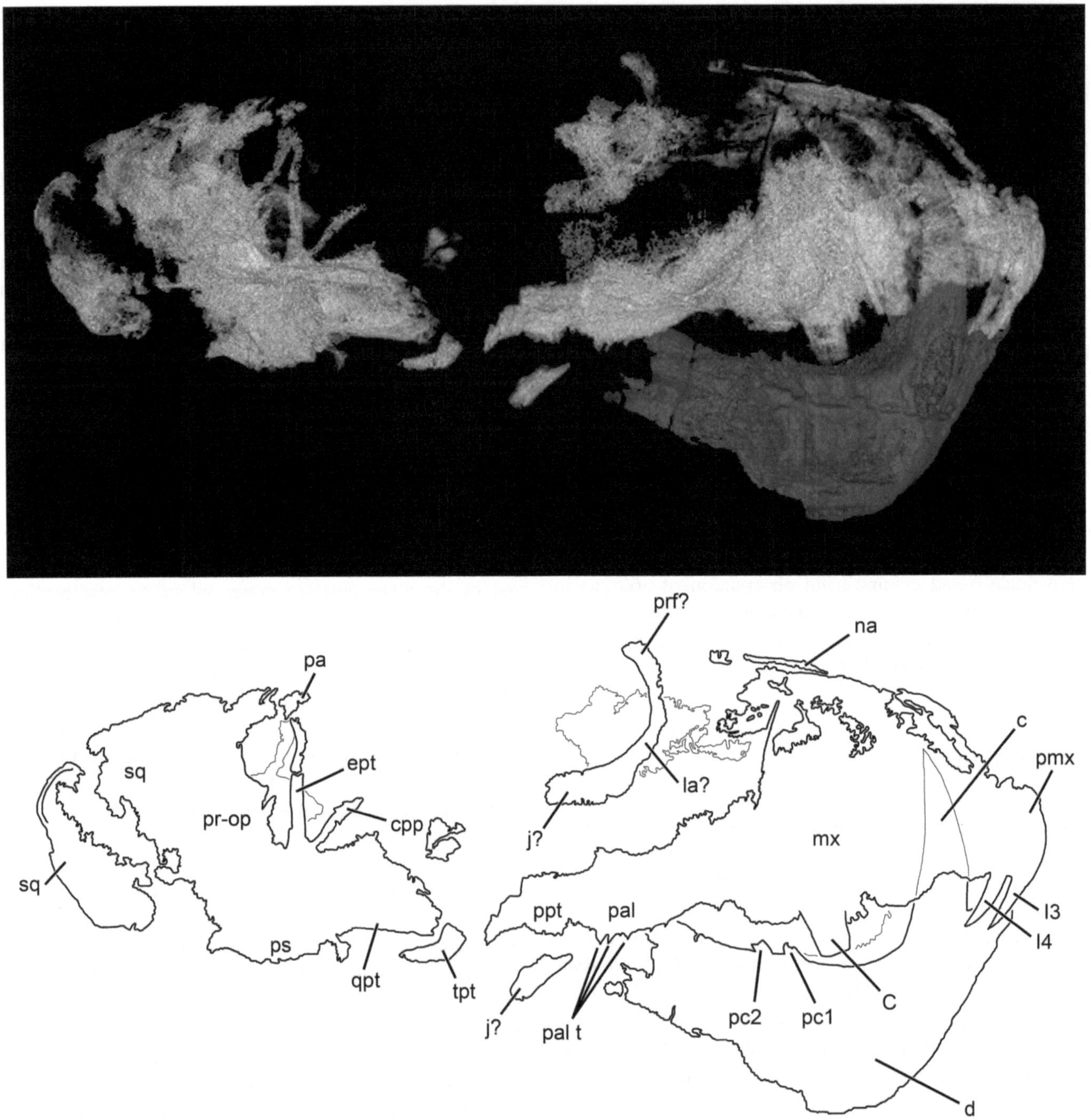

Fig. 11.2 CT rendering and interpretive drawing of AMNH FARB 5524 in right lateral view. In the rendering, the cranium is highlighted in *white* and the mandible in *gray*. In interpretive drawing, visible portions of *left side* of skull traced in *gray*. *c* Lower canine, *C* upper canine, *cpp* cultriform process of the parasphenoid, *ept* epipterygoid, *I* upper incisor, *j* jugal, *la* lacrimal, *mx* maxilla, *na* nasal, *pa* parietal, *pal* palatine, *pal t* palatine boss teeth, *pc* lower postcanine, *pmx* premaxilla, *ppt* palatal boss of the pterygoid, *pr-op* fused prootic-opisthotic, *prf* prefrontal, *ps* parasphenoid, *qpt* quadrate ramus of the pterygoid, *sq* squamosal, *tpt* transverse process of the pterygoid

The presence of only four upper incisors has also been suggested for another *Tapinocephalus* AZ gorgonopsian, *Eoarctops vanderbyli* (Sigogneau 1970; Gebauer 2007), although the dentition of that specimen is also poorly preserved. Additional preparation of the holotype of *E. vanderbyli*, SAM-PK-5598, will be necessary to confirm the presence of only four incisors in that taxon. Two postcanines are present, separated from the canine by a diastema. The upper dental formula for AMNH FARB 5524 is thus I4–5?/C1/PC2.

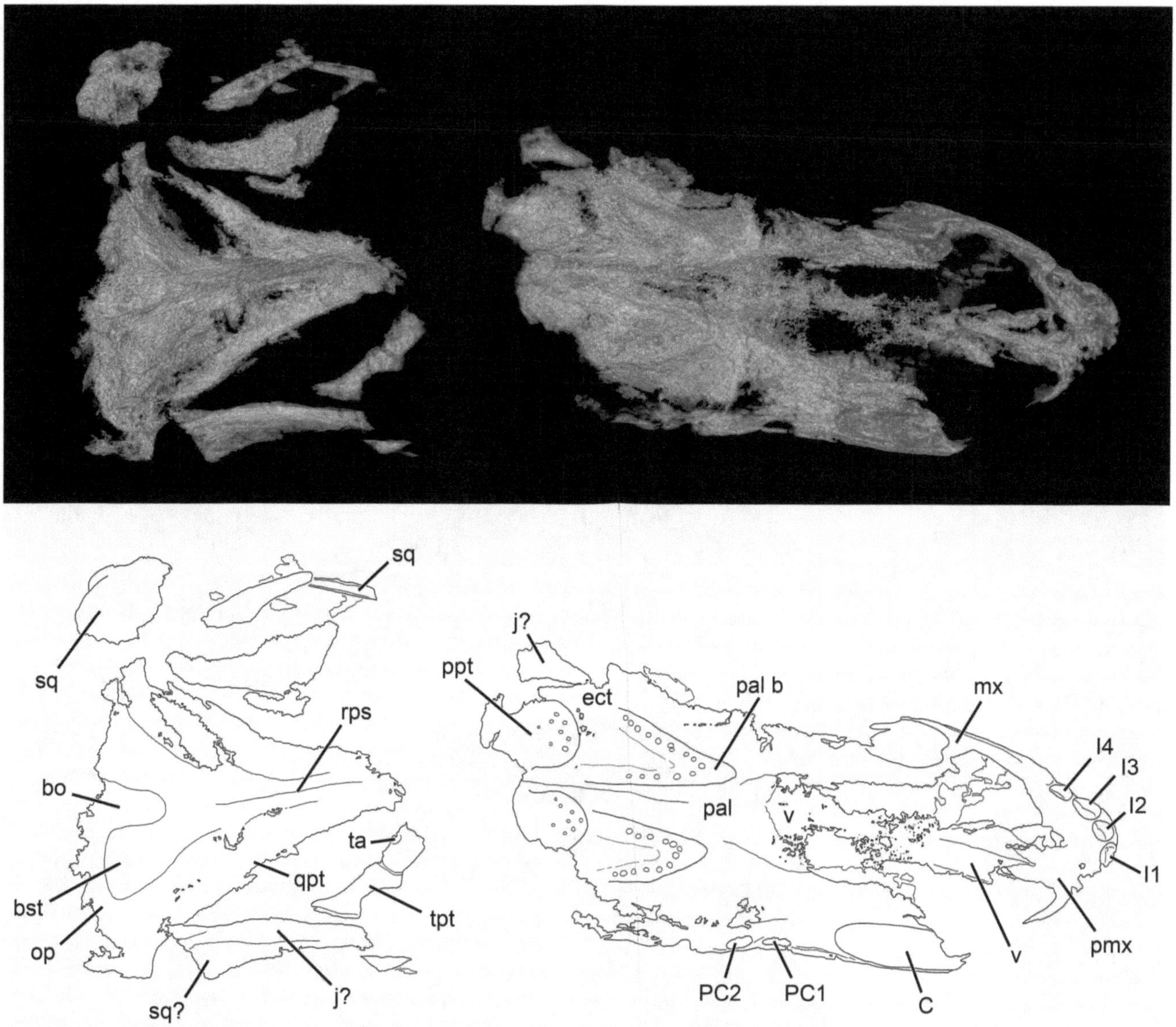

Fig. 11.3 CT rendering and interpretive drawing of AMNH FARB 5524 in palatal view (mandible digitally removed to better illustrate the palate). *bo* Basioccipital, *bst* basisphenoid tuber, *C* upper canine, *ect* ectopterygoid, *I* upper incisor, *j* jugal, *mx* maxilla, *op* opisthotic, *pal* palatine, *pal b* palatine boss, *PC* upper postcanine alveoli, *pmx* premaxilla, *ppt* palatal boss of the pterygoid, *rps* rostrum of the parasphenoid, *qpt* quadrate ramus of the pterygoid, *sq* squamosal, *ta* tooth alveolus on transverse process of the pterygoid, *tpt* transverse process of the pterygoid, *v* vomer

Palatally, the premaxilla is anteroposteriorly short (Fig. 11.3). It is thinnest at the lateral edge of the internal choana (where it houses the lower canine) and expands slightly towards the premaxillary midline suture to accommodate the incisors. A short, broad vomerine process connects the dentigerous portion of the premaxilla to the vomer. The vomer is poorly resolved in the scan, but clearly displays the typical gorgonopsian morphology, with an expanded anterior portion bearing three ridges: one on each edge and one down the midline. The vomer terminates posteriorly between the two palatine bosses. These bosses are elongate and delta-shaped, with pointed anterior tips. Both the lateral and medial rami of these bosses bear numerous small, conical teeth. The medial rami are longer than their lateral counterparts, extending posteriorly to a position between the palatal bosses of the pterygoid. The mid-palatine suture bears a prominent ridge running down the middle of the trough separating the palatine bosses.

The palatal bosses of the pterygoid are low and rounded and also bear numerous small, conical teeth (Fig. 11.3).

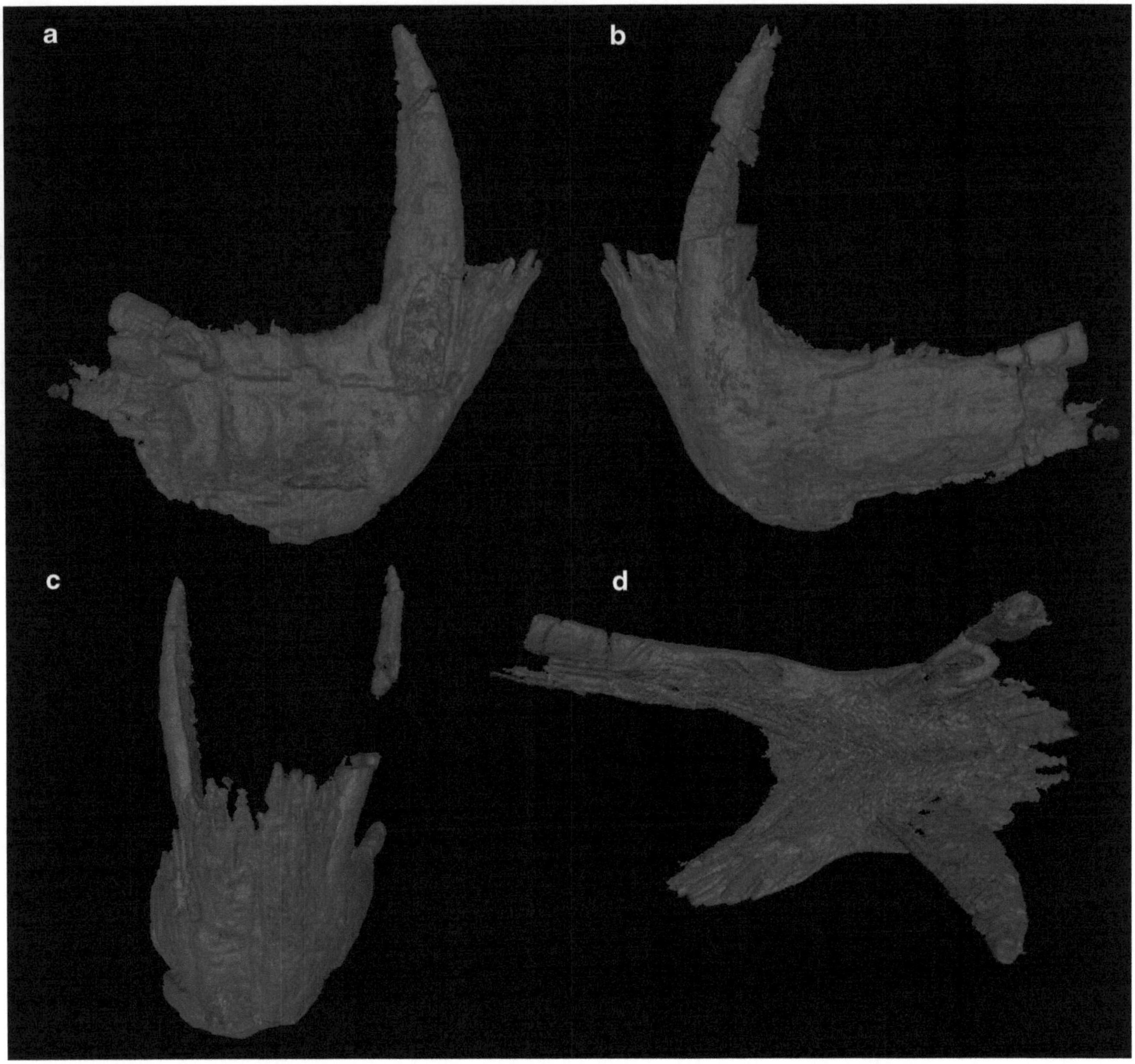

Fig. 11.4 CT rendering of the preserved portion of the mandible of *Eriphostoma microdon* in **a** right lateral, **b** left lateral, **c** anterior, and **d** dorsal views

These bosses are separated anteromedially from the palatine bosses and medially from each other by low troughs. The lateral region of the left transverse process of the pterygoid is preserved on the rear portion of the skull. This process is backswept, with a concave posterior margin. At least one tooth alveolus is visible on the ventral surface of the transverse process. The quadrate rami of the pterygoids broadly border the lateral edges of the parasphenoid, curving out laterally and separating at the level of the parasphenoid-basisphenoid suture. They attenuate posterolaterally towards the (unpreserved) quadrate.

Although parts of the basioccipital and paroccipital process of the opisthotic are preserved in AMNH FARB 5524, they are badly eroded and of indistinct morphology in the scan. The best-preserved basicranial elements are the parasphenoid and basisphenoid. The dorsal surface of the parasphenoid bears a narrow cultriform process extending anterodorsally. Ventrally, the parasphenoid rostrum is extremely thin and blade-like anteriorly, broadening slightly posteriorly near the junction with the basisphenoid. The basisphenoid tubera are expanded posterolaterally, but relatively narrow compared to most other gorgonopsians

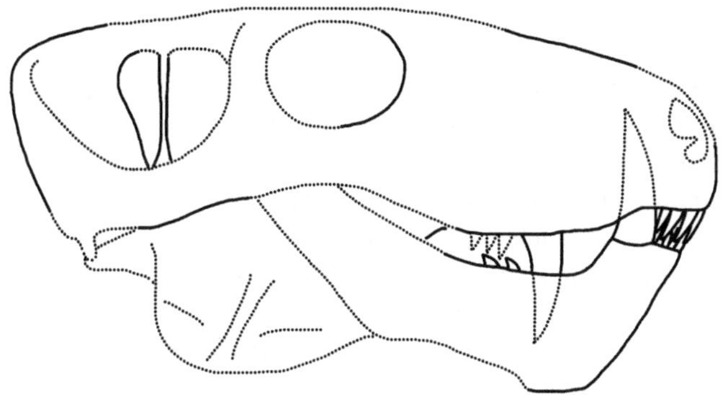

Fig. 11.5 Reconstruction of the skull of *Eriphostoma microdon* in lateral view. Preserved portions indicated by *solid line*, *dotted lines* are speculative but based on comparisons with other *Tapinocephalus* AZ gorgonopsians (particularly *Eoarctops vanderbyli*)

(see "Discussion"). The tubera are constricted near their union at the midpoint of the basisphenoid, with a marked ventral ridge on the constricted portion. The 'keel' of the parasphenoid rostrum continues onto the basisphenoid, terminating in a small, rounded process overhanging the parabasisphenoid fossa.

The epipterygoid (Fig. 11.2) originates in a dorsal groove on the quadrate ramus of the pterygoid, near the posterior terminus of the ramus. Although broad at its base, the epipterygoid is for most of its length a thin, strap-like, vertically-directed bone. It attenuates dorsally but then expands slightly where it contacts the ventral surface of the parietal (this region is poorly-preserved in AMNH FARB 5524, but fragments of the parietal-epipterygoid contact area are present on the left side). The epipterygoid is separated from the prootic posteriorly by a wide gap in the braincase wall. The prootic, opisthotic, supraoccipital, and postparietal form a broad, plate-like element bordering the basisphenoid ventrally, squamosal posteriorly, and parietal dorsally. The occipital portion of the squamosal is poorly preserved, but a large fragment of the squamosal portion of the left temporal arch is intact. The squamosal undercuts a posterior extension of the jugal anteriorly, terminating in a triangular process.

Mandible

The dentary is well preserved but broken off shortly behind the postcanines on both sides (Fig. 11.4). Most of the marginal teeth are poorly preserved. The lower incisors are all in place but most are missing the anterior edge of the crown. The right lower canine is missing the lingual edge of its root but is otherwise intact; the left lower canine is missing part of the crown at midheight, but has the crown tip intact. The left lower postcanines are lost, although distinct alveoli can be seen on the scan. The right lower postcanines are partially preserved, missing the crown tips and lingual sides.

The mandible of AMNH FARB 5524 has a tall, unfused symphysis with a steeply-angled anterior face (Fig. 11.4). The precanine portion of the dentary is very short, forming a thin wedge in lateral view. The lateral surface of the dentary bulges out slightly at the base of the lower canine, accommodating the massive canine root. The canine is separated from the postcanine tooth row by a diastema and a distinct 'step down' in the dorsal alveolar margin. Additionally, a prominent depression for the reception of the upper canine is present on the lateral surface of the mandible between the lower canine and the postcanine tooth row. Although the dentary is broken off behind the postcanine tooth row on both sides (immediately behind it on the right), it appears that lower jaw height decreased posteriorly, as in other gorgonopsians. The anterior edge of the coronoid process is preserved on the left dentary at the terminus of the postcanine tooth row, indicating that *Eriphostoma* had a short lower jaw in concordance with its relatively short snout.

The dentary of AMNH FARB 5524 has the dental formula i4/c1/pc2. The lower incisors are elongate and conical. The first two incisors are relatively longer and narrower than i3 and i4. It is uncertain whether serrations were present, based on the scan. The lower canines are blade-like: very tall, mesiodistally thin, and slightly recurved with anterior and posterior serrated edges. Although only the labial sides of the bases of the right lower postcanine crowns are preserved, they demonstrate that these teeth were strongly posteriorly canted, as in many gorgonopsians.

Discussion

Taxonomic Identity of Eriphostoma microdon

The new morphological information provided by the scan of AMNH FARB 5524 finally allows *Eriphostoma microdon* to be definitively identified as a gorgonopsian. Previous arguments for the gorgonopsian identification of this specimen (e.g., Boonstra 1935) have hinged on the presence of a tall, steep mandibular symphysis. Unfortunately, poor preparation of the specimen, leaving parts of the symphysis still covered with matrix, made this character somewhat questionable and subject to interpretation. However, the scan clearly shows the presence of a characteristically gorgonopsian symphysis in AMNH FARB 5524, as well as other gorgonopsian mandibular characters: a ventral 'step' in the alveolar margin behind the canine and few, strongly posteriorly canted postcanines. Previously hidden regions of the skull exhibit further gorgonopsian synapomorphies: the palatine bones meet on the ventral midline, separating the vomer from the pterygoids; the anterior portion of the vomer is broadly expanded with triple ridges; and the fossa for reception of the lower canine is confluent with the internal naris [although this feature also occurs in some therocephalians (van den Heever 1994)].

Gorgonopsians are a morphologically conservative group, rendering comparisons difficult at the specific level, especially in the case of a fragmentary specimen like AMNH FARB 5524. Fortunately, the palate of AMNH FARB 5524 is preserved well enough to permit detailed comparisons with other members of the group. Within Gorgonopsia, the elongate, delta-shaped palatine bosses exhibited by *Eriphostoma* are otherwise known only in *Gorgonops* itself. Most gorgonopsian taxa have either reniform palatine bosses with few, large teeth or elongate bosses with a single tooth row running in an anteromedial direction. The palatine bosses of *Aelurosaurus felinus* (based on the holotype, NHMUK R339) are also delta-shaped, but are proportionally much shorter than those of *Eriphostoma* and *Gorgonops*. Furthermore, in *A. felinus* the lateral rami of the palatine bosses are longer than the medial rami, opposite to the condition in *Eriphostoma* and *Gorgonops*. As the holotypes of *E. microdon* and *A. felinus* are similar in size (snout length 5.3 and 5.8 cm, respectively), these distinctions are unlikely to represent ontogenetic variation.

Like most gorgonopsian genera, *Gorgonops* is taxonomically problematic. Sigogneau (1970) placed many species formerly housed in separate genera (e.g., *Chiwetasaurus dixeyi*, *Gorgonognathus longifrons*, *Leptotracheliscops eupachygnathus*, *Pachyrhinos kaiseri*, *Scymnognathus whaitsi*) within an expanded *Gorgonops*, based on similar cranial proportions. However, these species have never been demonstrated to form a monophyletic unit, either within a formal phylogenetic analysis or on a discrete synapomorphy basis. The disparate palatal morphologies of these supposed *Gorgonops* species suggest that '*Gorgonops*' sensu Sigogneau (1970; Sigogneau-Russell 1989) may not represent a clade, but this must be tested within the context of a broader-scale revision and phylogenetic analysis of the Gorgonopsia. For the purposes of comparison with *Eriphostoma microdon*, *Gorgonops* will be represented solely by the type species, *G. torvus*, and specifically the specimens NHMUK R1647 (the holotype), AMNH FARB 5515, and BP/1/4089.

The skulls of *Eriphostoma microdon* and *Gorgonops torvus* are generally similar, although *Eriphostoma* has a proportionally shorter snout. Several characters allow these taxa to be distinguished, however. In *Eriphostoma*, the palatal bosses of the pterygoid are separated by a median trough, whereas these bosses are tightly appressed in *Gorgonops*. The maxilla of *Eriphostoma* is proportionally taller than in *Gorgonops*, and the snout curves more ventrally in lateral view. The portion of the choana that receives the lower canine is relatively larger in *Eriphostoma* than *Gorgonops*, both in mediolateral and anteroposterior dimensions. In *Eriphostoma*, the premaxilla is highly attenuate posteriorly in ventral view, with only a narrow strip of bone rimming the lower canine receptacle. In *Gorgonops* this region of the choana does not extend so far laterally, with a relatively thick premaxillary rim. *Gorgonops* has more postcanine teeth than *Eriphostoma* (3–4 vs. 2), although this character should not be considered particularly strong given that theriodont postcanine counts often vary ontogenetically. Similarly, *Eriphostoma* may have had only four incisors (as opposed to the usual gorgonopsian five), although it is possible that the absence of the fifth tooth is a preservational artifact in this specimen, so this character should also be considered questionable.

Tapinocephalus *Assemblage Zone Gorgonopsians*

Although few specimens of gorgonopsians are known from the *Tapinocephalus* AZ, the majority of them have been named as separate species. The characters differentiating these nominal species are vague at best, and for the most part their holotypes are not well enough prepared for thorough comparisons. A review of published *Tapinocephalus* AZ gorgonopsian specimens is instructive in this regard. It should be noted that the precise stratigraphic origin of most of these specimens is also vague, and in several cases is not sufficient to establish that they are truly from the *Tapinocephalus* AZ. Specimens labeled as from 'the Gouph' (also 'Gough' or 'Koup') or 'Gouph Tract, vicinity of Beaufort West' in particular could be from either the *Tapinocephalus* or *Pristerognathus* AZs (Angielczyk et al. 2005).

NHMUK 49419—NHMUK 49419 is a weathered snout with the anterior portion of the mandible preserved in occlusion. Lydekker (1890) referred this specimen to *Aelurosaurus* sp., which he distinguished from the type species *A. felinus* by the shorter snout and steeper mandibular symphysis. Broom referred this specimen to *Ictidosaurus angusticeps*, the holotype of which (SAM-PK-630) is a scylacosaurid therocephalian. Handwritten notes by L. D. Boonstra, J. A. Hopson, and J. A. van den Heever found with the specimen all note that NHMUK 49419 is clearly a gorgonopsian, a conclusion with which the present author agrees. The very steep mandibular symphysis of this specimen in particular is classically gorgonopsian in morphology. As with several other supposed *Tapinocephalus* AZ gorgonopsians, the locality data for this specimen is vague: 'the Gouph'. This specimen is actually one of two gorgonopsians Broom misidentified as *Ictidosaurus* *angusticeps*. The other, AMNH FARB 5527, is also an isolated snout and mandible [referred to *I. angusticeps* by Broom (1915b)], albeit from *Tropidostoma* AZ sediments near Beaufort West and thus beyond the scope of this chapter.

Scylacognathus parvus—Following *Eriphostoma*, the next named *Tapinocephalus* AZ gorgonopsians were *Scylacognathus parvus* (holotype AMG 3751) and *Scymnorhinus planiceps* (holotype AMG 3752) from Hottentots River, Beaufort West (Broom 1913). AMG 3751 is a small, somewhat crushed but generally well-preserved skull missing the left postorbital bar and right zygomatic arch. *Scylacognathus* has been treated as a valid taxon of gorgonopsian by most subsequent workers (e.g., Haughton and Brink 1954; Watson and Romer 1956; Boonstra 1963; Sigogneau 1970). Broom (1935) described a second species of *Scylacognathus*, *S. major* (holotype TM 256) from the *Cistecephalus* AZ. Sigogneau (1970) synonymized *S. major* with the type species *S. parvus* and referred an incomplete skull (BP/1/857) from the *Dicynodon* AZ to this species as well. If correct, these referrals would give *S. parvus* a remarkably long range for a Karoo therapsid (*Tapinocephalus-Dicynodon* AZs), equaled only by a few dicynodonts [e.g., *Diictodon feliceps* (Angielczyk and Sullivan, 2008)]. She also included the small gorgonopsians *Cynariops robustus*, *Cynarioides grimbeeki*, and *Cynarioides laticeps* within *Scylacognathus* in the new combinations *Scylacognathus robustus* and *S. grimbeeki* (including both *C. grimbeeki* and *C. laticeps*). Sigogneau-Russell (1989) noted that the holotype of *S. parvus* and some of the referred species represent immature gorgonopsians. Gebauer (2007) took this a step further, arguing that *Arctops* represents the adult morphology of *Scylacognathus* and accordingly synonymizing the two genera (with *Arctops willistoni*, the type species, and *A. watsoni* subsumed into *Scylacognathus parvus*).

Broomisaurus planiceps—AMG 3752, the holotype of *Scymnorhinus planiceps*, is a badly weathered snout and lower jaws. The lower jaws have been ground down from below to show the tooth roots, but the palate is still encased in matrix. Broom (1913) initially expressed doubt as to whether this specimen was a therocephalian or gorgonopsian, but later (Broom 1932, 1940) supported a gorgonopsian identification. The generic name *Scymnorhinus* was preoccupied by a shark (*Scymnorhinus* Bonaparte, 1846), leading Joleaud (1920) to rename the gorgonopsian *Broomisaurus*. Broom (1940) later described a second species of *Broomisaurus*, *B. rubidgei*, based on a complete skull (RC 19) from *Dicynodon* AZ rocks in New Bethesda. More recent studies of *B. rubidgei* conclude that it has no particular similarity to *B. planiceps*: Sigogneau (1970; also Sigogneau-Russell 1989) tentatively referred it to *Leontocephalus*, whereas Gebauer (2007) considered *B. rubidgei* indeterminate at the species level, and referred RC 19 to *Sycosaurus* sp. As for *B. planiceps*, Sigogneau (1970) retained it as a separate species (see also Sigogneau-Russell 1989), but expressed doubt as to its validity, and suggested that it could be synonymous with the better-known *Eoarctops*. Gebauer (2007) considered it a *nomen dubium*.

Cerdodon tenuidens—Broom (1915a) named two additional possible gorgonopsian taxa believed to be from the *Tapinocephalus* AZ (which at the time was called the '*Pareiasaurus* zone'): *Cerdodon tenuidens* (holotype NHMUK 49420) and *Cyniscodon lydekkeri* (holotype NHMUK 49409). Of the two, *Cerdodon* is known from more complete material, but like *Eriphostoma*, this taxon has variously been considered a gorgonopsian, therocephalian, or indeterminate theriodont. NHMUK 49420 (from Welterveden in the 'Gouph') consists of the left half of a partial skull and lower jaw preserving the area anterior to the postorbital bar. This specimen is preserved in a nodule, of which only the left side has been prepared to show surface detail. At the cranial midline the nodule has been cut and polished; unfortunately, no palatal details are visible in section. In his initial description, Broom (1915a) stated that this specimen had only three upper incisors, but subsequent studies (Broom 1932; Boonstra 1934) argued that four incisors are present. Broom (1915a) originally described this taxon as a therocephalian, and considered it to be most closely related to *Ictidosuchus primaevus*. Haughton (1924a) retained this placement, including it in the Ictidosuchidae with *Ictidosuchus* and *Arnognathus*, whereas Haughton and Brink (1954) still considered *Cerdodon* a therocephalian but placed it in Pristerognathidae [used at the time for all 'basal therocephalians', currently broken into Lycosuchidae and Scylacosauridae (van den Heever, 1994)]. Watson and Romer (1956) considered this specimen to be a gorgonopsian, and placed *Cerdodon* in their family Galesuchidae with the other *Tapinocephalus* AZ taxa (*Eoarctops*, *Galesuchus*, and *Scylacognathus*). Kitching (1977) included *Cerdodon* as a pristerognathid therocephalian, but considered the taxon a *nomen dubium*. Unfortunately no definitive synapomorphies of Gorgonopsia are visible in the specimen as currently prepared. However, in general gestalt it does appear to be a gorgonopsian. The lower jaw has a fairly steep chin, and the postcanines are small, close-packed, backswept, and separated from the canine by a lengthy diastema. The snout is short, the dorsal margin of the snout is straight, and the large orbit is level with the naris. It should be noted that although the symphyseal profile of the mandible is not as steep as in many large gorgonopsians, that of *Cerdodon* compares favorably with that of *Aelurosaurus felinus* (the type specimen, NHMUK R339), an undoubted gorgonopsian.

Cyniscodon lydekkeri—NHMUK 49409, the holotype of *Cyniscodon lydekkeri*, was originally described as a specimen of the Late Permian cynodont *Cynosuchus* (currently *Cynosaurus*) *suppostus* by Lydekker (1890). Broom (1915a) recognized that this specimen represents a gorgonopsian, and described it as a new taxon diagnosed by its relatively small size and unfused symphysis. Although its gorgonopsian status is not in doubt, *Cyniscodon lydekkeri* has been considered a *nomen dubium* by nearly all subsequent authors (Boonstra 1969; Sigogneau 1970; Sigogneau-Russell 1989), and it is unlikely that further study will overturn this assessment. NHMUK 49409 consists solely of a poorly-preserved fragment of right dentary, preserving the symphysis and postcanine tooth row but broken anterior to the coronoid region. Most of the bone surface has been lost to erosion or overpreparation and all of the teeth are broken off. However, the tooth roots are visible in section and show that this specimen had a dental formula of i4/c1/pc4. The symphysis is broken on its medial edge. The symphysis is typically gorgonopsian, with a steeply-angled anterior face and a distinct 'step' posterior to the canine, which is separated from the postcanines by a diastema. The provenance of this specimen is somewhat dubious—the original locality data is simply "Palmietfontein, Cape Colony", and as Broom (1915a, p. 167) noted, "there are many Palmietfonteins in the Karroo", although he considered Palmietfontein, Beaufort West (*Tapinocephalus* AZ) to be the most probable.

Galesuchus gracilis—Haughton (1915) described a partial, poorly-preserved skull missing the anterior portion of the snout (SAM-PK-2754, from Abrahamskraal, Prince Albert) as a new taxon of *Tapinocephalus* AZ gorgonopsian, *Galesuchus gracilis*. He did not provide a differential diagnosis in the original description, but later (Haughton 1924b) noted that *Galesuchus* can be distinguished from *Gorgonops* by a smaller participation of the frontal in the orbital margin. *Galesuchus* has been treated as valid by most subsequent authors (Haughton and Brink 1954; Watson and Romer 1956; Sigogneau 1970), but Gebauer (2007) considered it to be a *nomen dubium* based on poor preservation of the type and lack of any visible diagnostic features. Sigogneau-Russell (1989) referred several additional *Tapinocephalus* AZ gorgonopsian specimens to *Galesuchus* sp., including SAM-PK-11846 (from Veldmann Ween, Prince Albert), SAM-PK-11849 (from Veldmansrivier, Prince Albert), SAM-PK-K208 (from Lammerskraal, Prince Albert), and SAM-PK-K230 (from Dalajodon, Beaufort West). These specimens are highly incomplete and poorly preserved, and the character used to refer them to *Galesuchus* rather than *Eoarctops* (narrower intertemporal region; see Sigogneau-Russell 1989) is particularly prone to taphonomic distortion, making these referrals suspect.

Eoarctops vanderbyli—Haughton (1929) described the first nearly complete gorgonopsian skull from Abrahamskraal in the *Tapinocephalus* AZ (SAM-PK-5598) as the new taxon *Eoarctops vanderbyli*. He diagnosed *E. vanderbyli* as having fewer postcanines than *Scylacognathus parvus* [three vs. five, respectively, although 'five' is just Broom's (1913) estimate—only three postcanine roots per side are preserved in the holotype of *S. parvus*] and lacking the laterally compressed snout of *Eriphostoma microdon*. Unlike most of the previously mentioned taxa, *Eoarctops vanderbyli* has consistently been considered a valid gorgonopsian taxon since its description (Haughton and Brink 1954; Watson and Romer 1956; Sigogneau 1970; Sigogneau-Russell 1989), and was the main reference for Boonstra's (1969) 'generalized *Tapinocephalus* zone gorgonopsian' reconstruction. Most recently, Gebauer (2007) recognized *Eoarctops* as the only valid *Tapinocephalus* AZ-restricted gorgonopsian genus. Sigogneau (1970) referred a second skull (SAM-PK-12220, from Skoppelmaaikraal, Laingsburg) to this genus as an uncertain species (*Eoarctops* sp.). Even though it is represented by the most complete material for a *Tapinocephalus* AZ gorgonopsian, *Eoarctops* remains very difficult to diagnose, because of the poor preservation of the skull roof and unprepared nature of the palatal surface of the skull.

Pachyrhinos kaiseri—Broili and Schröder (1934) described a partial skull from La-de-da, south of Beaufort West, as a new gorgonopsian taxon, *Pachyrhinos kaiseri*. Uniquely among purported *Tapinocephalus* AZ gorgonopsians (but typical of Broili and Schröder's efforts), this specimen was thoroughly described, well prepared, and extensively compared with other known gorgonopsians. Broili and Schröder (1934) considered *Pachyrhinos* to be most similar to *Gorgonops* and *Scymnognathus*, and indeed Sigogneau (1970) synonymized these three genera [although Watson and Romer (1956) considered *Pachyrhinos* distinct enough to warrant the monotypic family Pachyrhinidae]. Subsequent studies (Sigogneau-Russell 1989; Gebauer 2007) have retained *P. kaiseri* within *Gorgonops*, but as a valid species (*Gorgonops kaiseri*). Although this record may indicate the presence of the genus *Gorgonops* in the *Tapinocephalus* AZ, La-de-da also includes *Pristerognathus* AZ exposures, so the stratigraphic position of this specimen should be considered uncertain (see also Sigogneau 1970).

Aelurosauroides watsoni—The most recently described *Tapinocephalus* AZ gorgonopsian, *Aelurosauroides watsoni*, was named by Boonstra (1934) for a partial skull (NHMUK R855, a snout and orbital region) collected by Thomas Bain in 'the Gouph' and initially referred to *Aelurosaurus felinus* (Lydekker 1890). Sigogneau (1970) synonymized this taxon with the otherwise *Cistecephalus* AZ-occurring taxon *Aelurosaurus felinus*, although Gebauer (2007) considered *A. watsoni* indeterminate. This specimen is generally poorly preserved but has been acid prepared, so unlike most *Tapinocephalus* AZ gorgonopsian specimens its palatal morphology is visible. Five upper incisors on each side are clearly visible. The roots of four postcanines are present in the left maxilla and five on the right; the left maxillary tooth row is well preserved, so the difference in postcanine count cannot be ascribed to taphonomic damage. The presence of varying postcanine counts on different sides of a single specimen underlines the problems inherent in historical diagnoses of gorgonopsian taxa based on minor variations in tooth count.

Summary—As should be indicated by the review above, rarity of specimens, poor preservation, and above all poor preparation have seriously hindered accurate alpha taxonomic determination of the gorgonopsians of the *Tapinocephalus* AZ. Of the described specimens, only two (the holotypes of *Aelurosauroides watsoni* and *Pachyrhinos kaiseri*) have been adequately prepared. Both of these specimens have been referred to longer-ranging taxa (*Aelurosaurus* and *Gorgonops*, respectively) and in both cases their origin in the *Tapinocephalus* AZ is questionable (they could also be from the *Pristerognathus* AZ). With regards to the other *Tapinocephalus* AZ taxa, it should be noted that even by gorgonopsian standards the nominal species are extremely homomorphic. All of these specimens exhibit the same general cranial morphology (where preserved): short, bulbous snouts and relatively large canines and orbits. Sigogneau-Russell (1989, p. 87) mentioned that *Broomisaurus*, *Eoarctops*, and *Scylacognathus* were very similar (perhaps congeneric), and even though she retained *Galesuchus* as separate from *Eoarctops*, she noted that they "may not be as different… as the two type-specimens would let one believe, both being crushed in opposite ways." Also, although she eventually concluded that the taxon was an indeterminate theriodont, Sigogneau (1970) noted a suspicious resemblance between *Eriphostoma microdon* and *Scylacognathus parvus*. Another potentially important characteristic of these specimens is the recurring report of only four upper incisors, which would be unique among gorgonopsians. This count has been listed or suggested for *Cerdodon* (Broom 1932; Boonstra 1934), *Eoarctops* (Haughton 1929; Sigogneau 1970; Gebauer 2007), and herein for *Eriphostoma*. Most of these accounts (including the present one) tend to consider the presence of four upper incisors in these specimens more likely attributable to taphonomic artifact than biological reality, but the recurrence of this feature is suggestive. However, *Broomisaurus* does appear to have the standard upper incisor count of five. Additional preparation of these specimens is desperately needed to resolve this issue.

The high degree of similarity between *Tapinocephalus* AZ gorgonopsian specimens (other than *Aelurosauroides watsoni* and *Pachyrhinos kaiseri*) suggests that they may all be synonymous, in which case the oldest-named taxon *Eriphostoma microdon* would have precedence. However, with the possible exception of the unusual upper incisor count (which is most likely a preservational artifact), these specimens do not show any clear autapomorphies that allow them to be diagnosed as a single distinct species. As shown above, *Eriphostoma microdon* is a valid taxon that can be differentiated from *Gorgonops* and other post-*Tapinocephalus* AZ gorgonopsians, but unfortunately the key palatal characters of *Eriphostoma* are not visible in its fellow *Tapinocephalus* AZ specimens. This is despite the fact that several of these taxa (*Scylacognathus parvus* and *Eoarctops vanderbyli*) are known from nearly complete skulls, and most of the others (*Broomisaurus planiceps*, *Cerdodon tenuidens*, and *Galesuchus gracilis*) include the most pertinent regions of the skull for comparison. Only through additional preparation or CT study can the conspecifity or distinction of these nominal gorgonopsians be determined.

If, as suspected here, most Middle Permian gorgonopsian specimens eventually prove referable to *Eriphostoma microdon*, it would imply very low species diversity in this group relative to coeval therapsid clades. The bulk of gorgonopsian diversification did not occur until the Late Permian (Sigogneau-Russell 1989), and prior to this time gorgonopsians were relatively small and rare. At the same time early therocephalians were filling the large-bodied, sabre-toothed predator niche later occupied by Late Permian gorgonopsians (Kemp 2005). It is possible that the extinction of these therocephalians (lycosuchids and scylacosaurids) at the end of the Middle Permian provided an adaptive release for gorgonopsians, but more precise biostratigraphic and abundance data for therocephalians and gorgonopsians in the *Pristerognathus* and *Tropidostoma* AZs is required to corroborate this hypothesis.

Palatal Evolution and Early Radiation of the Gorgonopsia

The discovery of *Gorgonops*-style palatal morphology (with elongate, delta-shaped palatine bosses with numerous teeth) in a *Tapinocephalus* AZ gorgonopsian (*Eriphostoma*) and its shared presence in only the next-oldest gorgonopsian taxa [*Aelurosaurus* ('*Aelurosauroides watsoni*') and *Gorgonops* ('*Pachyrhinos kaiseri*')] have important implications for palatal evolution and potentially phylogeny in the group. Most post-*Tapinocephalus* AZ gorgonopsians have discrete palatine bosses (either reniform or a thin ridge) with few teeth (Fig. 11.6d). In basal therapsids such as *Biarmosuchus*, the palatine dentition is similar to that of sphenacodont 'pelycosaurs', with numerous small teeth covering most of the palatal exposure of the palatine over a broad triangular expanse (Fig. 11.6a). Unlike in sphenacodontids, this dentition is raised on a palatine boss in

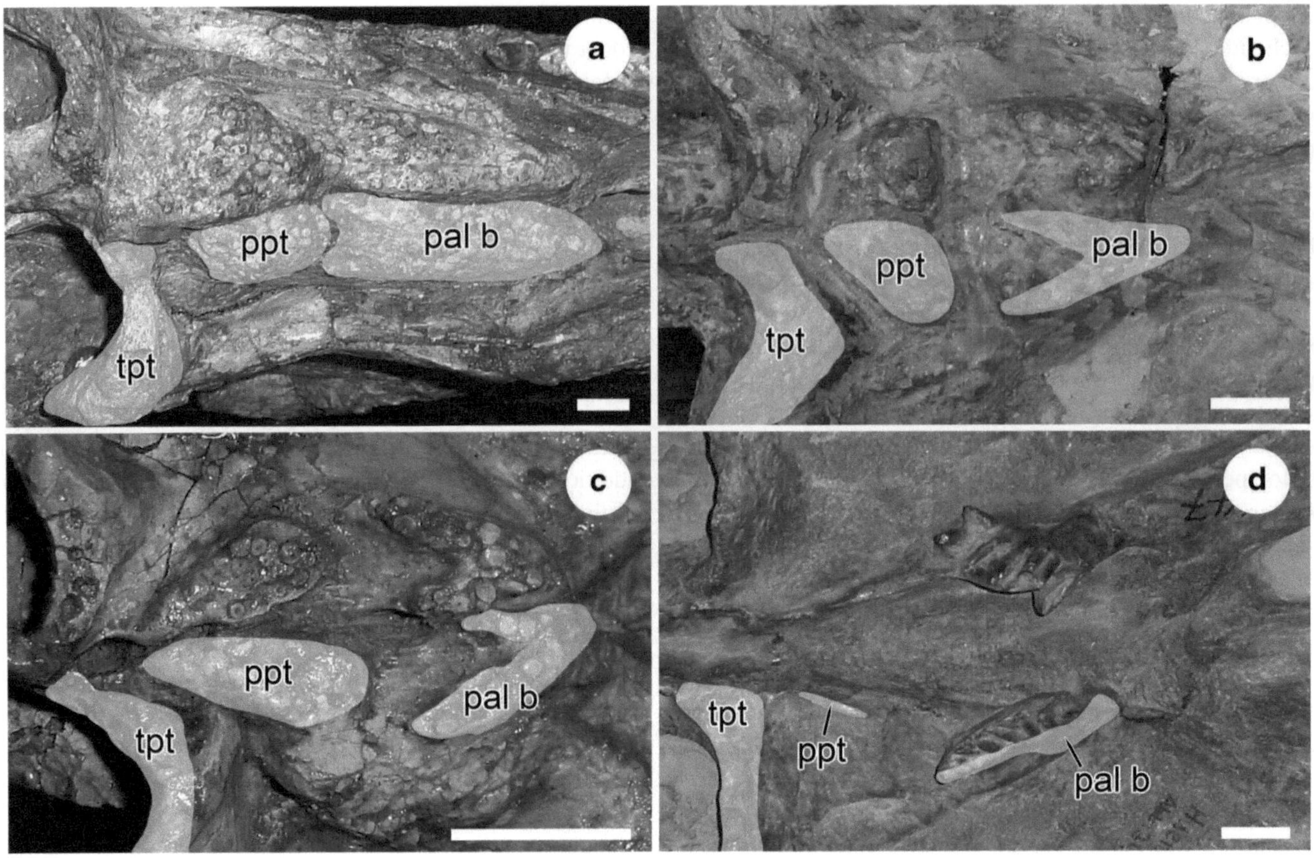

Fig. 11.6 Palatal boss morphology in gorgonopsians and an early therapsid. All specimens are in palatal view, anterior right. The left transverse process of the pterygoid, palatal boss of the pterygoid, and palatine boss are highlighted in each specimen. **a** *Biarmosuchus tener* (PIN 1758/1, holotype of *Eotitanosuchus olsoni*, from Ezhovo, Middle Permian of Russia), representing the primitive therapsid condition. The palatine bosses and palatal bosses of the pterygoid are nearly confluent and both bosses and the transverse process of the pterygoid are densely covered with teeth. **b** *Gorgonops torvus* (AMNH FARB 5515, from Beaufort West, Late Permian *Tropidostoma* AZ of South Africa). A single row of large teeth is present on the transverse process of the pterygoid. The palatine bosses and palatal bosses of the pterygoid are clearly separated. Both bosses bear relatively fewer, larger teeth than in *Biarmosuchus*. **c** *Aelurosaurus felinus* (NHMUK R339, holotype, from 'the Gouph', ?Late Permian of South Africa). A single row of large teeth is present on the transverse process of the pterygoid. The palatine and palatal pterygoid bosses are similar to those of *Gorgonops*, but the medial ramus of the palatine boss is shorter in *Aelurosaurus*. In both taxa the palatine bosses are relatively tightly appressed compared to the condition in *Eriphostoma microdon* (Fig. 11.3), in which these bosses are broadly separated by a trough. **d** *Cephalicustriodus kingoriensis* (GPIT/RE/7116, holotype, from Kingori Mountain, Late Permian Usili Formation of Tanzania), representing a Late Permian gorgonopsian. As in most Late Permian gorgonopsians, no dentition is present on the transverse processes of the pterygoid. The palatal bosses of the pterygoid are reduced to thin, toothless ridges in this specimen. Various palatine boss morphologies are present in Late Permian gorgonopsians; in this specimen the palatine boss is a narrow ridge angled anteromedially, bearing a single tooth row. *pal b* Palatine boss, *ppt* palatal boss of the pterygoid, *tpt* transverse process of the pterygoid. Scale bars equal 1 cm

Biarmosuchus, and in all therapsid groups the trend is towards more prominently-developed, discrete palatine bosses (albeit with a restricted area) with fewer, larger teeth. In most of these groups, the morphology of the palatine boss in the basalmost taxa is that of a delta- or horseshoe-shape, indicative of their origin from the ancestral triangular tooth patch. In basal dinocephalians [e.g., *Archaeosyodon* within Anteosauria and *Estemmenosuchus* within Tapinocephalia (Kammerer 2011)], the palatine boss is horseshoe-shaped, with numerous small teeth. Within anteosaurs, later species (e.g., *Titanophoneus*, *Anteosaurus*) are characterized by a prominent, reniform boss with few, large teeth. Similar bosses are present in the basal tapinocephalid tapinocephalian *Tapinocaninus*, although later tapinocephalids lose palatal teeth entirely. Finally, although the palatal dentition of the basalmost anomodont, *Biseridens qilianicus*, is in general highly aberrant for a therapsid, its palatine bosses are also elongate delta-shaped with numerous teeth (Liu et al. 2010). Only in the eutheriodonts is this pattern not observed—the earliest known therocephalians and cynodonts lack palatine dentition altogether, so the stages of evolution leading to this condition are unknown.

The status of *Eriphostoma*, *Aelurosaurus*, and *Gorgonops* as the stratigraphically earliest known diagnosable gorgonopsians, combined with their possession (unique among gorgonopsians) of a palatine boss morphology typical of basal members of other major therapsid clades (Figs. 11.3, 11.6b, c), suggests that these taxa may represent the most basal known gorgonopsians (and a possible vindication of the general use of *Gorgonops* as 'representative gorgonopsian' in higher-level analyses of therapsid phylogeny). Admittedly, this is a vague and altogether inadequate basis on which to base phylogenetic conclusions, but this is at least an initial hypothesis of gorgonopsian phylogeny to be tested. Our understanding of gorgonopsian phylogeny is largely nonexistent, the result of a general lack of interest, problems of outgroup choice, and above all the crippling difficulties of choosing operational taxonomic units among the unchecked wilds of gorgonopsian alpha taxonomy. Despite monographic revisions by Sigogneau (1970) and Gebauer (2007), gorgonopsian species diagnoses remain highly problematic, due in large part to the inadequate nature of their holotypes. New preparation and redescription of the types of most nominal gorgonopsian taxa (which are probably still greatly oversplit at the species level) will be necessary in order to produce an alpha taxonomic framework for the group approaching biological reality, as has been done for the other major therapsid clades (e.g., Hopson and Kitching 1972; King and Rubidge 1993; Kammerer et al. 2011). In addition to recognizing further synonymies among well-known taxa, it is probable that such research, when completed, will lead to the resurrection of several long-disused gorgonopsian taxa, as done here for *Eriphostoma*. Beyond admittedly scrappy holotypes like AMNH FARB 5524, several gorgonopsian species that have been considered *nomina dubia* since Boonstra's days are represented by nearly complete skulls, with their undiagnostic status solely attributable to lack of adequate preparation. Until we have a firm understanding of gorgonopsian morphospecies, it will be impossible to test phylogenetic hypotheses within the group.

Acknowledgments Special thanks are due to Matt Frenkel and Becky Rudolph of the Microscopy and Imaging Facility at the AMNH for their aid in the scanning of this specimen. Thanks are also due to the following curators and collection managers for permission to study the gorgonopsian specimens in their care: Billy DeKlerk (Albany Museum, Grahamstown), Carl Mehling (American Museum of Natural History, New York), Paul Barrett and Sandra Chapman (Natural History Museum, London), Bruce Rubidge, Michael Raath, and Fernando Abdala (Bernard Price Institute, Johannesburg), Johann Neveling and Ellen de Kock (Council for Geosciences, Pretoria), Bill Simpson (Field Museum of Natural History, Chicago), Jennifer Botha-Brink and Elize Butler (National Museum, Bloemfontein), Andrey Kurkin and Mikhail Ivakhnenko (Paleontological Institute, Moscow), Sheena Kaal (Iziko, the South African Museum, Cape Town), Stephany Potze (Ditsong, Pretoria), Pat Holroyd (University of California Museum of Paleontology, Berkeley), and Matt Carrano (National Museum of Natural History, Washington, DC). Tom Kemp, Eva Gebauer, and Christian Sidor are thanked for their helpful reviews of the original manuscript.

References

Abdala, F., Rubidge, B. S., & van den Heever, J. (2008). The oldest therocephalians (Therapsida, Eutheriodontia) and the early diversification of Therapsida. *Palaeontology, 51*, 1011–1024.

Angielczyk, K. D., & Sullivan, C. (2008). *Diictodon feliceps* (Owen, 1876), a dicynodont (Therapsida, Anomodontia) species with a Pangaean distribution. *Journal of Vertebrate Paleontology, 28*, 788–802.

Angielczyk, K. D., Fröbisch, J., & Smith, R. M. H. (2005). On the stratigraphic range of the dicynodont taxon *Emydops* (Therapsida: Anomodontia) in the Karoo Basin, South Africa. *Palaeontologia Africana, 41*, 23–33.

Bonaparte, C. L. (1846). Catalogo metodico dei pesci europei. *Atti della Settima Adunanza degli Scienziati Italiani, Napoli, Pt., 2*, 1–95.

Boonstra, L. D. (1934). Additions to our knowledge of the South African Gorgonopsia, preserved in the British Museum (Natural History). *Annals of the South African Museum, 31*, 175–213.

Boonstra, L. D. (1935). On the South African gorgonopsian reptiles preserved in the American Museum of Natural History. *American Museum Novitates, 772*, 1–14.

Boonstra, L. D. (1953). A report on a collection of fossil reptilian bones from Tanganyika Territory. *Annals of the South African Museum, 42*, 5–18.

Boonstra, L. D. (1963). Early dichotomies in the therapsids. *South African Journal of Science, 54*, 176–195.

Boonstra, L. D. (1969). The fauna of the *Tapinocephalus* Zone (Beaufort beds of the Karoo). *Annals of the South African Museum, 56*, 1–73.

Broili, F., & Schröder, J. (1934). Beobachtungen an Wirbieltieren der Karrooformation. IV. Ein neuer Gorgonopside aus den unteren Beaufort-Schichten. *Sitzungsberichte der mathematisch-naturwissenschaftlichen Abteilung der Bayerischen Akademie der Wissenschaften zu München, 1934*, 209–223.

Broili, F., & Schröder, J. (1935). Über die Skelettrests eines Gorgonopsiers aus den unteren Beaufort-Schichten. *Sitzungsberichte der mathematisch-naturwissenschaftlichen Abteilung der Bayerischen Akademie der Wissenschaften zu München, 1935*, 279–330.

Broom, R. (1911). On some new South African Permian reptiles. *Proceedings of the Zoological Society of London, 1911*, 1073–1082.

Broom, R. (1913). On four new fossil reptiles from the Beaufort series, South Africa. *Records of the Albany Museum, 2*, 397–401.

Broom, R. (1915a). On some new carnivorous therapsids in the collection of the British Museum. *Proceedings of the Zoological Society of London, 1915*, 163–173.

Broom, R. (1915b). Catalogue of types and figured specimens of fossil vertebrates in the American Museum of Natural History. II. Permian, Triassic and Jurassic reptiles of South Africa. *Bulletin of the American Museum of Natural History, 25*, 105–164.

Broom, R. (1930). On the structure of the mammal-like reptiles of the sub-order Gorgonopsia. *Philosophical Transactions of the Royal Society of London B, 218*, 345–371.

Broom, R. (1932). *The mammal-like reptiles of South Africa and the origin of mammals*. London: H. F. & G. Witherby.

Broom, R. (1935). On some new genera and species of Karroo fossil reptiles. *Annals of the Transvaal Museum, 18*, 55–72.

Broom, R. (1940). Some new Karroo reptiles from the Graaff-Reinet District. *Annals of the Transvaal Museum, 20*, 71–87.

Colbert, E. H. (1948). The mammal-like reptile *Lycaenops*. *Bulletin of the American Museum of Natural History, 89*, 353–404.

Gebauer, E. V. I. (2007). *Phylogeny and evolution of the Gorgonopsia with a special reference to the skull and skeleton of GPIT/RE/7113* ('Aelurognathus' parringtoni). Unpublished Ph.D. thesis, Eberhard-Karls Universität Tübingen.

Gebauer, E. V. I. (2013). Re-assessment of the taxonomic position of the specimen GPIT/RE/7113 (*Sauroctonus parringtoni* comb. nov., Gorgonopsia). In C. F. Kammerer, K. D. Angielczyk, & J. Fröbisch (Eds.), *Early evolutionary history of the Synapsida* (pp. 185–207). Dordrecht: Springer.

Haughton, S. H. (1915). Investigations in South African fossil reptiles and Amphibia. 7. On some new Gorgonopsians. *Annals of the South African Museum, 12*, 82–90.

Haughton, S. H. (1924a). A bibliographic list of pre-Stormberg Karroo Reptilia, with a table of horizons. *Transactions of the Royal Society of South Africa, 12*, 51–104.

Haughton, S. H. (1924b). Investigations in South African fossil reptiles and Amphibia. 12. On some Gorgonopsian skulls in the collection of the South African Museum. *Annals of the South African Museum, 12*, 499–517.

Haughton, S. H. (1929). On some new therapsid genera. *Annals of the South African Museum, 28*, 55–78.

Haughton, S. H., & Brink, A. S. (1954). A bibliographical list of Reptilia from the Karroo beds of Africa. *Palaeontologia Africana, 2*, 1–187.

Hopson, J. A., & Kitching, J. W. (1972). A revised classification of cynodonts (Reptilia; Therapsida). *Palaeontologia Africana, 14*, 71–85.

Huene, F. von. (1950). Die Theriodontier des ostafrikanischen Ruhuhu-Gebietes in der Tübinger Sammlung. *Neues Jahrbuch für Geologie und Paläontologie Abhandlungen, 92*, 47–136.

Ivakhnenko, M. F. (2008). Cranial morphology and evolution of Permian Dinomorpha (Eotherapsida) of Eastern Europe. *Paleontological Journal, 42*, 859–995.

Joleaud, M. L. (1920). Rectifications de nomenclature. *Revue Critique de Paléozoologie et de Paléophytologie, 24*, 36.

Kammerer, C. F. (2011). Systematics of the Anteosauria (Therapsida: Dinocephalia). *Journal of Systematic Palaeontology, 9*, 1–44.

Kammerer, C. F., Angielczyk, K. D., & Fröbisch, J. (2011). A comprehensive taxonomic revision of *Dicynodon* (Therapsida, Anomodontia) and its implications for dicynodont phylogeny, biogeography, and biostratigraphy. *Society of Vertebrate Paleontology Memoir, 11*, 1–158.

Kemp, T. S. (1969). On the functional morphology of the gorgonopsid skull. *Philosophical Transactions of the Royal Society of London Series B Biological Sciences, 256*, 1–83.

Kemp, T. S. (2005). *The origin and evolution of mammals*. Oxford: Oxford University Press.

King, G. M., & Rubidge, B. S. (1993). A taxonomic revision of small dicynodonts with postcanine teeth. *Zoological Journal of the Linnean Society, 107*, 131–154.

Kitching, J. W. (1977). The distribution of the Karroo vertebrate fauna. *Bernard Price Institute for Palaeontological Research Memoir No. 1*, 1–131.

Laurin, M. (1998). New data on the cranial anatomy of *Lycaenops* (Synapsida, Gorgonopsidae), and reflections on the possible presence of streptostyly in gorgonopsians. *Journal of Vertebrate Paleontology, 18*, 765–776.

Liu, J., Rubidge, B., & Li, J. (2010). A new specimen of *Biseridens qilianicus* indicates its phylogenetic position as the most basal anomodont. *Proceedings of the Royal Society of London B, 277*, 285–292.

Lydekker, R. (1890). *Catalogue of the Fossil Reptilia and Amphibia in the British museum (Natural History). Part IV. Containing the orders Anomodontia, Ecaudata, Caudata, and Labyrinthodontia; and supplement*. London: Trustees of the British Museum (Natural History).

Owen, R. (1881). On the Order Theriodontia, with a description of a new genus and species (*Ælurosaurus felinus*, Ow.). *Quarterly Journal of the Geological Society of London, 37*, 261–265.

Pravoslavlev, P. A. (1927). Gorgonopsidae from the North Dvina Expedition of V. P. Amalitzki. *Severo-Dvinskie raskopki Prof. V. P. Amalitskogo, 3*, 1–117 [in Russian].

Ray, S., & Bandyopadhyay, S. (2003). Late Permian vertebrate community of the Pranhita-Godavari valley, India. *Journal of Asian Earth Sciences, 21*, 643–654.

Rubidge, B. S. (Ed.) (1995). Biostratigraphy of the Beaufort Group (Karoo Supergroup). *South African Committee for Stratigraphy Biostratigraphic Series No. 1*, 1–46.

Sigogneau, D. (1970). *Révision systématique des Gorgonopsiens sud-africains. Cahiers de Paléontologie*. Paris: Centre National de la Recherche Scientifique.

Sigogneau-Russell, D. (1989). Theriodontia I. In P. Wellnhofer (Ed.), *Handbuch der Paläoherpetologie* (Vol. 17 B/I). Stuttgart: Gustav Fischer Verlag.

Smiley, T. M., Sidor, C. A., Maga, A., & Ide, O. (2008). The vertebrate fauna of the upper Permian of Niger. VI. First evidence of a gorgonopsian therapsid. *Journal of Vertebrate Paleontology, 28*, 543–547.

van den Heever, J. A. (1987). *The comparative and functional cranial morphology of the early Therocephalia (Amniota: Therapsida)*. Unpublished Ph.D. thesis, University of Stellenbosch.

van den Heever, J. A. (1994). The cranial anatomy of the early Therocephalia (Amniota: Therapsida). *Universiteit van Stellenbosch Annale, 1994*, 1–59.

Watson, D. M. S., & Romer, A. S. (1956). A classification of therapsid reptiles. *Bulletin of the Museum of Comparative Zoology, 114*, 35–89.

Wyllie, A. (2003). A review of Robert Broom's therapsid holotypes: Have they survived the test of time? *Palaeontologia Africana, 39*, 1–19.

Chapter 12
Re-assessment of the Taxonomic Position of the Specimen GPIT/RE/7113 (*Sauroctonus parringtoni* comb. nov., Gorgonopsia)

Eva V. I. Gebauer

Abstract The nearly complete gorgonopsian specimen GPIT/RE/7113, holotype of *Scymnognathus* (later *Aelurognathus*?) *parringtoni*, is redescribed. Comparisons with the type species of *Aelurognathus* (*A. tigriceps*) reveal that GPIT/RE/7113 is not referable to that genus. GPIT/RE/7113 shares a number of features with the Russian gorgonopsian *Sauroctonus progressus*, including a weakly flared zygomatic arch, interorbital and intertemporal regions of nearly equal width, broad nasal, naso-frontal suture situated anterior to the orbit and somewhat bow-shaped, prefrontal long and extremely low, terminating in a narrow anterior process, narrow vomer, parietal contribution to the occipital rim, and somewhat sloping dentary symphysis. Based on these characters, GPIT/RE/7113 is referred to *Sauroctonus* as *S. parringtoni* comb. nov. This represents the first instance of commonality in a gorgonopsian genus between the Eastern European and East African therapsid faunas.

Keywords Permian • Russia • Tanzania • Postcranium • Theriodontia • *Aelurognathus*

Introduction

The therapsid group Gorgonopsia includes the dominant carnivores of the Upper Permian. Within Therapsida, the gorgonopsians constitute the most basal group of the Theriodontia. The Gorgonopsia are characterized by a temporal opening which is larger than the orbit, the presence of a preparietal, a reduced and mobile quadrate and quadratojugal, enlarged upper canines, and a symphysis with a 'chin' (Sigogneau-Russell 1989). The individual gorgonopsian taxa are comparatively homogenous in their overall appearance, but they do differ in size, even if ontogenetic variation (which most probably explains part of this variability) is taken into account. Gorgonopsian alpha taxonomy has been somewhat complicated in the past, with some specimens being reassigned to different taxa through the years and depending on the author. One such specimen is GPIT/RE/7113 (former collection number, IGP U 28), which is housed in the collection of the Institut für Geowissenschaften in Tübingen, Germany. Although it is one of only a few nearly complete gorgonopsian skeletons in the world and is exceptionally well-preserved, its generic allocation is still uncertain. The specimen was first described by von Huene (1950) and allocated to the then-still-valid genus *Scymnognathus*. However, the genus *Scymnognathus* was never given a clear diagnosis with respect to other, similar genera. Sigogneau (1970) considered the type species *Scymnognathus whaitsi* to belong to the genus *Gorgonops* and distributed the remaining species between the genera *Lycaenops* and *Aelurognathus*. She then allocated GPIT/RE/7113 to the genus *Aelurognathus*, but still with a somewhat dubious position concerning actual affiliation to this genus. Sigogneau (1970) saw GPIT/RE/7113 as a specialized species of the genus *Aelurognathus*, since it differed from the typical *Aelurognathus* species by its narrower postorbital bars and longer temporal opening. She also discussed a possible relationship with the genus *Lycaenops*.

A thorough re-examination of the specimen has revealed that GPIT/RE/7113 does not fit well into the genus *Aelurognathus* (exemplified by the type species *A. tigriceps*). Both forms differ considerably, particularly in respect to the massive appearance of *Aelurognathus*, with its heavy skull arches, downturned zygomatic arch, more enlarged posterior region of the skull, and heavy mandible, which are not present to as great an extent in GPIT/RE/7113. Further investigation of gorgonopsian material by the author demonstrates that GPIT/RE/7113 differs from all other African taxa in most diagnostic features, which makes it difficult to

E. V. I. Gebauer (✉)
Staatliches Museum für Naturkunde Karlsruhe,
Erbprinzenstraße 13, 76133 Karlsruhe, Germany
e-mail: eva.gebauer@smnk.de

Christian F. Kammerer, Kenneth D. Angielczyk and Jörg Fröbisch (eds.), *Early Evolutionary History of the Synapsida*, Vertebrate Paleobiology and Paleoanthropology, DOI: 10.1007/978-94-007-6841-3_12,

allocate it to any South or East African genus. However, there is a close resemblance with the Russian taxon *Sauroctonus progressus*.

Institutional Abbreviations: AMG, Albany Museum, Grahamstown, South Africa; AMNH FARB, American Museum of Natural History, New York City, NY, USA; BSP, Bayerische Staatssammlung für Paläontologie und Geologie, Munich, Germany; BP, Bernard Price Institute for Palaeontological Research, University of the Witwatersrand, Johannesburg, South Africa; CAMZM, University Museum of Zoology, Cambridge, UK; FMNH, Field Museum of Natural History, Chicago, IL, USA; IGP/GPIT Institut und Museum für Geologie und Paläontologie, Tübingen, Germany; NHMUK, Natural History Museum, London, UK; PIN, Paleontological Institute, Russian Academy of Sciences, Moscow, Russia; RC, Rubidge Collection, Wellwood, Graaff-Reinet, South Africa; SAM, Iziko, the South African Museum, Cape Town, South Africa.

Anatomical Abbreviations: ang, angular; an (lr), reflected lamina of angular; ar, articular; bo, basioccipital; btub, basisphenoidal tubera; co, coronoid; de, dentary; ep, ectopterygoid; eo, exoccipital; f, frontal; fm, foramen magnum; fp, parietal foramen; fpt, fenestra posttemporalis; ip, interparietal; ivac, interpterygoid vacuity; j, jugal; l, lacrimal; mr, maxillary ridge; mx, maxilla; n, nasal; p, parietal; pal, palatine; par, paroccipital process; pbfos, parabasisphenoid fossa; pfos, palatal fossa; pmx, premaxilla; po, postorbital; pof, postfrontal; pp, preparietal; pra, prearticular; prf, prefrontal; pt, pterygoid; ptub, palatal tuberosities; rlr, ridge on reflected lamina; sa, surangular; smx, septomaxilla; soc, supraoccipital; sp, splenial; sq, squamosal; tab, tabular; v, vomer.

Systematic Paleontology

Therapsida Broom, 1905
Gorgonopsia Seeley, 1894
Gorgonopidae Lydekker, 1890
Sauroctonus Bystrow, 1955

Type species: *Arctognathus progressus* Hartmann-Weinberg, 1938.

Generic diagnosis: Posterior part of skull somewhat expanded but zygomatic arch does not flare laterally to a great extent; interorbital and intertemporal spaces nearly of the same width, snout narrow with sloping dorsal profile, nasal broad but slightly constricted in the middle, nasofrontal suture anteriorly situated to the orbit and somewhat bow-shaped, prefrontal long and extremely low, terminates in a narrow anterior process; postfrontal rather narrow and of the same width throughout its length, posterior margin extended in a narrow process; bone surface of maxilla and nasal strongly sculptured, skull arches rather slender, vomer narrow throughout its entire length, palatine tuberosities well developed and separated from each other, both with numerous teeth, parietal contributes to occipital rim, dentary symphysis somewhat sloping.

Included species: *Sauroctonus progressus* (Hartmann-Weinberg, 1938); *Sauroctonus parringtoni* (von Huene, 1950) comb. nov.

Sauroctonus parringtoni (von Huene, 1950) comb. nov. (Figs. 12.1, 12.2, 12.3, 12.4, 12.5, 12.6, 12.7, 12.8, 12.9, 12.10, 12.11, 12.12, 12.13)

1950 *Scymnognathus parringtoni* von Huene:48

1970 *Aelurognathus*? *parringtoni* Sigogneau:181.

Holotype: GPIT/RE/7113, an almost complete skull and skeleton.

Referred material: None.

Specific diagnosis: Interorbital and intertemporal spaces wider than in the type species, postfrontal narrow and long, dentary symphysis massive, supraoccipital low, paroccipital process massive.

Description

Preservation

The skull and mandible are nearly complete but somewhat compressed laterally (Fig. 12.1). The left side of the dorsal skull roof is considerably depressed, such that the zygomatic arch, squamosal, tabular, quadrate and the posterior parts of the postorbital and parietal are missing. Additionally, both epipterygoids, stapes, quadrate rami, the sphenethmoid region, and the vomer are not present. Although the lower jaw is less deformed, the right side is again compressed laterally and somewhat displaced posteriorly. Here the lower parts of the right reflected lamina and the left posterior ramus are missing, except for small parts of the coronoid process and the articular.

Cranium and Mandible

General Features of the Skull—The skull is rather low and the snout is almost as wide as high (Table 12.1). The dorsal profile of the snout is slightly sloping whereas the dorsal skull roof is straight. The curvature of the ventral margin of the maxilla and that of the zygomatic arch is only slight. All three skull arches are comparatively slender, with the suborbital arch being the thickest. The orbit is round and medium-sized and the temporal opening is long. The palate shows a narrow palatine and palatal fossa and the transverse apophyses are situated posteriorly. The occiput appears strongly concave;

Fig. 12.1 Photographs of the skull of GPIT/RE/7113 in **a** right lateral; **b** occipital; **c** ventral; **d** dorsal view. Scale bar equals 50 mm

however, this is mostly due to deformation. Thus, the occiput might originally have been only slightly concave but somewhat sloping antero-posteriorly. The median ridge is narrow but well developed, and terminates in a bulbous thickening above the foramen magnum. Dorsally, the occipital surface is considerably concave, forming two round depressions, one on each side of the median ridge.

Dorsal Surface of the Skull—The premaxilla is relatively low, since the external nares are situated ventrally. Posteriorly the bone is overlapped by the maxilla on the

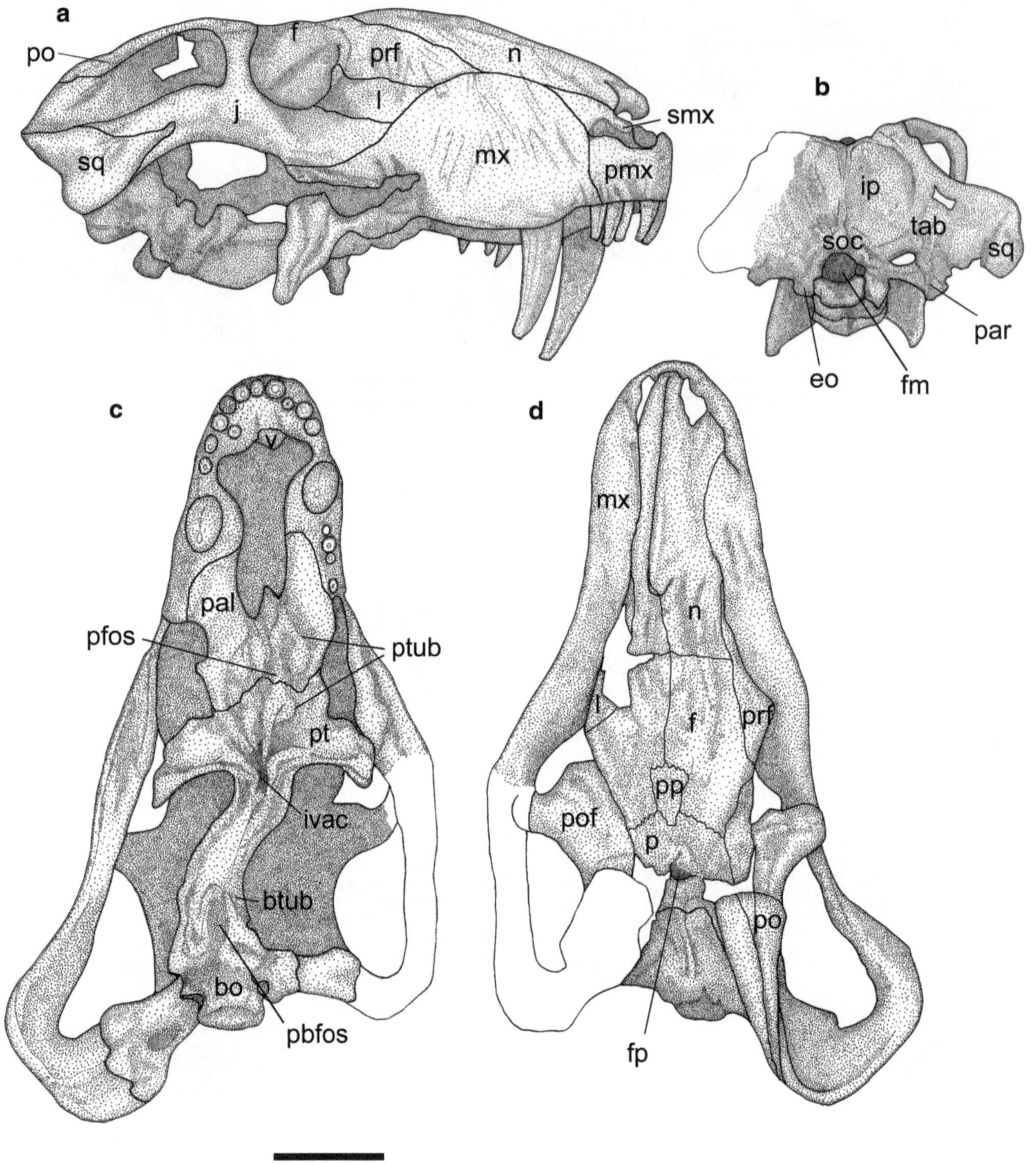

Fig. 12.2 Drawings of the skull of GPIT/RE/7113 in **a** right lateral; **b** occipital; **c** ventral; **d** dorsal view. Scale bar equals 50 mm

external surface of the skull, but on the internal face the premaxilla persists slightly further posteriorly, forming a scarf joint. The septomaxilla extends far posteriorly as a narrow process; however, it is rather low (Fig. 12.2a). There is no recess in the area of the septomaxilla foramen and thus the suture with the maxilla is regularly curved. The septomaxillary foramen perforates the bone at the dorsal end of a distinct oval fossa.

The maxilla is strongly sculptured with radial grooves and small pits. Its dorsal and posterior suture establishes a straight contact with the nasal, prefrontal, lacrimal and jugal. Postero-ventrally it forms a scarf joint with the

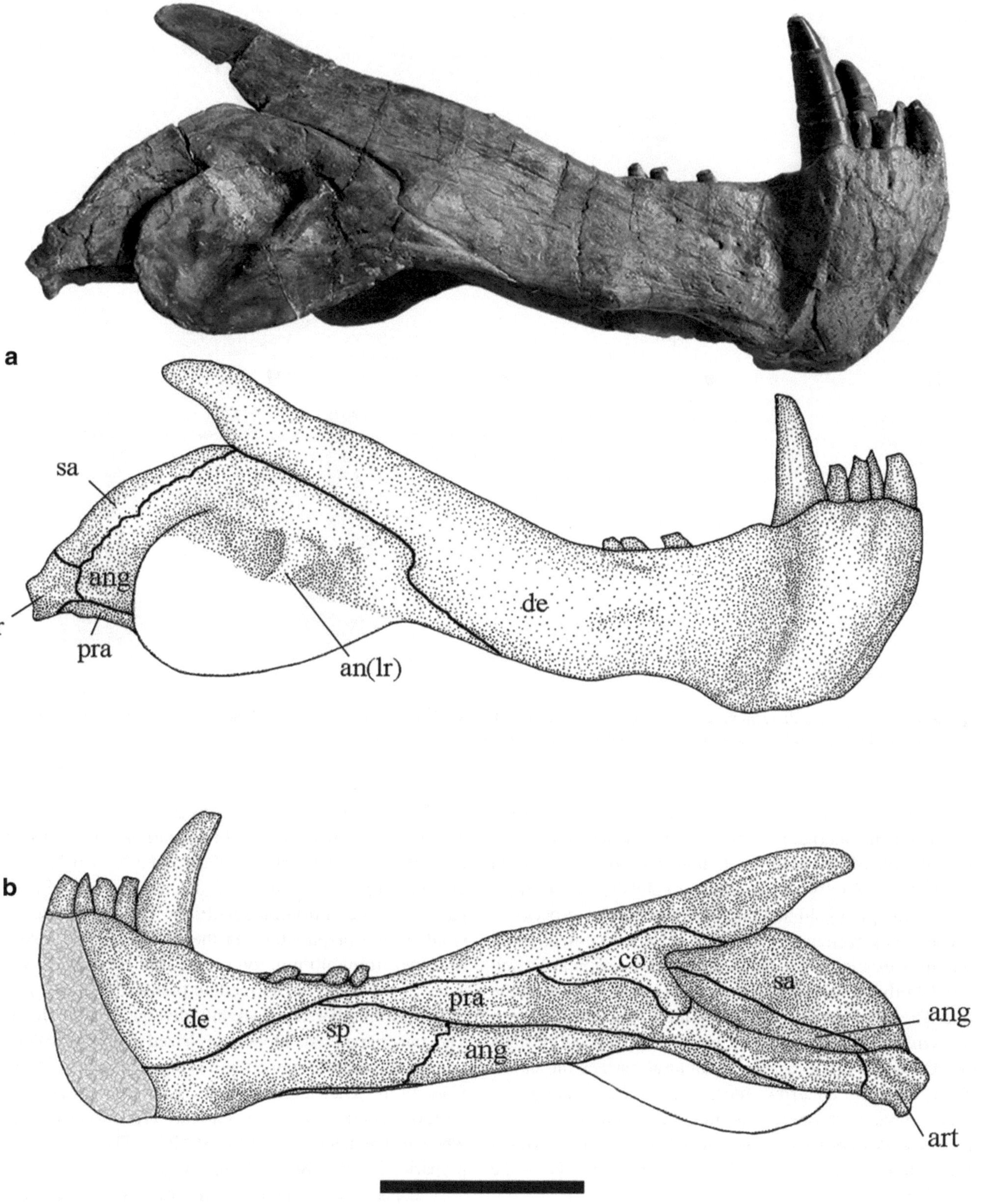

Fig. 12.3 Drawings and photographs of the lower jaw of GPIT/RE/7113 in **a** right lateral; **b** internal view. Scale bar equals 50 mm

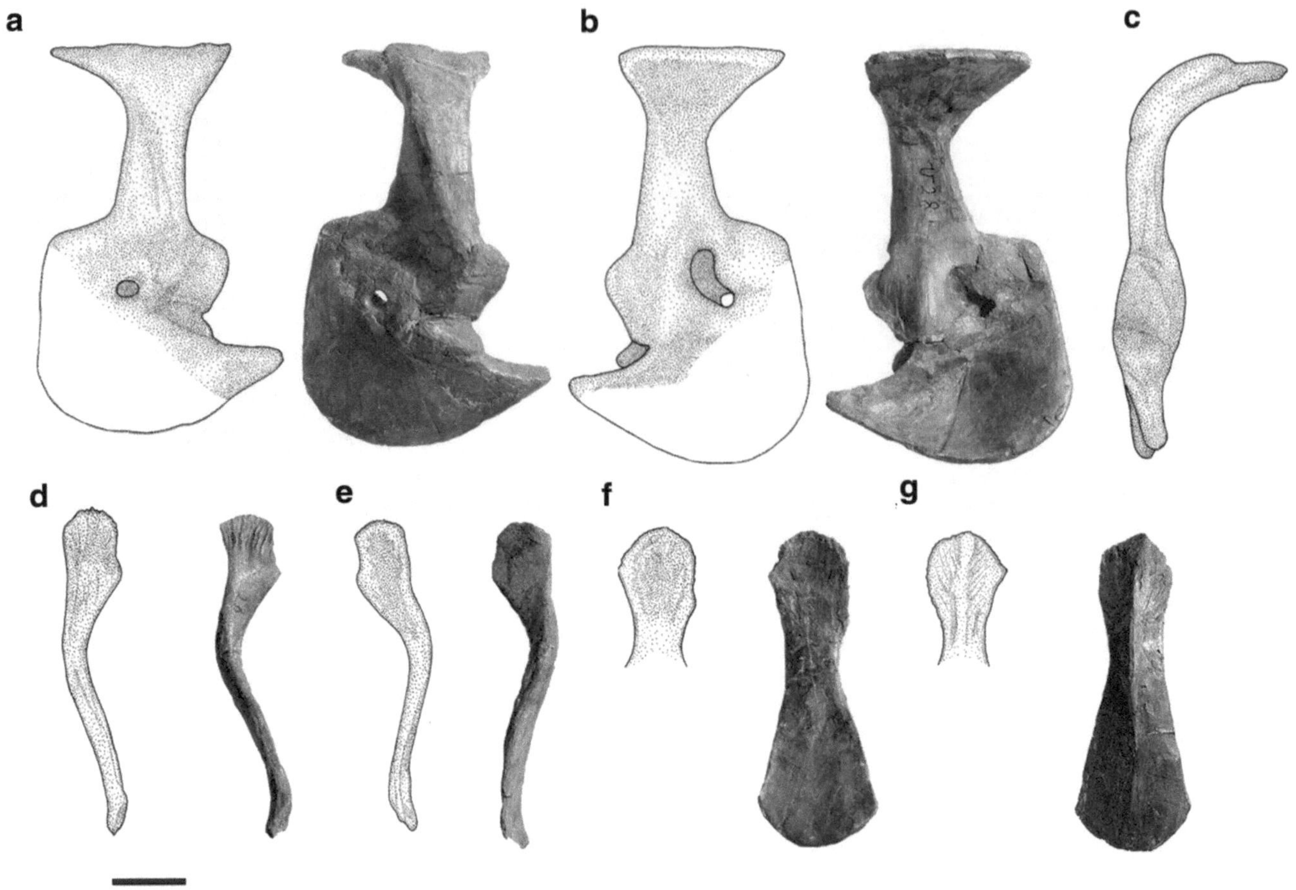

Fig. 12.4 Drawings and photographs of the left pectoral girdle of GPIT/RE/7113 in **a** lateral; **b** medial; and **c** posterior view; left clavicle in **d** ventral; **e** dorsal view; interclavicle in **f** ventral; **g** dorsal view. Scale bar equals 25 mm

jugal, sending a narrow process posteriorly on its external surface. On the internal face the maxilla constitutes the steep internal wall of the snout containing the postcanine tooth row postero-laterally and continuing caudally to terminate between the pterygoid and the jugal as a narrow process.

The small, rectangular lacrimal is somewhat sculptured with grooves and pits. It shows no antorbital depression and only the orbital margin is slightly off set from the rest of the bone. The jugal forms a moderately thick suborbital bar with an extensive double scarf joint with the squamosal posteriorly. On the external face the squamosal reaches far anteriorly by means of a narrow process that overlaps the jugal.

The nasal is strongly sculptured with oblong furrows anteriorly. Posteriorly the bow-shaped and strongly serrated naso-frontal suture is situated just in front of a boss-like elevation.

The large prefrontal reaches far anteriorly compared to other gorgonopsians. The contribution of the prefrontal to the dorsal margin of the orbit is, however, rather small. The surface of the prefrontal is covered with grooves and knobs but less sculptured than the maxilla and nasal.

The long frontal forms a considerable part of the orbital margin. A serrated suture with the parietal runs transversely from the medial margin of the frontal in a posterolateral direction up to the level of the anterior margin of the temporal fossa. Posteromedially the two frontals are separated by the preparietal. On the internal side of the skull roof two strong 40 mm long ridges are established on both sides of the anterodorsal median ossification at the level of the postorbital bar. They slightly diverge at an angle of 20°. The fan-shaped and moderately-sized preparietal does not reach the parietal foramen and is situated almost 10 mm in front of it. The anterior suture with the frontal and the posterior suture with the parietal are strongly serrated, whereas the lateral suture is straight. The surface of the preparietal is covered with narrow striations.

The parietal foramen is surrounded by a narrow ridge and situated on a slight elevation. It is well separated from the preparietal and is situated well in front of the occipital crest. The parietal does not contribute to the occipital surface and thus the suture with the interparietal (postparietal) directly forms the margin of the occipital crest.

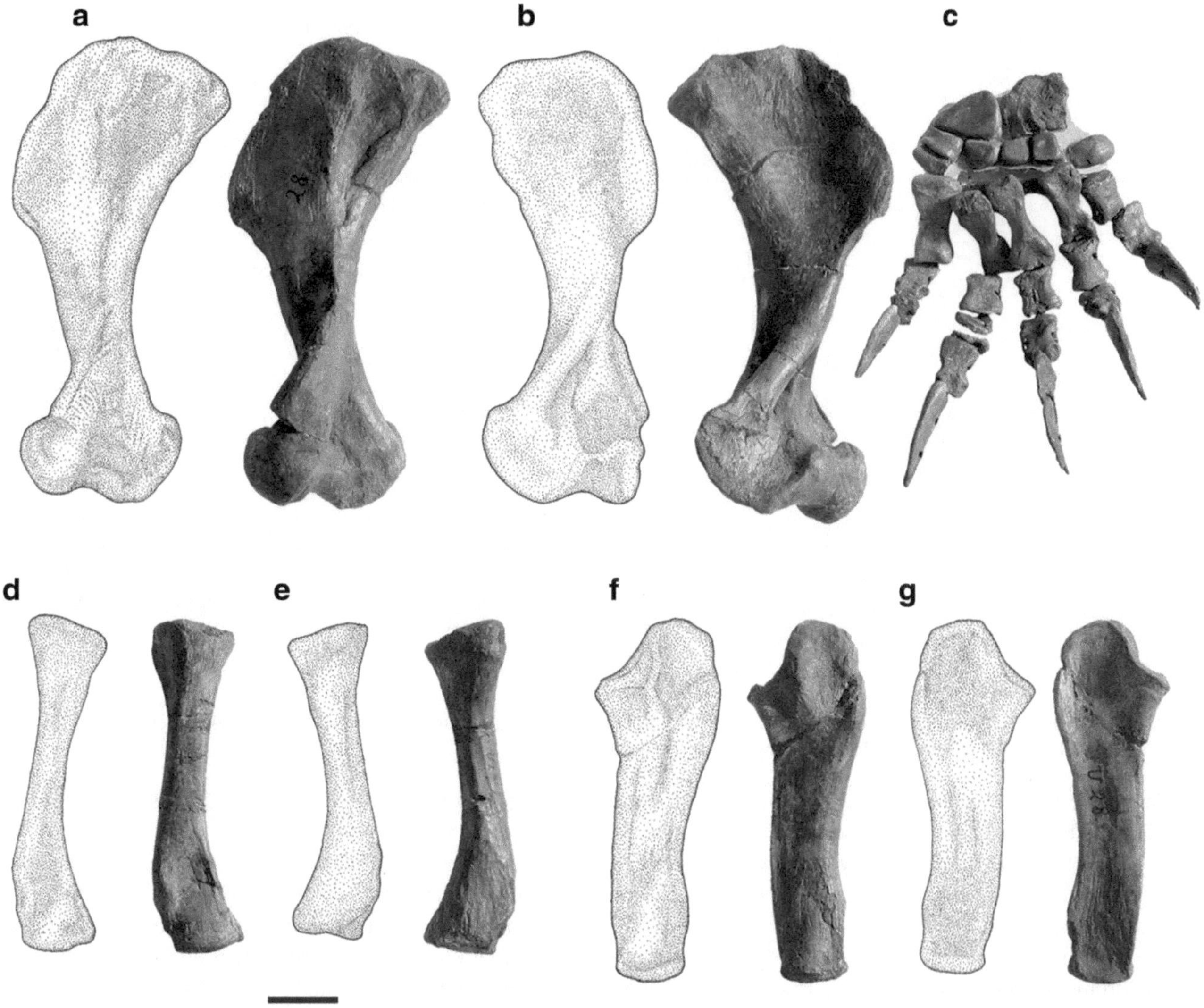

Fig. 12.5 Drawings and photographs of the anterior limb of GPIT/RE/7113. Left humerus in **a** dorsal; **b** ventral view; **c** left manus in dorsal view; left radius in **d** medial; **e** lateral view; left ulna in **f** medial; **g** lateral view. Scale bar equals 25 mm

The postfrontal is rather narrow compared to other gorgonopsians and retains the same width throughout its entire length. Dorsally it forms a long narrow process that extends posteriorly, establishing a scarf joint with the squamosal laterally.

Palatal Surface of the Skull—Most of the vomer is missing although von Huene (1950) described and figured it as if complete. However, inadequate preparation or cast production subsequent to von Huene's description damaged this part of the specimen irreparably. Thus, today only the anteriormost and posteriormost portions of the vomer are preserved. The entire vertical blade is missing and thus no information on its width can be given. However, the posterior portion is narrower than the anterior portion. The latter borders the premaxilla with a strongly serrated suture. Posteriorly the vomer continues between the palatines and extends posteriorly for almost 25 mm, separating the palatines and forming a deep groove between them.

The rather narrow and elongate palatine overlaps the maxilla anteriorly. Posterolaterally it is bordered by the ectopterygoid. The ectopterygoid does not reach far ventrally on the transverse apophyses of the pterygoid and thus seems to be wider than long. Posteriorly, the palatine is connected with the pterygoid by a strongly serrated suture. The pterygoid is composed of the palatal part, the vertical standing transverse apophyses and the quadrate ramus, which is missing. Medially, the deep but rather narrow palatal fossa is made up of both the palatine and pterygoid. It is bordered by the palatal tuberosities, which are separated from each other by a groove. The tuberosities on the palatine are larger than those on the pterygoid but both have numerous small teeth.

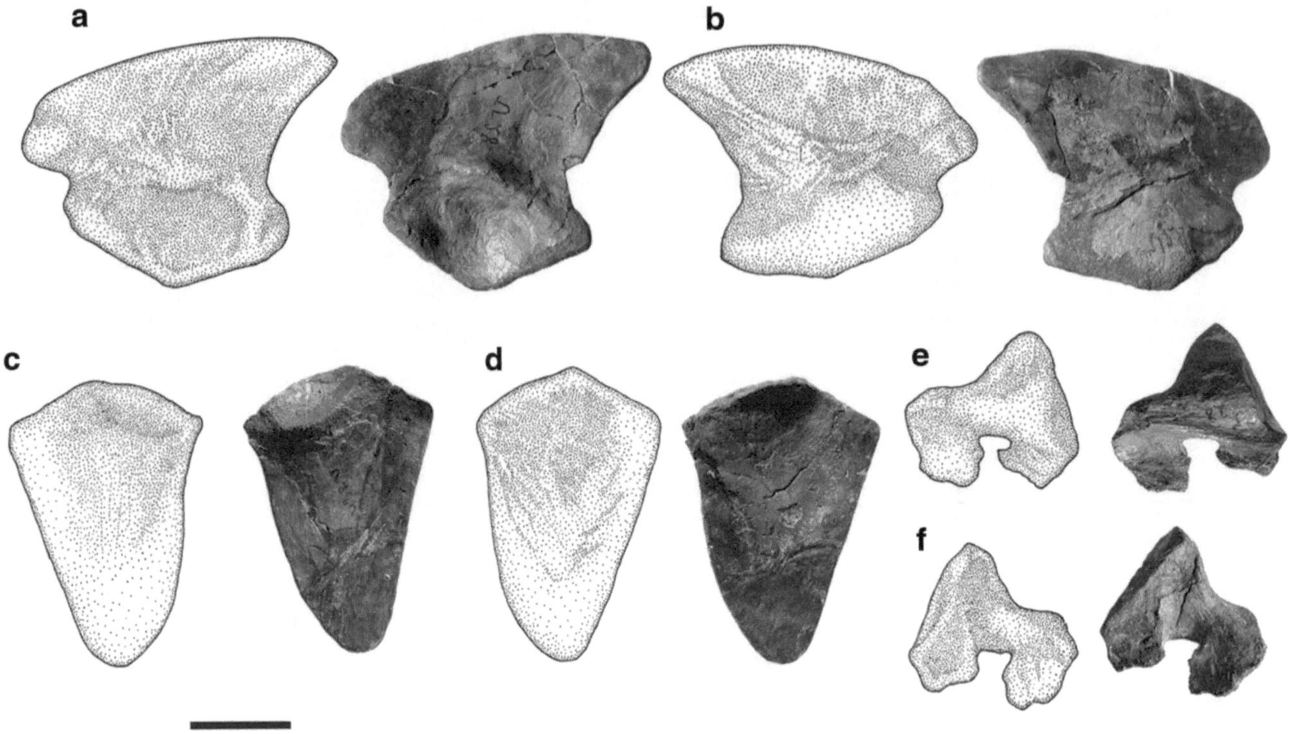

Fig. 12.6 Drawings and photographs of the pelvic girdle of GPIT/RE/7113. Right ilium in **a** lateral; **b** medial view; left ischium in **c** lateral; **d** medial view; left pubis in **e** lateral; **f** medial view. Scale bar equals 35 mm

The posteriorly situated interpterygoid vacuity is oval and deep. The transverse apophyses of the pterygoid are mostly directed ventrally and only somewhat posteriorly. The ventral margin forms a broad medial rim, which contains a few teeth.

Dorsally, only the ventralmost parts of the thin parasphenoid rostrum are preserved. Since the parasphenoid is heavily deformed in this area the usually hardly visible suture with the basisphenoid is recognizable as an oblique line. Dorsally, the basisphenoid forms the anteroventral margin of the foramen magnum and the posterior border of the hypophyseal fossa. Ventrally, the basisphenoidal tuberosities with the long and narrow para-basisphenoid fossa are narrow, elongate and strongly rugose. The basisphenoidal tuberosities and the basioccipital tubera are separated by a narrow trench. The occipital condyle is undivided and reniform with a smoothly rounded lower margin. Dorsally, there is no sutural separation from the exoccipital but the latter is distinguishable by its knobbly and tuberous appearance.

The prootic forms the mid-part of the anterior wall of the braincase and is laterally fused with the paroccipital process of the opisthotic. Dorsally it establishes the upper margin of the fenestra posttemporalis and anteriorly the prootic forms the anterolateral wall of the foramen magnum. The prootic is fused with the opisthotic ventrally. This bone is mainly composed of the paroccipital process, which is shifted somewhat anteriorly. It is comparatively low medially but widens laterally where it abuts against the tabular and squamosal with a massive facet.

Occiput—The squamosal is overlapped anterodorsally by the posterior process of the postorbital on the external side and the parietal dorsally. On the internal side, however, the squamosal reaches anteriorly with a small process, meeting the postorbital and parietal again but almost 35 mm farther anteriorly. On the lateral face, the squamosal sends a pointed process into the jugal and reaches anteriorly up to the middle of the temporal fossa. Posteriorly, the squamosal constitutes the lateral boundary of the occiput. It shares a long suture with the tabular and meets the paroccipital process of the opisthotic ventrally.

The quadrate is elongated in a dorsal direction and measures 40 mm in height. The posterior face is convex, whereas the anterior face is slightly concave. Medioventrally the posterior face shows two pronounced fossae, which are separated from each other by a thin ridge. The lower fossa is more elongate and served for the insertion of the stapes. The upper fossa is shallower and might have housed the quadrate ramus of the pterygoid and the epipterygoid. On the ventral margin the facet for articulation with the articular is concave in the middle and convex anteriorly and posteriorly. Posterolaterally, a small part of the quadratojugal is preserved. The recess on the ventrolateral margin might be the medial border of the quadratojugal foramen.

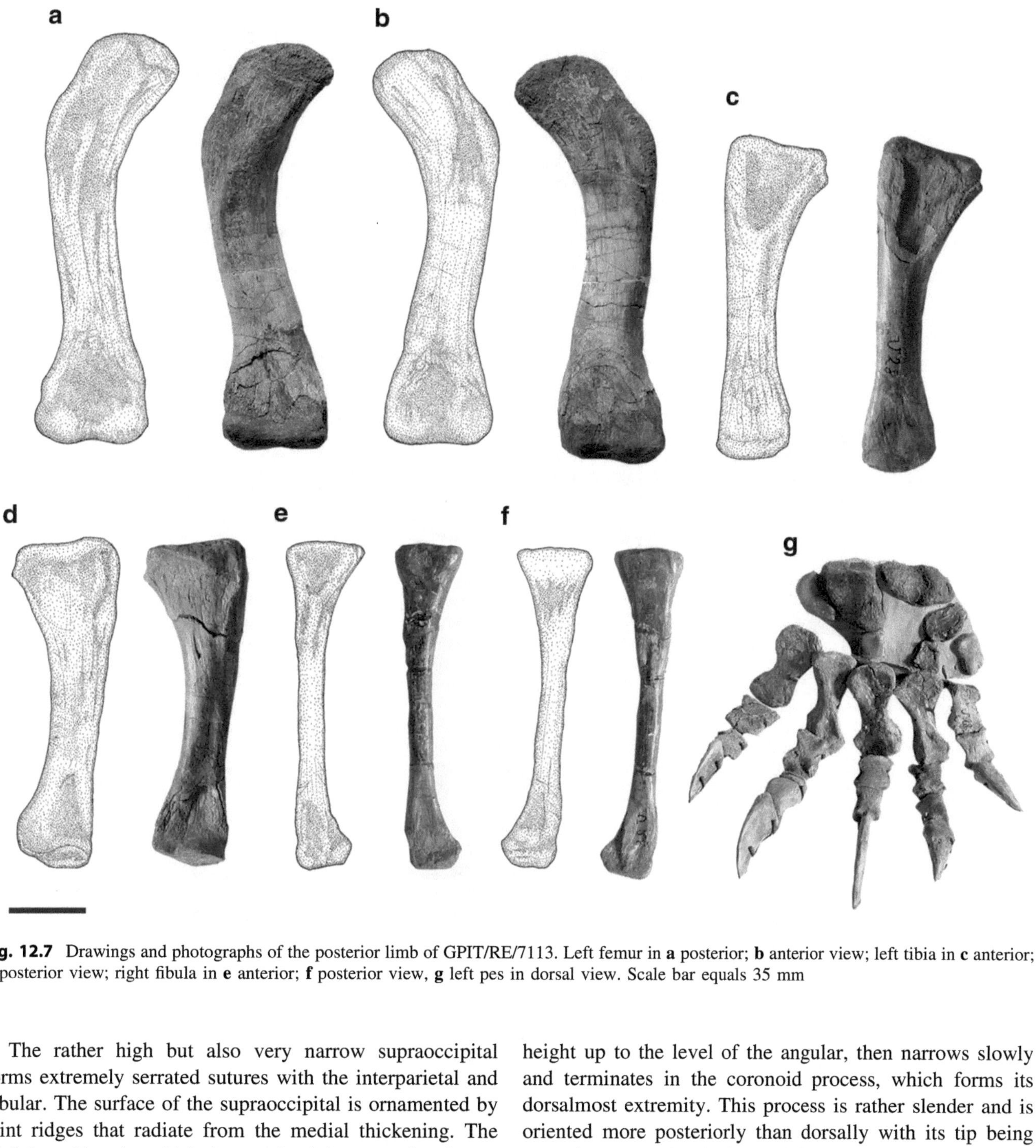

Fig. 12.7 Drawings and photographs of the posterior limb of GPIT/RE/7113. Left femur in **a** posterior; **b** anterior view; left tibia in **c** anterior; **d** posterior view; right fibula in **e** anterior; **f** posterior view, **g** left pes in dorsal view. Scale bar equals 35 mm

The rather high but also very narrow supraoccipital forms extremely serrated sutures with the interparietal and tabular. The surface of the supraoccipital is ornamented by faint ridges that radiate from the medial thickening. The slightly wider than high interparietal is somewhat larger than the supraoccipital and exhibits strongly serrated sutures as well. The tabular reaches the paroccipital process of the opisthotic ventrally, the posterior extensions of the parietal and postorbital dorsally, and the squamosal ventrolaterally.

Mandible—The symphyseal part of the dentary, which is covered with numerous small foramina, is more massive than the rest of the bone. Posteriorly, the dentary retains its height up to the level of the angular, then narrows slowly and terminates in the coronoid process, which forms its dorsalmost extremity. This process is rather slender and is oriented more posteriorly than dorsally with its tip being strongly rugose. On the internal face, the dentary forms only the upper two-thirds of the symphysis because it is internally covered by the splenial ventrally. Posteriorly, the medial exposure of the dentary narrows rapidly with a somewhat undulating suture and forms only the alveolar border and the coronoid process, since it is overlapped by the splenial, coronoid, and prearticular ventrally.

The splenial-dentary contact on the symphysis is marked by a posterolaterally oriented depression. In the area of the

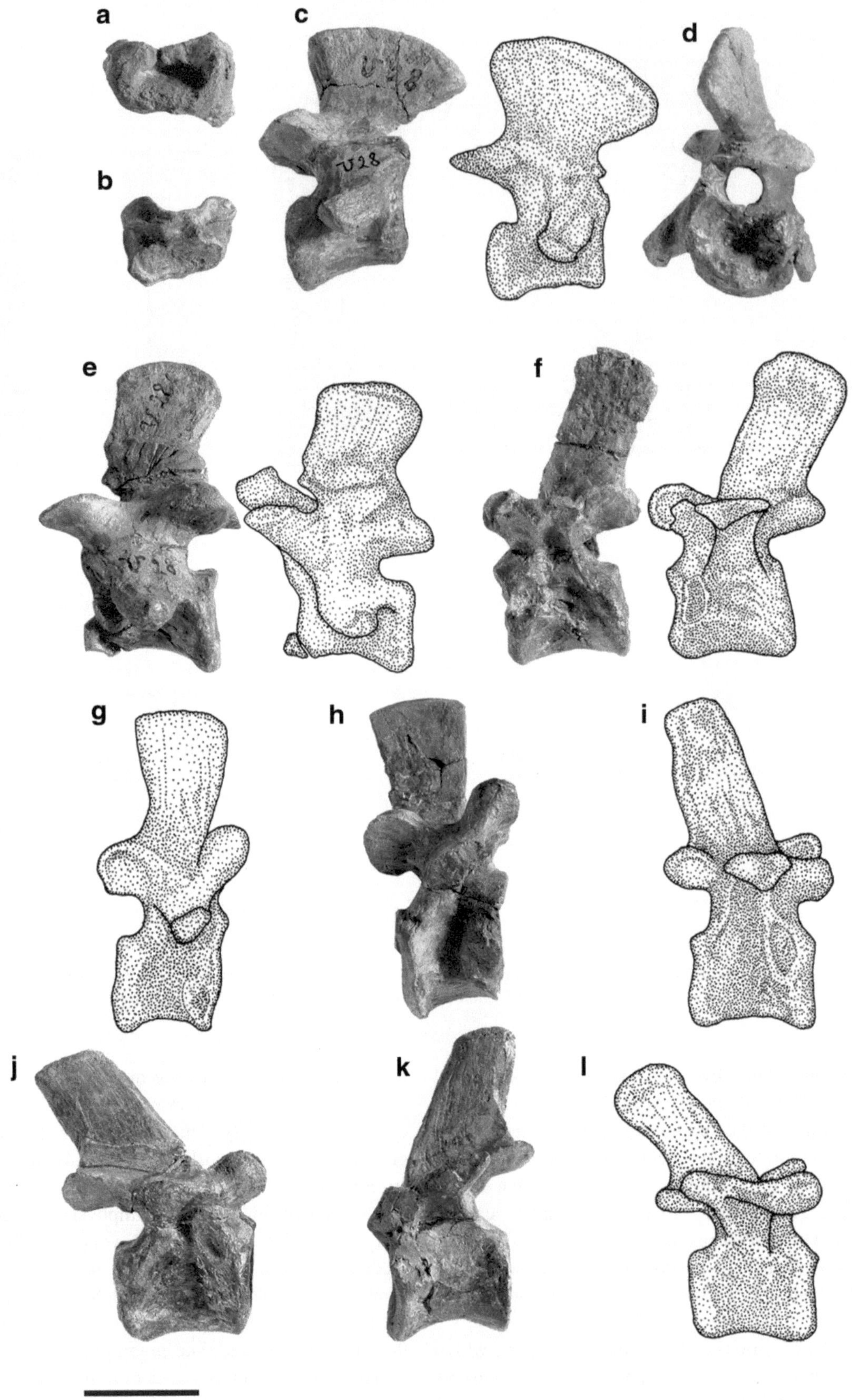

Fig. 12.8 Drawings and photographs of cervical and dorsal vertebrae of GPIT/RE/7113. Atlas in **a** posterior; **b** anterior view; Axis in **c** right lateral; **d** anterior view; **e** third cervical in, left lateral view; **f** seventh cervical in left lateral view; **g** first dorsal in right lateral view; **h** second dorsal in right lateral view; **i** sixth dorsal in right lateral view; **j** tenth dorsal in right lateral view; **k** fourteenth dorsal in left lateral view; **l** sixteenth dorsal in right lateral view; Scale bar equals 30 mm

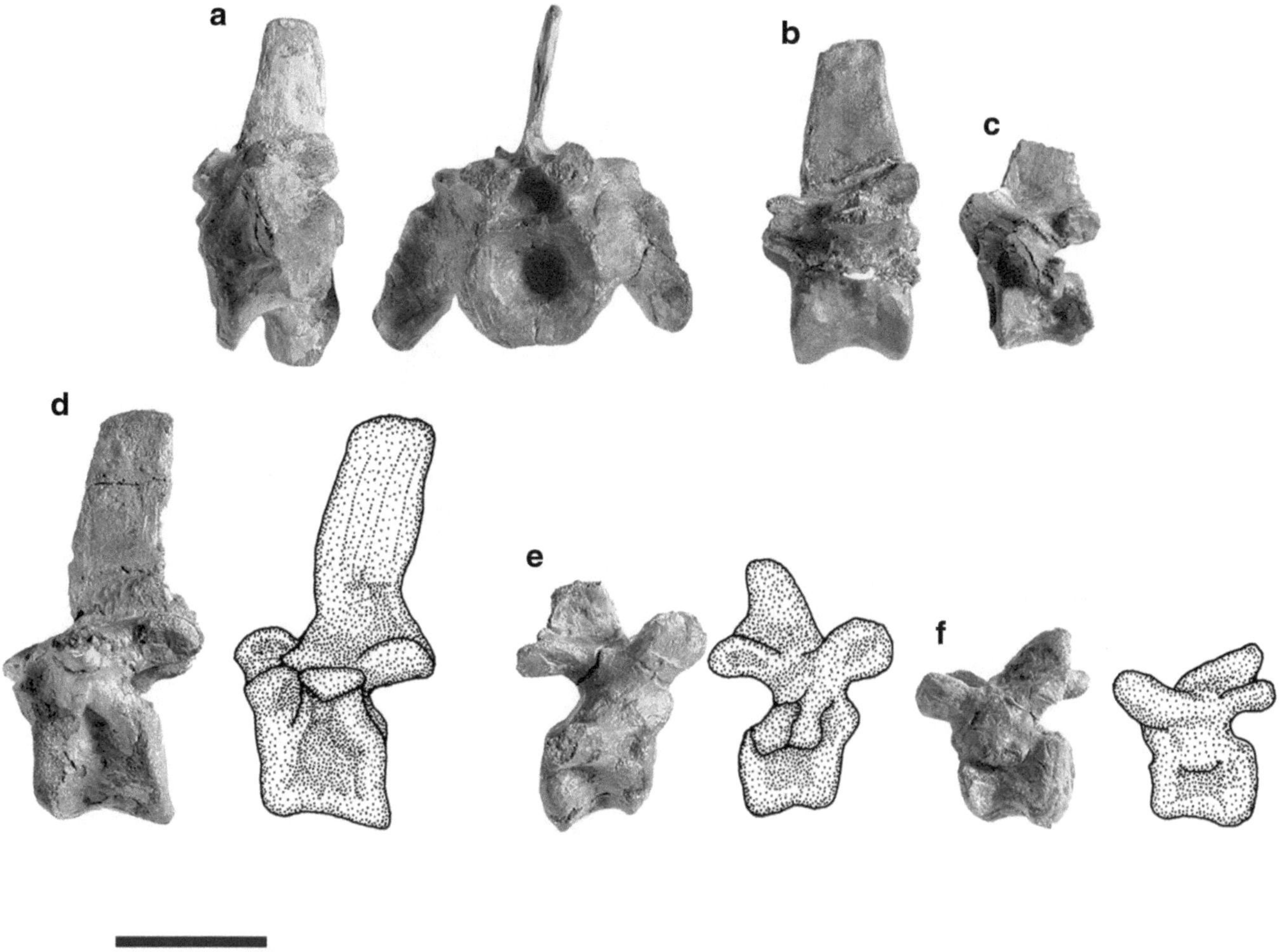

Fig. 12.9 Drawings and photographs of sacral and caudal vertebrae of GPIT/RE/7113. **a** First sacral in left lateral and anterior view; **b** second sacral in left lateral view; **c** third sacral in left lateral view; **d** first caudal in left lateral view; **e** third caudal in right lateral view; **f** sixth caudal in left lateral view. Scale bar equals 30 mm

postcanines, the splenial extends dorsally before again bending ventrally and sending a broad triangular process to contact the prearticular and angular, covering these two bones.

Laterally, the angular borders the dentary anteriorly with a narrow process that widens in a posterior direction and forms the reflected lamina. The anterodorsal ridge of the reflected lamina is well developed with a rounded contour and with an elongate depression in front and behind. Posteriorly, the curved suture with the surangular runs posteroventrally until the angular reaches the prearticular ventrally. On the internal face the reflected lamina is covered medially by a rod-like structure that is mainly composed of the prearticular.

The prearticular reaches up to the level of the last postcanine tooth anteriorly until it is covered by the dentary and splenial. Posteriorly, it widens and meets the angular ventrally. On the internal face the prearticular is covered by the small triradiate-shaped coronoid dorsally at the level of the anterior margin of the reflected lamina.

The surangular forms the dorsal margin of the posteroventrally directed posterior portion of the upper jaw. Posteroventrally it reaches the articular, but both bones are somewhat fused and thus no suture is discernible.

The articular has two articulating surfaces for the quadrate. The anterior fossa, which is medially situated, is smaller and almost round in shape. It lies entirely on the same level and is surrounded by a distinct ridge. The second fossa, which is laterally situated, is elongated mediolaterally and declines steeply posteroventrally. The dorsal and ventral margins as well as the lateral and medial ones bear ridges.

Dentition—The dentition is almost complete though most tips of the incisors are broken off. The upper incisors show the typical gorgonopsian condition, with the first three incisors having the same length, the fourth being the longest, and the fifth the shortest. Both canines show serrations on the

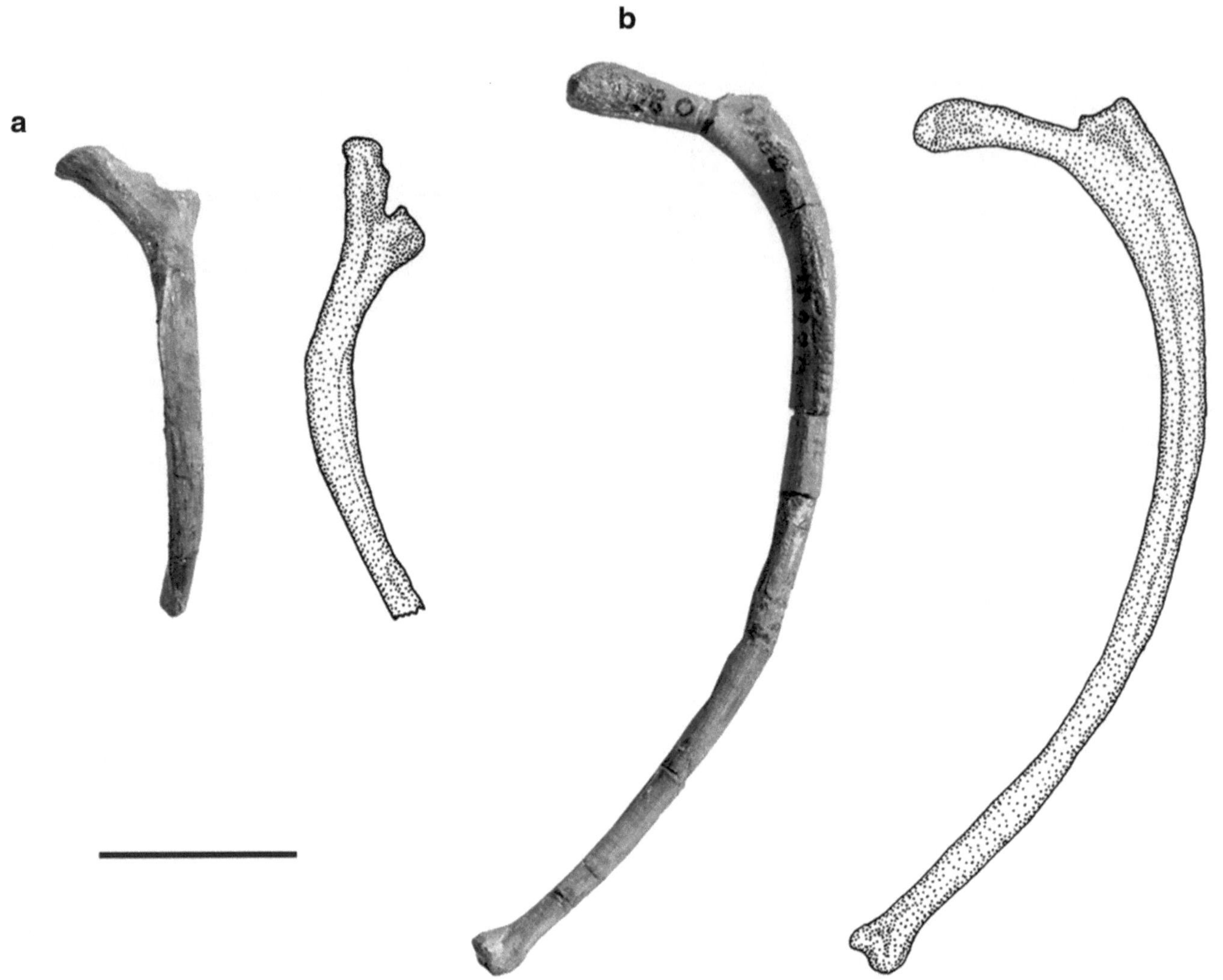

Fig. 12.10 Drawings and photographs of ribs of GPIT/RE/7113. **a** A left cervical rib in anterolateral and lateral view; **b** a left mid-dorsal rib in anterolateral view. Scale bar equals 30 mm

posterior margin and wear facets anterocoronally. There are at least four to five moderately sized postcanine teeth observable on the right side; however, the fifth might be a replacement tooth since it is situated medial to the fourth.

The lower jaw has the usual four incisors with recognizable wear facets and serrations on both margins. The first incisor is the largest whereas the remaining three have approximately the same length. The canine directly follows the last incisor. It is less crescent-shaped than the upper canine but also has serrations on both margins and wear facets on the coronal half of the anterior margin. On the right side three postcanine teeth are visible, although their tips are broken off. In front of the first postcanine tooth and between the first and second postcanine tooth, there remains an empty space 5 mm in width, but no alveolus is visible. On the right side, two apical parts of postcanine teeth are present and three alveolar roots are also visible. Thus it is possible that the number of postcanines in the lower jaw was five.

Postcranium

Pectoral Girdle and Forelimb—The pectoral girdle (Fig. 12.4; Table 12.2) is almost complete except for the cleithrum and sternum. The three endochondral bones, the scapula, coracoid and procoracoid, are fused, whereas the interclavicle and the two clavicles are disarticulated.

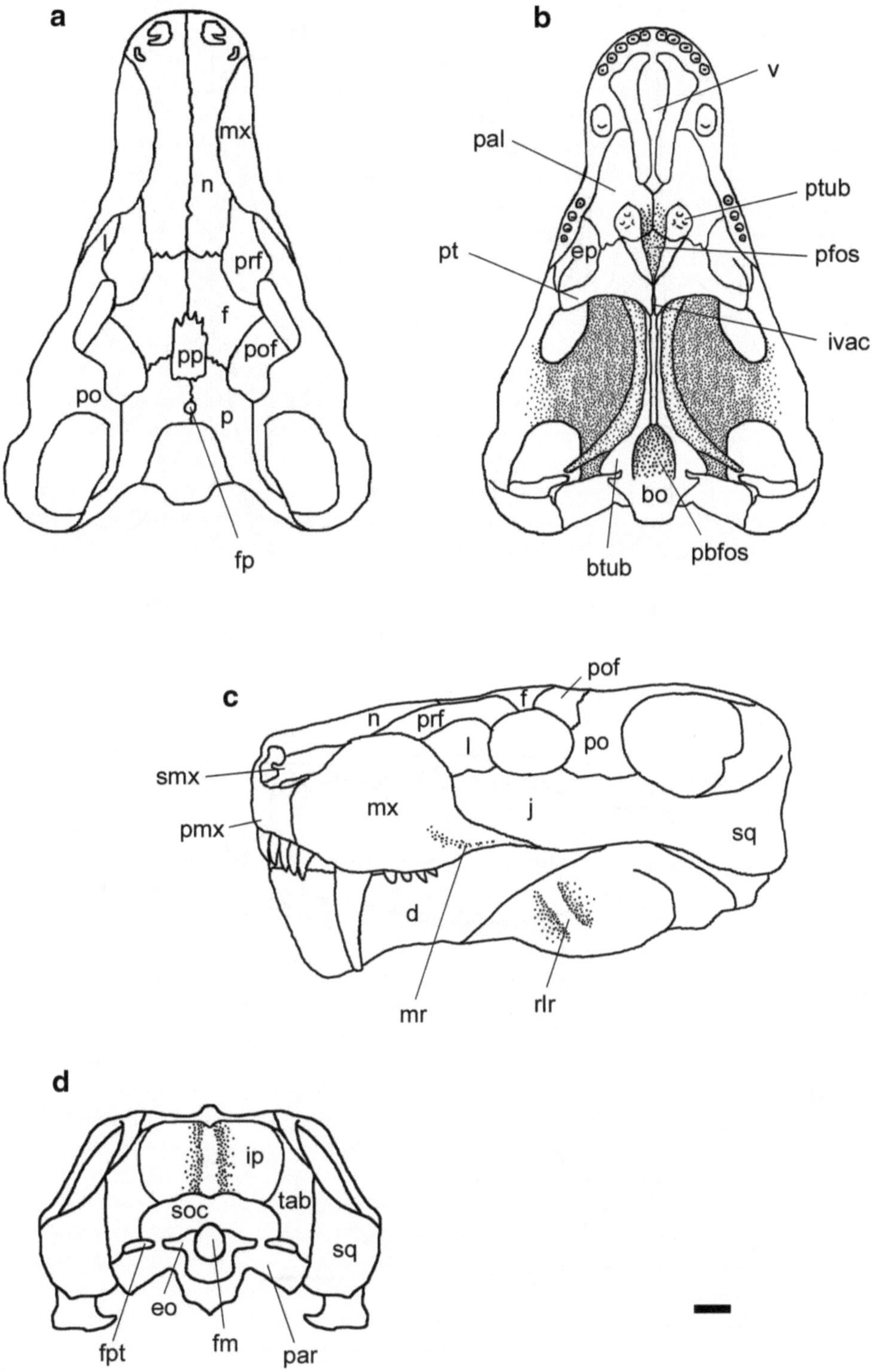

Fig. 12.11 Illustration of the characters in the genus *Aelurognathus* Haughton, 1924 based on the holotype of the type species, *Aelurognathus tigriceps* (SAM-PK-2342). **a** dorsal; **b** ventral; **c** left lateral; **d** occipital view. Scale bar equals 20 mm

The dorsal blade of the scapula is of medium thickness and only slightly curved medially. The glenoid is well preserved and forms two articulating facets that meet almost perpendicularly. The dorsal facet is higher than wide and is mostly formed by the scapula. The lower facet is wider than high with a rounded ventral margin and is exclusively formed

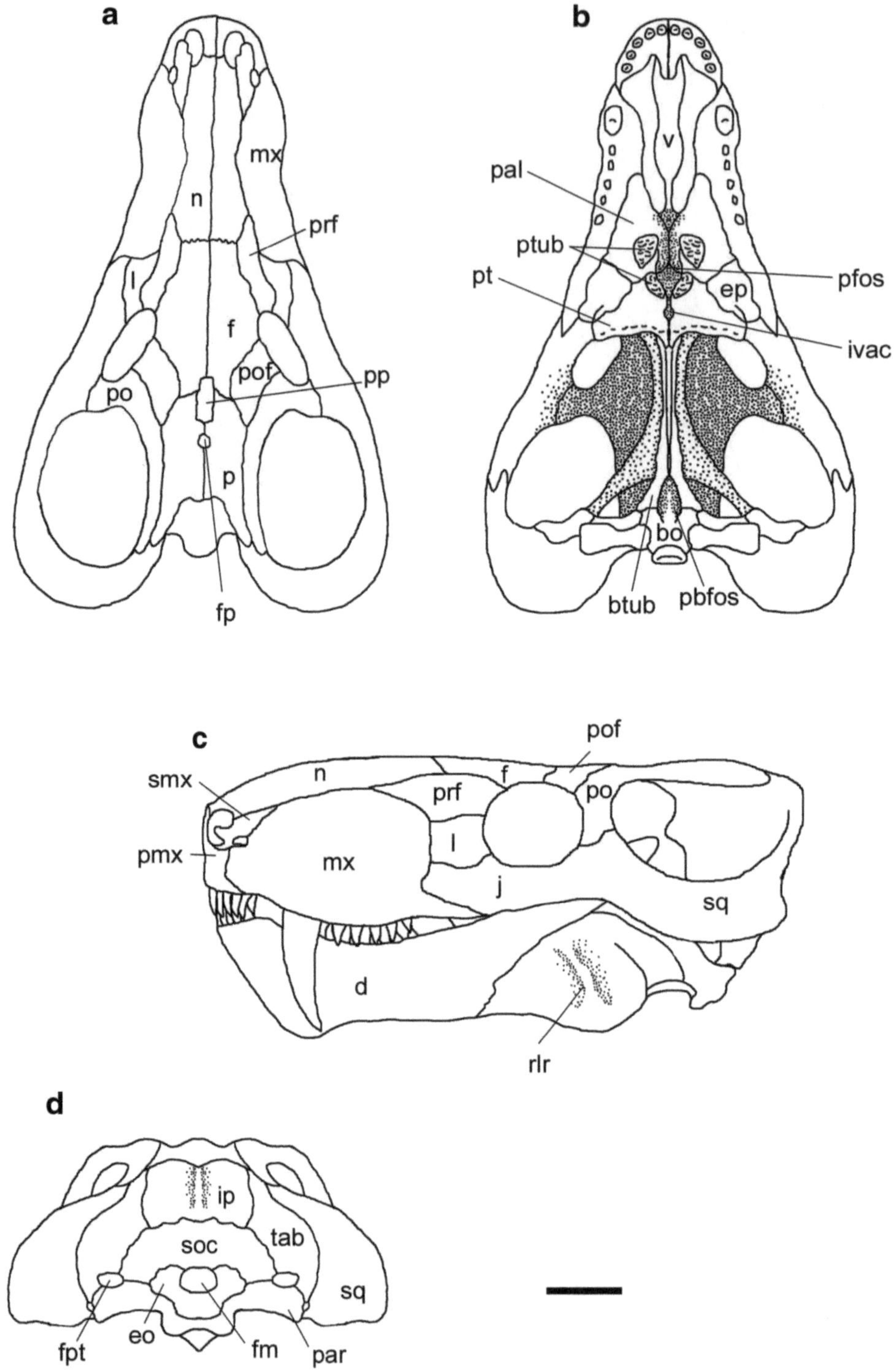

Fig. 12.12 Skull of *Sauroctonus progressus* (PIN 156/5) in **a** dorsal; **b** ventral; **c** left lateral; **d** occipital view. Scale bar equals 30 mm (Redrawn from Tatarinov 1974; Ivakhnenko 2001)

by the coracoid. The procoracoid foramen is entirely situated on the procoracoid. The medial face of the girdle is somewhat convex but posterior to the procoracoid foramen a broad elevation is established. The median keel of the interclavicle is well developed whereas the lateral parts, which are covered by the curved clavicles, are strongly rugose. The ventral face of the proximal extremity is covered with numerous ridges and grooves whereas the dorsal face is smooth.

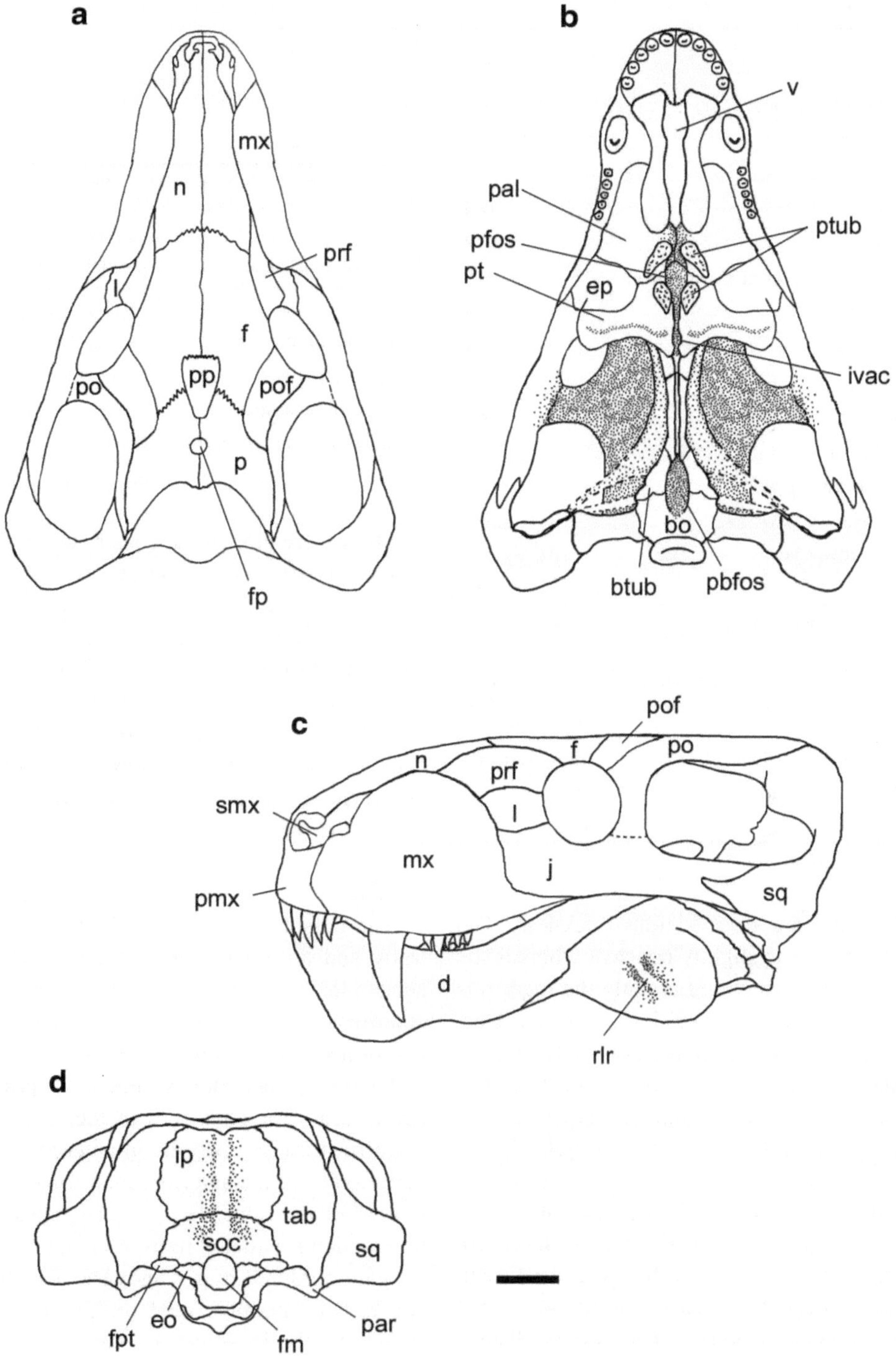

Fig. 12.13 Reconstruction of the skull of GPIT/RE/7113 in **a** dorsal; **b** ventral; **c** left lateral; **d** occipital view. Scale bar equals 30 mm

The anterior limb (Fig. 12.5) is completely preserved, at least on the left side. Only some components of the manus are missing on both sides.

The humerus is rather slender with the proximal and distal extremities only moderately expanded (Table 12.3). Both extremities are rotated at an angle of about 40°. The diaphysis shows an anteroposteriorly oval cross-section and is thus somewhat flattened dorsoventrally. The humeral head is convex, narrow and declines slightly posterodistally. Anteroventrally the deltopectoral crest is strongly rugose but comparatively weakly developed. Distally, the entepicondyle is flattened whereas the ectepicondyle is well rounded. The intercondylar fossa between them is rather shallow, but extends comparatively far medially.

Table 12.1 General measurements of the skull of GPIT/RE/7113

Skull length	250
Snout length	135
Length of posterior part of skull (from anterior margin of orbit to basioccipital condyle)	115
Length of palatine	150
Length of cranial basis	100
Height of snout	70
Width of snout (taken at the level of the canines)	45
Orbital width	65
Temporal width	75
Width of occiput	140
Height of the occiput	85
Length of a mandibular ramus	240
Height of symphysis of lower jaw	55
Length of symphysis of lower jaw	25
Width of symphysis of lower jaw	43
Height of dentary	37

All measurements in millimeters (mm)

Table 12.2 Measurements of the shoulder girdle of GPIT/RE/7113

Height of scapular blade	70 (deformed)
Width of posterior margin of scapular blade	65 (deformed)
Width of articular facet of glenoid	25
Length of interclavicle	105

All measurements in millimeters (mm)

The ulna and radius are of moderate thickness. The anterior surface of the ulna is slightly concave whereas the posterior surface is slightly convex. Laterally the margin is smoothly rounded, but medially a keel is established. Dorsally, the articular facet with the humerus is only slightly convex whereas the olecranon is broad but rather low. The ventral extremity is strongly rugose and less expanded.

The radius curves towards the ulna, especially in its ventral part. It is comparatively stout with massive proximal and distal extremities. The proximal extremity shows an oval cross-section whereas the distal one is more flattened.

The bones of the right manus are well-represented (with the exception of the claws). The ulnare is rather elongated and slightly constricted in the middle, whereas the radiale is square. Centrale 2 is situated between these two bones and has pointed proximal and distal ends. The intermedium is missing. The flat and wider than high centrale 1 is situated ventral to the radiale. Distal to centrale 1 is the mediolaterally oriented row of distal carpals. The fused fourth and fifth distal carpals are only slightly wider than the others (Sigogneau-Russell 1989). The metacarpals measure between 15 and 35 mm in length with the fourth and fifth being the longest. The first metacarpal is short and stout, the second longer and less constricted in the middle, the third and fourth are elongated, constricted in the middle and have expanded extremities, whereas the fifth, though also long, is relatively broad and flattened in the middle. The phalanges (as far as they are preserved) have the numbers: 2-3-4-5-3. Only the third claw is preserved: it is mediolaterally compressed and thus less distinctive.

Table 12.3 Measurements of the forelimb bones of GPIT/RE/7113

Length of humerus	170
Width of diaphysis of humerus	27
Length of ulna	138
Length of radius	125

All measurements in millimeters (mm)

Table 12.4 Measurements of the pelvis of GPIT/RE/7113

Length of dorsal margin of iliac blade	100
Height of ilium	90
Length of pubis	55
Length of ischium	100

All measurements in millimeters (mm)

Pelvic Girdle and Hind Limb—The pelvic girdle (Fig. 12.6; Table 12.4) is almost complete. Only the right pubis, the anteroventral and posteroventral parts of the left pubis, the ventral margin of the right ischium, and the dorsal, posterior and ventral margins of the left ischium are missing.

The dorsal margin of the iliac blade slopes posteroventrally and terminates in a broad posterior expansion. The acetabular crest is weakly developed. The posterior and anterior margins of the large acetabulum have the same length and form a broad triangle.

On the medial side, the iliac blade is slightly convex with strong scars for contact with the sacral ribs. A prominent ridge runs transversely in an anterodorsal to posteroventral direction from the anterior extension of the iliac blade to the middle of the bone. Here it terminates in a pointed elevation, which is situated in front of an elongate depression that is again bordered by a ridge. The acetabular part is convex and rather smooth except for the strongly rugose contact areas with the ischium and pubis.

The ischium is the most flattened of the three pelvic bones and only the acetabular part is slightly expanded. The blade narrows distinctly in a posteroventral direction. Again, the acetabular part is comparatively small and the ventral elevation is less well developed.

In the hind limb (Fig. 12.7; Table 12.5) both femora are completely preserved, as well as the right tibia, the proximal part of the left fibula and the distal part of the right fibula. The bones of the pes are incomplete, with only a few elements of the right preserved; these are, however, strongly weathered.

Table 12.5 Measurements of the hindlimb bones of GPIT/RE/7113

Length of femur	185
Width of diaphysis of femur	32
Length of tibia	142
Length of fibula	148

All measurements in millimeters (mm)

As in all other gorgonopsian taxa for which postcrania are known, the curved femur is longer and more slender than the humerus. The strongly rugose femoral head is only slightly expanded and rather flat, which may be due to preservation. The diaphysis is strongly curved and terminates in the two distal condyles, which are again rather flat. Both condyles have nearly the same size, which is in contrast to all other gorgonopsians where the medial condyle is usually somewhat larger.

The tibia and fibula are again rather flattened. The tibia is curved towards the fibula proximally, whereas the lower part is rather straight. Corresponding to this condition, the fibula is curved towards the tibia only with its distal extremity, which is in contrast to other taxa, where both bones are more strongly curved towards each other.

The components of the tarsus are also embedded in plaster and their arrangement is questionable. However, the bones are strongly weathered and as von Huene (1950) stated, all bones were found disarticulated so that he could only guess their correct arrangement. The largest bone, which has a concave dorsal surface, might be the calcaneum; medially another large bone, which is expanded more in a mediolateral direction, is probably the astragalus. Distal to this is a smaller bone that is wider than high and might be the centrale. None of the other elements of the tarsus are preserved.

The metatarsalia are complete. The first is short and wide, the second is narrower and has an expanded proximal extremity, the third has expanded proximal and distal extremities, and the fourth is long and slim, whereas the fifth is wide and not constricted in the middle. These are between 20 and 45 mm long, with the fourth being the longest. The phalanges are too incompletely preserved as to give any information about their numbers. The claws are not preserved.

Vertebral Column—The vertebrae are more or less completely preserved except for certain parts of some dorsals, and the caudals from the eighth caudal onwards (Figs. 12.8, 12.9).

The atlas intercentrum is a small and narrow crescentic element, and is somewhat more compact than the axis intercentrum (Sigogneau-Russell 1989). The atlas centrum is short and low in comparison to other taxa. The neural arch is somewhat T-shaped, with the horizontal bar forming the postzygapophysis. On the internal side a concave depression faces anteromedially (Sigogneau-Russell 1989). This facet might have embraced the dorsal parts of the basioccipital condyle.

The cervicals do not differ much in their morphology, except for the axis and the seventh cervical. The axis shows some characters of its own, whereas the latter more resembles the following dorsal vertebrae.

All cervicals, except for the seventh, are 25 mm long and 20 mm high, the seventh cervical being somewhat shorter and lower. These relations are observable in all other gorgonopsians. The anterior-posteriorly sloping centra are strongly amphicoelous. The lateral face of the axis is least depressed, whereas the anterior cervicals have an elongated depression on their ventral half, which becomes deeper and more rectangular in the posterior cervicals. In the last cervical, which is already like a dorsal, this depression is oriented in a dorsoventral direction.

The prezygapophyses are oval in an antero-posterior direction for all cervicals and do not meet each other in the middle. The articulating facets are almost horizontal in all cervicals, but get slightly steeper posteriorly. The postzygapophyses of the first five cervicals are fused in the middle.

The diapophysis merges into the ventral part of the prezygapophysis anterodorsally. The parapophysis is visible on the ventralmost edge of the centrum in the sixth cervical but presumably already existed in the fifth. The articular facet is surrounded by a low ridge, which leaves a slightly depressed area in the middle.

The neural spine measures 25 mm in the axis and gets somewhat higher in the following cervicals (30–35 mm). The fan-shaped neural spine of the axis is the most massive of the cervical spines. The neural spines of the following cervicals are steeply inclined postero-dorsally. In the third cervical the spine is shorter and more massive than in the following cervicals. All cervicals have a keel of variable sharpness on the anterior and posterior margins of their neural spines and the dorsal margin of the neural spine is slightly broadened. Compared to other gorgonopsian taxa, the general height of the spines in GPIT/RE/7113 seems to be rather short.

The dorsals again do not differ much from each other. The length of the vertebral centrum varies between 20 and 27 mm with most being 23 or 25 mm long; the height of the centrum varies between 20 and 25 mm and the neural spines measure between 30 and 40 mm in height.

The characteristic spool-shaped appearance is even more pronounced than in the cervicals. The zygapophyses are steep and both the pre- and postzygapophyses are oval and oriented in an antero-posterior direction, the posterior ones being slightly larger. The prezygapophyses reach freely in an anterodorsal direction whereas the postzygapophyses are directly attached laterally to the neural spine, which emerges in the middle.

The insertion area of the diapophysis is smaller than in the cervicals. The parapophysis, which inserts into the first dorsal approximately at mid-height of the vertebral body, has reached the dorsal margin of the vertebral centrum in the fifth dorsal and stretches beyond the dorsal margin of the body at latest at the tenth. In the seventeenth dorsal the parapophysis is still visible. It is not clear if the next two dorsals, which are the last of the series, have parapophyses since these two vertebrae are heavily weathered.

The steeply inclined neural spines are slender with a rounded dorsal margin and measure approximately 40 mm in the first nine dorsals. From the tenth dorsal onwards the neural spine broadens anteriorly and tapers in a dorsal direction. The anterior and posterior margins of all neural spines show a keel. Again, the spines of GPIT/RE/7113 are rather short but also massive.

The massive centrum of the first sacral measures 25 mm in length and 20 mm in height. The prezygapophyses are less steep than in the dorsals, whereas the oval postzygapophyses are again rather steep and far apart from each other. The transverse process is strongly expanded laterally and forms a massive sacral rib. The insertion area with the sacral rib covers almost the entire centrum and the ventral part of the neural arch between the zygapophyses. The neural spine is 30 mm high and narrow. Its anterior margin is inclined less steeply than the almost vertical posterior margin but has the same shape as in the last dorsals.

The centrum of the second sacral vertebra is as large as the first but the transverse process is smaller, less massive and exclusively oriented in a lateral direction. The insertion area for the sacral rib covers only the dorsal part of the vertebral centrum and the ventralmost part of the neural arch. The oval zygapophyses are oriented at a nearly 45° angle. The third sacral vertebra is much smaller than the preceding two. The transverse process resembles the shape of the first sacral almost perfectly, it is only more slender and smaller.

Only the first seven caudals are preserved, some of which are partially incomplete. The first caudal differs noticeably from the others. The vertebral centrum is square as in the sacrals, and it is larger and more massive than the following caudals. The neural spine is again as high and slender as in the anterior dorsals whereas the zygapophyses are less steep and the transverse process is massive but shortened and still of triangular section. The second caudal is almost completely restored in plaster, the third is fairly weathered but the vertebral centrum is smaller and more flattened than the others, which might be due to deformation. The transverse process remains massive but short. The following caudals are much smaller. The zygapophyses remain oval but get more steeply oriented further posteriorly. The postzygapophyses reach beyond the level of the posterior margin of the neural spine. The oval transverse processes become more slender and shorter until they only form small knots before they disappear completely at the tenth caudal. The pointed neural spines rapidly decrease in height.

Ribs—Although the skeleton of GPIT/RE/7113 is mounted with ribs, it is difficult to tell if these are reconstructed in the correct places since they were probably found disarticulated. Furthermore they are often broken and incomplete. Nevertheless the cervical and the anterior dorsal ribs are more complete and all have two articular heads. The posterior dorsal ribs are restored in plaster so that there is no information about the point when they become single headed. All ribs are strongly curved medially and markedly shorter in the cervical region (Fig. 12.10). Fractures and missing ends in the lumbar region prevent useful information on their length.

Discussion

Comparison with Aelurognathus *Haughton, 1924*

In order to clarify the unresolved taxonomic position of GPIT/RE/7113, it is necessary to look at the taxon *Aelurognathus* in more detail. The type species of this genus is *Aelurognathus tigriceps* (Broom and Haughton, 1913), with SAM-PK-2342 as the holotype (Fig. 12.11). Compared to this specimen, GPIT/RE/7113 displays many characters that clearly distinguish it from this and all other species of *Aelurognathus*. This is shown by Sigogneau's (1970) diagnosis for the genus *Aelurognathus*, which does not apply for GPIT/RE/7113 in many points, including: heavy skull (more gracile in GPIT/RE/7113), high temporal fossa (lower and rectangular in GPIT/RE/7113), wide interorbital space (more narrow in GPIT/RE/7113), thick suborbital and postorbital bar (more gracile in GPIT/RE/7113), narrow supraorbital portion of the frontal (wide in GPIT/RE/7113), high occiput (low in GPIT/RE/7113), massive and thick dentary (more gracile in GPIT/RE/7113).

Personal observation of the specimen indicates that GPIT/RE/7113 lacks all synapomorphies of *Aelurognathus* and is clearly distinguishable from this taxon. There are clear differences in palatal morphology between GPIT/RE/7113 and *Aelurognathus tigriceps*: the palatal tuberosities are confluent and V-shaped in *Aelurognathus*, but oval in GPIT/RE/7113, with the tuberosities on the palate separated from the ones on the pterygoid. GPIT/RE/7113 has teeth on the transverse process of the pterygoid, whereas *Aelurognathus* has none, and in GPIT/RE/7113 the vomer is narrow anteriorly but broad in *Aelurognathus*.

The holotype of *Aelurognathus tigriceps* (SAM-PK-2342) also differs significantly from GPIT/RE/7113 in other regards. The skull of SAM-PK-2342 is more massive and higher, the snout is heavier, and the posterior skull region is broader compared with GPIT/RE/7113. Almost all diagnostic bones differ as well. Thus, in *A. tigriceps* the posterior process of the maxilla is markedly elongated, the prefrontal is short, high and raised, the naso-frontal suture is situated posteriorly and is straight, the supraorbital portion of the frontal is small, the postfrontal is broad and large, the palatal fossa is wide, and the posterior para-basisphenoidal area is broad. Furthermore, the teeth are larger and more massive, the symphysis of the lower jaw is straighter and heavier, and the reflected lamina is stronger and situated more anteriorly than in GPIT/RE/7113. Additionally the genus *Aelurognathus* shows some characters such as the extremely convex ventral margin of the maxilla, the ridge on the maxilla postero-dorsal to the postcanine teeth, and the dorso-laterally constricted snout, which are absent in GPIT/RE/7113.

Comparison of GPIT/RE/7113 with Other Taxa

Cranium—The comparison is based on both literature and personal observation. First of all, thorough study of the descriptive literature provided a sound base for further studies. In a second step I personally investigated a large number of gorgonopsian specimens in collections worldwide (Gebauer 2007). The list below notes holotypes and referred material examined for specific comparisons; unless specified otherwise original specimens were examined firsthand:

Aelurosaurus felinus NHMUK R339

Aloposaurus gracilis AMNH FARB 5317

Cyonosaurus longiceps FMNH UC 1515 (cast); referred material: BP/1/137, BP/1/735, BP/1/2598

Scylacognathus parvus AMG 3751

Arctops willistoni NHMUK R4099; *Arctops watsoni* BP/1/698

Gorgonops torvus NHMUK R1647; referred material: AMNH FARB 5515, BP/1/1992

Eoarctops vanderbyli SAM-PK-5598

Lycaenops ornatus AMNH FARB 2240

Sycosaurus laticeps SAM-PK-4022

Clelandina rubidgei RC 57

Rubidgea atrox RC 13

Inostrancevia alexandri PIN 2005/1587 (literature); PIN 2005/1858 (personally)

Sauroctonus progressus PIN 156/5 (literature)

Suchogorgon golubevi PIN 4548/1, PIN 4548/10 (literature)

Viatkogorgon ivakhnenkoi PIN 2212/61 (literature)

The small-sized genera such as *Aelurosaurus*, *Aloposaurus* and *Cyonosaurus* differ in size and proportions from GPIT/RE/7113. Comparison with *Aelurosaurus* is further hampered by the fact that specimens of that taxon might be immature or sub-adult, which is in contrast to Owen (1881). Although the holotype and other specimens were referred to by several authors (Broom 1910, 1932; Watson 1912; Boonstra 1934), only Sigogneau (1970) remarked on the probable immaturity of the specimens. I concur with that view because the holotype in particular shows many features such as the short snout, large eyes, tooth replacement and an open symphysis which indicate an immature state. Overall the taxon differs greatly from GPIT/RE/7113 by its narrow posterior part of the skull, the straight occiput and the slender mandible. *Aloposaurus* shows a large preparietal and confluent parietal tuberosities, which are not present in GPIT/RE/7113. Although a probable immature state could be supposed for *Aloposaurus* as well, as already suggested by Broom (1932), the long snout, small orbits, small parietal foramen and the relatively high dentary are opposed to this. *Cyonosaurus* differs from GPIT/RE/7113 by its remarkably long snout (Olson 1937) and the postorbital which reaches far ventrally on the postorbital bar.

The snout in *Scylacognathus* and *Arctops* is sloping with a distinct 'bend' at the level of the anterior margin of the prefrontal, which distinguishes them from all other gorgonopsians. Additionally their interorbital and intertemporal widths are proportionally greater than in GPIT/RE/7113 (Sigogneau 1970).

The skull in *Gorgonops* is lower (Watson 1921) and the palatal tuberosities are elongate V-shaped with the tips pointing anteriorly, which distinguishes this taxon from all gorgonopsians other than Eriphostoma (Kammerer 2013). Additionally the contribution of the frontal to the orbital rim is much narrower than in GPIT/RE/7113.

The taxon *Eoarctops* differs from GPIT/RE/7113 by its shorter snout and broader temporal region. Further dissimilarity is the large and almost perfectly round preparietal which is a form that is truly unique in all gorgonopsians.

Lycaenops shows a remarkably deep skull (Broom 1925; Colbert 1948) with a short and high septomaxilla. The taxon differs from GPIT/RE/7113 as well by its quadrangular temporal opening and the confluent palatal tuberosities.

The rubidgeine genera are differentiated by the downturned zygomatic arch, the frontal not reaching the orbital rim, palatal teeth only present on the tuberosities of the palatine, no preparietal, strongly concave occiput and other characters that constitute this subfamily. The skull of *Sycosaurus* is generally more robust and is furthermore distinguished by a low supraoccipital and an interparietal that is considerably wider than high.

These comparative studies on a large number of specimens show that an allocation to the South African genera is precluded because GPIT/RE/7113 shares an insufficient number of characters with any described genus.

There is little information about the geographic dispersal of Russian and East African gorgonopsian forms. This is most likely due to the rarity of terrestrial Permian deposits in these parts of the world. However, over the years, some close relationships between a Russian form and a taxon from the eastern part of Africa or from South Africa have been reported such as the dinocephalian genera *Ulemosaurus* and *Moschops* (Battail and Surkov 2000) and the cosmopolitan genus *Lystrosaurus* with the Russian species *L. georgi* (Surkov et al. 2005).

In this study, the Russian taxa *Inostrancevia, Suchogorgon, Viatkogorgon* and *Sauroctonus* are considered. *Inostrancevia* has an elongated lacrimal, a narrow vomer, a sloping symphysis of the lower jaw and no supraorbital thickening. Furthermore, *Inostrancevia* differs remarkably from GPIT/RE/7113 with its large and massive skull with entirely dissimilar proportions and considerably larger canines. The two Russian taxa *Suchogorgon* and *Viatkogorgon* display more similarity to some extent but each taxon differs from GPIT/RE/7113 nevertheless: *Suchogorgon* has small orbits, the vomer is not narrowed in the middle part, and the postcanine teeth are extremely small (Tatarinov 2000). The upper canine of *Viatkogorgon* is directed in an anterior direction, the snout is high and narrow and the lacrimal wedges into the maxilla to a great extent (Tatarinov 1999). This leaves *Sauroctonus* as the only gorgonopsian genus matching the condition in GPIT/RE/7113, as will be detailed further below.

Postcranium—Postcranial material is rather limited in gorgonopsians in general and it is therefore questionable that the material at hand covers a range of taxa large enough to provide a sound basis for comparison. The available postcranial material shows that the differences observed between taxa are even less discernable than distinguishing characters of the skull.

A list of this material is given below:

Aelurognathus tigriceps SAM-PK-2342—pectoral girdle, humerus, hand

Arctognathus breviceps SAM-PK-9345—pectoral girdle, forelimb, vertebrae

Inostrancevia alexandri PIN 2005/1578—almost complete postcranial skeleton

Lycaenops? *microdon* SAM-PK-9344—humerus, pelvic girdle, hindlimb, vertebrae

Lycaenops ornatus AMNH FARB 2240—almost complete postcranial skeleton

Scylacops capensis SAM-PK-2343—pectoral girdle, humerus, anterior caudals

Gorgonops cf. *G. whaitsi* BSP 1934 VIII 28—almost complete postcranial skeleton

Gorgonopsia indet. CAMZM 883—almost complete postcranial skeleton

The scapular blade of the pectoral girdle in GPIT/RE/7113 is somewhat comparable with that of *Lycaenops ornatus*. Thus it is broader than in *Scylacops capensis* or *Arctognathus breviceps*, but narrower than in *Aelurognathus tigriceps* or *Inostrancevia*. The glenoid is larger than in *Scylacops capensis* but not as large as in *Lycaenops ornatus*. The length of the posterior extension of the coracoid is intermediate between *Arctognathus breviceps* and BSP 1934 VIII 28. The curved clavicle is more slender than in *Lycaenops ornatus* and *Inostrancevia*. The humerus is less stout in GPIT/RE/7113 than in other gorgonopsians, except *Scylacops capensis*, *Lycaenops ornatus* and SAM-PK-9344. The expansion of the dorsal extremity is intermediate between SAM-PK-9344 and *Lycaenops ornatus* in one instance and *Inostrancevia alexandri* and *Aelurognathus tigriceps* in the other. The diaphysis, which has an oval cross-section, is again comparable in shape with SAM-PK-9344 and *Lycaenops ornatus*. The distal condyles are less developed than in other taxa except for *Scylacops capensis*. The ulna and radius of GPIT/RE/7113 do not differ greatly from other taxa but are stouter than in *Lycaenops ornatus* and less massive than in *Arctognathus breviceps*.

The dorsal margin of the iliac blade slopes to a greater extent than in CAMZM 883 or SAM-PK-9344, its posterior expansion is, however, broader and less restricted ventrally than in SAM-PK-9344. The blade of the ischium narrows distinctly in a postero-ventral direction, which is in contrast to *Lycaenops ornatus* and CAMZM 883, but comparable to SAM-PK-9344. The femur of GPIT/RE/7113 is broader than that of *Lycaenops ornatus* and SAM-PK-9344 but not as massive as in CAMZM 883, BSP 1934 VIII 28 or *Inostrancevia alexandri*. The greater trochanter is offset only slightly from the rest of the bone and thus only comparable with BSP 1934 VIII 28. The tibia and fibula do not differ much from the other taxa in general shape with two exceptions for the tibia: in SAM-PK-9344 it is more slender and in *Inostrancevia alexandri* the proximal and distal extremities are extremely massive.

The shape of the atlas-axis complex of GPIT/RE/7113 is comparable with that observed in CAMZM 883, whereas the spine is lower and more expanded posteriorly in *Arctognathus breviceps* and higher in SAM-PK-9344. The bodies of all vertebrae do not differ much in the various taxa, but the length and orientation of the spine does. In GPIT/RE/7113 the spine is of intermediate height and the direction changes from vertical to posterodorsal. In *Arctognathus breviceps* and CAMZM 883 the spine is rather short and sloping. SAM-PK-9344 and *Inostrancevia alexandri* on the other hand have a long and slender spine, which is directed more dorsally than posterodorsally. In BSP 1934 VIII 28 and CAMZM 883 the vertebral body of the first sacral is also larger and more massive than the following caudals, whereas the broadness of the sacral rib is even more accentuated in those two specimens than in GPIT/RE/7113.

The comparison of postcranial material shows that GPIT/RE/7113 holds an intermediate position between the available taxa concerning features of the postcranial skeleton. It is obvious that its postcranial skeleton is more slender in its overall appearance than in the large species *Aelurognathus tigriceps* and *Inostrancevia alexandri*. Concerning the other taxa, however, there is no clear distinction possible. At best, considering only the available postcranial material, GPIT/RE/7113 may be most closely related to SAM-PK-9344 and *Lycaenops ornatus*.

Systematic Position

History of GPIT/RE/7113—The specimen GPIT/RE/7113 was first described by von Huene (1950) who placed it in the genus '*Scymnognathus*' as the new species *S. parringtoni*, mentioning a close relationship with the taxa *Aelurognathus* and '*Pachyrhinos*' (=*Gorgonops kaiseri*). von Huene (1950) saw the "shape of a typical *Scymnognathus*" because of the convex and ascending snout profile. However, the genus *Scymnognathus* never was clearly diagnosed with respect to other similar forms. This is evident by the fact that Sigogneau (1970) in her taxonomic reassessment of the Gorgonopsia placed the type species *Scymnognathus whaitsi* in the genus *Gorgonops* while distributing the remaining species between the genera *Lycaenops* and *Aelurognathus*. Furthermore von Huene's (1950) allocation was not based on specific characteristics and thus was poorly documented. In his description, von Huene (1950) remarked that the anterior margin of the orbit was situated in the middle of the skull length, the maxilla was high, the canine long and slender and the five to six postcanines were only slightly smaller than the incisors. Von Huene (1950) was also of the opinion that the frontal did not reach the orbit, which is clearly not the case. Indeed the contribution of the frontal to the suborbital rim is large. Von Huene (1950) stated that there was no step in the ventral maxillary border; that the maxillary, frontals and prefrontals were intensely sculptured; that the orbits were not covered with tuberosities; the postorbital bar and suborbital arches were slender; and the transverse process of the pterygoid was long. In the following detailed account of the single bones he mentioned the large and steeply oriented basioccipital tubera, the high interparietal, the posteriorly long parietal, the small parietal foramen which is surrounded by a ridge and is well separated from both the occipital crest and the preparietal. Furthermore, von Huene (1950) stated that the postorbital was long posteriorly, the septomaxilla narrow, the choanae elongate, the vomer narrow and the dentary symphysis relatively massive with a chin.

Sigogneau (1970) placed GPIT/RE/7113 into the genus *Aelurognathus*, although conditionally. Sigogneau (1970) did not see a close connection of GPIT/RE7113 with the genus *Gorgonops* because of the lower snout and the heavier skull arches in that genus and thus discussed a possible relationship with the genera *Aelurognathus* and *Lycaenops*. According to Sigogneau (1970), GPIT/RE/7113 shared with *Lycaenops* the slender skull arches and the convex profile of the snout. But the interorbital and intertemporal widths were the same in *Aelurognathus* as well as the small size of the orbits and the shape of the dentary. Although Sigogneau (1970) admitted that the temporal fossa was rather elongate, which would not fit the definition for the genus *Aelurognathus*, she emphasized the close resemblance in the postcranial material with *Aelurognathus tigriceps*. The cervicals had high neural arches in both GPIT/RE/7113 and *Aelurognathus tigriceps* and the anterior limbs were similar in proportions. Sigogneau (1970) considered these features sufficient for allocating GPIT/RE/7113 to the genus *Aelurognathus* as the species *A. parringtoni*.

History of the taxon *Sauroctonus progressus* (Hartmann-Weinberg, 1938)—The holotype PIN 156/5 was first described by Hartmann-Weinberg (1938) and allocated to the genus *Arctognathus* as a new species, *A. progressus*. Bystrow (1955) redescribed PIN 156/5 after re-preparation, and placed *A. progressus* in the new genus *Sauroctonus*. He mentioned the strongly concave occiput with a strong median keel and a small contribution of the parietal to the occipital rim, high maxilla, small prefrontal, massive paroccipital process, small ectopterygoid, teeth on the palatal tuberosities, broad and long splenial, and triangular coronoid. Tatarinov (1974) worked on the taxon again and contributed a particularly detailed account in his monograph on 'Theriodonts of the USSR', although much of this concerned nerve opening and blood vessel positions of little comparative value. He made some important new comparative contributions, however; he mentioned a preparietal, which was not recognized by either Hartmann-Weinberg (1938) or Bystrow (1955). Furthermore, Tatarinov (1974) stated that the supraorbital portion of the frontal was comparatively large, the prefrontal was longer at the orbit than figured by Bystrow (1955), the vomer was long and narrow, the palatal fossa was rather narrow, the supraoccipital was broad, and the interparietal was broader ventrally than dorsally. Finally he gave some measurements of the skull: total length 225 mm, preorbital length 110 mm, broadest width (in the temporal region) 150 mm, snout height 68 mm.

Sigogneau-Russell (1989) listed the genus and species as *Sauroctonus progressus* and noted the posteriorly narrow skull, the elongate temporal fossa, the small orbits and narrow skull arches, the very narrow interorbital and intertemporal spaces and the moderately high dentary. Finally, Ivakhnenko (2003) remarked on the high palatal tuberosities that have numerous teeth, the only slightly sculptured bones, and incisors that were only slightly larger than the postcanine teeth. Each author additionally provided a number of illustrations which are shown here as a composite in Fig. 12.12.

Conclusions—Although none of the previous authors ever mentioned GPIT/RE/7113 in connection with the taxon *Sauroctonus*, the similarity between the two specimens is overwhelming. For comparison see Figs. 12.12 and 12.13.

In both forms the skull is slender and the posterior part of the skull is only moderately enlarged, which means that the squamosal is flaring less laterally than in *Aelurognathus*. The sloping snout is narrow and somewhat higher than wide with the external nares situated ventrally. Posteriorly the skull roof is straight. The round orbit is rather small but well exposed in dorsal view and the temporal fossa is clearly elongate. In GPIT/RE/7113 the interorbital and intertemporal spaces are wider than in *Sauroctonus progressus*. They are, however, unusually narrow in this taxon, which may well be subject to deformation. Nevertheless, the difference in the width exists but this might be specific. In both GPIT/RE/7113 and *Sauroctonus* the ventral border of the maxilla is only slightly convex and the septomaxilla is rather narrow. The nasal is somewhat constricted in the middle with the naso-frontal suture situated anteriorly and slightly bow-shaped. The prefrontal is distinctly elongate and low and terminates anteriorly in a narrow process. The elongate lacrimal has no antorbital depression. On the dorsal skull roof the contribution of the frontal to the supraorbital margin is rather large whereas the preparietal is of medium size. In *Sauroctonus progressus* it is figured as extremely small and narrow by Tatarinov (1974) but somewhat larger in Ivakhnenko (2002). It seems, however, that the skull surface is rather weathered in this area and thus the delineation of the bone might be rather difficult and/or beyond recognition. The postfrontal is narrow in both forms but seems to be shorter in *Sauroctonus progressus*; its posterior margin is straight. Laterally the anterior squamosal process on the zygomatic arch reaches only to the middle of the temporal opening. All three skull arches are comparatively slender with the suborbital and zygomatic arch only slightly curved. The palates of the two taxa exhibit considerable similarities to one another as well. The vomer is slender throughout its entire length, a character that is shared only with *Aloposaurus*, *Cyonosaurus* and *Aelurognathus*. This indicates a rather plesiomorphic condition which is established by comparison with *Biarmosuchus* and *Herpetoskylax*. The palatine is small and the ectopterygoid is wider than long. The moderately broad palatal fossa is bordered by well-developed tuberosities that are separated from each other by a groove. Both tuberosities have numerous teeth with the palatal ones being much larger than those of the pterygoid. The tooth-bearing transverse processes of the pterygoid are posteriorly situated, somewhat more in GPIT/RE/7113 than in *Sauroctonus progressus*. The basisphenoid tubera are elongate and narrow, as is the basisphenoid fossa. The quadrangular occiput is rather convex with a well-developed median ridge and a massive paroccipital process. The lower jaw differs somewhat in the two forms since the symphysis is more massive in GPIT/RE/7113 than in *Sauroctonus progressus*. Nevertheless it is somewhat sloping in both taxa. In conclusion, it can be stated confidently that despite the geographic separation of *Sauroctonus* and GPIT/RE/7113 the latter should be included in this genus as the species *Sauroctonus parringtoni*.

Acknowledgments Sincere thanks go to the following who have supported this study by providing access to specimens in their care, discussion and assistance: Jenny Clack (Cambridge), Roger Smith (Cape Town), Bruce Rubidge and Mike Raath (Johannesburg), Carl Mehling (New York), Tom Kemp (Oxford), Denise Sigogneau-Russell (Paris), and Wolf-Ernst Reif and Frank Westphal (Tübingen). Financial support of the Landesgraduiertenförderung Baden-Württemberg is gratefully acknowledged.

References

Battail, B., & Surkov, M. V. (2000). Mammal-like reptiles from Russia. In M. J. Benton, M. A. Shishkin, D. M. Unwin, & E. N. Kurochkin (Eds.), *The age of dinosaurs in Russia and Mongolia* (pp. 86–119). Cambridge: Cambridge University Press.

Boonstra, L. D. (1934). A contribution to the morphology of the Gorgonopsia. *Annals of the South African Museum, 31*, 137–174.

Broom, R. (1905). On the use of the term Anomodontia. *Records of the Albany Museum, 1*, 266–269.

Broom, R. (1910). Observations on some specimens of South African fossil reptiles preserved in the British Museum. *Transactions of the Royal Society London, 2*, 19–25.

Broom, R., & Haughton, S. H. (1913). On a new species of Scymnognathus (S. tigriceps). *Annals of the South African Museum, 12*, 26–35.

Broom, R. (1925). On some carnivorous therapsids. *Records of the Albany Museum, 3*, 309–326.

Broom, R. (1932). *The mammal-like reptiles of South Africa and the origin of mammals*. London: H. F. & G. Witherby.

Bystrow, A. P. (1955). A gorgonopsian from the Upper Permian beds of the Volga. *Voprosy Paleontologii, 2*, 7–18.

Colbert, E. H. (1948). The mammal-like reptile *Lycaenops*. *Bulletin of the American Museum of Natural History, 89*, 357–404.

Gebauer, E. V. I. (2007). *Phylogeny and evolution of the Gorgonopsia with a special reference to the skull and skeleton of GPIT/RE/7113* ('Aelurognathus' parringtoni). Unpublished Ph.D. thesis, Eberhard-Karls Universität Tübingen.

Hartmann-Weinberg, A. (1938). Gorgonopsians as time indicators. *Problemi Paleontologii, 4*, 47–123.

Haughton, S. H. (1924). Investigations in South African fossil reptiles and Amphibia. 12. On some gorgonopsian skulls in the collection of the South African Museum. *Annals of the South African Museum, 12*, 499–518.

Huene, F. von. (1950). Die Theriodontier des ostafrikanischen Ruhuhu-Gebietes in der Tübinger Sammlung. *Neues Jahrbuch der Geologie und Paläontologie, Beilagen- Band, 92*, 47–136.

Ivakhnenko, M. F. (2001). Tetrapods from the East-European Placket – Late Paleozoic natural territorial complex. *Trudy Paleontologicheskogo Instituta, Akademiya Nauk* SSSR, *283*, 1–200.

Ivakhnenko, M. F. (2003). The features of lower jaw articulation in the gorgonopian *Suchogorgon* (Therapsida). *Paleontological Journal, 37*, 48–52.

Ivakhnenko, M. F. (2002). Taxonomy of East European gorgonopians (Therapsida). *Paleontological Journal, 36*, 283–292.

Kammerer, C. F. (2013). A redescription of Eriphostoma microdon Broom, 1911 (Therapsida, Gorgonopsia) from the *Tapinocephalus* Assemblage Zone of South Africa and a review of Middle Permian gorgonopsians. In C. F. Kammerer, K. D. Angielczyk, & J. Fröbisch (Eds.), *Early evolutionary history of the Synapsida* (pp. 171–184). Dordrecht: Springer.

Lydekker, R. (1890). *Catalogue of the fossil Reptilia and Amphibia in the British Museum (Natural History). Part IV. Containing the orders Anomodontia, Ecaudata, Caudata, and Labyrinthodontia; and supplement*. London: Trustees of the British Museum (Natural History).

Olson, E. C. (1937). The cranial morphology of a new gorgonopsian. *Journal of Geology, 45*, 511–527.

Owen, R. (1881). On the order Theriodontia with a description of a new genus and species (*Ælurosaurus felinus*, Ow.). *Quarterly Journal of the Geological Society London, 37*, 261–265.

Seeley, H. G. (1894). Researches on the structure, organisation and classification of the fossil Reptilia.—Part IX., Section 1. On the Therosuchia. *Philosophical Transactions of the Royal Society London, B, 185*, 987–1018.

Sigogneau, D. (1970). *Révision systematique des Gorgonopsiens sud-africains. Cahiers de Paléontologie*. Paris: Centre National de la Recherche Scientifique.

Sigogneau-Russell, D. (1989). Theriodontia I. In P. Wellnhofer (Ed.), *Handbuch der Paläoherpetologie* (Vol. 17 B/I). Stuttgart: Gustav Fischer Verlag.

Surkov, M. V., Kalandadze, N. N., & Benton, M. J. (2005). *Lystrosaurus georgi* a dicynodont from the Lower Triassic of Russia. *Journal of Vertebrate Paleontology, 25*, 402–413.

Tatarinov, L. P. (1974). Terriodont of USSR. *Trudy Paleontologicheskogo Instituta, Akademiya Nauk SSSR, 143*, 1–226.

Tatarinov, L. P. (1999). A new gorgonopid (Reptilia, Theriodontia) from the Upper Permian of the Vologda region. *Paleontologicheskii Zhurnal, 1*, 70–78.

Tatarinov, L. P. (2000). New Theriodonts (Reptilia) from the Late Permian Fauna of the Kotel'nich locality of the Kirov Region. *Paleontologicheskii Zhurnal, 5*, 76–82.

Watson, D. M. S. (1912). On some reptilian lower jaws. *Annals and Magazine of Natural History, 9*, 293–330.

Watson, D. M. S. (1921). The bases of classification of the Theriodontia. *Proceedings of the Zoological Society of London, 1*, 34–98.

Chapter 13
New Material of *Microgomphodon oligocynus* (Eutherapsida, Therocephalia) and the Taxonomy of Southern African Bauriidae

Fernando Abdala, Tea Jashashvili, Bruce S. Rubidge, and Juri van den Heever

Abstract An exceptionally well-preserved specimen of the bauriid therocephalian *Microgomphodon oligocynus* from the Burgersdorp Formation (Early-Middle Triassic, *Cynognathus* Assemblage Zone) of the South African Karoo is described. In addition, a taxonomic revision of bauriid therocephalians from southern Africa, based on firsthand examination of almost all know specimens, is presented. *Microgomphodon oligocynus* and *Bauria cynops* are recognized as the only valid species of southern African bauriids. *Microgomphodon oligocynus* is differentiated from *B. cynops* on the basis of clear-cut morphological features such as the presence of a complete postorbital bar, pineal foramen, contribution of the vomer to the osseous secondary palate, comparatively large orbits, presence of a lateral fossa on the posterior portion of the horizontal ramus and on the coronoid process of the dentary, and reduced number of postcanines. Procrustes analysis of the two best-preserved specimens of these species allowed recognition of further shape differences: *M. oligocynus* has a taller but narrower cranium, taller snout, temporal opening more expanded laterally, pterygoid process located more anteriorly, and smaller suborbital vacuity. The mandible of *M. oligocynus* has a higher symphysis, relatively short corpus, and more laterally-directed coronoid process. *Microgomphodon oligocynus* is known from the Olenekian to what are probably late Anisian levels in South Africa and Namibia, whereas *B. cynops* is restricted to the early Anisian of South Africa.

Keywords Karoo • Namibia • Theriodontia • Triassic • Procrustes analysis

Introduction

Therocephalians are a morphologically varied group of advanced therapsids that are well-represented in the Permo-Triassic South African Karoo Basin. This group was taxonomically very diverse during the Late Permian but became less varied in the Triassic, with ~30 genera in the South African Late Permian and only 10 from the Triassic *Lystrosaurus* and *Cynognathus* assemblage zones (LAZ and CAZ, respectively). Currently there is discussion as to the monophyly of the Therocephalia. Abdala (2007) and Botha et al. (2007) proposed that the taxon was paraphyletic, as *Theriognathus* was found to be the sister group of cynodonts and basal therocephalians such as Lycosuchidae and Scylacosauridae were recovered outside of the group consisting of all other therocephalians. Conversely, Huttenlocker (2009), using a large data matrix and more therocephalian terminal taxa, found the group to be monophyletic as previously proposed by Hopson and Barghusen (1986).

Bauriids from the CAZ were the last surviving therocephalians. Originally they were included among cynodonts (Broom 1911, 1913) as they share several common features (e.g., similar number of upper and lower incisors and the

F. Abdala (✉) · B. S. Rubidge
Evolutionary Studies Institute, University of the Witwatersrand, Private Bag 3, Wits, Johannesburg 2050, South Africa
e-mail: nestor.abdala@wits.ac.za

B. S. Rubidge
e-mail: bruce.rubidge@wits.ac.za

T. Jashashvili
Institute for Human Evolution, University of the Witwatersrand, Private Bag 3, Wits, Johannesburg 2050, South Africa
e-mail: tjashashvili@yahoo.fr

and

Anthropological Institute and Museum, University of Zurich, 8057 Zurich, Switzerland

and

Department of Geology and Paleontology, Georgian National Museum, 0105 Tbilisi, Georgia

J. van den Heever
Department of Botany and Zoology, University of Stellenbosch, Private Bag X1, Matieland 7602, South Africa
e-mail: javdh@maties.sun.ac.za

Christian F. Kammerer, Kenneth D. Angielczyk and Jörg Fröbisch (eds.), *Early Evolutionary History of the Synapsida*, Vertebrate Paleobiology and Paleoanthropology, DOI: 10.1007/978-94-007-6841-3_13,

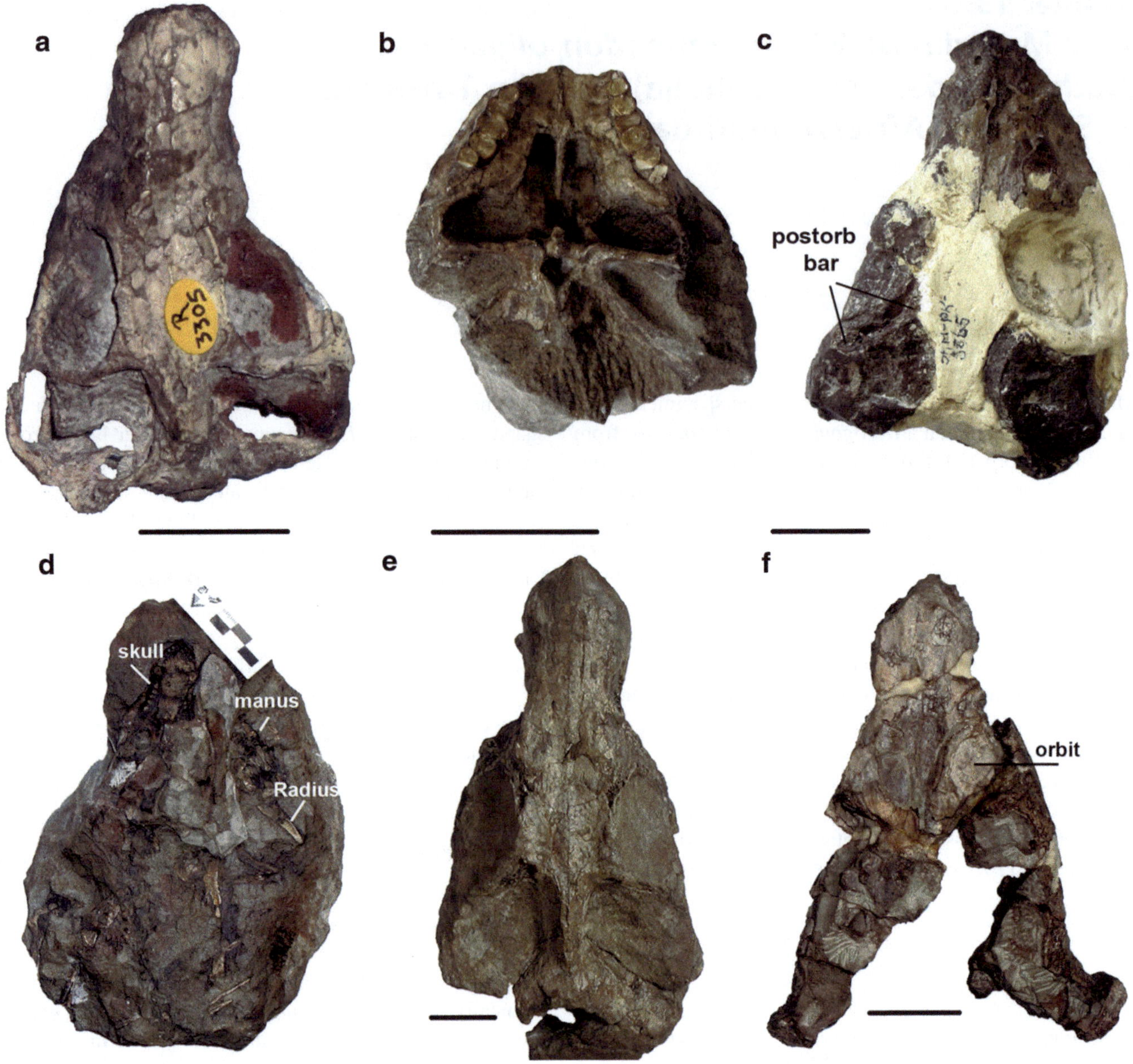

Fig. 13.1 Holotypes of bauriid species: **a** dorsal view of the skull of *Microgomphodon oligocynus*; **b** ventral view of the skull of *Melinodon simus*; **c** dorsal view of the skull of *Sesamodon browni*; **d** view of the block with skull and partial postcranium of *Aelurosuchus browni*; **e** dorsal view of the skull of *Bauria cynops*; **f** dorsal view of the skull of *Baurioides watsoni*; **g** dorsal view of the skull of *Watsoniella breviceps*; **h** block showing the partial skull and mandible, partial mandible and partial skeleton (*right side*) of *Sesamodontoides pauli*; **i** dorsal view of snout of *Bauria robusta*; **j** dorsal view of *Herpetogale marsupialis*. Scale bar in all figures except (**d**) is 2 cm

presence of an osseous secondary palate). However, many important features indicate that they were more closely related to therocephalians (Hopson and Barghusen 1986) and convergence is the best explanation for features in common with cynodonts (Watson 1921).

The first member of the Bauriidae in South Africa was described in the 1890s, followed by additional discoveries that took place until the middle of the 1970s. Here we present a historical review of the taxonomy of this family.

Microgomphodon oligocynus [represented by a small complete skull (Fig. 13.1a; Table 13.1)] and *Microgomphodon eumerus* (represented by a cranial fragment and part of the skeleton) were the first described representatives of what would be called bauriids, although they were originally reported as gomphodont reptiles by Seeley (1895). In a brief account lacking illustrations, Broom (1905) described the skulls of *Sesamodon browni* and *Melinodon simus* (Fig. 13.1b, c; Table 13.1). He allocated these forms to the

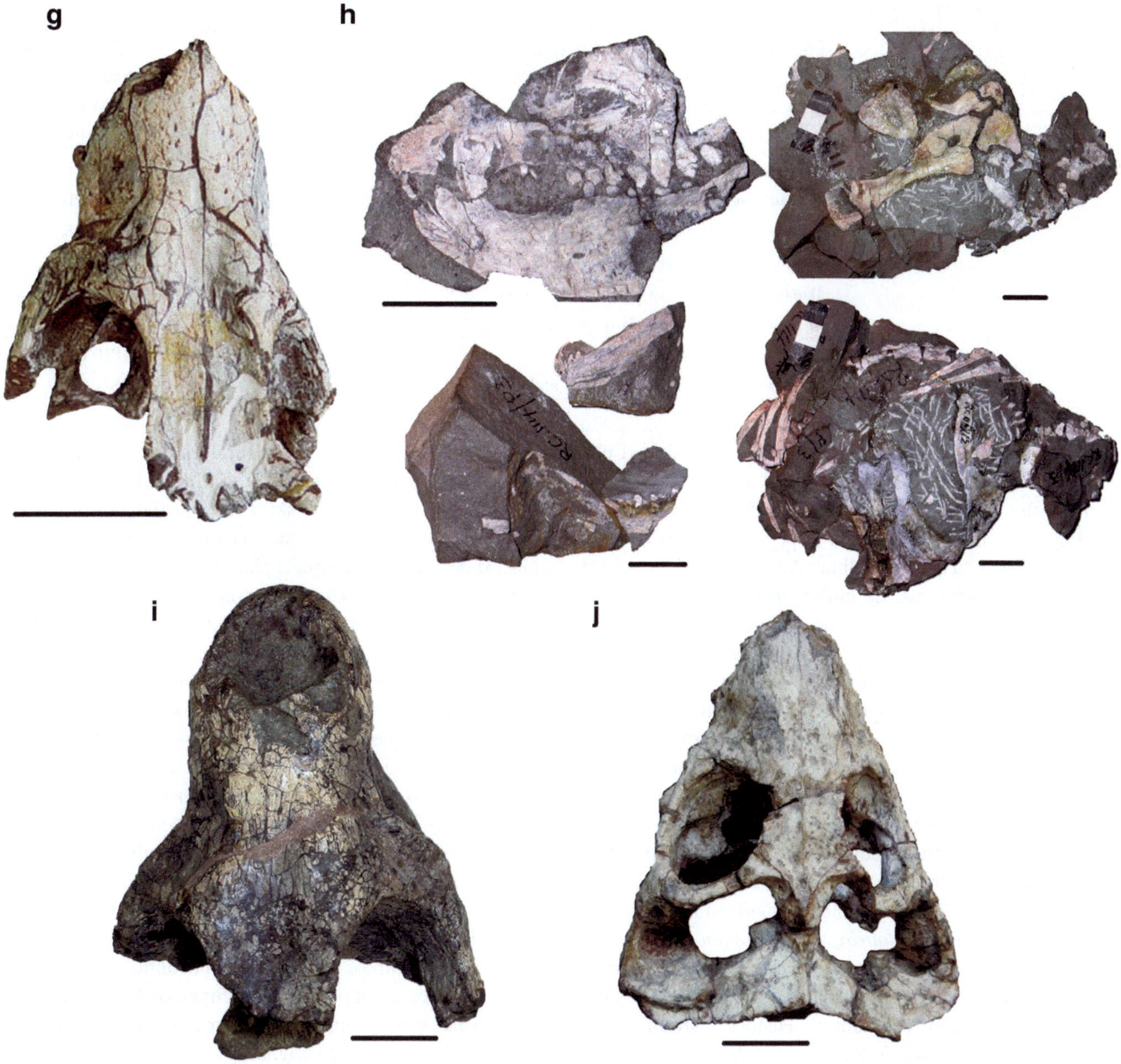

Fig. 13.1 (continued)

new family Sesamodontidae and placed these new taxa with "*Theriodonts* (or *Cynodonts*)..." but noted that they "may ultimately prove to be the type of a new Suborder connecting *Theriodonts* and Mammals" (Broom 1905, p. 273). Broom (1906) described 'a new cynodont' *Aelurosuchus browni* (Fig. 13.1d; Table 13.1) that he thought similar to *Microgomphodon oligocynus*, and later described *Bauria cynops* (Fig. 13.1e; Table 13.1), which he considered a primitive cynodont (Broom 1909). Broom (1911) briefly redescribed *B. cynops*, provided more illustrations, and proposed it as the type of the new family Bauridae (Broom 1911; the spelling of this family name was later correctly emended to Bauriidae). He recognized *B. cynops* as a cynodont that "retains many of the Therocephalian characters" (Broom 1911, p. 898), and redescribed and illustrated *Sesamodon browni* and *Melinodon simus* (Broom 1911, pp. 913–916). These two species were considered as closely allied and "pretty certainly" belonging to the same family (Broom 1911, p. 916).

Table 13.1 Nominal species of Bauriidae (see Fig. 13.1)

Taxon	Type specimen
Microgomphodon oligocynus Seeley 1895	NHMUK R3305
Microgomphodon eumerus Seeley 1895	NHMUK R3581
Sesamodon browni Broom 1905	SAM-PK-5865
Melinodon simus Broom 1905	SAM-PK-5866
Aelurosuchus browni Broom 1906	SAM-PK-5875
Bauria cynops Broom 1909	SAM-PK-1333
Baurioides watsoni Broom 1925	NHMUK R4095
Microhelodon eumerus Seeley 1895	NHMUK R3581
Watsoniella breviceps Broili and Schröder 1935	BSP 1934-VIII-13
Sesamodontoides pauli Broom 1950	RC 114
Bauria robusta Brink 1965	BP/1/1685
Herpetogale marsupialis Keyser and Brink 1978–1979	GSN R337

Although Broom (1911) recognized many shared characters between *Sesamodon* and *Bauria*, he did not explicitly include the former taxon in the family Bauriidae. In his phylogenetic tree (Broom 1911, p. 923), *Bauria* appears close to the "Therocephalian Ancestor", followed by *Aelurosuchus*, whereas the closely related *Melinodon* and *Sesamodon* appear close to the "Mammalian Ancestor". Watson (1913) considered *Microgomphodon oligocynus* and *Bauria cynops* as members of a common (innominate) family. Watson (1914) described additional specimens of *Bauria* and *Sesamodon* in the Natural History Museum, London. Referring to the latter taxon, Watson (1914, p. 1025; italics ours) states: "This type is *probably* represented in the British Museum by the anterior part of a skull broken off through the middle of the orbits" (FA was unable to locate this specimen in the Natural History Museum). In addition, Watson (1914, p. 1038) recognized Bauridae (sic), in which he included *Bauria*, *Microgomphodon*, and *Sesamodon*, as a distinct suborder separated from cynodonts.

Among the cynodonts housed in the collection of the American Museum of Natural History, Broom (1915) mentioned and figured a specimen of *Bauria cynops* and another of *Sesamodon browni*. Haughton (1922) described the palate and basicranium of the holotype of *Aelurosuchus browni* and assigned it to the Bauriamorpha. Broom (1925) proposed the new species *Baurioides watsoni* (Fig. 13.1f; Table 13.1) for the skull that was previously described as *Bauria cynops* by Watson (1914), based on differences in postcanine number. Broom (1931) argued that Seeley's (1895) *Microgomphodon eumerus* comprises a cranial fragment of a bauriid and the skeleton of a cynognathid cynodont and proposed the new combination *Microhelodon eumerus* (Seeley 1895) to name the cranial fragment. Broili and Schröder (1935) described the new species *Watsoniella breviceps* (Fig. 13.1g) based on a partial skull. Successive contributions by Broom (1937) and Boonstra (1938) described the same specimen of *Bauria cynops* housed at the American Museum of Natural History. Broom (1950) described *Sesamodontoides pauli* (Fig. 13.1h; Table 13.1) based on a fragmentary skull and partial postcranium of which he only illustrated the mandible. Brink and Kitching (1953) described a complete skull and mandible with part of the skeleton of *Bauria cynops* and mentioned an isolated left dentary of a second individual, which was not described. Brink (1963) provided a detailed description of *Bauria cynops*, using the two specimens reported previously by Brink and Kitching (1953) and two new specimens from the collection of the Bernard Price Institute. The last new species to be recognized from the South African Karoo was a partial skull named *Bauria robusta* (Fig. 13.1g; Table 13.1) described by Brink (1965). Finally, Keyser (1973a, b) found a *Sesamodon*-like skull in the Middle Triassic Omingonde Formation of Namibia that was later named *Herpetogale marsupialis* (Keyser and Brink 1977–1978) (Fig. 13.1h; Table 13.1).

The major aim of the present contribution is to provide a detailed description of a new specimen of *Microgomphodon oligocynus*, the best-preserved bauriid specimen yet known, recovered from the *Cynognathus* AZ, Subzone B of the Karoo Basin. We also present a taxonomic review of the southern Africa Bauriidae after firsthand examination of most of the existing specimens. To this end, and as an alternative of linear measurement comparison, we employed Procrustes analysis between the new specimen of *Microgomphodon oligocynus* (SAM-PK-K10160) and the best-preserved specimen of *Bauria cynops* (BP/1/1180). This analysis highlighted several differences between these "morphotypes" that were not perceived by visual inspection. Finally, we present details of the temporal and geographic distribution of these late therocephalian survivors in the Karoo Basin of southern Africa.

Institutional abbreviations: AMNH, American Museum of Natural History, New York City, NY, USA; BP, Bernard Price Institute for Palaeontological Research, University of the Witwatersrand, Johannesburg, South Africa; BSP, Bayerische Staatssammlung für Paläontologie und historische Geologie, Munich, Germany; GSN, Geological Survey of Namibia, Windhoek, Namibia; NHMUK, The Natural History Museum, London, UK; NMQR, National Museum, Bloemfontein, South Africa; SAM, Iziko, the South African Museum, Cape Town, South Africa; RC, Rubidge Collection, Wellwood, Graaff-Reinet, South Africa. UCMP, University of California Museum of Paleontology, Berkeley, CA, USA; USNM, National Museum of Natural History, Washington, D.C., USA

Materials and Methods

The new specimen, SAM-PK-K10160, consists of a cranium with mandible in occlusion, four cervical vertebrae (including ribs), and a manus. This specimen was found at the farm Lemoenfontein 44, Rouxville District, South Africa. For this study all holotypes of the nominal southern African bauriid species were examined (see Fig. 13.1; Table 13.1). Non-holotype material examined includes: AMNH 5517, 5622; BP/1/1180, 1679, 1685, 2523, 2837, 3770, 4655, 4678; NMQR 3183 and 3596.

The comparative approach of this work was twofold: (a) detailed inspection of the studied specimens to recognize qualitative variables that can be used for taxonomic purposes; and (b) a Procrustes analysis of two selected specimens representing *Microgomphodon oligocynus* (SAM-PK-K10160) and *Bauria cynops* (BP/1/1180), which are the best-preserved individuals referred to these taxa. The main intention of this approach was to highlight shape differences that may not be recognizable on a discrete character basis to provide further elements for taxonomical distinction. This is part of a study in progress by FA and TJ that will enlarge the sample of specimens studied with this methodology and explore some functional implications related to morphological differences among bauriid skulls.

Virtual Reconstruction

Computer tomographic scanning was used for virtual preparation of specimens SAM-PK-K10160 and BP/1/1180 in order to digitally separate the cranium and mandible. The specimens were CT-scanned at the Helen Joseph Hospital (Johannesburg, South Africa) on a Philips Brilliance 16 medical CT scanner (under 140 kV, tube current 165 mAs, beam collimation 0.5 cm, interstice distance 0.4 cm). 5.224 pixels per mm and depth 16 bits (unsigned) images were reconstructed using the sharp construction algorithm. The image stack was resampled to 20.898 pixels per mm and semi-automated routines of segmentation were undertaken using Avizo 6.2.1 software (Visualization Sciences Group, Mérignac, France). Threshold was defined between 2,200 and 4,000 HU, lower and upper level respectively, for automated segmentation routines. Manual segmentation was undertaken in situations where specimen and sediment density, or joint areas between mandible and cranium, were similar. 3D rendered surfaces of the cranium and mandible are shown in Fig. 13.2.

The description of the new material is supplemented with a 3D reconstruction of the specimen (Figs. 13.2, 13.3) which facilitates description of the anterior portion of the palate (concealed by the occluded mandible) (Fig. 13.2b) and reveals details of the postdentary bones in the mandible (Fig. 13.2e, g).

Correction of Deformation

During fossilization, SAM-PK-K10160 was distorted mostly on its left side (Fig. 13.2). In order to perform Procrustes analysis, correction of the deformation was undertaken by mirroring the less distorted side. Translation and rotation of the less deformed side of the cranium was performed until we achieved anatomical continuation (Zollikofer et al. 1998). Additionally, displacement of the zygomatic arch was corrected using translation and rotation of a portion of the arch until the anterior and posterior ends of this portion achieved anatomical continuation with the suborbital bar anteriorly and the posterior portion of the zygoma, posteriorly. In the mandible, the undistorted right side was mirrored to the left side using the same technique (Fig. 13.3).

Procrustes Analysis

In order to understand shape correspondence between the two genera of Bauriidae we preformed a Procrustes analysis (Gower 1975) using the most complete and best-preserved representative of each taxon. The least squares method was used to find the "best fit" of matrix A (in this case *Microgomphodon oligocynus*—SAM-PK-K10160) to matrix B (*Bauria cynops*—BP/1/1180) under scaling, rotation, and translation (Rohlf and Slice 1990; Dryden and Mardia 1998). Because landmarks provide the foundation of shape in Procrustes analysis, a series of landmarks were defined for the cranium and mandible (Fig. 13.3). Three types of landmarks (anatomical, mathematical, and pseudo-landmarks; Bookstein 1991) were used.

Procrustes analysis was performed using Morphologika (O'Higgins and Jones 2006). The configuration matrix after Procrustes analysis was used to explore the relationships

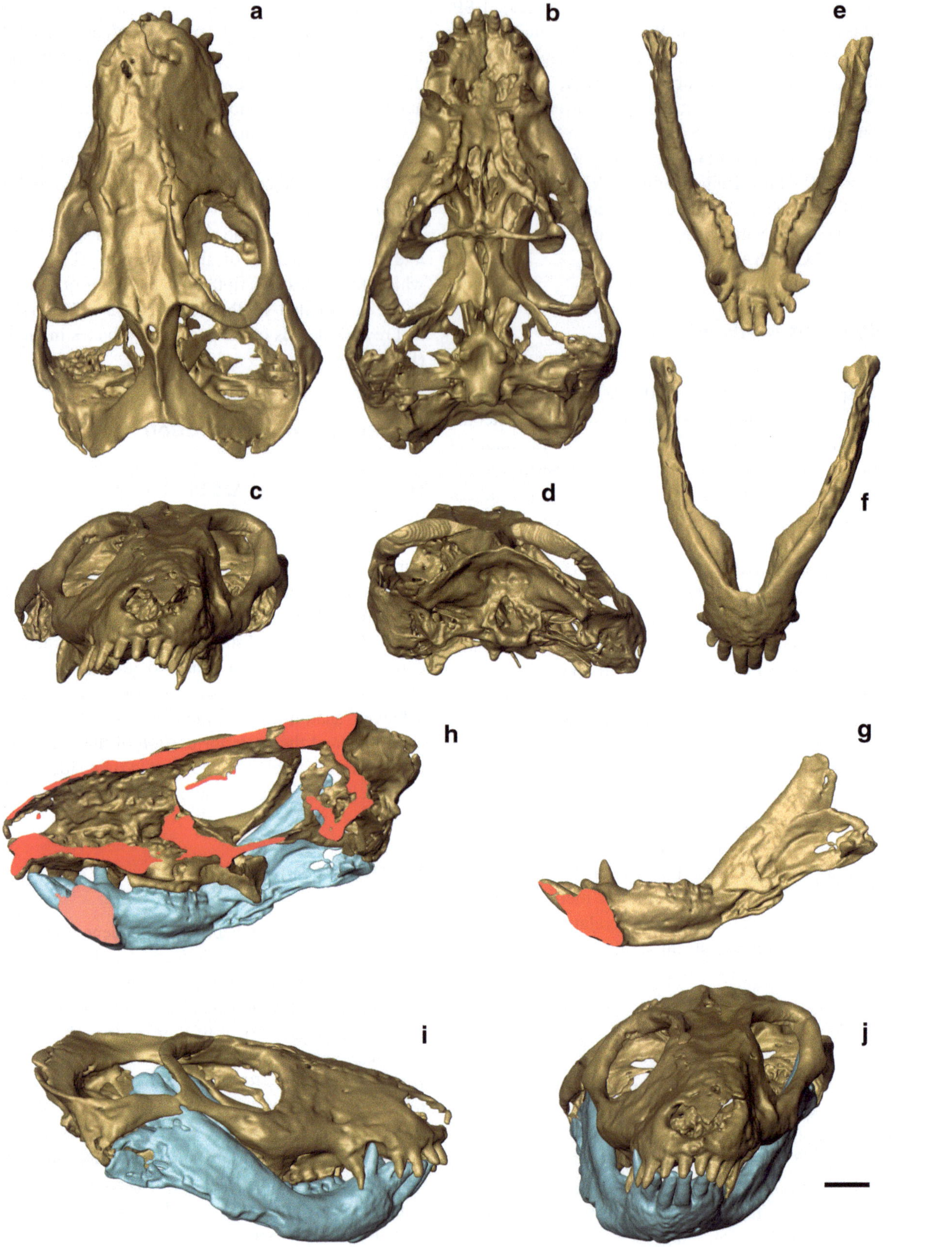

Fig. 13.2 Virtual reconstruction of *Microgomphodon oligocynus* (SAM-PK-K10160) cranium and mandible from different views. Cranium: **a** dorsal; **b** ventral; **c** anterior; **d** posterior. Mandible: **e** dorsal; **f** ventral; **g** posterior. Cranium and mandible together: **h** medial; **i** lateral; **j** anterior. Scale 1 cm

between these two genera (Fig. 13.4). Landmark surface warp rigid scaled module was used on the Avizo 6.2.1 software to rotate, translate, and scale 3D surfaces and a configuration matrix was produced for each specimen. Surface distance was measured between warped triangulated surfaces. For each vertex of a particular surface, Morphologika computes the closest point on the other surface. The colored map with shading from red to white (Fig. 13.4) represents surface differences between *Microgomphodon oligocynus* and *Bauria cynops*. Dark red indicates greater difference between the two specimens (Fig. 13.4a–d). Results of this analysis were also presented comparing shape configuration matrices for the cranium (Fig. 13.4a′, b′) and mandible (Fig. 13.4c′, d′) of *Microgomphodon oligocynus* (SAM–PK–K10160) represented in red and *Bauria cynops* (BP/1/1180) in grey/black.

Systematic Paleontology

Therapsida Broom, 1905
Therocephalia Broom, 1903
Bauriidae Broom, 1911
Microgomphodon oligocynus Seeley, 1895
1905 *Sesamodon browni* Broom: 272
1905 *Melinodon simus* Broom: 273
1935 *Watsoniella breviceps* Broili and Schröder: 23, Fig. 1
1977–1978 *Herpetogale marsupialis* Keyser and Brink: 91, Fig. 1
1977–1978 *Herpetogale saccatus* Keyser and Brink: 103, Table 5.1 (in error)

Holotype: NHMUK R3305, complete skull and lower jaws preserved in occlusion, from an unknown locality in Aliwal North District.

Referred Specimens: BSP 1934-VIII-13; SAM-PK-5865, SAM-PK-5866, SAM-PK-K10160; NMQR 3183; NMQR 3596; BP/1/4655; GSN R337.

Localities: See Table 13.1.

Horizon and Age: Levels of the Burgersdorp Formation corresponding to the faunas of the Subzone A and B of the CAZ, Karoo Basin, South Africa; upper Omingonde Formation, Otjiwarongo Basin, Namibia. Late Olenekian (Subzone A) to Anisian. Specimens collected from Subzone A of the CAZ are the oldest record of the taxon. The material from Namibia was collected high in the upper Omingonde Formation, near the contact with the Etjo Formation (Keyser and Brink 1977–1978) from levels probably of Late Anisian age (Abdala and Smith 2009).

Diagnosis: Small member of the Bauriidae presenting a relatively short snout and large orbits of almost equal size as the temporal openings; postorbital bar completed by an ascending process of the jugal; presence of parietal/pineal foramen; restricted exposure of the frontal on the dorsal orbital margin; presence of suborbital foramen oriented dorsally; short choanae with the vertical keel of the vomer ending near the base of the pterygoid flanges; well-developed, fan-shaped basisphenoidal keel; maxillary shelf curved immediately behind the canine; platform lateral to the postcanine row in the mandible extended anteriorly; vomer participates in the posterior margin of the secondary osseous palate; pterygoid process located anteriorly; presence of a lateral fossa on the posterior portion of the dentary horizontal ramus and on the coronoid process; first lower incisors remarkably large and procumbent; canine placed approximately at mid-length of the snout; postcanine number variable from 5/5 to 7/?7; postcanines oval and almost equally wide labially and lingually, and appear broad when observed in labial view.

Comments: The diagnostic characters mentioned above separate *Microgomphodon oligocynus* from *Bauria cynops*. These two taxa share a suite of characters indicating a close taxonomic affinity, including presence of a complete osseous secondary palate, dentary processes or shelves lateral to the postcanine series (clearly better developed in *M. oligocynus*), and expanded postcanine crowns which manifest occlusion.

Description

General Preservation

The preservation of SAM-PK-K10160 is excellent, with almost all sutures of the skull clearly visible (Fig. 13.5). The mandible is preserved in occlusion and the crowns of the postcanines are visible labially and lingually. Postdentary bones of the mandibles are preserved with the dorsal portion of the reflected lamina of the angular present on both sides, the ventral parts of the laminae being missing. The right lateral wall of the braincase and interorbital regions are completely preserved. The palate is preserved with all bones in their natural position, and in the basicranium the central para-basisphenoid keel and both stapes are well-preserved. Four cervical vertebrae, including the atlas-axis complex, first complete, and the second and third partial cervical ribs are preserved in situ.

Size and General Proportions

The basal skull length of SAM-PK-K10160 is 87.4 mm, similar to that of GSN R337 which is 88.57 mm. These two specimens are the largest representatives of the species. The

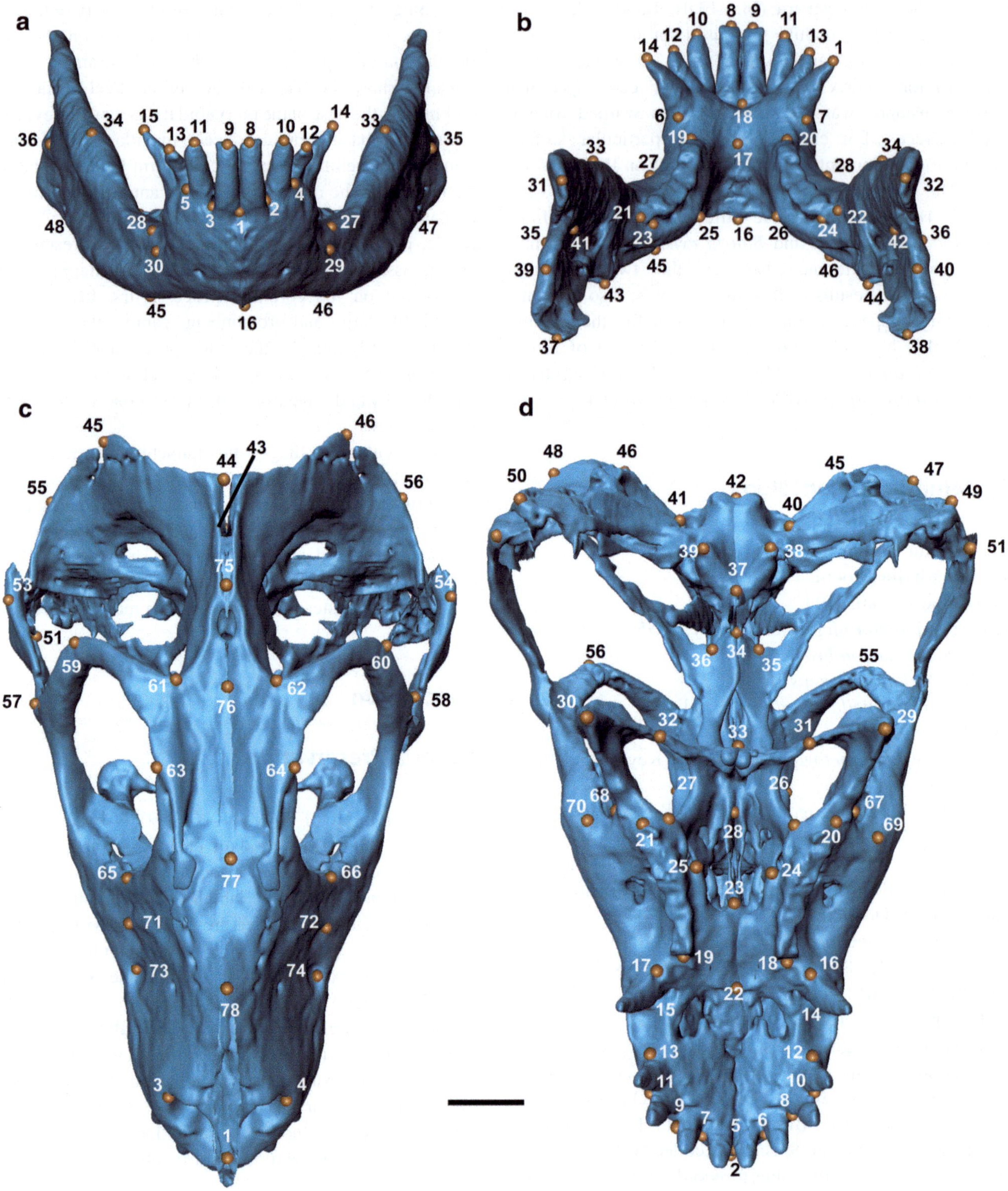
a
36 34 15 13 11 9 8 10 12 14 33 35
5 3 1 2 4
48 28 27 47
30 29
45 16 46
b
10 8 9 11
14 12 13 1
18
6 7
19 20
33 17 34
27 28
31 32
21 22
35 41 23 25 16 26 24 42 36
39 45 46 40
43 44
37 38
c
45 46
43
44
55 56
75
53 54
51
59 60
61 62
57 76 58
63 64
65 77 66
71 72
73 74
78
3 4
1
d
48 46 45 47
50 42 49
41 40
51
39 38
37
56 34 55
36 35
30 29
32 33 31
27 26
70 68 67 69
21 28 20
25 24
23
17 19 18 16
22
15 14
13 12
11 10
9 8
7 5 6
2

◀ **Fig. 13.3** Location of cranio-mandibular landmarks defined for the Procrustes analysis. **Mandibular landmarks**: **(1)** Anterior symphysis, below the two anterior incisors of both sides. **(2;3)** Between the base of first and second incisors; **(4;5)** between the base of second and third incisors; **(6;7)** behind the canine; **(8;9)** tip of the first incisor; **(10;11)** tip of the second incisor; **(12;13)** tip of the third incisor; **(14;15)** tip of the canine; **(16)** ventral symphysis. **(17)** symphysis where the dentaries contact each other posteriorly. **(18)** Symphysis behind the first incisors of both sides. **(19;20)** Anterior margin of the postcanine series (towards the center of the tooth); **(21;22)** posterior margin of the postcanine series (towards the center of the tooth); **(23;24)** medial corner of the last postecanine; **(25;26)** maximum internal curvature point of the dentary (observed in ventral view); **(27;28)** anterior base of the coronoid process; **(29;30)**. Anterior margin of the lateral canal of the dentary; **(31;32)** tip of the coronoid process; **(33;34)** half distance between landmark 27 and 31; **(35;36)** junction between the dentary and surangular in lateral view; **(37;38)** cranio-mandibular joint; **(39;40)** half distance between landmark 35 and 37; **(41;42)** end of the dorsal margin of the meckelian canal (represented as a bulged expansion of the dentary below the teeth); **(43;44)** angular process (where finish the embracing Meckelian canal of the dentary); **(45;46)** half distance between landmark 16 and 43; **(47;48)** contact between dentary and angular in lateral view. **Cranial landmarks**: **(1)** Tip of the snout; **(2)** distal middle point of the snout. **(3;4)** dorsoposterior margin of nasal opening; **(5)** between first incisors of each side; **(6;7)** between base of incisors 1st and 2nd; **(8;9)** between base of incisors 2nd and 3rd; **(10;11)** between base of incisor 3rd and 4th; **(12;13)** posterior margin of incisors; **(14;15)** anterior margin of canine; **(16;17)** posterior margin of canine; **(18;19)** anterior margin of postcanine row; **(20;21)** posterior extension of postcanine row; **(22)** central portion of the osseous palate at the level of the canine; **(23)** posterior end of the palate; **(24;25)** most internal point of maxillary curvature; **(26;27)** anterior portion of suborbital vacuity; **(28)** posterior margin of the vomer; **(29;30)** tip of pterygoid flange; **(31;32)** posterior portion of suborbital vacuity; **(33)** anterior margin of interpterygoid vacuity; **(34)** posterior margin of the interpterygoid vacuity; **(35;36)** opening of the quadrate ramus of pterygoid in the basicranial grider; **(37)** basal tubera anterior margin of basisphenoid; **(38;39)** basal tubera; **(40;41)** opening of the jugular foramina; **(42)** occipital condyle; **(43)** dorsal central point of foramen magnum; **(44)** end of the sagittal crest; **(45;46)** point between landmarks 45 and 53, 46 and 54 respectively; **(47;48)** contact of the posterior corner of temporal fossa; **(49;50)** squamosal lateral surface that contact quadrate; **(51;52)** squamoso–jugale suture in the ventral margin of the zygoma; **(53;54)** zygoma dorsal margin; **(55;56)** point between landmarks 45and 44, 46 and44 respectively; **(57;58)** posteroinferior margin of the orbit; **(59;60)** tip of postorbital process; **(61;62)** union postorbital bar and skull; **(63;64)** centre of dorsal margin of orbit; **(65;66)** front of the orbit; **(67;68)** base of pterygoidal process; **(69;70)** start of ventral margin of zygoma; **(71;72)** anterior border of lacrimal bone; **(73;74)** infraorbital foramen on the snout; **(75)** point between landmarks 44 and 76; **(76)** anterior margin of temporal region (no repetitive); **(77)** translation of the anterior portion of the orbit in dorsal view of the skull; **(78)** point between landmarks 1 and 77

skull has a triangular outline in dorsal view, with the maximum width of the skull (61.0 mm) at the posterior portion of the temporal region (Figs. 13.2a–d, 13.3a, 13.4a). Although the snout is short compared to other therocephalians, it is the longest region of the skull, representing 45 % of the basal skull length (Fig. 13.6). The orbital and temporal regions are subequal, respectively representing 30 and 29 % of the BSL. In GSN R237 the cranium proportions are slightly different, particularly in the snout (Fig. 13.6) (Table 13.2).

Snout and Orbits

The premaxilla forms the anterior margin of the snout and has a well-developed, dorsally directed ascending process. As the anterior portions of both nasals are preserved as an internal cast it is possible to observe the protrusion of the dorsal portion of the premaxillary ascending process between the nasals (Fig. 13.5a; contra Keyser and Brink 1977–1978). The anterior premaxillary foramen is located at the base of the ascending process near the suture with the other premaxilla. An elliptical external naris is directed anterolaterally, with its antero-posterior axis being longer than the dorsomedial axis (Figs. 13.2c, i, 13.5a, b). The anterior facial exposure of the premaxilla is restricted as it is covered by part of the facial extension and the intranarial process of the septomaxilla (Fig. 13.5b). The lateral exposure of the premaxilla is highest at its posterior end where the facial process of the septomaxilla is directed dorsoposteriorly (Fig. 13.5b) and the posterior margin of the premaxilla is slightly behind the level of the posterior margin of the external naris. The facial process of the septomaxilla is interposed for a long distance between the maxilla and the nasal, with a septomaxillary foramen placed at the base of the process and limited by the maxilla posteriorly (Fig. 13.5b). The intranarial process of the septomaxilla is reduced, in comparison with that of the scylacosaurid *Glanosuchus* (van den Heever 1994: Figs. 1, 2), and they do not contact the process of the opposite side. The dorsal projection of the intranarial process is small (Fig. 13.2i).

The maxilla forms a shelf lateral to the postcanine teeth that extends from immediately behind the canine to below the anterior margin of the orbit (Figs. 13.2b, i, 13.5b). In lateral view the maxillary shelf is somewhat concave behind the canine and slightly convex below the orbits. A series of anteriorly oriented nutritive foramina are present at the base of the canines and anterior to this tooth (Fig. 13.5b). A small dorsally oriented infraorbital foramen is present at the level of the canine buttress, and a small nutritive foramen is positioned immediately ventral to the "infraorbital foramen" on both sides of the snout (Figs. 13.2a, i, 13.5a, b). The posterior extension of the maxilla extends below the jugal reaching the middle of the orbit (Fig. 13.5b). The surface texture of the maxilla on the posterior extension is different from the rest of the bone,

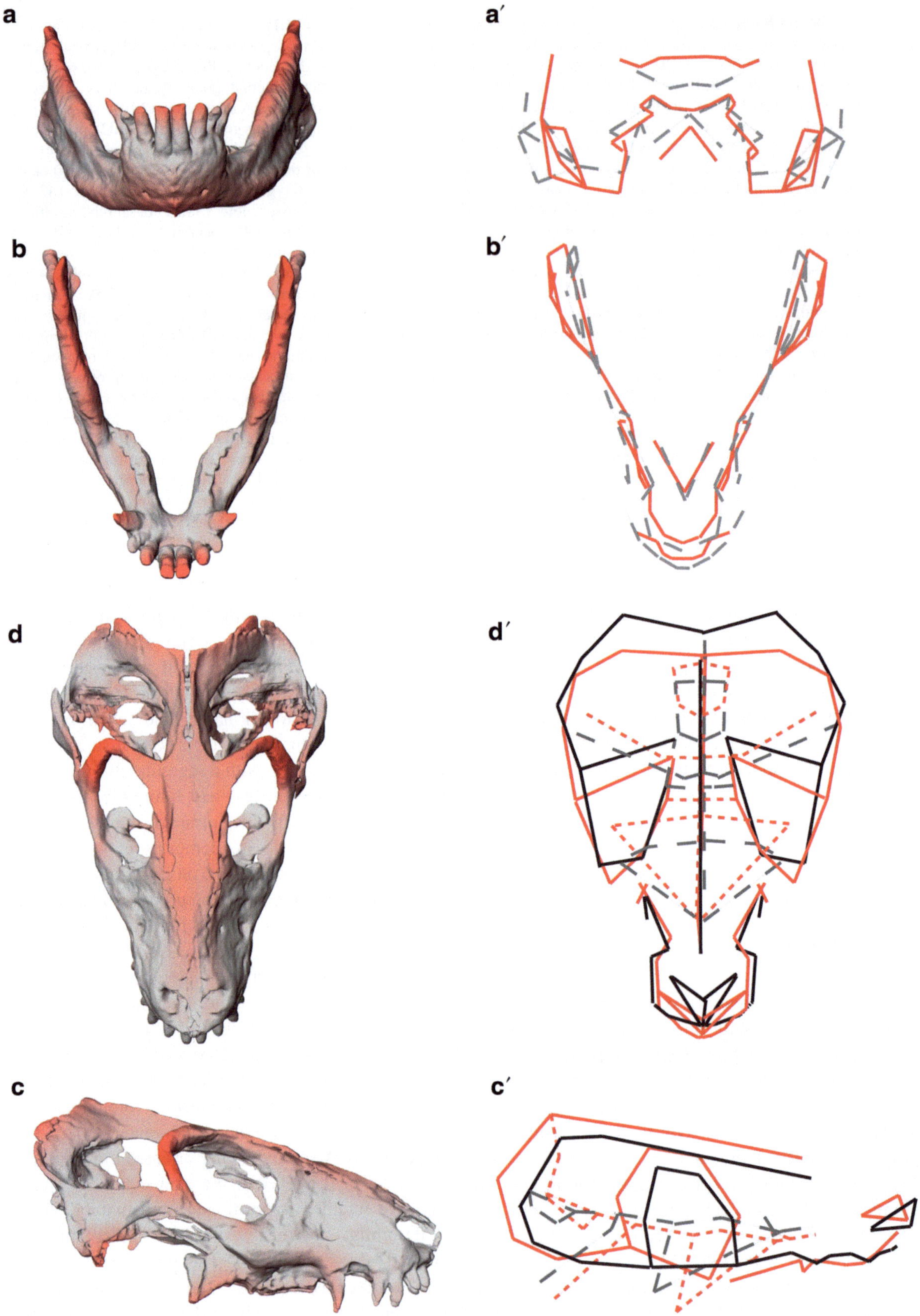
a
a′
b
b′
d
d′
c
c′

◀ **Fig. 13.4** Surface distance colour map expressed in *whitish* and *reddish* indicating lower and higher differences respectively, between the cranium (**a**, **b**) and mandible (**c**, **d**) of *Microgomphodon oligocynus* (SAM-PK-K10160) and *Bauria cynops* (BP/1/1180). Shape configuration matrix for the cranium (**a**′, **b**′) and mandible (**c**′, **d**′) of *Microgomphodon oligocynus* (SAM-PK-K10160) in *red* and *Bauria cynops* (BP/1/1180) in *grey/black*. In **a**′ and **b**′, *solid lines* represents dorsal view of cranium and *dots/dash lines* the ventral view of cranium. In **c**′ and **d**′, *solid red lines* represents *M. oligocynus*, whereas the *dashed grey lines* corresponds to *B. cynops*

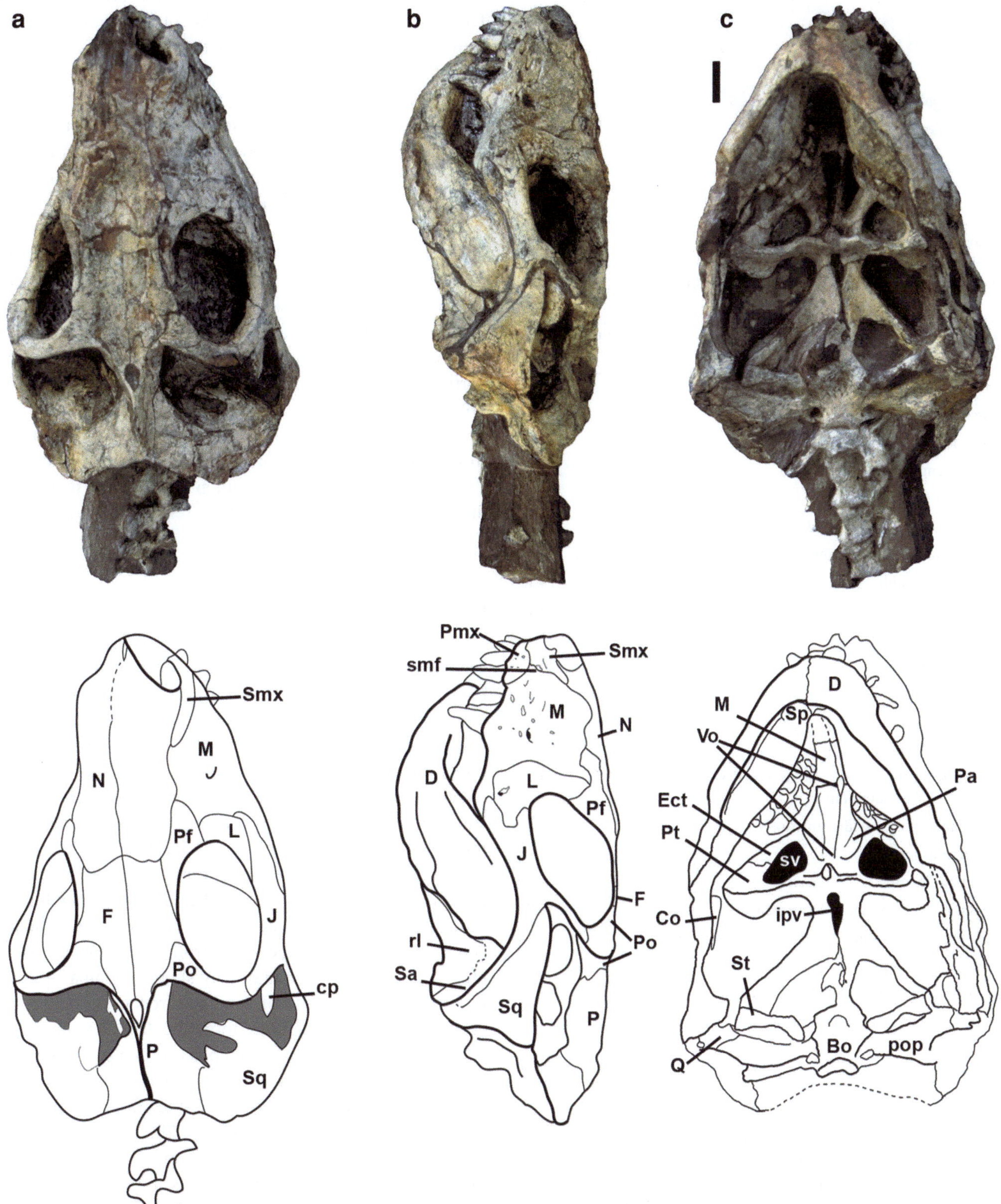

Fig. 13.5 SAM–PK–K10160. Photo and drawing in **a** dorsal, **b** lateral and **c** ventral views. *Bo* Basioccipital, *Co* coronoid, *cp* coronoid process, *d* dentary; *Ect* ectopterygoid, *f* frontal, *ipv* interpterygoid vacuity, *J* jugal, *L* lacrimal, *M* maxilla, *N* nasal, *P* parietal, *Pf* prefrontal, *Po* postorbital, *pop* paraoccipital process, *Pt* pterygoid, *Q* quadrate, *rl* reflected lamina, *Sa* surangular, *smf* septomaxillary foramen, *Smx* septomaxilla, *Sp* splenial, *Sq* squamosal, *St* stapes, *sv* suborbital vacuity, *Vo* vomer. Scale 1 cm

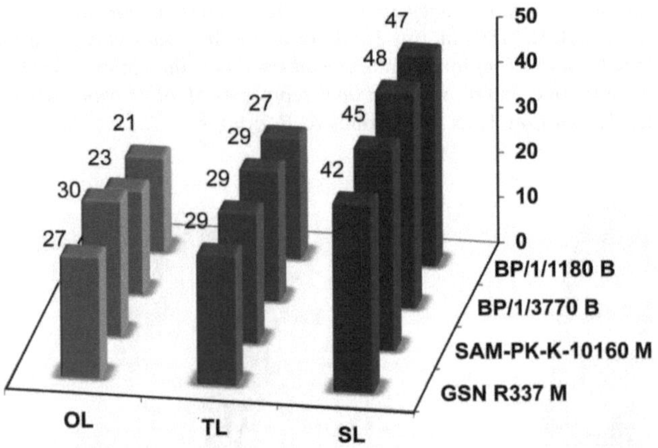

Fig. 13.6 Cranial proportions of selected specimens of Bauriidae: Proportion of the orbital length (*OL*), snout length (*SL*) and temporal length (*TL*) in relation to the basal cranial length. *B* are specimens of *Bauria cynops* and *M* of *Microgomphodon oligocynus*

Table 13.2 Measurements of SAM-PK-K10160 (in mm, except for postcanine number)

Basal cranial length	87.36
Dorsal cranial length	89.43
Snout length	39.22
Orbital length	26.53
Temporal length	25.18
Snout width at canine level	28.6
Maximum width of cranium	62.68
Interorbital width	19.19
Orbit diameter	25.17
Occiput width	35.44
Occiput height	24.27
Posterior root of the zygoma height	16.32
Palate length	34.65
Suborbital vacuity length	5.67
Length from tip of snout to pterygoid wings in the middle of the cranium	54.13
Basicranial girder width anteriorly	8.97
Basicranial length	14.07
Mandible length	74.24
Dentary length on the inferior margin	51.59
Zygoma length from orbit to posterior margin	42.54
Epipterygoid height	17.2
Epipterygoid width at the dorsal margin	9.16
Incisor–canine extension (until posterior margin of the canine)	28.36
Postcanine number	6/6

having faint striations directed posteriorly, probably related to adductor muscle attachment. The broadest portion of the wide nasal contacts the anterior margin of the prefrontal (Figs. 13.2a, 13.5a). The suture between the nasal and frontal forms an obtuse angle (approximately 150°; Figs. 13.2a, 13.5a).

The anterior margin of the lacrimal forms a convex suture with the maxilla which continues ventrally to the jugal-maxilla suture (Fig. 13.5a, b). Dorsoventrally, the lacrimal extends from the middle of the orbit to its base. The lacrimal foramen has a variable placement on the bone, being on the facial surface on the right side and on the orbital surface on the left.

The prefrontal is triangular on the facial region, with the anterior tip of the bone interposed between nasal and maxilla (Fig. 13.5a). A narrow posterior projection of the prefrontal extends half way along the dorsal rim of the orbit, thus restricting participation of the frontal in the dorsal margin of the orbit (Fig. 13.5a).

In dorsal view the frontal is broad but tapers posteriorly from its contact with the postorbital (Fig. 13.5a). The frontal has a small exposure on the dorsal orbital rim behind the prefrontal and ends at the level of the postorbital bar, in front of the pineal foramen. The frontals and the posterior portion of the nasals form a depressed area between the orbits (Fig. 13.2a, c).

The sutural contact between the postorbitals and the frontals is oriented anterolaterally and extends posteriorly slightly beyond the pineal foramen (Fig. 13.5a). Posteriorly, the parietals form a short sagittal crest (Figs. 13.2a, 13.5a).

Palate

There is a well-developed palatal foramen located medial to the paracanine fossa and limited by the premaxilla anteriorly and by the maxilla in the remaining margin. The paracanine fossa is a well-developed concavity located anterior to the upper canine (Figs. 13.2b, 13.5b). The osseous secondary palate is formed by a large extension of the maxilla and a small contribution of the vomer, posteriorly and in the middle of palate (Figs. 13.2b, 13.5c, 13.7a). The maxilla forms a long posterior and lateral projection which extends beyond the postcanine series, whereas the vomer forms a posteriorly directed keel that bisects the choanae (Figs. 13.5c, 13.7a). The area from the choanae to the pterygoid processes is remarkably short in comparison with that of *Bauria*, because of a notable reduction of the sub-orbital fossa. The palatines form the lateral walls of the choanae and the suture between palatine and pterygoid is not visible. A well-developed ventromedian crest is present at the base of the transverse process of the pterygoid in front of the interpterygoid vacuity (Figs. 13.2b, 13.5c, 13.7a). The circular suborbital fossa is small and the postero-lateral margin of the transverse process has a rounded projection directed posteriorly (Figs. 13.5c, 13.7a), which is of variable

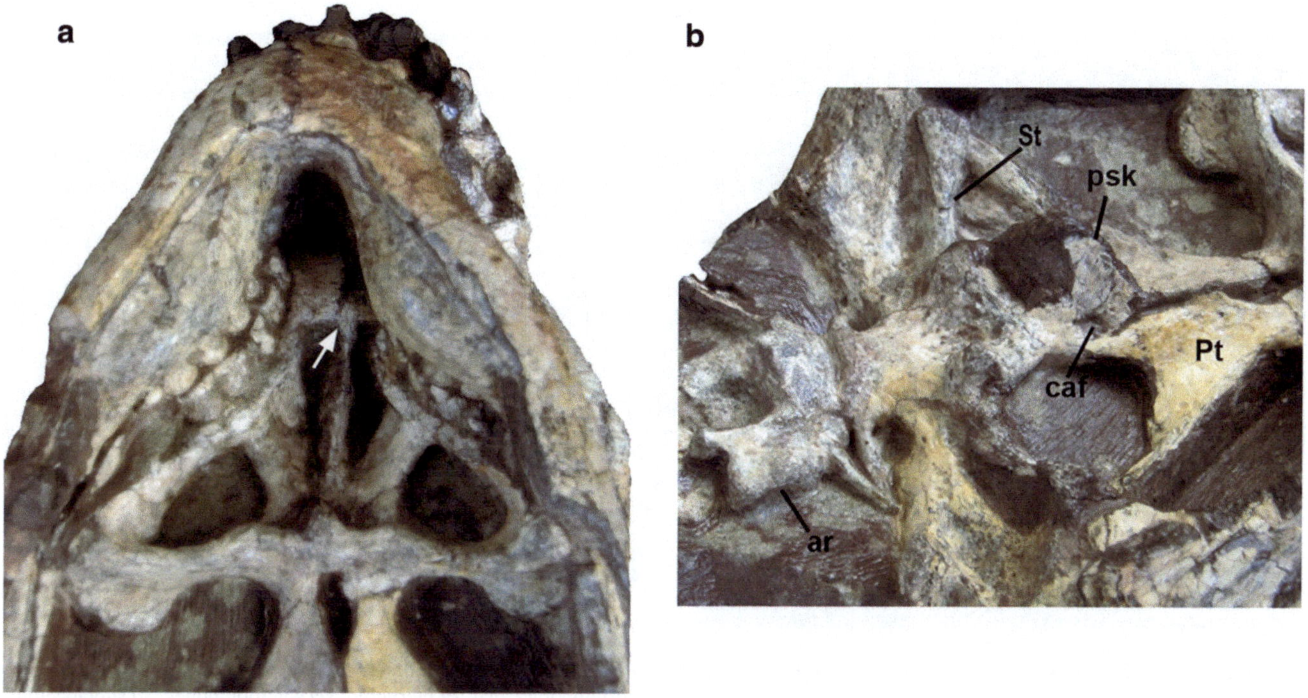

Fig. 13.7 SAM-PK-K10160. **a** Detail of the palate. *Arrow* indicates vomer contribution in the formation of the osseous palate. **b** Detail of the basicranium. Anterior to the right. *ar* atlantal rib, *caf* carotid foramen, *psk* parasphenoidal keel, *Pt* pterygoid, *St* stapes

proportions in the different specimens (that of GSN R337 is much less developed).

Zygoma and Temporal Region

The temporal opening is approximately rectangular, wider transversely. The zygomatic arch, including the suborbital and temporal bars, is formed by the posterior portion of the maxilla, the jugal, and the squamosal (Fig. 13.5b). The suborbital bar is slightly longer than the temporal bar and the ventral edge of the arch is convex anteriorly and concave in its central and posterior portions (Fig. 13.5b). At the anterior part of the orbit, the arch (comprising maxilla and jugal) is very high and robust. Posteriorly, the arch decreases to only half of its height at the level of the middle of the orbit and comprises only jugal (Fig. 13.5b). The posterior projection of the jugal extends almost as far as the end of the zygomatic arch. Posteriorly, the zygomatic arch becomes higher and is formed by the squamosal (Fig. 13.5b). The postorbital projection of the jugal is directed dorsally and contacts the postorbital halfway up the orbit (Fig. 13.5a, b).

The squamosal, which forms the posterior portion of the zygomatic arch, expands posterodorsally to form part of the occipital crest (Fig. 13.5b). This overhangs the ventrally directed squamosal sulcus in the area where the squamosal contacts the tabular and paroccipital process. In posterior view the squamosal forms a shallow ‘V’-shaped groove contacting the quadrate laterally and the paroccipital process medially.

The quadrate is only visible posteriorly as a rectangular bone having a wide horizontal (and presumably also anterior) contact with the articular (Fig. 13.2b).

Braincase and Basicranial Girder

The basicranial girder is wide anteriorly, where it is positioned between the quadrate rami of the pterygoids, and narrows posteriorly (Figs. 13.2b, 13.5c). A triangular interpterygoid vacuity is present at the anterior margin of the girder, behind the base of the central roots of the pterygoid lateral processes. The basioccipital-basisphenoid plate is pentagonal and no sutures are recognizable between these bones (Figs. 13.5c, 13.7b). A large, semilunar occipital condyle is located posteriorly and a well-developed, ventrally oriented jugular foramen is present anterolateral to the condyle. The robust basisphenoidal tubera are positioned on the anterior portion of the basisphenoid (Figs. 13.5c, 13.7b) and have a deep concave area between them. A remarkably high, fan-shaped parasphenoidal rostrum (Figs. 13.2b, 13.7b) is present anterior to the tubera, and the right carotid foramen is preserved at the base of the

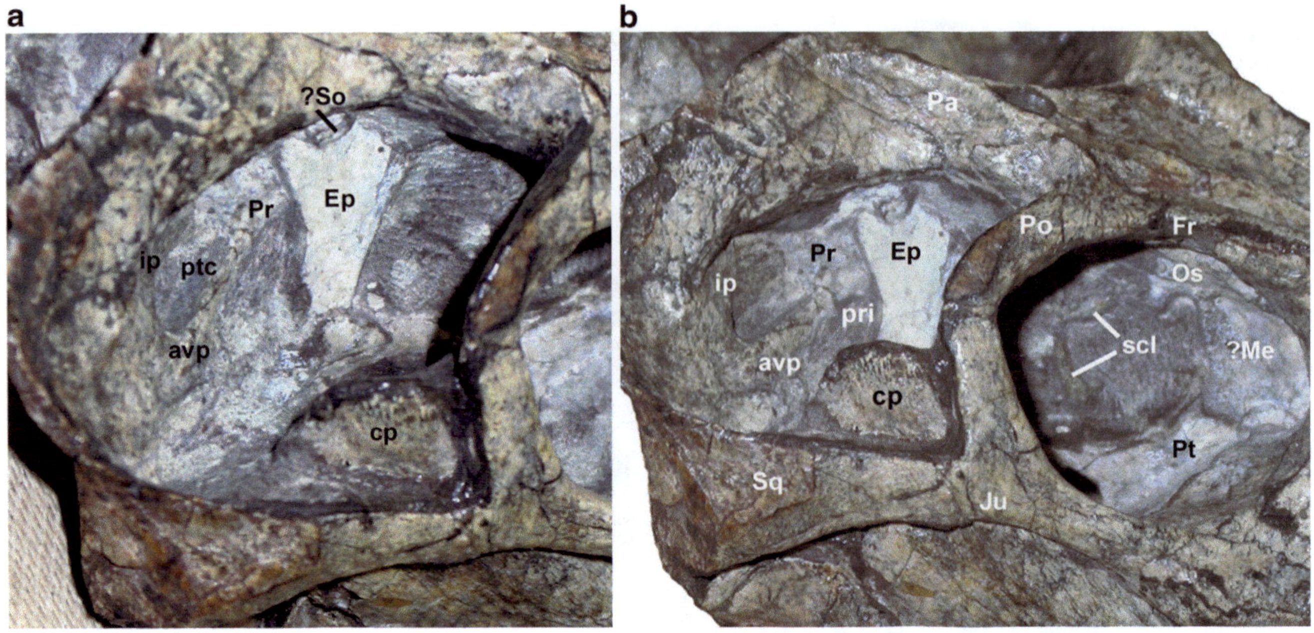

Fig. 13.8 SAM-PK-K10160. Detail of the **a** lateral wall of the skull; **b** interorbital region. *avp* antero–ventral process of the squamosal, *cp* coronoid process, *Ep* epipterygoid, *Fr* frontal, *ip* intermediate process of the squamosal, *Ju* jugal, *M* mesethmoid, *Os* orbitosphenoid, *Pa* parietal, *Po* postorbital, *Pr* prootic, *pri* prootic incisure, *ptc* postemporal canal, *Pt* pterygoid, *scl* sclerotic bones, *So* supraoccipital, *Sq* squamosal

rostrum. The paroccipital process of the opisthotic is narrow proximally and remarkably expanded distally, with the quadrate and occipital portion well-separated by a slight concavity (Fig. 13.7b).

The stapes, preserved in situ on both sides of the skull, is a cylindrical bone with both extremities expanded equally, and lacking a stapedial foramen (Fig. 13.7b). Its distal end is in contact with the paroccipital process dorsally and, visible only on the left side, with the quadrate distally.

Lateral and Interorbital Walls

In lateral view, the epipterygoid has a thin base and progressively increases in anteroposterior length toward its dorsal margin (Fig. 13.8a). On the left side, a short, anterior projection is present at its base. On the right side of the cranium, the posterodorsal portion of the epipterygoid seems to be superposed by the supraoccipital (Fig. 13.8a). The prootic is visible with its anterior margin in contact with the epipterygoid behind the dorsal third of the lamina and has short, paired posterolateral projections which contact a long projection of the squamosal laterally. The latter bone has two large and important processes: the intermediate process, in contact with the posterodorsal process of the prootic; and the anteroventral process, in contact with the central process of the prootic (Fig. 13.8a). An elliptical, well-developed post-temporal foramen is limited by these processes. The prootic incisure is located behind the epipterygoid, ventral to the body of the prootic and anterior to the connection between the anteroventral process of the squamosal and the central process of the prootic.

An ovoid interorbital vacuity is located in front of the anterior margin of the lamina of the epipterygoid and extends anteriorly to the ossified interorbital septum (Fig. 13.8b). Small and fragile bones in the right interorbital vacuity are interpreted as sclerotic ossicles. A thin orbitosphenoid is located dorsally and extends the length of the orbit. The left and right orbitosphenoids together form a 'V'-shaped structure, which is continued ventrally by an interorbital septum located in the anterior third of the orbit. Ventrally, this septum contacts the palatine and pterygoid.

Occiput

Most of the occipital region is masked by the cervical vertebrae (Fig. 13.5). Description of this region is based on digital preparation of the specimen (Fig. 13.2d). The occiput is triangular, with the foramen magnum having half of the height of the occipital plate (Fig. 13.2d). The dorsally situated interparietal bears a well-developed median ridge that continues ventrally onto the supraoccipital. The large tabular occupies a quarter of the occipital plate and meets the interparietal, supraoccipital, and exoccipital medially by means of a suture which extends obliquely in a ventrolateral direction. The tabular does not form part of the margin of the post-temporal fenestra. The ventral portion of the occiput comprises the paroccipital process which forms the base of the post-temporal fenestra and has a ventromedial projection on the occipital ventral margin. The lateral margin of this process forms a faint ridge that continues dorsally as a well-defined ridge on the suture between the lateral margin of the tabular and the squamosal.

Mandible

The dentary has a boomerang-shaped outline. The ventral margin of the bone is convex (Figs. 13.2i, g, 13.5b). The anterior portion of the dentary is high and has a very strong symphysis that maintains the anterior mandibular rami in its natural placement (Fig. 13.5c). A poorly defined dentary angle is present below the level of the middle of the orbit. The coronoid process is oriented obliquely and extends dorsally behind the postorbital bar (Figs. 13.2i, 13.5b), so that the process is located in the anterior portion of the temporal opening and near to the zygomatic arch. A fossa, delimited by the posteroventral margin of the dentary and the dorsal margin of the coronoid process, extends anteriorly on the coronoid and part of the horizontal processes of the dentary (Fig. 13.5b).

Behind the dentary symphysis a laminar splenial is present and meets its counterpart (Fig. 13.5c) in the midline. In SAM-PK-K10160, only the anterior part of the splenial is preserved and extends posteriorly as far as the level of the diastema between the lower canine and first postcanine. Further posteriorly, just below the internal expansion at the implantation of the postcanines, there is evidence of the original extent of the splenial: a shallow, horizontal canal close to the ventral margin of the dentary (Figs. 13.2g, h, 13.5c). The angular and prearticular form a stout bar, which is covered medially by a large coronoid bone, which was only possible to observe in CT images. This bone is preserved in contact with the pterygoid process of the cranium. Only the dorsal-most portion of the reflected lamina is preserved on both sides of the mandible (Fig. 13.5b). It is a thin plate which has a small, rounded area overhanging the remaining lamina on the right side. The medial portion of the bar formed by the prearticular is lateromedially expanded in the region of the craniomandibular articulation (Fig. 13.2f). No suture is visible between this element and the articular. The surangular is a strip of bone more developed in height posteriorly. On the medial side of the mandible the surangular forms a semicircle and reaches its greatest height at a level just below the top of the coronoid process (Fig. 13.2g). In lateral view, the surangular forms an overhanging strip of bone dorsal and posterior to the margin of the reflected lamina.

Dentition

The dental formula of the specimen is I4/3-C1/1-PC6/6. Only the most distal parts of the incisors have enamel as evidenced by their light brown colour, whereas the white basal portions of the exposed 'crowns' probably comprise only dentine. The upper incisors are smoothly conical, directed ventrally (Fig. 13.2c, i), and progressively increase in size posteriorly. The lower incisors are large and procumbent, with the first one being remarkably large, more than the double of the size of the upper incisors (Fig. 13.2e, i, j). Lower incisors reduce in size posteriorly; the second lower incisor is smaller than the first, but larger than the upper incisors (Fig. 13.2c). The third lower incisor is smaller than the upper incisors (Fig. 13.2e, j). The upper canine is located half way along the length of the snout behind a well-defined paracanine fossa that accommodates the lower canine in occlusion (Fig. 13.2a, i). The upper canine is almost twice the size of the first upper incisor, but smaller than the first lower incisor (Fig. 13.2i). Serrations are absent on the canine and all incisors.

All postcanines are of similar size with the exception of the smaller last lower postcanine (Fig. 13.5c). As the teeth are tightly occluded, postcanine crown morphology is only partially visible and only their convex labial and lingual sides are evident. On the left side (the less distorted in relation to the mandible, but more distorted relative to the cranium) each upper postcanine occludes with two teeth of the mandible. On the right, each upper postcanine occludes with only one postcanine (Fig. 13.5c).

Cervical Vertebrae

Four cervical vertebrae are preserved in articulation with the cranium (Fig. 13.9a–c). The axial centrum is larger than subsequent ones and all of them show a marked ventral midline keel (Fig. 13.9b). The atlas neural arch has a short dorsal plate with a small posterior projection representing the postzygapophysis and an additional ventral projection. The atlantal intercentrum is a large rectangular plate. Both the lateral projection of the intercentrum and the ventral projection of the neural arch are in contact with the atlantal rib (Figs. 13.7b, 13.9b) which is a large quadrangular plate with a short distal fusiform projection. From the right side

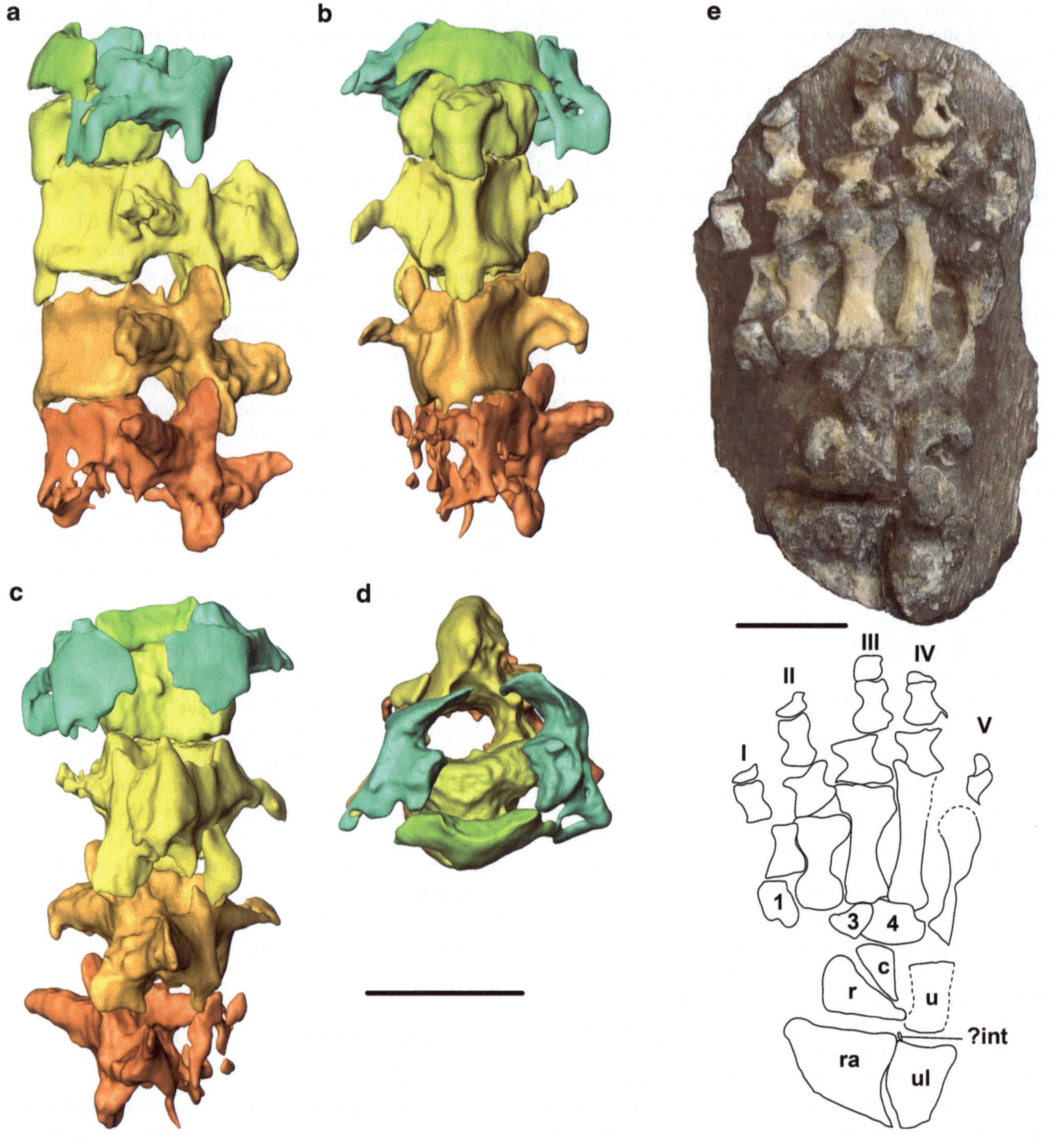

Fig. 13.9 SAM-PK-K10160. Virtual reconstruction of cervical vertebrae; **a** lateral, **b** ventral, **c** dorsal; and **d** anterior views; **e** right manus in dorsal view. *c* lateral centrale, *int* intermedium, *r* radiale, *ra* distal end of the radius, *u* ulnare, *ul* distal end of the ulna, *1–4* indicate distal carpals; *I–V* indicate digits. *Dashed lines* indicate lack of bone in the metacarpals and margin of the bone observed in ventral view in the case of the ulnare. Scale 1 cm

there is a contact between the atlas arch and the axis through the atlantal postzygapophysis (Fig. 13.9a).

The neural spine of the axis, which is expanded distally and forms a bulbous structure at its distal end, is only slightly larger than those of the remaining cervical vertebrae (Fig. 13.9c). The postzygapophyses of the axis are well-developed and articulate horizontally with the prezygapophyses of the third cervical. An axial transverse process is located at the level of the base of the prezygapophysis and articulates with the proximal portion of the rib. Neural arches of the two remaining cervicals (C3 and C4?) are the same height as the axis, but they are progressively shorter posteriorly (Fig. 13.9a). As in the axis, articular facets between zygapophyses are horizontal in these vertebrae.

Manus

The distal portions of radius and ulna are poorly preserved, with the radial larger than the ulnar portion. Seven carpal bones are visible in dorsal view (Fig. 13.9e). The triangular radiale shows a convex medial margin and a slight depression toward its medial side. This bone is more exposed dorsally than ventrally. The presence of the intermedium is inferred from a small osseous surface preserved close to the distal margins of the ulna and radius. The lateral centrale is also triangular and its pointed end is positioned between the radiale and the ulnare. The ulnare does not have a clear morphology in dorsal view, but ventrally is a quadrangular bone. Three distal carpals are preserved (Fig. 13.9e). The larger, interpreted as the third carpal is in contact with the ulnare, intermedium, and second distal. In ventral view the third carpal shows a rectangular morphology (wider than long) and a visible depression occupying two-thirds of the ventral surface of the bone. The smaller, triangular second carpal contacts parts of the third and second metacarpals. First distal carpal is larger than the second and smaller than the third. The first distal carpal is roughly quadrangular and has a low medial portion separated from the higher lateral portion by a well-marked concavity. Along its distal margin, it is in contact with the entire proximal surface of the first metacarpal; laterally, it meets the medial portion of the proximal surface of the second metacarpal (Fig. 13.9e). In addition, two ex situ bones located below the fifth, fourth and third metacarpals are probably remains of the other carpals.

The fourth and fifth metacarpals are long and the fourth is more robust than the fifth. Metacarpals three to one are successively shorter. The entire lateral margin of the first metacarpal is in contact with the central to distal portion of the lateral margin of the second metacarpal (Fig. 13.9e). The proximal portions of the second and third metacarpals are in contact, and the distal projection of the second metacarpal meets part of the diaphysis of the third metacarpal. This pattern of contact is also present between the third and fourth metacarpals. The second metacarpal is expanded at both the proximal and distal ends with the distal end being the largest. In contrast, the first metacarpal is much more expanded on the proximal end (Fig. 13.9e). The preserved phalanges are 2, 3, 3, 2(?), 2(?) (Fig. 13.9e). All are robust and quadrangular, with the proximal end usually more transversely expanded than the distal. No ungual phalanges are completely preserved (Fig. 13.9e).

Comparsion of Bauriid Species by Procrustes Analysis

The cranium of *Microgomphodon oligocynus* (based on SAM-PK-K10160) differs from that of *Bauria cynops* (based on BP/1/1180) in the following features: (1) cranium is higher and slightly larger, but less wide (Fig. 13.4a, a′, b, b′); (2) snout is higher (Fig. 13.4b′); (3) nasal opening ovoid compared to the elliptical and more anteriorly oriented opening of *B. cynops* (Fig. 13.4b, b′, a′); (4) orbit rounded, relatively larger, and more anteriorly oriented (Fig. 13.4b′, a′); (5) posterior angle of the zygoma much more pronounced (Fig. 13.4b, b′); (6) projection of the pterygoid process placed further anteriorly (half way along the cranial length) (Fig. 13.4b, b′); (7) lateral margin of the pterygoid process more robust and lateral margin of the suborbital vacuity directed inward (these two trends probably play an important role in the reduction of the size of the suborbital vacuity in *M. oligocynus*); (8) internal choanae less expanded antero-posteriorly (Fig. 13.4a′); (9) anterior and posterior margins of the postcanine series more expanded and more outwardly directed (Fig. 13.4a, a′); (10) canines more posteriorly curved (Fig. 13.4b); (11) incisors located more inwardly (Fig. 13.4b′).

In the mandible, the major differences between *Microgomphodon oligocynus* (SAM-PK-K10160) and *Bauria cynops* (BP/1/1180) are: (1) higher symphysis, less elongated in cranial view (Fig. 13.4c, c′); (2) horizontal ramus much more elevated and the corpus of the mandible relatively short (Fig. 13.4c′, d′); (3) higher and more laterally directed coronoid process (Fig. 13.4c, c′, d, d′); (4) depression of the posterior portion of the dentary deeper (Fig. 13.4c); (5) angular bone posterior to the dentary located more ventrally (Fig. 13.4c′); (6) shelf lateral to postcanines more concave (Fig. 13.4d); (7) postcanine series more curved laterally (Fig. 13.4d, d′); (8) incisors and canines relatively higher and more developed (Fig. 13.4c, d); (10) canines directed dorsally, whereas they are anteriorly oriented in *B. cynops* (Fig. 13.4d); (11) presence of diastema between the first postcanine and the canine (Fig. 13.4d, d′).

Table 13.3 Cranial length and selected characters of specimens studied

Taxa	Cranial length (mm)	Postcanine number	Postorbital bar	Pineal foramen
Microgomphodon oligocynus	65	6	Complete	Present
Sesamodon browni	*80	6	Complete	?
Melinodon simus	?	At least 7	?	?
Watsoniella breviceps	*85	6	?	?
Herpetogale marsupialis	89	5	Complete	Present
Bauria cynops (holotype)	*122	10/11	Incomplete	Absent
Aelurosuchus browni	92	8–?9	?	?
Sesamodontoides pauli	*80–85	8 (lp)	?	?
Bauria cynops (AMNH FARB 5622)	130	11	Incomplete	Absent
Bauria cynops (BP/1/3770) (Brink, 1963: 5th specimen)	117	9	Incomplete	Absent
Bauria cynops (BP/1/1180) (Brink, 1963: 3rd specimen)	114	At least 9	Incomplete	Absent
Bauria cynops (BP/1/2523) (Brink, 1963: 6th specimen)	?	10 (lp)	?	?
Bauria robusta	*132	11	?	?

Postcanine number refers to upper teeth excepts where lp is indicated
*Inferred

Discussion

Taxonomy of Southern African Bauriidae

The bauriids, which include the last therocephalian survivors, are well-represented in the Karoo Basin of South Africa and are also known from Namibia (Keyser 1973a, b; Keyser and Brink 1977–1978), China (Sun 1981, 1991) and Russia (Tatarinov 1973, 1974; Battail and Surkov 2000). The bauriids from southern Africa are small-sized taxa that attain a skull length of up to 133 mm. They can be easily recognized by the characteristic postcanines, as they are the only therocephalian group with buccolingually expanded postcanines (Crompton 1962). In addition, they show an osseous secondary palate comprised of the premaxilla, maxilla, and in some cases, a small contribution of the vomer. Another trait not known in other therocephalians is a maxillary shelf lateral to the tooth row series, a feature which is similar to that of most gomphodont cynodonts (i.e., cynodonts with buccolingually expanded postcanines). In Bauriidae, a similar expansion is also present on the dentary lateral to the lower postcanines and is a clear autapomorphy for the family.

We distinguish two morphotypes of Bauriidae from the CAZ of southern Africa, which represent different species. One morphotype is characterized by an incomplete postorbital bar, absence of parietal/pineal foramen, a large number of postcanines (from eight to 11), and the canine usually located anteriorly in the snout (Table 13.3). This type is herein assigned to the species *Bauria cynops* Broom (1909), of which we regard *Aelurosuchus browni*, *Baurioides watsoni*, *Microhelodon eumerus*, *Sesamodontoides pauli*, and *Bauria robusta* as synonyms. The inclusion of *A. browni* in this species is based on the anterior placement of the canine and the weak development of the maxillary shelf. *S. pauli* is tentatively included in *B. cynops* based on the postcanine crown that appears to be thin in lateral view. *Microhelodon eumerus* is also tentatively included in this species based on the location of the postcanines near the lateral margin of the maxilla (suggesting little development of the maxillary shelf) and the overall morphology of the preserved postcanines. Other specimens included within this species are shown in Table 13.4. *Bauria robusta* was originally described as being 20 % larger than the largest recognized specimen of *B. cynops* (AMNH FARB 5622); with 11 postcanines, small canines, powerful cheek bulges with deep depressions below and no interpterygoid vacuity. In Brink's (1965) description of the new species, the large size was recognized as the most important trait. Brink (1965, p. 123) noted that the specimen is represented by two-thirds of the skull and that the total preserved skull length is 100 mm. His estimated total skull length of 168 mm appears exaggerated in comparison with our estimation of around 133 mm for this specimen, which is similar to the size of the specimen AMNH FARB 5622. It is clear that size per se is not enough to differentiate this species. Considering the high number of postcanines and the location of the canine well anterior in the snout, we consider this species a junior synonym of *Bauria cynops*. One important character suggesting a difference between *B. robusta* and *B. cynops* was the absence of the interpterygoid vacuity. Further preparation of Brink's holotype indicates that the

Table 13.4 Geographic distribution of Bauriidae specimens from South Africa

Taxon	Specimen	Locality	District
Microgomphodon oligodens	NHMUK R3305	Near Aliwal North	Aliwal North
	AMNH FARB 5517		Aliwal North
	SAM-PK-5865	Erf 1 Commonage	Aliwal North
	SAM-PK-5866	Erf 1 Commonage	Aliwal North
	BP/1/4655	Hugoskop 620	Rouxville
	SAM-PK-K10160	Lemoenfontein 44	Rouxville
	BSP 1934-VIII-13	Kaaimansgat	Rouxville
	USNM PAL 289115	Matyantya	Cacadu
	USNM PAL 412433[a]	Matyantya	Cacadu
	USNM PAL 412401[a]	Matyantya	Cacadu
	NMQR 3189	Eerstegeluk 131	Bethlehem
	NMQR 3183	Jisreel 419	Harrismith
	NMQR 3596		Wepener
Bauria cynops	SAM-PK-5875	Melkspruit 12	Aliwal North
	BP/1/4678	Betjieskraal 36	Rouxville
	UCMP 4284	Bethel	Rouxville
	RC 114	Lady Frere Area	Cacadu
	BP/1/1180	Matyantya	Cacadu
	USNM PAL 23331	Lady Frere Commonage	Cacadu
	SAM–PK-1333	Vaal Bank 134	Albert
	NHMUK R4095	Essex	Albert
	BP/1/3770	Cragievar	Albert
	BP/1/1685	Grootdam	Albert
	BP/1/1679[b]	Grootdam	Albert
	AMNH FARB 5622	Winnaarsbaken	Albert
	BP/1/2523	Lady Frere	Albert
	BP/1/2837	Bersheba	Albert
	NHMUK R3581	Near Burgersdorp	Albert

[a] Specimens not observed, they are included here based on comments by Hotton in the museum catalog identifying them as *Sesamodon*
[b] Specimen lost, observed in photos, sketches and notes of Dr Mendrez–Carroll

vacuity is indeed present, however. It is reduced in size relative to that of BP/1/1180, but is similar in size to that of BP/1/3770. This suggests that the tendency toward reduction of the interpterygoid vacuity known in several therocephalians and some cynodonts (e.g., *Thrinaxodon*), appears to also be the case for Bauriidae. It should be mentioned that the interpterygoid vacuity in the large skull AMNH FARB 5622 is described as small by Broom (1937, p. 3) and as large by Boonstra (1938, p. 172), but the vacuity of this specimen is indeed smaller than that of smaller specimens of *B. cynops*.

The second type, herein assigned to *Microgomphodon oligocynus* Seeley (1895), has a complete postorbital bar, a pineal foramen, a small number of postcanines and the canine usually located posteriorly in the snout (Table 13.3). Another important trait is the presence of an extended fossa on the lateral surface of the dentary, in a similar placement as the masseteric fossa in cynodonts. This trait is clearly represented in SAM-PK-K10160 as well as in GSN R337, but it is absent in BSP 1934-VIII-13. In addition to the new specimen described here, *M. oligocynus* includes all material previously referred to *Sesamodon browni*, *Melinodon simus*, *Watsoniella breviceps*, and *Herpetogale marsupialis*. *Sesamodon browni* is included in this species as it has a complete postorbital bar (Fig. 13.1c), the canine is located posteriorly on the snout, and it has a reduced number of postcanines (Table 13.3). *Melinodon simus* is included based on a posterior keel of the vomer which extends posteriorly almost as far as the base of the pterygoid processes. Several features, including reduced number of postcanines, the posterior location of the canine, placement of the canine well offset from the postcanine series and restricted exposure of the frontal on the dorsal margin of the orbit, allow the inclusion of *W. breviceps* in *M. oligocynus*. NMQR 3189, described by King (1996) as *Bauria* sp., is also included in

a b

c d

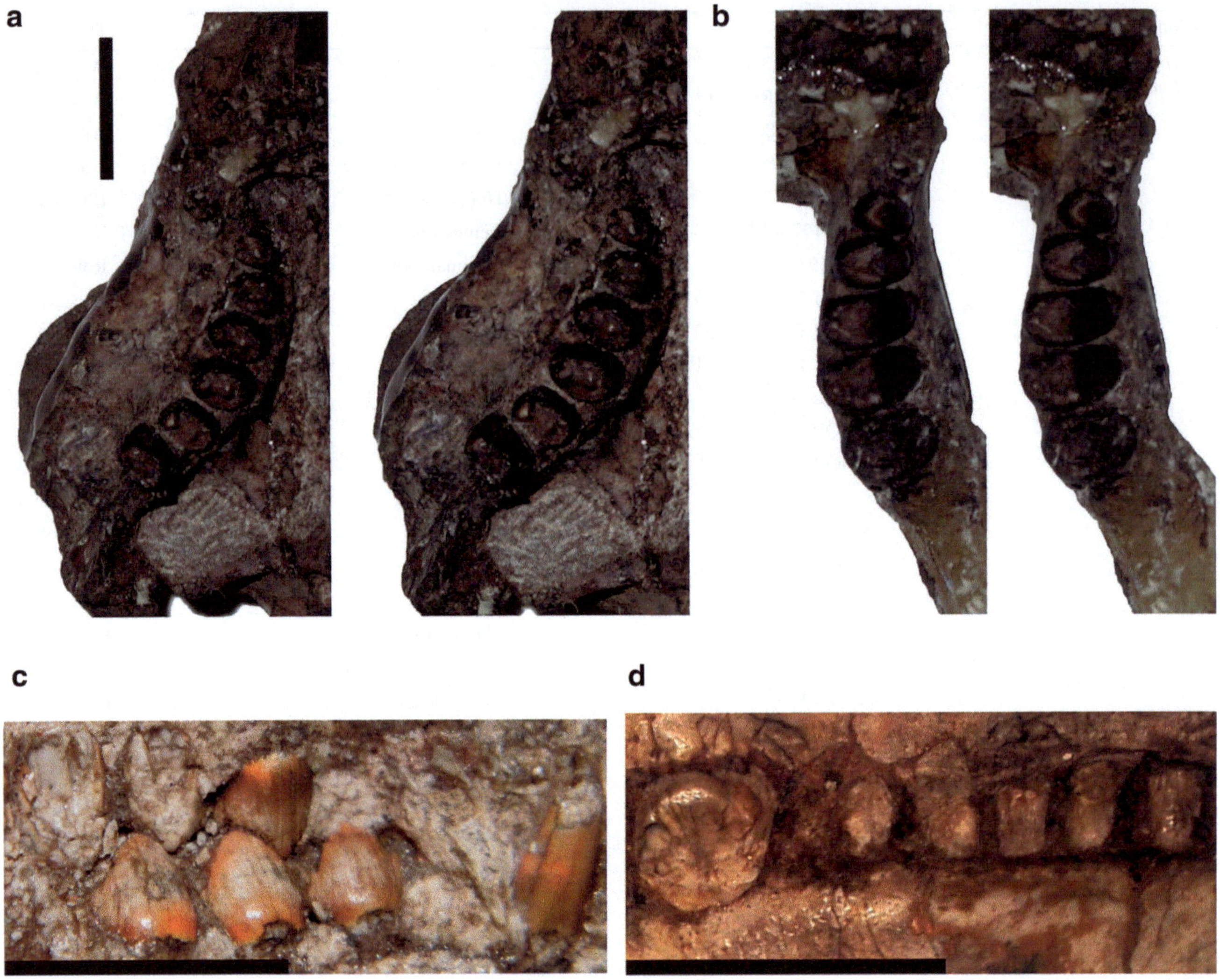

Fig. 13.10 Postcanines of Bauriidae. Stereopair in dorsal view of **a** upper and **b** lower postcanines of *Microgomphodon oligocynus* (BP/1/4655); **c** lateral view of the right upper and lower postcanines of *Microgomphodon oligocynus* (BMNH R3305); **d** lateral view of the left upper postcanines of *Bauria cynops* (BP/1/1180). Scale 1 cm

M. oligocynus, based on the morphology of the postcanines. The postcanine crowns of this taxon have an ovoid outline (Fig. 13.10a, b) with a long curved margin in lateral view, in contrast to the more restricted lateral exposure in the postcanines of *B. cynops* (compare Fig. 13.10c, d).

In addition to the above mentioned features, the Procrustes analysis allowed us to recognize the following additional differences: *M. oligocynus* has a taller but less wide cranium, taller snout, temporal opening more expanded laterally and smaller suborbital vacuity. The mandible of *M. oligocynus* has a higher symphysis, relatively shorter corpus, and more laterally directed coronoid process. It is interesting to note that the snout length, which intuitively appears as a noteworthy difference between these taxa, is not represented in the variations highlighted by the Procrustes analysis. This result is probably related to the condition of traits in the specimens analysed. The proportion of the length of the snout in relation to the basal cranial length in SAM-PK-K10160 and BP/1/1180 is the closest between representatives of the two different species (Fig. 13.6).

The presence of an ossified interorbital wall between the orbitosphenoid and the palatine in SAM-PK-K10160 is a feature infrequently found in therocephalians and cynodonts. Further preparation of BP/1/1180 enabled recognition of this structure in *Bauria cynops*. This condition was previously reported for whaitsiids from Tanzania (Kemp 1972) and among cynodonts was observed only in a specimen of *Diademodon tetragonus* (BSP 1934-VIII-20). We interpret this lamina as an ossified mesethmoid and the recovery of this delicate feature in both bauriid species indicates that this character is indeed present in the group.

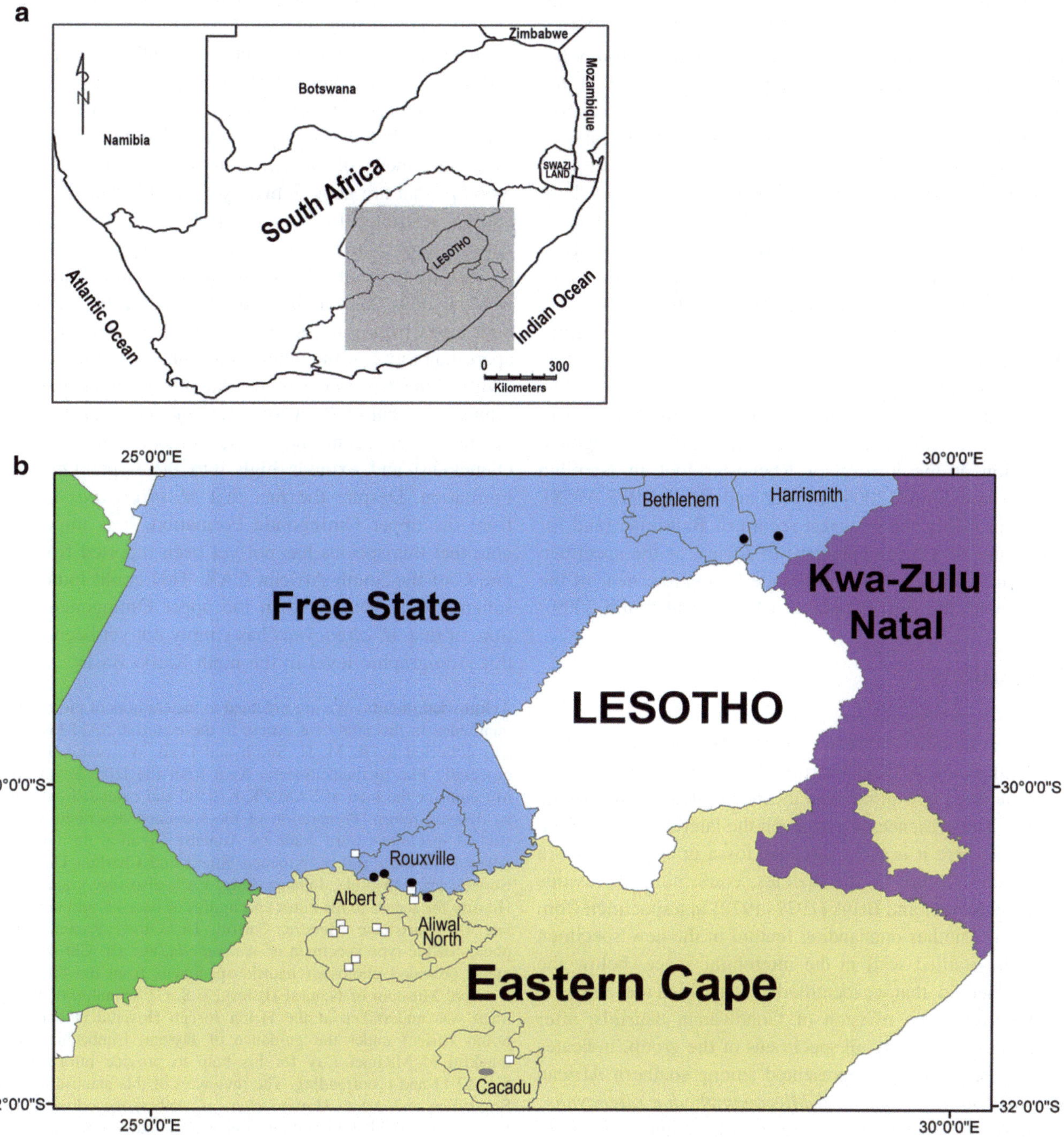

Fig. 13.11 Distribution of Bauriidae in the South African Karoo. *Grey area* in (**a**) is enlarged in (**b**). Only the districts with bauriid records are represented. *Round black, Microgomphodon oligocynus*; *square white Bauria cynops*; *ellipse grey*, the two species

Temporal and Geographic Distribution of Bauriidae in Southern Africa

Bauriids from the CAZ are represented by at least 25 specimens and their geographic distribution in the Karoo is restricted to the southwestern part of the basin, with only two exceptions from the northeastern portion (Table 13.4; Fig. 13.11). In fact, most of the bauriid localities are concentrated in the middle of the Eastern Cape and southern Free State provinces, with the greatest distance between localities being around 150 km.

Three locality clusters have been recognized in the Karoo. The northeastern cluster from outcrops in the Harrismith and Bethlehem districts have produced only

Microgomphodon oligocynus (Fig. 13.11), and are from Subzone A (i.e., Olenekian) of the CAZ (Neveling 2004). The other two clusters are in the districts of Rouxville, Albert, and Aliwal North on the one hand, and in the District of Cacadu on the other (Fig. 13.11). A relative abundance of *Bauria cynops* have come from the first of these two clusters, whereas in Cacadu, the locality of Matyantya is the only case in which remains of both species are found together. Most of the outcrops of these two clusters fall within Subzone B (i.e., lower Anisian) of the CAZ. However the two localities recording *M. oligocynus*, Hugokop 620 and Kaaimansgat Rouxville, also have remains of the trirachodontid *Langbergia modisei* (Abdala et al. 2006) and therefore are best interpreted as representing the Subzone A.

The only record in Africa of a CAZ-equivalent bauriid outside of South Africa is from the upper Omingonde Formation at the locality of Rhenosterkloof in Namibia (Abdala and Smith 2009; Keyser and Brink 1977–1978). This is probably the youngest record of Bauriidae (and for that matter of a therocephalian) in Africa, as the specimen was found in the upper Omingonde Formation, near of the contact with the Etjo Formation (Abdala and Smith 2009).

Conclusions

An exceptionally preserved new specimen (SAM-PK-K10160) of the African bauriid *Microgomphodon oligocynus* provides a new glimpse of morphological features of the species. The presence of a fossa on the lateral surface of the dentary (where it is the masseteric fossa in cynodonts) is a remarkable feature of this species, confirming a previous report by Keyser and Brink (1977–1978) in a specimen from Namibia. Another outstanding feature in the new specimen is a thin ossified wall in the interobital space, below the orbitosphenoid, that we identified as an ossified mesethmoid.

Our taxonomic revision of Gondwanan bauriids, after examination of nearly all specimens of the group, indicates that two species were represented among southern African bauriids: *Bauria cynops* and *Microgomphodon oligocynus*.

Microgomphodon oligocynus is differentiated from *B. cynops* on the basis of clear-cut morphological features such as smaller size (89 vs. ~130 mm of basal skull length in the largest specimens of each species, respectively) and the presence of a complete postorbital bar, pineal foramen, and comparatively large orbits. In addition, *M. oligocynus* features fewer postcanine teeth, with a long curved margin in lateral view, that contrast with the numerous teeth presenting more restricted lateral exposure observed in *B. cynops*.

Further putative differences between these species are provided by a Procrustes analysis of the two best preserved specimens of these species: *M. oligocynus* has a taller but less wide cranium, taller snout, temporal opening more expanded laterally, and smaller suborbital vacuity. The mandible of *M. oligocynus* has a higher symphysis, relatively short corpus, and more laterally directed coronoid process.

Microgomphodon oligocynus appears to have been widely distributed geographically and temporally. In contrast, specimens of *Bauria cynops* are known only from outcrops that are geographically close together (~150 km) and are restricted to the subzone B of the CAZ (probably Anisian-aged). *Microgomphodon oligocynus* is the only species from the northeastern portion of the Karoo Basin and the only African bauriid recorded from Namibia, in both cases from outcrops which are geographically further apart than those of the other representatives of the group. In addition the age range of the latter species appears to be longer than that of *B. cynops*. *Microgomphodon oligocynus* has its earliest occurrence in Subzone A of the CAZ (Late Olenekian) and extends high into the upper Omingonde Formation. Despite the fact that *M. oligocynus* is known from the upper Omingonde Formation, it is important to note that this species has not yet been reported from subzone C of the South African CAZ. This could indicate that subzone C is younger than the upper Omingonde Formation, or that *M. oligocynus* has simply not yet been found at this stratigraphic level in the main Karoo Basin.

Acknowledgments We are indebted to the curators of the collections mentioned in this study for access to the material. SAM-PK-K10160 was collected by R. M. H. Smith and Derik Wolvaardt. FA would especially like to thank Sheena Kaal from the Iziko-South African Museum for the loan of SAM-PK-K10160 and extended patience for its delayed return. Preparation of the specimen described and other studied specimens were made by Annelise Crean at the Iziko-South African Museum and A. Nthaopa Ntheri and Charlton Dube at the Bernard Price Institute, Johannesburg. FA is also very grateful to Jim Hopson for access to his notes and figures of bauriids and those of the late Dr. Christiane Mendrez-Carroll. Dr. Oliver Rauhut supplied photos of the type specimen of *Watsoniella*, and Dr. Christian Kammerer provided photos of bauriid specimens from the Smithsonian National Museum of Natural History, U.S. CT-scanning of the specimen was undertaken at the Helen Joseph Hospital (Johannesburg, South Africa) under the guidance of Jaymati Limbachia. We are thankful to Michael Day for his help to provide information for Fig. 13.11 and proofreading. The reviewers of this manuscript, Trond Sirgurdsen and Adam Huttenlocker, as well as the editor Christian Kammerer provided important suggestions that improved the final result of this work. The research of FA and BR research was funded by DST, NRF and PAST. TJ's research was supported by the Claude Leon Foundation.

References

Abdala, F. (2007). Redescription of *Platycraniellus elegans* (Therapsida, Cynodontia) from the Lower Triassic of South Africa, and the cladistic relationships of Eutheriodontia. *Palaeontology, 50*, 591–618.

Abdala, F., & Smith, R. M. H. (2009). A middle Triassic cynodont fauna from Namibia and its implications for the biogeography of Gondwana. *Journal of Vertebrate Paleontology, 29*, 837–851.

Abdala, F., Neveling, J., & Welman, J. (2006). A new trirachodontid cynodont from the lower levels of the Burgersdorp Formation (Lower Triassic) of the Beaufort Group, South Africa and the cladistic relationships of Gondwanan gomphodonts. *Zoological Journal of the Linnean Society, 147*, 383–413.

Battail, B., & Surkov, M. V. (2000). Mammal-like reptiles from Russia. In M. J. Benton, E. N. Kurochkin, M. A. Shishkin, & D. Unwin (Eds.), *The age of dinosaurs in Russia and Mongolia* (pp. 86–119). New York: Cambridge University Press.

Bookstein, F. L. (1991). *Morphometric tools for landmark data: Geometry and biology*. Cambridge: Cambridge University Press.

Boonstra, L. D. (1938). On a South African mammal-like reptile, *Bauria cynops*. *Palaeobiologica, 6*, 164–183.

Botha, J., Abdala, F., & Smith, R. (2007). The oldest cynodont: New clues on the origin and early diversification of the Cynodontia. *Zoological Journal of the Linnean Society, 149*, 477–492.

Brink, A. S. (1963). On *Bauria cynops* Broom. *Palaeontologia Africana, 8*, 39–56.

Brink, A. S. (1965). A new large bauriamorph from the *Cynognathus*-Zone of South Africa. *Palaeontologia Africana, 9*, 123–127.

Brink, A. S., & Kitching, J. W. (1953). On some new *Cynognathus* zone specimens. *Palaeontologia Africana, 1*, 29–48.

Broili, F., & Schröder, J. (1935). Beobachtungen an Wirbieltieren der Karrooformation. VII. Ein neuer Bauriamorphe aus der Cynognathus-zone. *Sitzungsberichte der mathematisch-naturwissenschaftlichen Abteilung der Bayerischen Akademie der Wissenschaften zu München*, 1935, 21–63

Broom, R. (1903). On the classification of the theriodonts and their allies. *Report of the South African Association for the Advancement of Science, 1*, 286–294.

Broom, R. (1905). Preliminary notice of some new fossil reptiles collected by Mr. Alfred Brown at Aliwal North, South Africa. *Records of the Albany Museum, 1*, 269–275.

Broom, R. (1906). On a new cynodont reptile (*Aelurosuchus browni*). *Transactions of the South African Philosophical Society, 16*, 376–378.

Broom, R. (1909). Notice of some new South African fossil amphibians and reptiles. *Annals of the South African Museum, 7*, 270–278.

Broom, R. (1911). On the structure of the skull in cynodont reptiles. *Proceedings of the Zoological Society of London, 81*, 893–925.

Broom, R. (1913). South African fossil reptiles. *The American Museum Journal, 13*, 334–346.

Broom, R. (1915). Catalogue of types and figured specimens of fossil vertebrates in the American Museum of Natural History. II.- Permian, Triassic and Jurassic reptiles of South Africa. *Bulletin of the American Museum of Natural History, 25*, 105–164.

Broom, R. (1925). On some carnivorous therapsids. *Records of the Albany Museum, 3*, 309–326.

Broom, R. (1931). Notices of some new genera and species of Karroo fossil reptiles. *Records of the Albany Museum, 4*, 161–166.

Broom, R. (1937). On the palate, occiput and hind foot of *Bauria cynops* Broom. *American Museum Novitates, 946*, 1–6.

Broom, R. (1950). Some fossil reptiles from the Karroo beds of Lady Frere. *South African Journal of Science, 47*, 86–88.

Crompton, A. W. (1962). On the dentition and tooth replacement in two bauriamorph reptiles. *Annals of the South African Museum, 46*, 231–255.

Dryden, I. L., & Mardia, K. V. (1998). *Statistical shape analysis*. London: Wiley.

Gower, J. C. (1975). Generalized Procrustes analysis. *Psychometrika, 40*, 33–50.

Haughton, S. H. (1922). On some Upper Beaufort Therapsida. *Transactions of the Royal Society of South Africa, 10*, 299–307.

Hopson, J. A., & Barghusen, H. R. (1986). An analysis of therapsid relationships. In N. Hotton, P. D. MacLean, J. J. Roth, & E. C. Roth (Eds.), *The ecology and biology of mammal-like reptiles* (pp. 83–106). Washington, DC: Smithsonian Institution Press.

Huttenlocker, A. (2009). An investigation into the cladistic relationships and monophyly of therocephalian therapsids (Amniota: Synapsida). *Zoological Journal of the Linnean Society, 157*, 865–891.

Kemp, T. S. (1972). Whaitsiid Therocephalia and the origin of cynodonts. *Philosophical Transactions of the Royal Society of London B, 264*, 1–54.

Keyser, A. W. (1973a). New Triassic vertebrate fauna from south west Africa. *South African Journal of Science, 69*, 113–115.

Keyser, A. W. (1973b). A new Triassic vertebrate fauna from south west Africa. *Palaeontologia Africana, 16*, 1–15.

Keyser, A. W., & Brink, A. S. (1977–1978). A new bauriamorph (*Herpetogale marsupialis*) from the Omingonde Formation (Middle Triassic) of South West Africa. *Annals of the Geological Service, 12*, 91–105.

King, G. (1996). A description of the skeleton of a bauriid therocephalian from the Early Triassic of South Africa. *Annals of the South African Museum, 104*, 379–393.

Neveling, J. (2004). Stratigraphic and sedimentological investigation of the contact between the *Lystrosaurus* and the *Cynognathus* assemblage zones (Beaufort Group: Karoo Supergroup). *Bulletin of the Council for Geosciences, 137*, 1–165.

O'Higgins, P., & Jones, N. (2006). Morphologika—tools for statistical shape analysis. York: Hull York Medical School. http://www.york.ac.uk/res/fme/resources/software.htm

Rohlf, F. J., & Slice, D. E. (1990). Extensions of the Procrustes method for the optimal superimposition of landmarks. *Systematic Zoology, 39*, 40–59.

Seeley, H. G. (1895). Researches on the structure, organization, and classification of the fossil Reptilia. Part IX, Section 4. On the Gomphodontia. *Philosophical Transactions of the Royal Society of London B, 186*, 1–57.

Sun, A.-L. (1981). Reidentification of *Traversodontoides wangwuensis* Young. *Vertebrata Palasiatica, 19*, 1–4.

Sun, A.-L. (1991). A review of Chinese therocephalian reptiles. *Vertebrata Palasiatica, 29*, 85–94.

Tatarinov, L. P. (1973). Cynodonts of Gondwanan habit in the Middle Triassic of the USSR. *Paleontological Journal, 1973*, 200–205.

Tatarinov, L. P. (1974). Theriodonts of the USSR. *Transactions of the Palaeontological Institute, Academy of Sciences of the USSR, 143*, 1–250.

van den Heever, J. A. (1994). The cranial anatomy of the early Therocephalia (Amniota: Therapsida). *Annals of the University of Stellenbosch, 1*, 1–59.

Watson, D. M. S. (1913). On some features of the structure of therocephalian skull. *Annals and Magazine of Natural History, 8*, 65–79.

Watson, D. M. S. (1914). Notes on some carnivorous therapsids. *Proceedings of the Zoological Society of London, 1914*, 1021–1038.

Watson, D. M. S. (1921). The bases of classification of the Theriodontia. *Proceedings of the Zoological Society of London, 1921*, 35–98.

Zollikofer, C. P. E., Ponce De León, M. S., & Martin, R. D. (1998). Computer-assisted paleoanthropology. *Evolutionary Anthropology: Issues, News, and Reviews, 6*, 41–54.

Chapter 14
The Traversodontid Cynodont *Mandagomphodon hirschsoni* from the Middle Triassic of the Ruhuhu Valley, Tanzania

James A. Hopson

Abstract *Mandagomphodon hirschsoni* (gen. nov., comb. nov.) is one of three species of traversodontid cynodont placed in *Scalenodon* [type species *S. angustifrons* (Parrington, 1946)] by Crompton (1972). It is based on a partial skull and lower jaws from the Middle Triassic of the Ruhuhu Valley, southwestern Tanzania. The upper postcanine teeth were used to diagnose species of *Scalenodon*, but newer traversodontid material indicates that the three species represent distinct genera. Material of "*S.*" *hirschsoni*, except for the postcanines, has not been described. It is unusual among traversodontids in having only three upper incisors, which are enlarged and procumbent. Three enlarged, procumbent anterior lower teeth are interpreted as two incisors and a canine. Analysis of postcanine wear facets indicates that the power stroke of the lower teeth was entirely in a posterior direction, including a slightly downward and backward grinding movement at the end of the stroke.

Keywords Africa • Dental function • Gomphodontia • Manda beds • Therapsida • Theriodontia

This chapter includes one or more new nomenclatural-taxonomic actions, registered in Zoobank, and for such purposes the official publication date is Sep 2013.

J. A. Hopson
Department of Organismal Biology, University of Chicago, 1027 East 57th Street, Chicago, IL 60637, USA

Present Address:
J. A. Hopson (✉)
3051 Piney Ridge Road, Ludington, MI 49431, USA
e-mail: jhopson@uchicago.edu

Introduction

Cynodonts are a clade of synapsids that includes Mammalia as its most derived subgroup. The non-mammalian component of the Cynodontia comprises a series of taxa mainly of Late Permian and Triassic age that includes, in addition to the persistently carnivorous ancestral lineage of mammals, several specialized subgroups of which one became highly modified for an herbivorous diet. This plant-eating clade of cynodonts, the Gomphodontia, is characterized by the transverse expansion of the postcanine teeth, which develop a precise crown-to-crown occlusion. Of the three families of Triassic gomphodonts, two—the Gomphognathidae (=Diademodontidae) and Trirachodontidae—show limited taxonomic and morphological diversity and are restricted to the Early and Middle Triassic, primarily of south and east Africa [though *Diademodon* has recently been described from the Early to Middle Triassic of Argentina (Martinelli et al. 2009)]. The supposed trirachodontids *Sinognathus* and *Beishanodon* (Gao et al. 2010) from China are more likely to be probainognathians close to *Aleodon brachyrhamphus* (Hopson in preparation). In contrast, the third family, the Traversodontidae, is taxonomically and morphologically much more diverse and occupies a much wider geographic range, including North and South America, Europe, Africa, Madagascar, and India.

Sues and Hopson (2010) have recently summarized the history of traversodontid discovery and description. The family Traversodontidae was established by Huene (1936) for *Gomphodontosuchus* and *Traversodon* from the Middle or Late Triassic of Brazil. Crompton (1955) named the genus *Scalenodon*, based on *Trirachodon angustifrons* Parrington, 1946, from the Middle Triassic Manda Beds of Tanzania. Though not placing *Scalenodon* in the family Traversodontidae, Crompton noted that the mandibular postcanines of *S. angustifrons* are "remarkably similar" to those of *Traversodon*. Since then, approximately 20 new species of traversodontid have been described, indicating a wide range

Christian F. Kammerer, Kenneth D. Angielczyk and Jörg Fröbisch (eds.), *Early Evolutionary History of the Synapsida*, Vertebrate Paleobiology and Paleoanthropology, DOI: 10.1007/978-94-007-6841-3_14,

of dental morphologies evolving from an ancestral pattern similar to that of *S. angustifrons*. An even more primitive (more like that of *Diademodon*) postcanine tooth pattern occurs in the South American traversodontids *Pascualgnathus* (Bonaparte 1966) and *Andescynodon* (Bonaparte 1967), which probably gave rise to the pattern seen in *S. angustifrons* and all other traversodontids.

The main interest of traversodontids to students of the ancestry of mammals is the fact that they have developed a mammal-like complexity to their postcanine teeth, which involves precise occlusion between upper and lower teeth and a complex pattern of shearing and crushing/grinding seen in few other non-mammalian tetrapods (see Sues and Reisz 1998; Reisz and Sues 2000). Associated with the complex occlusal relations of the traversodontid dentition is the active movement of the lower jaw and teeth in a fore-aft (propalinal) rather than a transverse direction, as occurs in most mammals.

Crompton (1972) in a review of dental morphology and function in gomphodont cynodonts, named three new traversodontid species from the Manda Beds of the Ruhuhu Valley in southwest Tanzania. He included all three species in the genus *Scalenodon*. Two of the new species named by Crompton (1972), *S. hirschsoni* and *S. attridgei*, were collected in 1963 by the British Museum (Natural History)—University of London Joint Palaeontological Expedition; the third species, *S. charigi*, was based on a specimen collected by Parrington in 1933, which had earlier been referred by Crompton (1955) to cf. *Gomphodontosuchus brasiliensis*. In diagnosing the four species attributed to *Scalenodon*, Crompton (1972) utilized only the upper postcanine teeth.

Scalenodon angustifrons is now known to be a relatively primitive traversodontid in which the postcanine teeth display the main features of the traversodont cheek tooth pattern. The three species of *Scalenodon* named by Crompton in 1972 show most of these features, but, with knowledge of the broader range of traversodontid postcanine variation now available, it is clear that they are distinct enough to be placed in separate genera, as has been noted by Hopson (1984) and Abdala and Ribeiro (2003). "*S.*" *hirschsoni* has been included in cladistic analyses of cynodonts by Hopson and Kitching (2001, where it is misspelled as "*S.*" *hirschoni*) and Liu and Olsen (2010) and in cladistic analyses of gomphodonts by Abdala and Ribeiro (2003), Abdala et al. (2006), Kammerer et al. (2008), and Sues and Hopson (2010). Although characters of "*S.*" *hirschsoni* have been included in published data matrices, the type material, with the exception of the postcanines, has yet to be described.

The purpose of this study is to provide a detailed description of the skull and dentition of the known material of "*Scalenodon*" *hirschsoni* Crompton, 1972, and to provide a new generic name and an expanded diagnosis for this taxon. The other new species referred by Crompton (1972) to *Scalenodon*, i.e., *S. attridgei* and *S. charigi*, will be redescribed, subjected to a phylogenetic analysis, and, if necessary, provided with new generic names in a future paper. Liu and Abdala (2013) have already done this, including both species in the same genus as "*S.*." *hirschsoni*, but I intend to test their conclusions with my own analysis (Hopson in preparation).

The completeness and quality of preservation of the postcanine teeth of "*S.*" *hirschsoni* permit a detailed description of dental wear and a reinterpretation of the functional morphology of the postcanine dentition.

As noted by Sues and Hopson (2010), the monophyly of Traversodontidae has been questioned because some authors (e.g., Crompton and Ellenberger 1957; Hopson 1984, 1985; Sues 1985; Hopson and Kitching 2001; Hopson and Barghusen 1986; Hopson and Sues 2006) have argued that Tritylodontidae, specialized rodent-like herbivores of Early Jurassic to Early Cretaceous age, nests within Traversodontidae, rendering the latter family paraphyletic. Tritylodontidae Cope, 1884 has priority over Traversodontidae Huene, 1936, so if tritylodontids were indeed to fall among the traversodontids, the former name would become the valid name of the group. Despite these uncertainties, I have utilized the name Traversodontidae in this paper, recognizing the uncertainty of its validity, because it is almost universally used by workers on cynodonts.

Materials and Methods

The material described here as "*Scalenodon*" *hirschsoni* was originally considered to pertain to a single individual, catalogued as NHMUK R8577 (Crompton 1972). The holotype includes the snout and lower jaws, the teeth of which were originally in occlusion but were separated during preparation (Crompton 1972, p. 49). However, included among the dissociated skull parts is a fragment consisting of portions of the primary and secondary palates and the pterygoid flanges that duplicates parts preserved in the type skull. The presence of a second individual of "*S.*" *hirschsoni* among the type material raises the question of which incomplete pieces pertain to the individual represented by the holotypic snout (the teeth of which present the diagnostic features of the species) and lower jaws and which may pertain to a second individual. All are from the same locality, and though they vary in color from reddish-brown to dark gray-brown, some specimens (notably the fused dentaries) are reddish in some parts and dark gray in others, canceling the value of such color differences in distinguishing individuals.

For purposes of description of the skull and lower jaws of "*S.*" *hirschsoni*, all identifiable fragments, with the exception of the duplicated palatal elements, are herein considered to pertain to the type skull, which consists with absolute certainty only of the snout plus orbital region and the lower jaws. In addition, an isolated right jugal has a reasonably good fit with the rear of the right maxilla, once weathering and some breakage are taken into account. An isolated braincase with attached right zygomatic arch lacks an identifiable contact with the anterior part of the skull; however, in terms of size and non-duplication of parts, it is likely that the braincase pertains to the type specimen. A number of unidentified small fragments cannot be assigned to either individual and are not further discussed. The poorly preserved distal portion of a left humerus may pertain to either individual.

The second individual is about equal in size to the type specimen. Though less complete, it is better preserved than the type and possesses more of the primary palate and pterygoid flanges.

Systematic Paleontology

Therapsida Broom, 1905
Cynodontia Owen, 1861
Eucynodontia Kemp, 1982
Cynognathia Seeley, 1908
Gomphodontia Seeley, 1894
Traversodontidae Huene, 1936
Mandagomphodon gen. nov.

Etymology: *Manda*, for the Manda Beds from which the material was collected, and, from the Greek, *gomphios*, molar, and *odon*, tooth; *gomphodon* is a commonly used suffix in generic names of gomphodont cynodonts.

Holotype: The Natural History Museum, London, NHMUK R8577, anterior half of skull, most of right dentary and dentigerous portion of left dentary. Also referred to the type are a right jugal and a partial braincase with attached right zygomatic arch.

Referred Specimen: The Natural History Museum, London, NHMUK: PV R11974, partial palate including primary and secondary palate and left pterygoid flange.

Horizon: Lifua Member of Manda Beds.

Locality: Locality U12 of the British Museum (Natural History)—University of London Joint Palaeontological Expedition of 1963, Ruhuhu Valley, southwest Tanzania, between the Hiasi and Njalila streams, just south of the Rutukira River; the most northerly of the Expedition's localities west of the Njalila (Crompton 1972, p. 43). Cox (1991, p. 768, Fig. 1) shows a map of these localities, including U12.

Age: The fossiliferous middle to upper parts of the Lifua Member of the Manda Beds are considered to be of Anisian (Middle Triassic) age (Wopfner 2002).

Diagnosis: Traversodontid cynodont with three upper and probably two lower incisors, all of which are enlarged (except possibly the first upper incisor) and markedly procumbent; the upper canine is separated from the incisors by a very shallow paracanine fossa; the postcanine tooth rows diverge only slightly from the midline axis; the presumed lower canine (third dentary tooth) inclines forward, nearly paralleling the procumbent incisors. Upper postcanines with rectangular crowns that are oriented perpendicularly, not obliquely, to the midline axis; transverse crest of uppers relatively low and anteroposteriorly broad, with three main cusps, the labial cusp separated from the central and lingual cusps by a deep U-shaped (rather than V-shaped) notch; prominent anterolabial and anterolingual cusps present; anterior and posterior cingula well-developed, labial cingulum absent. Lower postcanines with a robust anterior transverse ridge consisting of two large cusps that are relatively low and anteroposteriorly broad; large anterolabial and posterolabial accessory cusps. Suborbital flange of jugal appears to be absent.

Remarks: *Mandagomphodon hirschsoni* is a very unusual traversodontid because of its unique mixture of plesiomorphic and highly autapomorphic features. It differs from *Luangwa* and *Traversodon* in lacking a strong sigmoid curvature to the lower border of the dentary (this feature in *S. angustifrons* is uncertain) and from all three of these genera in lacking great height of the zygomatic arch above the tooth row. In both of these features it resembles *Massetognathus*. Its upper postcanine teeth lack an external cingulum, which occurs in the three apparently primitive genera noted above, but like *S. angustifrons*, it possesses an anterolabial accessory cusp in front of the lower transverse ridge. *Luangwa, Boreogomphodon*, and *Nanogomphodon* have a transverse cingulum on the anterior face of the lower transverse ridge, but its homology with the single anterolabial cusp of *Scalenodon* and *Mandagomphodon* is uncertain.

Description

Skull

This description is based on the type snout and the associated jugal, braincase, and zygomatic arch, which probably pertain to the same individual (Figs. 14.1, 14.2, 14.3). Details are added from the isolated palatal region pertaining to a second individual (Fig. 14.2).

The type snout of *Mandagomphodon hirschsoni* has been dorsoventrally compressed and its dorsal midline shifted to the right. The anterodorsal surface has suffered the greatest

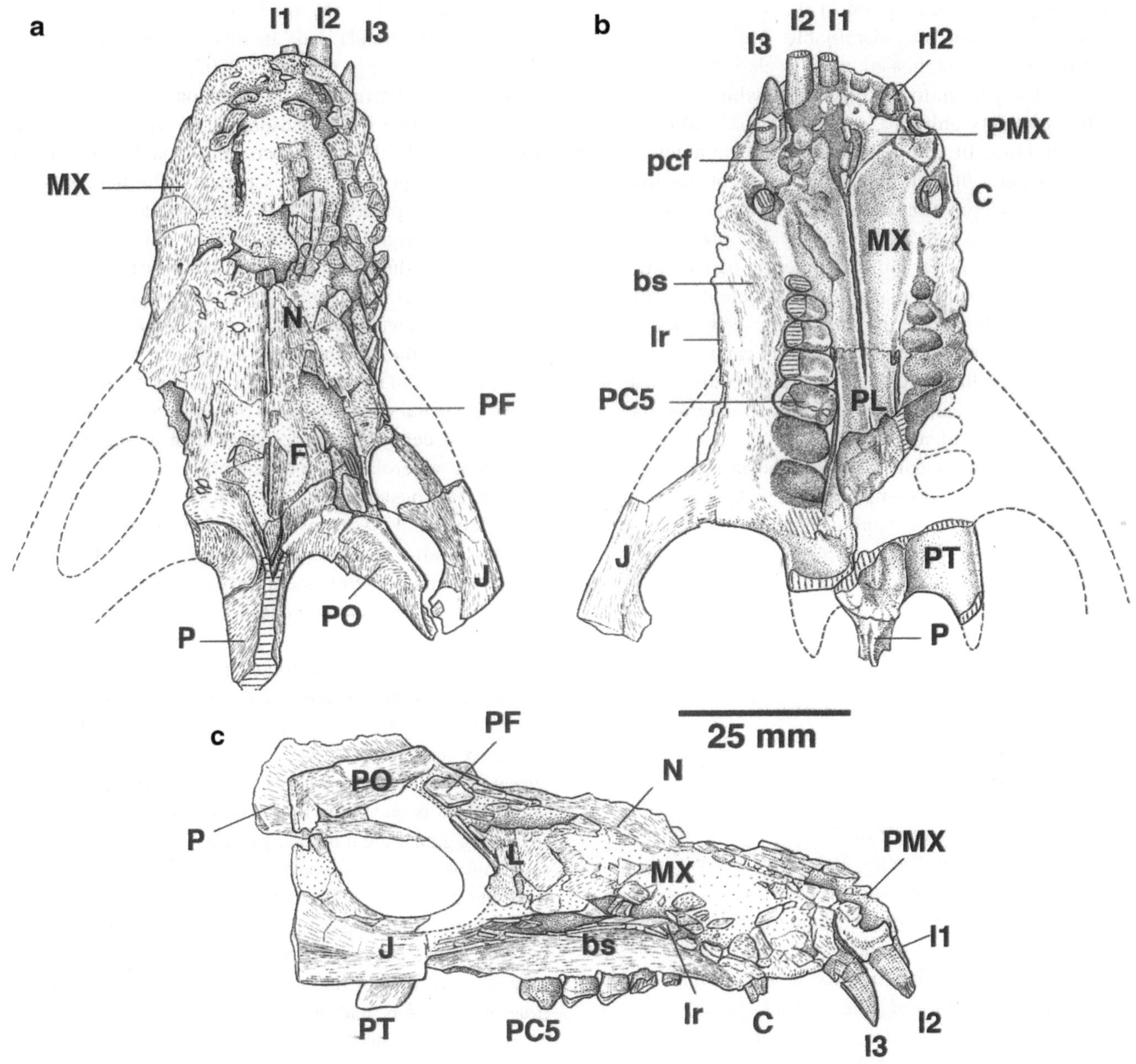

Fig. 14.1 *Mandagomphodon hirschsoni*, Crompton 1972. Anterior portion of holotype skull, NHMUK R8577. **a** dorsal view; **b** ventral view; **c** right lateral view. *bs* buccal shelf, *C* canine, *F* frontal, *I* incisor, *J* jugal, *L* lacrimal, *lr* lateral ridge, *MX* maxilla, *N* nasal, *P* parietal, *PC* postcanine, *pcf* paracanine fossa, *PF* prefrontal, *PL* palatine, *PMX* premaxilla, *PO* postorbital, *PT* pterygoid, *rI* replacing incisor

collapse, which has obscured the external nares except along their anteroventral margins. The bone on the left side of the snout is only moderately damaged, whereas that on the right side is shattered into numerous fragments. The anterior margin of the right orbit has been pushed down and back so that the restored orbit is an asymmetrical oval with its long axis oriented posterodorsally.

The ventral surface of the snout has suffered much less damage than the dorsal surface. Lateral compression has caused the rear of the secondary palate to buckle upward into the choanal passage, which has pulled the posterior halves of the maxillary postcanine tooth rows closer together. The undistorted secondary palate of the referred specimen is therefore wider than that of the type. Sutures cannot be distinguished on either specimen except on the palate.

The anterior end of the premaxilla, above the first two procumbent incisors, is very shallow; more posteriorly, the premaxilla curves back and up above the large third incisor. It forms a distinct edge on the margin of the narial opening, suggesting it was overlain by the septomaxilla, no part of

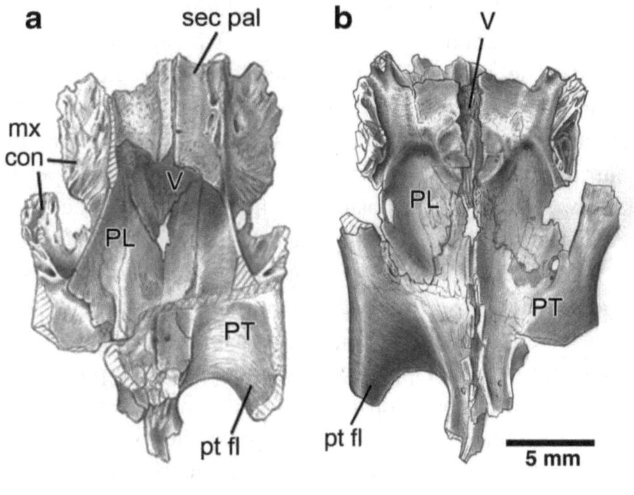

Fig. 14.2 *Mandagomphodon hirschsoni*, Crompton 1972. Ventral view of referred specimen, NHMUK: PV R11974, isolated portion of the palatal region. **a** ventral view; **b** dorsal view. *mx con* portions of palatine and pterygoid that overlie posterior part of maxilla, *PL* palatine, *PT* pterygoid, *pt fl* pterygoid flange, *sec pal* secondary palate, *V* vomer

which is identifiable. On the palatal surface, the premaxilla forms a short shelf behind the first two incisors and is pierced by a small foramen behind the first incisor. More laterally, the posterior margin of the premaxilla is uncertain, but it appears to form the medial border of the alveolus for the third incisor. It presumably forms the bar between the incisive foramina and definitely forms their lateral margins. Whether the paired premaxillae meet behind the incisive foramina, or whether the posterior rim is formed by the maxillae, cannot be determined with certainty.

The maxilla undoubtedly forms the greater part of the lateral surface of the snout, but it is heavily damaged dorsally and its sutures with other facial elements cannot be determined. Laterally, it supports a rounded longitudinal ridge that extends back from the slight swelling for the canine root to pass below the orbit, where it merges into the suborbital portion of the jugal. Above this ridge on the better-preserved left side, a shallow depression lies between the low canine swelling and the lacrimal region; such a depression is seen in most gomphodonts. Below the longitudinal ridge, the maxilla bends sharply inward to form a moderately broad, concave overhang lateral to the cheek teeth, termed the buccal shelf by Sues and Hopson (2010). On the right side, between the canine and fifth postcanine, the overhang is wider than on the left because an elongated piece of laterally displaced maxilla appears to form an outer extension of the right shelf (Fig. 14.1c). The less damaged

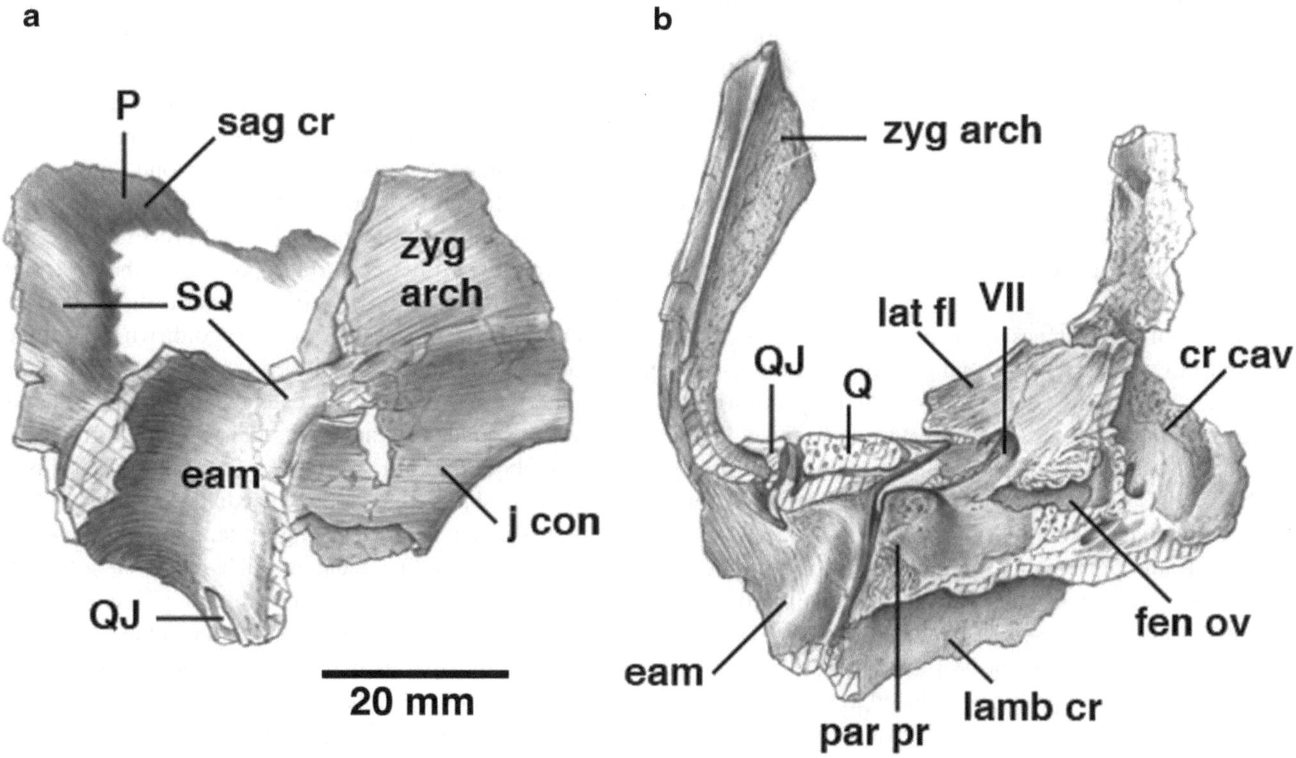

Fig. 14.3 *Mandagomphodon hirschsoni*, Crompton 1972. Fragment of braincase and right zygomatic arch referred to holotype skull, NHMUK R8577. **a** Lateral view, **b** ventral view. *cr cav* cranial cavity, *eam* external auditory meatus of squamosal, *fen ov*, presumed position of fenestra ovalis, *j con* contact surface on squamosal for jugal, *lamb cr* lambdoidal crest of squamosal, *lat fl* lateral flange of prootic, *P* parietal, *par pr* paroccipital process of the opisthotic, *Q* quadrate, *QJ* quadratojugal, *sag cr* sagittal crest, *SQ* squamosal, *zyg arch* zygomatic arch of squamosal, *VII* hollow containing foramen for facial nerve

surface on the left shows that the shelf was much narrower. The concavity of the overhang is seen best lateral to the diastema, where it forms a shallow but distinct depression behind the canine. Lateral to the last postcanine, the maxillary overhang expands outward to contact the jugal in the base of the zygomatic arch. Anterior to the left canine is a small foramen in the side of the maxilla, as is usual in cynodonts. The maxilla extends forward as far as the lateral margin of the alveolus of the third incisor. In palatal view, the maxilla forms a convex-outward ridge that extends forward from the canine alveolus. On the palatal side of the ridge is a very shallow oval depression, preserved on both sides, that lies anterior and slightly medial to the canine alveolus. As will be discussed below, the depression appears to be homologous to the paracanine fossa of other cynodonts, which houses the tip of the lower canine.

The palatal suture between premaxilla and maxilla appears to extend posterointernally from the lingual side of the third incisor to the posterior part of the incisive foramen. As noted above, it is unclear whether the premaxilla or the maxilla forms the posterior border of the foramen. In ventral view, the maxillae form a concave-upward secondary palate that lies between the canines and the postcanine tooth rows, with left and right maxillae meeting on the midline in a low narrow ridge. Behind the canine is a long diastema formed by a prominent rounded ridge that passes posteromedially to meet the first postcanine. The asymmetrical distortion of the palatal region has pushed the right postcanine dentition several millimeters closer to the midline than is the left. The maxilla meets the palatine in the secondary palate in a transverse, slightly interdigitating, suture at the level of the posterior third of the third postcanine. From the major palatine foramen in the lateral part of the palatine just behind the suture, a shallow groove extends forward on to the maxilla.

Portions of the left nasal are preserved anterior to the orbit, but no sutures with surrounding bones are preserved to indicate the limits of the nasal. On the midline, the medial edges of the nasals are raised as a ridge that is continuous with a prominent midline ridge on the frontals.

The frontals appear to be depressed between the orbits. They are bounded laterally by raised ridges formed by the postorbitals and presumably the prefrontals. The prominent midline ridge on the frontals extends into the narrow cleft between the posteriorly-converging postorbitals. The contact with the parietals cannot be seen.

The prefrontals are probably included in the raised ridges lateral to the frontals and also in the orbital wall, but they are too fragmentary for reliable identification.

Fragments of bone anterior to the right orbit and within the orbital rim undoubtedly pertain to the lacrimal, but no sutures or foramina are preserved to corroborate this.

Most of the right postorbital, including the postorbital bar, and the cranial portion of the left postorbital are preserved. The postorbital bar extends posterolaterally and slightly downward to form the posterodorsal quadrant of the orbital rim. Its dorsal surface is rounded in section, but ventrally it extends down and back as a deep tapering flange. Medially, this postorbital flange extends back as a vertical lappet overlying the lateral surface of the parietal. The postorbital of *Mandagomphodon hirschsoni* appears to be unusually deep, although a thorough comparison with other advanced cynodonts is not possible at present. A less prominent vertical lappet forming the rear of the postorbital bar was noted by Hopson and Kitching (2001) in the probainognathian *Lumkuia fuzzi*, which they interpreted as marking the area of attachment of the anteriormost portion of the temporalis muscle. Among gomphodonts, such a deepening of the postorbital bar is absent in *Massetognathus* (Romer 1967, Figs. 3, 10), but present in *Traversodon* (personal observation), where it is less prominent than that of *M. hirschsoni*.

The anterior end of the sagittal crest of the parietal is preserved between the cranial lappets of the postorbitals. The narrow crest continues on the isolated braincase back to the divergence of the lambdoidal crests. The dorsal part of the sagittal crest is missing, except for about 12 mm at its posterior end. Evidence of a parietal foramen is absent.

The right jugal is preserved as an isolated fragment consisting of the suborbital region and most of the postorbital process. It appears to have a contact with the posterolateral process of the maxilla lateral to the rear of the last postcanine. An elongate near-horizontal process on the maxilla fits into a groove on the anteromedial end of the jugal. In this orientation, the postorbital process of the jugal is directed toward the postorbital process of the postorbital and the anteromedial process of the jugal is directed toward the anterolateral process of the pterygoid, which it almost contacts behind the maxilla, as is usual in gomphodonts. The anterior end of the jugal extends about 4 mm laterally beyond the contact with the maxilla where it forms a vertical bar about 6.5 mm deep. The ventral margin of this bar is a natural sharp-edged ridge that underlies the posterior two-thirds of the orbit; it is slightly convex, but lacks evidence of a descending flange, which in traversodonts lies below the posterior half of the orbit. It is likely that the anterior extension of the jugal bar contacts the longitudinal ridge of the maxilla, as is usual in gomphodonts.

The palatines are preserved both on the type skull (Fig. 14.1c) and the isolated palatal region (Fig. 14.2). The latter specimen preserves the dorsal surface of the palatine, which is usually hidden by external skull bones. In both specimens, the palatine portion of the secondary palate is about 14 mm in length. In the little-distorted palatal specimen, the palatine portion of the secondary palate is 14 mm wide, whereas in the more laterally compressed type skull, in which it is buckled upward, it is 10.5 mm wide. On the lateral margin of the palatine, shortly behind the transverse suture with the maxilla, is a conspicuous foramen, the major palatine foramen, from which a shallow groove extends forward. Two smaller, elongate foramina lie behind this foramen and in the palatal specimen, a tiny foramen lies slightly medial to the posteriormost foramen. A pair of slender ridges extend back and slightly outward from the posterolateral margins of the secondary palate, bounding the posterior parts of the maxillae laterally and the lateral walls of the choanal passage medially. In the roof of the choana, the palatines curve medially above the fused vomers in the primary palate, nearly meeting on the midline. Posterior to the secondary palate, the palatines meet the pterygoids in the roof of the choanal trough and contribute to the anterior parts of the low palatal ridges that converge slightly toward the rear. In the isolated palatal specimen, the contact surfaces with the medial surfaces of the maxillae are large thin plates that extend dorsolaterally from the margins of the secondary palate and the more posterior palatine ridges. These plates are slightly concave laterally and bear ridges, grooves, rugosities, and small foramina on their contact surface with the maxillae. They extend above the level of the dorsal surface of the primary palate. Medial to the plates, the palatines in the roof of the primary palate bear a broad longitudinal trough that opens laterally through a notch or foramen, presumably into the nasal cavity.

The fused vomers are preserved only on the palatal specimen, where they form a broad plate in the roof of the air passage above the palatines in the secondary palate. They taper toward the rear but are missing from the roof of the choanal trough, where they are expected to contact the pterygoids between the more lateral palatines.

The pterygoids are best preserved on the palatal specimen, though the base of the right pterygoid flange and the anterolateral process that appears to contact the jugal behind the maxilla are also preserved on the type skull. On the primary palate, the pterygoids bear the rear portions of the low ridges in the choanal roof, which converge slightly toward the rear and end 5 mm apart at the junction of the basipterygoid rami of the pterygoids with the posteromedial margin of the pterygoid flanges. The pterygoid flanges are robust, with flat lateral surfaces that are oriented backwards about 35° from the horizontal. Continuing forward from the pterygoid flange is a robust process with a concave ventromedially-facing surface that overlies the rear of the maxilla dorsolaterally. In the type, this process of the pterygoid, which is exposed in the floor of the orbit, bends outward toward the anteromedial process of the jugal. On its ventral surface in the palatal specimen, this process has a prominent anterolaterally-directed foramen on both sides, immediately lateral to the concavity for the rear of the maxilla. It is not visible in the type skull where it may be covered by the suborbital flange of the maxilla. A short section of the basipterygoid rami of the pterygoids are preserved on the palatal fragment.

Much of the right squamosal is preserved on the isolated braincase (Fig. 14.3). The cranial process is a narrow triangular sheet of bone that lies against the parietal at the posterior end of the sagittal crest. It extends sharply back from the apex of the sagittal crest to form the lambdoidal crest; in the complete skull the diverging lambdoidal crests would overhang the occiput. The zygomatic process of the squamosal joins the lower end of the cranial process at the V-shaped notch. Although the posterior margin of the lambdoidal crest is damaged, enough is preserved to show that it extended behind and below the base of the zygomatic arch. Continuous ventrally with the lambdoidal crest is a narrow, slightly concave ridge that forms the lateral boundary of the middle ear cavity (and possibly supported a tympanum; Allin and Hopson 1992). On the medial side of this ridge, the squamosal has a vertical surface bounded in front by a medially-directed vertical lappet. The anterior surface of this lappet bears striations, indicating that it was overlain by another bone, most likely the lateral flange of the prootic (which is damaged in this specimen). In life, the space between the preserved distal end of the paroccipital process (which is formed by unfinished endochondral bone) and the distal and anterior squamosal surfaces probably held a cartilaginous distal and anterior extension of the paroccipital process. Lateral to the medial lappet, the squamosal forms a transverse plate that supports on its anterior face the preserved dorsal ends of the quadrate and quadratojugal. The latter bone lies in a deep notch in the squamosal so that it is visible from behind. Posterodorsal to the quadrate region is a broad shallow trough, the external auditory meatus, which is overhung by the outturned dorsal border of the squamosal. The zygomatic arch curves forward and outward from the level of the V-shaped notch. Its posterolateral surface bears a conspicuous sulcus, the external auditory meatus, which extends from the presumed tympanic ridge up, forward, and slightly laterally on the outer surface of the zygoma. The zygomatic arch is very deep, its incomplete dorsal margin perhaps extending as high as the sagittal crest. The lower margin of the squamosal portion of the zygomatic arch is preserved as a very thin, slightly concave, knife-edged plate. On the broken anterior end of the arch, the squamosal is seen to increase in width about 10 mm above its lower edge; this change in thickness is traceable for about 15 mm back on the damaged lateral surface of the squamosal. It is likely that the thinner lower portion of the squamosal was the contact surface for the rear of the jugal, which would have extended back nearly to the jaw joint. As preserved, the anterior end of the zygomatic arch angles medially too much to meet the suborbital portion of the jugal; it is restored as angled more laterally.

All but the posterior-most part of the sidewall of the braincase and none of the basicranium is preserved; hence the orbitosphenoid, epipterygoid, basisphenoid, parasphenoid, and basioccipital, are not represented. Most of the prootic and opisthotic and the right side of the occiput above the level of the foramen magnum are complete, though sutures are not visible.

The paroccipital process of the opisthotic is sufficiently preserved to show that its anteroventral surface slopes posteroventrally and is composed of finished bone (Fig. 14.3b). Comparison with other cynodonts indicates that this surface forms a partial roof for the middle ear cavity (Hopson 1966; Allin and Hopson 1992). More distally, this surface curves ventrally and slightly anteriorly to end in a ventrally-bulging process formed by unfinished bone. Behind this process is a shallow trough of finished bone that is directed toward the presumed "tympanic" ridge of the squamosal. As noted above, the unossified distal and anterodistal parts of the paroccipital process were probably finished in cartilage, which implies that the ventrally-bulging process was a much larger structure than indicated by its preserved size, perhaps resembling the ossified crista parotica of tritylodontids (Hopson 1966; Sues 1986). At the proximal end of the paroccipital process is a broken mass of spongy bone around an irregular gap where the fenestra ovalis would be expected to lie.

Though sutures are lacking, the prootic of cynodonts usually contributes a thin lappet of bone to the anterior face of the paroccipital process (Parrington 1946; Hopson and Kitching 2001). The concave anterior margin of the paroccipital process forms the posterior rim of the pterygoparoccipital foramen. This foramen is usually walled in front by the lateral flange of the prootic, which extends laterally to contact the lappet of squamosal covering the anterodistal end of the paroccipital process. Here, the lateral flange has been pushed upward and forward by compression so that it lies 3 mm anterior to the squamosal lappet; the latter, however, shows anterior striations presumably marking the contact surface with the lateral flange. A sulcus in the prootic anterior to the proximal end of the paroccipital process contains a small facial foramen.

The triangular occiput is bounded above by the posterior end of the sagittal (parietal) crest and laterally by the posterolaterally-directed lambdoidal crests of the squamosal. The paroccipital process is exposed ventrally, but the foramen magnum and adjacent elements are not preserved. Bulges at the proximal ends of the paroccipital processes probably represent the dorsal parts of the exoccipitals. From them on either side, rounded ridges extend dorsolaterally, dividing the occiput into a single dorsomedian depression and two lateral depressions (of which only the right is preserved). These depressions contain matrix, but the lateral one is undoubtedly floored by the tabular and is penetrated by the small posttemporal foramen. The dorsomedian depression is probably formed by the supraoccipital in its lower half and the postparietal above.

The upper parts of the quadrate and quadratojugal are preserved in place, but their distal, articular portions are broken off. The upper part of the quadrate lies in a sulcus on the anterior face of the squamosal adjacent to the squamosal lappet that abuts the paroccipital process. The preserved part is 5.5 mm wide and 7.8 mm long and 3.4 mm from front to back. The broken surface shows that the outer bone is dense but the inner bone is spongy with fairly large cavities. The medial side of the quadrate is rounded in section whereas the lateral side is flat to slightly concave. Its posterolateral corner lies in front of a vertical groove in the squamosal; this is just above the expected position of the inverted V-shaped emargination that houses the posterior process of the quadrate.

The quadratojugal lies in a deep notch in the squamosal (which, as preserved, is 6.5 mm long) that lies immediately lateral to the quadrate sulcus. The medial face of the notch is flat (matching the flat medial surface of the quadratojugal) whereas the lateral face is flat dorsally but narrows ventrally to form a longitudinal ridge that fits into a matching groove on the lateral side of the quadratojugal. As preserved, the quadratojugal is about 1.0 mm in transverse width and 4.5 mm from front to back.

Lower Jaw

The only preserved parts of the lower jaws are partial dentaries indistinguishably fused at the symphysis (Fig. 14.4). The more complete right dentary consists of the horizontal ramus, which lacks the angular region, and the vertical ramus preserving the lower part of the coronoid process to about the level of the articular process, the latter lacking a small portion of its posterior tip. The left dentary is broken vertically behind the last (seventh) postcanine. The surface bone of the left ramus is largely smooth and undamaged, whereas that of the right ramus has suffered fracturing and some distortion. The medial surface of the right ramus is covered with a layer of fragmented bone embedded in a thin layer of matrix; it is possible that the fragmented bone represents parts of the coronoid and splenial, even portions of the postdentary rod formed by surangular, angular, and prearticular, but this is uncertain.

The total preserved length of the more complete right dentary is 103.3 mm, with an estimated restored length of about 118.3 mm. The fused symphysis has a slightly convex anterior profile and extends a short distance below the ventral margin of the horizontal ramus at the level of

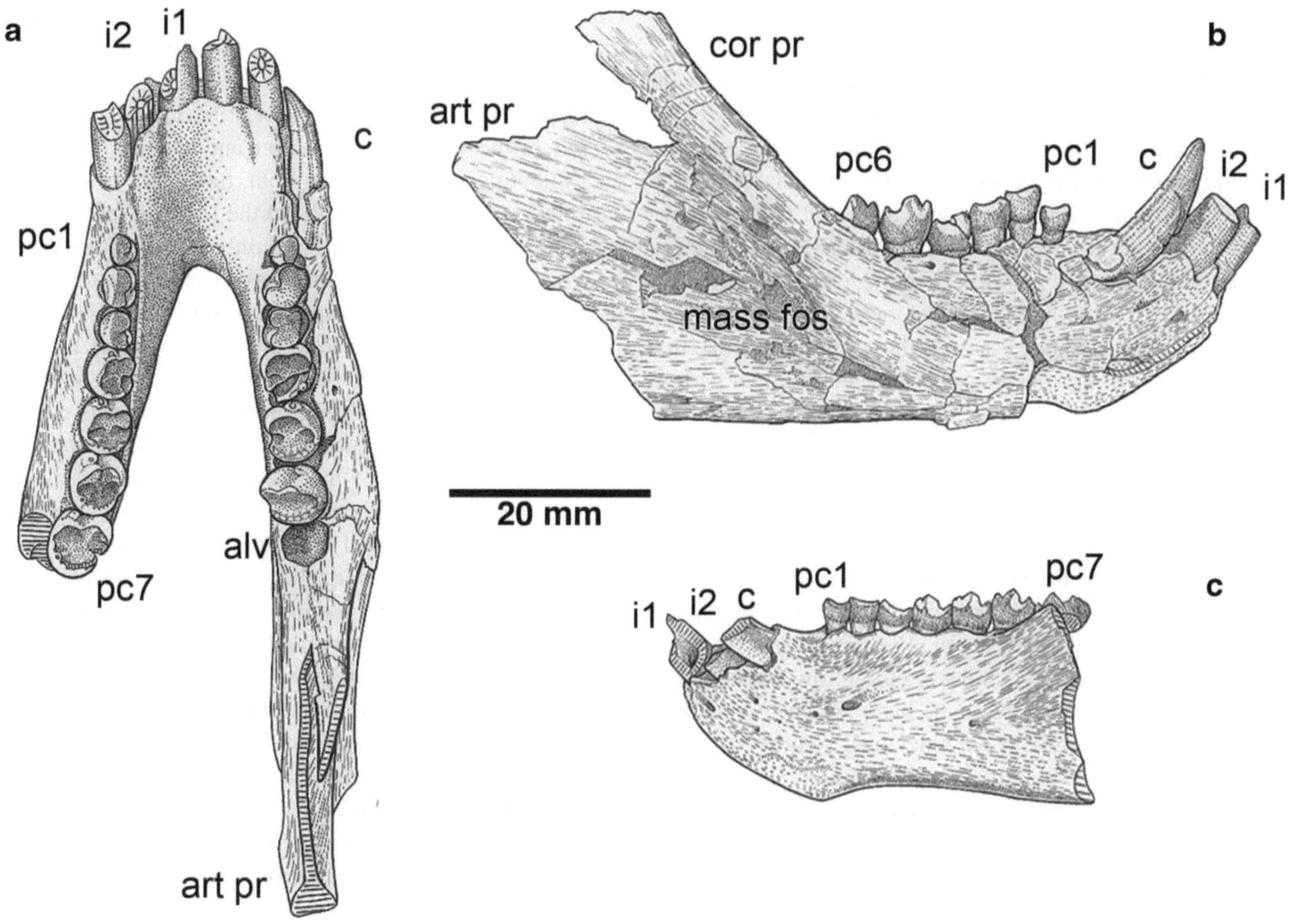

Fig. 14.4 *Mandagomphodon hirschsoni*, Crompton 1972. Holotype lower jaw, NHMUK R8577. **a** Dorsal view of fused dentaries; **b** lateral view of right dentary; **c** lateral view of left dentary. *alv* alveolus, *art pr* possible articular process, *c* canine, *cor pr* coronoid process, *i* incisor, *mass fos* masseteric fossa, *pc* postcanine

the canine/postcanine diastema. The horizontal length of the symphysis is 24.5 mm. The height of the jaw below the diastema is 20.0 mm. Behind the symphysis, the ventral profile of the dentary is slightly concave, with the shallowest part of the horizontal ramus lying below the rear of the second postcanine, where the dentary is 19.6 mm high. The dentary deepens very slightly toward the rear of the tooth row (well shown on the left side) and becomes straight or very slightly convex below the coronoid process (shown only on the slightly damaged right side). The deepest part of the ramus anterior to the rise into the coronoid process, at midlength of the fifth postcanine, is 20.2 mm.

Approximately the lower two-thirds of the horizontal ramus is broadly convex in cross section; but the dorsal third of the ramus forms a shallowly concave, medially inclined surface that extends from the swelling of the canine root back to the base of the coronoid process. This concave surface corresponds to the overhanging concave buccal shelf in the skull, though it is much less prominent. It is likely to have been covered by a fleshy cheek, as has been interpreted for *Arctotraversodon plemmyridon* (originally ?*Scalenodontoides plemmyridon*) (Hopson 1984).

The lateral surface of the well-preserved left ramus has a few mental foramina, the largest of which lies below the contact between the first and second postcanines and opens anteriorly into a short groove. Slightly anterior to this opening is a much smaller foramen, which also opens anteriorly. Below the rear half of the fifth postcanine, on the thickened anterior margin of the masseteric fossa is a tiny foramen that opens toward the rear. On the symphyseal part of the dentary are three very small mental foramina below the canine and a single larger oval foramen below the gap between the first and second incisors. The relative sparseness of mental foramina in the symphyseal region of *M. hirschsoni* is notable, suggesting that the skin in this region is less tightly connected to the underlying bone than in other cynodonts, in which the symphyseal region is often rugose and more densely covered with small foramina.

The base of the thickened ridge that forms the anterior margin of the coronoid process rises lateral to the sixth postcanine tooth and extends posterodorsally at an angle of about 40° from the horizontal postcanine tooth row. The coronoid process is preserved to about the height of the articular process of the dentary, some 17.0 mm above the alveolar margin of the tooth row. The lateral surface of the coronoid process forms a concave depression that is deepest just behind the thickened anterior margin and that is bounded below by the thickened ventral border of the dentary. Anteroventrally, this depression is bounded by the thickened masseteric crest, at about the level of the seventh postcanine. The lower part of this depression is the masseteric fossa, the attachment area of the masseter muscle, and the upper part of the depression, on the lateral surface of the coronoid process, is the attachment surface for the temporalis muscle.

Though the rear of the dentary is missing, the preserved portion above the missing angular region forms a thin posterodorsally-directed ridge, which is interpreted here as the lower margin of the articular process. If this is correct, then the missing part of the angular region of the dentary is about 20.0 mm long and about 17.0 mm high. On the medial side of the dentary behind the tooth row is the trough for the postdentary jaw bones, which are not preserved. The trough is bounded below by the thickened ventral border of the dentary. Its dorsal margin extends obliquely up and back above the missing angular region, behind which it continues back on the underside of the articular process as a transversely-widened surface that presumably supported the rear of the postdentary rod. This widened surface ends at a near-triangular break, the dorsal part of which represents the broken posterior end of the coronoid process. Neither the coronoid nor the splenial bone can be distinguished on the inside of the lower jaw. Below the tooth row a narrow groove extends forward to form a deep sulcus just above the posteroventral margin of the symphysis; this part of the trough was undoubtedly overlain by the splenial, which presumably met its counterpart in the symphyseal sulcus.

When viewed from above, the dentaries diverge at a low angle, thickening posteriorly lateral to the tooth row before rising into the coronoid process. The anterior end of the postcanine row lies slightly internal to the level of the canine, from which it is separated by a short diastema marked by a rounded, posteromedially-directed ridge. The tooth row is curved, concave-outward, and toward its rear overhangs the medial side of the dentary. On the lingual side of the jaw is a continuous shallow groove that lies just below the alveolar margins of the postcanine teeth. Such a groove in other cynodonts has been interpreted as housing the dental lamina.

Dentition

The dental formula of *Mandagomphodon hirschsoni* is I3, C1, PC7/i2, c1, pc7 (abbreviations for upper teeth in upper case, for lower teeth in lower case). The only possible uncertainty involves whether the third lower tooth is the third incisor or the canine.

Upper Incisors (Fig. 14.1)—Of the upper incisors preserved in the type skull, the first and second on the right lack the tips of the crowns, but the third is essentially complete and well-preserved. On the left, the second incisor is in the process of erupting and only its well-preserved tip is exposed. The first and third left incisors are represented by alveoli containing fragmentary roots.

The three upper incisors are all procumbent, being inclined forward about 37° from the vertical. The first upper incisor is the smallest of the three. The tip of its crown is broken off and the cross section on the break is a transverse oval. A small portion of the mesial marginal ridge is preserved. The posterolateral face of the crown appears to possess a flat longitudinal facet that is adjacent to a similar facet on the second incisor. The second incisor has a greater diameter, but is equally procumbent. Its tip is missing on the right, but the erupting I2 on the left has a spatulate tip, with mesial and distal ridges extending down the crown and a raised central area on its lingual face. Its labial face has a broadly rounded transverse surface. On the posteromesial and distal parts of the right I2 are what appear to be longitudinal planar wear surfaces. These presumed wear facets in I1 and I2 suggest that the procumbent first lower incisor occluded between the first and second upper incisors and the second lower incisor occluded against the posterodistal face of I2. The crown of the right third upper incisor is well-preserved and complete to its tip. It is shorter than the restored crown of I2 and slightly less procumbent. It resembles a canine in that its crown is recurved, with sharp mesial and distal cutting ridges that lie parallel to the curved margin of the premaxilla. However, it resembles an incisor in that the lingual side of the crown has slight concavities and a more rounded central surface internal to the marginal ridges, whereas the labial side of the crown is more broadly and smoothly convex than the lingual. Also, although the anterior cutting ridge lies on the mesial side of the crown apically, further toward the base it curves on to the lingual side of the crown and the mesial face of the tooth becomes smoothly rounded in section. This is seen in incisors but rarely in canines. The enamel of the labial face of the crown has a rugose surface of short longitudinal ridges; the enamel on the lingual side is smooth.

The presence in *M. hirschsoni* of three upper incisors that are procumbent and of which some (or all) are enlarged with respect to the primitive traversodont condition also characterizes *Exaeretodon*, *Menadon*, *Protuberum* and *Scalenodontoides*.

Lower Incisors (Fig. 14.4)—The anterior end of the lower jaw has three closely clustered teeth on each side, separated from the first postcanine tooth by a moderate diastema. Two interpretations of their homologies are possible: there are either three incisors and no canine or two incisors and a semi-incisiform canine. As discussed below, I interpret the third tooth to be a canine; thus, I believe there are only two lower incisors in *M. hirschsoni*. Until now, no gomphodont (excluding *Sinognathus* and tritylodontids, whose gomphodont relationships have been questioned) has been known to possess fewer than three lower incisors.

All of the lower incisors lack the apical portions of their crowns, but because the preserved portions do not taper apically, they appear to have been longer than the upper incisors. The better-preserved right incisors are inclined forward about 35° from the vertical. They are enlarged, about the same diameter as the second upper incisor, but are more robust than the uppers, having an oval cross-section that is longer labiolingually than mesiodistally. In lateral profile they appear to be nearly straight. The first left and two right incisors are broken off relatively low on the crown and show a long exposure of basal dentine, with very thin enamel adjacent to the breaks. The crown of the second left incisor is partially erupted and is broken off just above the level of the alveolus. The cross-section shows a rim of thick enamel, which on the labial face of the erupting crown is strongly rugose and longitudinally-ridged.

Upper Canine (Fig. 14.1)—Both upper canines are in the process of erupting so that the crowns are only partially visible in their much larger alveoli. The apices of both crowns are broken off. The mesial (anterior) part of the right canine appears to be broken off, but the distal (posterior) part of the crown has slightly convex labial and lingual surfaces separated by a distal cutting ridge without serrations. The right canine is more heavily damaged. The preserved enamel on the lateral surfaces of both canines bears longitudinal rugosities. The right canine is slightly procumbent whereas the left is inclined more steeply forward.

Lower Canine (Fig. 14.4)—This tooth is strongly inclined forward, about 30° from the vertical, so that its crown lies very close to that of the second incisor; thus it appears functionally to be part of the incisor series. However, the lower part of the crown curves strongly back so that its embedded root is nearly horizontal in orientation, as indicated by a low swelling below the diastema external to the level of the first postcanine tooth. Furthermore, when the lower postcanines are occluded with the uppers, the tip of the presumed lower canine lies below the shallow depression identified above as the paracanine fossa. Finally, the presumed canine has sharp posterior and anterior cutting ridges that extend to the base of the crown, although this is not necessarily a clear distinction because the crowns of the lower incisors are broken off and show no trace of mesial and distal ridges. However, the third and only complete upper incisor does resembles the lower presumed canine in having sharp mesial and distal ridges, although its mesial ridge passes lingually to terminate on the medial face of the crown, as often occurs in incisors, though not canines.

The lower canine is strongly recurved, in contrast to the nearly straight crowns of the incisors, and it is interpreted here as being much shorter than the lower incisors, because the latter, though incomplete, are almost as long as the canine, yet show no sign of tapering apically.

Dimorphism between upper and lower canines is rare in cynodonts, though it occurs in *Exaeretodon, Protuberum*, and *Scalenodontoides*, where, however, it is the upper canine that lies adjacent to the incisors and the lower canine that is separated from the incisors by a long diastema; consequently, in these taxa the paracanine fossa, which receives the tip of the lower canine, lies behind rather than in front of the upper canine as it does in *Mandagomphodon*.

Upper Postcanines (Figs. 14.1, 14.5a, 14.6)—The upper postcanine dentition is preserved only on the right side, and consists of five well-preserved teeth implanted in their alveoli and two large alveoli at the end of the tooth row.

The preserved upper postcanines increase in transverse diameter from front to back. The anterior to posterior decrease in wear on the crowns indicates that they erupted in sequence from front to back, the usual eruption pattern in traversodontids. In addition, the color of the enamel darkens posteriorly, from an amber color in PC2-4 to a deep brown in PC5, indicating thicker enamel in the latter. The unworn enamel in PC5 has a pattern of fine dorsoventral ridges.

Postcanine five (PC5) is the least worn of the upper postcanines and thus best shows the main features of the crown. Crompton (1972, pp. 50, 51) presents excellent illustrations of this tooth, describing it through a comparison with the upper postcanines of *Scalenodon angustifrons*.

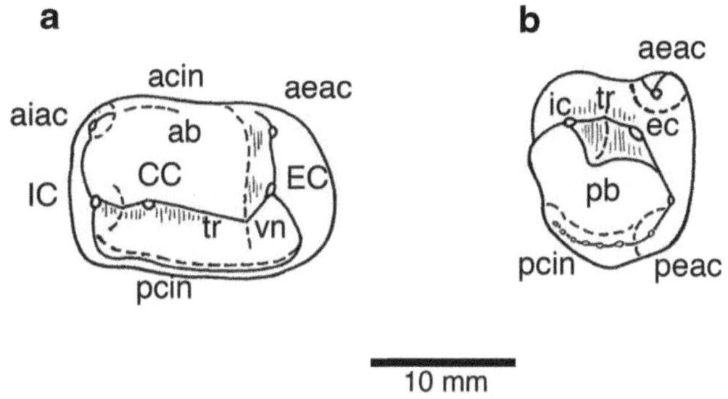

Fig. 14.5 *Mandagomphodon hirschsoni*, Crompton 1972. **a** Upper, and **b** lower postcanines with cusps and other features labeled. *ab* anterior basin, *acin* anterior cingulum, *aeac* anteroexternal accessory cusp, *aiac* anterointernal accessory cusp, *CC* main central cusp, *EC, ec* main external cusp, *IC, ic* main internal cusp, *pb* posterior basin, *pcin* posterior cingulum, *peac* posterior external accessory cusp, *tr* transverse ridge, *vn* V-shaped notch. Modified from Crompton (1972)

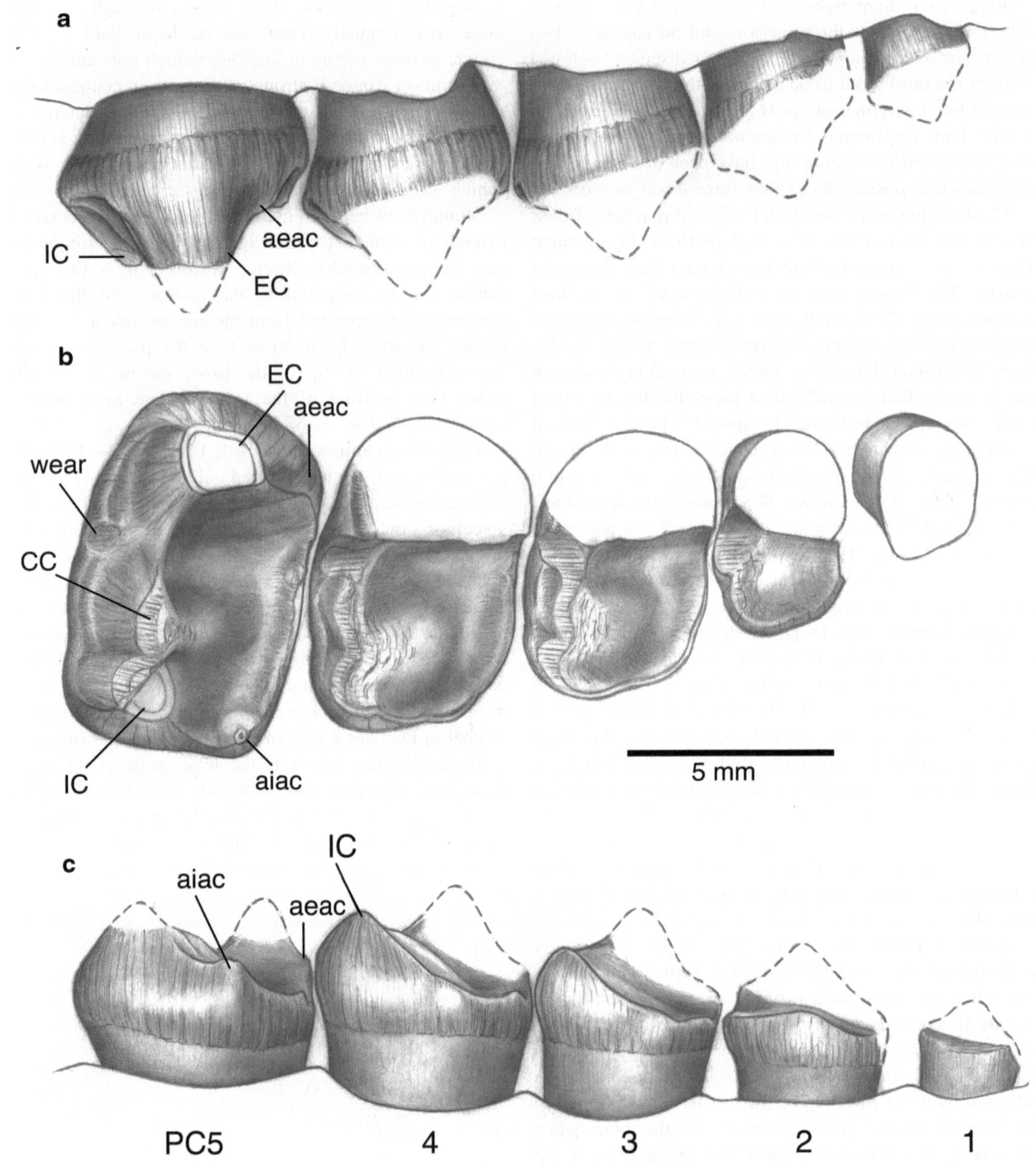

Fig. 14.6 *Mandagomphodon hirschsoni*, Crompton 1972. Right upper postcanines 1–5 of holotype skull, NHMUK R8577. **a** Buccal view; **b** crown view; **c** lingual view (rotated 180° so occlusal surface faces dorsally). *aeac* anteroexternal accessory cusp, *aiac* anterointernal accessory cusp, *CC* main central cusp, *EC* main external cusp, *IC* main internal cusp, *PC* postcanine, *wear* wear facet on posterior cingulum caused by main lower external cusp

I shall present a detailed description of PC5, including wear facets, and then shall describe the crowns of the preceding postcanines and how they have been modified by progressive wear. Functional interpretations will be discussed after the lower dentition is described. Identification of upper postcanine features are shown in Fig. 14.5a.

The fifth right upper postcanine is roughly rectangular in crown view (unlike the elongate oval shape of *S. angustifrons*). It is 9.1 mm in transverse diameter and 5.7 mm in anteroposterior diameter, thus being 1.6 times wider than long. This contrasts with *Scalenodon angustifrons* (Crompton 1955) in which the ratio of width to length of the largest posterior postcanines ranges from 1.6 to 2.1, indicating, as noted by Crompton (1972), an average greater length in *M. hirschsoni*. As also noted by Crompton (1972, pp. 50, 51), that portion of the *M. hirschsoni* crown anterior to the transverse ridge is considerably longer than in *S. angustifrons*, but this is also the case for the portion behind the transverse ridge, which bears a distinct posterior cingulum, as well as for the ridge itself. As in *S. angustifrons*, the transverse ridge is formed by three main cusps joined by a continuous crest that extends from the apex of the main external cusp to the apex of the internal cusp. The internal and central cusps are partly conjoined on the lingual half of the crown and are separated from the taller external cusp by a deep embayment with a broad, nearly semicircular, posterior profile (in contrast to the almost V-shaped profile of *S. angustifrons*). This embayment is referred to in *Boreogomphodon* as a "V-shaped notch" by Sues and Hopson (2010). The labial face of the main external cusp is convex, whereas its lingual face is a vertical, near longitudinal, planar surface. The anterior ridge on the external cusp is an oblique, nearly straight cutting edge that is notched toward its basal end by a small anterior accessory cusp lying on the ridge. Unlike in *S. angustifrons* and *Luangwa drysdalli*, the labial surface of the main external cusp lacks a cingulum.

The anterior basin of the upper postcanine is bounded anteriorly by a low transverse cingulum ridge. This ridge ends labially at the anteroexternal accessory cusp and lingually at a large anterointernal accessory cusp that lies at the junction of the transverse cingulum ridge and the prominent anterior ridge of the main internal cusp. The anterior basin is narrowest lingually, where the main central and internal cusps bulge forward from the transverse crest. Labial to the central cusp, the basin is deeper and extends further back, being bounded posteriorly by the slope up to the deep "V-shaped" notch and laterally by the planar longitudinal surface on the lingual side of the main external cusp. The posterior cingulum is a robust cuspidate ridge that bounds a narrow transverse trough behind the main transverse ridge. The posterior cingulum continues labially and lingually as faint ridges that pass down the rear of the main external and internal cusps toward their apices.

The little-worn enamel of the fifth upper postcanine is relatively thick on the peripheral parts of the crown and very thin within the anterior basin. Wear has removed the lingual enamel from both the main external cusp and its anterior accessory cusp and has worn the enamel on the ridge that extends down the lingual side of the main cusp to join the ridge of the "V-shaped" notch. The thin enamel on the anterior faces of the central and internal cusps appears to be worn through and the apex of the anterolingual accessory cusp shows slight abrasion wear. On the posterior side of the transverse ridge, the central and internal cusps and the posterior slope of the "V-shaped notch" have oblique facets on the thick enamel. The posterior cingulum is notched by wear in the enamel behind the valley separating the central and internal cusps and behind the lingual end of the "V-shaped" notch.

In all of the more anterior upper postcanines, the external cusp has been broken off. The topography of the preserved parts of the crowns becomes progressively lower and more broadly rounded than in PC5. In PC4, wear has truncated the anterior and lingual walls of the anterior basin, leaving a rim of worn enamel around the exposed dentine in the basin, and has entirely obliterated the anterolingual accessory cusp. The anterior faces of the central and internal main cusps are also heavily worn and their apices are smoothly rounded, with dentine extensively exposed on both cusps. On its lingual side, the internal cusp has a raised rim of enamel and adjacent dentine. The posterobasal parts of these cusps preserve enamel with oblique posterodorsally-sloping wear facets bearing anteroposteriorly-oriented striations. The posterior cingulum is worn posteriorly and is notched by wear behind the junction of the central and internal cusps. A deep longitudinal trough truncates the slope and cingulum behind the "V-shaped" notch.

The third postcanine has a more subdued topography than PC4, the central and internal cusps being very low and rounded, with steeper, enamel-covered posterior faces and worn, more gradually-sloping, anterior faces. The lateral part of the anterior basin is deeper than the medial part and is continuous with a deep wear surface with a V-shaped cross-section on the preceding tooth. The enamel across the rear of the transverse ridge bears a continuous oblique wear facet and, more basally, the entire posterior cingulum is worn off, leaving a posterodorsally-sloping facet.

The second postcanine has a featureless surface of exposed dentine surrounded by a rim of enamel lingual to the base of its broken external cusp.

The first postcanine is much smaller than the following tooth and its crown appears to be both worn and damaged, so that no reliable details can be determined.

Lower Postcanines (Figs. 14.4, 14.5b, 14.7)—The lower postcanine dentition is complete on both sides, with seven teeth in place on the left and six in place on the right, plus an isolated right crown that was presumably lost from the empty seventh alveolus and which has been glued back in place. In crown view, the tooth rows are seen to diverge toward the rear, so the labial profile of the tooth rows is slightly concave. In addition, the fourth to seventh teeth incline progressively inward toward the midline. As in the upper postcanines, the

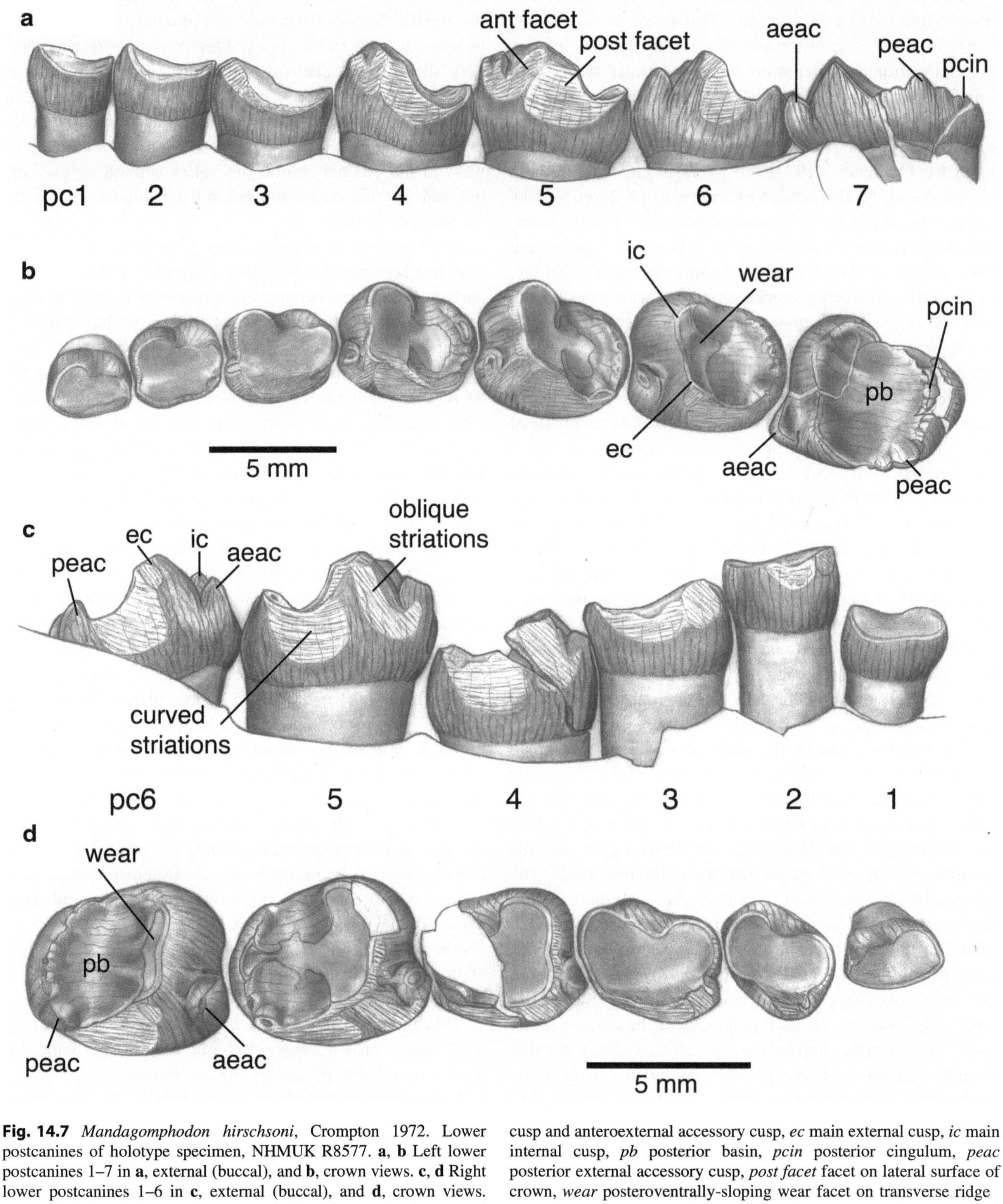

Fig. 14.7 *Mandagomphodon hirschsoni*, Crompton 1972. Lower postcanines of holotype specimen, NHMUK R8577. **a**, **b** Left lower postcanines 1–7 in **a**, external (buccal), and **b**, crown views. **c**, **d** Right lower postcanines 1–6 in **c**, external (buccal), and **d**, crown views. *aeac* anteroexternal accessory cusp, *ant facet* facet on main external cusp and anteroexternal accessory cusp, *ec* main external cusp, *ic* main internal cusp, *pb* posterior basin, *pcin* posterior cingulum, *peac* posterior external accessory cusp, *post facet* facet on lateral surface of crown, *wear* posteroventrally-sloping wear facet on transverse ridge

enamel darkens, therefore, is thicker, in more posterior teeth. The unworn enamel on the posterior teeth is rugose, with a pattern of short irregular dorsoventral ridges.

The description of the crown pattern (see Fig. 14.5b for identification of features) is based on both sixth postcanines, which are little worn, as well as the unworn but damaged left pc7. As with the upper postcanines, the lowers are compared with those of *S. angustifrons*, as was done by Crompton (1972). The crown pattern in *M. hirschsoni*, as in *S. angustifrons*, is typically traversodont, with a tall anterior transverse ridge formed by two cusps and a low posterior basin walled laterally by a narrow concave-upward ridge descending from the anterolabial cusp and behind by a cuspidate cingulum. A large posterolabial accessory cusp forms the posterolateral margin of the rear cingulum, as it does in *S. angustifrons*. Internal to it, the cingulum descends to the posterolabial side of the crown, where it joins a very low, broadly-rounded ridge that bounds the lingual side of the basin. The floor of the basin slopes obliquely downward toward this low ridge, which is soon obliterated in more worn teeth so that the basin becomes open internally.

In both *M. hirschsoni* and *S. angustifrons*, as is usual in Middle Triassic traversodontids, the internal cusp on the transverse ridge is labiolingually wider than the external cusp; the latter cusp, however, is taller and more robust. Both species possess a prominent cusp, very large in *M. hirschsoni*, on the anterolabial face of the transverse ridge. The transverse ridge in *M. hirschsoni* is lower and anteroposteriorly broader than that of *S. angustifrons*, with an oblique rather than nearly vertical posterior slope. The posterior basin in *M. hirschsoni* is consequently proportionately shorter than that of *S. angustifrons*. In *M. hirschsoni*, a low ridge passes from the apex of the lingual cusp down the posterior slope into the basin; its presence is uncertain in *S. angustifrons*.

Wear on the crowns varies slightly on the right and left lower postcanines, with teeth on the right being less worn than the matching teeth on the left. The descriptions of wear will concentrate on the left side, with some observations made from the right. The newly erupted left pc7 shows no signs of wear, but in pc6 a thin posteroventrally-sloping facet extends across both cusps of the anterior transverse ridge. The facet on the labial cusp extends steeply downward from the apex of the cusp into the notch between the two cusps, truncating the thick enamel, which bears longitudinal striations, and extending across a narrow band of exposed dentine. On the near-horizontal labial ridge of the main internal cusp, between notch and apex, the enamel and adjacent dentine are also truncated by wear. Crompton (1972, Fig. 8F) illustrates these wear facets in the right pc6. Wear also extends into the dentine in a broad V down the posterior slope of the transverse ridge below the notch between the main cusps. Though not totally clear, the enamel on the anterodorsal face of the transverse ridge appears to have faint wear surfaces. As noted below, anterior wear on the transverse ridge is clearly present in more anterior postcanines.

The external face of pc6 shows two planar wear facets in the enamel, a smaller one high on the anterolabial face of the main lateral cusp and a larger one that covers much of the lateral surface of the crown below the concave ridge that descends from the main cusp. The small anterolabial facet, when viewed from above, is angled slightly inward anteriorly; posteriorly it truncates the larger, more longitudinal wear facet. It bears parallel striations that are angled posterodorsally between 20 and 25° from the horizontal. The more posterior facet fades out at the level of the large posterolabial cusp on the heel. It bears striations that tend to be oriented upward posteriorly at a low angle to the horizontal.

The wear on pc5 is much more extensive than that on pc6. The notch between the main anterior cusps is totally obliterated, leaving a single posteriorly-sloping transverse facet with thick enamel in front of the transverse ridge and a long posteroventrally-sloping wear facet on the dentine that extends down into the posterior basin. The wear facet on the posterior slope of the transverse ridge of the right pc5 is illustrated by Crompton (1972, Fig. 7B). The floor of the basin appears to have separate wear surfaces separated by a low anteroposterior ridge that bisects the the basin, a more labial and shallower depression behind the main external cusp and a more lingual and deeper depression behind the main internal cusp.

The enamel on the anterodorsal face of the transverse ridge bears a distinct anteroventrally-sloping wear facet near the crest of the ridge. The facet is anteroposteriorly slightly curved, rather than being planar, and striations are not evident.

On the outer wall of the crown of pc5 are two much larger wear facets than seen on pc6. The anterior facet is roughly V-shaped, with one limb extending anteroventrally and slightly medially down the anterolabial side of the main external cusp and the second limb truncating the entire lateral face of the anterolabial accessory cusp. The tip of this cusp is also worn by apical abrasion. The parallel striations on both limbs of the facet are oriented steeply up and back. A narrow vertical strip of unworn enamel separates this anterior facet from the larger, more posterior, wear facet in the enamel that covers most of the lateral surface of the crown. This facet is planar and longitudinally-oriented, lying below the concave lateral ridge and extending back and upward to truncate the side of the posterolabial accessory cusp. The striations on the anterior part of the facet slope only slightly up and back, whereas on the posterolabial cusp they slope more steeply up and back.

The fourth lower postcanine possesses the main features of pc5, but wear is more extensive. Truncated enamel extends across the crest of the transverse ridge, which is worn down lower on the crown than in pc5, lying only slightly above the tip of the anterolabial accessory cusp. Enamel wraps around to the posterior side of the internal cusp and merges ventrally into a thin surface of enamel on the lingual margin of the posterior basin. The worn dentine on the posterior slope of the transverse ridge merges into the worn dentine on the floor of the posterior basin. The posterior cingulum ridge bears an elongate oval wear facet that truncates the enamel and an enclosed patch of dentine. The wear on the enamel on the anterior face of the transverse ridge forms a thin facet that extends across both cusps; the facets bear clear anteroposteriorly-oriented striations.

The main labial cusp of the fourth postcanine has on its anterolateral face a facet that bears surface striations that are oriented posterodorsally at an angle of about 35°. It contacts the more posterior lateral facet, forming in crown view a slight angulation with it. The anterolabial accessory cusp has a truncated apex, with an abraded wear facet that slopes down and forward. The lateral surface of the cusp is polished and may bear a few faint striations.

The lateral face of pc4 shows a wear feature not seen on more posterior teeth: the concave ridge passing down and back from the anterolateral main cusp and the lateral face of the posterolabial accessory cusp are so heavily worn that a large exposure of deeply-incised dentine is present. This feature is shown well on the left pc4, but is not ascertainable on the damaged right pc4. Below this wear surface, the enamel on the lateral surface of the crown bears a large planar wear facet that extends from the apex of the main labial cusp to the base of the posterolabial accessory cusp. There are several prominent concave-upward striations low on the facet and fainter striations higher on the facet that appear to extend up and back at low angles. On the anterolateral face of the main labial cusp is a second facet that contacts the lateral facet, forming in crown view a slight angulation between them. It bears surface striations that are oriented posterodorsally at an angle of about 35°.

The crowns of the anterior three lower postcanines are much more worn than pc4 and lack distinct cusps, including the anterolabial and posterolabial accessory cusps. A rim of enamel forms the circumference of each crown, with the entire central portion formed by worn dentine. Though greatly truncated, the transverse ridge remains the highest part of the crown. Its posterior face slopes at a low angle back to the posterior basin, the rear of which rises slightly to form the basin's posterior wall. The dentine-covered floor of the posterior basin slopes lingually and merges with the lingual embayment of enamel, the surface of which slopes medially at a low angle.

Wear on the labial side of pc1-3 varies between the two tooth rows. On the right side, the labial enamel is less truncated by wear, so less dentine is exposed along the concave cutting ridge; thus, the lateral enamel wear facet is dorsoventrally deeper than on the left. On the right pc3, the lateral enamel facet has striations that show a less consistent trend than on more posterior teeth, but the majority of the striations appear to slope slightly anterodorsally. The smaller facet on the enamel of the anterolabial face of the main labial cusp has striations that are more parallel and slope relatively steeply up and back. These two facets differ from those on more posterior teeth in that they meet at a worn, rounded, surface rather than at a sharp angle. On the left side, where the dorsolabial surface of the crown is truncated by exposed dentine, the lateral enamel facet is narrow dorsoventrally and the orientation of the lateral striations is not clear. On the more anterior facet on the main labial cusp, the parallel striations slope strongly posterodorsally. The linear contact between the two facets forms a distinct angle.

The lateral facets on pc2 on both sides of the jaw are similar to those of pc3, except that the lateral facets are very narrow and the anterolabial facets cannot be distinguished on the smoothly polished enamel. The right pc2 is rotated anterolabially in its alveolus, which could be considered a postmortem event except that the lateral wear facet is restricted to the enamel on the outer side of the anterolabial cusp and the enamel on the more posterior part of the lateral surface is unworn. The shear surface on the main labial cusp of the matching upper postcanine is broken off, so an unusual wear pattern here cannot be determined. As noted above, the crown of pc1 on both sides is so truncated that the labial wear facets are no longer present. However, the enamel on the anterior face of the crown is more rounded and polished than that of pc3, which in turn is more rounded than that of pc4, which still bears traces of planar, sharp-edged wear facets.

Interpretation of Postcanine Occlusion in *Mandagomphodon hirschsoni*

Several features of the occlusion of upper and lower postcanines of *M. hirschsoni* are evident from the description of the wear facets. These are essentially as noted by Crompton (1972).

1. Resting occlusion in unworn postcanines is with the transverse ridge of the lower postcanine fitted into the anterior basin of the matching upper postcanine and

the transverse ridge of the upper postcanine fitted into the posterior basin of the lower postcanine.

2. The initial movement of the lower jaw for chewing is anteroventrally so that the transverse ridge of the lower postcanine comes to lie immediately below the transverse ridge of the preceding upper postcanine (Fig. 14.8a). The transverse ridge of the lower tooth moves up the posterior side of the transverse ridge of the preceding upper tooth. At the same time, the anterolabial face of the main lower external cusp contacts the main upper external cusp on its posterolingual face, behind the extension of the transverse ridge on to the inner face of this cusp (Fig. 14.8b).
3. The path of the tip of the main labial cusp of the lower postcanine at the beginning of the retractive stroke is to move up and back on the posterior slope of the "V-shaped" notch in the upper transverse ridge. As the tip of the cusp passes up and back into the basin of the following upper postcanine, the anterolabial accessory cusp of the lower also slides up and back on the posterior slope of the "V-shaped" notch. The main labial cusp continues up and back across the anterior part of the basin of its matching upper tooth (Fig. 14.8c); it then shifts down and back along the anterior slope of the "V-shaped notch" of the upper transverse ridge (Fig. 14.8d).
4. The path of the wide main lingual cusp of the lower postcanine at the beginning of the closing and retractive stroke is to move up and back on the posterior faces of the medial and lingual cusps of the upper transverse ridge of the preceding tooth (positions in Fig. 14.8a, b). It continues up and back across the more medial part of the upper basin of its matching tooth (position of Fig. 14.8c), then moves down and back along the anterior slope of the medial and lingual cusps of the upper transverse ridge (position of Fig. 14.8d).

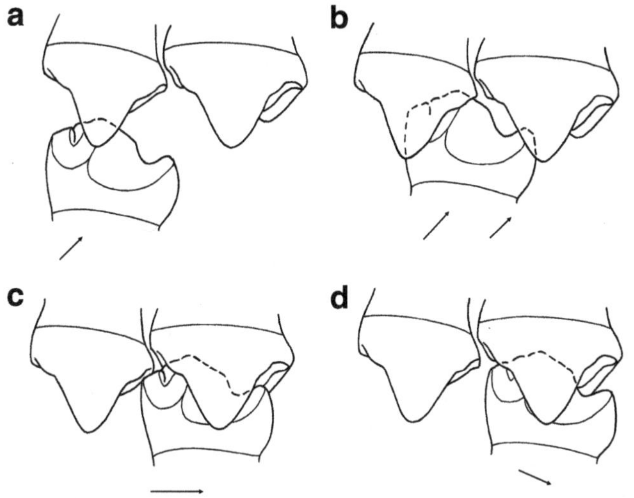

Fig. 14.8 *Mandagomphodon hirschsoni*, Crompton 1972. Stages in the retractive power stroke of a lower postcanine across two adjacent upper postcanines. *Arrows* below each figure show direction of movement at each stage of contact, as indicated by striations on lower tooth. **a** At beginning of the power stroke, the lower postcanine moves posterodorsally so that main external lower cusp contacts internal face of the main external cusp of the anterior upper postcanine. **b** At later stage, the lower tooth is near end of posterodorsal movement in which main external cusp and anteroexternal accessory cusp contact the rear surface of the transverse ridge of the anterior upper postcanine; at the same time, the posteroexternal cusp of the lower postcanine contacts the internal shear surface of the main external cusp of the posterior upper postcanine, creating posterodorsally-sloping striations on the lower cusp. **c** The lower postcanine moves posteriorly, as indicated by near-horizontal striations in the center of the posterolateral wear facet, to reach resting occlusion, at which stage, the main lower cusps occlude in the anterior basin of the upper postcanine. **d** The lower postcanine continues posteriorly and somewhat ventrally beyond resting occlusion so that its truncated transverse ridge (*broken line*) grinds against the anterior face of the medial portion of the upper transverse ridge that is formed by the central and internal main cusps. This movement is indicated by posteroventrally-sloping striations on the anterior portion of the posterolateral wear facet

The functional interpretation of these movements is much as described by Crompton in *Scalenodon angustifrons* and *Scalenodon* (now *Mandagomphodon*) *hirschsoni* (1972, Figs. 5–9). The principal shearing surfaces are as illustrated by Crompton (1972, Fig. 6C): (1) a longitudinal shear surface on the inner face of the main labial cusp in the upper postcanine that is paired with a longitudinal shear surface on the labial face of the matching lower crown; and (2) a transverse shear surface on the transverse ridge of the upper postcanine that is paired with a transverse surface on the transverse ridge of the lower postcanine. The functional unit associated with a single lower postcanine is more complex than this because it includes the transverse ridges of two upper postcanines, the one on the matching upper tooth that the lower tooth contacts during resting occlusion and that of the preceding upper postcanine that the lower tooth contacts during dynamic occlusion. That portion of the upper shear surface that lies posterior and somewhat labial to the upper transverse ridge contacts that portion of the lower shear surface that lies on the anterolabial face of the lower transverse ridge of the succeeding lower tooth (for example, the anterior part of pc5 shears against the posterior part of PC4). Only the portion of the upper longitudinal shear surface that lies anterior to the upper transverse ridge shears against the longitudinal labial face of its matching lower postcanine (for example, the lateral face of pc5 shears against the large longitudinal shear surface on the lingual face of the main upper external cusp of PC5).

With respect to the transverse shear surfaces, the wear surface on the rear of the transverse ridge of the preceding upper tooth shears against the anterior face of the transverse ridge of the following lower tooth (for example, the transverse ridge of pc5 shears against the transverse ridge of PC4). The posterior face of the lower transverse ridge also possesses a distinct wear facet that matches wear on the anterior face of the transverse ridge of its matching upper tooth. Crompton (1972, p. 51) noted that "it is difficult to account for these facets if the power stroke of the lower jaw was directed dorso-posteriorly," although "they could have resulted from the postero-dorsal surface of the transverse ridge of the lower tooth being drawn backwards and downwards across the antero-ventral surface of the transverse ridge of the upper" (Crompton 1972, p. 51, Fig. 9D). However, Crompton pointed out, apparently disapprovingly, that this movement "would have required that the lower postcanines be dragged down an inclined plane". Consequently, he proposed that the same wear facets "would have been formed if the mandible had moved forwards and upwards during dynamic occlusion... so that the leading edge of the transverse ridge of the lowers sheared past the trailing edge of the transverse ridge of the uppers" (Crompton 1972, p. 51, Fig. 9E).

Can any of the wear facets on the postcanine teeth be used to test whether the wear on the rear of the lower transverse ridge and on the front of the upper transverse ridge were made by a posteroventral or an anterodorsal movement of the lower tooth across the upper? It would seem that either direction of movement would produce similar oblique striations. The parallel striations on the anterolateral wear facet of the main labial cusp of the lower postcanines slope strongly up and back, clearly demonstrating that the initial movement of the lower crown was in a posterodorsal direction (Fig. 14.7b, d). Thus, to demonstrate anterodorsal movement of the lower postcanines, one would expect to see equally strong striations on their external wear facets (made by contact with the lingual side of the main upper external cusp) sloping up and forward, parallel to the truncated dorsal surface of the lower transverse ridge (Fig. 14.8a). Variably present and usually faint striations that slope obliquely anterodorsally-posteroventrally do occur on the external longitudinal wear facet of lower crowns, but they are mixed with more horizontally-oriented striations and even posterodorsally-oriented striations. This mixture of differently-oriented, overlapping striations can be interpreted more plausibly as being caused by the lower crown consistently moving in a generally posterior direction, but with an initial posterodorsal direction (indicated by striations on the posterolabial accessory cusp), then a more horizontal direction (indicated by striations on the middle of the lateral surface of the crown), and, finally, a posteroventral direction (indicated by striations high on the lateral surface of the main external cusp). What is lacking among these overlapping striations is a clear set of parallel striations superimposed on the fainter ones but discontinuous from them. This strongly suggests that the reversal in direction of the lower teeth from a backward and downward movement to a powerful forward and upward movement lacks support in the wear facets and striations on the lower postcanines.

I interpret the tooth movements that caused the external striation patterns as follows (Fig. 14.8): (1) during the oblique posterodorsal movement of the lower crown (which is recorded in the striations on the anterolabial main and accessory cusps) the posterolabial accessory cusp of the lower crown contacted the upper longitudinal shearing surface on the inner face of the main external cusp (see Crompton 1972, Fig. 9B), resulting in anteroventrally sloping striations on the posterolabial cusp (Fig. 14.8b); (2) further posterodorsal movement of the lower crown, to a position where the lower transverse ridge moves into the deepest part of the anterior basin of the upper, is represented by the forward continuation of the lateral striations, which now slope slightly anteroventrally to horizontally, on to the middle of the large external wear facet (Fig. 14.8c) on the lower crown (see Crompton 1972, Fig. 9C); (3) the final movement of the rear face of the lower transverse ridge down and back against the anterior face of the upper transverse ridge is represented by the anterodorsal/posteroventral slope (Fig. 14.8d) of the anterior- and dorsal-most striations on the lower crown (Crompton 1972, Fig. 9D). Thus, the evidence of the lateral striations on the lower postcanines suggests a posteroventral crushing/grinding movement, continuous with the posterodorsal and subsequent horizontal and backward movement of the lower crown, rather than a reversal of direction of the lower crown to create an anterodorsal shearing movement of the lower transverse ridge on the upper ridge.

That an oblique downward movement of the lower postcanine can occur toward the end of the occlusal stroke, where it moves down an inclined plane formed by the upper crown, is demonstrated by a variety of living (usually herbivorous) mammals. In Fig. 14.9, the proposed movement of the lower postcanine in relation to the upper postcanine in *Mandagomphodon* is compared with the movement of the lower molar across the upper in the primate *Saimiri* (Kay and Hiiemae 1974, Fig. 9C). In both, the initial phase of upward movement (Phase I) causes shear between tall lower cusps passing internal (or posterior in *Mandagomphodon*) to tall upper cusps. This phase is followed by Phase II, in which the tall lower cusps move obliquely downward and medially (posteriorly in *Mandagomphodon*) across an inclined plane formed by low upper cusps. Kay and Hiiemae (1974, p. 243) argue that "Phase I and Phase II [in primates] are parts of a continuous movement, distinguished only by a change of direction from anteromedial and upward to anteromedial and downward." I believe this interpretation is applicable to *Mandagomphodon*, which differs only in that movement of the lower tooth is posterior rather than medial. The gomphodont also differs from mammals in that it utilizes two upper molariform teeth to achieve the functional results that

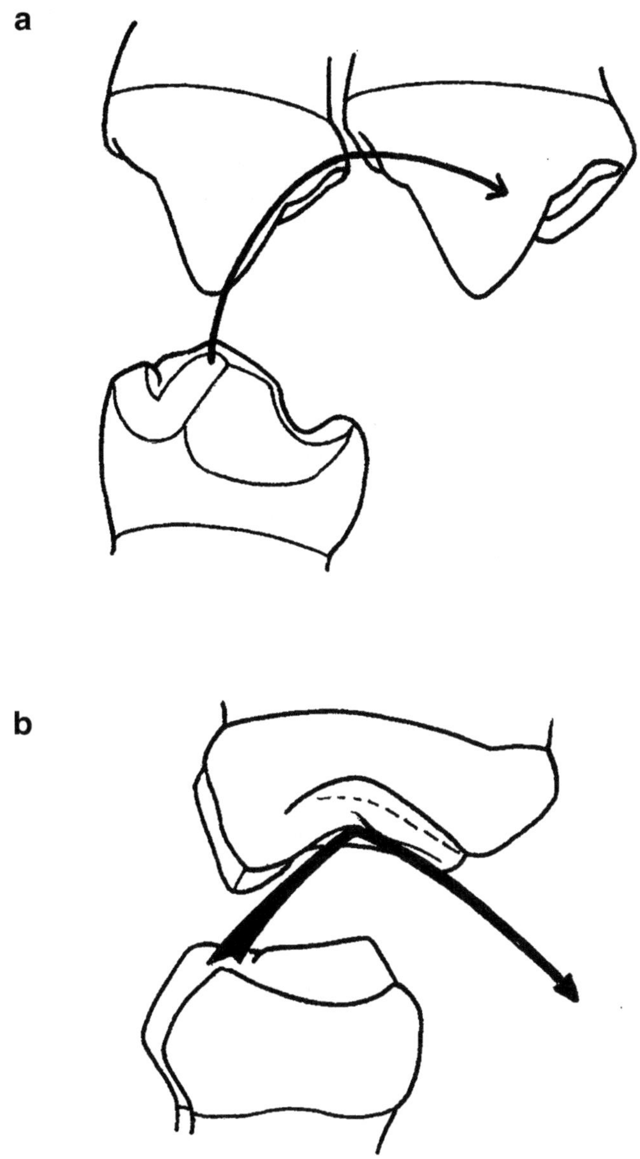

◀Fig. 14.9 Comparison of the curved trajectory of a lower molariform tooth across the matching upper molariform(s) in **a** the gomphodont *Mandagomphodon hirschsoni* Crompton, 1972, and **b** the living cebid primate *Saimiri sciureus* (modified from Kay and Hiiemae 1974). Movement of the lower tooth, as indicated by the *arrows*, is in a posterior direction in **a** and in a medial direction in **b**. Thus the cutting component in the gomphodont (*phase* I of Kay and Hiiemae) is due to movement of the anterior face of the transverse ridge of the lower postcanine up and back against the rear of the upper transverse ridge of the preceding postcanine. (The longitudinal shear also contributes to cutting the food.) In the primate mammal, where the lower jaws move transversely, the shear component of *phase* I is due to movement of the external surfaces of the outer cusps of the lower molar up and inward across the inner surfaces of the external cusps of the upper molar. The grinding component (*phase* II of Kay and Hiiemae) occurs in the gomphodont as the lower postcanine moves down and back beyond centric occlusion at the end of the power stroke, so that the posterodorsal faces of the lower main cusps grind against the anteroventral faces of the central and internal main upper cusps. In the primate, the grinding in phase II occurs as the dorsomedial surfaces of the outer cusps of the lower molar pass medially across the ventrolaterally-facing surfaces of the inner cusps of the upper molar. **a** and **b** not to scale

mammals achieve with one upper molar. Other implications of these differences are discussed below.

As noted above, as wear of the postcanine crowns of *Mandagomphodon* increases from posterior to anterior, the transverse ridges become increasingly lower in height and all of the tooth cusps are eliminated; furthermore, the distinctness of the wear facets is lost on the anterior postcanines, indicating a loss of direct contact between upper and lower teeth. Also, as noted above, the worn enamel on anterior teeth is more rounded and polished than that of more posterior teeth and wear facets are less clear. The increase in smoothing and polishing of the enamel of anterior teeth is undoubtedly due to abrasion by food between the teeth rather than tooth-on-tooth contact. The occlusal surfaces of these teeth, which are formed entirely by dentine, are also smoothly rounded, with only slight topography remaining. This nearly complete destruction of the crown topography in anterior postcanines is characteristic of all traversodontids, indeed of all gomphodonts.

The power stroke of the lower jaws of gomphodonts during food processing is in a posterior direction, as attested by the positions of shearing edges and grinding surfaces on the crowns and by the orientation of wear facets and striations on the enamel. Thus, unworn or little worn postcanines appear to function optimally as the lower crown moves backwards across the matching upper crown. Another characteristic of gomphodont postcanines is that they increase in size from front to back, which is a consequence of their serial eruption as the skull and lower jaws elongate posteriorly during ontogeny and consequently create space at the back of the tooth row for eruption of newly-formed teeth. Thus, for optimal function of the postcanine dentition, the length and shape of the retractive power stroke must be determined by the least worn and largest (i.e., longest) teeth, which lie at or near the posterior end of the tooth row. A consequence of the length of the power stroke being adapted to the functional requirements of the longest postcanines is that for shorter more anterior postcanines the power stroke becomes incompatible to optimal functioning. It is now much too long for these teeth to perform useful shearing or grinding. Thus, the distinct occlusal wear facets seen in more posterior teeth become rounded and polished in more anterior teeth because these teeth no longer possess useful tooth-on-tooth contact. It is also likely that the greater length of movement of lower teeth relative to uppers anteriorly along the tooth row would have increased the rate of wear, and hence of obliteration of the crown surfaces, in comparison to more posterior teeth. As noted above, enamel thickness increases posteriorly along the tooth row, so that the thinner enamel of more anterior teeth probably contributed to an increased rate of wear. However, rather than being a disadvantage, the rapid elimination of the original topography of anterior teeth meant that these teeth could not interfere with the optimal functioning of the little-worn posterior teeth.

The postcanine dentition of *Mandagomphodon hirschsoni*, which shares its main characteristics with most other gomphodonts, functions optimally only in the posterior-most part of the tooth row. The traversodontid postcanine dentition in its functional complexity can be compared with that of mammals. However, the dentitions of most mammals (with the exception of those with planar occlusal surfaces, such as rodents and elephants) function with a transverse rather than fore-aft (propalinal) movement of the lower jaws. Thus, whereas mammalian molars (and premolars) can maintain functional occlusion along the entire tooth row throughout most of an animal's life, and the length of the transverse power stroke can be adjusted to the degree of wear of the tooth crowns, gomphodonts must adjust the length of the longitudinal power stroke, at most, to the last few teeth in the jaw, with the more anterior teeth progressively losing optimal function. Thus, the functional surface and lifetime of use in an herbivorous mammal's cheek dentition is much greater than that of a gomphodont cynodont, which is restricted to a few posterior teeth that appear to lose optimal function fairly rapidly. It also follows that replacement of small, heavily-worn anterior postcanines by unworn teeth of comparable length would not increase the functionality of the dentition if the length of the power stroke remained adapted to optimizing function of the longer posterior teeth.

Tritylodontids, which may have been derived from within Traverodontidae, appear to have circumvented this "functional constraint" by eliminating features that functioned in transverse shear; thus, the postcanine crowns consisted of multiple parallel rows of shearing cusps, with functional boundaries between individual teeth no longer present. Thus, the power stroke no longer needed to be adjusted to the posterior-most teeth and the whole tooth row functioned as a single functional unit.

Acknowledgments My thanks to Dr. Angela Milner and Ms. Sandra Chapman of the Natural History Museum, London, for the loan of the material of *Mandagomphodon hirschsoni*. I also thank Ms. Claire Vanderslice for her carefully and beautifully rendered drawings of these fossils. Thanks also to Christian Kammerer and the anonymous reviewers for helping to improve the manuscript. I wish also to acknowledge the important contribution of Dr. A. W. "Fuzz" Crompton in shaping my approach to the study of vertebrate paleontology and functional biology. My research has over the years been generously supported by the National Science Foundation.

References

Abdala, F., & Ribeiro, A. M. (2003). A new traversodontid cynodont from the Santa Maria Formation (Ladinian-Carnian) of southern Brazil, with a phylogenetic analysis of Gondwanan traversodontids. *Zoological Journal of the Linnean Society, 139*, 529–545.

Abdala, F., Neveling, J., & Welman, J. (2006). A new trirachodontid cynodont from the lower levels of the Burgersdorp Formation (Lower Triassic) of the Beaufort Group, South Africa and the cladistic relationships of Gondwanan gomphodonts. *Zoological Journal of the Linnean Society, 147*, 383–413.

Allin, E. F., & Hopson, J. A. (1992). Evolution of the auditory system in Synapsida ("mammal-like reptiles" and primitive mammals) as seen in the fossil record. In D. B. Webster, R. F. Fay, & A. N. Popper (Eds.), *The evolutionary biology of hearing*. (pp. 587–614). New York: Springer-Verlag.

Bonaparte, J.-F. (1966). Una nueva "fauna" Triásica de Argentina. (Therapsida: Cynodontia–Dicynodontia). Consideraciones filogeneticas y paleobiogeográficas. *Ameghiniana, 4*, 243–296.

Bonaparte, J.-F. (1967). Dos nuevas ''faunas'' de reptiles triásicos de Argentina *Gondwana Stratigraphy* (IUGS symposium) (pp. 283–306) (published in 1969).

Broom, R. (1905). On the use of the term Anomodontia. *Records of the Albany Museum, 1*, 266–269.

Cope, E. D. (1884). The Tertiary Marsupialia. *American Naturalist, 18*, 686–697.

Cox, C. B. (1991). The Pangaea dicynodont *Rechnisaurus* and the comparative biostratigraphy of Triassic dicynodont faunas. *Palaeontology, 34*, 767–784.

Crompton, A. W. (1955). On some Triassic cynodonts from Tanganyika. *Proceedings of the Zoological Society of London, 1125*, 617–669.

Crompton, A. W. (1972). Postcanine occlusion in cynodonts and tritylodontids. *Bulletin of the British Museum (Natural History): Geology, 21*, 27–71.

Crompton, A. W., & Ellenberger, F. (1957). On a new cynodont from the Molteno beds and the origin of the tritylodontids. *Annals of the South African Museum, 44*, 1–14.

Gao, K.-Q., Fox, R. C., Zhou, C.-F., & Li, D.-Q. (2010). A new nonmammalian eucynodont (Synapsida: Therapsida) from the Triassic of Northern Gansu Province, China, and its biostratigraphic and biogeographic implications. *American Museum Novitates, 3685*, 1–25.

Hopson, J. A. (1966). The origin of the mammalian middle ear. *American Zoologist, 6*, 437–450.

Hopson, J. A. (1984). Late Triassic traversodont cynodonts from Nova Scotia and southern Africa. *Palaeontologia Africana, 25*, 181–201.

Hopson, J. A. (1985). Morphology and relationships of *Gomphodontosuchus brasiliensis* von Huene (Synapsida, Cynodontia, Tritylodontoidea) from the Triassic of Brazil. *Neues Jahrbuch für Geologie und Paläontologie, Monatshefte, 1985*, 285–299.

Hopson, J. A., & Barghusen, H. R. (1986). An analysis of therapsid relationships. In N. Hotton, P. D. MacLean, J. J. Roth, & E. C. Roth (Eds.), *The ecology and biology of mammal-like reptiles* (pp. 83–106). Washington, DC: Smithsonian Institution Press.

Hopson, J. A., & Kitching, J. W. (2001). A probainognathian cynodont from South Africa and the phylogeny of nonmammalian cynodonts. *Bulletin of the Museum of Comparative Zoology, Harvard University, 156*, 5–35.

Hopson, J. A., & Sues, H.-D. (2006). A traversodont cynodont from the Middle Triassic (Ladinian) of Baden-Württemberg (Germany). *Paläontologische Zeitschrift, 80*, 124–129.

Huene, F. von. (1936). *Die Fossilen Reptilien des Südamerikanischen Gondwanalandes. Ergebnisse der Sauriergrabungen in Südbrasilien*

1928/29. Lieferung 2 (Cynodontia, pp. 93–159). Tubingen: Franz F. Heine.

Kammerer, C. F., Flynn, J. J., Ranivoharimanana, L., & Wyss, A. R. (2008). New material of *Menadon besairiei* (Cynodontia: Traversodontidae) from the Triassic of Madagascar. *Journal of Vertebrate Paleontology, 28*, 445–462.

Kay, R. F., & Hiiemae, K. M. (1974). Jaw movement and tooth use in recent and fossil primates. *American Journal of Physical Anthropology, 40*, 227–256.

Kemp, T. S. (1982). *Mammal-like reptiles and the origin of mammals*. London: Academic Press.

Liu, J., & Olsen P. (2010). The phylogenetic relationships of Eucynodontia (Amniota: Synapsida). *Journal of Mammalian Evolution, 17*, 151–176.

Liu, J., & Abdala, F. (2013). Phylogeny and taxonomy of the Traversodontidae. In C. F. Kammerer, K. D. Angielczyk, & J. Fröbisch (Eds.), *Early evolutionary history of the Synapsida* (pp. 255–279). Dordrecht: Springer.

Martinelli, A. G., de la Fuente, M., & Abdala, F. (2009). *Diademodon tetragonus* Seeley, 1894 (Therapsida: Cynodontia) in the Triassic of South America and its biostratigraphic implications. *Journal of Vertebrate Paleontology, 29*, 852–862.

Owen, R. (1861). *Palaeontology, or a systematic summary of extinct animals and their geologic relations* (2nd ed.). Edinburgh: Adam and Charles Black.

Parrington, F. R. (1946). On the cranial anatomy of cynodonts. *Proceedings of the Zoological Society of London, 116*, 181–197.

Reisz, R. R., & Sues, H.-D. (2000). Herbivory in Late Paleozoic and Triassic terrestrial vertebrates. In H.-D. Sues (Ed.), *Evolution of herbivory in terrestrial vertebrates* (pp. 9–41). Cambridge: Cambridge University Press.

Romer, A. S. (1967). The Chañares (Argentina) Triassic reptile fauna: 3. Two new gomphodonts. *Massetognathus pascuali and M. teruggii. Breviora, 264*, 1–25.

Seeley, H. G. (1894). Researches on the structure, organization, and classification of the fossil Reptilia. Part IX, Section 3. On *Diademodon. Philosophical Transactions of the Royal Society of London B, 185*, 1029–1041.

Seeley, H. G. (1908). On the dentition of the palate in the South African fossil reptile genus *Cynognathus. Geological Magazine, 5*, 486–491.

Sues, H.-D. (1985). The relationships of the Tritylodontidae (Synapsida). *Zoological Journal of the Linnean Society, 85*, 205–217.

Sues, H.-D. (1986). The skull and dentition of two tritylodontid synapsids from the Lower Jurassic of western North America. *Bulletin of the Museum of Comparative Zoology, Harvard University, 151*, 217–268.

Sues, H.-D., & Hopson, J. A. (2010). Anatomy and phylogenetic relationships of *Boreogomphodon jeffersoni* (Cynodontia: Gomphodontia) from the Upper Triassic of Virginia. *Journal of Vertebrate Paleontology, 30*, 1202–1220.

Sues, H.-D., & Reisz, R. R. (1998). Origins and early evolution of herbivory in tetrapods. *TREE, 13*, 141–145.

Wopfner, H. (2002). Tectonic and climatic events controlling deposition in Tanzanian Karoo basins. *Journal of African Earth Sciences, 34*, 167–177.

Chapter 15
Phylogeny and Taxonomy of the Traversodontidae

Jun Liu and Fernando Abdala

Abstract We review the taxonomic history of traversodontid cynodont genera and species and the previous classifications and phylogenetic analyses of the group. 17 genera and 22 species are accepted as valid taxa within Traversodontidae. The phylogenetic relationships of traversodontids and other gomphodonts (including 6 trirachodontid species and *Diademodon tetragonus*) are analyzed based on 77 characters. *Scalenodon angustifrons* and (*Andescynodon+Pascualgnathus*) are found to be the most basal traversodontids. The three species of *Scalenodon* included in the analysis (*Scalenodon angustifrons*, *S. attridgei*, and *S. hirschsoni*) do not form a clade, supporting recognition of a separate genus for the latter two. A monophyletic Gomphodontosuchinae is recovered, as suggested by previous analyses. The distribution of traversodontids is discussed, with Gondwana and especially Africa suggested as the ancestral area for Gomphodontia and Traversodontidae.

Keywords Cynodontia • Gomphodontia • Gondwana • Laurasia • Triassic

Introduction

The Family Traversodontidae was established by von Huene (1936) to include Brazilian Triassic cynodonts with labiolingually expanded postcanines (von Huene 1928, 1936). Representatives of this group were recognized in the Triassic of Argentina shortly thereafter; these taxa were represented by partial skulls, some of them erroneously allied with *Belesodon magnificus*, a Brazilian taxon with sectorial dentition (Cabrera 1943). Crompton (1955) recognized the first representative of this group in Africa, *Scalenodon angustifrons* from the Manda beds of Tanzania. He also recognized the presence of additional traversodontids in this fauna with postcanines resembling those of the Brazilian taxon *Gomphodontosuchus brasiliensis* (Crompton 1955, p. 659). Another African traversodontid, *Scalenodontoides macrodontes*, was described based on a large partial lower jaw including dentition found in Lesotho (Crompton and Ellenberger 1957).

Several important new traversodontid discoveries occurred during the 1960s, most of them in Argentina. Bonaparte (1962, 1963a, 1966a) provided extensive descriptions of the skull, postcranium, and endocranial cavities of *Exaeretodon frenguellii*, based on rich, new material found in successive expeditions to the Upper Triassic Ischigualasto Formation. Bonaparte (1963b, c) also proposed two new traversodontid taxa, *Proexaeretodon vincei* and *Ischignathus sudamericanus*, found in the same deposits as *E. frenguellii*. Several new traversodontid taxa were discovered in earlier rocks in western Argentina: *Massetognathus* (represented by three species, *M. pascuali*, *M. teruggii*, and *M. major*) and the closely related *Megagomphodon oligodens* from the Middle Triassic Chañares Formation (Romer 1967, 1972); *Andescynodon mendozensis* and *Rusconiodon mignonei* from the Cerro de las Cabras Formation (Bonaparte 1969, 1970); and *Pascualgnathus polanskii* from the Rio Seco de la Quebrada Formation, originally considered a trirachodontid but later reinterpreted as a basal traversodontid (Bonaparte 1966b, 1970; see also Martinelli 2010a). During this period, new discoveries also occurred in Africa, with the description of *Luangwa drysdalli* from the upper Ntawere Formation of Zambia (Brink 1963).

J. Liu (✉)
Key Laboratory of Vertebrate Evolution and Human Origin of Chinese Academy of Sciences, Institute of Vertebrate Paleontology and Paleoanthropology, Chinese Academy of Sciences, 100044 Beijing, People's Republic of China
e-mail: liujun@ivpp.ac.cn

F. Abdala
Evolutionary Studies Institute, Private Bag 3, Wits, Johannesburg, 2050, South Africa
e-mail: nestor.abdala@wits.ac.za

Christian F. Kammerer, Kenneth D. Angielczyk and Jörg Fröbisch (eds.), *Early Evolutionary History of the Synapsida*, Vertebrate Paleobiology and Paleoanthropology, DOI: 10.1007/978-94-007-6841-3_15,

Crompton (1972, see also Hopson 2013) described three new species from the Manda beds of Tanzania (*Scalenodon attridgei*, *S. charigi*, and *S. hirschsoni*) and also developed a detailed study of postcanine occlusion in traversodontid cynodonts, based mainly on evidence from *Scalenodon* species. Barberena (1974, 1981a, b) produced an extensive review of South American traversodontid cynodonts, including the description of additional material of the poorly known *Traversodon stahleckeri* and the new species *Massetognathus ochagaviae* (see also Teixeira 1987; Liu et al. 2008) from southern Brazil. Chatterjee (1982) described the first Indian traversodontid, the Late Triassic *Exaeretodon statisticae*. Hopson (1984) redescribed the holotype of *Scalenodontoides macrodontes*, adding new information about the snout of this taxon. Hopson (1985) provided a redescription of the type and only specimen of *Gomphodontosuchus brasiliensis* from the Middle Triassic of Brazil and proposed a close relationship with the Late Triassic traversodontids *Exaeretodon* and *Scalenodontoides*.

Goñi (1986; see also Goñi and Goin 1988) presented a detailed description of the dentition and an analysis of dental replacement in the Argentinean traversodontid *Andescynodon mendozensis* and subsequently postulated a hypothesis about the origin of gomphodont morphology in the postcanines of this taxon (Goñi and Goin 1987). Goñi and Abdala (1989) restudied the Argentinean traversodontid *Rusconiodon mignonei*, including the holotype and additional unpublished material. Goñi and Goin (1990) described unpublished postcanine material of *Exaeretodon frenguellii*, mainly to produce a biomechanical analysis of mastication in this taxon.

Two new traversodontids, *Menadon besairiei* and *Dadadon isaloi*, were recovered from the 'Isalo II' beds of southwestern Madagascar (Flynn et al. 1999, 2000; Kammerer et al. 2008). Recently, new species of Brazilian traversodontids have been described, including *Exaeretodon riograndensis* (Abdala et al. 2002; Oliveira 2006; Oliveira et al. 2007), *Santacruzodon hopsoni* (Abdala and Ribeiro 2003), and *Luangwa sudamericana* (Abdala and Teixeira 2004). The latter taxon represents the first commonality of a traversodontid genus between South America and Africa. The latest addition to the Brazilian traversodontid record is *Protuberum cabralense* (Reichel et al. 2009), represented by a fairly complete skeleton exhibiting bizarre rib morphology with a series of protuberances along their shafts.

Traversodontid records from Laurasia are much rarer than on the southern continents. The first record of traversodontids from the northern hemisphere was reported by Tatarinov (1973, 1988): *Antecosuchus ochevi*, represented by a partial maxilla and postcanines, and *Scalenodon boreus*, represented by isolated teeth, both from the Middle Triassic of Russia.

Hopson (1984) described the first traversodontid from North America: ?*Scalenodontoides plemmyridon*, represented by a couple of mandibles without postcanines, an isolated large canine and postcanine. Important additions to the North American traversodontid record occurred during the 1990s, with the discovery of the small traversodontid *Boreogomphodon jeffersoni* from the Upper Triassic of Virginia and North Carolina (Sues and Olsen 1990; Sues et al. 1994; Liu and Sues 2010; Sues and Hopson 2010). In addition, a new upper postcanine tooth of ?*Scalenodontoides plemmyridon* was described by Sues et al. (1992), who reassessed the taxonomic identity of this species, placing it in the new genus *Arctotraversodon*. Sues et al. (1999) described another small traversodontid cynodont, *Plinthogomphodon herpetairus*, represented by a fragment of the snout with dentition and postcranial bones from the Upper Triassic of North Carolina.

Hahn et al. (1988) reported the first record of traversodontids in western Europe (see also Sigogneau-Russell and Hahn 1994): two isolated postcanines of *Microscalenodon nanus* from southern Belgium. Although subsequent discoveries have increased the number of nominal traversodontid species from Europe, the more recent finds are also limited to dental records. Godefroit and Battail (1997) described several new cynodonts represented by isolated teeth from the Upper Triassic of Saint-Nicolas-de-Port, France. They reported two new species of traversodontids, *Maubeugia lotharingica* and *Rosieria delsatei*, and also described unnamed traversodontid taxa as genus aff. *Rosieria*, genus aff. *Microscalenodon*, and gen. et sp. indet., each of them represented by one isolated postcanine. Godefroit (1999) described an isolated tooth interpreted as a traversodontid upper postcanine of the new species *Habayia halbardieri* from Upper Triassic levels of southern Belgium. More recently, Hopson and Sues (2006) described an isolated lower postcanine of the new species *Nanogomphodon wildi* from the Middle Triassic of Germany.

Recent revisions of traversodontids have proposed synonymy of several of the nominal species. Abdala and Giannini (2000) studied all the traversodontid material from the Argentinean Chañares Formation (originally assigned to three species of *Massetognathus* and *Megagomphodon oligodens*) and supported previous conclusions (e.g., Hopson and Kitching 1972; Battail 1991) that there was only one species represented (*Massetognathus pascuali*), and that most of the diagnostic characters proposed to differentiate these taxa can be explained as ontogenetic variation. Liu and Powell (2009) considered *Rusconiodon mignonei* to be a junior synonym of *Andescynodon mendozensis*, and Liu (2007) regarded *Ischignathus sudamericanus* as a junior synonym of *Exaeretodon argentinus*. More recently, Liu and Sues (2010) proposed *Plinthogomphodon* as a junior synonym of *Boreogomphodon*, but a specific differentiation was tentatively proposed mainly because of their occurrences in different ages (i.e., Carnian and Norian).

In the current study, 22 traversodontid species are recognized as valid: *Andescynodon mendozensis*, *Arctotraversodon plemmyridon*, *Boreogomphodon herpetairus*, *Boreogomphodon jeffersoni*, *Dadadon isaloi*, *Exaeretodon argentinus*, *Exaeretodon riograndensis*, *Gomphodontosuchus brasiliensis*, *Luangwa drysdalli*, *Luangwa sudamericana*, *Mandagomphodon attridgei*, *Mandagomphodon hirschsoni*, *Massetognathus pascuali*, *Massetognathus ochagaviae*, *Menadon besairiei*, *Nanogomphodon wildi*, *Pascualgnathus polanskii*, *Protuberum cabralense*, *Santacruzodon hopsoni*, *Scalenodon angustifrons*, *Scalenodontoides macrodontes*, and *Traversodon stahleckeri* (see discussion below and Table 15.1).

Systematic and Phylogenetic History of Traversodontidae

The Family Traversodontidae originally included three cynodont species from the Brazilian Triassic: *Traversodon stahleckeri*, ?*Traversodon major*, and *Gomphodontosuchus brasiliensis* (von Huene 1936). This group was characterized by von Huene (1936) as possessing expanded molars, a skull with proportions similar to that of galesaurids and chiniquodontids, and the anterior root of the zygoma located high on the maxilla, above the level of the teeth. Later, von Huene (1948) included taxa from the Argentinean Triassic described by Cabrera (1943) in this family. Romer (1956) and Watson and Romer (1956) included *T. stahleckeri* in

Table 15.1 List of worldwide traversodontid cynodonts

Taxon	Stratum	Age	Size
South America (9 genera)			
Andescynodon mendozensis	Cerro de las Cabras Formation	Late Anisian/Early Ladinian	Medium
Pascualgnathus polanskii	Río Seco de la Quebrada Formation	Late Anisian/Early Ladinian	Medium
Luangwa sudamericana	Santa Maria Formation	?Early Carnian	Medium
Massetognathus pascuali	Chañares Formation	Early Carnian	Medium
Massetognathus ochagaviae	*Dinodontosaurus* Assemblage Zone, Santa Maria Formation	Early Carnian	Medium
Traversodon stahleckeri	*Dinodontosaurus* Assemblage Zone, Santa Maria Formation	Early Carnian	Large
Santacruzodon hopsoni	*Santacruzodon* Assemblage Zone, Santa Maria Formation	Late Carnian	Medium
Protuberum cabralense	*Dinodontosaurus* Assemblage Zone, Santa Maria Formation	Early Carnian	Large
Gomphodontosuchus brasiliensis	*Hyperodapedon* Assemblage Zone, Santa Maria Formation	Early Norian	Medium
Exaeretodon riograndensis	*Hyperodapedon* Assemblage Zone, Santa Maria Formation	Early Norian	Large
Exaeretodon argentinus	Ischigualasto Formation	Early Norian	Large
Africa (6 genera)			
Luangwa drysdalli	Upper Ntawere Formation	Late Anisian/Early Ladinian	Medium
Luangwa sp.	Upper Omingonde Formation	Middle Triassic	Large
Scalenodon angustifrons	Manda Beds	Late Anisian/Early Ladinian	Medium
Mandagomphodon attridgei	Manda Beds	Late Anisian/Early Ladinian	–
Mandagomphodon hirschsoni	Manda Beds	Late Anisian/Early Ladinian	Medium
Dadadon isaloi	'Isalo II' (=Makay Formation)	Late Carnian	Medium
Menadon besairiei	'Isalo II' (=Makay Formation)	Late Carnian	Large
Scalenodontoides macrodontes	Lower Elliot Formation	Late Norian	Large
India (1 genus)			
Exaeretodon sp.	Lower Maleri Formation	Early Norian	–
North America (2 genera)			
Arctotraversodon plemmyridon	Wolfville Formation, Fundy Group	Carnian	Large
Boreogomphodon jeffersoni	Tomahawk Creek Member, Turkey Branch Formation; Pekin Formation	Carnian	Small
Boreogomphodon herpetairus	Muddy sandstone of Lithofacies Association II, Newark Supergroup	? Early Norian	Small
Europe (1 genus)			
Nanogomphodon wildi	Lower Keuper or Erfurt Formation	Ladinian	Small

Diademodontidae, and erected the new family Gomphodontosuchidae for *G. brasiliensis*. After the discovery of numerous additional South African specimens, Romer (1966) modified his proposal, including all of these forms in Traversodontidae in his third edition of *Vertebrate Paleontology*.

Bonaparte (1963d) presented an extensive review of traversodontids, with a large list of characters typical for the family that included *Gomphodontosuchus* and *Traversodon* from Brazil; *Exaeretodon*, *Proexaeretodon*, and *Ischignathus* from Argentina; and *Scalenodon* and *Scalenodontoides* from Africa. He proposed that Traversodontidae originated from procynosuchids, independent of the lineage that give rise to thrinaxodontids, cynognathids, and diademodontids. Bonaparte (1963d), based mainly on evidence presented by Crompton and Ellenberger (1957), also considered the possibility that tritylodontids originated from basal traversodontids.

Hopson and Kitching (1972) introduced modifications to traversodontid classification, including all the taxa under the Subfamily Traversodontinae, which along with Diademodontinae and Trirachodontinae were members of the Family Diademodontidae. This group and the Family Tritylodontidae were included under the Superfamily Tritylodontoidea.

Tatarinov (1974) presented a classification of therapsids in which he included the Family Traversodontidae in the Gomphognathoidea. He recognized two subfamilies of traversodontids, Scalenodontinae (including *Scalenodon boreus*) and Traversodontinae (including *Antecosuchus ochevi*).

Hopson (1984, 1985) suggested that *Scalenodon angustifrons* derived from *Andescynodon* and proposed that *Luangwa*, *Traversodon*, and possibly *Scalenodon attridgei* formed a group more derived than *S. angustifrons*. He interpreted *Massetognathus*, *Scalenodon hirschsoni*, *Gomphodontosuchus*, *Exaeretodon*, *Scalenodontoides*, and probably *Scalenodon charigi* as representing a group of the most derived traversodontids. Hopson (1985) presented a cladogram in which *Exaeretodon* and *Scalenodontoides* were sister taxa, whereas *Gomphodontosuchus* and, tentatively, *S. charigi* were basal to the (*Exaeretodon*+*Scalenodontoides*) clade.

Brink (labeled 1982, but published in 1986) listed 15 genera and 17 species in Traversodontidae: nine from South America (*Andescynodon mendozensis*, *Rusconiodon mignonei*, *Pascualgnathus polanskii*, *Massetognathus pascuali*, *Massetognathus teruggii*, *Traversodon stahleckeri*, *Gomphodontosuchus brasiliensis*, *Exaeretodon argentinus*, and *Ischignathus sudamericanus*); three from Africa (*Luangwa drysdalli*, *Scalenodon angustifrons*, and *Scalenodontoides macrodontes*); two from Russia (*Antecosuchus ochevi* and *Scalenodon boreus*); and one from China (*Traversodontoides wangwuensis*). He also considered two taxa (*Colbertosaurus muralis* from Argentina and *Theropsodon njalilus* from Tanzania) as Traversodontidae *incertae sedis*. Brink's (1986) catalogue was not exhaustive in its coverage of traversodontids and several taxa were not included in his work.

Battail (1991, Fig. 8) considered traversodontids to be a monophyletic group closely related to tritylodontids. He recognized the following valid species: *Andescynodon mendozensis*, *Exaeretodon frenguellii*, *Exaeretodon vincei*, *Gomphodontosuchus brasiliensis*, *Ischignathus sudamericanus*, *Scalenodontoides macrodontes*, ?*Scalenodontoides plemmyridon*, *Massetognathus pascuali*, *Pascualgnathus polanskii*, *Rusconiodon mignonei*, *Scalenodon angustifrons*, *Scalenodon hirschsoni*, *Scalenodon attridgei*, ?*Scalenodon charigi*, *Scalenodon drysdalli*, ?*Scalenodon boreus*, *Theropsodon njalilus*, *Traversodon stahleckeri* and perhaps *Microscalenodon nanus*. The species *E. vincei* presented as a new combination was based on a single specimen originally described as *Proexaeretodon vincei* by Bonaparte (1963). Another new combination was *Scalenodon drysdalli* (originally *Luangwa drysdalli*), based on the similarity of dentition of a second specimen of *L. drysdalli* described by Kemp (1980) with species of *Scalenodon*. Finally, Battail (1991) also proposed the synonymy of the three Argentinean species of *Massetognathus* and *Megagomphodon oligodens*.

The monophyly of Traversodontidae is the subject of current debate, with some scholars (Sues 1985; Hopson and Barghusen 1986; Hopson and Kitching 2001; Sues and Hopson 2010) arguing that tritylodontids are derived from traversodonts, a hypothesis put forward by Crompton and Ellenberger (1957; see also Crompton 1972). Other researchers support the hypothesis of a monophyletic Traversodontidae within Cynognathia (Luo 1994; Abdala 1998, 2007; Liu and Olsen 2010), where tritylodontids are members of Probainognathia, closely related to mammaliaforms.

Flynn et al. (1999, p. 765) proposed a stem-based definition of Traversodontidae as "the clade consisting of all cynodont species sharing a more recent common ancestor with *Exaeretodon* than with *Probainognathus* or Mammalia". This definition is problematic because it would include *Cynognathus, Diademodon*, and trirachodontids within this group based on recent phylogenetic hypotheses and is therefore much more inclusive than the traditional meaning of the taxon. More recently, Kammerer et al. (2008, p. 446) considered Traversodontidae to represent the most inclusive clade containing *Traversodon stahleckeri* von Huene 1936, p. 132 but not *Trirachodon kannemeyeri* Seeley 1895, p. 48 or *Diademodon tetragonus* Seeley 1894, p. 1030. We follow the definition of Kammerer et al. (2008) here.

Godefroit and Battail (1997) presented a phylogeny of Traversodontidae based on a reduced set of dental characters mapped on a tree including 14 genera. Abdala (1998) analyzed the interrelationships of nine traversodontid genera, using a parsimony-based program to generate shortest trees from a data matrix of 11 dental characters. All the following contributions explored the relationships of traversodontids using data sets analyzed by parsimony programs. Flynn et al. (2000) studied the interrelationships of six traversodontid genera using 16 characters. While presenting the phylogeny of non-mammalian cynodonts, Hopson and Kitching (2001) included seven genera and species of traversodontids. Abdala and Ribeiro (2003) presented a phylogeny of 13 traversodontids with 21 dental and seven craniomandibular characters. Abdala et al. (2006) expanded the data set to 42 characters in an analysis designed to recover gomphodont cynodont relationships. Kammerer et al. (2008) presented a phylogeny with a revised version of the data set of Abdala et al. (2006), whereas Reichel et al. (2009) added a new taxon (*Protuberum*) to the analysis of Abdala and Ribeiro (2003). Recently, Sues and Hopson (2010) presented the phylogeny of 17 traversodontid taxa, including for the first time taxa from Laurasia, and recovered a clade composed of *Boreogomphodon*, *Arctotraversodon* and *Nanogomphodon* from the Northern Hemisphere as the sister group to most other known Middle and Late Triassic traversodontids from Gondwana.

In previous phylogenetic analyses, researchers have obtained consensus on the relationships of some Middle and Late Triassic Gondwanan taxa (e.g., *Exaeretodon*, *Gomphodontosuchus*, *Menadon*, *Protuberum*, and *Scalenodontoides* have been consistently found to be more closely related to each other than to other traversodontid taxa). However, important differences exist between these previous analyses regarding the interrelationships of basal traversodontids. To help resolve this issue, we assembled an enhanced dataset including postcranial characters to reexamine the phylogeny of traversodontids, and, based on this result, revise the taxonomy of the group.

Institutional Abbreviations: AMNH FARB, Fossil Amphibians, Reptiles, and Birds Collection, American Museum of Natural History, New York City, NY, USA; BP, Bernard Price Institute for Palaeontological Research, University of the Witwatersrand, Johannesburg, South Africa; BSP, Bayerische Staatssammlung für Paläontologie und historische Geologie, Munich, Germany; CAMZM, University Museum of Zoology, Cambridge, UK; CGP, Council for Geosciences, Pretoria, South Africa; CM, Carnegie Museum of Natural History, Pittsburgh, Pennsylvania, USA; FMNH, Field Museum of Natural History, Chicago, IL, USA; GPIT, Institut und Museum für Geologie und Paläontologie der Universität Tübingen, Tübingen, Germany; GSN, Geological Survey of Namibia, Windhoek, Namibia; IRSNB, Institut Royal des Sciences Naturelles de Belgique, Brussels, Belgium; MACN, Museo Argentino de Ciencias Naturales "Bernardino Rivadavia", Buenos Aires, Argentina; MCN, Museu de Ciências Naturais, Fundação Zoobotânica do Rio Grande do Sul, Porto Alegre, Brazil; MCP, Laboratório de Paleontologia, Museu de Ciências e Tecnologia, Pontifícia Universidade Católica do Rio Grande do Sul, Porto Alegre, Brazil; MCZ, Museum of Comparative Zoology, Harvard University, Cambridge, MA, USA; MGB, Museu Guido Borgomanero, Mata, Brazil; MLP: Museo de La Plata, La Plata, Argentina; MNHN, Muséum National d'Histoire Naturelle, Paris, France; NCSM, North Carolina State Museum, Raleigh, NC, USA; NHMUK, Natural History Museum, London, UK; NMQR, National Museum, Bloemfontein, South Africa; NSM, Nova Scotia Museum, Halifax, Canada; PIN, Paleontological Institute, Russian Academy of Sciences, Moscow, Russia; PULR, Museo de Antropología, Universidad Nacional de La Rioja, La Rioja, Argentina; PVL, Colección de Palaeontología de Vertebrados, Instituto Miguel Lillo, Universidad Nacional de Tucumán, Argentina; PVSJ, Museo de Ciencias Naturales, Universidad Nacional de San Juan, San Juan, Argentina; SAM, Iziko, the South African Museum, Cape Town, South Africa; SMNS, Staatliches Museum für Naturkunde Stuttgart, Stuttgart, Germany; UA, Université d'Antananarivo, Antananarivo, Madagascar; TSK, collection of Prof. Thomas Kemp, Oxford University, Oxford, UK; UFRGS-PV, Setor de Paleovertebrados, Instituto de Geociências, Universidade Federal do Rio Grande do Sul, Porto Alegre, Brazil; UNC, Collections of the Department of Geology, University of North Carolina, Chapel Hill, USA; USNM, National Museum of Natural History, Washington D.C., USA; VMNH, Virginia Museum of Natural History, Martinsville, VA, USA; YPM-PU, former Princeton University collection, now housed in the Peabody Museum of Natural History, Yale University, New Haven, CT, USA.

Phylogenetic Analysis

All currently accepted valid traversodontid species were included in this analysis except for *Boreogomphodon* (originally *Plinthogomphodon*) *herpetairus*, whose postcanine morphology cannot be differentiated from that of *Boreogomphodon jeffersoni*. The sectorial-toothed cynodonts *Thrinaxodon liorhinus* and *Cynognathus crateronotus* and the gomphodonts *Beishanodon youngi*, *Cricodon metabolus*, *Diademodon tetragonus*, an unnamed South African taxon (CGP JSM100), *Langbergia modisei*, *Sinognathus gracilis*, and *Trirachodon berryi* were also included in this analysis; *Thrinaxodon liorhinus* was used to root the most parsimonious trees (MPTs). The character list includes

18 craniomandibular characters, 16 tooth position characters, 32 dental morphology characters, and 10 postcranial characters (Appendix 15.1).

The homology between sectorial and gomphodont teeth remains an unsolved problem in phylogenetic analyses of cynodont relationships. Abdala and Ribeiro (2003; see also Hopson 2005) accepted the homology of the sectorial border of gomphodont tooth with the sectorial teeth, and considered the lingual border of the former a new structure. However Abdala and Ribeiro (2003; see also Martinelli 2010b) also proposed an explanation of homology based on the rotation of sectorial teeth to constitute the gomphodont postcanines in *Trirachodon*. In this case the aligned main cusps of the sectorial tooth would be homologous with the transverse crest of gomphodont postcanines. In this study, the gomphodont tooth is interpreted as a neomorphic dental structure and the cusps of the gomphodont tooth are not considered homologous with the cusps of the sectorial tooth.

The data matrix (Appendix 15.2) was run in TNT (Goloboff et al. 2008). All characters were equally weighted and multistate characters were treated as unordered. Ten random addition sequences with 10 trees retained per replication in TNT produced seven most parsimonious trees (MPTs) with tree length of 212 steps. The number of MPTs increased to 13 with increase to 50 addition sequences and 50 trees retained per replication. The number of MPTs did not vary with continued increase of these settings.

Traversodontidae and the clade (*Exaeretodon+Scalenodontoides*) are the best supported monophyletic groups in our analysis, with a Bremer support of three. Advanced traversodontids (including node Y on Fig. 15.1 and Gomphodontosuchinae, see below) have a Bremer support of two and remaining groups have a support of one.

The base of Traversodontidae in the strict consensus is represented by a large polytomy including seven terminals and five monophyletic groups. In the majority consensus tree (Fig. 15.1), *Nanogomphodon* is placed as the most basal traversodontid. This placement is mainly based on the presence of one or more cuspules located anterolabially in the lower postcanines and the lack of score of most characters. The fact that *Nanogomphodon* is only known by an isolated lower postcanine is therefore the main reason of this basal placement of the taxon. Because of its incompleteness, we produced another analysis removing *Nanogomphodon* from the dataset. This resulted in 26 MPTs retained of 211 steps, whose majority rule consensus is almost identical to Fig. 15.1, other than recovering a monophyletic Trirachodontidae (Fig. 15.2).

In the results of this analysis, traversodontids exhibit a basal polytomy formed by *Scalenodon angustifrons* from Tanzania, a clade formed by the Argentinean taxa *Pascualgnathus* and *Andescynodon*, and another clade including the remaining traversodontids. The latter clade includes a monophyletic group formed by *Luangwa drysdalli* and *L. sudamericana*, from Zambia and Brazil respectively. A second polytomy follows formed by *Traversodon* from Brazil, a clade formed by *Mandagomphodon hirschsoni* and *M. attridgei* and a clade including the remaining traversodontids. The latter clade is composed of the South American *Santacruzodon*, which is the sister taxon of a monophyletic *Massetognathus*. The next-diverging clade (N in Fig. 15.1) is composed of the Laurasian taxa *Arctotraversodon* and *Boreogomphodon*. The Malagasy traversodontid *Dadadon* represents the sister taxon to Gomphodontosuchinae (sensu Kammerer et al. 2008), a group composed of (*Gomphodontosuchus* (*Menadon* (*Protuberum* (*Scalenodontoides+Exaeretodon*)))). The two species of *Exaeretodon* and *Scalenodontoides* form a polytomy.

A monophyletic Trirachodontidae is only recovered in the majority consensus tree of Fig. 15.2, and consequently there is no unambiguous synapomorphy for this group. The specimen CGP JSM100 [interpreted by Hopson (2005) as a probable juvenile *Trirachodon* of uncertain species] is the most basal trirachodontid followed by a polytomy including *Trirachodon, Cricodon*, and a monophyletic group formed by *Langbergia* and the Laurasian *Beishanodon* and *Sinognathus*. This result is slightly different from that of Gao et al. (2010), in which they recognized two monophyletic groups for Laurasian and African forms.

Major interrelationships revealed here are mostly consistent with the previous phylogenetic hypotheses of Abdala and Ribeiro (2003), Abdala et al. (2006), and Kammerer et al. (2008), but differ from those of Sues and Hopson (2010). The primary similarity with previous cladistic analyses of traversodontids is the recovery of a monophyletic Gomphodontosuchinae, but, differing from the recent phylogeny by Reichel et al. (2009), we did not recover a monophyletic group composed of *Menadon* and *Protuberum*. The recovery of a monophyletic group composed of *Santacruzodon* and *Massetognathus* is similar to results by Kammerer et al. (2008) and Sues and Hopson (2010).

As mentioned by Abdala et al. (2006), the placement of basal forms continues to be the most variable area of traversodontid phylogeny. Unlike in Abdala et al. (2006), *Luangwa* never appears as the most basal traversodontid in this study. The most basal traversodontids in the analysis of Sues and Hopson (2010) form a monophyletic group of the Argentinean taxa *Andescynodon* and *Pascualgnathus*, but our result indicates that the African *Scalenodon angustifrons* alone is the most basal traversodontid on 6 of 13 MPTs and together with *Andescynodon* and *Pascualgnathus* on 4 of 13 MPTs that do not consider *Nanogomphodon*. *Andescynodon* plus *Pascualgnathus* occupies the most basal position on 3 of 13 MPTs, is slightly more advanced than *S.*

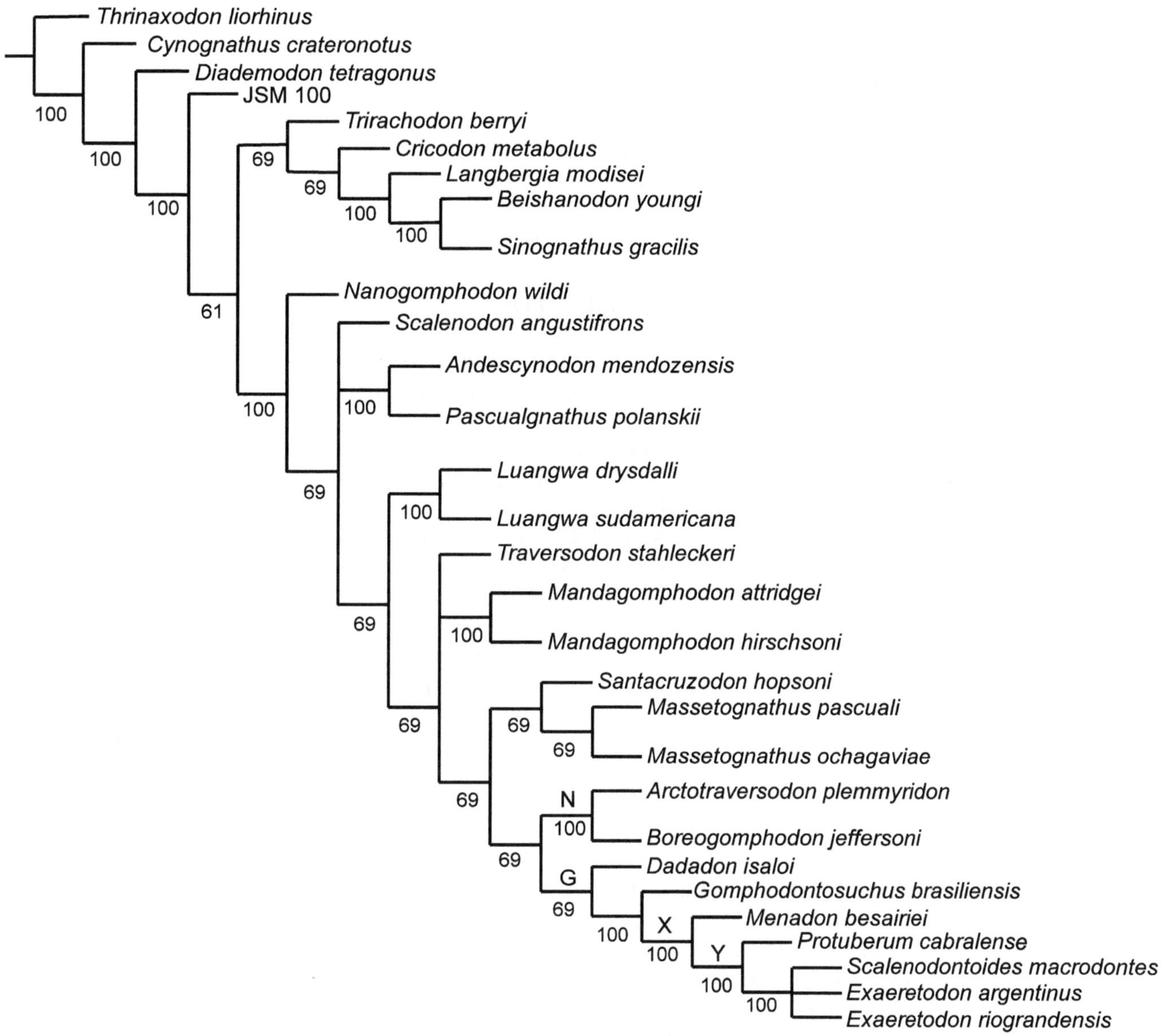

Fig. 15.1 Majority rule consensus tree of 13 most parsimonious trees (tree length = 212) for the data set provided in Appendix 15.2. The *numbers* indicate the frequency of clades in the fundamental trees. *Letters* are used to refer to unnamed clades discussed in the text

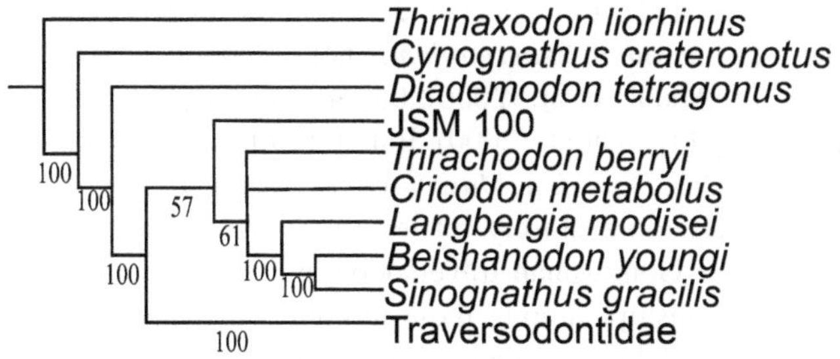

Fig. 15.2 Majority rule consensus tree from the parsimony analysis for the date set provided in Appendix 15.2, after removal of the 'wildcard' taxon *Nanogomphodon*. The interrelationships among traversodontids are the same as those in Fig 12.1 (other than exclusion of *Nanogomphodon wildi*). The *numbers* indicate the frequency of clades in the fundamental trees

angustifrons on 2 of 13 MPTs, and groups with the Laurasian clade (*Arctotraversodon+Boreogomphodon*) on 4 of 13 MPTs. *Scalenodon angustifrons* formed a monophyletic group with *Luangwa* in Sues and Hopson's (2010) hypothesis, and the clade has a derived placement relative to (*Andescynodon+Pascualgnathus*) and *Scalenodon hirschsoni* in their majority-rule consensus (Sues and Hopson 2010, Fig. 12B). North American traversodontids occupy a basal position in the cladogram of Sues and Hopson (2010), contra the results of our analysis.

Most multispecific genera other than *Scalenodon* are recovered as monophyletic in our analysis. *Scalenodon hirschsoni* and *S. attridgei* should be placed in a different

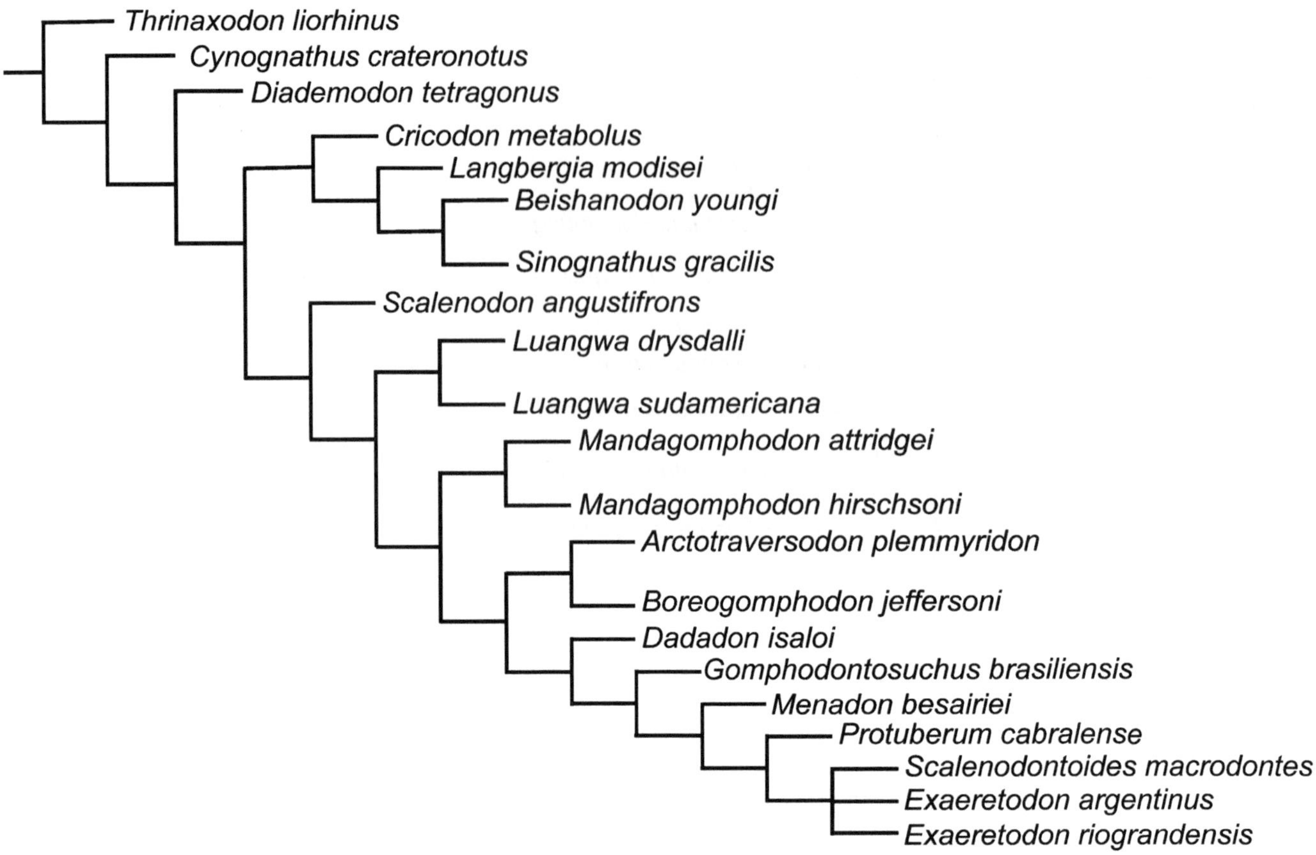

Fig. 15.3 Agreement subtree of 13 most parsimonious trees for the data set provided in Appendix 15.2 without wildcard taxa

genus than *Scalenodon angustifrons*, as proposed by Hopson and Kitching (2001) and Hopson (2013). Following the latter contribution, they are placed in the genus *Mandagomphodon*. The position of *M. hirschsoni* has varied in different phylogenetic hypotheses (Hopson and Kitching 2001; Abdala and Ribeiro 2003; Abdala et al. 2006), and is recovered in yet a different placement here (Figs. 15.1, 15.2). The relationships of *M. attridgei* within traversodontids have not been analyzed in previous phylogenies; here, it forms a clade with *M. hirschsoni*. Battail (1991) synonymized *Luangwa* with *Scalenodon*, but this analysis does not support his hypothesis.

From the agreement subtrees (Fig. 15.3), the wildcard taxa in Traversodontidae include *Traversodon*, *Massetognathus*, and *Santacruzodon* in addition to the previously mentioned (*Andescynodon+Pascualgnathus*) and *Nanogomphodon*.

Arctotraversodon plemmyridon was at first tentatively referred to *Scalenodontoides* (Hopson 1984), but Sues et al. (1992) suggested it could possibly be linked to *Boreogomphodon jeffersoni* by the presence of three anterior cusps on the lower postcanines. They indicated the existence of a distinct lineage of traversodontid cynodonts in Europe and eastern North America, which includes *Arctotraversodon*, *Boreogomphodon*, *Plinthogomphodon*, and *Nanogomphodon* (Hopson and Sues 2006; Sues and Hopson 2010). *Plinthogomphodon* was recently proposed as a synonym of *Boreogomphodon* (Liu and Sues 2010); in our analysis *Arctotraversodon* forms a clade with *Boreogomphodon*.

Hopson (1984, 1985) proposed close relationships between *Gomphodontosuchus*, *Exaeretodon*, and *Scalenodontoides*. Flynn et al. (2000) added *Menadon* to this clade and Kammerer et al. (2008) applied the name Gomphodontosuchinae to this group. Later, *Protuberum* was also included in the clade (Reichel et al. 2009). In the current analysis, the sister taxon of *Exaeretodon* is not *Menadon* as hypothesized by Abdala et al. (2006), but *Scalenodontoides* as suggested by Flynn et al. (2000), Battail (2005), and Kammerer et al. (2008). *Dadadon*, *Massetognathus*, and *Santacruzodon* are always more closely related to Gomphodontosuchinae than other Gondwanan traversodontids, although the exact interrelationships among these terminals differ in previous studies (Abdala and Ribeiro 2003; Abdala et al. 2006; Reichel et al. 2009; Gao et al. 2010; Sues and Hopson 2010).

Based on the current phylogenetic hypothesis presented in Figs. 15.1 and 15.2, the pattern of dental evolution in traversodontids is more irregular than proposed by Abdala and Ribeiro (2003), although patterns are partially obscured by unresolved relationships at the base of the tree. The presence of three upper incisors is a synapomorphy of clade Y, but also convergent with *Pascualgnathus*, *Mandagomphodon hirschsoni*, and *M. attridgei*. The presence of two lower incisors is convergent *in M. hirschsoni* and *Sinognathus*. Enlarged incisors evolved independently in four groups: clade X, *Arctotraversodon*, *M. hirschsoni*, and *Cricodon*. The diastema between the last upper incisor and the upper canine disappears in *Massetognathus* and clade G (Gomphodontosuchinae+*Dadadon*), suggesting parallel evolution in these two groups. The diastema between the upper canine and postcanines is remarkably lengthened in Gomphodontosuchinae. The paracanine fossa is anteromedial with respect to the upper canine in basal traversodontids and changed to a posteromedial position in clade Y, passing through a medial position in *Massetognathus*, *Dadadon*, *Gomphodontosuchus*, and *Menadon*. Because clade N and *Santacruzodon* have the primitive state, parallel evolution may have occurred in *Massetognathus* compared to *Menadon* and *Gomphodontosuchus*. Sectorial teeth present until adulthood is a derived character of *Boreogomphodon*, possibly indicating paedomorphosis in this taxon. A transverse cusp row positioned centrally on the crown in both upper and lower postcanines is the primitive state for gomphodonts. Derived states, transverse cusp row on posterior part of the crown of upper postcanines and on anterior margin of the lower postcanines, are both acquired in Traversodontidae, while a convergent evolution appeared in *Trirachodon* for a posterior position of the transverse cusp row on the crown of some upper postcanines, e.g., BSP 1934 VIII 21 (Broili and Schröder 1935, Fig. 3). Another derived state, transverse cusp row on anterior part of the crown of upper postcanines, is present in *Andescynodon* and *Pascualgnathus*. A slightly developed shouldering [shouldering is defined as the extension of the anterolabial margin of the upper postcanine forward, producing a 'shoulder-like' process over the preceding tooth, following Romer (1967) and Abdala et al. (2006)] of the upper gomphodont dentition is a synapomorphy of clade G. The posteriormost postcanines are inclined posteromedially in relation to the longitudinal axis of the skull in many species, e.g., *Exaeretodon*, *Traversodon*, and *Mandagomphodon attridgei*, and Abdala and Ribeiro (2003) suggested that this feature is possibly related to the shouldering; however, the correlation of these two features is low for known traversodontids.

Systematic Paleontology

The classification of Traversodontidae summarized here is based on the phylogenetic analysis resulting from this contribution. Diagnosis, stratigraphy and geographic distribution, and specimen representation for each taxon is presented. Most diagnoses are represented by a distinctive combination of characters and, in several cases, autapomorphies. Ages follow the Triassic Time Scale recently advanced by Walker et al. (2009).

Family Traversodontidae von Huene, 1936

Definition: (revised from Kammerer et al. 2008) The most inclusive clade containing *Traversodon stahleckeri* von Huene, 1936 but not *Trirachodon kannemeyeri* Seeley, 1895.

Diagnosis: Cynognathian cynodonts characterized by the absence of the ectopterygoid; presence of an epipterygoid-quadrate contact; maxillary platform lateral to the postcanine series (convergent in *Trirachodon* and bauriid therocephalians); labiolingually expanded upper postcanines with a deep occlusal basin; upper gomphodont teeth wider than lowers with outline varying from ellipsoid to rectangular; lower gomphodont teeth quadrangular in shape, with anteriorly positioned transverse crest.

Genus *Andescynodon* Bonaparte, 1969

Synonym: *Rusconiodon* Bonaparte, 1970.

Type species: *Andescynodon mendozensis* Bonaparte, 1969.

Species *Andescynodon mendozensis* Bonaparte, 1969

Synonym: *Rusconiodon mignonei* Bonaparte, 1970.

Holotype: PVL 3833, incomplete skull with poorly preserved teeth.

Stratum typicum: Cerro de las Cabras Formation.

Locus typicus: 5 km west of Villa de Potrerillos, Mendoza Province, Argentina.

Referred material: PVL 3834–3836, 3840 (holotype of *Rusconiodon mignonei*), 3890, 3891, 3892(a, b, c, d), 3894–3900, 3894-1, 3895–3900, 3903, 3907, 4069–4072, 4390, 4423–4432; PVL unnumbered, incomplete pelvis (Abdala 2000; Liu and Powell 2009).

Age: Late Anisian/Early Ladinian, Middle Triassic.

Diagnosis: Small to medium-sized traversodontid, characterized by 9–11 upper postcanines in adults (fewer in larger individuals); transverse crest of upper postcanines located anteriorly; paracanine fossa perforating the snout dorsolaterally in adults; lower postcanines with a cingulum formed by one small cusp anterior to the transverse crest; differentiated from *Pascualgnathus* by the relatively flat skull, having a shorter and narrower temporal region, fewer upper postcanines, and shorter upper tooth row in adults (modified from Liu and Powell 2009).

Comments: All the known specimens of *Andescynodon mendozensis* and *Rusconiodon mignonei* come from the same locality, and their only significant difference is the dorsolabial perforation of the paracanine fossa in *Rusconiodon mignonei* (Goñi and Abdala 1989; Abdala 2000). This feature was recently explained as ontogenetic variation and forms showing the dorsolabially open paracanine fossa are interpreted as adult specimens of *A. mendozensis* (Liu and Powell 2009). FA, however, entertains the possibility that they represent different taxa.

Genus *Pascualgnathus* Bonaparte, 1966

Type species: *Pascualgnathus polanskii* Bonaparte, 1966.

Species *Pascualgnathus polanskii* Bonaparte, 1966

Holotype: MLP 65-VI-18-1, skull and mandibles, and partial postcranial skeleton.

Stratum typicum: Río Seco de la Quebrada Formation, Puesto Viejo Group.

Locus typicus: "Puesto Viejo" locality, west of Colonia Las Malvinas Mendoza, Departamento de San Rafael, Mendoza, Argentina.

Referred material: MLP 65-VI-18-2, skull, mandibles and partial postcranial skeleton; PVL 3466, poorly preserved, dorsoventrally crushed skull and mandible (Bonaparte 1966b); PVL 4416, skull and mandibles articulated but remarkably crushed, connected to the first three cervical vertebrae (Abdala 2000).

Age: Late Anisian/Early Ladinian, Middle Triassic.

Diagnosis: Medium-sized traversodontid with the following combination of characters: three upper incisors; hypertrophied canines; lingual cusp of upper postcanines connected to the transverse ridge located in the anterior part of the crown; one main labial cusp followed by a posterior cusp forming the labial margin; posterior labial cusp distinct and persistent in the last upper postcanines; constricted and large (about 30 % of the skull length) snout with a dorsal perforation of the paracanine fossa; high, long, and sharp parietal crest (modified from Martinelli 2010a).

Genus *Scalenodon* Crompton, 1955

Type species: *Trirachodon angustifrons* Parrington, 1946.

Comments: Five species have been referred to this genus, four from Tanzania (Crompton 1972) and one more recently proposed from Russia (Tatarinov 1973). The Russian species, represented by isolated teeth, is not a traversodontid but a bauriid therocephalian (Sues and Hopson 2010). Two of the Tanzanian species (*S. attridgei*, including *S. charigi* as a junior synonym, and *S. hirschsoni*) do not form a clade with the type species and are here placed in a separate genus.

Species *Scalenodon angustifrons* (Parrington, 1946)

Synonym: *Trirachodon angustifrons* Parrington, 1946.

Holotype: CAMZM T907 (Field Catalogue no. 120B).

Stratum typicum: Lifua Member of the Manda Beds.

Locus typicus: Ruhuhu Valley, Tanzania; Stockley's bone locality B29 between Gingama and Tschikonge.

Referred material: Many specimens from the same locality as the holotype, including CAMZM T908–918, T925, T946 (Field catalogue nos. 120A, 111B, 110A, 134B, 112B, 111C, 131, 129A, 113D, 112C, 119B) (Crompton 1955, 1972).

Age: Late Anisian/Early Ladinian, Middle Triassic.

Diagnosis: Medium-sized traversodontid with transverse crest formed by three cusps, close to the middle of the crown on most upper postcanines; absence of anterior cingulum or anterolabial accessory cusp on upper postcanines; absence of zygomatic process on the jugal.

Comments: Specimens from the Upper Omingonde Formation of Namibia were referred to this species by Brink (1986), some of these specimens were recently redescribed as *Luangwa* sp. (Abdala and Smith 2009). There is no current evidence for the presence of *Scalenodon angustifrons* in the Middle Triassic of Namibia.

Genus *Luangwa* Brink, 1963

Type species: *Luangwa drysdalli* Brink, 1963.

Age: Middle to ?Late Triassic.

Distribution: Upper Ntawere Formation, Zambia; Upper Omingonde Formation, Namibia; Santa Maria Formation, Brazil.

Diagnosis: Small to medium-sized traversodontid with a rounded margin of the zygomatic process of jugal; short snout; quadrangular temporal opening of the same size or slightly larger than orbits; 7–10 gomphodont postcanines with the last upper teeth inclined obliquely; posterior projection of the angular process of the dentary; posterior cingulum behind the transverse crest in upper postcanines, labial cingulum on anterior portion of upper postcanines.

Comments: Specimens recognized as *Luangwa* sp. were recently described from the Upper Omingonde Formation of Namibia (Abdala and Smith 2009). One particular specimen, CGP R572, is remarkably larger than remaining representatives of this genus.

Species *Luangwa drysdalli* Brink, 1963

Holotype: BP/1/3731 (Karoo Field Catalogue 3727), complete skull and lower jaw.

Stratum typicum: Upper Ntawere Formation.

Locus typicus: Northern part of the Luangwa Valley, Zambia.

Referred material: BP/1/3733, poorly preserved anterior portion of the skull with dentition (Abdala and Teixeira 2004); TSK 121, partial skull with lower jaws and some postcranial bones (Kemp 1980).

Age: Late Anisian/Early Ladinian, Middle Triassic.

Distribution: Upper Ntawere Formation; northern part of the Luangwa Valley, Zambia.

Diagnosis: Larger species of *Luangwa* with a posterior cingulum extending along the entire posterior border of the upper postcanines; absence of a posterior accessory cusp behind the main cusp on the labial crest of the upper postcanines; four cusps in the transverse crest of the upper postcanines, with a small additional cusp between the central and external cusps, cingulum composed of several tiny cusps in front of the transverse crest of the lower postcanines; tiny cusps forming a labial cingulum on the lower postcanines.

Comments: This species was referred to the genus *Scalenodon* by Battail (1991, p. 59), however, these taxa are clearly distinct (Abdala and Teixeira 2004).

Species *Luangwa sudamericana* Abdala and Teixeira, 2004

Holotype: MCP 3167PV, a well-preserved partial skull and mandible.

Stratum typicum: Santa Maria Formation.

Locus typicus: Exact locality unknown, Rio Grande do Sul, Brazil.

Referred material: UFRGS-PV 0267T, a partial skull and a left lower jaw with three postcanines.

Age: ?Early Carnian, Late Triassic.

Diagnosis: Smaller species of *Luangwa* differing from *L. drysdalli* by the presence of a posterior cingulum extending along part of the posterior border of the upper postcanines; well defined posterior accessory cusp behind the main cusp on the labial crest of the upper postcanines; three cusps in the transverse crest of the upper postcanines; cingulum composed of two tiny cuspules in front of the transverse crest of the lower postcanines; absence of a labial cingulum on the lower postcanines.

Genus *Traversodon* von Huene, 1936

Type species: *Traversodon stahleckeri* von Huene, 1936.

Species *Traversodon stahleckeri* von Huene, 1936

Holotype: GPIT/RE/7170, an incomplete skull, jaw, and some postcranial bones.

Stratum typicum: *Dinodontosaurus* Assemblage Zone, Santa Maria Formation.

Locus typicus: Cynodont Sanga, west of Chiniquá, Paraná Basin, Rio Grande do Sul, Brazil (von Huene 1936).

Referred material: GPIT 1045, 1063, 1069; UFRGS-PV 0224T (specimen described by Barberena 1981a).

Age: Early Carnian, Late Triassic.

Diagnosis: Large traversodontid with short snout; sharp, powerful angular process of the dentary; nine to ten ovoid upper postcanines with a well-developed posterior cingulum; eight lower postcanines lacking cingulum anterior to the transverse crest; with the labial cusp lower than lingual cusp.

Comments: A partial skull lacking lower jaw was described and referred to this species by Barberena (1974, 1981a). Unfortunately, preservation of the dentition and other important traits are poor in this specimen and we are not totally confident that this material is representative of the species *Traversodon stahleckeri*. We consider that material representing this taxon in the Tübingen collection (particularly a maxilla with a well preserved last postcanine in eruption) may prove to be more closely related (i.e., in the same monophyletic group) to *Luangwa* than is portrayed by the phylogeny presented here.

Genus *Mandagomphodon* Hopson, 2013

Type species: *Scalenodon hirschsoni* Crompton, 1972.

Age: Late Anisian/Early Ladinian, Middle Triassic.

Distribution: Lifua Member of Manda Beds, Tanzania.

Diagnosis: Traversodontid with three enlarged upper incisors; well-developed anterior and posterior cingulum in upper postcanines.

Species *Mandagomphodon hirschsoni* (Crompton, 1972)

Holotype: NHMUK R8577, partial skull and mandible with well-preserved postcanine teeth.

Stratum typicum: Lifua Member of Manda Beds.

Locus typicus: Ruhuhu Valley, southwest Tanzania locality U12 of the BM(NH)-University of London Joint Palaeontological Expedition, 1963. Between the Hiasi and Njalila streams, just south of the Rutukira River; the most northerly of the expedition's localities west of the Njalila.

Age: Late Anisian/Early Ladinian, Middle Triassic.

Diagnosis: Medium-sized *Mandagomphodon* traversodontid characterized by two lower procumbent incisors; an anterior wall in the posterior postcanines; cingulum in front of the transverse crest of the lower postcanines formed by an enormous isolated cusp.

Species *Mandagomphodon attridgei* (Crompton, 1972) comb. nov.

Synonym: *Scalenodon charigi* Crompton, 1972.

Holotype: NHMUK R8578, a well preserved snout with complete upper dentition of a young individual.

Stratum typicum: Lifua Member of Manda Beds.

Locus typicus: Ruhuhu Valley, S.W. Tanzania; locality U2 of the BM(NH)-University of London Joint Palaeontological Expedition, 1963. Next to the Peramiho-Litumba dirt road, on its southwestern side.

Referred material: CAMZM 922 (Ruhuhu Field catalogue no. 136), partial left maxilla with two postcanines (holotype of *Scalenodon charigi*).

Locality: Ruhuhu Valley, locality B26 of Stockley (1932, p. 620). Gingama south of the Ruhuhu River.

Age: Late Anisian/Early Ladinian, Middle Triassic.

Diagnosis: *Mandagomphodon* traversodontid characterized by upper postcanines with ovoid outline; high anterior transverse ridge on upper postcanines.

Comments: Specimen CAMZM 922 ("*Scalenodon charigi*") is from an individual double the size of the holotype and the two preserved postcanines are worn out. Crompton (1972) provided several diagnostic characters for *Scalenodon charigi*. However, the differences between

NHMUK R8578 and CAMZM 922 may possibly be related to the poor preservation of CAMZM 922 or represent cases of individual variation. For example, the central cusp of the posterior transverse ridge and tiny cuspules of the cingulum are hardly preserved in worn postcanines. The overall similar crown shape of postcanines, the oblique orientation of the last postcanines, and the size of the external anterior accessory cusp in relation to the external cusp support the synonymy between *M. attridgei* and *S. charigi*.

Genus *Massetognathus* Romer, 1967

Synonym: *Megagomphodon* Romer, 1972.

Type species: *Massetognathus pascuali* Romer, 1967.

Age: Early Carnian, Late Triassic.

Distribution: Chañares Formation, La Rioja Province, Argentina; Santa Maria Formation, Rio Grande do Sul, Brazil.

Diagnosis: Medium-sized traversodontid with the posterior extension of the secondary palate beyond the anterior border of orbit; extreme development of the maxillary platform lateral to the postcanine series; absence of the zygomatic process of jugal; mesiodistally enlarged incisors with denticulated cutting margins; upper incisors close to the canine; small canines, the upper ones lateral to the paracanine fossa; incipient shouldering between upper postcanines; two anterolabial accessory cusps on the nearly rectangular upper postcanines; high and sharp transverse crest of lower postcanines; the anterolabial cusp wider than the anterolingual in lower postcanines.

Species *Massetognathus pascuali* Romer, 1967

Synonyms: *Massetognathus teruggii* Romer, 1967; *Massetognathus major* Romer, 1972; *Megagomphodon oligodens* Romer, 1972.

Holotype: PULR 10 (former MLP No. 65-XI-14-1), a complete well preserved skull with lower jaw.

Stratum typicum: Chañares Formation; Ischigualasto-Villa Union Basin.

Locus typicus: About two miles north of the point where the Chañares River debouches into the Campo de Talampaya, in western La Rioja Province, Argentina.

Referred materials: PULR 13 (former MLP No. 65-XI-14-2), skull and jaws (holotype of *M. teruggii*) (Romer 1966); PULR 11 (former MLP No. 65-XI-14-15), skull (holotype of *M. major*), PULR unnumbered (former MLP No. 65-XI-14-16), skull and lower jaws (holotype of *Megagomphodon oligodens*) (Romer 1972); PVL 3901–3906, 4014, 4016, 4168, 4439–4443, 4613, 4614, 4676, 4726–4729, 5441, 5443–5445 (Abdala 2000), PVL 5683; MCZ 3691, 3786, 3798, 3801, 3804, 3806, 3807, 4021, 4138, 4208, 4215, 4216, 4258, 4265; NHMUK R8430 (Abdala and Giannini 2000); BP/1/4245; MCP 3284 (Teixeira 1995).

Age: Early Carnian, Late Triassic.

Distribution: Chañares Formation, Ischigualasto-Villa Union Basin, Argentina and tentatively in the Santa Maria Formation, Rio Grande do Sul, Brazil (Liu et al. 2008).

Diagnosis: Snout is subequal to temporal region; small canines; absence of anterior cingulum on upper postcanines; presence of posterior cingulum on lower postcanine; dorsal margin of the coracoid equal to that of procoracoid in medial view; 'T'-shaped lumbar ribs with restricted contact between successive ribs.

Comments: The synonymy of *Massetognathus pascuali* and *M. teruggii* was proposed by Hopson and Kitching (1972). Battail (1991) recognized *Massetognathus pascuali* as the only valid traversodontid species from the Chañares Formation, a hypothesis confirmed by the analysis of Abdala and Giannini (2000).

Species *Massetognathus ochagaviae* Barberena, 1981

Holotype: UFRGS-PV 0255T (G), a skull without mandible.

Stratum typicum: *Dinodontosaurus* Assemblage Zone, Santa Maria Formation, Paraná Basin.

Locus typicus: 3.5 km from Prof. Parreira train station, southeast of Melos, General Câmara District, Rio Grande do Sul, Brazil.

Neotype: MCP 3871 PV, a skull without mandible.

Locus neotypicus: Rincão do Pinhal, Agudo Municipality, Rio Grande do Sul, Brazil.

Referred material: UFRGS-PV 0070T, 0071T, 0125T, 0239T, 0241T, 0242T, 0243T, 0245T, 0265T, 0273T, 0397T, 1064T; AMNH FARB 7802, 7803, 21400–21410 (Liu et al. 2008).

Age: Early Carnian, Late Triassic.

Diagnosis: Snout shorter than temporal region; skull and mandible taller than *Massetognathus pascuali*; postcanine number less variable than in *M. pascuali*; isosceles triangular shape of the maxillary platform lateral to the postcanines in ventral view; short lingual ridge forming a nearly triangular basin on upper postcanines; robust lower canines; subrectangularly shaped lower postcanines (modified from Liu et al. 2008).

Comments: The holotype described by Barberena (1981b) could not be located in the UFRGS collection. MCP 3871 PV was recommended as neotype in case of definitive loss of the holotype (Liu et al. 2008).

Genus *Santacruzodon* Abdala and Ribeiro, 2003

Type species: *Santacruzodon hopsoni* Abdala and Ribeiro, 2003.

Species *Santacruzodon hopsoni* Abdala and Ribeiro, 2003

Holotype: MCN PV 2768, fragmentary skull with lower jaws.

Stratum typicum: *Santacruzodon* Assemblage Zone (Soares et al. 2011), Santa Maria Formation.

Locus typicus: Suburbs of the city of Santa Cruz do Sul (S 29° 44′ 25″, W 52° 27′ 01″), Rio Grande do Sul, Brazil (Abdala et al. 2001).

Referred material: MCN PV 2751, MCN PV 2752, MCP 4044 PV three lower jaws; MCN PV 2770 incomplete maxilla with postcanines; MCP 4034 PV fragmentary skull and lower jaw with postcanines (Abdala and Ribeiro 2003).

Age: Late Carnian, Late Triassic.

Diagnosis: Traversodontid characterized by a ball-shaped ventrally projecting suborbital process; incisors flattened labiolingually with a series of 7–9 marginal cuspules; upper postcanines present an anterior small crest formed by a series of cingular cuspules; three labial cusps in the upper postcanines with the posterior cusp very large, representing more than half the length of the labial crest; anterolingual cusp strongly inclined posteriorly on lower gomphodont teeth (modified from Abdala and Ribeiro 2003).

Genus *Arctotraversodon* Sues, Hopson, and Shubin, 1992

Type species: ?*Scalenodontoides plemmyridon* Hopson, 1984.

Species *Arctotraversodon plemmyridon* (Hopson, 1984)

Synonym: ?*Scalenodontoides plemmyridon* Hopson, 1984.

Holotype: YPM-PU 19190, the horizontal ramus of a right dentary with a small portion of the left dentary.

Stratum typicum: Wolfville Formation, Fundy Group.

Locus typicus: Northeast corner of Burntcoat headland, 1.5 miles northwest of Noel, Hants County, Nova Scotia, Canada.

Age: Carnian, Late Triassic.

Diagnosis: Very large traversodontid with greatly enlarged posterior mental foramen on dentary; broad, chin-like symphyseal region of lower jaw; lower incisors with large mesial and distal marginal cuspules; crowns of post-canine teeth distinctly compressed anteroposteriorly; upper gomphodont teeth with prominent central cusp; three cusps on anterior crest of lower gomphodont teeth; posterior heel of lower gomphodont teeth without raised rim and labial accessory cusp placed high on labial ridge, rather than on margin of heel (after Sues et al. 1992).

Referred material: From Burntcoat shore: YPM-PU 19190-A, a partial canine tooth; YPM-PU 21693, a small partial dentary lacking teeth; NSM 983GF2.1, isolated right lower postcanine tooth from type locality; NSM 990GF89.1, left upper postcanine tooth. From Evangeline Beach (west outcrop), Kings County, Nova Scotia: YPM-PU 22343, fragmentary small dentary without teeth (Hopson 1984; Sues et al. 1992).

Genus *Boreogomphodon* Sues and Olsen, 1990

Synonym: *Plinthogomphodon* Sues, Olsen, and Carter, 1999.

Type species: *Boreogomphodon jeffersoni* Sues and Olsen, 1990.

Age: Late Triassic.

Distribution: North Carolina and Virginia, United States.

Species *Boreogomphodon jeffersoni* Sues and Olsen, 1990

Holotype: USNM 437632, a left maxilla with three teeth.

Stratum typicum: Tomahawk Creek Member of the Turkey Branch Formation, Newark Supergroup.

Locus typicus: Tomahawk locality, near Midlothian, Chesterfield County, Virginia, United States.

Referred material: From Virginia, cranial remains: CM 20050, 76800, 76801, 76803; USNM 437635, 437636, 448562, 448570, 448593, 448599, 448632, 448633; VMNH 3575, 3578. Isolated teeth: CM 76805, 76807, 76808, 76810, 76812, 76815, 76818; USNM 448563–448569, 448571–448573, 448575, 448576, 448578, 448597, 448601, 448625, 448629. Postcranial bones: USNM 448598, 448602; VMNH 3577 (Sues and Hopson 2010). From North Carolina, cranial remains: NCSM 11466, 15295, 16292, 16297, 16358, 16364, 18300, 19587, 20660, 20662, 20692, 20698, 20700, 20704, 20712, 21370, 21371 (Liu and Sues 2010).

Age: Carnian, Late Triassic.

Distribution: Turkey Branch Formation, Virginia and Pekin Formation, North Carolina, United States.

Diagnosis: Small traversodontid with the dorsal surface of the snout presenting pronounced, irregular sculpturing composed of ridges and grooves; jugal without distinct suborbital process; zygomatic arches bowed laterally at about mid-length rather than reaching greatest width posteriorly; presence of sectorial postcanines in adults; upper gomphodont teeth nearly triangular in crown view and posterior cingulum absent; three cusps in the transverse crest of the upper gomphodont teeth, the central and lingual confluent, and separated from the labial cusp by a valley; the transverse crest of lower gomphodont teeth is formed by three cusps in all but the smallest individuals; anterolabial cingular cuspules in front of the transverse crest (modified from Sues and Hopson 2010).

Comments: The North Carolina materials differ from those from Virginia because their lower gomphodont teeth are mostly represented by specimens having two cusps on the anterior ridge. Assessment of a taxonomic distinction must await completion of the study of the cranial material.

Species *Boreogomphodon herpetairus* (Sues, Olsen, and Carter, 1999)

Holotype: UNC 15576, partial snout preserved in two pieces.

Stratum typicum: Muddy sandstone of Lithofacies Association II, Durham sub-basin of Deep River Basin, Newark Supergroup.

Locus typicus: Genlee, Durham County, North Carolina, United States.

Referred material: UNC 15656, a few ingested postcranial elements including a complete humerus.

Age: ?Early Norian, Late Triassic.

Comments: The specimens were found associated with a large "rauisuchian", and by the fossilization features, interpreted as ingested by this archosaur (Peyer et al. 2008). There is no clear diagnostic feature for this taxon. The species is tentatively accepted because of its different stratigraphic occurrence in relation to *Boreogomphodon jeffersoni* (Liu and Sues 2010).

Genus *Nanogomphodon* Hopson and Sues, 2006

Type species: *Nanogomphodon wildi* Hopson and Sues, 2006.

Species *Nanogomphodon wildi* Hopson and Sues, 2006

Holotype: SMNS 51962, left lower postcanine tooth lacking apical portion of the root.

Stratum typicum: Lower Keuper or Erfurt Formation.

Locus typicus: Housing development "Leitenäcker II" in Michelbach an der Bilz, Baden-Württemberg, Germany.

Age: Ladinian, Middle Triassic.

Diagnosis: Traversodontid having lower postcanine with anterior transverse ridge composed of three principal cusps and long basined "heel" bordered posteriorly by a ridge bearing three small cusps; distinct ridge bounding lingual side of basin; anterior cingulum with five cuspules (from Hopson and Sues 2006).

Genus *Dadadon* Flynn et al., 2000

Type species: *Dadadon isaloi* Flynn, Parrish, Rakotosamimanana, Ranivoharimanana, Simpson, and Wyss, 2000.

Species *Dadadon isaloi* Flynn et al., 2000

Holotype: UA 10606, partial skull with complete right upper dentition, excepting the first left incisor and right incisors.

Stratum typicum: Basal 'Isalo II' beds (Makay Formation sensu Razafimbelo 1987)

Locus typicus: East of Sakaraha, northern Morondava Basin, Madagascar.

Referred material: UA 10605, edentulous rostrum including the right orbit and nasal chamber, with four incisors, one canine, and nine postcanine alveoli.

Age: Late Carnian, Late Triassic.

Diagnosis: Medium-sized traversodontid with a large, rounded, ventrally projecting suborbital process; the upper tooth row extends posteriorly beyond the anterior margin of the subtemporal fossa; posterior upper postcanines obliquely oriented and with incipient shouldering; a posterolabial projection of the ectoloph forms a metastyle-like structure; strong labial cingulum on the upper postcanines (modified from Flynn et al. 2000).

Subfamily Gomphodontosuchinae Watson and Romer, 1956

Type genus: *Gomphodontosuchus* von Huene, 1928.

Composition: *Exaeretodon*, *Gomphodontosuchus*, *Menadon*, *Protuberum*, and *Scalenodontoides*.

Age: Late Triassic.

Distribution: South Africa, Lesotho, Madagascar, India, Argentina, and Brazil.

Diagnosis: A group of traversodontids characterized by high position of anterior root of zygomatic arch; posterior extension of the jugal well-developed dorsally above the squamosal in the zygomatic arch; well-developed angular process of dentary; enlarged incisors, the lowers being procumbent; three upper incisors; last upper incisor close to upper canine and the latter far from the first upper postcanine; reduced lower canine; very oblique orientation of upper gomphodont teeth; anterior wall in upper postcanines; lack of central cusp on posterior transverse crest; presence of distinct anterolingual accessory cusp of upper gomphodont teeth; absence of posterior cingulum on the lower postcanines.

Genus *Gomphodontosuchus* von Huene, 1928

Type species: *Gomphodontosuchus brasiliensis* von Huene, 1928.

Species *Gomphodontosuchus brasiliensis* von Huene, 1928

Holotype: GPIT unnumbered, anterior half of a skull and associated lower jaws.

Stratum typicum: *Hyperodapedon* Assemblage Zone, Santa Maria Formation.

Locus typicus: Santa Maria city, Rio Grande do Sul, Brazil.

Age: Early Norian, Late Triassic.

Diagnosis: Small traversodontid with a short and high rostrum; short and broad secondary palate with its posterior margin located anterior to the orbit; five or six upper and lower postcanines; paracanine fossa placed medially to the upper canine; massive chin-like dentary symphysis; forward position of the anterior border of the coronoid process (at the level of the fourth postcanine); upper postcanine outline varying from triangular anteriorly to quadrangular in the posterior teeth.

Comments: The only known specimen was interpreted as a juvenile, particularly resembling juvenile specimens of *Exaeretodon* (Hopson 1985). FA considers it possible that *Gomphodontosuchus* may be the sister taxon of *Menadon*, or even a juvenile of the latter taxon, although they do not form a monophyletic group in the current analysis.

Genus *Menadon* Flynn et al., 2000

Type species: *Menadon besairiei* Flynn, Parrish, Rakotosamimanana, Ranivoharimanana, Simpson, and Wyss, 2000.

Species *Menadon besairiei* Flynn et al., 2000

Holotype: UA 10601, skull and mandibles with eroded left side.

Stratum typicum: Basal 'Isalo II' beds (Makay Formation); Madagascar.

Locus typicus: East of Sakaraha, northern Morondava Basin, Madagascar.

Referred material: FMNH PR 2104, an isolated mandible; FMNH PR 2444, a partial skull and postcranium; field number 8-31-98-387, an isolated right pelvis (Kammerer et al. 2008).

Locality: Drainage of the Malio River, Morondava Basin, Madagascar.

Age: Late Carnian, Late Triassic.

Diagnosis: Large traversodontid characterized by four large upper incisors; lower incisors procumbent; third and fourth upper incisors caniniform, with fourth upper incisor strongly recurved, presenting serrated margins; upper canines small, of equal height to incisors, distinctly canted forwards; canine alveolus proportionally narrower than incisor alveoli, with a much greater anteroposterior than labiolingual length; low number of postcanines (8 uppers and 6–7 lowers); upper and lower postcanines quadrangular, roughly trapezoidal in outline; axial spine with concave dorsal profile and elongate posterior process overhanging shortened neural spine of subsequent cervical vertebra; expanded ribs present; caudal neural spines very tall; anterior edge of the iliac blade sloping upwards at a $\sim 45°$ angle; acetabular buttresses of the three pelvic bones largely confluent; posterior process of the iliac blade short and directed away from the dorsal edge of the ischium; ischium and pubis extremely constricted, resulting in a large obturator foramen (modified from Kammerer et al. 2008).

Genus *Protuberum* Reichel, Schultz, and Soares, 2009

Type species: *Protuberum cabralensis* Reichel, Schultz, and Soares, 2009.

Species *Protuberum cabralense* Reichel, Schultz, and Soares, 2009

Holotype: MGB 368-100, skull without lower jaw, and several elements of the postcranium.

Stratum typicum: *Dinodontosaurus* Assemblage Zone, Santa Maria Formation.

Locus typicus: Outcrop in the Municipality of Novo Cabrais, Rio Grande do Sul, Brazil.

Referred material: UFRGS-PV 0981T, a proximal fragment of a right cervical rib; UFRGS-PV 0983T, an isolated vertebra; UFRGS-PV 0985T, an isolated vertebra; UFRGS-PV 0986T, an isolated vertebra; UFRGS-PV 1009T, a left cervical rib; UFRGS-PV 1010T, a left thoracic rib; UFRGS-PV 1011T, a fragment of a thoracic rib (Reichel et al. 2009).

Locality: Some vertebrae and ribs of the referred material were found in Rincão do Pinhal, Municipality of Agudo, Rio Grande do Sul, Brazil.

Age: Early Carnian, Late Triassic.

Diagnosis: Large traversodontid with upper postcanines having two main cusps (one labial and one lingual) connected by a sharp transverse crest and lacking shouldering; anteroposteriorly elongated paracanine fossa posteriorly placed in relation to the upper canine; short parietal crest; well-developed descending process of the jugal; bifurcated paroccipital process; incisive foramina totally enclosed by the maxillae; a bony thickening that forms wide crests on the dorsal surface of the skull; small upper canine contiguous with the last incisor; ribs with very pronounced processes on their dorsal border, the most proximal of these is generally the largest and the others become smaller distally; the iliac blade has a series of rugosities along its dorsal border (modified from Reichel et al. 2009).

Genus *Exaeretodon* Cabrera, 1943

Synonyms: *Theropsis* Cabrera, 1943; *Ischignathus* Bonaparte, 1963; *Proexaeretodon* Bonaparte, 1963.

Type species: *Exaeretodon frenguellii* Cabrera, 1943, a junior synonym of *E. argentinus* (Cabrera, 1943).

Age: Early Norian, Late Triassic.

Distribution: Ischigualasto Formation, San Juan and La Rioja Provinces, Argentina; Santa Maria Formation, Brazil; lower Maleri Formation, India.

Diagnosis: Very large traversodontids lacking internarial bar; upper postcanines with a well-developed posterolabial accessory cusp and extensive shouldering resulting in a separation between a labial lobe and a lingual one (including the occlusal basin); anterolingual cusp of the lower postcanines strongly inclined posteriorly; divergent zygomatic arches; well-developed descending process of the jugal; three lower incisors; large upper canines but reduced lowers; ribs lacking costal plates.

Species *Exaeretodon argentinus* (Cabrera, 1943)

Synonyms: *Belesodon argentinus* Cabrera, 1943; *Exaeretodon frenguellii* Cabrera, 1943; *Theropsis robusta* Cabrera, 1943; *Ischignathus sudamericanus* Bonaparte, 1963b; *Proexaeretodon vincei* Bonaparte, 1963c.

Holotype: MLP 43-VII-14-2, incomplete left mandibular ramus.

Stratum typicum: Ischigualasto Formation.

Locus typicus: Hoyada de Ischigualasto, San Juan, Argentina.

Referred material: MLP 43-VII-14-1, 43-VII-14-3, 43-VII-14-4, and some specimens beginning with 61-VIII-2; MACN 18063, skull with lower jaws and some postcranial bones; MACN 18114, skull and articulated lower jaw and some postcranial bones (Bonaparte 1966a); several specimens in PVL (see Abdala 2000), including PVL 2564, skull, mandible, an atlantal arch, and a dorsal vertebra, holotype of *Ischignathus sudamericanus*; MCZ 7047, a complete skull with lower jaws (Chatterjee 1982); MCZ 3779, 4493, 111-64A, 377-58 M; MACN 18114, 18125; MCP 1522PV, PVSJ 157.

Age: Early Norian, Upper Triassic.

Distribution: Ischigualasto Formation, San Juan and La Rioja Provinces, Argentina.

Diagnosis: Variable number of postcanines (from 6 to 10); anterolabial cusp wider than the anterolingual on lower postcanines; absence of crest in the lateral flange of the prootic.

Species *Exaeretodon riograndensis* Abdala, Barberena, and Dornelles, 2002

Holotype: MCP 1522PV, complete skull plus proatlas and atlas arches on the occiput, dislocated from their anatomical position.

Stratum typicum: *Hyperodapedon* Assemblage Zone, Santa Maria Formation.

Locus typicus: Kilometer 136 of the railroad Br 287, 6 km west of the Botucarai hill, in the Candelaria district, Rio Grande do Sul State, Brazil.

Referred material: MCP 2361 PV, skull lacking the anterior portion of the rostrum, same locality as holotype; MCP 3843 PV, skull and lower jaw in occlusion, lacking the right temporal region, right dentary also incomplete, same locality as holotype; UFRGS PV 0715T, skull, lower jaws and some postcranial bones, Sitio Janner (53°17′30″W, 29°39′09.68″S), near Agudo city, Rio Grande do Sul, Brazil (Oliveira et al. 2007).

Age: Early Norian, Late Triassic.

Diagnosis: Presence of a series of crests in the lateral flange of the prootic anterior to the fenestra ovalis, the number of postcanines less variable in ontogeny than in *Exaeretodon argentinus* (from Abdala et al. 2002).

Comments: Several additional, mostly unpublished, specimens of *Exaeretodon riograndensis* have been collected in recent years, making it the most abundant cynodont known from the *Hyperodapedon* Assemblage Zone.

Exaeretodon sp. indet.

Composition: *Exaeretodon statisticae* Chatterjee, 1982.

Material: ISIR 303, a few mandibular fragments; ISIR 304, a partial skull (holotype of *E. statisticae*).

Stratum typicum: Lower Maleri Formation.

Locus typicus: Venkatapur village, District of Adilabad, Andhra Pradesh, south India (Chatterjee 1982).

Age: Early Norian, Late Triassic.

Comments: Although clearly a specimen of *Exaeretodon*, this Indian taxon does not have clear diagnostic characters, therefore it is not possible to place it in any of the recognized species. We entertain the hypothesis that this may represent a different species because of geographical distribution.

Genus *Scalenodontoides* Crompton and Ellenberger, 1957

Type species: *Scalenodontoides macrodontes* Crompton and Ellenberger, 1957.

Species *Scalenodontoides macrodontes* Crompton and Ellenberger, 1957

Holotype: MNHN 1957-23, paired dentaries lacking the region behind the postcanines.

Stratum typicum: Base of the Lower Elliot Formation.

Locus typicus: "Site A" of Crompton and Ellenberger (1957), Morobong Hill, Mohale's Hoek district, Lesotho.

Referred material and distribution: SAM-PK-K336, right half of a large snout, collect at "Site B, approximately 100 yards west of Site A", Morobong Hill, Mohale's Hoek district, Lesotho (Hopson 1984); BP/1/5395, a skull with lower jaw, farm Norwood, Sterkstroom, Eastern Cape Province (Gow and Hancox 1993); MNHN 1955-25, a skull lacking lower jaw, Leribe district, Lesotho (Battail 2005); NMQR 3053, paired premaxillae and maxillae with associated skull fragments, farm Patriotsklip, Jamestown, Eastern Cape Province (Kammerer et al. 2008).

Age: Late Norian, Late Triassic.

Diagnosis: Very large traversodontid with a very robust skull, approximately as broad as long; snout short and broad; temporal region remarkably short with the temporal opening being wider than long; overhanging nuchal table in the dorsal portion of the cranium composed mostly of the parietal; lower portion of the symphysis extends ventrally as a chin-like projection; occipital condyles widely separated; labial and lingual lobes of the last upper postcanines clearly distinct and demarcated by a constriction; anterolabial cusp of lower postcanines much larger than the anterolingual cusp in crown view; the ridge which passes back from the apex of the lingual cusp to the heel describes a distinct, lingually concave curve (revised from Battail 2005).

Comments: This is the latest traversodontid from Gondwana and quite likely the world. The absence of the nuchal table from a specimen recently described by Battail (2005) represents a remarkable difference in relation to the only previously known cranium of this taxon (Gow and Hancox 1993). This has been interpreted as being related to sexual dimorphism (Battail 2005).

Family Traversodontidae *incertae sedis*

?Traversodon major von Huene 1936

Material: GPIT unnumbered.

Stratum: *Dinodontosaurus* Assemblage Zone, Santa Maria Formation.

Locality: Sanga north of house of Theotônio Béles Xavier and Sanga 'of the tree on the road', Chiniquá, Rio Grande do Sul, Brazil.

Comments: Six specimens coming from two different "sangas" in the locality of Chiniquá were included in this taxon, including a fragmentary mandibular symphysis, a partial maxilla with empty alveoli and postcranial remains. Von Huene (1936) distinguished this taxon from *Traversodon stahleckeri* primarily based on the larger size of the remains. The maxillary fragment was later transferred to *Exaeretodon major* (Barberena 1974), a conclusion cautiously supported by Abdala et al. (2002, pp. 320–321). We believe that there is no clear diagnostic character that allows a definitive inclusion of this maxilla in *Exaeretodon* and prefer to consider these remains, as well as the other five specimens described by von Huene (1936), as *incertae sedis* and *Exaeretodon major* as a *nomen dubium*.

Theropsodon njalilus von Huene 1950

Holotype: GPIT/RE/7162, a complete but poorly preserved skull with lower jaws in occlusion.

Stratum typicum: Lifua Member of the Manda Beds.

Locus typicus: Ruhuhu Valley, Tanzania.

Age: Late Anisian/Early Ladinian, Middle Triassic.

Distribution: Known only from the holotype.

Comments: This specimen cannot be allocated to a particular traversodontid taxon because the postcanines are not readily available, and it is considered a *nomen dubium* (Hopson and Kitching 1972).

Colbertosaurus muralis (Minoprio, 1954)

Synonym: *Colbertia muralis* Minoprio, 1954

Holotype: Cast of partial pair of lower jaws, AMNH FARB 7610.

Stratum typicum: Potrerillos Formation.

Locus typicus: Obligación Quarry, Cubhilla de las Vacas, 20 km to the west of the city of Mendoza, Argentina.

Age: Middle Triassic.

Distribution: Known only from the holotype.

Comments: The holotype and only specimen of *Colbertosaurus muralis* includes only incomplete lower jaws with a few broken teeth (Minoprio 1954, 1957). Not enough diagnostic information can be found at the genus level, and the name should be considered as a *nomen dubium* (Hopson and Kitching 1972).

Unnamed traversodontid

Material: GSN OM-5.

Stratum: Upper Omingonde Formation.

Locus: Farm Omingonde 96, Etjo Mountain, Namibia.

Age: Ladinian, Middle Triassic.

Comments: This taxon is based on a medium-sized specimen recently described by Abdala and Smith (2009), presenting as the most remarkable feature a long and thin basicranial girder and long sagittal crest. The specimen shows occluding jaws, but the internal view of the postcanines allows the specimen to be recognized as a traversodontid.

Taxa of Uncertain Taxonomic Position

Genus ***Microscalenodon*** Hahn, Lepage, and Wouters, 1988

Species ***Microscalenodon nanus*** Hahn, Lepage, and Wouters, 1988

Holotype: IRSNB R405, an upper postcanine.

Stratum typicum: Bonebed Habay-le-Vieille-2, Sables de Mortinsart Formation.

Locus typicus: Habay-le-Vieille, Gaume, Belgium.

Referred material: IRSNB R406, a lower postcanine.

Age: Rhaetian, Late Triassic.

Genus ***Maubeugia*** Godefroit and Battail, 1997

Species ***Maubeugia lotharingica*** Godefroit and Battail, 1997

Holotype: IRSNB R172, a left upper postcanine, with its root nearly completely preserved.

Stratum typicum: "Rhaetian" sandstone.

Locus typicus: Quarry at Rosières-aux-Salines, Saint-Nicolas-de-Port (Meurthe-et-Moselle), France.

Age: Late Norian-Early Rhaetian, Late Triassic.

Genus ***Rosieria*** Godefroit and Battail, 1997

Species ***Rosieria delsatei*** Godefroit and Battail, 1997

Holotype: IRSNB R173, an upper left postcanine.

Stratum typicum: "Rhaetian" sandstone

Locus typicus: Quarry at Rosières-aux-Salines, Saint-Nicolas-de-Port (Meurthe-et-Moselle), France.

Age: Late Norian-Early Rhaetian, Late Triassic

Comments: A further possible specimen has also been identified, IRSNB R174.

Genus *Habayia* Godefroit, 1999

Species *Habayia halbardieri* Godefroit, 1999

Holotype: IRSNB R203, a right upper postcanine teeth.

Stratum typicum: Bonebed Habay-le-Vieille-3, Grès de Mortinsart.

Locus typicus: Habay-la-Vieille, southern Belgium, side of the speedway E25–E411.

Age: Rhaetian, Late Triassic.

Comments: The four taxa mentioned above are represented by tiny isolated teeth (1 mm or less). Although they resemble traversodontid postcanines in being labiolingually expanded, they also show a morphology that departs quite considerably from indisputable traversodontid teeth. Godefroit and Battail (1997, p. 604), who described two of these species, recognized the possibility that the taxonomy of these forms might be reevaluated with more complete material. In addition, Hopson and Sues (2006, p. 125) were not convinced that these Upper Triassic European teeth represent traversodontids. We are uncertain whether these forms belong in Traversodontidae, as they represent the only Rhaetian evidence of labiolingually expanded tooth forms (not considering the clearly different morphology of tritylodontids and haramyids). Therefore, we leave these four species in limbo until more material can help in reassessing their taxonomy.

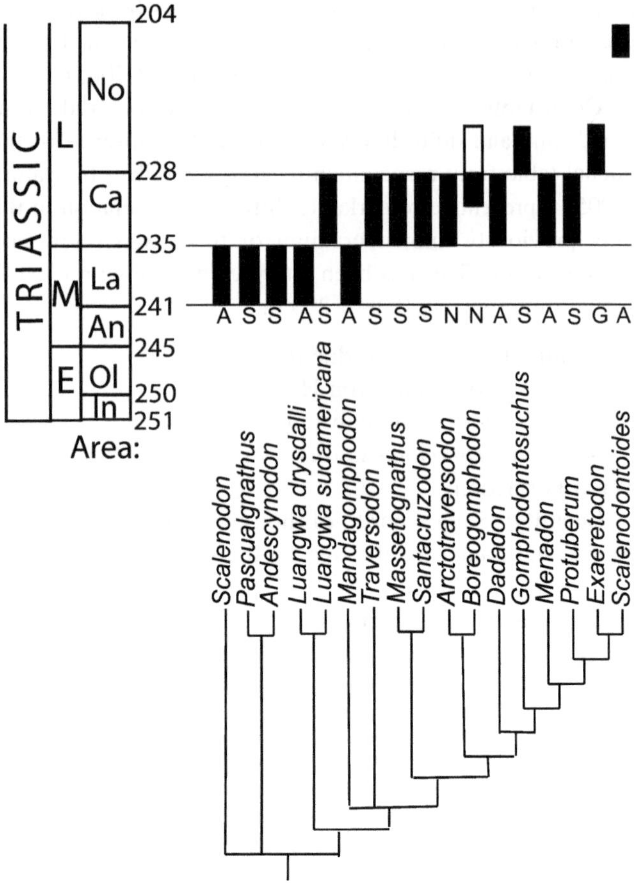

Fig. 15.4 A simplified cladogram from Fig. 15.1 (excluding *Nanogomphodon wildi*) to indicate temporal and spatial distribution of the traversodontid taxa included in the analysis. Time-scale based on Walker and Geissman (2009). The *blank box* means uncertain time range. *An* Anisian, *La* Ladinian, *Ca* Carnian, *In* Induan, *No* Norian, *Ol* Olenekian. *A* Africa, *E* Eurasia, *G* Gondwana, *M* Madagascar, *N* North America, *S* South America

Non-traversodontid Taxa

Species *Scalenodon boreus* Tatarinov, 1973

Holotype: PIN 2973/1, left upper postcanine tooth.

Stratum typicum: Donguz Formation.

Locus typicus: Southern Cisuralia, Orenburg Province, Karagachka, Russia.

Referred material: PIN 2973/2, isolated upper postcanine from the same site.

Age: Anisian-Ladinian, Middle Triassic.

Distribution: Only known from the type locality.

Comments: This taxon is based on two ovoid postcanines with morphology reminiscent of the circular teeth of *Neotrirachodon exspectatus*, also from the Donguz Formation. The latter was considered a trirachodontid by Tatarinov (2002), but it is in fact a bauriid therocephalian (Abdala and Smith 2009). Therefore we believe that the tooth of *Scalenodon boreus* is also more likely a bauriid, as concluded by Sues and Hopson (2010).

Distribution

Currently, 22 species and 17 genera of traversodontids are considered valid. The distribution of these animals is uneven in space and time (Table 15.1; Fig. 15.4): 11 species in South America, seven in Africa (including Madagascar), three in North America, and one each in Eurasia and India. Therefore, this diverse group of non-mammaliaform cynodonts is clearly predominantly Gondwanic (with 19 species), with only four species found in Laurasia.

The older traversodontids are known from Anisian/Ladinian localities in Africa and South America and their sudden emergence produced a diversity peak around the end of the Anisian (Abdala and Ribeiro 2010). This group remains well-represented in the Carnian of Gondwana. The oldest record in Laurasia is Ladinian and is restricted to an isolated tooth, reflecting a poor record of this group in comparison with contemporaneous deposits from Gondwana. Small isolated teeth from Norian-Rhaetian deposits of Europe described as traversodontids are not considered here, as we believe that additional material is necessary before these taxa can be included unquestionably in the group (see taxonomic section above). The only, rather small, radiation of Laurasian traversodontids occurred in the Late Triassic of North America, with one ?Late Ladinian-Early Carnian or Early-Middle Carnian species (Sues and Olsen 1990), one from the Late Carnian (Rayfield et al. 2005), and one from the Late Carnian or Early Norian (Peyer et al. 2008).

A general pattern of size increase in traversodontids from the Middle to the Upper Triassic is recognized (e.g., Battail 2001), with the largest forms being known from the Norian. This trend is clearer in South America, with a variation of skull length from 10 cm in Anisian/Ladinian forms to 45 cm in the Norian (these being the largest traversodontids known). In Africa, some Anisian/Ladinian traversodontids have a skull length of at least 20 cm (Abdala and Smith 2009), larger than that of the Carnian *Menadon* (skull length approximately 16 cm; Flynn et al. 2000), but smaller than the Norian *Scalenodontoides* (skull length 28 cm; Battail 2005) (and some partial remains clearly indicate even larger size in *Scalenodontoides*). It is more difficult to follow this trend in Laurasia because of the restricted record, but it is clear that traversodontids also attained large size at the end of the Carnian, at least in North America (e.g., *Arctotraversodon*).

Every monophyletic group must have a center of origin, or ancestral area (Bremer 1992). The hypothesis in vicariance biogeography that ancestral distribution was identical to the present distribution loses value for a cosmopolitan group. For example, the distribution of the common ancestor of all modern humans is considered to have been restricted to Africa rather than cosmopolitan. For a widespread species, dispersal out of the ancestral area must have occurred after speciation. Some methods have been proposed to find the ancestral area, such as the Progression Rule (Hennig 1966) and Ancestral Area Analysis (Bremer 1992). The Progression Rule assumes that basal members of a monophyletic group will be found near that center of the ancestral area. This method has been criticized, especially because of the bias caused by missing basal taxa in the fossil record (Humphries 1992; Ebach 1999). However, this is not a flaw but rather shows the importance of basal taxa; we can only build our hypothesis on current data, regardless of whether it will be falsified by future discoveries. The distribution of early, basal taxa on a phylogenetic tree is a good indicator of ancestral area. The basal placement of *Scalenodon angustifrons* from Africa suggests that traversodontids originated in Africa. However, considering the weak support of the basal nodes in our phylogeny, this hypothesis requires of additional corroboration. In any case, the origin of traversodontids is indeed circumscribed to Gondwana based on current phylogenetic results. This analysis includes almost all known species of Gomphodontia and the current phylogenetic results strongly indicate that gomphodont cynodonts originated in Africa.

In our results, *Nanogomphodon* does not group with the other two Laurasian taxa *Boreogomphodon* and *Arctotraversodon*, suggesting that traversodontids migrated at least twice from Gondwana to Laurasia. This, however should be considered with extreme caution, as *Nanogomphodon* represents a wildcard in our phylogeny. During the Triassic, major continents were connected as the Pangaean supercontinent. Based on the faunal assemblages of the Early Triassic, early Middle Triassic, and Early Jurassic, dispersal across the Pangaean land mass must have been possible for tetrapods during the Triassic. However, only three Early to Middle Triassic genera of Gomphodontia (*Sinognathus*, *Beishanodon*, and *Nanogomphodon*) are known in Eurasia. The poor representation of gomphodonts and traversodontids in particular in the Late Triassic of Eurasia could be the result of bias in preservation and study, e.g., up to now only one possible Late Triassic terrestrial tetrapod has been reported from China (Liu et al. 2001). We hope that further discoveries in this region will improve our understanding of gomphodont evolution in the northern hemisphere.

Acknowledgments We thank Christian F. Kammerer, Agustín G. Martinelli, and an anonymous reviewer whose comments greatly improved this paper. The cooperation and hospitality of the staff of various museums and institutions greatly facilitated our comparative studies. We would like to thank Michael W. Maisch (GPIT), Sandra Chapman (NMHUK), Tom Kemp (Oxford), Ray Symonds (CAMZM), Johann Neveling (CGP), Jennifer Botha and Elize Butler (NMQR), Roger Smith and Sheena Kaal (SAM), Alejandro Kramarz and Agustín G. Martinelli (MACN), Ana Maria Ribeiro (MCN), Marcelo Reguero and Rosendo Pascual (MLP), Maria C. Malabarba (MCP), Guillermo F. Vega (PULR), Jaime E. Powell (PVL), Ricardo Martinez (PVSJ), Marina B. Soares and Cesar L. Schultz (UFRGS), John Flynn (AMNH), Olivier Rieppel, Elaine Zeiger, and William F. Simpson (FMNH), Charles R. Schaff and Wu Shaoyuan (MCZ), Hans-Dieter Sues and Matthew Carrano (USNM), James A. Hopson (University of Chicago), and Vince Schneider (NCSM). Financial support for this project was provided by the Chinese Academy of Sciences (KZCX2-YW BR-07), Columbia University through a Faculty Fellowship, the Climate Center of Lamont-Doherty Earth Observatory, and the Theodore Roosevelt Memorial Fund of AMNH. The Field Museum provided grants making research visits to Chicago possible. FA is supported by the National Research Foundation, South Africa.

Appendix 15.1: List of Morphological Characters

The following abbreviations are used to identify authors that previously used a particular character in data matrices: R, (Rowe 1988); M, (Martínez et al. 1996); F, (Flynn et al. 2000); H, (Hopson and Kitching 2001); A, (Abdala and Ribeiro 2003); Ab, (Abdala et al. 2006); SH, (Sidor and Hancox 2006); K, (Kammerer et al. 2008); S, (Sues and Hopson 2010); L, (Liu and Olsen 2010).

1. Adult maximum skull size: large (greater than 25 cm) (0), medium to small (1).
2. Snout (preorbit) in adults in relation to temporal region: longer (0), subequal (1), shorter (2) [Ab28].
3. Two side of temporal fenestra: divergent posteriorly (0), nearly parallel (1), bulge in the middle (2) [H39, Ab33].
4. Premaxilla forms posterior border incisive foramen: absent (0), present (1) [H1, Ab29].
5. Vomer exposure in incisive foramen (at anterior ends of maxillae on palate): present (0), absent (1) [M21].
6. Vomer: with (0) or without (1) vertical septum extending posteriorly beyond level of secondary palate [SH65].
7. Internarial bar: present (0), absent (1) [F5, A22, Ab21].
8. Parietal foramen in adults: present (0), absent (1) [H7, F6, A24, Ab23].
9. Ectopterygoid: present (0), absent (1) [H9, Ab30].
10. The posterior extension of secondary palate relative to anterior border of orbit: shorter (0), subequal (1), longer (2) [H15, S2].
11. Posterior extension of the jugal dorsally above the squamosal in the zygomatic arch: absent or with a small extension (0), well-developed (1) [F15, A26, Ab25].
12. The position of anterior root of the zygomatic arch relative to the ventral margin of the maxilla: nearly at same level or slightly higher (0), remarkable higher (1).
13. Zygomatic process of the jugal: little projected (0), conspicuously projected (1), absent (2), a ball-like process (3) [F16, H21, A25, Ab24].
14. Diameter of suborbital bar below center of orbit (anterior to suborbital process, where present): greater than 1/2 diameter of bar below posterior part of orbit (posterior to suborbital process) (0), less than 1/2 diameter of bar below posterior part of orbit (1) [S35].
15. Maxilla in the margin of the subtemporal fenestra: excluded (0), included (1) [Ab31].
16. Epipterygoid-quadrate contact: present (0), absent (1) [Ab32].
17. Frontal-epipterygoid contact: present (0), absent (1) [R39, H35].
18. Palatine: does not meet frontal (0), meets frontal but neither element contributes significantly to medial orbit wall (1) [H23].
19. Notch separating lambdoidal crest from zygomatic arch: shallow (0); deep, V-shaped (1) [H43].
20. Lower jaw symphysis as a chin-like process in adult: absent or little developed (0), well developed (1).
21. Dentary with sigmoid ventral curvature: absent (0), present (1) [S34].
22. Dentary angular process: not or very weakly projected posteriorly (0), projected posteriorly as distinct process (1) [A28, Ab27, S32].
23. Elongated mental foramen below postcanine tooth row and above coronoid ridge: absent (0), present (1) [S29].
24. Coronoid ridge anterior to masseteric fossa: absent to low (0), very strong, outturned (1) [S30].
25. Position of the upper canine in relation to paracanine fossa: posterolateral (0), lateral (1), anterolateral (2) [A6, Ab5, S1].
26. Diastema between upper incisors and canine: present (0), absent (1) [A3, Ab2].
27. Diastema between canine and maxillary postcanines in adult: short (0), long (1) [F1].
28. Diastema between canine and dentary postcanines: long (0), absent or very short (less than one tooth length) (1).
29. Maxillary labial platform lateral to the postcanine series: absent (0), present (1) [H77, A23, Ab22].
30. Posteromedial inclination of the last few upper gomphodont postcanines: absent or small (0), oblique (1) [F9, H72, A9, Ab8].
31. Axis of posterior part of maxillary tooth row: directed lateral to subtemporal fossa (0), directed towards center of fossa (1), directed toward medial rim of fossa (2) [H78, Ab36].
32. Maxillary tooth row extent relative to anterior margin of the subtemporal fossa in adult: anterior (0), at the same level (1), posterior (2).
33. Coronoid process of the mandible: covers the last postcanine (0), does not cover (1) [A27, Ab26].
34. Postcanine occlusion: absent (0), present (1) [Ab41].
35. Shearing planes between the outer surface of the main cusp of the lower and the inner surfaces of the main cusps of the uppers postcanines: present (0), absent (1) [Ab42].
36. Upper incisor number: four (0), three (1) [H53, F4, A1, Ab0].
37. Lower incisor number: three (0), two (1) [H54].
38. Incisor procumbency: (0) absent; (1) present [K43].
39. Incisor cutting margins: serrated (0), smoothly ridged (1), denticulated (2) [H55, Ab34].
40. Incisor size: small (0), enlarged (1) [F3, H56, A2, Ab1].
41. Upper canine size: large (0), reduced in size (1) [H57, A4, Ab3].
42. Lower canine size: large (0), reduced in size (1) [H58, A5, Ab4].
43. Canine serrations: present (0), absent (1) [H59, Ab35].
44. Postcanine tooth row in adults: formed by sectorial (0), conical, gomphodont and sectorial (1), gomphodont and sectorial (2), gomphodont (3) [H80, Ab37].
45. Overall morphology of the upper gomphodont postcanines in occlusive view: ovoid-ellipsoid (0), nearly rectangular (1), nearly triangular (2) [A7, Ab6].
46. Labial cingulum on anterior portion of the upper postcanines (external to the sectorial ridge): absent (0), present (1) [H61, A14, Ab13].
47. Posterior cingulum on upper postcanines: present (0), absent (1) [F7, A13, Ab12].

48. Shouldering in the posterior margin of upper postcanines: absent (0), slightly developed (1), well developed (2). [F2, A8, Ab7]
49. Anterior profile of principal labial cusp: convex (0), concave (1) [S31].
50. Number of cusps in the transverse crest of the upper postcanines: two (0), three or more (1) [F8, H63, A11, Ab10].
51. Central cusp of upper transverse row: midway between labial and lingual cusps (0), closer to lingual cusp (1) [H65, A12, Ab11].
52. Anterolabial accessory cusp on upper postcanines: one (0), absent (1), two or more (2) [H67].
53. Posterolabial accessory cusp on upper postcanines: present (0), absent (1) [H68].
54. Position of upper transverse cusp row on crown: central (0), anterior half of crown (1), at posterior part (2) [H64, A10, Ab9].
55. Distinct anterolingual accessory cusp on upper postcanines: absent (0), present (1) [H69, A15, Ab14].
56. Anterior cingulum in the upper postcanines: present (0), absent (1) [Ab39].
57. Anterior transverse crest on upper postcanines: absent or low (0), high (1) [F14, H70].
58. Lingual ridge on upper postcanines: absent (0), present (1) [F10, H71].
59. Overall morphology of the lower gomphodont postcanines in occlusal view: circular (0), ovoid-ellipsoid (1), quadrangular (2) [H62, A17, Ab16].
60. Transverse crest in lower postcanines: central (0), anterior (1) [A18, Ab17].
61. Number of cusps in the transverse crest of the lower postcanines: two (0), three or more (1) [H73, A19, Ab18].
62. Anterior cingulum in the lower postcanines: cuspules disposed on the entire margin (0), one or more cuspules located anterolabially (1), absent (2) [H74, A21, Ab20].
63. Posterior cingulum on the lower gomphodont postcanines: present (0), absent (1) [Ab38].
64. Size of the anterior cusps in the lower postcanines: labial lower than lingual (0), labial higher than lingual (1) [A20, Ab19].
65. Widest lower cusp in transverse row of lower postcanines: lingual (0), labial (1), middle (2) [F11, H76].
66. Anterolingual cusp of lower postcanines: nearly vertical (0); strongly inclined posteriorly (1).
67. Deep occlusal basins in the postcanines: absent (0), present (1) [H75, Ab40].
68. Anapophysis: absent (0), present (1) [L124].
69. Expanded costal plates on ribs: present (0), absent (1) [H82].
70. Lumbar costal plates with ridge overlapping preceding rib: present (0), absent (1) [H83].
71. Procoracoid in glenoid: present (0), barely present or absent (1) [H88].
72. The dorsal margin of the coracoid in medial view in relation to that of the procoracoid: shorter (0), equal or longer than (1).
73. Cranial margin of the procoracoid: convex (0), nearly straight (1), obviously concave (2).
74. Angle between ventral margin on anterior and posterior process of iliac blade: small (e.g., less than 140°) (0), large (1).
75. Length of anterior process of ilium anterior to acetabulum (relative to diameter of acetabulum): less than 1.5 (0), greater than 1.5 (1) [H94].
76. Dorsal profile of ilium: strongly convex (0), flat to concave (1) [H96].
77. The trochanter major position relative to the femoral head: distal (0), close, major part in same height (1).

Appendix 15.2: Character Matrix

Thrinaxodon	1020000000	0020010000	000000000?	0000?00010	0010???00?	-??-??????	???????101	0000000
Cynognathus	0000000000	0000001000	000000010?	0000?00000	0000???00?	-??-??????	???????000	0000000
Diademodon	0000000000	0010001010	0000000000	00(01)1000000	0001000000	-000?00000	000???0100	0001000
Beishanodon	0210?00101	002000??1?	????000?10	10???0?0??	0????0000?	??00?000??	??????0???	???????
Cricodon	1??0??0???	0???0????0	000000?010	1011100001	0002000001	1000?00010	100???0100	??????0
Langbergia	1110100001	001000?010	0000000000	1011100010	0002000001	0000?00010	100???0???	???????
Sinognathus	121??1?101	00200???10	0000000010	1011101010	00?2000001	0110??0010	10????0???	???????
Trirachodon	1110100(01)01	0010001010	0000000010	10(01)1100000	0002000001	(01)11(02)?00010	100???0100	000000?
JSM100	1????00??1	?0?0?????0	0000??0000	?0?1???0?0	0012100001	1110?00020	100???0???	???????
Andescynodon	11001101?1	00000???10	0000000010	2101000010	0003101000	-101?11021	011101110?	?111111
Arctotraversodon	0?????????	?????????1	0?11???0??	???10?0021	???3?01011	0?120?0?21	1212001???	???????
B.jeffersoni	112??101?2	002001??10	0111000010	2111000010	001220(01)111	1(02)12010121	(01)111(01)0101?	112???1
Dadadon	11101?01?1	?0300???1?	????110?11	21?100?010	0?13111111	10120101??	??????1???	???????
E. argentinus	0001111111	111111?110	1100211011	2201010111	011310121?	-00-111121	021111101?	1?11111
E. riograndensis	0001111111	111111??10	1100211?11	22?1010111	011310121?	-00-111121	021111101?	????11?
Gomphodontosuchus	????1?0??0	?1??0????0	1?00110011	2001000?10	111310111?	-??-011121	02111?1???	???????
L. drysdalli	101???00??	00110??010	1100000111	2101000000	0003010011	1(02)12000121	0101001101	0001110
L. sudamericana	1?????00?1	?001?????1	1100000111	2101000000	0003010011	1002000121	01010?1???	???????
M. ochagaviae	1211110112	002001?01?	?000110011	2(12)010000?0	0013101111	12120?0121	02?1101??1	???????
M. pascuali	1111110111	0020010010	000011001(01)	2(12)01000020	1113101111	1212000121	020010100?	1111111
Menadon	10111?0??0	11210???10	1100111011	2001000111	110310111?	-01-011121	021???101?	1?0111?
Pascualgnathus	110011?1?1	00000??010	0000000010	2101010010	0013000000	-101?11021	0?????1101	00?0011
Protuberum	101?111011	111111??1?	????211?11	22?101?111	1?131?1110	-0121111??	?????110?	???????
Santacruzodon	1?????0??0	?03??????0	1100000011	2101000020	??13201111	1212000121	0201011???	???????
S. angustifrons	110???00??	?0200???10	??00000010	2001000000	0003011011	111(02)010021	0100001???	???00?0
S. attridgei	???0110??1	????0?????	????000?11	21?101?010	0?13100011	10121011??	??????1???	???????
S. hirschoni	1??0110?11	?0?001?1?0	0?00000011	2101011011	1013100011	1012100121	0100001???	???????
Scalenodontoides	012???11?1	11?11???11	?10021101?	2201010111	?1?310121?	-00-?11121	0211111???	???????
Traversodon	1???1?00?1	?1110????1	1100000011	2101000010	0013110011	1?02000121	02?0?01?01	11????1
Nanogomphodon	??????????	??????????	??????????	???10?????	??????????	????????21	1101101???	???????

References

Abdala, F. (1998). An approach to the phylogeny of gomphodont cynodonts based on dental characters. *Journal of African Earth Sciences, 27*, 1–2.

Abdala, F. (2000). Catalogue of non-mammalian cynodonts in the Vertebrate Paleontology Collection of the Institute Miguel Lillo, Universidad Nacional de Tucumán, with comments on species. *Ameghiniana, 37*, 463–476.

Abdala, F. (2007). Redescription of *Platycraniellus elegans* (Therapsida, Cynodontia) from the Early Triassic of South Africa, and the cladistic relationships of eutheriodonts. *Palaeontology, 50*, 591–618.

Abdala, F., & Giannini, N. P. (2000). Gomphodont cynodonts of the Chañares Formation, the analysis of an ontogenetic sequence. *Journal of Vertebrate Paleontology, 20*, 501–506.

Abdala, F., & Ribeiro, A. M. (2003). A new traversodontid cynodont from the Santa Maria Formation (Ladinian-Carnian) of southern Brazil, with a phylogenetic analysis of Gondwanan traversodontids. *Zoological Journal of the Linnean Society, 139*, 529–545.

Abdala, F., & Ribeiro, A. M. (2010). Distribution and diversity patterns of Triassic cynodonts (Therapsida, Cynodontia) in Gondwana. *Palaeogeography, Palaeoclimatology, Palaeoecology, 286*, 202–217.

Abdala, F., & Smith, R. M. H. (2009). A Middle Triassic cynodont fauna from Namibia and its implications for the biogeography of Gondwana. *Journal of Vertebrate Paleontology, 29*, 837–851.

Abdala, F., & Teixeira, A. M. S. (2004). A traversodontid cynodont of African affinity in the South American Triassic. *Palaeontologia Africana, 40*, 11–22.

Abdala, F., Ribeiro, A. M., & Schultz, C. L. (2001). A rich cynodont fauna of Santa Cruz do Sul, Santa Maria Formation (Middle-Upper Triassic), in southern Brazil. *Neues Jahrbuch für Geologie und Paläontologie, Monatshefte, 2001*, 669–687.

Abdala, F., Barberena, M. C., & Dornelles, J. (2002). A new species of the traversodontid cynodont *Exaeretodon* from the Santa Maria Formation (Middle/Late Triassic) of southern Brazil. *Journal of Vertebrate Paleontology, 22*, 313–325.

Abdala, F., Neveling, J., & Welman, J. (2006). A new trirachodontid cynodont from the lower levels of the Burgersdorp Formation (Lower Triassic) of the Beaufort Group, South Africa and the cladistic relationships of Gondwanan gomphodonts. *Zoological Journal of the Linnean Society, 147*, 383–413.

Barberena, M. C. (1974). *Contribuição ao Conhecimento dos Cinodontes Gonfodontes (Cynodontia, Tritylodontoidea) do Brasil.* Tese Livre Docência: Universidade Federal do Rio Grande do Sul.

Barberena, M. C. (1981a). Novos materiais de *Traversodon stahleckeri* da Formação Santa Maria (Triássico do Rio Grande do Sul). *Pesquisas, 14*, 149–162.

Barberena, M. C. (1981b). Uma nova espécie de *Massetognathus* (*Massetognathus ochagaviae*, sp. nov.) da Formação Santa Maria, Triássico do Rio Grande do Sul. *Pesquisas, 14*, 181–195.

Battail, B. (1991). Les cynodontes (Reptilia, Therapsida); une phylogenie. *Bulletin du Muséum National d'Histoire Naturelle, Section C, Sciences de la terre, paléontologie, géologie, minéralogie, 13*, 17–105.

Battail, B. (2001). A short review of studies of cynodonts. In L. A. Hector (Ed.), *Publicacion especial, VII International Symposium on Mesozoic Terrestrial Ecosystems* (pp. 29–38). Buenos Aires: Asociacion Paleontológica Argentina.

Battail, B. (2005). Late Triassic traversodontids (Synapsida, Cynodontia) in South Africa. *Palaeontologia Africana, 41*, 67–80.

Bonaparte, J. F. (1962). Descripción del cráneo y mandíbula de *Exaeretodon frenguellii* Cabrera, y su comparación con Diademodontidae, Tritylodontidae y los cinodontes sudamericanos. *Publicaciones del Museo Municipal de Ciencias Naturales y Tradicional de Mar del Plata, 1*, 135–202.

Bonaparte, J. F. (1963a). Descripción del esqueleto postcraneano de *Exaeretodon* (Cynodontia—Traversodontidae). *Acta Geológica Lilloana, 4*, 5–52.

Bonaparte, J. F. (1963b). Descripción de *Ischignathus sudamericanus* n. gen. n. sp., nuevo cindonte gonfodonte del Triásico Medio Superior de San Juan, Argentina (Cynodontia-Traversodontidae). *Acta Geológica Lilloana, 4*, 111–128.

Bonaparte, J. F. (1963c). Un nuevo cinodonte gonfodonte del Triásico Medio Superior de San Juan, *Proexaeretodon vincei* n. gen., n. sp. (Cynodontia—Traversodontidae). *Acta Geológica Lilloana, 4*, 129–133.

Bonaparte, J. F. (1963d). La familia Traversodontidae (Therapsida-Cynodontia). *Acta Geológica Lilloana, 4*, 163–194.

Bonaparte, J. F. (1966a). Sobre las cavidades cerebral, nasal y otras estructuras del cráneo de *Exaeretodon* sp. (Cynodontia, Traversodontidae). *Acta Geológica Lilloana, 8*, 5–31.

Bonaparte, J. F. (1966b). Una nueva 'fauna' triásica de Argentina (Therapsida, Cynodontia, Dicynodontia); consideraciones filogenéticas y paleobiogeográficas. *Ameghiniana, 4*, 243–294.

Bonaparte, J. F. (1969). Dos nuevas "faunas" de reptiles triásicos de Argentina. *Gondwana Stratigraphy, I.U.G.S.* (pp. 283–306). Mar del Plata.

Bonaparte, J. F. (1970). Annotated list of the South American Triassic tetrapods. In S. H. Haughton (Ed.), *Second Gondwana symposium, proceedings and papers (South Africa 1970)* (pp. 665–682). Pretoria: CSIR.

Bremer, K. (1992). Ancestral areas—a cladistic reinterpretation of the center of origin concept. *Systematic Biology, 41*, 436–445.

Brink, A. S. (1963). Two cynodonts from the Ntawere Formation in the Luangwa Valley of Northern Rhodesia. *Palaeontologia Africana, 8*, 77–96.

Brink, A. S. (1986). *Illustrated bibliographical catalogue of the Synapsida*. Handbook 10, Part 1, South African Geological Survey.

Broili, F., & Schröder, J. (1935). Beobachtungen an Wirbeltieren der Karrooformation. X. Über die Bezahnung von *Trirachodon* Seeley. *Sitzungsberichte der Bayerischen Akademie der Wissenchaften, Mathematisch-Naturwissenschaftliche, Abteilung, 1935*, 189–198.

Cabrera, A. (1943). El primer hallazgo de terápsidos en la Argentina. *Notas del Museo de La Plata, 8*, 317–331.

Chatterjee, S. (1982). A new cynodont reptile from the Triassic of India. *Journal of Paleontology, 56*, 203–214.

Crompton, A. W. (1955). On some Triassic cynodonts from Tanganyika. *Proceedings of the Zoological Society of London, 125*, 617–669.

Crompton, A. W. (1972). Postcanine occlusion in cynodonts and tritylodontids. *Bulletin of the British Museum (Natural History), Geology, 21*, 29–71.

Crompton, A. W., & Ellenberger, F. (1957). On a new cynodont from the Molteno beds and the origin of the tritylodontids. *Annals of the South African Museum, 44*, 1–13.

Ebach, M. C. (1999). Paralogy and the centre of origin concept. *Cladistics, 15*, 387–391.

Flynn, J. J., Parrish, J. M., Rakotosamimanana, B., Simpson, W. F., Whatley, R. L., & Wyss, A. R. (1999). A Triassic fauna from Madagascar, including early dinosaurs. *Science, 286*, 763–765.

Flynn, J. J., Parrish, J. M., Rakotosamimanana, B., Ranivoharimanana, L., Simpson, W. F., & Wyss, A. R. (2000). New traversodontids (Synapsida, Eucynodontia) from the Triassic of Madagascar. *Journal of Vertebrate Paleontology, 20*, 422–427.

Gao, K.-Q., Fox, R. C., Zhou, C.-F., & Li, D.-Q. (2010). A new nonmammalian eucynodont (Synapsida, Therapsida) from the Triassic of northern Gansu Province, China, and its biostratigraphic and biogeographic implications. *American Museum Novitates, 3685*, 1–25.

Godefroit, P. (1999). New traversodontid (Therapsida, Cynodontia) teeth from the Upper Triassic of Habay-la-Vieille (southern Belgium). *Palaeontologische Zeitschrift, 73*, 385–394.

Godefroit, P., & Battail, B. (1997). Late Triassic cynodonts from Saint Nicolas de Port (northeastern France). *Geodiversitas, 19*, 567–631.

Goloboff, P. A., Farris, J. S., & Nixon, K. C. (2008). TNT, a free program for phylogenetic analysis. *Cladistics, 24*, 774–786.

Goñi, R. (1986). Reemplazo de dientes postcaninos en *Andescynodon mendozensis* Bonaparte (Cynodontia, Traversodontidae). *Actas del IV Congreso Argentino de Paleontología y Bioestratigrafía* (pp. 7–14), Mendoza.

Goñi, R., & Abdala, F. (1989). Consideraciones sobre la morfología craneodentaria de *Rusconiodon mignonei* Bonaparte (Cynodontia, Traversodontidae); diagnosis, afinidades y variaciones ontogenéticas. *Ameghiniana, 25*, 237–244.

Goñi, R., & Goin, F. J. (1987). El origen de los postcaninos gonfodontes de *Andescynodon mendozensis* Bonaparte (Cynodontia, Traversodontidae). *Ameghiniana, 24*, 235–239.

Goñi, R., & Goin, F. J. (1988). Morfología dentaria y biomecánica masticatoria de los cinodontes (Reptilia, Therapsida) del Triásico Argentino; I, *Andescynodon mendozensis* Bonaparte (Traversodontidae). *Ameghiniana, 25*, 139–148.

Goñi, R., & Goin, F. J. (1990). Morfologia dentaria y biomecánica masticatoria de los cinodontes (Reptilia, Therapsida) del Triasico argentino; II, *Exaeretodon frenguellii* Cabrera; Traversodontidae. *Ameghiniana, 27*, 327–336.

Gow, C. E., & Hancox, P. J. (1993). First complete skull of the Late Triassic *Scalenodontoides* (Reptilia, Cynodontia) from southern Africa. *New Mexico Museum of Natural History and Science Bulletin, 3*, 161–168.

Hahn, G., Lepage, J. C., & Wouters, G. (1988). *Traversodontiden zaehne* (Cynodontia) aus der ober Trias von Gaume (Sued Belgien). *Bulletin de l'Institut Royal des Sciences Naturelles de Belgique. Sciences de la Terre, 58*, 177–186.

Hennig, W. (1966). *Phylogenetic systematics*. Urbana: University of Illinois Press.

Hopson, J. A. (1984). Late Triassic traversodont cynodonts from Nova Scotia and southern Africa. *Palaeontologia Africana, 25*, 181–201.

Hopson, J. A. (1985). Morphology and relationships of *Gomphodontosuchus brasiliensis* von Huene (Synapsida, Cynodontia, Tritylodontoidea) from the Triassic of Brazil. *Neues Jahrbuch für Geologie und Paläontologie Monatshefte, 1985*, 285–299.

Hopson, J. A. (2005). A juvenile gomphodont cynodont specimen from the *Cynognathus* Assemblage Zone of South Africa, implications for the origin of gomphodont postcanine morphology. *Palaeontologia Africana, 41*, 53–66.

Hopson, J. A. (2013). The traversodontid cynodont *Mandagomphodon hirschsoni* from the Middle Triassic of the Ruhuhu Valley, Tanzania. In C. F. Kammerer, K. D. Angielczyk, & J. Fröbisch (Eds.), *Early evolutionary history of the Synapsida* (pp. 233–253). Dordrecht: Springer.

Hopson, J. A., & Barghusen, H. R. (1986). An analysis of therapsid relationships. In N. Hotton III, P. D. MacLean, J. J. Roth, & E. C. Roth (Eds.), *The ecology and biology of mammal-like reptiles* (pp. 83–106). Washington DC: Smithsonian Institution Press.

Hopson, J. A., & Kitching, J. W. (1972). A revised classification of cynodonts (Reptilia; Therapsida). *Palaeontologia Africana, 14*, 71–85.

Hopson, J. A., & Kitching, J. W. (2001). A probainognathian cynodont from South Africa and the phylogeny of nonmammalian cynodonts. *Bulletin of the Museum of Comparative Zoology, 156*, 5–35.

Hopson, J. A., & Sues, H.-D. (2006). A traversodont cynodont from the Middle Triassic (Landian) of Baden-Württemberg (Germany). *Paläontologische Zeitschrift, 80*, 124–129.

Huene, F. von. (1928). Ein Cynodontier aus der Trias Brasiliens. *Centralblatt für Mineralogie, Geologie und Paläontologie, Abt. B, 1928*, 251–270.

Huene, F. von. (1936). Die fossilen Reptilien des südamerikanischen Gondwanalandes. *Ergebnisse der Sauriergrabungen in Südbrasilien 1928–1929. Lieferung 2* (pp. 93–159). Tübingen: Verlag Franz F. Heine.

Huene, F. von (1948). Gleiche Cynodontier in der Obertrias Nordargentiniens und Südbrasiliensis. *Neues Jahrbuch für Mineralogie, Geologie und Paläontologie Monatshefte, 1945*(48), 378–382.

Huene, F. von (1950). Die Theriodontier des ostafrikanischen Ruhuhu-Gebietes in der Tübinger Sammlung. *Neues Jahrbuch für Geologie und Paläontologie Abhandlungen, 92*, 47–136.

Humphries, C. J. (1992). Cladistic biogeography. In P. L. Forey, I. J. Kitching, & C. J. Humphries (Eds.), *Cladistics—a practical course in systematics* (pp. 137–159). Oxford: Clarendon Press.

Kammerer, C. F., Flynn, J. J., Ranivoharimanana, L., & Wyss, R. R. (2008). New material of *Menadon besairiei* (Cynodontia, Traversodontidae) from the Triassic of Madagascar. *Journal of Vertebrate Paleontology, 28*, 445–462.

Kemp, T. S. (1980). Aspects of the structure and functional anatomy of the Middle Triassic cynodont *Luangwa*. *Journal of Zoology (London), 191*, 193–239.

Liu, J. (2007). The taxonomy of the traversodontid cynodonts *Exaeretodon* and *Ischignathus*. *Revista Brasileira de Paleontologia, 10*, 133–136.

Liu, J., & Olsen, P. E. (2010). The phylogenetic relationships of Eucynodontia (Amniota, Synapsida). *Journal of Mammalian Evolution, 17*, 151–176.

Liu, J., & Powell, J. (2009). Osteology of *Andescynodon* (Cynodontia, Traversodontidae) from the Middle Triassic of Argentina. *American Museum Novitates, 3674*, 1–19.

Liu, J., & Sues, H.-D. (2010). Dentition and tooth replacement of *Boreogomphodon* (Cynodontia, Traversodontidae) from the Upper Triassic of North Carolina, USA. *Vertebrata PalAsiatica, 48*, 169–184.

Liu, J., Wu, X. C., & Li, J. L. (2001). The first reptile from the Tongchuan Formation and its stratigraphical significance. *Vertebrata PalAsiatica, 39*, 67–71.

Liu, J., Soares, M. B., & Reichel, M. (2008). *Massetognathus* (Cynodontia, Traversodontidae) from the Santa Maria Formation of Brazil. *Revista Brasileira de Paleontologia, 11*, 27–36.

Luo, Z. X. (1994). Sister-group relationships of mammals and transformations of diagnostic mammalian characters. In N. C. Fraser & H.-D. Sues (Eds.), *In the shadow of the dinosaurs, Early Mesozoic tetrapods* (pp. 98–128). Cambridge: Cambridge University Press.

Martinelli, A. G. (2010a). On the postcanine dentition of *Pascualgnathus polanskii* Bonaparte (Cynodontia: Traversodontidae) from the Middle Triassic of Argentina. *Geobios, 43*, 629–638.

Martinelli, A. G. (2010b). Trirachodontids (Therapsida, Cynodontia) as non-gomphodont cynodonts: Testing a hypothesis. *Congreso Argentino de Paleontologia y Bioestratigrafia y Congreso Latinoamericano de Paleontologia. La Plata* (p. 182).

Martínez, R. N., May, C. L., & Forster, C. A. (1996). A new carnivorous cynodont from the Ischigualasto Formation (Late Triassic, Argentina), with comments on eucynodont phylogeny. *Journal of Vertebrate Paleontology, 16*, 271–284.

Minoprio, J. L. (1954). Theriodonte en el Triássico de Mendoza. *Anales de la Sociedad Cientifica Argentina, 157*, 31–37.

Minoprio, J. L. (1957). Nota aclaratoria sobre *Colbertia muralis* (Cambio de denominación). *Ameghiniana, 1*, 113.

Oliveira, T. V. (2006). *Descrição osteológica de materiais póscranianos de dois cinodontes não-mamalianos do Meso/Neotriássico (Formação Santa Maria, Bacia do Paraná) do Rio Grande do Sul, Brasil*. Dissertação (Mestrado), Universidade Federal do Rio Grande do Sul.

Oliveira, T. V., Schultz, C. L., & Soares, M. B. (2007). O esqueleto pós-craniano de *Exaeretodon riograndensis* Abdala et al., (Cynodontia, Traversodontidae), Triássico do Brasil. *Revista Brasileira de Paleontologia, 10*, 79–94.

Parrington, F. R. (1946). On the cranial anatomy of cynodonts. *Proceedings of the Zoological Society of London, 116*, 181–197.

Peyer, K., Carter, J. G., Sues, H.-D., Novak, S. E., & Olsen, P. E. (2008). A new suchian archosaur from the Upper Triassic of North Carolina. *Journal of Vertebrate Paleontology, 28*, 363–381.

Rayfield, E. J., Barrett, P. M., McDonnell, R. A., & Willisi, K. J. (2005). A geographical information system (GIS) study of Triassic vertebrate biochronology. *Geological Magazine, 142*, 327–354.

Razafimbelo, E. (1987). Le basin de Morondava (Madagascar). *Synthèse géologique et structurale*. Ph.D. thesis, University Louis Pasteur, Strasbourg.

Reichel, M., Schultz, C., & Soares, M. B. (2009). A new traversodontid cynodont (Therapsida, Eucynodontia) from the Middle Triassic Santa Maria Formation of Rio Grande do Sul, Brazil. *Palaeontology, 52*, 229–250.

Romer, A. S. (1956). *Osteology of the reptiles*. Chicago: University of Chicago Press.

Romer, A. S. (1966). *Vertebrate paleontology*. Chicago: University of Chicago Press.

Romer, A. S. (1967). The Chañares (Argentina) Triassic reptile fauna. III. Two new gomphodonts, *Massetognathus pascuali and M. teruggii*. *Breviora, 264*, 1–25.

Romer, A. S. (1972). The Chañares (Argentina) Triassic reptile fauna. XVII. The Chañares gomphodonts. *Breviora, 396*, 1–9.

Rowe, T. (1988). Definition, diagnosis and origin of Mammalia. *Journal of Vertebrate Paleontology, 8*, 241–264.

Seeley, H. G. (1894). Researches on the structure, organization, and classification of the fossil Reptilia. Part IX, section 3. On *Diademodon*. *Philosophical Transactions of the Royal Society of London, Series B, Biological Sciences, 185*, 1029–1041.

Seeley, H. G. (1895). Researches on the structure, organization, and classification of the fossil Reptilia. Part IX, section 4. On the Gomphodontia. *Philosophical Transactions of the Royal Society of London, Series B, Biological Sciences, 186*, 1–57.

Sidor, C. A., & Hancox, P. J. (2006). *Elliotherium kersteni*, a new tritheledontid from the Lower Elliot Formation (Upper Triassic) of South Africa. *Journal of Paleontology, 80*, 333–342.

Sigogneau, R. D., & Hahn, G. (1994). Late Triassic microvertebrates from Central Europe. In N. C. Fraser & H. -D. Sues (Eds.), *In the shadow of dinosaurs: Early Mesozoic tetrapods* (pp. 197–213). Cambridge: Cambridge University Press.

Soares, M. B., Abdala, F., & Bertoni, C. M. (2011). A sectorial toothed cynodont from the Triassic Santa Cruz do Sul fauna, Santa Maria Formation, Southern Brazil. *Geodiversitas, 33*, 265–278.

Sues, H.-D. (1985). The relationships of the Tritylodontidae (Synapsida). *Zoological Journal of the Linnean Society, 85*, 205–217.

Sues, H.-D., & Hopson, J. A. (2010). Anatomy and phylogenetic relationships of *Boreogomphodon jeffersoni* (Cynodontia, Gomphodontia) from the Upper Triassic of Virginia. *Journal of Vertebrate Paleontology, 30*, 1202–1220.

Sues, H.-D., & Olsen, P. E. (1990). Triassic vertebrates of Gondwanan aspect from the Richmond Basin of Virginia. *Science, 249*, 1020–1023.

Sues, H.-D., Hopson, J. A., & Shubin, N. H. (1992). Affinities of ?*Scalenodontoides plemmyridon* Hopson, 1984 (Synapsida, Cynodontia) from the Upper Triassic of Nova Scotia. *Journal of Vertebrate Paleontology, 12*, 168–171.

Sues, H.-D., Olsen, P. E., & Kroehler, P. A. (1994). Small tetrapods from the Upper Triassic of the Richmond basin (Newark Supergroup Virginia). In N. C. Fraser & H.-D. Sues (Eds.), *In the shadow of dinosaurs: Early Mesozoic tetrapods* (pp. 161–170). Cambridge: Cambridge University Press.

Sues, H.-D., Olsen, P. E., & Carter, J. G. (1999). A late Triassic traversodont cynodont from the Newark Supergroup of North Carolina. *Journal of Vertebrate Paleontology, 19*, 351–354.

Tatarinov, L. P. (1973). Cynodonts of Gondwanan habit in the middle Triassic of the USSR. *Paleontological Journal, 7*, 200–205.

Tatarinov, L. P. (1974). Terriodont of USSR. *Trudy Paleontologicheskogo Instituta, Akademiya Nauk SSSR, 143*, 1–226.

Tatarinov, L. P. (1988). On the morphology and systematic position of the gomphodont cynodont *Antecosuchus ochevi*. *Paleontological Journal, 22*, 82–90.

Tatarinov, L. P. (2002). Gomphodont cynodonts (Reptilia, Theriodontia) from the Middle Triassic of the Orenburg Region. *Paleontologicheskii Zhurnal, 2002*, 58–61.

Teixeira, A. M. S. (1987). Novas observações osteologicas e taxonômicas sobre *Massetognathus ochagaviae* Barberena, 1981 (Reptilia, Therapsida, Cynodontia). *Paula-Coutiana, 1*, 39–49.

Teixeira, A. M. S. (1995). A família Traversodontidae (Therapsida, Cynodontia) no sul do Brasil e suas relações com formas afins no domínio gondwânico. Unpublished Ph.D. thesis, Universidade Federal do Rio Grande do Sul.

Walker, J. D., & Geissman, J. W., compilers. (2009). *Geologic time scale, Geological Society of America*. doi:10.1130/2009.CTS004R2C.

Watson, D. M. S., & Romer, A. S. (1956). A classification of therapsid reptiles. *Bulletin of the Museum of Comparative Zoology, 114*, 37–89.

Part IV
Therapsid Diversity Patterns and the End-Permian Extinction

Chapter 16
Introduction

Kenneth D. Angielczyk

The end-Permian mass extinction looms large in synapsid research because of the magnitude of the event and the fact that non-mammalian synapsids were the dominant tetrapods in terrestrial communities immediately before and after the extinction. Given that synapsids were such diverse and abundant components of latest Permian tetrapod communities, it is perhaps not surprising that nearly all synapsid subclades were strongly affected by the extinction (e.g., Irmis and Whiteside 2011). At the same time, it's noteworthy that at least some synapsids were able to rapidly recover and rediversify following the extinction and others were very successful in what were presumably degraded environmental conditions in its immediate aftermath. For example, even though there is a complete turnover among cynodonts at the Permo-Triassic boundary, cynodont species diversity is essentially unchanged between the latest Permian and the earliest Triassic, and cynodonts rapidly diversify during the Middle Triassic (Abdala and Ribeiro 2010; Botha-Brink et al. 2012). The dicynodont *Lystrosaurus* is similarly famous for its high local abundance (Nicolas and Rubidge 2010; Smith et al. 2012) and its global geographic range in the Early Triassic (Rubidge 2005; Fröbisch 2009).

The past two decades have witnessed an explosion of interest in the end-Permian extinction in the terrestrial realm, largely coinciding with the general renewed interest in the event within the paleontology and Earth science communities. During this time, much work has focused on documenting fine-scale data needed to reconstruct the timing and sequence of biological and environmental events that occurred during the extinction (e.g., Smith 1995; MacLeod et al. 2000; Ward et al. 2000, 2005; Smith and Ward 2001; Tverdokhlebov et al. 2002, 2005; Retallack et al. 2003; Benton et al. 2004; Taylor et al. 2009; Newell et al. 2010). However, an increasing number of papers now present and test hypotheses about potential extinction mechanisms and reasons for survivorship in the terrestrial (e.g., Angielczyk et al. 2005; Angielczyk and Walsh 2008; Botha-Brink and Angielczyk 2010; Fröbisch et al. 2010) and marine realms (e.g., Renne et al. 1995; Becker et al. 2001; Knoll et al. 2007; Bottjer et al. 2008), examine diversity patterns of individual clades near the Permo-Triassic boundary, particularly in the light of how potential biases might influence our picture of the event (e.g., Fröbisch 2008; Ruta and Benton 2008; Bernard et al. 2010, Botha-Brink et al. 2012; Huttenlocker et al. 2011; Irmis and Whiteside 2011; Ruta et al. 2011), reassess previously-presented data (Gastaldo et al. 2005, 2009; Gastaldo and Rolerson 2008; Pace et al. 2009), and investigate patterns of recovery in the extinction's aftermath (e.g., Smith and Botha 2005; Botha and Smith 2006; Roopnarine et al. 2007; Sahney and Benton 2008; Roopnarine and Angielczyk 2012). There has also been a debate over whether tetrapod extinctions near the end of the Middle Permian, particularly the extinction of dinocephalians, were contemporaneous with the end-Guadalupian mass extinction observed in the marine realm (e.g., Retallack et al. 2006; Lucas 2009; Smith et al. 2012). Synapsids, of course, have figured prominently in all of this work.

Much of our picture of the end-Permian extinction in the terrestrial realm and its effect on tetrapods comes from two geographic areas, the Fore-Ural region of Russia (e.g., Benton et al. 2004; Newell et al. 2010) and the Karoo Basin of South Africa (e.g., Ward et al. 2000, 2005). The Karoo Basin has received particular attention because of its wide exposures of fossiliferous rocks, its likely preservation of a continuous sedimentological record through the Permo-Triassic boundary (Smith 1995; MacLeod et al. 2000; Ward et al. 2000, 2005; Smith and Ward 2001; Retallack et al. 2003; Smith and Botha 2005; Botha and Smith 2006; although see Gastaldo et al. 2009; Pace et al. 2009), and a long history of paleontological research that has resulted in

K. D. Angielczyk (✉)
Department of Geology, Field Museum of Natural History, 1400 South Lake Shore Drive, Chicago, IL 60605, USA
e-mail: kangielczyk@fieldmuseum.org

Christian F. Kammerer, Kenneth D. Angielczyk and Jörg Fröbisch (eds.), *Early Evolutionary History of the Synapsida*, Vertebrate Paleobiology and Paleoanthropology, DOI: 10.1007/978-94-007-6841-3_16,

an unmatched fossil record of Permo-Triassic tetrapods (Nicolas and Rubidge 2009, 2010). The two chapters in this section both present new results from the Karoo record of the Permo-Triassic transition, and in some ways reflect the two main trends in research on the subject: one presents detailed new data on the stratigraphic occurrences of latest Permian and earliest Triassic tetrapods in one boundary section, whereas the other examines changes in Permo-Triassic synapsid diversity throughout the Karoo and whether these patterns are biased by available rock outcrop areas.

Botha-Brink et al. (2013) provide an introduction to the lithostratigraphy and vertebrate paleontology of a Permo-Triassic boundary section located on the farm Nooitgedacht 68, a locality that is interesting for several reasons. For example, it has produced the largest sample by far of the Permian dicynodont *Lystrosaurus maccaigi*. This is the largest species of *Lystrosaurus* (Grine et al. 2006), and the first to appear in the Karoo fossil record (Botha and Smith 2007), but it is also rare and relatively poorly known. In particular, juvenile specimens of *L. maccaigi* are nearly unknown, and there has long been uncertainty regarding whether *L. maccaigi* simply represents large individuals of other *Lystrosaurus* species. Botha-Brink et al. document 50 specimens from the locality, almost doubling the available sample of *L. maccaigi*. Moreover, the specimens range in size from juvenile to large adult, making it possible for the ontogeny of *L. maccaigi* to be studied in detail for the first time, particularly if some or all of the specimens can be incorporated into a morphometric framework similar to those of Ray (2005), Grine et al. (2006), or Camp (2010).

In addition to the *L. maccaigi* specimens, Botha-Brink et al. document the occurrences of a number of additional tetrapods in their measured stratigraphic section. These data are the most detailed yet available for some of the taxa, and are key for providing insight into long-standing questions such as where the last appearances of *Dicynodon lacerticeps*, *Daptocephalus leoniceps*, and *Dinanomodon gilli* lie relative to the Permo-Triassic boundary (e.g., see Kammerer et al. 2011). Indeed, explicitly linking stratigraphic ranges for Karoo taxa to voucher specimens collected in measured stratigraphic sections is a major step forward compared to the much more informally documented stratigraphic ranges previously available (e.g., Rubidge 1995), and echoes work being done elsewhere in the basin (e.g., Angielczyk and Rubidge 2009, 2013).

Finally, the Nooitgedacht 68 section is an important new data point for comparison with other Karoo Permo-Triassic boundary sections. There has been debate recently concerning the continuity of deposition across the Permo-Triassic boundary in the Karoo Basin, as well as the placement of the boundary itself, and the nature of lithostratigraphic changes across the boundary and their environmental implications (compare Smith 1995; MacLeod et al. 2000; Ward et al. 2000, 2005; Smith and Ward 2001; Retallack et al. 2003; Smith and Botha 2005; Botha and Smith 2006; Gastaldo et al. 2005, 2009; Pace et al. 2009). The lithostratigraphic section and faunal changes described by Botha-Brink et al. (2013) generally fit well with the interpretation developed previously by Roger Smith and colleagues elsewhere in the Karoo. As the Nooitgedacht 68 section becomes more thoroughly studied, it will be interesting to see whether this interpretation holds, as well as whether the section can help resolve questions such as whether the deposition of the thinly-laminated (heterolithic) mudrocks typically used to mark the Permo-Triassic boundary in the Karoo were deposited synchronously across the basin.

Fröbisch (2013) examines broad patterns of Permian and Triassic synapsid taxonomic diversity in the Karoo Basin. There has been renewed interest in patterns of tetrapod taxonomic diversity at a range of geographic and temporal scales (e.g., Lloyd et al. 2008; Barrett et al. 2009; Butler et al. 2009, 2011; Marx 2009; Benson et al. 2010; Marx and Uhen 2010; Benton et al. 2011; Benson and Mannion 2012; Lloyd 2012), although data from the Permian and Triassic of South Africa have mostly received attention as part of larger, regional compilations (e.g., Fröbisch 2008; Abdala and Ribeiro 2010; although see Irmis and Whiteside 2011). Fröbisch's paper helps to address this shortcoming by focusing in detail on the Karoo synapsid record, and he investigates possible biases resulting from the nature of the fossil record by applying the modeling approach of Smith and McGowan (2007) and McGowan and Smith (2008).

His results suggest several interesting patterns. First, the Karoo fossil record of synapsids does appear to be correlated with the amount of available rock outcrop area, although only when the *Lystrosaurus* Assemblage Zone is excluded from consideration. This is significant because it implies that the low diversity of the *Lystrosaurus* zone is unlikely to be simply an artifact of limited exposures, but instead reflects real changes in diversity caused by the end-Permian extinction. It also generally fits well with the conclusions of a specimen-based analysis that also considered the effects of outcrop area on diversity patterns in the Karoo (Irmis and Whitside 2011), although that analysis found the *Dicynodon* Assemblage Zone to be an outlier that was more diverse than expected given its outcrop area. Nevertheless both studies suggest that it may be inadvisable to take diversity patterns in the Karoo at face value, despite relatively even geographic sampling in the basin (Nicolas and Rubidge 2009).

Second, when potential sampling biases are taken into account, different synapsid clades show distinct patterns of diversity change over time. Given their diversity, it is perhaps not surprising that the pattern for anomodonts closely resembles the pattern for all synapsids, but they are not the sole driver of the pattern. Moreover, the fact that each clade has a distinct diversity pattern is encouraging because it implies that although outcrop area does overprint diversity signals to some degree, it is still possible to extract biologically-relevant information from the data. The distinct clade-level diversity patterns also corroborate the observations of Irmis and Whiteside (2011), and extends them by also considering the *Eodicynodon*, *Tapinocephalus*, and *Pristerognathus* assemblage zones.

Third, there is evidence for a mid-Permian extinction among synapsids in addition to the well-defined end-Permian event, but it appears to be strongly driven by signals in a few subclades, particularly dinocephalians. This calls into question whether a severe end-Guadalupian event occurred on land as it did in the marine realm. In particular, dinocephalian taxonomy is still undergoing extensive revision (Atayman et al. 2009; Kammerer 2011), and the stratigraphic ranges of individual dinocephalian taxa generally are not well constrained. If the number of valid dinocephalian taxa is significantly reduced, as seems likely (Atayman et al. 2009), and/or the improved stratigraphic range data show that dinocephalian taxa do not all disappear more or less simultaneously, the terrestrial mid-Permian extinction may turn out to be a non-event, at least in the Karoo Basin.

A final important thing to keep in mind when considering papers such as Fröbisch (2013) or Irmis and Whiteside (2011) is that the best way to account for potential biases in fossil diversity data is still very much an open question (Benton et al. 2011), and new techniques continue to be proposed (e.g., Benson and Mannion 2012; Lloyd 2012). Therefore, although these works provide the most detailed and nuanced working hypotheses of diversity changes in the Karoo to date, it will be interesting to see how they fare as they are tested with other methods and as new data become available. A particularly important step in this regard will be checking and updating specimen identifications in the Karoo database of Nicolas and Rubidge (2009, 2010; also see Smith et al. 2012), a project that is already underway for the anomodont specimens (J. Fröbisch, personal communication).

Taken together, Botha-Brink et al.'s and Fröbisch's chapters are important contributions to our understanding of diversity patterns through time and the end-Permian extinction in the Karoo Basin. At the same time, when put into a broader context, both chapters emphasize the number of unanswered questions that remain and the additional research needed to fully address these topics. That this is the case despite over 150 years of paleontological research in the Karoo Basin, including nearly two decades of intense scrutiny of its record of the end-Permian extinction, highlights the potential insights that will be possible as the number of synthetic studies grows.

References

Abdala, F., & Ribeiro, A. M. (2010). Distribution and diversity patterns of Triassic cynodonts (Therapsida, Cynodontia) in Gondwana. *Palaeogeography, Palaeoclimatology, Palaeoecology, 286*, 202–217.

Angielczyk, K. D., & Rubidge, B. S. (2009). The Permian dicynodon *Colobodectes cluveri* (Therapsida, Anomodontia), with notes on its ontogeny and stratigraphic range in the Karoo Basin, South Africa. *Journal of Vertebrate Paleontology, 29*, 1162–1173.

Angielczyk, K. D., & Rubidge, B. S. (2013). Skeletal morphology, phylogenetic relationships, and stratigraphic range of *Eosimops newtoni* Broom, 1921, a pylaecephalid dicynodont (Therapsida, Anomodontia) from the Middle Permian of South Africa. *Journal of Systematic Palaeontology, 11*, 189–229.

Angielczyk, K. D., & Walsh, M. L. (2008). Patterns in the evolution of nares size and secondary palate length in anomodont therapsids (Synapsida): Implications for hypoxia as a cause for end-Permian terrestrial vertebrate extinctions. *Journal of Paleontology, 82*, 528–542.

Angielczyk, K. D., Roopnarine, P. D., & Wang, S. C. (2005). Modeling the role of primary productivity disruption in end-Permian extinctions, Karoo Basin, South Africa. *New Mexico Museum of Natural History and Science Bulletin, 30*, 16–23.

Atayman, S., Rubidge, B. S., & Abdala, F. (2009). Taxonomic re-evaluation of tapinocephalid dinocephalians. *Palaeontologia Africana, 44*, 87–90.

Barrett, P. M., McGowan, A. J., & Page, V. (2009). Dinosaur diversity and the rock record. *Proceedings of the Royal Society B, 276*, 2667–2674.

Becker, L., Poreda, R. J., Hunt, A. G., Bunch, T. E., & Rampino, M. R. (2001). Impact event at the Permian-Triassic boundary: Evidence from extraterrestrial noble gases in fullerenes. *Science, 291*, 1530–1533.

Benson, R. B. J., & Mannion, P. D. (2012). Multi-variate models are essential for understanding vertebrate diversification in deep time. *Biology Letters, 8*, 127–130.

Benson, R. B. J., Butler, R. J., Lindgren, J., & Smith, A. S. (2010). Palaeodiversity of Mesozoic marine reptiles: Mass extinctions and temporal heterogeneity in geologic megabiases affecting vertebrates. *Proceedings of the Royal Society B, 277*, 829–834.

Benton, M. J., Tverdokhlebov, V. P., & Surkov, M. V. (2004). Ecosystem remodelling among vertebrates at the Permian-Triassic boundary in Russia. *Nature, 432*, 97–100.

Benton, M. J., Dunhill, A. M., Lloyd, G. T., & Marx, F. G. (2011). Assessing the quality of the fossil record: Insights from vertebrates. In A. J. McGowan & A. B. Smith (Eds.), *Comparing the geological and fossil records: Implications for biodiversity studies* (Vol. 358, pp. 63–94). London: Geological Society (Special publications).

Bernard, E. L., Ruta, M., Tarver, J. E., & Benton, M. J. (2010). The fossil record of early tetrapods: Worker effort and the end-Permian mass extinction. *Acta Palaeontologia Polonica, 55*, 229–239.

Botha, J., & Smith, R. M. H. (2006). Rapid vertebrate recuperation in the Karoo Basin of South Africa following the end-Permian extinction. *Journal of African Earth Sciences, 45*, 502–514.

Botha, J., & Smith, R. M. H. (2007). *Lystrosaurus* species composition across the Permo-Triassic boundary of South Africa. *Lethaia, 40*, 125–137.

Botha-Brink, J., & Angielczyk, K. D. (2010). Do extraordinarily high growth rates in Permo-Triassic dicynodonts (Therapsida, Anomodontia) explain their success before and after the end-Permian extinction? *Zoological Journal of the Linnean Society, 160*, 341–365.

Botha-Brink, J., Abdala, F., & Chinsamy-Turan, A. (2012). The radiation and osteohistology of nonmammaliaform cynodonts. In A. Chinsamy-Turan (Ed.), *Forerunners of mammals: Radiation, histology, biology* (pp. 223–246). Bloomington: Indiana University Press.

Botha-Brink, J., Huttenlocker, A. K., & Modesto, S. P. (2013). Vertebrate paleontology of Nooigedacht 68: A *Lystrosaurus maccaigi*-rich Permo-Triassic boundary locality in South Africa. In C. F. Kammerer, K. D. Angielczyk, & J. Fröbisch (Eds.), *Early evolutionary history of the Synapsida* (pp. 289–304). Dordrecht: Springer.

Bottjer, D. J., Clapham, M. E., Fraiser, M. L., & Powers, C. M. (2008). Understanding mechanisms for the end-Permian mass extinction and the protracted Early Triassic aftermath and recovery. *GSA Today, 18*, 4–10.

Butler, R. J., Barrett, P. M., Nowbath, S., & Upchurch, P. (2009). Estimating the effects of the rock record on pterosaur diversity patterns: Implications for hypotheses of bird/pterosaur competitive replacement. *Paleobiology, 35*, 432–446.

Butler, R. J., Benson, R. J., Carrano, M. T., Mannion, P. D., & Upchurch, P. (2011). Sea level, dinosaur diversity and sampling biases: Investigating the 'common cause' hypothesis in the terrestrial realm. *Proceedings of the Royal Society B, 278*, 1165–1170.

Camp, J. A. (2010). *Morphological variation and disparity in Lystrosaurus (Therapsida: Dicynodontia)*. Unpublished MS thesis, University of Iowa.

Fröbisch, J. (2008). Global taxonomic diversity of anomodonts (Tetrapoda, Therapsida) and the terrestrial rock record across the Permian-Triassic boundary. *PLoS One*, *3*, e3733. doi: 223 10.1371/journal.pone.0003733.

Fröbisch, J. (2009). Composition and similarity of global anomodont-bearing tetrapod faunas. *Earth Science Reviews, 95*, 119–175.

Fröbisch, J. (2013). Synapsid diversity and the rock record in the Permian-Triassic Beaufort Group (Karoo Supergroup), South Africa. In C. F. Kammerer, K. D. Angielczyk, & J. Fröbisch (Eds.), *Early evolutionary history of the Synapsida* (pp. 305–319). Dordrecht: Springer.

Fröbisch, J., Angielczyk, K. D., & Sidor, C. A. (2010). The Triassic dicynodont *Kombuisia* (Synapsida, Anomodontia) from Antarctica, a refuge for terrestrial tetrapods during and after the end-Permian extinction. *Naturwissenschaften, 97*, 187–196.

Gastaldo, R. A., & Rolerson, M. W. (2008). *Katbergia* gen. nov., a new trace fossil form Upper Permian and Lower Triassic rocks of the Karoo Basin: Implications for palaeoenvironmental conditions at the P/Tr extinction event. *Palaeontology, 51*, 215–229.

Gastaldo, R. A., Adendorff, R., Bamford, M., Labandeira, C. C., Neveling, J., & Sims, H. (2005). Taphonomic trends of macrofloral assemblages across the Permian-Triassic boundary, Karoo Basin, South Africa. *Palaios, 20*, 479–497.

Gastaldo, R. A., Neveling, J., Clark, C. K., & Newbury, S. S. (2009). The terrestrial Permian-Triassic boundary event bed is a nonevent. *Geology, 37*, 199–202.

Grine, F. E., Forster, C. A., Cluver, M. A., & Georgi, J. A. (2006). Cranial variability, ontogeny and taxonomy of *Lystrosaurus* from the Karoo Basin of South Africa. In M. T. Carrano, T. J. Gaudin, R. W. Blob, & J. R. Wible (Eds.), *Amniote paleobiology: Perspectives on the evolution of mammals, birds, and reptiles* (pp. 403–503). Chicago: The University of Chicago Press.

Huttenlocker, A. K., Sidor, C. A., & Smith, R. M. H. (2011). A new specimen of *Promoschorhynchus* (Therapsida: Therocephalia: Akidnognathidae) from the Lower Triassic of South Africa and its implications for theriodont survivorship across the Permo-Triassic boundary. *Journal of Vertebrate Paleontology, 31*, 405–421.

Irmis, R. B., & Whiteside, J. M. (2011). Delayed recovery of non-marine tetrapods after the end-Permian mass extinction tracks global carbon cycle. *Proceedings of the Royal Society B, 279*, 1310–1318.

Kammerer, C. F. (2011). Systematics of the Anteosauria (Therapsida: Dinocephalia). *Journal of Systematic Palaeontology, 9*, 261–304.

Kammerer, C. F., Angielczyk, K. D., & Fröbisch, J. (2011). A comprehensive taxonomic revision of *Dicynodon* (Therapsida, Anomodontia) and its implications for dicynodont phylogeny, biogeography, and biostratigraphy. *Society of Vertebrate Paleontology Memoir, 11*, 1–158.

Knoll, A. H., Bambach, R. K., Payne, J. L., Pruss, S., & Fischer, W. W. (2007). Paleophysiology and end-Permian mass extinction. *Earth and Planetary Science Letters, 256*, 295–313.

Lloyd, G. T. (2012). A refined modelling approach to assess the influence of sampling on palaeobiodiveristy curves: New support for declining Cretaceous dinosaur richness. *Biology Letters, 8*, 123–126.

Lloyd, G. T., Davis, K. E., Pisani, D., Tarver, J. E., Ruta, M., Sakamoto, M., et al. (2008). Dinosaurs and the Cretaceous Terrestrial Revolution. *Proceedings of the Royal Society B, 275*, 2483–2490.

Lucas, S. G. (2009). Timing and magnitude of tetrapod extinctions across the Permo-Triassic boundary. *Journal of Asian Earth Sciences, 36*, 491–502.

MacLeod, K. G., Smith, R. M. H., Koch, P. L., & Ward, P. D. (2000). Timing of mammal-like reptile extinctions across the Permian-Triassic boundary in South Africa. *Geology, 28*, 227–230.

Marx, F. (2009). Marine mammals through time: When less is more in studying palaeodiversity. *Proceedings of the Royal Society B, 276*, 887–892.

Marx, F., & Uhen, M. D. (2010). Climate, critters, and cetaceans: Cenozoic drivers of the evolution of modern whales. *Science, 327*, 993–996.

McGowan, A. J., & Smith, A. B. (2008). Are global Phanerozoic marine diversity curves truly global? A study of the relationship between regional rock records and global Phanerozoic marine diversity. *Paleobiology, 34*, 80–103.

Newell, A. J., Sennikov, A. G., Benton, M. J., Molostovskaya, I. I., Golubev, V. K., Minikh, A. V., et al. (2010). Disruption of playa-lacustrine depositional systems at the Permo-Triassic boundary: Evidence from Vyazniki and Gorokhovets on the Russian Platform. *Journal of the Geological Society, London, 167*, 695–716.

Nicolas, M., & Rubidge, B. S. (2009). Assessing content and bias in South African Permo-Triassic Karoo tetrapod fossil collections. *Palaeontologia Africana, 44*, 13–20.

Nicolas, M., & Rubidge, B. S. (2010). Changes in Permo-Triassic terrestrial tetrapod ecological representation in the Beaufort Group (Karoo Supergroup) of South Africa. *Lethaia, 43*, 45–59.

Pace, D. W., Gastaldo, R. A., & Neveling, J. (2009). Early Triassic aggradational and degradational landscapes of the Karoo Basin and evidence for climate oscillation following the P-Tr event. *Journal of Sedimentary Research, 79*, 316–331.

Ray, S. (2005). *Lystrosaurus* (Therapsida, Dicynodontia) from India: taxonomy, relative growth, and cranial dimorphism. *Journal of Systematic Palaeontology, 3*, 203–221

Renne, P. R., Zichao, Z., Richards, M. A., Black, M. T., & Basu, A. R. (1995). Synchrony and causal relations between Permian-Triassic boundary crises and Siberian flood volcanism. *Science, 269*, 1413–1416.

Retallack, G. J., Smith, R. M. H., & Ward, P. D. (2003). Vertebrate extinction across Permian-Triassic boundary in Karoo Basin, South Africa. *Geological Society of America Bulletin, 115*, 1133–1152.

Retallack, G. J., Metzgar, C. A., Greaver, T., Jahren, A. H., Smith, R. M. H., & Sheldon, N. D. (2006). Middle-Late Permian extinction on land. *Geological Society of America Bulletin, 118*, 1398–1411.

Roopnarine, P. D., & Angielczyk, K. D. (2012). The evolutionary palaeoecology of species and the tragedy of the commons. *Biology Letters, 8*, 147–150.

Roopnarine, P. D., Angielczyk, K. D., Wang, S. C., & Hertog, R. (2007). Trophic network models explain instability of Early Triassic terrestrial communities. *Proceedings of the Royal Society B-Biological Sciences, 274*, 2077–2086.

Rubidge, B. S. (Ed.) (1995). Biostratigraphy of the Beaufort Group (Karoo Supergroup). *South African Committee for Stratigraphy Biostratigraphic Series, 1*, 1–46.

Rubidge, B. S. (2005). Reuniting lost continents—fossil reptiles from the ancient Karoo and their wanderlust. *South African Journal of Geology, 108*, 135–172.

Ruta, M., & Benton, M. J. (2008). Calibrated diversity, tree topology and the mother of mass extinctions: The lesson of temnospondyls. *Palaeontology, 51*, 1261–1288.

Ruta, M., Cisneros, J. C., Liebrecht, T., Tsuji, L. A., & Müller, J. (2011). Amniotes through major biological crises: Faunal turnover among parareptiles and the end-Permian mass extinction. *Palaeontology, 54*, 1117–1137.

Sahney, S., & Benton, M. J. (2008). Recovery from the most profound mass extinction of all time. *Proceedings of the Royal Society B-Biological Sciences, 275*, 759–765.

Smith, A. B., & McGowan, A. J. (2007). The shape of the Phanerozoic marine palaeodiversity curve: How much can be predicted from the sedimentary rock record of western Europe? *Palaeontology, 50*, 765–774.

Smith, R. M. H. (1995). Changing fluvial environments across the Permian-Triassic boundary in the Karoo Basin, South Africa and possible causes of tetrapod extinctions. *Palaeogeography, Palaeoclimatology, Palaeoecology, 117*, 81–104.

Smith, R. M. H., & Botha, J. (2005). The recovery of terrestrial vertebrate diversity in the South African Karoo Basin after the end-Permian extinction. *Comptes Rendus Palevol, 4*, 555–568.

Smith, R. M. H., & Ward, P. D. (2001). Pattern of vertebrate extinction across an event bed at the Permian-Triassic boundary in the Karoo Basin of South Africa. *Geology, 29*, 1147–1150.

Smith, R., Rubidge, B., & van der Walt, M. (2012). Therapsid biodiversity patterns and paleoenvironments of the Karoo Basin, South Africa. In A. Chinsamy-Turan (Ed.), *Forerunners of mammals: Radiation histology biology* (pp. 31–62). Bloomington: Indiana University Press.

Taylor, G. K., Tucker, C., Twitchett, R. J., Kearsey, T., Benton, M. J., Newell, A. J., et al. (2009). Magnetostratigraphy of Permian/Triassic boundary sequences in the Cis-Urals, Russia: No evidence for a major temporal hiatus. *Earth and Planetary Science Letters, 281*, 36–47.

Tverdokhlebov, V. P., Tverdokhlebova, G. I., Surkov, M. V., & Benton, M. J. (2002). Tetrapod localities from the Triassic of the SE of European Russia. *Earth-Science Reviews, 60*, 1–66.

Tverdokhlebov, V. P., Tverdokhlebova, G. I., Minikh, A. V., Surkov, M. V., & Benton, M. J. (2005). Upper Permian vertebrates and their sedimentological context in the South Urals, Russia. *Earth-Science Reviews, 69*, 27–77.

Ward, P. D., Montgomery, D. R., & Smith, R. (2000). Altered river morphology in South Africa related to the Permian-Triassic extinction. *Science, 289*, 1740–1743.

Ward, P. D., Botha, J., Buick, R., De Kock, M. O., Erwin, D. H., Garrison, G. H., et al. (2005). Abrupt and gradual extinction among Late Permian land vertebrates in the Karoo Basin, South Africa. *Science, 307*, 709–714.

Chapter 17
Vertebrate Paleontology of Nooitgedacht 68: A *Lystrosaurus maccaigi*-rich Permo-Triassic Boundary Locality in South Africa

Jennifer Botha-Brink, Adam K. Huttenlocker, and Sean P. Modesto

Abstract The farm Nooitgedacht 68 in the Bethulie District of the South African Karoo Basin contains strata that record a complete Permo-Triassic boundary sequence providing important new data regarding the end-Permian extinction event in South Africa. Exploratory collecting has yielded at least 14 vertebrate species, making this locality the second richest Permo-Triassic boundary site in South Africa. Furthermore, fossils include 50 specimens of the otherwise rare Late Permian dicynodont *Lystrosaurus maccaigi*. As a result, Nooitgedacht 68 is the richest *L. maccaigi* site known. The excellent preservation, high concentration of *L. maccaigi*, presence of relatively rare dicynodonts such as *Dicynodontoides recurvidens* and *Dinanomodon gilli*, and the large size of many of these Permian individuals makes Nooitgedacht 68 a particularly interesting site for studying the dynamics of the end-Permian extinction in South Africa.

Keywords Late Permian • Early Triassic • End-Permian extinction • Karoo Basin • Dicynodontia • Therocephalia • Gorgonopsia

J. Botha-Brink (✉)
Department of Karoo Palaeontology, National Museum, Bloemfontein, South Africa

and

Department of Zoology and Entomology, University of the Free State, Bloemfontein, South Africa
e-mail: jbotha@nasmus.co.za

A. K. Huttenlocker
Department of Biology, University of Washington, Seattle, WA 98195, USA
e-mail: huttenla@u.washington.edu

S. P. Modesto
Department of Biology, Cape Breton University, Sydney, NS, Canada
e-mail: seanmodesto@yahoo.ca

Introduction

The end-Permian extinction, which occurred 252.6 Ma ago (Mundil et al. 2004), is widely regarded as the most catastrophic mass extinction in Earth's history (Erwin 1994). Much research has focused on the cause(s) of the extinction (e.g., Renne et al. 1995; Wignall and Twitchett 1996; Knoll et al. 1996; Isozaki 1997; Krull et al. 2000; Hotinski et al. 2001; Becker et al. 2001, 2004; Sephton et al. 2005), the paleoecology and paleobiology of the flora and fauna prior to and during the event (e.g., Ward et al. 2000; Smith and Ward 2001; Wang et al. 2002; Gastaldo et al. 2005) and the consequent recovery period (Benton et al. 2004; Smith and Botha 2005; Botha and Smith 2006; Sahney and Benton 2008). Although the most complete Permo-Triassic boundary (PTB) sequences are located in marine deposits (e.g., Clark et al. 1986; Gruszczynski et al. 1989; Jin et al. 2000; Miller and Foote 2003; Racki 2003; Wignall and Newton 2003; Payne et al. 2004; Lehrmann et al. 2006; Fraiser and Bottjer 2007; Chen et al. 2007), recent intense research at nonmarine Permo-Triassic localities has revealed well-preserved, complete boundary sequences in the South Urals Basin in Russia (Benton 2003; Benton et al. 2004; Sahney and Benton 2008), Sydney Basin in Australia (Morante et al. 1994; Morante 1996; Retallack et al. 1998; Retallack 1999), central Transantarctic Mountains in Antarctica (Retallack and Krull 1999; Retallack et al. 2007), Raniganj Basin in eastern India (Sarkar et al. 2003), and especially the Karoo Basin of South Africa (MacLeod et al. 2000; Ward et al. 2000; Smith and Ward 2001; Retallack et al. 2003; Smith and Botha 2005; Botha and Smith 2006).

The southern portion of the Beaufort Group in the South African Karoo Basin (Fig. 17.1a) preserves the most complete and best-preserved record of the end-Permian extinction on land (Ward et al. 2000, 2005; Retallack et al. 2003). Several localities in the Beaufort Group preserve complete terrestrial Permo-Triassic sequences, but the most widely known are the section on the farm Bethal (a.k.a. "Bethel

Christian F. Kammerer, Kenneth D. Angielczyk and Jörg Fröbisch (eds.), *Early Evolutionary History of the Synapsida*, Vertebrate Paleobiology and Paleoanthropology, DOI: 10.1007/978-94-007-6841-3_17,

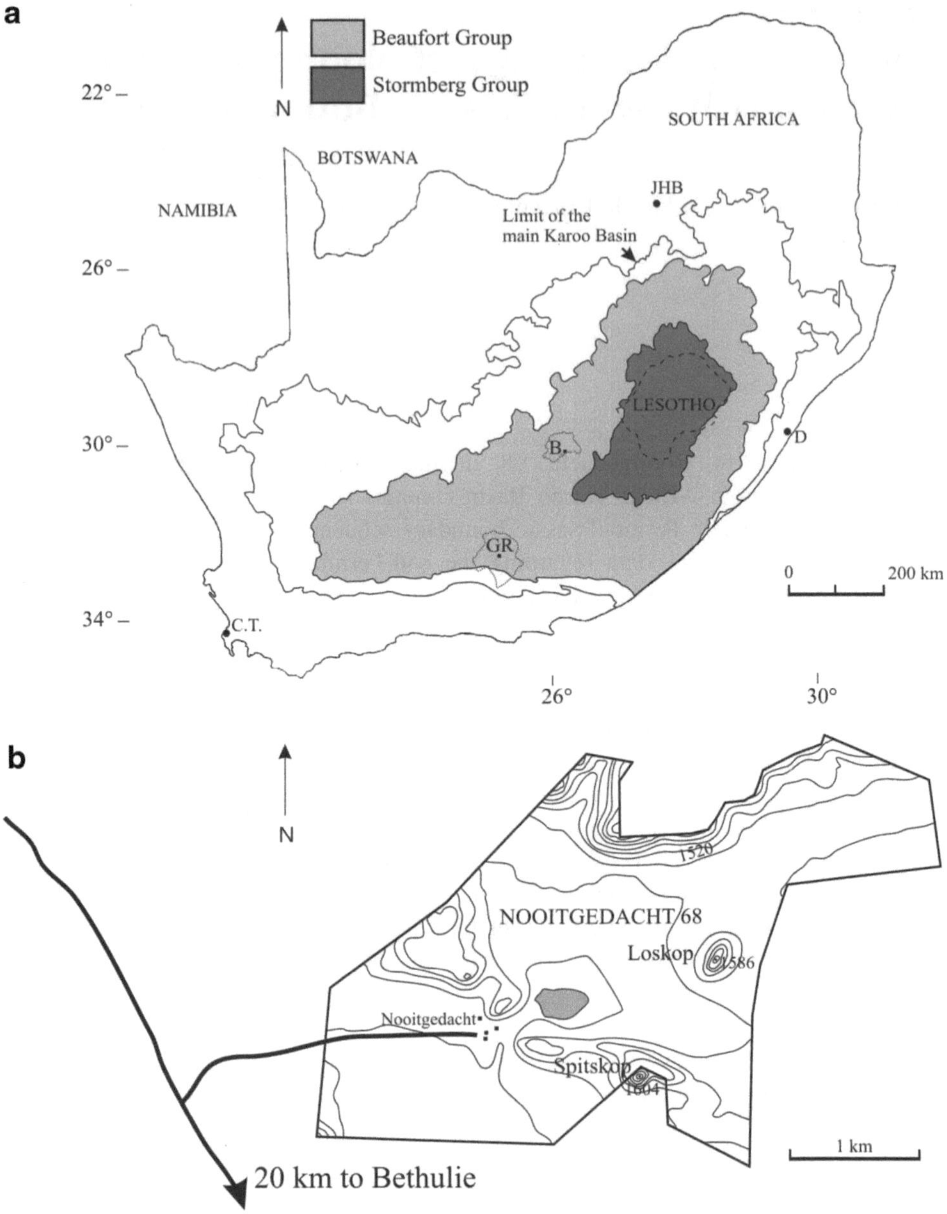

Fig. 17.1 **a** Map of South Africa showing the location of the town Bethulie (*square*) in the Bethulie District (*B*) and the town Graaff-Reinet (*square*) in the Graaff-Reinet District (*GR*). The farm Bethal (Bethel 763) is located approximately 20 km to the south–east of Bethulie, and the Lootsberg and Old Wapadsberg passes are located 58 km to the north–east of Graaff-Reinet. Nooitgedacht 68 is located 20 km north of the town Bethulie in the Bethulie District. **b** Detailed map showing the positions of Loskop and Spitskop on Nooitgedacht 68, Bethulie District. Numbers are elevation in meters. *C.T.* Cape Town, *D* Durban, *JHB* Johannesburg

763" on map 3026AD Tampasfontein) in the Bethulie District (Smith 1995) and sections on the Lootsberg and Old Wapadsberg passes in the Graaff-Reinet District (Ward et al. 2000). These sections are well known not only for their complete preservation of the PTB, but also for their highly fossiliferous rocks, which have facilitated detailed regional studies on the paleoecology and paleobiology of Permo-Triassic tetrapods associated with the extinction on land (Smith and Ward 2001; Botha and Smith 2006, 2007).

The farm Nooitgedacht 68 in the Bethulie District was first discovered as a potentially rich fossil locality by the late James Kitching in the 1970s (Kitching 1977). Johann Eksteen, a former assistant at the National Museum, Bloemfontein, also collected several specimens from this locality during the 1970s. Kitching noted that both Permian and Triassic strata were preserved on this farm, and he collected typical Permian vertebrates, including large dicynodonts he identified as *Daptocephalus leoniceps* and *Lystrosaurus platyceps*. Subsequently, the genus *Daptocephalus* was proposed to be a synonym of *Dicynodon* (Cluver and Hotton 1981), but recent work has confirmed that *Daptocephalus* is distinct from *Dicynodon* (Kammerer et al. 2011). *Lystrosaurus platyceps* is now considered a synonym of *Lystrosaurus curvatus* (Grine et al. 2006). He also collected typical Triassic taxa comprising the dicynodonts *Lystrosaurus murrayi* and *L. declivis*, and the archosauromorph *Proterosuchus fergusi* (formerly *Proterosuchus vanhoepeni*: Welman 1998). Kitching also recovered the therocephalian *Moschorhinus kitchingi* (a possible subjective junior synonym of *Tigrisuchus simus* according to Mendrez 1974; Kammerer 2008) from *Dicynodon* Assemblage Zone strata (Kitching's "*Daptocephalus* Zone"), and *Lystrosaurus curvatus* as well as apparent "*Ictidosuchops*-like" therocephalians from *Lystrosaurus* Assemblage Zone strata (Kitching's "*Lystrosaurus* Zone"). One of these *Ictidosuchops*-like therocephalians was re-identified by Fourie and Rubidge (2007) as *Regisaurus jacobi*, a therocephalian restricted to the lowermost Triassic *Lystrosaurus* Assemblage Zone (Rubidge 1995; Nicolas and Rubidge 2010).

Moschorhinus kitchingi and *Lystrosaurus curvatus* are taxa that have been recovered from both sides of the Permo-Triassic boundary and are thus termed PTB markers (Botha and Smith 2006, 2007). The presence of these taxa, as well as that of *Proterosuchus*, which is the first taxon to appear above the PTB (Smith and Botha 2005), prompted the authors to revisit Nooitgedacht 68 in order to determine if a complete PTB sequence was preserved at this locality. New collecting efforts by the authors during the 2008, 2010 and 2011 field seasons allowed for the measuring of precise stratigraphic logs and the acquisition of 73 positively identified new vertebrate fossil occurrences of which 38 were in situ, across a complete PTB sequence. The discovery of any new complete terrestrial PTB sequences is important as each locality provides fresh insight into the dynamics of the end-Permian extinction in the South African Karoo Basin.

Institutional Abbreviations: BP, Bernard Price Institute for Palaeontological Research, University of the Witwatersrand, Johannesburg, South Africa; NMQR, National Museum, Bloemfontein, South Africa.

Field Techniques

The farm Nooitgedacht 68 is situated approximately 20 km to the north of the small town of Bethulie in the Bethulie District, Free State Province, South Africa. The farm contains two fossiliferous localities; an isolated butte named Loskop and a large pinnacle named Spitskop, ca. 800 m to the south-west of Loskop (Fig. 17.1b). Approximately 93 m of section on Loskop and 42 m on Spitskop were measured using standard field methods, including Jacob's staff and sight level as well as a Munsell geological rock-color chart (2009 revision). A stratigraphic log was recorded on Loskop to an accuracy of 10 cm and measured 54 m in total, to the base of the Katberg Formation (Fig. 17.2a, b). The uppermost 39 m of Loskop comprises coarse sheet sandstone units exhibiting massive and horizontal bedding typical of the Katberg Formation and was not studied in detail, as it was not the focus of the current study. The stratigraphic positions of all in situ vertebrate fossils were mapped onto the log in order to assess the faunal turnover across the PTB interval at this site. A detailed log was also recorded in the same manner on Spitskop and measured 42 m in total, also to the base of the Katberg Formation. Spitskop revealed similar geological features to those observed on Loskop. In general, both hills reveal an overall lithology that transitions upwards from olive-gray siltstones and mudrock interbedded with well-sorted, fine-grained sandstone bodies into progressively more dusky red, heterolithic mudrock alternating with siltstones and fine-grained sandstones. Higher in the succession the lithology becomes more arenaceous, with sandstones becoming coarser-grained and conglomeratic in places.

Lithostratigraphy of the Permo-Triassic Boundary Section and In Situ Vertebrate Records

Ward et al. (2000), MacLeod et al. (2000), Smith and Ward (2001), Retallack et al. (2003), Smith and Botha (2005) and Botha and Smith (2006) define the PTB as an interval that marks the last occurrence of the Late Permian dicynodonts *Dicynodon* and *Lystrosaurus maccaigi*, and associated facies that comprise "rhythmically-bedded" laminated dark reddish-brown (2.5YR 3/4) and olive-gray (5Y 5/2) siltstone-mudstone couplets (but see Gastaldo et al. 2009 and Pace et al. 2009 for differing interpretations). The laminite beds coincide with a negative excursion in $\delta^{13}C$ at the Bethal (Bethulie District) and Lootsberg (Graaff-Reinet District) sections (Macleod et al. 2000; Ward et al. 2005). Smith and Ward (2001) initially placed the PTB at the base of the laminite beds as no fossils had been found within the beds themselves. Further collecting, however, revealed the presence of in situ Permian taxa such as *Dicynodon* and *Lystrosaurus maccaigi* and the absence of Triassic taxa. Consequently, the PTB has been placed at the top of the laminite beds in more recent studies (e.g., Retallack et al. 2003; Smith and Botha 2005; Botha and Smith 2006). Following the lithologic descriptions and stratigraphic interpretations of Retallack et al. (2003), Smith and Botha (2005), and Botha and Smith (2006), we place the Permo-Triassic boundary at Nooitgedacht 68 at the contact between a thick (8 m at Loskop, 4 m at Spitskop) rubified, thinly laminated heterolithic mudrock section and a 50 cm thick siltstone body (Fig. 17.2), below which the last occurrence of the locally abundant *L. maccaigi* is documented (Botha and Smith 2007).

Lateral continuity of the strata and continuous sedimentation at Loskop and Spitskop suggests that the slopes of Loskop and Spitskop are equivalent in age. The similar faunal assemblages further support this interpretation. To

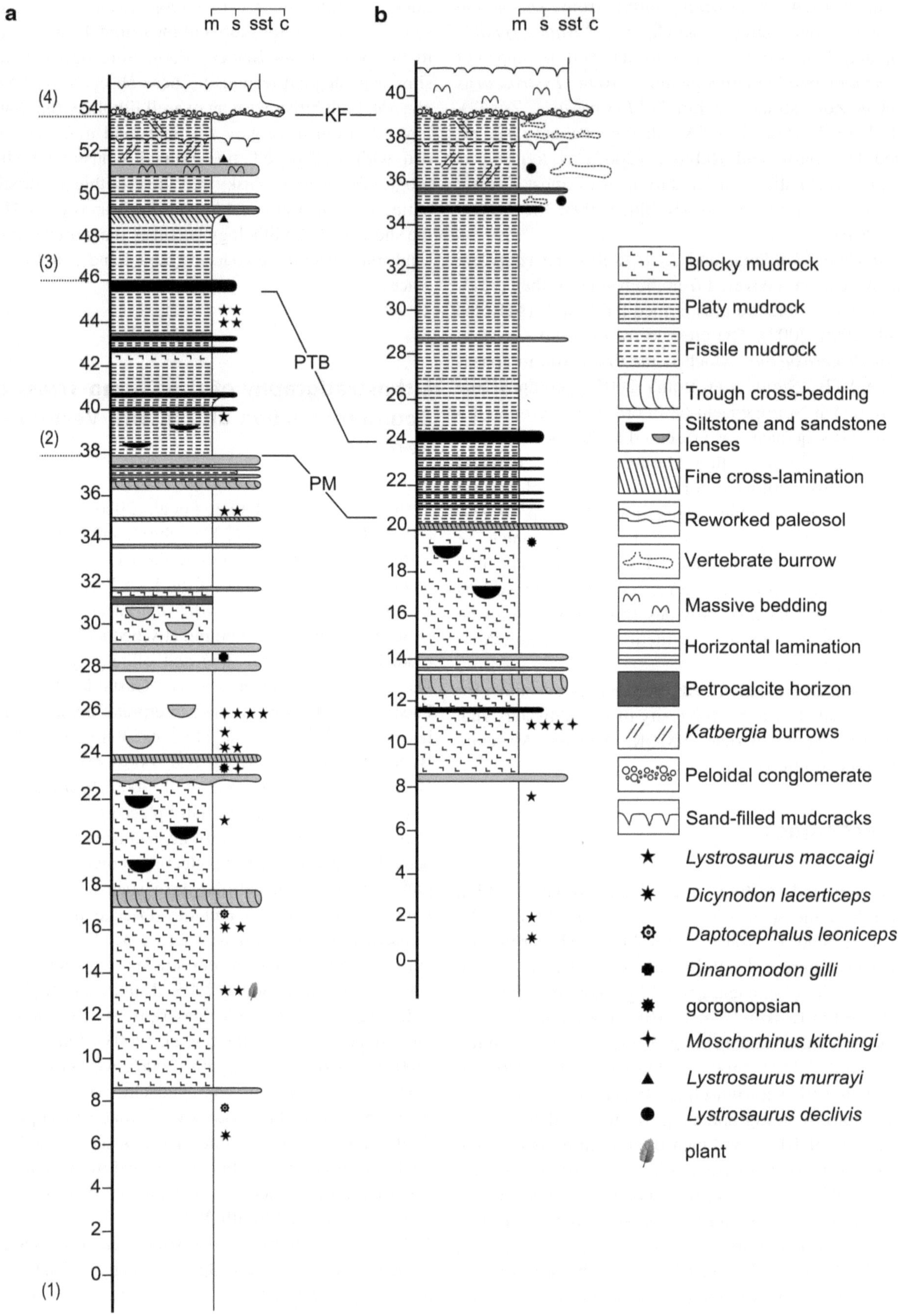
a
m s sst c
(4)
(3)
(2)
(1)
54
52
50
48
46
44
42
40
38
36
34
32
30
28
26
24
22
20
18
16
14
12
10
8
6
4
2
0
KF
PTB
PM
b
m s sst c
40
38
36
34
32
30
28
26
24
22
20
18
16
14
12
10
8
6
4
2
0
Blocky mudrock
Platy mudrock
Fissile mudrock
Trough cross-bedding
Siltstone and sandstone lenses
Fine cross-lamination
Reworked paleosol
Vertebrate burrow
Massive bedding
Horizontal lamination
Petrocalcite horizon
Katbergia burrows
Peloidal conglomerate
Sand-filled mudcracks
Lystrosaurus maccaigi
Dicynodon lacerticeps
Daptocephalus leoniceps
Dinanomodon gilli
gorgonopsian
Moschorhinus kitchingi
Lystrosaurus murrayi
Lystrosaurus declivis
plant

◀ **Fig. 17.2** Sedimentological log of Upper Permian and Lower Triassic strata on **a**, Loskop and **b**, Spitskop, Nooitgedacht 68, Bethulie District, and stratigraphic positions of in situ specimens. Vertical scale *tick marks* are in meters. The measured section begins at 0 m. *Numbers* in parentheses refer to sections as in text. *c* conglomerate, *KF* Katberg Formation, *m* mudstone, *s* siltstone, *sst* sandstone, *PM* Palingkloof Member of the Balfour Formation, *PTB* Permo-Triassic Boundary

avoid repetition, and as the strata on Loskop and Spitskop revealed similar strata with complete PTB sections (as identified by the description above), the following geological description is based on Loskop as it revealed better exposures of the PTB.

−46 to −7.7 m Interval Below the PTB (Upper Balfour Formation to Base of Palingkloof Member)

This facies comprises alternating blocky grayish brown (5YR 3/2) and very dusky red (10R 2/2) mudrock interspersed with fine-grained, medium to well-sorted sandstone bodies, one of which contains large-scale trough cross-bedding 8.3 m into the section. Rhizocretions are present in the lower levels of this unit. Further up, a fine-grained, medium-sorted lenticular sandstone body with a basal scour surface containing calcareous siltstone clasts interpreted as reworked pedogenic material, is present on the south-western slope of Loskop, but this feature is absent from the rest of the slope. At this horizon very dusky red (10R 2/2) mudrock alternates with minor olive-gray (5Y 3/2) sandstone lenses, and dark brown weathering calcareous nodules are present. The facies grades upwards into blocky dusky red (10R 2/2) and olive-gray (5Y 4/1) siltstone representing a petrocalcic horizon. Very fine cross laminae were observed in the fine-grained olive-gray sandstone in the upper levels (approximately 35 m into the section), above which a 25 cm thick, moderately well-sorted (fine- to medium-grained) unit with large scale trough cross-bedding was observed. The uppermost levels of this unit at approximately 36 m into the section (10 m below the PTB) comprise a fissile, light olive-gray siltstone (5Y 5/2) periodically interrupted by lenses of fine-grained, light olive-gray (5Y 5/2) sandstone.

Vertebrate fossils are abundant, consisting of both isolated loose fragments and well-preserved in situ skulls and postcrania of the dicynodonts *Dicynodon lacerticeps*, *Daptocephalus leoniceps*, *Dinanomodon gilli*, *Lystrosaurus maccaigi* and the therocephalian *Moschorhinus kitchingi*. An articulated skull and anterior skeleton of the therocephalian *Ictidosuchoides longiceps* (NMQR 3686) was recovered loose from the lower slope of Loskop approximately 26 m below the PTB. The lithified impression of a fallen tree trunk (Fig. 17.3b) was preserved in a well-sorted, very fine-grained sandstone in the lowermost level of this

Fig. 17.3 **a** Fossilized tree impression. **b** Close up of fossilized tree impression. **c** Fossilized burrow cast found as float on the lowermost slopes of Loskop, Nooitgedacht 68. Geological hammer = 32 cm

unit, and was associated with an in situ *Lystrosaurus maccaigi* skull, NMQR 3706 (approximately 33 m below the PTB). A fossilized burrow infill was also observed on the lowermost slopes (Fig. 17.3c). The stratigraphically lowest vertebrate fossils from Loskop were an in situ *Lystrosaurus maccaigi* skull (NMQR 3706) from 33 m below the PTB, and the lowermost in situ *Dicynodon lacerticeps* specimen (NMQR 3943), recorded at 40 m below the PTB.

−7.7 to 0 m (Palingkloof Member, Balfour Formation to PTB)

This interval marks the beginning of a succession of thinly laminated (heterolithic), dusky red (10R 2/2) mudrock containing thin olive-gray (5Y 5/2) siltstone lenses overlying a 25 cm thick sandstone bed. Several in situ *Lystrosaurus maccaigi* skulls were recorded and collected approximately 6.8 m into this heterolithic succession, within a dusky red laminated, fissile mudrock overlain by a 50 cm thick olive-gray, planar-bedded siltstone, the base of which is interpreted as the position of the PTB (Fig. 17.4). The uppermost occurrence of *Lystrosaurus maccaigi* at this locality occurs within this dusky red mudrock, approximately 1.5 m below the base of the thick siltstone bed.

0–8 m Above PTB (Lower Palingkloof Member, Balfour Formation)

Dusky red and olive-gray laminated mudrock and siltstone beds continue upward to a 50 cm thick platy siltstone body. Fine cross-laminae are seen in the uppermost section of this unit, which also contains in situ bone fragments of *Lystrosaurus*. Many of these fragments consist of jumbled-up, disarticulated skeletons. They cannot be identified to species level with any certainty, but most likely represent either *L. murrayi* or *L. declivis* based on the strong ventral elongation of the snout, a short basicranial axis, widely exposed parietals on the skull roof and the absence of teeth apart from maxillary tusks. The material lacks the gentle curving snout seen in *L. curvatus* and overly large orbits that characterise both *L. curvatus* and *L. maccaigi* (Grine et al. 2006). Furthermore the angle of the orbits is not consistent with that of *L. maccaigi*, but appears to be more similar to that of *L. murrayi* or *L. declivis* (Botha and Smith 2007).

This interval is overlain by a very fine-grained, horizontally laminated sandstone. The fissile mudrock extends further upwards and encompasses a thick massively-bedded sandstone body, above which in situ *Katbergia* (Gastaldo and Rolerson 2008) burrows (probable decapod crustacean, referred to as callianassid or *Macanopsis* burrows in previous publications; Smith and Ward 2001; Retallack et al. 2003), ripples, and infilled mudcracks are present.

Fig. 17.4 **a** The Permo-Triassic boundary (PTB) sequence exposed on the farm Nooitgedacht 68, Bethulie District of South Africa. **b** The PTB is positioned at the top of a basin-wide section containing thick rubified, thinly laminated dusky red mudrock. *Fm* Formation, *PTB* Permo-Triassic boundary

Approximately 4.5 m into this interval an in situ burrow cast (field number JB021044) was observed on Spitskop (Fig. 17.5). The burrow is preserved as a 164 cm long, siltstone-filled branched tunnel measuring up to 8 cm in diameter. Most of the main tunnel is preserved only in outline, but the smaller branch preserves some three-

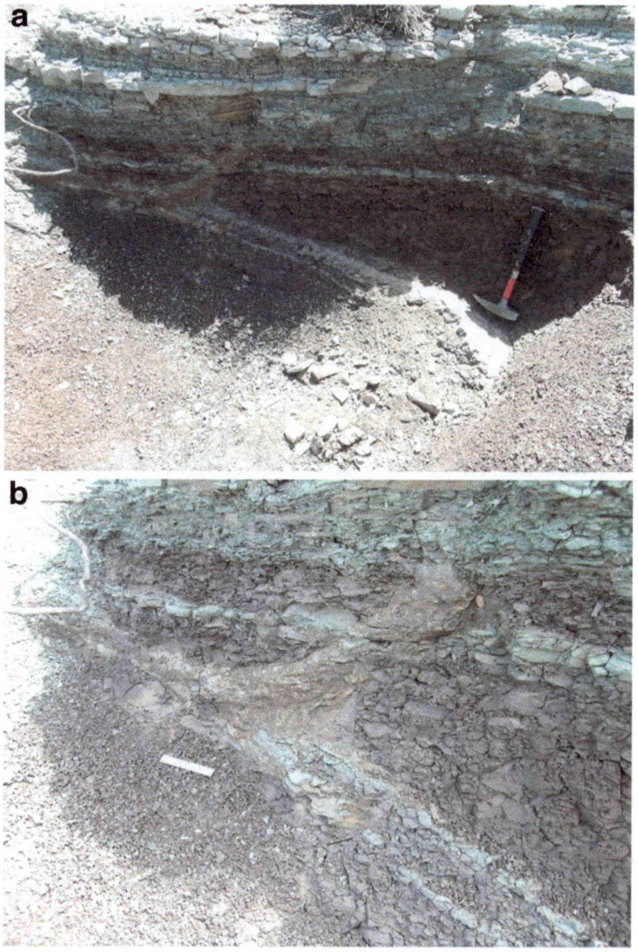

Fig. 17.5 **a** In situ fossilized vertebrate burrow cast on Spitskop, Nooitgedacht 68. Geological hammer 32 cm, **b** Close up of fossilized burrow cast. Scale bar 7 cm

dimensionality suggesting that the cross-sectional shape of the tunnel is oval. The architectural morphology comprises a downward-oriented, straight, sub-horizontal burrow formed at a low angle of 30° and penetrates a stratigraphic distance of 61 cm. A branch extends a distance of 30 cm from the main tunnel at about 30° from the horizontal plane. The surface morphology comprises thin elongate ridges, which are interpreted as scratch marks. They measure up to 3 mm in width and are oriented longitudinally or tangentially to the long axis of the tunnel. A longitudinal medial groove running along the base of the tunnel, noted in other burrow casts of similar size (Miller et al. 2001; Damiani et al. 2003), was not observed. The burrow signatures, which include a low-angle (30°), diagonally-oriented tunnel with prominent scratch marks approximately 3 mm in diameter, suggest that the tracemaker was most likely a small vertebrate (Hasiotis et al. 1999). Another four in situ burrows of similar size and morphology were observed on Spitskop at approximately 7 m into the section, just below the first sandstone of the Katberg Formation and were also most likely created by small vertebrates. A large (approximately 500 mm diameter) burrow, similar to that described by Modesto and Botha-Brink (2010), which contained juvenile *Lystrosaurus* bones, was also observed approximately 2 m below the first sandstone of the Katberg Formation.

8–40 m Above PTB (Lower Katberg Formation)

Approximately 8 m above the PTB the fissile mudrock unit is replaced by a sheet sandstone-dominated succession, which continues to the top of Loskop. The base of this sandstone succession is characterized by a pebble lag. The horizontally bedded, medium-grained, grayish olive (10Y 4/2) sandstone bodies are interspersed with olive-gray and dusky red mudrocks, which contain loose and in situ *Lystrosaurus murrayi* and *L. declivis* fossils. Small broken fragments of bone within the sandstone bodies and loose *Lystrosaurus* sp. preserved in nodules are also observed. Ripples and infilled mudcracks are present in the lower levels and a steady increase in occurrence of conglomerates is observed throughout this unit.

Systematic Paleontology

This section provides a brief description of vertebrate fossils collected from the study area. Many of these were collected by the authors during the course of fieldwork for the present study, but others were collected during previous expeditions by Kitching and Eksteen. Vertebrate collections from Nooitgedacht 68 are housed primarily in the collections of the Bernard Price Institute, Johannesburg (BP) and National Museum, Bloemfontein (NMQR).

Synapsida Osborn, 1903
Therapsida Broom, 1905
Dicynodontia Owen, 1860a
Dicynodontoides recurvidens (Owen, 1876)

Material: BP/1/4027, skull and postcranial elements collected by J. W. Kitching during the 1970s from Nooitgedacht 68, uppermost Permian Balfour Formation, *Dicynodon* Assemblage Zone.

Remarks: This material was originally identified as Dicynodontia indet. by Kitching (1977), but has recently been assigned to the kingoriid *Dicynodontoides* (previously *Kingoria*) *recurvidens* by Angielczyk et al. (2009). The skull contains several features characteristic of *D. recurvidens* including a large lateral dentary shelf and closed mandibular fenestra on the lower jaw, a postcaniniform

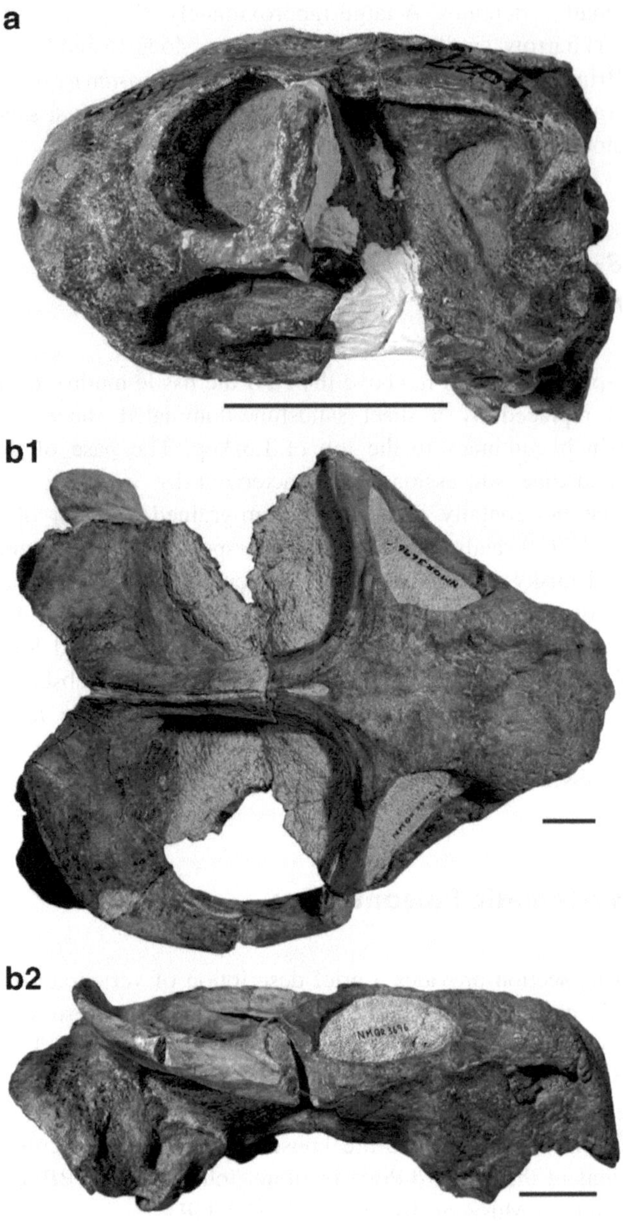

Fig. 17.6 **a** BP/1/4027, *Dicynodontoides recurvidens*, Permian. **b** NMQR 3696, *Dinanomodon gilli*, Permian in dorsal (**b1**) and right lateral (**b2**) view. Scale bars 5 cm

keel, an embayment anterior to the caniniform process, a parietal foramen located near the middle of the intertemporal bar and raised above the level of the surrounding bones, as well as the absence of postfrontals and postcanines (Angielczyk et al. 2009) (Fig. 17.6a).

Dicynodon lacerticeps Owen, 1845

Material: BP/1/4025, 4026, 4028, cranial and postcranial elements collected by J. W. Kitching in the 1970s; NMQR 1644, 3644, 3664, 3701, 3943, several relatively complete skulls, collected by J.P. Eksteen in the 1970s and more recently by the authors from the lowermost slopes of Loskop and Spitskop, Nooitgedacht 68, uppermost Permian Balfour Formation, *Dicynodon* Assemblage Zone.

Remarks: Kitching (1977) collected *Daptocephalus leoniceps* from Nooitgedacht 68. *Daptocephalus leoniceps* was synonymized with *Dicynodon lacerticeps* by Cluver and Hotton (1981), but has recently been resurrected by Kammerer et al. (2011). However, the material listed above was identified as *Dicynodon lacerticeps* on the basis of an anteriorly directed caniniform process, narrow intertemporal region, the absence of nasal bosses and the presence of tusks (Brink 1986; Angielczyk and Kurkin 2003; Kammerer et al. 2011).

Daptocephalus leoniceps (Owen, 1876)

Material: BP/1/3985, skull collected by J. W. Kitching in the 1970s; NMQR 3645, 3942, skulls collected by the authors from the lowermost slopes of Loskop and Spitskop, Nooitgedacht 68, uppermost Permian Balfour Formation, *Dicynodon* Assemblage Zone.

Remarks: The material was identified as *Daptocephalus leoniceps* on the basis of a steeply angled sloping snout, an anteroposteriorly short premaxilla, a ventrally directed caniniform process, and a particularly long, narrow intertemporal bar with an extremely narrow median exposure of the parietals (Kammerer et al. 2011).

Dinanomodon gilli (Broom, 1932)

Material: BP/1/5287; NMQR 3696, almost complete skull, with only left zygomatic arch missing, collected in situ by the authors from the lowermost slopes of Loskop, Nooitgedacht 68, uppermost Permian Balfour Formation, *Dicynodon* Assemblage Zone.

Remarks: *Dinanomodon gilli* (formerly *Dinanomodon rubidgei*) is similar in appearance to *Dicynodon lacerticeps* and *Daptocephalus leoniceps*. However, NMQR 3696 is identified as *D. gilli* on the basis of a ridge on the premaxilla, an extended dorsal process of the premaxilla that nearly contacts the frontals, a convex dorsal edge on the external naris, and a long intertemporal region. The premaxilla also tapers more sharply than it does in *Dicynodon* (Brink 1986; Kammerer et al. 2011) (Fig. 17.6b).

Lystrosaurus maccaigi (Seeley, 1898)

Material: BP/1/3972, 4052; NMQR 1637, 1639, 1640a, 1641, 1643, 1648, 1650, 1653, 3208, 3641, 3642, 3646, 3647, 3648, 3658, 3663, 3684a, 3687, 3689, 3690, 3693, 3694, 3695, 3699, 3700, 3702, 3703, 3705, 3706, 3708, 3711, 3712, 3713, 3919, 3922a, 3934, 3935, 3936, 3938, 3940. Skulls and associated skeletons in varying stages of completeness and at various ontogenetic ages, collected by J. W. Kitching and J. P. Eksteen during the 1970s and more recently by the authors from the lowermost slopes of Loskop and Spitskop, Nooitgedacht 68, uppermost Permian Balfour Formation, *Dicynodon* Assemblage Zone.

Remarks: The material is assigned to *Lystrosaurus* on the basis of a strong ventral elongation of the snout, a short basicranial axis, widely exposed parietals on the skull roof and the absence of teeth apart from maxillary tusks (Cluver 1971; Grine et al. 2006; Angielczyk 2007; Fröbisch 2007; Fröbisch and Reisz 2008; Kammerer and Angielczyk 2009). It is assigned to *L. maccaigi* on the basis of relatively large orbits (compared to *L. murrayi* and *L. declivis*) and markedly prominent pre- and postorbital bosses in larger individuals (Brink 1951; Cluver 1971). The facial surface in all individuals, regardless of size, slopes forward and down from its junction with the frontal plane, similar to *L. declivis*, but differs from the latter species in that the premaxillary plane lies at a sharper angle and the orbits of *L. maccaigi* are more upward and forward facing compared to *L. declivis* (Grine et al. 2006). *L. maccaigi* is a notably more robust animal compared to the other *Lystrosaurus* species, regardless of body size (Fig. 17.7a, b).

Lystrosaurus curvatus (Owen, 1876)

Material: BP/1/3976, skull collected by J. W. Kitching during the 1970s from Nooitgedacht 68, uppermost Permian Balfour Formation, *Dicynodon* Assemblage Zone.

Remarks: Kitching (1977) listed *Lystrosaurus platyceps* as one of the *Lystrosaurus* species collected from Nooitgedacht 68. However, *Lystrosaurus platyceps* is now considered a junior synonym of *L. curvatus* (Grine et al. 2006). The material is assigned to *Lystrosaurus* for the same reasons given above and to the species *L. curvatus* on the basis of particularly large orbits, a gently curved snout and the absence or weak development of a frontonasal ridge, longitudinal, premaxillary ridge and postorbital and prefrontal bosses (Brink 1951; Cluver 1971; Grine et al. 2006) (Fig. 17.7c).

Lystrosaurus declivis (Owen, 1860b)

Material: BP/1/3990, skull and complete skeleton collected by J. W. Kitching during the 1970s; NMQR 1655, 3209, skulls collected by J. P. Eksteen during the 1970s and NMQR 3688, 3925, 3926, 3929, 3933, skulls collected by the authors from the uppermost slopes of Loskop and Spitskop, Nooitgedacht 68, Lower Triassic Palingkloof Member, Balfour Formation, *Lystrosaurus* Assemblage Zone.

Remarks: The material is assigned to *Lystrosaurus* for the same reasons given above and to the species *L. declivis* on the basis of a snout that is longer than the length of the skull roof, prefrontal bosses, a prominent frontonasal ridge, a longitudinal, premaxillary ridge and the absence of postorbital bosses (Cluver 1971; Grine et al. 2006; Botha and Smith 2007) (Fig. 17.7d).

Lystrosaurus murrayi (Huxley, 1859)

Material: BP/1/3974, 3975, 3977, 3978, 3979, 4030, 4039, 4040, skulls and skeletons collected by J. W. Kitching during the 1970s; NMQR 3212, collected by J. P. Eksteen during the 1970s and NMQR 3649, 3932, 3937, skulls and associated postcrania collected by the authors from the uppermost slopes of Loskop and Spitskop, Nooitgedacht 68, Lower Triassic Palingkloof Member, Balfour Formation, *Lystrosaurus* Assemblage Zone.

Remarks: The material is assigned to *Lystrosaurus* for the same reasons given above and to the species *L. murrayi* on the basis of a shorter, curved snout compared to other *Lystrosaurus* species, the absence of postorbital bosses, the anterior surface of the snout lying at right angles to the parieto-preparietal plane and weakly developed prefrontal bosses, frontonasal ridge and longitudinal, premaxillary ridge (Cluver 1971; Grine et al. 2006; Botha and Smith 2007) (Fig. 17.7e).

Gorgonopsia Seeley, 1894

Material: BP/1/3982, partial maxilla collected by J. W. Kitching during the 1970s from Nooigedacht 68; NMQR 3707, almost complete, laterally compressed skull, with associated humerus, rib and vertebrae found in situ by the authors from the lowermost slopes of Spitskop, and NMQR 4000, complete skull, articulated vertebral column, pelvic girdle and hind limbs, and disarticulated forelimbs found in situ by the authors from the lowermost slopes of Loskop, Nooitgedacht 68, Upper Permian Balfour Formation, *Dicynodon* Assemblage Zone.

Remarks: Specimen BP/1/3982 consists only of a partial maxilla and was identified as an indeterminate gorgonopsian by J. W. Kitching. Specimens NMQR 3707 and NMQR 4000 are identified as gorgonopsians on the basis of the enlarged, serrated canines, absence of precanines, well-developed incisors, reduced number of postcanines, deep snout, presence of a preparietal bone, far anterior position of the reflected lamina of the angular which bears a single lateral, non-radiating ridge and lacks a free dorsal margin, deepened dentary with a well-developed mental protuberance and a dorsally tapering coronoid process (Brink 1986; Hopson and Barghusen 1986). The specimens are currently under study and have not yet been assigned to a genus (Fig. 17.8a).

Therocephalia Broom, 1903a

Akidnognathidae Nopcsa, 1923

Moschorhinus kitchingi Broom, 1920

Material: BP/1/3983, large weathered skull collected by J. W. Kitching during the 1970s; NMQR 1640b, fragmentary skull and postcrania collected by J. P. Eksteen in the 1970s; NMQR 3684, isolated femur; NMQR 3939, complete skull, partially articulated anterior skeleton and disarticulated pelvis and hind limbs, found in situ by the authors associated with NMQR 4000, a gorgonopsian skeleton, on the lower slopes of Loskop; and NMQR 3921, complete skull found in situ by the authors on the lower slopes of Spitskop, Nooitgedacht 68, Upper Permian Balfour Formation, *Dicynodon* Assemblage Zone.

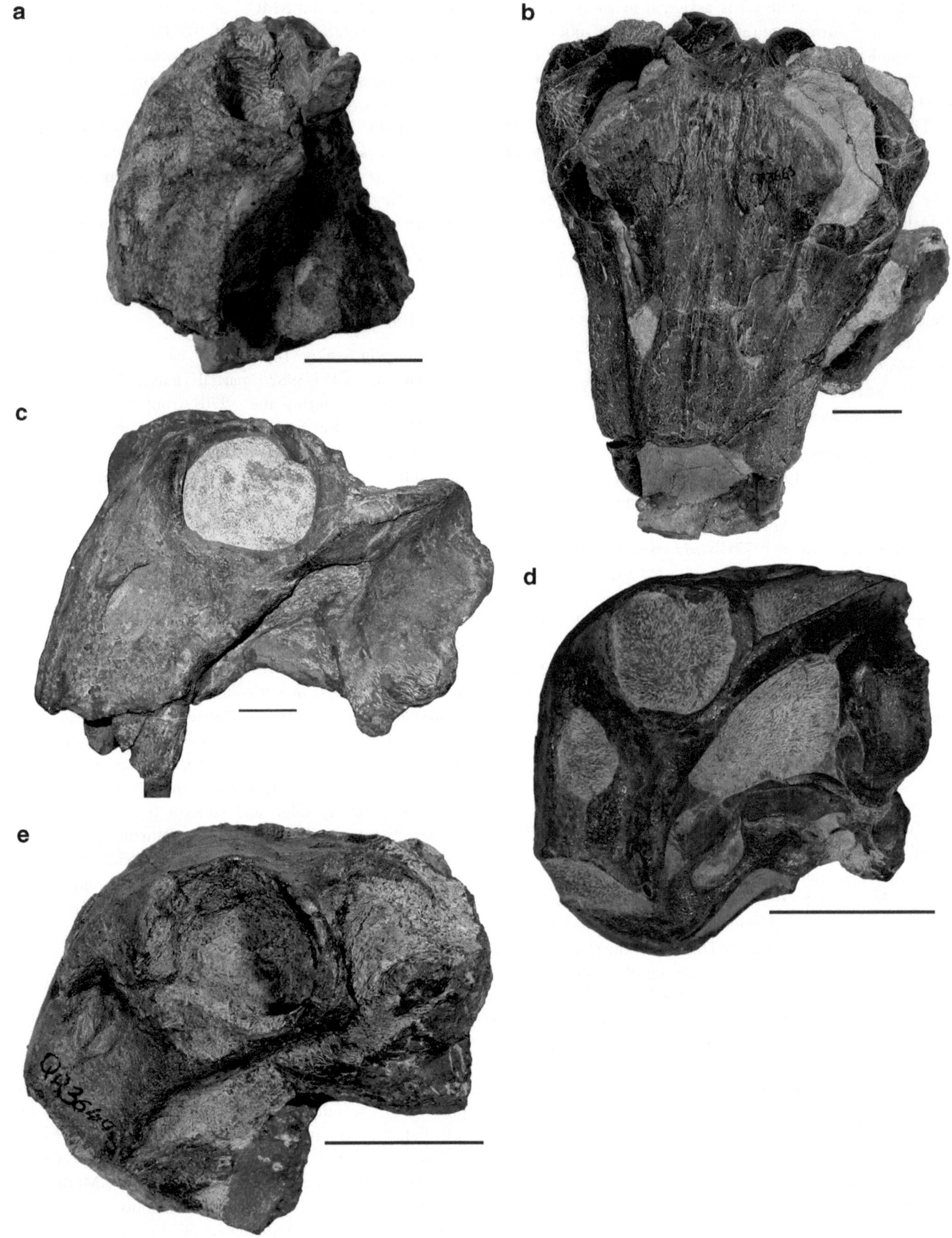

Fig. 17.7 **a** NMQR 3642, juvenile *Lystrosaurus maccaigi*, Permian. **b** NMQR 3663, adult *Lystrosaurus maccaigi*, Permian. **c** BP/1/3976, *Lystrosaurus curvatus*, Permo-Triassic. **d** NMQR 3209, *Lystrosaurus declivis*, Triassic. **e** NMQR 3649, *Lystrosaurus murrayi*, Triassic. Specimen in **b** in anterior view, all other specimens in left lateral view. Scale bars 5 cm

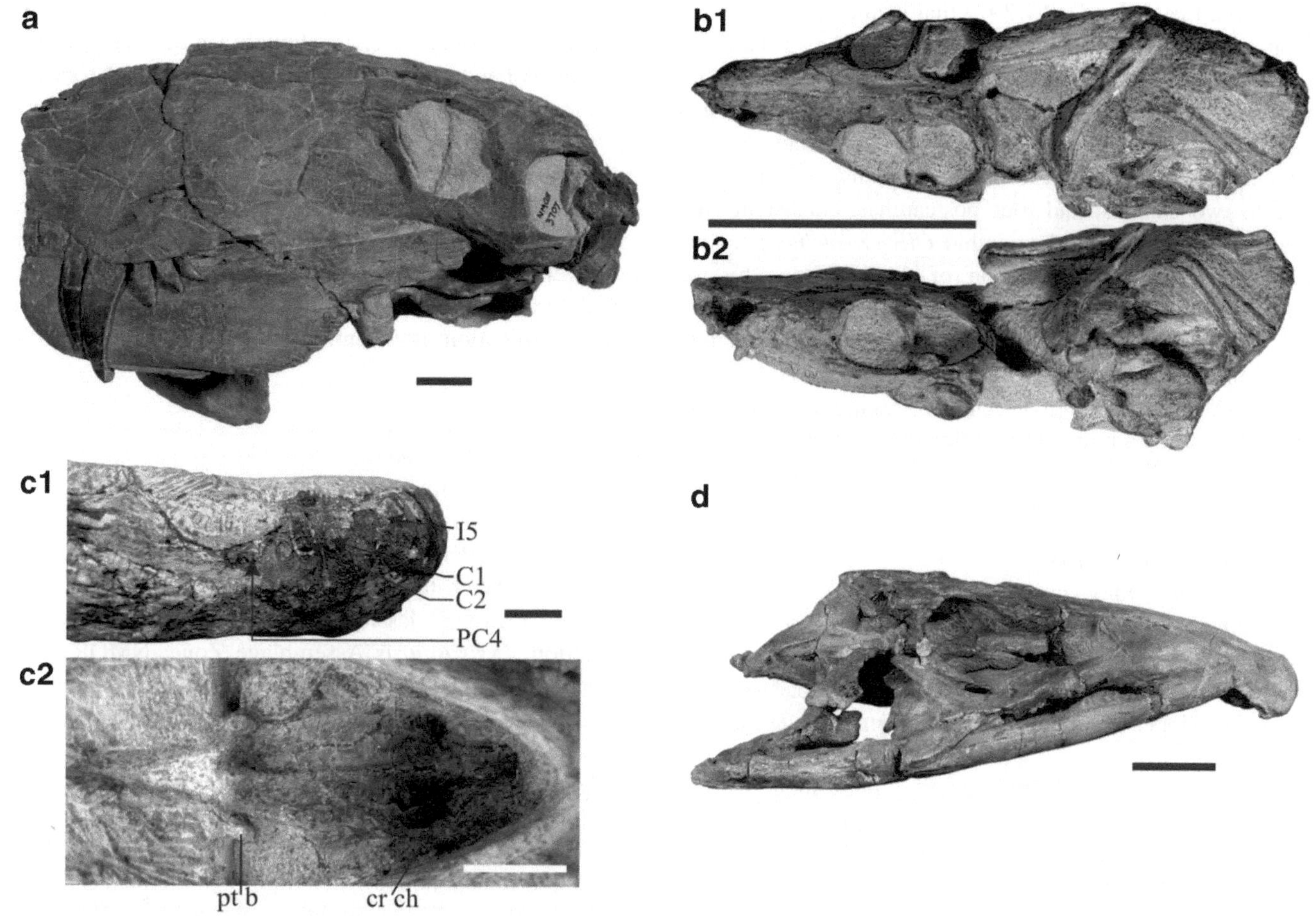

Fig. 17.8 **a** NMQR 3707, Gorgonopsia., Permian. **b** NMQR 3686, baurioid therocephalian *Ictidosuchoides longiceps*, Permian, in dorsal (**b1**) and left lateral (**b2**) views. **c** BP/1/3973, skull of the akidnognathid therocephalian *Olivierosuchus parringtoni*, Triassic, in right lateral (**c1**) and ventral (**c2**) views (scale bars 10 mm). **d** BP/1/3993, archosauromorph reptile *Proterosuchus fergusi*, Triassic. *cr ch* crista choanalis, *C1* first canine, *C2* second canine, *I5* fifth incisor, *PC4* fourth postcanine, *pt b* pterygoid boss. Scale bars 5 cm

Remarks: The skulls are identified as *Moschorhinus kitchingi* on the basis of enlarged, anteriorly facing external nares, a vomer that broadly overlaps the vomerine process of the premaxilla, the contribution of the premaxilla and maxilla to a fossa for the lower canine on the palatal surface, a blunt crista choanalis and an upper dental formula of I5:pC1:C1:PC3/4 (5 incisors, one precanine, one canine and 3/4 postcanines) (Mendrez 1974; Durand 1991). Additional disarticulated theriodont postcranial material, NMQR 3684, includes a large proximal end of a femur (shaft diameter ~27 mm) and was also found associated with a specimen of *Dicynodon lacerticeps* (NMQR 3701). The overall size of this femur and comparisons with an articulated hind limb of *Moschorhinus kitchingi* (NMQR 3351) allowed a positive identification of the specimen as *M. kitchingi*. The presence of *Moschorhinus* at Nooitgedacht 68 is particularly important, as it is the only tetrapod, apart from *Lystrosaurus* and *Promoschorhynchus* (Huttenlocker et al. 2011), to have been recovered from both sides of the Permo-Triassic boundary (Smith and Botha 2005). It is restricted to the uppermost Permian Balfour Formation (*Dicynodon* Assemblage Zone) and lowermost Triassic Palingkloof Member, Balfour Formation (*Lystrosaurus* Assemblage Zone).

Olivierosuchus parringtoni (Brink, 1965)

Material: BP/1/3973 (Fig. 17.8c), partial skull and relatively complete articulated skeleton, lacking the right manus, left pes, left femur, left ilium, part of the left tibia, the right tibia and fibula, collected by J. W. Kitching in 1964 from Nooitgedacht 68.

Remarks: BP/1/3973 was initially identified as *Ictidosuchops* by Kitching (1977) and Fourie (2001), but was later reassigned to *Regisaurus jacobi* by Fourie and Rubidge (2007). These authors based this identification on features such as the presence of a particularly long lacrimal, the absence of precanines, the absence of a pineal foramen, the presence of palatal teeth on the pterygoid boss and an upper dental formula of I6:C1:PC10 (Fourie and Rubidge 2007). However, re-examination of the specimen has found that the portion of the skull of BP/1/3973 that would have preserved a pineal foramen is absent (largely weathered away) and the tooth count has been inflated. Impressions of only four postcanine roots are visible and there are two subequal canines positioned directly next to one another (see Fig. 17.8c1). The correct tooth count is, thus, five incisors, one to two canines, and four postcanines, consistent with that of the Triassic akidnognathid *Olivierosuchus parringtoni*. Likewise, re-examination of the specimen by the authors could not independently confirm the presence of pterygoid boss teeth (Fig. 17.8c2), barring an assignment to *Regisaurus*, but consistent with *Olivierosuchus*. Thus, as the features diagnosing the material as *Regisaurus* are found to be equivocal, and no discrete features distinguish BP/1/3973 from *Olivierosuchus*, we tentatively reassign the specimen to *O. parringtoni*.

Baurioidea Broom, 1911
Ictidosuchoides longiceps (Broom, 1920)

Material: NMQR 3686 (Fig. 17.8b), complete skull and articulated pectoral girdle and forelimbs collected by the authors loose from the lower slopes of Loskop, Nooitgedacht 68, Upper Permian Balfour Formation, *Dicynodon* Assemblage Zone.

Remarks: Although two small "*Ictidosuchops*-like" baurioids were reported from the *Lystrosaurus* Assemblage Zone by Kitching (1977, p. 96), *Ictidosuchoides longiceps* (NMQR 3686) represents the second Permian therocephalian genus, and the first small baurioid, discovered at Nooitgedacht 68. The material is identified as *I. longiceps* on the basis of its long, narrow rostrum, bearing numerous precanine and postcanine maxillary teeth, concave precanine diastema, subtle rounded antorbital buttress (as in all baurioids), apparent absence of pterygoid boss teeth and retention of a pineal foramen [the latter two features distinguishing it from immature regisaurid baurioids in younger (*Lystrosaurus* Assemblage Zone) deposits] (Brink 1986). The range of *I. longiceps* is likely restricted to the Upper Permian, as most Triassic specimens referred to the genus were misidentified subadult specimens belonging to other families (Kammerer 2008; Huttenlocker et al. 2011). Though found as float, the provenance of this specimen on the distal slopes of Loskop is likely and consistent with a Late Permian age, as it could not have drifted to these outer, low-grade gullies without having been carried.

Baurioidea indet.

Material: BP/1/4021, two almost complete skulls with articulated anterior skeletons and one articulated hind limb collected by J. W. Kitching during the 1970s.

Remarks: Aside from *Olivierosuchus parringtoni* (above), a second therocephalian was reported from the *Lystrosaurus* Assemblage Zone by Kitching (1977). Recent study and preparation of the specimen has revealed its identity as probable baurioid, though a more precise identification is pending. Large orbits compared to the anteroposteriorly short temporal fenestra and a wide intertemporal region suggest that the specimens represent immature individuals. The lower jaw, however, is extremely straight and slender with numerous (8+) small, closely spaced postcanines. The high tooth count coupled with an apparently well-developed crista choanalis, apparent absence of a pineal foramen, and a sharp mastoid process may support their identification as juvenile regisaurid baurioids (Mendrez 1972).

Reptilia Laurenti, 1768
Diapsida Osborn, 1903
Archosauromorpha Huene, 1946
Proterosuchus fergusi Broom, 1903b

Material: BP/1/3993, 4016, skulls with lower jaw collected by J. W. Kitching during the 1970s from Nooigedacht 68, Lower Triassic Palingkloof Member, Balfour Formation, *Lystrosaurus* Assemblage Zone; NMQR 3924, partial skull found loose on Spitskop, Nooitgedacht 68, Lower Triassic Palingkloof Member, Balfour Formation, *Lystrosaurus* Assemblage Zone.

Remarks: The Bernard Price material was originally identified as the proterosuchid archosauriform *Proterosuchus vanhoepeni* by Kitching (1977), but that taxon is now considered to be a junior synonym of *Proterosuchus fergusi* (Welman 1998). The material is assigned to *P. fergusi* (Fig. 17.8d) on the basis of a narrow and relatively long pointed snout, a premaxilla which overhangs the lower jaw, the presence of a large oval antorbital fenestra and a second antorbital fenestra between the premaxilla and maxilla, a small posttemporal fenestra and a small lateral mandibular fenestra on the lower jaw (Welman 1998). NMQR 3924 is assigned to *Proterosuchus* on the basis of the skull roof and tooth morphology. *Proterosuchus* plays an important role in Pangaean-scale reconstructions of Early Triassic paleobiogeography as proterosuchids are the first terrestrial vertebrates to appear above the PTB in Karoo-aged basins of South Africa and Russia (Smith and Botha 2005; Benton et al. 2004).

Stratigraphic and Sedimentological Interpretations

The facies succession and fossil vertebrate associations at Nooitgedacht 68 are similar to those at other PTB sites in the South African Karoo Basin (e.g., Bethal, Bethulie District; Old Wapadsberg Pass, Graaff-Reinet District; Smith 1995; Smith and Ward 2001; Smith and Botha 2005). The sequence changes from drab olive-gray massive siltstone beds interbedded with well-sorted, fine grained sandstone bodies to more dusky red heterolithic mudrock and is capped by a facies dominated by massive and horizontally bedded sheet sandstone bodies with prominent pebble lag deposits. This transition is interpreted as a change in fluvial style from an alluvial plain with large, meandering rivers and expansive lowland floodplains to a network of ephemeral, braided streams (Smith 1995). We interpret the interval containing the laminated dusky red and olive-gray mudrock and siltstone couplets to be the Palingkloof Member, which also preserves the PTB (as identified by Smith 1995; Smith and Ward 2001; Smith and Botha 2005). We place the PTB at the top of the laminated section. *Dicynodon* is not as abundant as *Lystrosaurus maccaigi* at Nooitgedacht 68. However, similar to *Dicynodon*, *L. maccaigi* is restricted to the uppermost Permian and is also considered to be a biostratigraphic marker of the continental PTB (Botha and Smith 2007). Significantly, this pattern remains true even given the local abundance of *Lystrosaurus maccaigi* at Nooitgedacht 68, which should be expected to reduce the Signor–Lipps Effect (Signor and Lipps 1982). In spite of this local abundance, which has fostered greater sampling (almost doubling the known sample size of *L. maccaigi*), this taxon is still not found above the lithostratigraphic marker beds. We thus place the PTB at the base of a siltstone body within this interval just above the Last Appearance Datum of *Lystrosaurus maccaigi*.

Discussion

To date, 112 positively identifiable (38 in situ specimens from our collecting efforts) vertebrate fossils have been recovered from Nooitgedacht 68. The presence of numerous in situ and loose skulls and skeletons of the dicynodonts *Dicynodontoides*, *Dicynodon*, *Daptocephalus*, *Dinanomodon* and *Lystrosaurus maccaigi* on the lowermost slopes of Loskop and Spitskop, Nooitgedacht 68 confirms that these strata lie in the uppermost *Dicynodon* Assemblage Zone, which is Late Permian in age (Rubidge 1995). Furthermore, two large, yet-to-be-described specimens NMQR 4000 and NMQR 3707 (skull lengths both 500 mm), are assignable to Gorgonopsia, a therapsid group that is restricted to the Permian, were recovered in situ from the lower slopes of Loskop and Spitskop, respectively.

Taxa recovered from the upper slopes of Loskop and Spitskop comprise the Triassic *Lystrosaurus murrayi* and *L. declivis*. Although these species were recovered in situ during our field expeditions, Kitching (1977) also found *L. murrayi*, *L. declivis*, *Proterosuchus fergusi*, and a therocephalian that has now been identified as *Olivierosuchus parringtoni* (previously *Regisaurus jacobi*, Fourie and Rubidge 2007) from the uppermost slopes of Loskop. Our latest field expedition also produced a *Proterosuchus fergusi* skull from the Lower Triassic Palingkloof Member of the *Lystrosaurus* Assemblage Zone from Spitskop. Both *Olivierosuchus* and *Proterosuchus* are restricted to the lowermost Triassic *Lystrosaurus* Assemblage Zone, and along with the presence of Triassic *Lystrosaurus* species, confirm an Early Triassic age for the uppermost slopes of Loskop and Spitskop (Botha and Smith 2006; Botha-Brink and Modesto 2011).

Due to the continuous sedimentation and gradual transition from typical Permian (greenish gray to olive-gray siltstone beds with fine-grained sandstone bodies) to Triassic (sandstone rich succession of massive fine-grained sandstone bodies with gullied basal scours and dark reddish brown or dusky red siltstone beds) strata as has been noted at other Permo-Triassic boundary sections in the Karoo Basin (Smith 1995; Smith and Ward 2001; Smith and Botha 2005), we propose that a similarly complete Permo-Triassic boundary sequence is preserved at Loskop and Spitskop on Nooitgedacht 68. Approximately 16 m of the Palingkloof Member, Balfour Formation is preserved at Loskop, and an 8 m thick sequence of dusky red laminites, previously referred to as the 3.7 m 'event bed' (e.g., Smith and Ward 2001; Retallack et al. 2003), preserves the PTB interval itself. Approximately 19 m of the Palingkloof Member, with a 4 m thick PTB sequence, is preserved at Spitskop.

In the southern Karoo Basin, *Dicynodon* is usually the most abundant tetrapod recovered from the uppermost *Dicynodon* Assemblage Zone, whereas other taxa, including *Lystrosaurus maccaigi*, are relatively rare. The genus *Lystrosaurus* comprises four species in South Africa, viz. *Lystrosaurus maccaigi*, *L. curvatus*, *L. murrayi*, and *L. declivis* (Grine et al. 2006). *Lystrosaurus curvatus* is the rarest species, with only a handful of specimens known, closely followed by *L. maccaigi* (with only 35 identified specimens in South African fossil collections). In contrast, thousands of *L. murrayi* and *L. declivis* specimens have been recovered from the lowermost Triassic *Lystrosaurus* Assemblage Zone (J. Botha-Brink personal observation of museum collections, 2007). To date, Nooitgedacht 68 has yielded 50 positively identifiable *Lystrosaurus maccaigi* specimens, ranging from juvenile (70 mm snout length) to adult (230 mm snout length), and is consequently the richest *L. maccaigi* site known in South Africa. Interestingly, four in situ specimens have been found within the heterolithic PTB interval itself, which is more than previously recorded from other PTB sites.

The presence of *Dicynodontoides recurvidens* (BP/1/4027) at this site is also noteworthy. A recent study on this taxon confirmed its First Appearance Datum as being in the lower Upper Permian *Pristerognathus* Assemblage Zone (Angielczyk et al. 2009). However, although its range extends into the Upper Permian *Dicynodon* Assemblage Zone, the upper limit of *D. recurvidens* has yet to be confirmed as it is a relatively rare taxon, and field notes on previously collected specimens are not detailed enough to pinpoint its Last Appearance Datum (Angielczyk et al. 2009). Consequently, it is currently thought to have disappeared before the onset of the end-Permian extinction and is not considered to be a victim of the event (Angielczyk et al. 2009). Although we do not know the exact stratigraphic level at which specimen BP/1/4027 was collected, only 46 m of the Upper Permian *Dicynodon* Assemblage Zone are exposed at Loskop, with most of the fossiliferous outcrops comprising no more than 34 m, thus the specimen that J. W. Kitching collected must have been recovered from somewhere in these strata, which indicates that *Dicynodontoides* was a victim of the end-Permian extinction.

The recovery of an in situ *Dinanomodon gilli* specimen (NMQR 3696) is also significant. Earlier collecting efforts recovered *Dinanomodon* from 25 m below the PTB in the upper Permian Balfour Formation, *Dicynodon* Assemblage Zone (Ward et al. 2005; R. Smith personal communication, 2011). However, the specimen collected at Nooitgedacht 68 was found 17 m below the PTB and thus slightly extends the stratigraphic range of this taxon to closer to the PTB and confirms that *Dinanomodon* was also a victim of the end-Permian extinction.

The presence of at least 14 species at Nooitgedacht 68 makes this locality the second richest PTB site in South Africa. The diversity at this site is surpassed only by that at the well-known PTB locality at Bethal, Bethulie District, from where at least 18 species have been recovered. The excellent preservation, presence of rare taxa and overall abundance of Permian tetrapods at Nooitgedacht 68 indicates that it will be an excellent site to confirm previous observations made at other PTB sites in South Africa and will provide insight into the disappearance of Permian taxa during this mass extinction event.

Acknowledgments Thank you to Elize Butler, Emily Byington, John Nyaphuli, Joël Mohoi, Nthaopa Ntheri and Paul October for their very hard work in the field. A grant to JBB from the National Research Foundation (GUN 65681) and African Origins Programme (GUN 65244) of South Africa and a Discovery Grant (no. 288126-07) to SPM from the Natural Sciences and Engineering Research Council (NSERC) of Canada made this research possible. Helpful comments from Bruce Rubidge, Fernando Abdala, Jörg Fröbisch and an anonymous reviewer are gratefully acknowledged.

References

Angielczyk, K. D. (2007). New specimens of the Tanzanian dicynodont '*Cryptocynodon*' *parringtoni* Von Huene, 1942 (Therapsida, Anomodontia), with an expanded analysis of Permian dicynodont phylogeny. *Journal of Vertebrate Paleontology, 27*, 116–131.

Angielczyk, K. D., & Kurkin, A. A. (2003). Phylogenetic analysis of Russian Permian dicynodonts (Therapsida: Anomodontia): Implications for Permian biostratigraphy and Pangaean biogeography. *Zoological Journal of the Linnean Society, 139*, 157–212.

Angielczyk, K. D., Sidor, C. A., Nesbitt, S. J., Smith, R. M. H., & Tsuji, L. A. (2009). Taxonomic revision and new observations on the postcranial skeleton, biogeography, and biostratigraphy of the dicynodont genus *Dicynodontoides*, the senior subjective synonym of *Kingoria* (Therapsida, Anomodontia). *Journal of Vertebrate Paleontology, 29*, 1174–1187.

Becker, L., Poreda, R. J., Hunt, A. G., Bunch, T. E., & Rampino, M. (2001). Impact event at the Permian-Triassic boundary: Evidence from extraterrestrial noble gases in fullerenes. *Science, 291*, 1530–1533.

Becker, L., Poreda, R. J., Basu, A. R., Pope, K. O., Harrison, T. M., Nicholson, C., et al. (2004). Bedout: A possible end-Permian impact crater offshore of Northwestern Australia. *Science, 304*, 1469–1476.

Benton, M. J. (2003). *When life nearly died. The greatest mass extinction of all time*. London: Thames and Hudson.

Benton, M. J., Tverdokhlebov, V. P., & Surkov, M. V. (2004). Ecosystem remodelling among vertebrates at the Permian-Triassic boundary in Russia. *Nature, 432*, 97–100.

Botha, J., & Smith, R. M. H. (2006). Rapid vertebrate recovery in the Karoo Basin of South Africa following the end-Permian extinction. *Journal of African Earth Sciences, 45*, 502–514.

Botha, J., & Smith, R. M. H. (2007). *Lystrosaurus* species composition across the Permo-Triassic boundary of South Africa. *Lethaia, 40*, 125–137.

Botha-Brink, J., & Modesto, S. P. (2011). A new skeleton of the therocephalian synapsid *Olivierosuchus parringtoni* from the Lower Triassic South African Karoo Basin. *Palaeontology, 54*, 591–606.

Brink, A. S. (1951). On the genus *Lystrosaurus* Cope. *Transactions of the Royal Society of South Africa, 33*, 107–120.

Brink, A. S. (1965). A new ictidosuchid (Scaloposauria) from the *Lystrosaurus*-Zone. *Palaeontologia Africana, 9*, 129–138.

Brink, A. S. (1986). *Illustrated bibliographical catalogue of the Synapsida*. Handbook 10, Part 1. Pretoria: Geological Survey of South Africa.

Broom, R. (1903a). On the classification of the theriodonts and their allies. *Report of the South African Association for the Advancement of Science, 1*, 286–294.

Broom, R. (1903b). On a new reptile (*Proterosuchus fergusi*) from the Karroo beds of Tarkastad, South Africa. *Annals of the South African Museum, 4*, 159–164.

Broom, R. (1905). On the use of the term Anomodontia. *Albany Museum Records, 1*, 266–269.

Broom, R. (1911). On the structure of the skull in cynodont reptiles. *Proceedings of the Zoological Society of London, 81*, 893–925.

Broom, R. (1920). On some new therocephalian reptiles from the Karoo Beds of South Africa. *Proceedings of the Zoological Society of London, 1920*, 343–355.

Broom, R. (1932). *The mammal-like reptiles of South Africa and the origin of mammals*. London: H. F. and G. Witherby.

Chen, Z. Q., Tong, J., Kaiho, K., & Kawahata, H. (2007). Onset of biotic and environmental recovery from the end-Permian mass extinction within 1–2 million years: A case study of the Lower Triassic of the Meishan section, South China. *Palaeogeography, Palaeoclimatology, Palaeoecology, 252*, 176–187.

Clark, D. L., Cheng-Yuan, W., Orth, C. S., & Gilmore, J. S. (1986). Conodont survival and low iridium abundances across the Permian-Triassic boundary. *Science, 233*, 984–986.

Cluver, M. A. (1971). The cranial morphology of the dicynodont genus *Lystrosaurus*. *Annals of the South African Museum, 56*, 155–274.

Cluver, M. A., & Hotton, N., III. (1981). The genera *Dicynodon* and *Diictodon* and their bearing on the classification of the Dicynodontia (Reptilia, Therapsida). *Annals of the South African Museum, 83*, 99–146.

Damiani, R., Modesto, S., Yates, A., & Neveling, J. (2003). Earliest evidence of cynodont burrowing. *Proceedings of the Royal Society B, 270*, 1747–1752.

Durand, J. F. (1991). A revised description of the skull of *Moschorhinus* (Therapsida, Therocephalia). *Annals of the South African Museum, 99*, 381–413.

Erwin, D. H. (1994). The Permo-Triassic extinction. *Nature, 367*, 231–236.

Fourie, H. (2001). *Morphology and function of the postcrania of selected genera of Therocephalia* (Amniota: Therapsida). Unpublished Ph.D. thesis, University of the Witwatersrand.

Fourie, H., & Rubidge, B. S. (2007). The postcranial skeletal anatomy of the therocephalian *Regisaurus* (Therapsida: Regisauridae) and its utilization for biostratigraphic correlation. *Palaeontologia Africana, 42*, 1–16.

Fraiser, M. L., & Bottjer, D. J. (2007). Elevated atmospheric CO_2 and the delayed biotic recovery from the end-Permian mass extinction. *Palaeogeography, Palaeoclimatology, Palaeoecology, 252*, 164–175.

Fröbisch, J. (2007). The cranial anatomy of *Kombuisia frerensis* Hotton (Synapsida, Dicynodontia), and a new phylogeny of anomodont therapsids. *Zoological Journal of the Linnean Society, 150*, 117–144.

Fröbisch, J., & Reisz, R. R. (2008). A new species of *Emydops* (Synapsida, Anomodontia) and a discussion of dental variability and pathology in dicynodonts. *Journal of Vertebrate Paleontology, 28*, 770–787.

Gastaldo, R. A., & Rolerson, M. W. (2008). *Katbergia* gen. nov., a new trace fossil from the Late Permian and Early Triassic of the Karoo Basin: Implications for paleoenvironmental conditions at the P/Tr extinction event. *Palaeontology, 51*, 215–229.

Gastaldo, R. A., Adendorff, R., Bamford, M., Labandeira, C. C., Neveling, J., & Sims, H. (2005). Taphonomic trends of macrofloral assemblages across the Permian-Triassic boundary, Karoo Basin, South Africa. *Palaios, 20*, 479–497.

Gastaldo, R. A., Neveling, J., Clark, C. K., & Newbury, S. S. (2009). The terrestrial Permian-Triassic boundary event bed is a nonevent. *Geology, 37*, 199–202.

Grine, F. E., Forster, C. A., Cluver, M. A., & Georgi, J. A. (2006). Cranial variability, ontogeny and taxonomy of *Lystrosaurus* from the Karoo Basin of South Africa. In M. Carrano, R. W. Blob, T. J. Gaudin, & J. R. Wible (Eds.), *Amniote paleobiology: Perspectives on the evolution of mammals, birds, and reptiles* (pp. 403–503). Chicago: University of Chicago Press.

Gruszczynski, M., Halas, S., Hoffman, A., & Malkowksi, K. (1989). A brachiopod calcite record of the oceanic carbon and oxygen isotope shifts at the Permian/Triassic transition. *Nature, 337*, 64–68.

Hasiotis, S. T., Miller, M. F., Isbell, J. L., Babcock, L. E., & Collinson, J. W. (1999). Is Triassic crayfish fossil evidence from Antarctica really burrow evidence of mammal-like reptiles? Resolving vertebrate from invertebrate burrows. *Freshwater Crayfish, 12*, 71–81.

Hopson, J. A., & Barghusen, H. R. (1986). An analysis of therapsid relationships. In N. Hotton III, P. D. Maclean, J. J. Roth, & E. C. Roth (Eds.), *The ecology and biology of mammal-like reptiles* (pp. 83–106). Washington, D.C.: Smithsonian Institution Press.

Hotinski, R. M., Bice, K. L., Kump, L. R., Najjar, R. G., & Arthur, M. A. (2001). Ocean stagnation and end-Permian anoxia. *Geology, 29*, 7–10.

Huene, F. von (1946). Die grossen Stämme der Tetrapoden in den geologischen Zeiten. *Biologisches Zentralblatt, 65*, 268–275.

Huttenlocker, A. K., Sidor, C. A., & Smith, R. M. H. (2011). A new specimen of *Promoschorhynchus* (Therapsida: Therocephalia: Akidnognathidae) from the Lower Triassic of South Africa and its implications for theriodont survivorship across the Permo-Triassic boundary. *Journal of Vertebrate Paleontology, 31*, 405–421.

Huxley, T. H. (1859). On a new species of dicynodont (*D. murrayi*) from near Colesburg, South Africa; and on the structure of the skull in dicynodonts. *Quarterly Journal of the Geological Society of London, 15*, 649–659.

Isozaki, Y. (1997). Permo-Triassic boundary superanoxia and stratified superocean: Records from lost deep sea. *Science, 276*, 235–238.

Jin, Y. G., Wang, Y., Shang, Q. H., Cao, C. Q., & Erwin, D. H. (2000). Pattern of marine mass extinction near the Permian-Triassic boundary in South China. *Science, 289*, 432–436.

Kammerer, C. (2008). A new therocephalian from the *Cistecephalus* Assemblage Zone of South Africa and new information on therocephalian systematics. *Journal of Vertebrate Paleontology, 28*, 98A–99A.

Kammerer, C. F., & Angielczyk, K. D. (2009). A proposed higher taxonomy of anomodont therapsids. *Zootaxa, 2018*, 1–24.

Kammerer, C. F., Angielczyk, K. D., & Fröbisch, J. (2011). A comprehensive taxonomic revision of *Dicynodon* (Therapsida, Anomodontia) and its implications for dicynodont phylogeny, biogeography, and biostratigraphy. *Society of Vertebrate Paleontology Memoir, 11*, 1–158.

Kitching, J. W. (1977). Distribution of the Karoo vertebrate fauna. *Bernard Price Institute for Palaeontological Research Memoir 1*, 1–131.

Knoll, A. H., Bambach, R. K., Canfield, D. E., & Grotzinger, J. P. (1996). Comparative earth history and Late Permian mass extinction. *Science, 273*, 452–457.

Krull, E. S., Retallack, G. J., & Campbell, H. J. (2000). $d^{13}C_{org}$ chemostratigraphy of the Permian-Triassic boundary in the Maitai Group, New Zealand: Evidence for high-latitudinal methane release. *New Zealand Journal of Geology and Geophysics, 43*, 21–32.

Laurenti, J. N. (1768). *Classis Reptilium. Specimen medicum, exhibens synopsis Reptilium emendatum, cum experimentis circa venena et antidote Reptilium Austriacorum*. J. Thom., Nob, et Trattnern, Vienna.

Lehrmann, D. J., Ramezani, J., Bowring, S. A., Martin, M. W., Montgomery, P., Enos, P., et al. (2006). Timing of recovery from the end-Permian extinction: Geochronologic and biostratigraphic constraints from south China. *Geology, 34*, 1053–1056.

MacLeod, K. G., Smith, R. M. H., Koch, P. L., & Ward, P. D. (2000). Timing of mammal-like reptile extinctions across the Permian-Triassic boundary in South Africa. *Geology, 28*, 227–230.

Mendrez, C. H. (1972). On the skull of *Regisaurus jacobi*, a new genus of Bauriamorpha Watson and Romer 1956 (=Scaloposauria Boonstra 1953), from the *Lystrosaurus*-Zone of South Africa. In K. A. Joysey & T. S. Kemp (Eds.), *Studies in vertebrate evolution* (pp. 191–212). New York: Winchester Press.

Mendrez, C. H. (1974). A new specimen of *Promoschorhynchus platyrhinus* Brink 1954 (Moschorhinidae) from the *Daptocephalus*-

Zone (Upper Permian) of South Africa. *Palaeontologia Africana, 17*, 69–85.

Miller, A. I., & Foote, M. (2003). Increased longevities of post-Paleozoic marine genera after mass extinctions. *Science, 302*, 1030–1032.

Miller, M. F., Hasiotis, S. T., Babcock, L. E., Isbell, J. L., & Collinson, J. W. (2001). Tetrapod and large burrows of uncertain origin in Triassic high paleolatitude floodplain deposits, Antarctica. *Palaios, 16*, 218–232.

Modesto, S. P., & Botha-Brink, J. (2010). A burrow cast with *Lystrosaurus* skeletal remains from the Lower Triassic of South Africa. *Palaios, 25*, 274–281.

Morante, R. (1996). Permian and early Triassic isotopic records of carbon and strontium in Australia and a scenario of events about the Permian-Triassic boundary. *Historical Biology, 11*, 289–310.

Morante, R., Veevers, J. J., Andrew, A. S., & Hamilton, P. J. (1994). Determining the Permian-Triassic boundary in Australia using C-isotope chemostratigraphy. *Australian Petroleum Exploration Association Journal, 34*, 330–336.

Mundil, R., Ludwig, K. R., Metcalf, I., & Renne, P. R. (2004). Age and timing of the Permian mass extinctions: U/Pb dating of closed-system zircons. *Science, 305*, 1760–1763.

Nicolas, M., & Rubidge, B. S. (2010). Changes in Permo-Triassic terrestrial tetrapod ecological representation in the Beaufort Group (Karoo Supergroup) of South Africa. *Lethaia, 43*, 45–59.

Nopcsa, F. (1923). *Die Familien der Reptilien*. Berlin: Verlag von Gebrüder Borntraeger.

Osborn, H. F. (1903). On the primary division of the Reptilia into two sub-classes, Synapsida and Diapsida. *Science, 17*, 275–276.

Owen, R. (1845). Report on the reptilian fossils of South Africa. Part I. Description of certain fossil crania, discovered by A. G. Bain, Esq., in sandstone rocks at the south-eastern extremity of Africa, referable to different species of an extinct genus of Reptilia (*Dicynodon*), and indicative of a new tribe or sub-order of Sauria. *Transactions of the Geological Society of London, Second Series, 7*, 59–84.

Owen, R. (1860a). On the orders of fossil and recent Reptilia and their distribution in time. *Report of the British Association for the Advancement of Science, 1859*, 153–166.

Owen, R. (1860b). On some reptilian fossils from South Africa. *Quarterly Journal of the Geological Society of London, 16*, 49–54.

Owen, R. (1876). *Descriptive and illustrated catalogue of the Fossil Reptilia of South Africa in the collection of the British Museum of Natural History*. London: British Museum.

Pace, D. W., Gastaldo, R. W., & Neveling, J. (2009). Early Triassic aggradational and degradational landscapes of the Karoo Basin and evidence for climate oscillation following the P-Tr event. *Journal of Sedimentary Research, 79*, 316–331.

Payne, J. L., Lehrmann, D. L., Wei, J., Orchard, M. J., Schrag, D. P., & Knoll, A. H. (2004). Large perturbations of the carbon cycle during recovery from the end-Permian extinction. *Science, 305*, 506–509.

Racki, G. (2003). End-Permian mass extinction: Oceanographic consequences of double catastrophic volcanism. *Lethaia, 36*, 171–173.

Renne, P. R., Zichao, Z., Richards, M. A., Black, M. A., & Basu, A. R. (1995). Synchrony and causal relations between Permian-Triassic boundary crises and Siberian flood volcanism. *Science, 269*, 1413–1416.

Retallack, G. J. (1999). Postapocalyptic greenhouse palaeoclimate revealed by earliest Triassic paleosols in the Sydney basin, Australia. *Geological Society of America Bulletin, 111*, 52–70.

Retallack, G. J., & Krull, E. S. (1999). Landscape ecological shift at the Permian-Triassic boundary in Antarctica. *Australian Journal of Earth Sciences, 46*, 786–812.

Retallack, G. J., Seyodolali, A., Krull, E. S., Holser, W. T., Ambers, C. P., & Kyte, F. T. (1998). Search for evidence of impact at the Permian-Triassic boundary in Antarctica and Australia. *Geology, 26*, 979–982.

Retallack, G. J., Smith, R. M. H., & Ward, P. D. (2003). Vertebrate extinction across Permian-Triassic boundary in Karoo Basin, South Africa. *Geological Society of America Bulletin, 115*, 1133–1152.

Retallack, G. J., Greaver, T., & Jahren, A. H. (2007). Return to Coalsack Bluff and the Permian-Triassic boundary in Antarctica. *Global Planetary Change, 55*, 90–108.

Rubidge, B. S. (Ed.). (1995). Biostratigraphy of the Beaufort Group (Karoo Supergroup). *South African Committee for Stratigraphy Biostratigraphic Series 1*, 1–46.

Sahney, S., & Benton, M. J. (2008). Recovery from the most profound mass extinction of all time. *Proceedings of the Royal Society B, 275*, 759–765.

Sarkar, A., Yoshioka, H., Ebihara, M., & Naraoka, H. (2003). Geochemical and organic carbon isotope studies across the continental Permo-Triassic boundary of Raniganj Basin, Eastern India. *Palaeogeography, Palaeoclimatology, Palaeoecology, 191*, 1–14.

Seeley, H. G. (1894). Researches on the structure, organisation and classification of the Fossil Reptilia. IX, section 1. On the Therosuchia. *Philosophical Transactions of the Royal Society London, B, 185*, 987–1018.

Seeley, H. G. (1898). On the skull of *Mochlorhinus platyceps* from Bethulie, Orange Free State, preserved in the Albany Museum, Grahamstown. *Annals and Magazine of Natural History, 1*, 164–176.

Sephton, M. A., Looy, C. V., Brinkhuis, H., Wignall, P. B., de Leeuw, J. W., & Visscher, H. (2005). Catastrophic soil erosion during the end-Permian crisis. *Geology, 33*, 941–944.

Signor, P. W., III, & Lipps, J. H. (1982). Sampling bias, gradual extinction patterns, and catastrophes in the fossil record. In L. T. Silver & P. H. Schultz (Eds.), *Geological implications of impacts of large asteroids and comets on the Earth* (pp. 291–296). Boulder: Geological Society of America Special Publication.

Smith, R. M. H. (1995). Changing fluvial environments across the Permian-Triassic boundary in the Karoo Basin, South Africa and possible causes of tetrapod extinctions. *Palaeogeography, Palaeoclimatology, Palaeoecology, 117*, 81–104.

Smith, R. M. H., & Botha, J. (2005). The recovery of terrestrial vertebrate diversity in the South African Karoo Basin after the end-Permian extinction. *Comptes Rendus Palevol, 4*, 555–568.

Smith, R. M. H., & Ward, P. D. (2001). Pattern of vertebrate extinctions across an event bed at the Permian-Triassic boundary in the Karoo Basin of South Africa. *Geology, 29*, 1147–1150.

Wang, X., Zhang, Z., Metcalfe, I., & Foster, C. (2002). Palynofloral assemblage across the terrestrial Permian-Triassic boundary at Dalongkou, Xinjiang, NW China. In G. A. Brock & J. A. Talent (Eds.), *First international palaeontological congress* (pp. 166–167). Sydney: Geological Society of Australia.

Ward, P. D., Montgomery, D. R., & Smith, R. (2000). Altered river morphology in South Africa related to the Permian-Triassic extinction. *Science, 289*, 1740–1743.

Ward, P. D., Botha, J., Buick, R., De Kock, M. O., Erwin, D. H., Garrison, G. H., et al. (2005). Abrupt and gradual extinction among Late Permian land vertebrates in the Karoo Basin, South Africa. *Science, 307*, 709–714.

Welman, J. (1998). The taxonomy of the South African proterosuchids (Reptilia, Archosauromorpha). *Journal of Vertebrate Paleontology, 18*, 340–347.

Wignall, P. B., & Newton, R. (2003). Contrasting deep-water records from the Upper Permian and Lower Triassic of South Tibet and British Columbia: Evidence for a diachronous mass extinction. *Palaios, 18*, 153–167.

Wignall, P. B., & Twitchett, R. J. (1996). Oceanic anoxia and the end-Permian mass extinction. *Science, 272*, 1155–1158.

Chapter 18
Synapsid Diversity and the Rock Record in the Permian-Triassic Beaufort Group (Karoo Supergroup), South Africa

Jörg Fröbisch

Abstract This study investigates diversity patterns of Synapsida in the Permian-Triassic sequence of the Karoo Basin, South Africa. Permian-Triassic synapsids represent the dominant terrestrial tetrapods of their time and play a central role in assessing the impact of the end-Permian mass extinction on terrestrial ecosystems. On the regional scale of the Karoo Basin, synapsid diversity shows a mid-Permian extinction and a pronounced extinction event at the end of the Permian, whereas the subclades of Synapsida exhibit clade-specific diversity patterns. Taxonomic diversity estimates (TDEs) of Synapsida and its subclades are not significantly correlated with outcrop area for the complete time series. However, after exclusion of the *Lystrosaurus* Assemblage Zone from all data series, the TDEs of the majority of synapsid subclades show statistically significant strong positive correlations with outcrop area. Nonetheless, diversity residuals, resulting from modeled diversity estimates, exhibit clade-specific patterns with varying support for a mid-Permian event and strong support for an end-Permian extinction. The results confirm studies at the global scale and imply that synapsid diversity in the Karoo Basin is at least partially biased by the Permian-Triassic terrestrial rock record. Moreover, Anomodontia, the most speciose clade of non-mammalian synapsids, is not the sole driver of the synapsid diversity signal. Instead, there seems to be a general synapsid pattern, with each subclade diverging from this pattern to varying degrees for clade-specific reasons. Thus, despite the obvious rock record bias, the end-Permian extinction maintains its major impact on synapsid diversity and therefore on the composition and structure of past and present terrestrial ecosystems.

J. Fröbisch (✉)
Museum für Naturkunde, Leibniz-Institut für Evolutions- und Biodiversitätsforschung, Humboldt-Universität, Invalidenstraße 43, 10115 Berlin, Germany
e-mail: joerg.froebisch@mfn-berlin.de

Keywords Karoo Basin • End-Permian extinction • Sampling bias

Introduction

The fossil record of Synapsida is exceptional in that it documents in much detail the transition from basal, pelycosaur-grade synapsids to derived synapsids and the acquisition of increasingly large numbers of mammalian features, such as the mammalian phalangeal formula, upright locomotion, bony secondary palate, the mammalian middle ear, and ecological specializations (e.g., Olson 1944, 1959; Allin 1975; Cluver 1978; Hopson 1995; Fröbisch 2006; Luo 2007; Fröbisch and Reisz 2009).

The initial diversification of synapsids marks their early success in terms of dominance of Paleozoic (in particular Permian) ecosystems on land (e.g., Olson 1966; Kissel and Reisz 2004). Yet, the greatest mass extinction in Earth's history at the end of the Permian had a major impact on the composition and structure of terrestrial ecosystems, including a significant decline in synapsid diversity and disparity (e.g., Benton et al. 2004; Ward et al. 2005; Roopnarine et al. 2007). Only three major synapsid clades survived the end-Permian extinction, namely dicynodont anomodonts, therocephalians, and cynodonts (e.g., Kemp 2005; Fröbisch 2007). Ultimately, only cynodonts survived to give rise to crown mammals. The latter experienced a major diversification during the Cenozoic, leading to the success of mammals in today's ecosystems (see Prothero 2006; Rose 2006).

Traditional studies of paleobiodiversity utilized raw data from the fossil record to reconstruct diversity curves throughout the Phanerozoic (e.g., Raup 1972; Sepkoski et al. 1981; Raup and Sepkoski 1984; Benton 1985, 1995). More recent contributions strongly emphasize the need to apply methods of sampling standardization to minimize existing sampling biases (e.g., Alroy et al. 2001, 2008). In

Christian F. Kammerer, Kenneth D. Angielczyk and Jörg Fröbisch (eds.), *Early Evolutionary History of the Synapsida*, Vertebrate Paleobiology and Paleoanthropology, DOI: 10.1007/978-94-007-6841-3_18,

particular, a wide range of studies on marine invertebrates have clearly demonstrated that the raw patterns of taxonomic diversity are strongly correlated with several proxies for rock availability, including the number of formations, number of localities, or outcrop area per time interval (e.g., Raup 1972, 1976; Peters and Foote 2001; Smith 2001; Smith et al. 2001; Crampton et al. 2003; Peters 2005, 2006; Smith and McGowan 2005, 2007; McGowan and Smith 2008). Recently, this correlation has also been documented for a number of vertebrate clades in the terrestrial realm, including anomodont therapsids, dinosaurs, and pterosaurs (Upchurch and Barrett 2005; Fröbisch 2008; Lloyd et al. 2008; Barrett et al. 2009; Butler et al. 2009, 2010; Mannion et al. 2011), as well as for marine reptiles in the oceans (Benson et al. 2010). Thus, it is possible that rock availability is driving the diversity patterns seen in the fossil record.

However, a correlation between these two variables does not necessarily imply a causal relationship. Alternatively, both variables may be influenced or controlled by a third factor, such as sea-level changes, known as the 'common cause' hypothesis (e.g., Sepkoski 1976; Peters and Foote 2001; Smith et al. 2001; Peters 2005; Benton and Emerson 2007; Benton 2009). However, there seem to be divergent signals in the marine and terrestrial environments. The 'common cause' scenario with sea-level changes as third variable appears to be well-supported in the shallow marine realm (e.g., Sepkoski 1976; Smith 2001; Peters 2005, 2006), whereas the influence of external factors on terrestrial diversity patterns is currently poorly understood (Benton and Emerson 2007; Butler et al. 2009). In fact, sea-level changes do not seem to be strongly correlated with the diversity on land (Fara 2002; Butler et al. 2010). The investigation of the 'common cause' hypothesis is beyond the scope of this study.

Instead, the present study thoroughly investigates the relationship of rock availability and the diversity of non-mammalian synapsids, as few previous studies have dealt with this topic explicitly (Fröbisch 2008; Abdala and Ribeiro 2010). The goal is to investigate for the first time diversity patterns of Synapsida as a whole and its subclades at the regional scale of the South African Karoo Basin. The latter contains an almost unbroken sequence of continental sedimentation from the Middle Permian to the Early Jurassic with exceptional records of amphibians, synapsids, parareptiles, and eureptiles alike (Rubidge 2005). Specifically, the continuous deposits of the Permian-Triassic Beaufort Group represent one of the best records of terrestrial biodiversity during that time frame (Rubidge 1995). Special focus of the present study will be given to rock availability, the significance of the end-Permian extinction in the terrestrial realm, and the extraction of biological signals from the diversity patterns observed in the fossil record by applying corrections for a geological sampling bias.

Materials and Methods

This study is based on a dataset of Karoo vertebrates compiled by Angielczyk et al. (2005; also see Roopnarine et al. 2007), which was subsequently updated using the subsequent literature and personal observations. The dataset represents a current list of vertebrate genus richness per assemblage zone (AZ) in the South African Karoo Basin, but has been modified to include synapsid genera only (see Appendix 18.1). Taxonomic diversity estimates (TDEs) were counted as genera present within one time bin (i.e. assemblage zone) not including phylogenetic ghost lineages.

King's (1991) compilation of outcrop area values of the eight well-established South African assemblage zones served as proxy for rock availability. For each AZ, King (1991) provided values of the total outcrop area (TO), the "unproductive" area (UO), and the adjusted "productive" outcrop area (PO), the latter representing the total outcrop area minus the "unproductive" area. In this framework, the "unproductive" values represent areas covered with soil or vegetation, areas with no relief, or unsampled areas (see King 1991 for details). There are potential problems with the outcrop data provided by King (1991): First, the total outcrop data was collected from a relatively low-resolution map with the assemblage zone boundaries of Keyser and Smith (1979) superimposed on it. Second, the total outcrop values include areas where the Beaufort Series is overlain by Cenozoic rocks or soil. Third, the "unproductive" area values that King (1991) provided to correct for uneven sampling were derived from multiple variables with different correction factors: areas with low relief (100 % unproductive; less than 100 m elevation difference with 10 km distance), areas with vegetation cover (50 % unproductive; Karoo exposures north of the Orange River), and areas with lack of sampling (100 % unproductive; graphically estimated). Nonetheless, the values provided by King (1991) represent the only available data on outcrop areas of the assemblage zones of the South African Beaufort Group. A new compilation of detailed outcrop area data from local-scale geological maps would have the potential to be more accurate. However, local-scale geological maps contain lithological units rather than biostratigraphic assemblage zone data; conversion from the former to the latter is not always simple and would introduce another potential source of error.

All data were log-transformed, applying the function [f(TDE) = $\log_{10}$(TDE + 1)]. In addition, the data were transformed using generalized differencing (McKinney 1990; see also Alroy 2000) rather than simple first difference transformation, to remove a potential short-term autocorrelation of the diversity data. Generalized differencing (GD) works by detrending the data by calculating residuals of the diversity data regressed against time, and then removing autocorrelation by taking differences between neighboring values, which are ultimately modulated by the strength of the correlation between the neighbors. In this weighting scheme, weak autocorrelation is treated as little or no change to the original curve, whereas generalized differencing approximates simple first differencing when autocorrelation is strong. As autocorrelation could be an important factor, the results of the generalized differenced data series should be regarded as most accurate from an analytical perspective.

To assess the impact of the terrestrial rock record on the diversity patterns of terrestrial vertebrates in the Karoo Basin, Pearson's product-moment correlation coefficient (r), Spearman's rank correlation coefficient (r_s) and Kendall's rank correlation coefficient (t) were calculated to test for a correlation between taxonomic diversity and outcrop area, the latter being used as a proxy for sedimentary rock volume per assemblage zone. Pearson's r is a measure of the strength of correlation between two variables, Spearman's r_s measures whether peaks in two data series are correlated via a rank ordered correlation, and Kendall's τ measures whether two curves simultaneously rise and fall. Significant correlations have a p value of less than 0.05. In addition, significance levels were adjusted for multiple comparisons (Bonferroni correction), which utilizes a p value that is divided by the number (n) of pairwise comparisons with the same data series, i.e. $0.05/n$ (e.g., see Quinn and Keough 2002). The number of pairwise comparisons with the same data series (family size) ranged from two to eleven. Both significance levels are discussed below.

The method of Smith and McGowan (2007; see also McGowan and Smith 2008) was used to model the diversity of synapsids, assuming that rock availability (outcrop area) is a perfect predictor of the TDE for each synapsid clade. The modeled diversity estimate (MDE) is constructed by independently rank-ordering the $\log_{10}$-transformed TDE and outcrop area values. A least-squares regression of the form $y = mx + c$ was calculated for the re-ordered data. This equation was then applied to the outcrop area values in their original order, representing the MDE for each time interval. The difference between TDE and MDE is the residual diversity, i.e. the component of diversity that cannot simply be explained by variation in rock availability.

To examine the effect of variable duration of the eight time bins (i.e. assemblage zones), the correlation between bin length and taxonomic diversity, as well as between bin length and outcrop area, also were examined. Moreover, data series were corrected for bin duration and reanalyzed for correlation. Stratigraphic correlation of the South African assemblage zones with the international marine stages is mainly based on Rubidge (2005; see also Fröbisch 2009).

Finally, previous analyses clearly showed that the Early Triassic time interval was a distinct outlier with particularly low taxon counts despite increased sampling (Benton et al. 2004; Fröbisch 2008). Therefore, the Lower Triassic *Lystrosaurus* AZ was excluded in an additional set of analyses to test the impact of the end-Permian extinction event on the relationship between synapsid diversity and outcrop area in the Beaufort Group of the South African Karoo Basin.

All statistical analyses were performed in the software package PAST version 2.0 (Hammer et al. 2001).

Results

Raw Taxonomic Diversity Estimates for Synapsida in the Karoo Basin

At the genus level, the unadjusted synapsid TDE in the Karoo Basin suggests a rapid initial diversification, a pronounced mid-Permian extinction event, a second but more gradual diversification in the Late Permian that peaks in the *Dicynodon* AZ, and a distinct and equally pronounced extinction event at the end of the Permian, followed by low diversity in the Lower and Middle Triassic assemblage zones (Fig. 18.1). In fact, the peak in synapsid richness in the *Tapinocephalus* AZ (45 genera) is identical that of the *Dicynodon* AZ (45 genera), whereas the number of known synapsid taxa after the mid-Permian extinction in the *Pristerognathus* AZ (11 genera) is even slightly lower than in the Lower Triassic *Lystrosaurus* AZ (13 genera). Thus, the drop in synapsid diversity in the South African Karoo Basin is slightly more pronounced at the mid-Permian event (about 75 %) than the corresponding diversity drop at the end-Permian event (about 71 %).

Individual synapsid clades exhibit quite divergent diversity patterns (Fig. 18.1). For example, anomodonts seem to most closely reflect the diversity pattern of Synapsida as a whole. This may not be surprising, as anomodonts represent the richest Permian-Triassic synapsid clade. Interestingly, however, anomodonts do not seem to exclusively drive the synapsid pattern. In fact, there seems to be a general synapsid pattern that is preserved even when

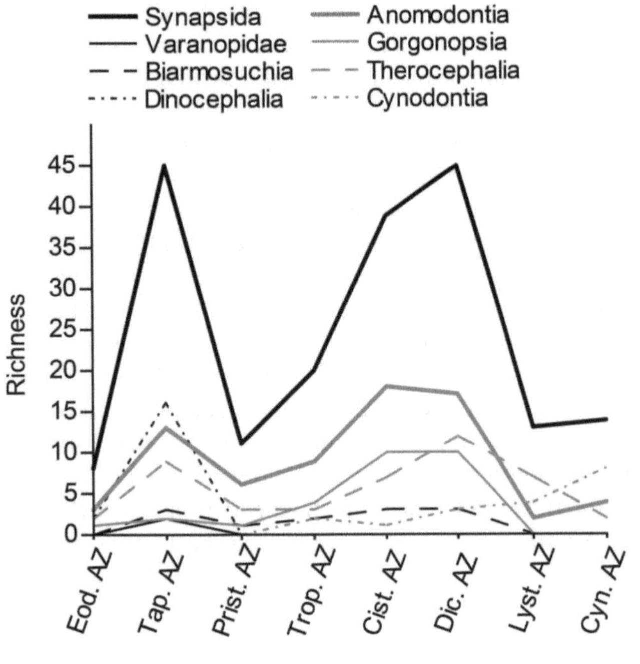

Fig. 18.1 Raw genus diversity curves (TDEs) of Synapsida and its subclades for the South African assemblage zones

anomodonts are excluded from the dataset (Fig. 18.1; Table 18.1). The main differences between the diversity curves of all synapsids and that of anomodonts are that in anomodonts the diversity peak in the *Tapinocephalus* AZ (13 genera) is lower than that in the Late Permian (18 genera), that the Late Permian diversity peak of anomodonts is actually slightly higher in the *Cistecephalus* AZ (18 genera) than in the *Dicynodon* AZ (17 genera), and that the drop in diversity is less pronounced at the mid-Permian event (54 %) compared to the end-Permian event (88 %). In contrast, only two varanopid genera are known from South Africa, both occurring in the *Tapinocephalus* AZ. Dinocephalians are also confined to the Middle Permian assemblage zones of the Karoo Basin, but they reach an enormous diversity peak in the *Tapinocephalus* AZ, before they disappear entirely shortly thereafter in the *Pristerognathus* AZ. Biarmosuchians display a relatively low and nearly constant diversity profile with a maximum of three genera per assemblage zone, although they are absent from the *Eodicynodon* and the Triassic assemblage zones. Gorgonopsian diversity does not reflect a mid-Permian event but increases to a maximum of ten genera before the clade's extinction at the end of the Permian. Therocephalians show a diversity pattern similar to anomodonts and Synapsida as a whole, with a distinct mid-Permian extinction (67 %) in therocephalians. However, their diversity profile continues to fall after the end-Permian event from the Early Triassic into the Middle Triassic, with diversity drops of 42 and 71 %, respectively. In contrast, members of the Cynodontia first appear in the *Tropidostoma* AZ and their diversity increases nearly continuously even across the end-Permian extinction event, reaching its maximum of eight genera in the Middle Triassic *Cynognathus* AZ (Fig. 18.1).

Influence of the Rock Record of the Karoo Basin on Synapsid Diversity

At first sight, the distribution of outcrop area through time appears similar to the diversity profile of Synapsida as a whole (Fig. 18.2). However, the peak in outcrop area in the Middle Triassic *Tapinocephalus* AZ is quite low and thereafter outcrop area seems to increase more or less continuously, reaching a maximum in the Lower Triassic *Lystrosaurus* AZ instead of the Upper Permian *Dicynodon* AZ, before decreasing significantly in the Middle Triassic *Cynognathus* AZ.

All three data series for outcrop area (TO, PO, UO) are significantly and strongly correlated with one another when log-transformed, even after Bonferroni correction (see Table 18.1 for details on all correlation values). For the generalized differenced (GD) data series, only the correlation between TO and PO remains significant. None of the data series for outcrop area are correlated with bin length. TDEs of Synapsida as a whole and its subclades are likewise not correlated with bin length. Moreover, for the complete data series, only the diversity profiles of Therocephalia and Synapsida minus anomodonts display a statistically significant positive correlation between log-transformed and generalized differenced data and outcrop area (TO), although Bonferroni correction renders these correlations non-significant. These results hold true when TDEs are compared with King's (1991) productive outcrop area (PO) only (results not shown), which is not surprising due to the significant positive correlation between total outcrop area (TO) and productive outcrop area (PO). Thus, it seems obvious that the effects of correction factor for the total outcrop area suggested by King (1991) are negligible. Furthermore, the results are consistent when the data series are standardized by time (results not shown).

The *Lystrosaurus* AZ has been shown to be a distinct outlier, due to its particularly low diversity compared to sampling (Fröbisch 2008). Interestingly, after exclusion of the *Lystrosaurus* AZ from all data series, the TDEs of selected synapsid subclades show a statistically significant and strong positive correlation with outcrop area (TO) for both log-transformed and generalized differenced (GD) data. For the comparison of GD values for the TDE and outcrop area excluding the *Lystrosaurus* AZ, the *Lystrosaurus* AZ values were on the one hand simply deleted from the already transformed data series and on the other hand

Table 18.1 Summary of values for Pearson's product-moment correlation coefficient (r), Spearman's rank correlation coefficient (r_s) and Kendall's rank correlation coefficient (t) for comparisons of the main data series used in this study, including raw data and data transformed using generalized differencing

Variable 1	Variable 2	N	Pearson's ρ	Spearman's r_s	Kendall's τ	p-value
logTO	logBL	8	0.072; p = 0.866	0.084; p = 0.845	0.000; p = 1.000	0.0045
logPO	logBL	8	−0.056; p = 0.895	0.024; p = 0.964	−0.036; p = 0.900	0.0045
logUO	logBL	8	0.322; p = 0.437	0.275; p = 0.508	0.255; p = 0.378	0.0045
logTO	logPO	8	**0.984; p = 0.00001**[*]	**0.994; p = 0.0001**[*]	**0.982; p = 0.0007**[*]	0.0045
logTO	logUO	8	**0.755; p = 0.030**	**0.898; p = 0.005**	**0.764; p = 0.008**	0.0045
logPO	logUO	8	0.673; p = 0.068	**0.881; p = 0.007**[*]	**0.714; p = 0.013**[*]	0.0167
GD TO	GD PO	7	**0.978; p = 0.0001**[*]	**0.964; p = 0.0004**[*]	**0.905; p = 0.004**[*]	0.0050
GD TO	GD UO	7	0.708; p = 0.075	0.643; p = 0.110	0.429; p = 0.176	0.0050
GD PO	GD UO	7	0.600; p = 0.155	0.571; p = 0.167	0.333; p = 0.293	0.0250
logTDE (S)	logBL	8	0.070; p = 0.868	0.398; p = 0.323	0.296; p = 0.305	0.0042
logTDE (Va)	logBL	8	0.314; p = 0.449	0.415; p = 0.500	0.364; p = 0.208	0.0042
logTDE (Bi)	logBL	8	−0.236; p = 0.573	−0.063; p = 0.907	−0.041; p = 0.887	0.0042
logTDE (Di)	logBL	8	0.216; p = 0.608	0.188; p = 0.786	0.160; p = 0.579	0.0042
logTDE (An)	logBL	8	−0.136; p = 0.748	−0.120; p = 0.779	−0.109; p = 0.705	0.0042
logTDE (Go)	logBL	8	−0.427; p = 0.292	−0.274; p = 0.507	−0.231; p = 0.424	0.0042
logTDE (Th)	logBL	8	−0.142; p = 0.738	0.110; p = 0.788	0.077; p = 0.790	0.0042
logTDE (Cy)	logBL	8	0.543; p = 0.165	0.491; p = 0.218	0.385; p = 0.182	0.0042
logTDE (S-An)	logBL	8	0.208; p = 0.621	0.340; p = 0.403	0.302; p = 0.335	0.0042
logTDE (S)	logTO	8	0.648; p = 0.082	0.584; p = 0.133	0.519; p = 0.072	0.0042
logTDE (Va)	logTO	8	0.099; p = 0.815	0.083; p = 1.000	0.073; p = 0.801	0.0042
logTDE (Bi)	logTO	8	0.397; p = 0.330	0.403; p = 0.304	0.410; p = 0.155	0.0042
logTDE (Di)	logTO	8	−0.182; p = 0.666	−0.314; p = 0.536	−0.267; p = 0.355	0.0042
logTDE (An)	logTO	8	0.352; p = 0.393	0.299; p = 0.470	0.400; p = 0.166	0.0042
logTDE (Go)	logTO	8	0.308; p = 0.458	0.281; p = 0.495	0.269; p = 0.351	0.0042
logTDE (Th)	logTO	8	**0.756; p = 0.030**	**0.774; p = 0.033**	**0.616; p = 0.033**	0.0042
logTDE (Cy)	logTO	8	0.532; p = 0.175	0.466; p = 0.243	0.385; p = 0.182	0.0042
logTDE (S-An)	logTO	8	**0.726; p = 0.041**	**0.727; p = 0.050**	**0.566; p = 0.050**	0.0042
logTDE (S)[a]	logTO[a]	7	**0.915; p = 0.004**[*]	**0.918; p = 0.008**	**0.850; p = 0.007**	0.0056
logTDE (Va)[a]	logTO[a]	7	0.195; p = 0.675	0.206; p = 0.857	0.183; p = 0.565	0.0056
logTDE (Bi)[a]	logTO[a]	7	**0.796; p = 0.032**	**0.906; p = 0.017**	**0.813; p = 0.010**	0.0056
logTDE (Di)[a]	logTO[a]	7	−0.102; p = 0.827	−0.225; p = 0.643	−0.202; p = 0.524	0.0056
logTDE (An)[a]	logTO[a]	7	**0.887; p = 0.008**	**0.955; p = 0.004**[*]	**0.878; p = 0.006**	0.0056
logTDE (Go)[a]	logTO[a]	7	0.695; p = 0.083	**0.844; p = 0.028**	**0.667; p = 0.035**	0.0056
logTDE (Th)[a]	logTO[a]	7	**0.757; p = 0.049**	**0.844; p = 0.027**	**0.718; p = 0.024**	0.0056
logTDE (Cy)[a]	logTO[a]	7	0.440; p = 0.324	0.374; p = 0.388	0.264; p = 0.406	0.0056
logTDE (S-An)[a]	logTO[a]	7	**0.878; p = 0.009**	**0.873; p = 0.018**	**0.750; p = 0.018**	0.0056
GD TDE (S)	GD TO	7	0.672; p = 0.098	0.643; p = 0.110	0.619; p = 0.051	0.0045
GD TDE (Va)	GD TO	7	−0.241; p = 0.603	−0.286; p = 498	−0.048; p = 0.881	0.0045
GD TDE (Bi)	GD TO	7	0.378; p = 0.403	0.250; p = 0.595	0.429; p = 0.176	0.0045
GD TDE (Di)	GD TO	7	−0.156; p = 0.739	0.000; p = 1.000	0.048; p = 0.881	0.0045

(continued)

Table 18.1 (continued)

Variable 1	Variable 2	N	Pearson's ρ	Spearman's r_s	Kendall's τ	p-value
GD TDE (An)	GD TO	7	0.286; p = 0.534	0.250; p = 0.595	0.429; p = 0.176	0.0045
GD TDE (Go)	GD TO	7	0.175; p = 0.707	0.179; p = 0.713	0.238; p = 0.453	0.0045
GD TDE (Th)	GD TO	7	**0.886; p = 0.008**	**0.786; p = 0.048**	**0.619; p = 0.050**	0.0045
GD TDE (Cy)	GD TO	7	0.352; p = 0.438	0.464; p = 0.302	0.333; p = 0.293	0.0045
GD TDE (S-An)	GD TO	7	0.748; p = 0.053	0.714; p = 0.088	0.524; p = 0.099	0.0045
GD TDE (S)[a]	GD TO[a]	6	**0.964; p = 0.002**[*]	**1.000; p = 0.003**[*]	**1.000; p = 0.005**[*]	0.0056
GD TDE (Va)[a]	GD TO[a]	6	−0.207; p = 0.694	−0.257; p = 0.564	−0.067; p = 0.851	0.0056
GD TDE (Bi)[a]	GD TO[a]	6	**0.957; p = 0.003**[*]	**1.000; p = 0.003**[*]	**1.000; p = 0.005**[*]	0.0056
GD TDE (Di)[a]	GD TO[a]	6	−0.163; p = 0.758	−0.086; p = 0.803	−0.067; p = 0.851	0.0056
GD TDE (An)[a]	GD TO[a]	6	**0.991; p = 0.0001**[*]	**1.000; p = 0.003**[*]	**1.000; p = 0.005**[*]	0.0056
GD TDE (Go)[a]	GD TO[a]	6	**0.857; p = 0.029**	**0.886; p = 0.017**	**0.733; p = 0.039**	0.0056
GD TDE (Th)[a]	GD TO[a]	6	**0.897; p = 0.015**	**0.886; p = 0.017**	**0.733; p = 0.039**	0.0056
GD TDE (Cy)[a]	GD TO[a]	6	0.204; p = 0.698	0.143; p = 0.714	0.067; p = 0.851	0.0056
GD TDE (S-An)[a]	GD TO[a]	6	**0.875; p = 0.022**	**0.886; p = 0.017**	**0.733; p = 0.039**	0.0056
GD TDE (S)[b]	GD TO[b]	6	**0.896; p = 0.020**	**0.829; p = 0.033**	**0.733; p = 0.039**	0.0056
GD TDE (Va)[b]	GD TO[b]	6	−0.265; p = 0.612	−0.257; p = 0.564	−0.067; p = 0.851	0.0056
GD TDE (Bi)[b]	GD TO[b]	6	**0.953; p = 0.003**[*]	**0.943; p = 0.003**[*]	**0.867; p = 0.015**	0.0056
GD TDE (Di)[b]	GD TO[b]	6	−0.198; p = 0.707	−0.314; p = 0.564	−0.200; p = 0.573	0.0056
GD TDE (An)[b]	GD TO[b]	6	**0.990; p = 0.0002**[*]	**1.000; p = 0.003**[*]	**1.000; p = 0.005**[*]	0.0056
GD TDE (Go)[b]	GD TO[b]	6	**0.931; p = 0.007**	**0.886; p = 0.017**	**0.733; p = 0.039**	0.0056
GD TDE (Th)[b]	GD TO[b]	6	0.767; p = 0.075	0.771; p = 0.102	0.600; p = 0.091	0.0056
GD TDE (Cy)[b]	GD TO[b]	6	0.230; p = 0.661	−0.257; p = 0.564	−0.200; p = 0.573	0.0056
GD TDE (S-An)[b]	GD TO[b]	6	0.746; p = 0.088	0.714; p = 0.136	0.467; p = 0.188	0.0056
logTDE (S)	MDE (S)	8	0.648; p = 0.082	0.599; p = 0.125	0.546; p = 0.059	0.0125
logTDE (Va)	MDE (Va)	8	0.099; p = 0.815	0.083; p = 1.000	0.071; p = 0.805	0.0125
logTDE (Bi)	MDE (Bi)	8	0.397; p = 0.330	0.401; p = 0.334	0.403; p = 0.163	0.0125
logTDE (Di)	MDE (Di)	8	−0.182; p = 0.666	−0.312; p = 0.500	−0.262; p = 0.364	0.0125
logTDE (An)	MDE (An)	8	0.352; p = 0.392	0.286; p = 0.462	0.357; p = 0.216	0.0125
logTDE (Go)	MDE (Go)	8	0.308; p = 0.458	0.279; p = 0.496	0.265; p = 0.359	0.0125
logTDE (Th)	MDE (Th)	8	**0.756; p = 0.030**	**0.800; p = 0.023**	**0.643; p = 0.026**	0.0125
logTDE (Cy)	MDE (Cy)	8	0.532; p = 0.175	0.488; p = 0.227	0.416; p = 0.150	0.0125
logTDE (S-An)	MDE (S-An)	8	**0.726; p = 0.041**	**0.735; p = 0.046**	**0.593; p = 0.040**	0.0125
logTDE (S)[a]	MDE (S)[a]	7	**0.915; p = 0.004**[*]	**0.937; p = 0.005**[*]	**0.878; p = 0.006**[*]	0.0167
logTDE (Va)[a]	MDE (Va)[a]	7	0.195; p = 0.675	0.204; p = 0.857	0.178; p = 0.574	0.0167
logTDE (Bi)[a]	MDE (Bi)[a]	7	**0.796; p = 0.032**	**0.898; p = 0.019**	**0.794; p = 0.012**[*]	0.0167
logTDE (Di)[a]	MDE (Di)[a]	7	−0.102; p = 0.827	−0.222; p = 0.667	−0.197; p = 0.534	0.0167
logTDE (An)[a]	MDE (An)[a]	7	**0.887; p = 0.008**[*]	**0.929; p = 0.007**[*]	**0.810; p = 0.011**[*]	0.0167
logTDE (Go)[a]	MDE (Go)[a]	7	0.695; p = 0.083	**0.837; p = 0.027**	**0.651; p = 0.040**	0.0167
logTDE (Th)[a]	MDE (Th)[a]	7	**0.757; p = 0.049**	**0.873; p = 0.017**	**0.751; p = 0.018**	0.0167
logTDE (Cy)[a]	MDE (Cy)[a]	7	0.440; p = 0.324	0.408; p = 0.371	0.309; p = 0.330	0.0167
logTDE (S-An)[a]	MDE (S-An)[a]	7	**0.878; p = 0.009**[*]	**0.883; p = 0.015**[*]	**0.781; p = 0.014**[*]	0.0167

(continued)

Table 18.1 (continued)

Variable 1	Variable 2	N	Pearson's ρ	Spearman's r_s	Kendall's τ	p-value
logTDE (S)	logTDE (An)	8	**0.893; p = 0.003***	**0.802; p = 0.022**	**0.618; p = 0.032**	0.0100
logTDE (S-An)	logTDE (S)	8	**0.964; p = 0.0001***	**0.946; p = 0.002***	**0.868; p = 0.003***	0.0100
logTDE (S-An)	logTDE (An)	8	**0.744; p = 0.034**	0.675; p = 0.077	0.445; p = 0.123	0.0100
GD TDE (S)	GD TDE (An)	7	**0.853; p = 0.015***	**0.893; p = 0.012***	**0.810; p = 0.011***	0.0167
GD TDE (S-An)	GD TDE (S)	7	**0.939; p = 0.002***	**0.857; p = 0.012***	**0.714; p = 0.024**	0.0167
GD TDE (S-An)	GD TDE (An)	7	0.637; p = 0.124	0.714; p = 0.088	0.524; p = 0.099	0.0167

The *p* value in the right column is adjusted for multiple comparisons (Bonferroni correction), as discussed in the Materials and Methods
An Anomodontia, *Bi* Biarmosuchia, *BL* bin length, *Cy* Cynodontia, *Di* Dinocephalia *GD* generalized differenced data, *Go* Gorgonopsia, *N* number of data points for each comparison, *log* log-transformed data, *PO* productive outcrop area, *S* Synapsida, *S-An* Synapsida excluding anomodonts, *Th* Therocephalia, *TO* total outcrop area, *UO* unproductive outcrop area, *Va* Varanopidae

[a] data series excluding the *Lystrosaurus* AZ

[b] recalculated data series after exclusion of the *Lystrosaurus* AZ

* statistically significant value after Bonferroni correction; bold values, statistically significant with *p* value of less than 0.05

the GD values were re-calculated after exclusion of the *Lystrosaurus* AZ. Using the first calculation scheme, the TDEs of Synapsida as a whole, Synapsida minus anomodonts, and the synapsid subclades Biarmosuchia, Anomodontia, Gorgonopsia, and Therocephalia (see below) show a strong positive correlation with outcrop area (TO) for both log-transformed and generalized differenced data. Of these clades, only the correlations between outcrop area and Synapsida TDE minus anomodonts and therocephalian TDE and outcrop area (TO) are non-significant when the GD values are recalculated after deletion of the *Lystrosaurus* AZ. Bonferroni correction of the *p* values renders most of the correlations of the log-transfromed data non-significant, except for Pearson's r for Synapsida and Spearman's r_s for Anomodontia. Nonetheless, for the GD data series the strong positive correlation between TDE and outcrop area remains robust for Synapsida, Biarmosuchia, Anomodontia, and Gorgonopsia.

Corrected Diversity Estimates for Synapsida in the Karoo Basin

The modeled diversity estimates (MDEs) are derived from the outcrop area and its correlation (based on least-squares regression) with the respective taxonomic diversity estimates (TDEs). MDEs of Synapsida and its subclades are not correlated with the respective TDEs for the complete data series (Table 18.1). Conversely, when the *Lystrosaurus* AZ values are excluded, there is again an obvious and statistically significant positive correlation between the MDE and TDE in Synapsida, Biarmosuchia, Anomodontia, Gorgonopsia, Therocephalia, and Synapsida minus anomodonts (Table 18.1), comparable to the relationship between outcrop area and TDE in these clades (see above).

Calculation of diversity residuals by subtracting the MDEs from the TDEs of Synapsida and its subclades results in clade-specific residual trends (Fig. 18.3). The trend in residual taxonomic diversity in Synapsida as a whole has a similar pattern of peaks and troughs when compared to that of raw or log-transformed diversity. The residual values are positive in the Middle Permian reaching a peak in the *Tapinocephalus* AZ, followed by values close to zero in the subsequent two assemblage zones and a continuous rise to a low peak in the Upper Permian *Dicynodon* AZ. Thereafter, synapsid residuals drop to high negative values in the earliest Triassic *Lystrosaurus* AZ, followed by low negative values in the Middle Triassic *Cynognathus* AZ.

Similar but slightly varying trends in residual taxonomic diversity are also seen in biarmosuchians, anomodonts, gorgonopsians, and therocephalians. In biarmosuchians, residual diversity is positive throughout the Permian assemblage zones of the Karoo Basin, increases towards a maximum in the lower Upper Permian *Pristerognathus* AZ, thereafter gradually decreases towards low positive values in the Late Permian, and finally drops abruptly to high negative values in the Triassic.

The trend in residual diversity of anomodonts is also positive throughout the entire Permian, exhibiting high positive values in the Middle and early Late Permian, values close to zero in the *Tropidostoma* AZ, and a small peak in the late Late Permian, before dropping to high negative values in the Early and Middle Triassic.

Gorgonopsian residual diversity is highest in the Middle Permian *Eodicynodon* AZ, drops to a low negative value in the *Tapinocephalus* AZ, thereafter increases to high positive values in the Late Permian, and drops to high negative values in the Triassic.

The residual diversity of therocephalians starts with high positive values in the Middle and early Late Permian with a peak in the *Tapinocephalus* AZ, drops abruptly to high

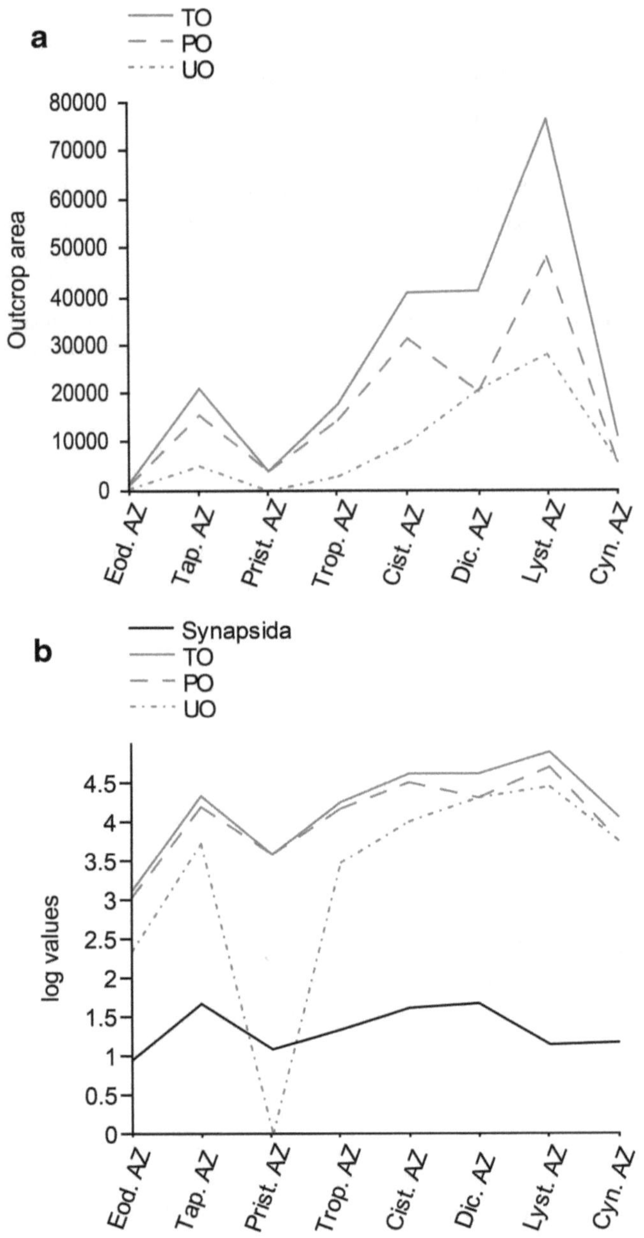

Fig. 18.2 Profiles of **a** raw outcrop area (total, productive, and unproductive), and **b** log-transformed genus diversity of Synapsida and outcrop area for the South African assemblage zones

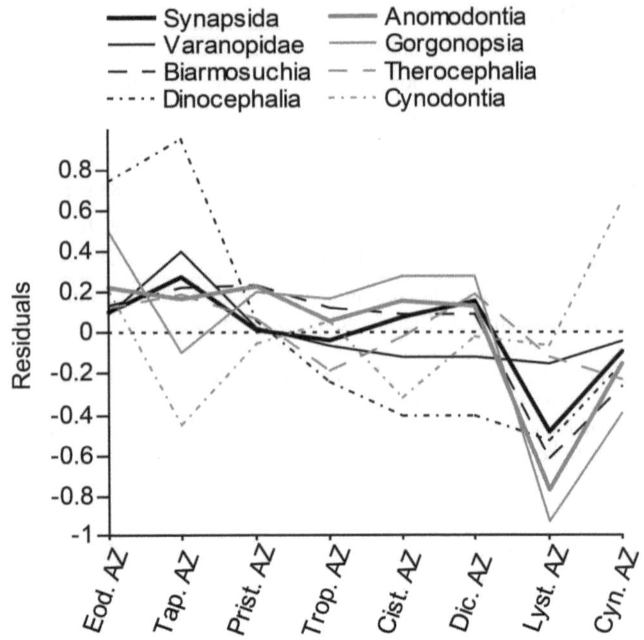

Fig. 18.3 Plots of residual taxonomic diversity after correction using a model in which rock availability (outcrop area) is a perfect predictor of the TDE for Synapsida and its subclades per South African assemblage zone

negative values in the *Tropidostoma* AZ, increases again to a high in the Upper Permian *Dicynodon* AZ, before it drops to high negative values in the Triassic and an overall minimum in the *Cynognathus* AZ.

Interestingly, cynodonts seem to show an almost inverse pattern compared to those seen in all synapsids and the already-discussed subclades. Their residual diversity profile exhibits a minimum in the Middle Permian *Tapinocephalus* AZ, reaches values close to zero in the early Late Permian before dropping to a second minimum in the Upper Permian *Cistecephalus* AZ, values close to zero in the latest Permian and Early Triassic and an overall maximum in the Middle Triassic *Cynognathus* AZ.

The residual diversity trends in varanopids and dinocephalians are very different from the general synapsid pattern, exhibiting high positive values in the Middle Permian only and dropping to negative or borderline positive values for all younger assemblage zones.

Discussion

Previous studies of diversity patterns of Permian-Triassic tetrapods (including synapsids) were undertaken at the global scale (Benton 1985, 1995; Maxwell 1992; Sahney and Benton 2008; Sahney et al. 2010) as well as on the regional scales of Russia (Benton et al. 2004) and South Africa (King 1990, 1991). These studies have primarily focused on ecological aspects of the extinction event (e.g., turnover rates of specific guilds, such as carnivores and herbivores), have employed coarse taxonomic levels such as families, or have targeted the differentiation between amphibians and amniotes only. Other studies focused explicitly on Permian-Triassic boundary sequences to investigate the detailed stratigraphic occurrences of relevant taxa and their significance for interpreting patterns of the end-Permian extinction in the terrestrial realm (Smith and Ward 2001; Smith and Botha 2005; Ward et al. 2005; Botha and Smith 2006, 2007; Botha-Brink et al. 2013). Selected studies specifically downplayed the influence of sampling biases on Permian-Triassic diversity patterns (e.g., Sahney and Benton 2008; Sahney et al. 2010). The only exception to this is an investigation of the global diversity of anomodont synapsids, which clearly demonstrated the impact of the rock record on diversity trends (Fröbisch 2008).

The present study thoroughly investigates the relationship of rock availability and synapsid diversity. For the first time, it addresses synapsid diversity in its entirety and focuses on the specific response of individual synapsid clades to the end-Permian event. By taking into account the rock record, the present analysis uncovers biological signals underpinning observed patterns of fluctuating taxonomic diversity. The results of this work emphasize the need for sampling standardization and specifically the control of rock availability, as highlighted in recent paleobiological contributions (e.g., Alroy et al. 2001, 2008; Peters and Foote 2001; Smith 2001; Smith et al. 2001; Crampton et al. 2003; Peters 2005, 2006; Fröbisch 2008; Lloyd et al. 2008; McGowan and Smith 2008; Barrett et al. 2009; Butler et al. 2009, 2010; Benson et al. 2010; Mannion et al. 2011). Whereas the effect of a 'common cause' for similar patterns in diversity and rock availability requires additional testing (e.g., Sepkoski 1976; Peters and Foote 2001; Smith et al. 2001; Peters 2005; Benton and Emerson 2007; Benton 2009), the role of sea-level changes seems to be doubtful with respect to diversity patterns in the terrestrial realm (Butler et al. 2010). This suggests that sampling biases should be regarded as the null hypothesis for explaining short-term fluctuations in diversity and that a correction for those biases is essential to extract genuine biological signals from raw diversity counts.

On the regional scale of the South African Karoo Basin, raw synapsid diversity shows two distinct and equally pronounced extinction events, namely a mid-Permian event, potentially reflecting an event synchronous with the marine end-Guadalupian extinction (but see Lucas 2009), as well as an end-Permian event (Fig. 18.1). It is notable that the individual synapsid clades portray divergent diversity patterns through time, suggesting the reflection of at least some genuine biological signals rather than a pure sampling bias imposed by the rock record.

Anomodontia, the most speciose and abundant clade within non-mammalian synapsids in the Karoo Basin, seems to most closely reflect the diversity pattern of all synapsids. Thus, due to the large amount of data shared between these two clades, anomodonts seem to drive synapsid diversity to a certain degree. However, the general synapsid pattern persists when anomodonts are excluded from the synapsid dataset (Table 18.1). In fact, the general synapsid pattern is furthermore approximated by biarmosuchians, gorgonopsians, and therocephalians. The lack of correlation between outcrop area and TDEs of selected subclades of Synapsida, in particular Varanopidae and Dinocephalia, most likely results from their limited occurrence in the eight assemblage zones. In contrast, the diversity profile of cynodonts, which are absent from the three oldest assemblage zones, displays a distinct pattern that is divergent from any other synapsid clade in that it continuously rises across the Permian-Triassic boundary. This pattern most likely represents a genuine signal, as this is also apparent in the residual diversity of this clade. The observed continuous diversity increase of cynodonts across the Permian-Triassic boundary has also been noted by previous studies on a broader geographic scale (Abdala and Ribeiro 2010). As cynodonts ultimately gave rise to mammals, this pattern matches well with Sepkoski's (1981) observation in the marine realm where the 'Modern fauna' was largely unaffected by the end-Permian extinction, which ultimately influenced the shaping of the modern diversity.

In general, the TDEs of Synapsida and its subclades are not significantly correlated with outcrop area for the complete Permian-Triassic time series in the Karoo Basin, the only exception being Therocephalia. These results conform to observation made by previous workers at the regional scales of South Africa and Russia (King 1991; Benton et al. 2004). However, Fröbisch (2008) pointed out that scatter plots of global anomodont diversity and anomodont-bearing tetrapod faunas per time interval show distinct outliers in the mid-Permian and in the Early Triassic, despite a statistically significant positive correlation between the two variables. The mid-Permian outlier resulted from particularly high taxon counts in a few faunas, whereas the Early Triassic outlier was based on very low taxon counts despite a high number of known faunas. Therefore, an exclusion of the Lower Triassic *Lystrosaurus* AZ values seems to be the obvious way to test the impact of the end-Permian extinction event on the relationship between synapsid diversity and outcrop area in the South African Karoo Basin. Hence, after exclusion of the *Lystrosaurus* AZ values from the datasets, the TDEs of selected synapsid subclades (Synapsida, Biarmosuchia, Anomodontia, and Gorgonopsia) show strong, statistically significant positive correlations with outcrop area. This implies that the pattern of synapsid diversity in the Karoo Basin is significantly biased by the Permian-Triassic terrestrial rock record, confirming previous observations of anomodont synapsids at a global scale (Fröbisch 2008). However, the rock record bias is not evident in all clades and is only apparent after removal of the *Lystrosaurus* AZ from the datasets. Thus, despite the obvious bias, the end-Permian extinction maintains a major impact on synapsid diversity.

The trends of residual taxonomic diversity, calculated by subtracting MDEs from TDEs, exhibit clade-specific patterns (see above), but the profiles of some clades show a clear similarity. This is the case for Synapsida,

Biarmosuchia, Anomodontia, Gorgonopsia, and Therocephalia, which altogether are quite similar to the raw and log-transformed diversity profile of Synapsida as a whole. However, they show only varying support for a mid-Permian event but strongly suggest a high impact for the end-Permian extinction. The clade-specific residual trends among the subclades of Synapsida most likely represent genuine biological signals. This is best reflected in the variation among all clades, but specifically in the entirely divergent pattern displayed by cynodonts, which is almost the inverse of that seen in most other clades.

A pressing question remains: how many extinction events are reflected in the Permian-Triassic synapsid record of the South African Karoo Basin? Previous studies documented a mid-Permian extinction, possibly coinciding with the marine end-Guadalupian event, and an end-Permian extinction (King 1991; Fröbisch 2008; Sahney and Benton 2008). However, the influence of the mid-Permian event does not seem to be reflected by the diversity trends of all individual subclades of Synapsida. Instead, it results mainly from the complete extinction of varanopids and the much more diverse dinocephalians, as well as from a decrease in diversity in therocephalians (see Fig. 18.3). Thus, a simultaneous extinction in the mid-Permian, involving multiple synapsid lineages, is currently not supported by the data, questioning its identification as a mass extinction event on land. Conversely, the end-Permian extinction is clearly reflected in the majority of diversity profiles of the individual synapsid subclades, except the Cynodontia (and the already extinct Varanopidae and Dinocephalia). Thus, the results of the present study do not question the importance of the end-Permian extinction event but rather emphasize its significance and impact on terrestrial ecosystems.

Finally, Uhen and Pyenson (2007) and Marx (2009) suggested based on their studies of cetaceans that smaller, well-defined groups at lower taxonomic levels may preserve a larger amount of detectable biological information and may not be as prone to bias as larger clades. This hypothesis is only to a certain degree supported by the present study on synapsid subclades. First, strong biological signals imposed my mass extinction events may obscure the apparent correlation between rock record and paleodiversity, but this doesn't mean that a bias is absent. Second, whereas the diversity of some clades doesn't correlate with outcrop area, others are strongly correlated with the latter, emphasizing the need for thorough investigations of this matter in a large range of taxonomic groups and subsequent corrections for existing biases.

Conclusion

The most severe extinction event in Earth's history at the end of the Permian had a major impact on terrestrial ecosystems. Synapsids, the dominant terrestrial tetrapods in the Permian-Triassic, are of particular importance for assessing the impact of the end-Permian mass extinction in the terrestrial realm. The present study focuses on the regional scale of the Karoo Basin, examines the relationship of rock availability and synapsid diversity, and for the first time presents diversity patterns for Synapsida as a whole and its major subclades to extract clade-specific trends and potential genuine biological signals. The raw TDE of Synapsida reflects two distinct and equally pronounced extinction events, namely a mid-Permian and an end-Permian event, whereas the individual subclades of Synapsida exhibit varying diversity profiles. Thereby, the diversity profile of anomodonts, the most speciose and abundant clade of Permian-Triassic synapsids, most closely reflects the pattern of all synapsids, whereas the profile of cynodont diversity describes an almost inverse course. Nonetheless, it is interesting to note that there seems to be a general synapsid diversity pattern that persists when anomodonts are excluded from the dataset. The TDEs of synapsids are not significantly correlated with outcrop area for the entire time series, the only exception being Therocephalia. However, when the values of the *Lystrosaurus* AZ are excluded from all datasets, the TDEs of Synapsida, Biarmosuchia, Anomodontia, and Gorgonopsia display a statistically significant strong positive correlation with outcrop area. Thus, the pattern of synapsid diversity in the Beaufort Group of the South African Karoo Basin is significantly biased by heterogeneity in the Permian-Triassic terrestrial rock record, confirming previous observations on a global scale. This suggests that regional and global patterns may not be that different after all. However, the rock record bias is not evident in all clades and only apparent after removal of the *Lystrosaurus* AZ from the dataset. Most notably, the trends of residual taxonomic diversity, calculated by subtracting MDEs from TDEs, exhibit clade-specific diversity patterns, which imply the reflection of potential genuine biological signals rather than a pure sampling bias. A simultaneous mid-Permian extinction event is not reflected in the corrected diversity profiles of all synapsid subclades, but results mainly from the complete extinction of varanopids and the much more diverse dinocephalians, as well as a diversity decrease in therocephalians. Thus, the presence of a simultaneous mid-Permian extinction event, involving multiple synapsid lineages, is not supported by the data. In contrast, the end-Permian extinction is clearly reflected in most of the individual diversity profiles within synapsids. Hence, the impact of the end-Permian mass extinction on the diversity of synapsids and its role in shaping the composition and structure of terrestrial ecosystems remains unquestioned.

Acknowledgments I would like to thank Ken Angielczyk and Christian Kammerer for organizing the 'Synapsid Symposium' at the 69th Annual Meeting of the Society of Vertebrate Paleontology in Bristol, UK. Ken Angielczyk further provided the majority of the taxonomic and stratigraphic data on Karoo vertebrates. This work greatly benefited from reviews by Roger Benson, Philip Mannion, and Marcello Ruta. This study was financially supported by the Deutsche Forschungsgemeinschaft (FR 2457/3-1) and a Sofja Kovalevskaja Award of the Alexander von Humboldt Foundation.

Appendix 18.1

List of synapsid genera per assemblage zone of the South African Karoo Basin

Eodicynodon Assemblage Zone

Australosyodon
Eodicynodon
"*Eodicynodon*" *oelofseni*
Glanosuchus
Ictidosaurus
Patranomodon
Tapinocaninus
indet. gorgonopsian

Tapinocephalus Assemblage Zone

Alopecodon
Anomocephalus
Anteosaurus
Avenantia
Brachyprosopus
Bullacephalus
Chelydontops
Colobodectes
Crapartinella
Criocephalosaurus
Delphinognathus
Diictodon
Elliotsmithia
Eoarctops
Eosimops
Galechirus
Galeops
Galepus
Glanosuchus
Heleosaurus
Hipposaurus
Ictidosaurus
Jonkeria
Keratocephalus
Lanthanostegus
Lycosuchus
Mormosaurus
Moschops
Pachydectes
Pardosuchus
Phocosaurus
Pristerodon
Pristerognathus
Prosictodon
Riebeeckosaurus
Robertia
Scylacognathus
Scylacosaurus
Simorhinella
Struthiocephalus
Struthiocephaloides
Styracocephalus
Tapinocephalus
Taurocephalus
Titanosuchus

Pristerognathus Assemblage Zone

Dicynodontoides
Diictodon
Emydops
Endothiodon
Eosimops
Glanosuchus
Hipposaurus
Hofmeyria
Pristerodon
Pristerognathus
Scylacognathus

Tropidostoma Assemblage Zone

Charassognathus
Cistecephalus
Cyonosaurus
Dicynodontoides
Diictodon
Emydops
Endothiodon
Gorgonops
Hofmeyria
Ictidosuchoides
Ictidosuchops
Lobalopex
Lophorhinus
Lycaenops
Oudenodon
Pristerodon
Procynosuchus
Rhachiocephalus
Scylacognathus
Tropidostoma

Cistecephalus Assemblage Zone

Aelurognathus
Aelurosaurus
Aloposaurus
Arctognathus
Aulacephalodon
Basilodon
Choerosaurus
Cistecephalus
Clelandina
Compsodon
Cyonosaurus
Dicynodon
Dicynodontoides
Diictodon
Dinanomodon
Emydops
Endothiodon
Euchambersia
Euptychognathus
Gorgonops
Herpetoskylax
Ictidostoma
Ictidosuchoides
Ictidosuchops
Kitchinganomodon
Lycaenodon
Lycaenops
Mirotenthes
Myosauroides
Notaelurodon
Odontocyclops
Oudenodon
Paraburnetia
Pristerodon
Procynosuchus
Rhachiocephalus
Rubidgea
Scylacognathus
Sintocephalus

Dicynodon Assemblage Zone

Aelurognathus
Aelurosaurus
Akidnognathus
Aloposaurus
Arctognathus
Aulacephalodon
Basilodon
Burnetia
Cerdosuchoides
Cistecephaloides
Clelandina
Cynosaurus
Cyonosaurus
Daptocephalus
Dicynodon
Dicynodontoides
Diictodon
Dinanomodon
Emydops
Ictidochampsa
Ictidorhinus
Ictidosuchoides
Ictidosuchops
Keyseria
Kwazulusaurus
Lemurosaurus
Lycaenops
Lycideops
Lystrosaurus

Myosauroides
Nanictidops
Nanictosaurus
Notaelurodon
Oudenodon
Pelanomodon
Polycynodon
Pristerodon
Procynosuchus
Propelanomodon
Rubidgea
Scaloporhinus
Scylacognathus
Sycosaurus
Theriognathus
Tigrisuchus

Lystrosaurus Assemblage Zone

Ericiolacerta
Galesaurus
Ictidosuchoides
Lystrosaurus
Myosaurus
Notaelurodon
Olivierosuchus
Platycraniellus
Progalesaurus
Regisaurus
Scaloposaurus
Thrinaxodon
Tigrisuchus

Cynognathus Assemblage Zone

Angonisaurus
Bauria
Bolotridon
Cistecynodon
Cricodon
Cynognathus
Diademodon
Kannemeyeria
Kombuisia
Langbergia
Lumkuia
Mircrogomphodon
Shansiodon
Trirachodon

References

Abdala, F., & Ribeiro, A. M. (2010). Distribution and diversity patterns of Triassic cynodonts (Therapsida, Cynodontia) in Gondwana. *Palaeogeography, Palaeoclimatology, Palaeoecology, 286*(3–4), 202–217.

Allin, E. F. (1975). Evolution of the mammalian middle ear. *Journal of Morphology, 147*, 403–438.

Alroy, J. (2000). New methods for quantifying macroevolutionary patterns and processes. *Paleobiology, 26*, 707–733.

Alroy, J., Marshall, C. R., Bambach, R. K., Bezusko, K., Foote, M., Fursich, F. T., et al. (2001). Effects of sampling standardization on estimates of Phanerozoic marine diversification. *Proceedings of the National Academy of Sciences of the United States of America, 98*, 6261–6266.

Alroy, J., Aberhan, M., Bottjer, D. J., Foote, M., Fursich, F. T., Harries, P. J., et al. (2008). Phanerozoic trends in the global diversity of marine invertebrates. *Science, 321*, 97–100.

Angielczyk, K. D., Roopnarine, P. D., & Wang, S. C. (2005). Modeling the role of primary productivity disruption in end-Permian extinctions, Karoo Basin, South Africa. *New Mexico Museum of Natural History and Science Bulletin, 30*, 16–23.

Barrett, P. M., McGowan, A. J., & Page, V. (2009). Dinosaur diversity and the rock record. *Proceedings of the Royal Society B: Biological Sciences, 276*, 2667–2674.

Benson, R. B. J., Butler, R. J., Lindgren, J., & Smith, A. S. (2010). Mesozoic marine tetrapod diversity: Mass extinctions and temporal heterogeneity in geological megabiases affecting vertebrates. *Proceedings of the Royal Society B-Biological Sciences, 277*, 829–834.

Benton, M. J. (1985). Mass extinction among non-marine tetrapods. *Nature, 316*, 811–814.

Benton, M. J. (1995). Diversification and extinction in the history of life. *Science, 268*, 52–58.

Benton, M. J. (2009). The fossil record: Biological or geological signal? In D. Sepkoski & M. Ruse (Eds.), *The paleobiological revolution: Essays on the growth of modern paleontology* (pp. 43–59). Chicago: University of Chicago Press.

Benton, M. J., & Emerson, B. C. (2007). How did life become so diverse? The dynamics of diversification according to the fossil record and molecular phylogenetics. *Palaeontology, 50*, 23–40.

Benton, M. J., Tverdokhlebov, V. P., & Surkov, M. V. (2004). Ecosystem remodelling among vertebrates at the Permian-Triassic boundary in Russia. *Nature, 432*, 97–100.

Botha, J., & Smith, R. M. H. (2006). Rapid vertebrate recuperation in the Karoo Basin of South Africa following the end-Permian extinction. *Journal of African Earth Sciences, 45*, 502–514.

Botha, J., & Smith, R. M. H. (2007). *Lystrosaurus* species composition across the Permo-Triassic boundary in the Karoo Basin of South Africa. *Lethaia, 40*, 125–137.

Botha-Brink, J., Huttenlocker, A. K., & Modesto, S. P. (2013). Vertebrate paleontology of Nooitgedacht 68: A *Lystrosaurus maccaigi*-rich Permo-Triassic boundary locality in South Africa. In C. F. Kammerer, K. D. Angielczyk, & J. Fröbisch (Eds.), *Early evolutionary history of the Synapsida* (pp. 289–304). Dordrecht: Springer.

Butler, R. J., Barrett, P. M., Nowbath, S., & Upchurch, P. (2009). Estimating the effects of sampling biases on pterosaur diversity patterns: Implications for hypotheses of bird/pterosaur competitive replacement. *Paleobiology, 35*, 432–446.

Butler, R. J., Benson, R. B. J., Carrano, M. T., Mannion, P. D., & Upchurch, P. (2010). Sea level, dinosaur diversity and sampling biases: Investigating the 'common cause' hypothesis in the terrestrial realm. *Proceedings of the Royal Society B-Biological Sciences, 278*, 1165–1170.

Cluver, M. A. (1978). The skeleton of the mammal-like reptile *Cistecephalus* with evidence of a fossorial mode of life. *Annals of the South African Museum, 76*, 213–246.

Crampton, J. S., Beu, A. G., Cooper, R. A., Jones, C. M., Marshall, B., & Maxwell, P. A. (2003). Estimating the rock volume bias in paleobiodiversity studies. *Science, 301*, 358–360.

Fara, E. (2002). Sea-level variations and the quality of the continental fossil record. *Journal of the Geological Society, 159*, 489–491.

Fröbisch, J. (2006). Locomotion in derived dicynodonts (Synapsida, Anomodontia): A functional analysis of the pelvic girdle and hind limb of *Tetragonias njalilus*. *Canadian Journal of Earth Sciences, 43*, 1297–1308.

Fröbisch, J. (2007). The cranial anatomy of *Kombuisia frerensis* Hotton (Synapsida, Dicynodontia) and a new phylogeny of anomodont therapsids. *Zoological Journal of the Linnean Society, 150*, 117–144.

Fröbisch, J. (2008). Global taxonomic diversity of anomodonts (Tetrapoda, Therapsida) and the terrestrial rock record across the Permian-Triassic boundary. *PLoS ONE, 3*(11), e3733. doi: 10.1371/journal.pone.0003733.

Fröbisch, J. (2009). Composition and similarity of global anomodont-bearing tetrapod faunas. *Earth-Science Reviews, 95*, 119–157.

Fröbisch, J., & Reisz, R. R. (2009). The Late Permian herbivore *Suminia* and the early evolution of arboreality in terrestrial vertebrate ecosystems. *Proceedings of the Royal Society B-Biological Sciences, 276*, 3611–3618.

Hammer, Ø., Harper, D. A. T., & Ryan, P. D. (2001). PAST: Paleontological statistics software package for education and data analysis. *Palaeontologia Electronica, 4*, 1–9.

Hopson, J. A. (1995). Patterns of evolution in the manus and pes of non-mammalian therapsids. *Journal of Vertebrate Paleontology, 15*, 615–639.

Kemp, T. S. (2005). *The origin and evolution of mammals*. Oxford: Oxford University Press.

Keyser, A. W., & Smith, R. M. H. (1979). Vertebrate biozonation of the Beaufort Group with special reference to the Western Karoo Basin. *Annals of the Geological Survey of South Africa, 12*, 1–35.

King, G. M. (1990). Life and death in the Permo-Triassic: The fortunes of the dicynodont mammal-like reptiles. *Sidney Haughton Memorial Lecture, 3*, 17.

King, G. M. (1991). Terrestrial tetrapods and the end Permian event: A comparison of analyses. *Historical Biology, 5*, 239–255.

Kissel, R. A., & Reisz, R. R. (2004). Synapsid fauna of the Upper Pennsylvanian Rock Lake shale near Garnett, Kansas and the diversity pattern of early amniotes. In G. Arratia, M. V. H. Wilson, & R. Cloutier (Eds.), *Recent advances in the origin and early radiation of vertebrates* (pp. 409–428). München: Pfeil Verlag.

Lloyd, G. T., Davis, K. E., Pisani, D., Tarver, J. E., Ruta, M., Sakamoto, M., et al. (2008). Dinosaurs and the Cretaceous terrestrial revolution. *Proceedings of the Royal Society B: Biological Sciences, 275*, 2483–2490.

Lucas, S. G. (2009). Timing and magnitude of tetrapod extinctions across the Permo-Triassic boundary. *Journal of Asian Earth Sciences, 36*, 491–502.

Luo, Z.-X. (2007). Transformation and diversification in early mammal evolution. *Nature, 450*, 1011–1019.

Mannion, P. D., Upchurch, P., Carrano, M. T., & Barrett, P. M. (2011). Testing the effect of the rock record on diversity: A multidisciplinary approach to elucidating the generic richness of sauropodomorph dinosaurs through time. *Biological Reviews, 86*, 157–181.

Marx, F. G. (2009). Marine mammals through time: When less is more in studying palaeodiversity. *Proceedings of the Royal Society B: Biological Sciences, 276*, 887–892.

Maxwell, D. (1992). Permian and Early Triassic extinction of non-marine tetrapods. *Palaeontology, 35*, 571–583.

McGowan, A. J., & Smith, A. B. (2008). Are global Phanerozoic marine diversity curves truly global? A study of the relationship between regional rock records and global Phanerozoic marine diversity. *Paleobiology, 34*, 80–103.

McKinney, M. L. (1990). Classifying and analyzing evolutionary trends. In K. J. McNamara (Ed.), *Evolutionary trends* (pp. 28–58). Tucson: University of Arizona Press.

Olson, E. C. (1944). Origin of mammals based upon the cranial morphology of the therapsid suborders. *Special Papers of the Geological Society of America, 55*, 1–136.

Olson, E. C. (1959). The evolution of mammalian characters. *Evolution, 13*, 344–353.

Olson, E. C. (1966). Community evolution and the origin of mammals. *Ecology, 47*, 291–302.

Peters, S. E. (2005). Geologic constraints on the macroevolutionary history of marine animals. *Proceedings of the National Academy of Sciences of the United States of America, 102*, 12326–12331.

Peters, S. E. (2006). Genus extinction, origination, and the durations of sedimentary hiatuses. *Paleobiology, 32*, 387–407.

Peters, S. E., & Foote, M. (2001). Biodiversity in the Phanerozoic: A reinterpretation. *Paleobiology, 27*, 583–601.

Prothero, D. R. (2006). *After the dinosaurs: The age of mammals*. Bloomington: Indiana University Press.

Quinn, G. P., & Keough, M. J. (2002). *Experimental design and data analysis for biologists*. Cambridge: Cambridge University Press.

Raup, D. M. (1972). Taxonomic diversity during the Phanerozoic. *Science, 177*, 1065–1071.

Raup, D. M. (1976). Species diversity in the Phanerozoic: An interpretation. *Paleobiology, 2*, 289–297.

Raup, D. M., & Sepkoski, J. J. (1984). Periodicity of extinctions in the geologic past. *Proceedings of the National Academy of Sciences of the United States of America, 81*, 801–805.

Roopnarine, P. D., Angielczyk, K. D., Wang, S. C., & Hertog, R. (2007). Trophic network models explain instability of Early Triassic terrestrial communities. *Proceedings of the Royal Society B-Biological Sciences, 274*, 2077–2086.

Rose, K. D. (2006). *The beginning of the age of mammals*. Baltimore: Johns Hopkins University Press.

Rubidge, B. S. (1995). Biostratigraphy of the Beaufort Group (Karoo Supergroup). *South African Committee for Stratigraphy Biostratigraphic Series, 1*, 1–46.

Rubidge, B. S. (2005). Re-uniting lost continents—Fossil reptiles from the ancient Karoo and their wanderlust. *South African Journal of Geology, 108*, 135–172.

Sahney, S., & Benton, M. J. (2008). Recovery from the most profound mass extinction of all time. *Proceedings of the Royal Society B-Biological Sciences, 275*, 759–765.

Sahney, S., Benton, M. J., & Ferry, P. A. (2010). Links between global taxonomic diversity, ecological diversity and the expansion of vertebrates on land. *Biology Letters, 6*, 544–547.

Sepkoski, J. J. (1976). Species diversity in the Phanerozoic: Species-area effects. *Paleobiology, 2*, 298–303.

Sepkoski, J. J. (1981). A factor analytic description of the Phanerozoic marine fossil record. *Paleobiology, 7*, 36–53.

Sepkoski, J. J., Bambach, R. K., Raup, D. M., & Valentine, J. W. (1981). Phanerozoic marine diversity and the fossil record. *Nature, 293*, 435–437.

Smith, A. B. (2001). Large-scale heterogeneity of the fossil record: Implications for Phanerozoic biodiversity studies. *Philosophical*

Transactions of the Royal Society of London Series B-Biological Sciences, 356, 351–367.
Smith, A. B., & McGowan, A. J. (2005). Cyclicity in the fossil record mirrors rock outcrop area. *Biology Letters, 1*, 443–445.
Smith, A. B., & McGowan, A. J. (2007). The shape of the Phanerozoic marine palaeodiversity curve: How much can be predicted from the sedimentary rock record of Western Europe? *Palaeontology, 50*, 765–774.
Smith, A. B., Gale, A. S., & Monks, N. E. A. (2001). Sea-level change and rock record bias in the Cretaceous: A problem for extinction and biodiversity studies. *Paleobiology, 27*, 241–253.
Smith, R. M. H., & Ward, P. D. (2001). Pattern of vertebrate extinction across an event bed at the Permian-Triassic boundary in the Karoo Basin of South Africa. *Geology, 29*, 1147–1150.
Smith, R. M. H., & Botha, J. (2005). The recovery of terrestrial vertebrate diversity in the South African Karoo Basin after the end-Permian extinction. *Comptes Rendus Palevol, 4*, 555–568.
Uhen, M. D., & Pyenson, N. D. (2007). Diversity estimates, biases, and historiographic effects: Resolving cetacean diversity in the Tertiary. *Palaeontologia Electronica, 10*, 11A: 22p.
Upchurch, P., & Barrett, P. M. (2005). A phylogenetic perspective on sauropod diversity. In K. A. Curry-Rogers & J. A. Wilson (Eds.), *The Sauropods: Evolution and paleobiology* (pp. 104–124). Berkeley: University of California Press.
Ward, P. D., Botha, J., Buick, R., De Kock, M. O., Erwin, D. H., Garrison, G. H., et al. (2005). Abrupt and gradual extinction among Late Permian land vertebrates in the Karoo Basin, South Africa. *Science, 307*, 709–714.

Subject Index

Note: Page numbers followed by "*f*" and "*t*" indicate figures and tables respectively.

Christian F. Kammerer, Kenneth D. Angielczyk and Jörg Fröbisch (eds.), *Early Evolutionary History of the Synapsida*,
Vertebrate Paleobiology and Paleoanthropology, DOI: 10.1007/978-94-007-6841-3,

Taxonomic Index

Christian F. Kammerer, Kenneth D. Angielczyk and Jörg Fröbisch (eds.), *Early Evolutionary History of the Synapsida*,
Vertebrate Paleobiology and Paleoanthropology, DOI: 10.1007/978-94-007-6841-3,

MIX
Papier aus verantwortungsvollen Quellen
Paper from responsible sources
FSC® C105338

If you have any concerns about our products,
you can contact us on
ProductSafety@springernature.com

In case Publisher is established outside the EU,
the EU authorized representative is:
Springer Nature Customer Service Center GmbH
Europaplatz 3, 69115 Heidelberg, Germany

Printed by Libri Plureos GmbH
in Hamburg, Germany